SHOP REFERENCE

FOR
STUDENTS AND APPRENTICES

SHOP REFERENCE
FOR STUDENTS AND APPRENTICES

SECOND EDITION

COMPILED BY EDWARD G. HOFFMAN

CHRISTOPHER J. MCCAULEY, SENIOR EDITOR
MUHAMMED IQBAL HUSSAIN, ASSOCIATE EDITOR

2000
INDUSTRIAL PRESS INC.
NEW YORK

Library of Congress Cataloging-in-Publication Data

Shop reference for students and apprentices / compiled by Edward G. Hoffman; Christopher J. McCauley, senior editor; Muhammed Iqbal Hussain, associate editor. —2nd ed.

520 p. 127 × 178 cm.

ISBN 0-8311-3079-2

I. Machine-shop practice—Handbook, manuals, etc.

I. Hoffman, Edward G.

II. McCauley, Christopher J,

III. Hussain, Muhammed Iqbal

TJ1065.S615 2000 670.42'3--dc21 00-057240

Previously published under the title
Student's Shop Reference Handbook

Industrial Press Inc.
989 Avenue of the Americas, New York, NY 10018
Tel: 212-889-6330 Toll-Free: 1-888-528-7852 Fax: 212-545-8327
www.industrialpress.com Email: info@industrialpress.com

Shop Reference for Students and Apprentices
Second Edition

6 7 8 9 0

FOREWORD

THE AIM OF THE EDITOR in compiling the *Shop Reference Handbook for Students and Apprentices* has been to provide an affordable and dependable handbook for students in the machine shop, the tool room, and the drafting room. The material included has been taken from *Machinery's Handbook* and other authoritative sources and is presented in as clear, accurate, and easy-to-follow form as possible. In it the reader will find a wide range of useful formulas and data together with an extensive text.

Data and information from many American National Standards Institute (ANSI) Standards will be found in this book and have been extracted with permission of the publisher, the American Society of Mechanical Engineers, Three Park Ave., New York, New York 10016. These Standards are revised periodically; ASME should be contacted for information concerning the current edition of any particular document.

The nineteen sections of this handbook are thoroughly described in the *Table Of Contents* cover those areas of interest commonly encountered by machinists, toolmakers, diemakers, drafters, and other shop and manufacturing personnel. From the ability to understand and use shop mathematics to the reading and interpreting of shop drawings, the editor's intent is to provide the information and know-how that students will need as they prepare themselves for jobs in the metalworking industries.

The sections *Conversion Factors* and *Mathematics* covers those aspects of applied mathematics that are needed on the job. From basic conversions between the inch and metric systems to the solving of problems involving arithmetic, geometry, algebra, and trigonometry, all of the essential information, formulas, and tables needed to make accurate calculations are provided.

The section *Engineering Drawings* outlines the standard methods of presentation and the conventions used in preparing engineering drawings and deals with the fundamentals of dimensioning, and the understanding of the application of tolerances, allowances, fits, and surface finish with respect to drawings.

The section *Inspection* covers the proper use of measuring instruments and methods, including the sine bar, calculating tapers, and measurement over pins.

The section *Allowances And Tolerances For Fits* deals in depth with inch and metric standard tolerances, allowances, limits and fits, preferred numbers and sizes, and contains a great deal of essential standard allowance, tolerance and fits data.

The five sections *Pins*, *Standard Tapers*, *Screw Thread Systems*, *Common Hardware*, and *Gears And Gearing* describe the sizes, forms, and dimensions of standard machine elements commonly encountered in and around the shop.

The remaining six sections, *Cutting Speeds and Feeds*, *Cutting Tools*, *Tool Wear*, *Cutting Fluids for Machining*, *Machining Nonferrous Metals*, and *Materials* cover machining methods and materials selection, specify the recommended speeds and feeds for various kinds of machining operations on different materials, and the types and compositions of metals commonly used in machine construction.

An extensive index has been prepared to enable the user to quickly and conveniently find the information and data that he or she requires.

Suggestions and comments concerning this Handbook are welcome.

TABLE OF CONTENTS

CONVERSION FACTORS

MATHEMATICS

MATHEMATICS (Cont.)

ENGINEERING DRAWINGS

TABLE OF CONTENTS

ENGINEERING DRAWINGS (Cont.)

INSPECTION

ALLOWANCES AND TOLERANCES FOR FITS

ALLOWANCES AND TOLERANCES FOR FITS (Cont.)

PINS

STANDARD TAPERS

KEYS AND KEYSEATS

TABLE OF CONTENTS

SCREW THREAD SYSTEMS

COMMON HARDWARE

GEARS AND GEARING

GEARS AND GEARING (Cont.)

INDEXING

CUTTING SPEEDS AND FEEDS

TABLE OF CONTENTS

CUTTING SPEEDS AND FEEDS (Cont.)

CUTTING TOOLS

TOOL WEAR

TOOL WEAR (Cont.)

CUTTING FLUIDS FOR MACHINING

MACHINING NONFERROUS METALS

MATERIALS

TABLE OF CONTENTS

MATERIALS (Cont.)

MATERIALS (Cont.)

HARDNESS TESTING

CONVERSION FACTORS

Table 1. Metric Conversion Factors

Multiply	By	To Obtain
	Length	
centimeter	0.03280840	foot
centimeter	0.3937008	inch
fathom	1.8288[a]	meter (m)
foot	0.3048[a]	meter (m)
foot	30.48[a]	centimeter (cm)
foot	304.8[a]	millimeter (mm)
inch	0.0254[a]	meter (m)
inch	2.54[a]	centimeter (cm)
inch	25.4[a]	millimeter (mm)
kilometer	0.6213712	mile [U.S. statute]
meter	39.37008	inch
meter	0.5468066	fathom
meter	3.280840	foot
meter	0.1988388	rod
meter	1.093613	yard
meter	0.0006213712	mile [U. S. statute]
microinch	0.0254[a]	micrometer [micron] (μm)
micrometer [micron]	39.37008	microinch
mile [U. S. statute]	1609.344[a]	meter (m)
mile [U. S. statute]	1.609344[a]	kilometer (km)
millimeter	0.003280840	foot
millimeter	0.03937008	inch
rod	5.0292[a]	meter (m)
yard	0.9144[a]	meter (m)
	Area	
acre	4046.856	meter^2 (m^2)
acre	0.4046856	hectare
centimeter^2	0.1550003	inch^2
centimeter^2	0.001076391	foot^2
foot^2	0.09290304[a]	meter^2 (m^2)
foot^2	929.0304[a]	centimeter^2 (cm^2)
foot^2	92,903.04[a]	millimeter^2 (mm^2)
hectare	2.471054	acre
inch^2	645.16[a]	millimeter^2 (mm^2)
inch^2	6.4516[a]	centimeter^2 (cm^2)
inch^2	0.00064516[a]	meter^2 (m^2)
meter^2	1550.003	inch^2
meter^2	10.763910	foot^2
meter^2	1.195990	yard^2
meter^2	0.0002471054	acre

Table 1. *(Continued)* **Metric Conversion Factors**

Multiply	By	To Obtain
mile^2	2.5900	kilometer^2
millimeter^2	0.00001076391	foot^2
millimeter^2	0.001550003	inch^2
yard^2	0.8361274	meter^2 (m^2)
Volume (including Capacity)		
centimeter^3	0.06102376	inch^3
foot^3	28.31685	liter
foot^3	28.31685	liter
gallon [U.K. liquid]	0.004546092	meter^3 (m^3)
gallon [U.K. liquid]	4.546092	liter
gallon [U. S. liquid]	0.003785412	meter^3 (m^3)
gallon [U.S. liquid]	3.785412	liter
inch^3	16,387.06	millimeter^3 (mm^3)
inch^3	16.38706	centimeter^3 (cm^3)
inch^3	0.00001638706	meter^3 (m^3)
liter	0.001[a]	meter^3 (m^3)
liter	0.2199692	gallon [U. K. liquid]
liter	0.2641720	gallon [U. S. liquid]
liter	0.03531466	foot^3
meter^3	219.9692	gallon [U. K. liquid]
meter^3	264.1720	gallon [U. S. liquid]
meter^3	35.31466	foot^3
meter^3	1.307951	yard^3
meter^3	1000.[a]	liter
meter^3	61,023.76	inch^3
millimeter^3	0.00006102376	inch^3
quart [U.S. Liquid]	0.946	liter
quart [U.K. Liquid]	1.136	liter
yard^3	0.7645549	meter^3 (m^3)
Velocity, Acceleration, and Flow		
centimeter/second	1.968504	foot/minute
centimeter/second	0.03280840	foot/second
centimeter/minute	0.3937008	inch/minute
foot/hour	0.00008466667	meter/second (m/s)
foot/hour	0.00508[a]	meter/minute
foot/hour	0.3048[a]	meter/hour
foot/minute	0.508[a]	centimeter/second
foot/minute	18.288[a]	meter/hour
foot/minute	0.3048[a]	meter/minute
foot/minute	0.00508[a]	meter/second (m/s)
foot/second	30.48[a]	centimeter/second
foot/second	18.288[a]	meter/minute
foot/second	0.3048[a]	meter/second (m/s)
foot/second^2	0.3048[a]	meter/second^2 (m/s^2)
foot^3/minute	28.31685	liter/minute

Table 1. *(Continued)* **Metric Conversion Factors**

Multiply	By	To Obtain
$foot^3$/minute	0.0004719474	$meter^3$/second (m^3/s)
gallon [U. S. liquid]/min.	0.003785412	$meter^3$/minute
gallon [U. S. liquid]/min.	0.00006309020	$meter^3$/second (m^3/s)
gallon [U. S. liquid]/min.	0.06309020	liter/second
gallon [U. S. liquid]/min.	3.785412	liter/minute
gallon [U. K. liquid]/min.	0.004546092	$meter^3$/minute
gallon [U. K. liquid]/min.	0.00007576820	$meter^3$/second (m^3/s)
inch/minute	25.4[a]	millimeter/minute
inch/minute	2.54[a]	centimeter/minute
inch/minute	0.0254[a]	meter/minute
inch/$second^2$	0.0254[a]	meter/$second^2$ (m/s^2)
kilometer/hour	0.6213712	mile/hour [U. S. statute]
liter/minute	0.03531466	$foot^3$/minute
liter/minute	0.2641720	gallon [U.S. liquid]/minute
liter/second	15.85032	gallon [U. S. liquid]/minute
mile/hour	1.609344[a]	kilometer/hour
millimeter/minute	0.03937008	inch/minute
meter/second	11,811.02	foot/hour
meter/second	196.8504	foot/minute
meter/second	3.280840	foot/second
meter/$second^2$	3.280840	foot/$second^2$
meter/$second^2$	39.37008	inch/$second^2$
meter/minute	3.280840	foot/minute
meter/minute	0.05468067	foot/second
meter/minute	39.37008	inch/minute
meter/hour	3.280840	foot/hour
meter/hour	0.05468067	foot/minute
$meter^3$/second	2118.880	$foot^3$/minute
$meter^3$/second	13,198.15	gallon [U. K. liquid]/minute
$meter^3$/second	15,850.32	gallon [U. S. liquid]/minute
$meter^3$/minute	219.9692	gallon [U. K. liquid]/minute
$meter^3$/minute	264.1720	gallon [U. S. liquid]/minute
	Mass and Density	
grain [$\frac{1}{7000}$ 1b avoirdupois]	0.06479891	gram (g)
gram	15.43236	grain
gram	0.001[a]	kilogram (kg)
gram	0.03527397	ounce [avoirdupois]
gram	0.03215074	ounce [troy]
gram/$centimeter^3$	0.03612730	pound/$inch^3$
hundredweight [long]	50.80235	kilogram (kg)
hundredweight [short]	45.35924	kilogram (kg)
kilogram	1000.[a]	gram (g)
kilogram	35.27397	ounce [avoirdupois]
kilogram	32.15074	ounce [troy]

Table 1. *(Continued)* **Metric Conversion Factors**

Multiply	By	To Obtain
kilogram	2.204622	pound [avoirdupois]
kilogram	0.06852178	slug
kilogram	0.0009842064	ton [long]
kilogram	0.001102311	ton [short]
kilogram	0.001[a]	ton [metric]
kilogram	0.001[a]	tonne
kilogram	0.01968413	hundredweight [long]
kilogram	0.02204622	hundredweight [short]
kilogram/meter^3	0.06242797	pound/foot^3
kilogram/meter^3	0.01002242	pound/gallon [U. K. liquid]
kilogram/meter^3	0.008345406	pound/gallon [U. S. liquid]
ounce [avoirdupois]	28.34952	gram (g)
ounce [avoirdupois]	0.02834952	kilogram (kg)
ounce [troy]	31.10348	gram (g)
ounce [troy]	0.03110348	kilogram (kg)
pound [avoirdupois]	0.4535924	kilogram (kg)
pound/foot^3	16.01846	kilogram/meter^3 (kg/m^3)
pound/inch^3	27.67990	gram/centimeter^3 (g/cm^3)
pound/gal [U. S. liquid]	119.8264	kilogram/meter^3 (kg/m^3)
pound/gal [U. K. liquid]	99.77633	kilogram/meter^3 (kg/m^3)
slug	14.59390	kilogram (kg)
ton [long 2240 lb]	1016.047	kilogram (kg)
ton [short 2000 lb]	907.1847	kilogram (kg)
ton [metric]	1000.[a]	kilogram (kg)
ton [Metric]	0.9842	ton[long 2240 lb]
ton [Metric]	1.1023	ton[short 2000 lb]
tonne	1000.[a]	kilogram (kg)
	Force and Force/Length	
dyne	0.00001[a]	newton (N)
kilogram-force	9.806650[a]	newton (N)
kilopound	9.806650[a]	newton (N)
newton	0.1019716	kilogram-force
newton	0.1019716	kilopond
newton	0.2248089	pound-force
newton	100,000.[a]	dyne
newton	7.23301	poundal
newton	3.596942	ounce-force
newton/meter	0.005710148	pound/inch
newton/meter	0.06852178	pound/foot
ounce-force	0.2780139	newton (N)
pound-force	4.448222	newton (N)
poundal	0.1382550	newton (N)
pound/inch	175.1268	newton/meter (N/m)
pound/foot	14.59390	newton/meter (N/m)

Table 1. *(Continued)* **Metric Conversion Factors**

Multiply	By	To Obtain
	Bending Moment or Torque	
dyne-centimeter	0.0000001[a]	newton-meter (N · m)
kilogram-meter	9.806650[a]	newton-meter (N · m)
ounce-inch	7.061552	newton-millimeter
ounce-inch	0.007061552	newton-meter (N · m)
newton-meter	0.7375621	pound-foot
newton-meter	10,000,000.[a]	dyne-centimeter
newton-meter	0.1019716	kilogram-meter
newton-meter	141.6119	ounce-inch
newton-millimeter	0.1416119	ounce-inch
pound-foot	1.355818	newton-meter (N · m)
	Moment of Inertia and Section Modulus	
moment of inertia [kg · m^2]	23.73036	pound-foot^2
moment of inertia [kg · m^2]	3417.171	pound-inch^2
moment of inertia [lb · ft^2]	0.04214011	kilogram-meter^2 (kg · m^2)
moment of inertia [lb · inch^2]	0.0002926397	kilogram-meter^2 (kg · m^2)
moment of section [foot^4]	0.008630975	meter^4 (m^4)
moment of section [inch^4]	41.62314	centimeter^4
moment of section [meter^4]	115.8618	foot^4
moment of section [centimeter^4]	0.02402510	inch^4
section modulus [foot^3]	0.02831685	meter^3 (m^3)
section modulus [inch^3]	0.00001638706	meter^3 (m^3)
section modulus [meter^3]	35.31466	foot^3
section modulus [meter^3]	61,023.76	inch^3
	Momentum	
kilogram-meter/second	7.233011	pound-foot/second
kilogram-meter/second	86.79614	pound-inch/second
pound-foot/second	0.1382550	kilogram-meter/second (kg · m/s)
pound-inch/second	0.01152125	kilogram-meter/second (kg · m/s)
	Pressure and Stress	
atmosphere [14.6959 lb/inch^2]	101,325.	pascal (Pa)
bar	100,000.[a]	pascal (Pa)
bar	14.50377	pound/inch^2
bar	100,000.[a]	newton/meter^2 (N/m^2)
hectobar	0.6474898	ton [long]/inch^2
kilogram/centimeter^2	14.22334	pound/inch^2
kilogram/meter^2	9.806650[a]	newton/meter^2 (N/m^2)
kilogram/meter^2	9.806650[a]	pascal (Pa)
kilogram/meter^2	0.2048161	pound/foot^2
kilonewton/meter^2	0.1450377	pound/inch^2
newton/centimeter^2	1.450377	pound/inch^2
newton/meter^2	0.00001[a]	bar
newton/meter^2	1.0[a]	pascal (Pa)

Table 1. *(Continued)* **Metric Conversion Factors**

Multiply	By	To Obtain
newton/meter^2	0.0001450377	pound/inch^2
newton/meter^2	0.1019716	kilogram/meter^2
newton/millimeter^2	145.0377	pound/inch^2
pascal	0.00000986923	atmosphere
pascal	0.00001[a]	bar
pascal	0.1019716	kilogram/meter^2
pascal	1.0[a]	newton/meter^2 (N/m^2)
pascal	0.02088543	pound/foot^2
pascal	0.0001450377	pound/inch^2
pound/foot^2	4.882429	kilogram/meter^2
pound/foot^2	47.88026	pascal (Pa)
pound/inch^2	0.06894757	bar
pound/inch^2	0.07030697	kilogram/centimeter^2
pound/inch^2	0.6894757	newton/centimeter^2
pound/inch^2	6.894757	kilonewton/meter^2
pound/inch^2	6894.757	newton/meter^2 (N/m^2)
pound/inch^2	0.006894757	newton/millimeter^2 (N/mm^2)
pound/inch^2	6894.757	pascal (Pa)
ton [long]/inch^2	1.544426	hectobar
	Energy and Work	
Btu [International Table]	1055.056	joule (J)
Btu [mean]	1055.87	joule (J)
calorie [mean]	4.19002	joule (J)
foot-pound	1.355818	joule (J)
foot-poundal	0.04214011	joule (J)
joule	0.0009478170	Btu [International Table]
joule	0.0009470863	Btu [mean]
joule	0.2386623	calorie [mean]
joule	0.7375621	foot-pound
joule	23.73036	foot-poundal
joule	0.9998180	joule [International U. S.]
joule	0.9999830	joule [U. S. legal, 1948]
joule [International U. S.]	1.000182	joule (J)
joule [U. S. legal, 1948]	1.000017	joule (J)
joule	0.0002777778	watt-hour
watt-hour	3600.[a]	joule (J)
	Power	
Btu [International Table]/hour	0.2930711	watt (W)
foot-pound/hour	0.0003766161	watt (W)
foot-pound/minute	0.02259697	watt (W)
horsepower [550 ft-lb/s]	0.7456999	kilowatt (kW)
horsepower [550 ft-lb/s]	745.6999	watt (W)
horsepower [electric]	746.[a]	watt (W)
horsepower [metric]	735.499	watt (W)
horsepower [U. K.]	745.70	watt (W)
kilowatt	1.341022	horsepower [550 ft-lb/s]
watt	2655.224	foot-pound/hour

Table 1. *(Continued)* **Metric Conversion Factors**

Multiply	By	To Obtain
watt	44.25372	foot-pound/minute
watt	0.001341022	horsepower [550 ft-lb/s]
watt	0.001340483	horsepower [electric]
watt	0.001359621	horsepower [metric]
watt	0.001341022	horsepower [U. K.]
watt	3.412141	Btu [International Table]/hour
	Viscosity	
centipoise	0.001[a]	pascal-second (Pa · s)
centistoke	0.000001[a]	meter²/second (m²/s)
meter²/second	1,000,000.[a]	centistoke
meter²/second	10,000.[a]	stoke
pascal-second	1000.[a]	centipoise
pascal-second	10.[a]	poise
poise	0.1[a]	pascal-second (Pa · s)
stoke	0.0001[a]	meter²/second (m²/s)
	Temperature	
To Convert From	To	Use Formula
temperature Celsius, t_C	temperature Kelvin, t_K	$t_K = t_C + 273.15$
temperature Fahrenheit, t_F	temperature Kelvin, t_K	$t_K = (t_F + 459.67)/1.8$
temperature Celsius, t_C	temperature Fahrenheit, t_F	$t_F = 1.8\ t_C + 32$
temperature Fahrenheit, t_F	temperature Celsius, t_C	$t_C = (t_F - 32)/1.8$
temperature Kelvin, t_K	temperature Celsius, t_C	$t_C = t_K - 273.15$
temperature Kelvin, t_K	temperature Fahrenheit, t_F	$t_F = 1.8\ t_K - 459.67$
temperature Kelvin, t_K	temperature Rankine, t_R	$t_R = 9/5\ t_K$
temperature Rankine, t_R	temperature Kelvin, t_K	$t_K = 5/9\ t_R$

[a] Where an asterisk is shown, the figure is exact.

Symbols of SI units, multiples and sub-multiples are given in parentheses in the right-hand column

Greek Letters and Standard Abbreviations.—The Greek letters are frequently used in mathematical expressions and formulas. The Greek alphabet is given below.

Α	α	Alpha	Η	η	Eta	Ν	ν	Nu	Τ	τ	Tau
Β	β	Beta	Θ	ϑ θ	Theta	Ξ	ξ	Xi	ϒ	υ	Upsilon
Γ	γ	Gamma	Ι	ι	Iota	Ο	ο	Omicron	Φ	φ	Phi
Δ	δ	Delta	Κ	κ	Kappa	Π	π	Pi	Χ	χ	Chi
Ε	ε	Epsilon	Λ	λ	Lambda	Ρ	ρ	Rho	Ψ	ψ	Psi
Ζ	ζ	Zeta	Μ	μ	Mu	Σ	σ ς	Sigma	Ω	ω	Omega

Table 2. Decimal Equivalents of Fractions of an Inch

1/64	0.015 625	11/32	0.343 75	43/64	0.671 875
1/32	0.031 25	23/64	0.359 375	11/16	0.687 5
3/64	0.046 875	3/8	0.375	45/64	0.703 125
1/16	0.062 5	25/64	0.390 625	23/32	0.718 75
5/64	0.078 125	13/32	0.406 25	47/64	0.734 375
3/32	0.093 75	27/64	0.421 875	3/4	0.750
7/64	0.109 375	7/16	0.437 5	49/64	0.765 625

Table 2. Decimal Equivalents of Fractions of an Inch

1/8	0.125	29/64	0.453 125	25/32	0.781 25
9/64	0.140 625	15/32	0.468 75	51/64	0.796 875
5/32	0.156 25	31/64	0.484 375	13/16	0.812 5
11/64	0.171 875	1/2	0.500	53/64	0.828 125
3/16	0.187 5	33/64	0.515 625	27/32	0.843 75
13/64	0.203 125	17/32	0.531 25	55/64	0.859 375
7/32	0.218 75	35/64	0.546 875	7/8	0.875
15/64	0.234 375	9/16	0.562 5	57/64	0.890 625
1/4	0.250	37/64	0.578 125	29/32	0.906 25
17/64	0.265 625	19/32	0.593 75	59/64	0.921 875
9/32	0.281 25	39/64	0.609 375	15/16	0.937 5
19/64	0.296 875	5/8	0.625	61/64	0.953 125
5/16	0.312 5	41/64	0.640 625	31/32	0.968 75
21/64	0.328 125	21/32	0.656 25	63/64	0.984 375

Table 3. Miscellaneous Conversion Factors (English Units)

Multiply	By	To Obtain
atmospheres	29.92	inches of mercury (32 deg. F.)
atmospheres	14.70	pounds/inch^2
British thermal units/hour	12.96	foot-pounds/minute
circular miles	0.7854	square miles
feet of water (60 deg. F.)	0.8843	inches of mercury (60 deg. F.)
feet of water (60 deg. F.)	0.4331	pounds/inch^2
feet/minute	0.01136	miles/hour
foot-pounds/second	0.07716	British thermal units/minute
gallons (U.S.) of water (60 deg. F.)	8.337	pounds of water (60 deg. F.)
gallons (U.S.)/second	8.021	feet^3/minute
inches of mercury (32 deg. F.)	0.03342	atmospheres
inches of mercury (60 deg. F.)	1.131	feet of water (60 deg. F.)
inches of mercury (60 deg. F.)	0.4898	pounds/inch^2
inches of water (60 deg. F.)	0.03609	pounds/inch^2
knots (International)	1.151	miles (statute)/hour
miles/hour	88	feet/minute
miles (statute)/hour	0.8690	knots (International)
ounces (avoirdupois)	0.9115	ounces (troy)
ounces (troy)	1.097	ounces (avoirdupois)
ounces (troy)	0.06857	pounds (avoirdupois)
pounds (avoirdupois)	14.58	ounces (troy)
pounds of water (60 deg. F.)	0.01603	feet^3
pounds of water (60 deg. F.)	0.1199	gallons (U.S.)
pounds/inch^2	0.06805	atmospheres
pounds/inch^2	2.309	feet of water (60 deg. F.)
pounds/inch^2	2.042	inches of mercury (60 deg. F.)
pounds/inch^2	27.71	inches of water (60 deg. F.)
square mils	1.273	circular miles

Use of Conversion Tables.—On this and following pages, tables are given that permit conversion from English to metric units and vice versa over a wide range of values. Where the desired value cannot be obtained directly from these tables, a simple addition of two or more values taken directly from the table will suffice as shown in the following examples:

Example 1: Find the millimeter equivalent of 0.4476 inch.

0.4	in	=	10.16000 mm
0.04	in	=	1.01600 mm
0.007	in	=	0.17780 mm
0.0006	in	=	0.01524 mm
0.4476	in	=	11.36904 mm

Example 2: Find the inch equivalent of 84.9 mm.

80.	mm	=	3.14961 in.
4.	mm	=	0.15748 in.
0.9	mm	=	0.03543 in.
84.9	mm	=	3.34252 in.

Table 4. Inch to Millimeters and Millimeters to Inch Conversion Table

Inches to Millimeters											
in.	mm	in.	mm	in.	mm	in.	mm	in.	mm	in.	mm
10	254.00000	1	25.40000	0.1	2.54000	0.01	0.25400	0.001	0.02540	0.0001	0.00254
20	508.00000	2	50.80000	0.2	5.08000	0.02	0.50800	0.002	0.05080	0.0002	0.00508
30	762.00000	3	76.20000	0.3	7.62000	0.03	0.76200	0.003	0.07620	0.0003	0.00762
40	1,016.00000	4	101.60000	0.4	10.16000	0.04	1.01600	0.004	0.10160	0.0004	0.01016
50	1,270.00000	5	127.00000	0.5	12.70000	0.05	1.27000	0.005	0.12700	0.0005	0.01270
60	1,524.00000	6	152.40000	0.6	15.24000	0.06	1.52400	0.006	0.15240	0.0006	0.01524
70	1,778.00000	7	177.80000	0.7	17.78000	0.07	1.77800	0.007	0.17780	0.0007	0.01778
80	2,032.00000	8	203.20000	0.8	20.32000	0.08	2.03200	0.008	0.20320	0.0008	0.02032
90	2,286.00000	9	228.60000	0.9	22.86000	0.09	2.2860	0.009	0.22860	0.0009	0.02286
100	2,540.00000	10	254.00000	1.0	25.40000	0.10	2.54000	0.010	0.25400	0.0010	0.02540
Millimeters to Inches											
mm	in.	mm	in.	mm	in.	mm	in.	mm	in.	mm	in.
100	3.93701	10	0.39370	1	0.03937	0.1	0.00394	0.01	0.000039	0.001	0.00004
200	7.87402	20	0.78740	2	0.07874	0.2	0.00787	0.02	0.00079	0.002	0.00008
300	11.81102	30	1.18110	3	0.11811	0.3	0.01181	0.03	0.00118	0.003	0.00012
400	15.74803	40	1.57480	4	0.15748	0.4	0.01575	0.04	0.00157	0.004	0.00016
500	19.68504	50	1.96850	5	0.19685	0.5	0.01969	0.05	0.00197	0.005	0.00020
600	23.62205	60	2.36220	6	0.23622	0.6	0.02362	0.06	0.00236	0.006	0.00024
700	27.55906	70	2.75591	7	0.27559	0.7	0.02756	0.07	0.00276	0.007	0.00028
800	31.49606	80	3.14961	8	0.31496	0.8	0.03150	0.08	0.00315	0.008	0.00031
900	35.43307	90	3.54331	9	0.35433	0.9	0.03543	0.09	0.00354	0.009	0.00035
1000	39.37008	100	3.93701	10	0.39370	1.0	0.03937	0.10	0.00394	0.010	0.00039

For inches to centimeters, shift decimal point in mm column one place to left and read centimeters, thus:

$$40 \text{ in} = 1016 \text{ mm} = 101.6 \text{ cm}$$

For centimeters to inches, shift decimal point of centimeter value one place to right and enter mm column, thus:

$$70 \text{ cm} = 700 \text{ mm} = 27.55906 \text{ inches}$$

Table 5. Fractional Inch — Millimeter and Foot — Millimeter Conversion Tables (Based on 1 inch = 25.4 millimeters, exactly)

Fractional Inch to Millimeters

in.	mm	in.	mm	in.	mm	in.	mm
1/64	0.397	17/64	6.747	33/64	13.097	49/64	19.447
1/32	0.794	9/32	7.144	17/32	13.494	25/32	19.844
3/64	1.191	19/64	7.541	35/64	13.891	51/64	20.241
1/16	1.588	5/16	7.938	9/16	14.288	13/16	20.638
5/64	1.984	21/64	8.334	37/64	14.684	53/64	21.034
3/32	2.381	11/32	8.731	19/32	15.081	27/32	21.431
7/64	2.778	23/64	9.128	39/64	15.478	55/64	21.828
1/8	3.175	3/8	9.525	5/8	15.875	7/8	22.225
9/64	3.572	25/64	9.922	41/64	16.272	57/64	22.622
5/32	3.969	13/32	10.319	21/32	16.669	29/32	23.019
11/64	4.366	27/64	10.716	43/64	17.066	59/64	23.416
3/16	4.762	7/16	11.112	11/16	17.462	15/16	23.812
13/64	5.159	29/64	11.509	45/64	17.859	61/64	24.209
7/32	5.556	15/32	11.906	23/32	18.256	31/32	24.606
15/64	5.953	31/64	12.303	47/64	18.653	63/64	25.003
1/4	6.350	1/2	12.700	3/4	19.050	1	25.400

Inches to Millimeters

in.	mm	in.	mm	in.	mm	in.	mm	in.	mm	in.	mm
1	25.4	3	76.2	5	127.0	7	177.8	9	228.6	11	279.4
2	50.8	4	101.6	6	152.4	8	203.2	10	254.0	12	304.8

Feet to Millimeters

ft	mm	ft	mm	ft	mm	ft	mm	ft	mm
100	30,480	10	3,048	1	304.8	0.1	30.48	0.01	3.048
200	60,960	20	6,096	2	609.6	0.2	60.96	0.02	6.096
300	91,440	30	9,144	3	914.4	0.3	91.44	0.03	9.144
400	121,920	40	12,192	4	1,219.2	0.4	121.92	0.04	12.192
500	152,400	50	15,240	5	1,524.0	0.5	152.40	0.05	15.240
600	182,880	60	18,288	6	1,828.8	0.6	182.88	0.06	18.288
700	213,360	70	21,336	7	2,133.6	0.7	213.36	0.07	21.336
800	243,840	80	24,384	8	2,438.4	0.8	243.84	0.08	24.384
900	274,320	90	27,432	9	2,743.2	0.9	274.32	0.09	27.432
1000	304,800	100	30,480	10	3,048.0	1.0	304.80	0.10	30.480

Example 1: Find millimeter equivalent of 293 feet, 5 47/64 inches.

200 ft	=	60,960.00	mm
90 ft	=	27,432.00	mm
3 ft	=	914.40	mm
5 in.	=	127.00	mm
47/64 in.	=	18.653	mm
293 ft 5 47/64 in.	=	89,452.053	mm

Example 2: Find millimeter equivalent of 71.86 feet.

70.00	ft	=	21,336.00	mm
1.00	ft	=	304.80	mm
0.80	ft	=	243.84	mm
0.06	ft	=	18.288	mm
71.86	ft	=	21,902.928	mm

Table 6. Decimals of an Inch to Millimeters
(Based on 1 inch = 25.4 millimeters, exactly)

Inches	0.000	0.001	0.002	0.003	0.004	0.005	0.006	0.007	0.008	0.009
	Millimeters									
0.000	…	0.0254	0.0508	0.0762	0.1016	0.1270	0.1524	0.1778	0.2032	0.2286
0.010	0.2540	0.2794	0.3048	0.3302	0.3556	0.3810	0.4064	0.4318	0.4572	0.4826
0.020	0.5080	0.5334	0.5588	0.5842	0.6096	0.6350	0.6604	0.6858	0.7112	0.7366
0.030	0.7620	0.7874	0.8128	0.8382	0.8636	0.8890	0.9144	0.9398	0.9652	0.9906
0.040	1.0160	1.0414	1.0668	1.0922	1.1176	1.1430	1.1684	1.1938	1.2192	1.2446
0.050	1.2700	1.2954	1.3208	1.3462	1.3716	1.3970	1.4224	1.4478	1.4732	1.4986
0.060	1.5240	1.5494	1.5748	1.6002	1.6256	1.6510	1.6764	1.7018	1.7272	1.7526
0.070	1.7780	1.8034	1.8288	1.8542	1.8796	1.9050	1.9304	1.9558	1.9812	2.0066
0.080	2.0320	2.0574	2.0828	2.1082	2.1336	2.1590	2.1844	2.2098	2.2352	2.2606
0.090	2.2860	2.3114	2.3368	2.3622	2.3876	2.4130	2.4384	2.4638	2.4892	2.5146
0.100	2.5400	2.5654	2.5908	2.6162	2.6416	2.6670	2.6924	2.7178	2.7432	2.7686
0.110	2.7940	2.8194	2.8448	2.8702	2.8956	2.9210	2.9464	2.9718	2.9972	3.0226
0.120	3.0480	3.0734	3.0988	3.1242	3.1496	3.1750	3.2004	3.2258	3.2512	3.2766
0.130	3.3020	3.3274	3.3528	3.3782	3.4036	3.4290	3.4544	3.4798	3.5052	3.5306
0.140	3.5560	3.5814	3.6068	3.6322	3.6576	3.6830	3.7084	3.7338	3.7592	3.7846
0.150	3.8100	3.8354	3.8608	3.8862	3.9116	3.9370	3.9624	3.9878	4.0132	4.0386
0.160	4.0640	4.0894	4.1148	4.1402	4.1656	4.1910	4.2164	4.2418	4.2672	4.2926
0.170	4.3180	4.3434	4.3688	4.3942	4.4196	4.4450	4.4704	4.4958	4.5212	4.5466
0.180	4.5720	4.5974	4.6228	4.6482	4.6736	4.6990	4.7244	4.7498	4.7752	4.8006
0.190	4.8260	4.8514	4.8768	4.9022	4.9276	4.9530	4.9784	5.0038	5.0292	5.0546
0.200	5.0800	5.1054	5.1308	5.1562	5.1816	5.2070	5.2324	5.2578	5.2832	5.3086
0.210	5.3340	5.3594	5.3848	5.4102	5.4356	5.4610	5.4864	5.5118	5.5372	5.5626
0.220	5.5880	5.6134	5.6388	5.6642	5.6896	5.7150	5.7404	5.7658	5.7912	5.8166
0.230	5.8420	5.8674	5.8928	5.9182	5.9436	5.9690	5.9944	6.0198	6.0452	6.0706
0.240	6.0960	6.1214	6.1468	6.1722	6.1976	6.2230	6.2484	6.2738	6.2992	6.3246
0.250	6.3500	6.3754	6.4008	6.4262	6.4516	6.4770	6.5024	6.5278	6.5532	6.5786
0.260	6.6040	6.6294	6.6548	6.6802	6.7056	6.7310	6.7564	6.7818	6.8072	6.8326
0.270	6.8580	6.8834	6.9088	6.9342	6.9596	6.9850	7.0104	7.0358	7.0612	7.0866
0.280	7.1120	7.1374	7.1628	7.1882	7.2136	7.2390	7.2644	7.2898	7.3152	7.3406
0.290	7.3660	7.3914	7.4168	7.4422	7.4676	7.4930	7.5184	7.5438	7.5692	7.5946
0.300	7.6200	7.6454	7.6708	7.6962	7.7216	7.7470	7.7724	7.7978	7.8232	7.8486
0.310	7.8740	7.8994	7.9248	7.9502	7.9756	8.0010	8.0264	8.0518	8.0772	8.1026
0.320	8.1280	8.1534	8.1788	8.2042	8.2296	8.2550	8.2804	8.3058	8.3312	8.3566
0.330	8.3820	8.4074	8.4328	8.4582	8.4836	8.5090	8.5344	8.5598	8.5852	8.6106
0.340	8.6360	8.6614	8.6868	8.7122	8.7376	8.7630	8.7884	8.8138	8.8392	8.8646
0.350	8.8900	8.9154	8.9408	8.9662	8.9916	9.0170	9.0424	9.0678	9.0932	9.1186
0.360	9.1440	9.1694	9.1948	9.2202	9.2456	9.2710	9.2964	9.3218	9.3472	9.3726
0.370	9.3980	9.4234	9.4488	9.4742	9.4996	9.5250	9.5504	9.5758	9.6012	9.6266
0.380	9.6520	9.6774	9.7028	9.7282	9.7536	9.7790	9.8044	9.8298	9.8552	9.8806
0.390	9.9060	9.9314	9.9568	9.9822	10.0076	10.0330	10.0584	10.0838	10.1092	10.1346
0.400	10.1600	10.1854	10.2108	10.2362	10.2616	10.2870	10.3124	10.3378	10.3632	10.3886
0.410	10.4140	10.4394	10.4648	10.4902	10.5156	10.5410	10.5664	10.5918	10.6172	10.6426
0.420	10.6680	10.6934	10.7188	10.7442	10.7696	10.7950	10.8204	10.8458	10.8712	10.8966
0.430	10.9220	10.9474	10.9728	10.9982	11.0236	11.0490	11.0744	11.0998	11.1252	11.1506
0.440	11.1760	11.2014	11.2268	11.2522	11.2776	11.3030	11.3284	11.3538	11.3792	11.4046
0.450	11.4300	11.4554	11.4808	11.5062	11.5316	11.5570	11.5824	11.6078	11.6332	11.6586
0.460	11.6840	11.7094	11.7348	11.7602	11.7856	11.8110	11.8364	11.8618	11.8872	11.9126
0.470	11.9380	11.9634	11.9888	12.0142	12.0396	12.0650	12.0904	12.1158	12.1412	12.1666
0.480	12.1920	12.2174	12.2428	12.2682	12.2936	12.3190	12.3444	12.3698	12.3952	12.4206
0.490	12.4460	12.4714	12.4968	12.5222	12.5476	12.5730	12.5984	12.6238	12.6492	12.6746
0.500	12.7000	12.7254	12.7508	12.7762	12.8016	12.8270	12.8524	12.8778	12.9032	12.9286

Table 6. Decimals of an Inch to Millimeters *(Continued)*
(Based on 1 inch = 25.4 millimeters, exactly)

Inches	0.000	0.001	0.002	0.003	0.004	0.005	0.006	0.007	0.008	0.009
	Millimeters									
0.510	12.9540	12.9794	13.0048	13.0302	13.0556	13.0810	13.1064	13.1318	13.1572	13.1826
0.520	13.2080	13.2334	13.2588	13.2842	13.3096	13.3350	13.3604	13.3858	13.4112	13.4366
0.530	13.4620	13.4874	13.5128	13.5382	13.5636	13.5890	13.6144	13.6398	13.6652	13.6906
0.540	13.7160	13.7414	13.7668	13.7922	13.8176	13.8430	13.8684	13.8938	13.9192	13.9446
0.550	13.9700	13.9954	14.0208	14.0462	14.0716	14.0970	14.1224	14.1478	14.1732	14.1986
0.560	14.2240	14.2494	14.2748	14.3002	14.3256	14.3510	14.3764	14.4018	14.4272	14.4526
0.570	14.4780	14.5034	14.5288	14.5542	14.5796	14.6050	14.6304	14.6558	14.6812	14.7066
0.580	14.7320	14.7574	14.7828	14.8082	14.8336	14.8590	14.8844	14.9098	14.9352	14.9606
0.590	14.9860	15.0114	15.0368	15.0622	15.0876	15.1130	15.1384	15.1638	15.1892	15.2146
0.600	15.2400	15.2654	15.2908	15.3162	15.3416	15.3670	15.3924	15.4178	15.4432	15.4686
0.610	15.4940	15.5194	15.5448	15.5702	15.5956	15.6210	15.6464	15.6718	15.6972	15.7226
0.620	15.7480	15.7734	15.7988	15.8242	15.8496	15.8750	15.9004	15.9258	15.9512	15.9766
0.630	16.0020	16.0274	16.0528	16.0782	16.1036	16.1290	16.1544	16.1798	16.2052	16.2306
0.640	16.2560	16.2814	16.3068	16.3322	16.3576	16.3830	16.4084	16.4338	16.4592	16.4846
0.650	16.5100	16.5354	16.5608	16.5862	16.6116	16.6370	16.6624	16.6878	16.7132	16.7386
0.660	16.7640	16.7894	16.8148	16.8402	16.8656	16.8910	16.9164	16.9418	16.9672	16.9926
0.670	17.0180	17.0434	17.0688	17.0942	17.1196	17.1450	17.1704	17.1958	17.2212	17.2466
0.680	17.2720	17.2974	17.3228	17.3482	17.3736	17.3990	17.4244	17.4498	17.4752	17.5006
0.690	17.5260	17.5514	17.5768	17.6022	17.6276	17.6530	17.6784	17.7038	17.7292	17.7546
0.700	17.7800	17.8054	17.8308	17.8562	17.8816	17.9070	17.9324	17.9578	17.9832	18.0086
0.710	18.0340	18.0594	18.0848	18.1102	18.1356	18.1610	18.1864	18.2118	18.2372	18.2626
0.720	18.2880	18.3134	18.3388	18.3642	18.3896	18.4150	18.4404	18.4658	18.4912	18.5166
0.730	18.5420	18.5674	18.5928	18.6182	18.6436	18.6690	18.6944	18.7198	18.7452	18.7706
0.740	18.7960	18.8214	18.8468	18.8722	18.8976	18.9230	18.9484	18.9738	18.9992	19.0246
0.750	19.0500	19.0754	19.1008	19.1262	19.1516	19.1770	19.2024	19.2278	19.2532	19.2786
0.760	19.3040	19.3294	19.3548	19.3802	19.4056	19.4310	19.4564	19.4818	19.5072	19.5326
0.770	19.5580	19.5834	19.6088	19.6342	19.6596	19.6850	19.7104	19.7358	19.7612	19.7866
0.780	19.8120	19.8374	19.8628	19.8882	19.9136	19.9390	19.9644	19.9898	20.0152	20.0406
0.790	20.0660	20.0914	20.1168	20.1422	20.1676	20.1930	20.2184	20.2438	20.2692	20.2946
0.800	20.3200	20.3454	20.3708	20.3962	20.4216	20.4470	20.4724	20.4978	20.5232	20.5486
0.810	20.5740	20.5994	20.6248	20.6502	20.6756	20.7010	20.7264	20.7518	20.7772	20.8026
0.820	20.8280	20.8534	20.8788	20.9042	20.9296	20.9550	20.9804	21.0058	21.0312	21.0566
0.830	21.0820	21.1074	21.1328	21.1582	21.1836	21.2090	21.2344	21.2598	21.2852	21.3106
0.840	21.3360	21.3614	21.3868	21.4122	21.4376	21.4630	21.4884	21.5138	21.5392	21.5646
0.850	21.5900	21.6154	21.6408	21.6662	21.6916	21.7170	21.7424	21.7678	21.7932	21.8186
0.860	21.8440	21.8694	21.8948	21.9202	21.9456	21.9710	21.9964	22.0218	22.0472	22.0726
0.870	22.0980	22.1234	22.1488	22.1742	22.1996	22.2250	22.2504	22.2758	22.3012	22.3266
0.880	22.3520	22.3774	22.4028	22.4282	22.4536	22.4790	22.5044	22.5298	22.5552	22.5806
0.890	22.6060	22.6314	22.6568	22.6822	22.7076	22.7330	22.7584	22.7838	22.8092	22.8346
0.900	22.8600	22.8854	22.9108	22.9362	22.9616	22.9870	23.0124	23.0378	23.0632	23.0886
0.910	23.1140	23.1394	23.1648	23.1902	23.2156	23.2410	23.2664	23.2918	23.3172	23.3426
0.920	23.3680	23.3934	23.4188	23.4442	23.4696	23.4950	23.5204	23.5458	23.5712	23.5966
0.930	23.6220	23.6474	23.6728	23.6982	23.7236	23.7490	23.7744	23.7998	23.8252	23.8506
0.940	23.8760	23.9014	23.9268	23.9522	23.9776	24.0030	24.0284	24.0538	24.0792	24.1046
0.950	24.1300	24.1554	24.1808	24.2062	24.2316	24.2570	24.2824	24.3078	24.3332	24.3586
0.960	24.3840	24.4094	24.4348	24.4602	24.4856	24.5110	24.5364	24.5618	24.5872	24.6126
0.970	24.6380	24.6634	24.6888	24.7142	24.7396	24.7650	24.7904	24.8158	24.8412	24.8666
0.980	24.8920	24.9174	24.9428	24.9682	24.9936	25.0190	25.0444	25.0698	25.0952	25.1206
0.990	25.1460	25.1714	25.1968	25.2222	25.2476	25.2730	25.2984	25.3238	25.3492	25.3746
1.000	25.4000	…	…	…	…	…	…	…	…	…

Use previous table to obtain whole inch equivalents to add to decimal equivalents above. All values given in this table are exact; figures to the right of the last place figures are all zeros.

Table 7. Millimeters to Inches
(Based on 1 inch = 25.4 millimeters, exactly)

Millimeters	0	1	2	3	4	5	6	7	8	9
	Inches									
0	...	0.03937	0.07874	0.11811	0.15748	0.19685	0.23622	0.27559	0.31496	0.35433
10	0.39370	0.43307	0.47244	0.51181	0.55118	0.59055	0.62992	0.66929	0.70866	0.74803
20	0.78740	0.82677	0.86614	0.90551	0.94488	0.98425	1.02362	1.06299	1.10236	1.14173
30	1.18110	1.22047	1.25984	1.29921	1.33858	1.37795	1.41732	1.45669	1.49606	1.53543
40	1.57480	1.61417	1.65354	1.69291	1.73228	1.77165	1.81102	1.85039	1.88976	1.92913
50	1.96850	2.00787	2.04724	2.08661	2.12598	2.16535	2.20472	2.24409	2.28346	2.32283
60	2.36220	2.40157	2.44094	2.48031	2.51969	2.55906	2.59843	2.63780	2.67717	2.71654
70	2.75591	2.79528	2.83465	2.87402	2.91339	2.95276	2.99213	3.03150	3.07087	3.11024
80	3.14961	3.18898	3.22835	3.26772	3.30709	3.34646	3.38583	3.42520	3.46457	3.50394
90	3.54331	3.58268	3.62205	3.66142	3.70079	3.74016	3.77953	3.81890	3.85827	3.89764
100	3.93701	3.97638	4.01575	4.05512	4.09449	4.13386	4.17323	4.21260	4.25197	4.29134
110	4.33071	4.37008	4.40945	4.44882	4.48819	4.52756	4.56693	4.60630	4.64567	4.68504
120	4.72441	4.76378	4.80315	4.84252	4.88189	4.92126	4.96063	5.00000	5.03937	5.07874
130	5.11811	5.15748	5.19685	5.23622	5.27559	5.31496	5.35433	5.39370	5.43307	5.47244
140	5.51181	5.55118	5.59055	5.62992	5.66929	5.70866	5.74803	5.78740	5.82677	5.86614
150	5.90551	5.94488	5.98425	6.02362	6.06299	6.10236	6.14173	6.18110	6.22047	6.25984
160	6.29921	6.33858	6.37795	6.41732	6.45669	6.49606	6.53543	6.57480	6.61417	6.65354
170	6.69291	6.73228	6.77165	6.81102	6.85039	6.88976	6.92913	6.96850	7.00787	7.04724
180	7.08661	7.12598	7.16535	7.20472	7.24409	7.28346	7.32283	7.36220	7.40157	7.44094
190	7.48031	7.51969	7.55906	7.59843	7.63780	7.67717	7.71654	7.75591	7.79528	7.83465
200	7.87402	7.91339	7.95276	7.99213	8.03150	8.07087	8.11024	8.14961	8.18898	8.22835
210	8.26772	8.30709	8.34646	8.38583	8.42520	8.46457	8.50394	8.54331	8.58268	8.62205
220	8.66142	8.70079	8.74016	8.77953	8.81890	8.85827	8.89764	8.93701	8.97638	9.01575
230	9.05512	9.09449	9.13386	9.17323	9.21260	9.25197	9.29134	9.33071	9.37008	9.40945
240	9.44882	9.48819	9.52756	9.56693	9.60630	9.64567	9.68504	9.72441	9.76378	9.80315
250	9.84252	9.88189	9.92126	9.96063	10.0000	10.0394	10.0787	10.1181	10.1575	10.1969
260	10.2362	10.2756	10.3150	10.3543	10.3937	10.4331	10.4724	10.5118	10.5512	10.5906
270	10.6299	10.6693	10.7087	10.7480	10.7874	10.8268	10.8661	10.9055	10.9449	10.9843
280	11.0236	11.0630	11.1024	11.1417	11.1811	11.2205	11.2598	11.2992	11.3386	11.3780
290	11.4173	11.4567	11.4961	11.5354	11.5748	11.6142	11.6535	11.6929	11.7323	11.7717
300	11.8110	11.8504	11.8898	11.9291	11.9685	12.0079	12.0472	12.0866	12.1260	12.1654
310	12.2047	12.2441	12.2835	12.3228	12.3622	12.4016	12.4409	12.4803	12.5197	12.5591
320	12.5984	12.6378	12.6772	12.7165	12.7559	12.7953	12.8346	12.8740	12.9134	12.9528
330	12.9921	13.0315	13.0709	13.1102	13.1496	13.1890	13.2283	13.2677	13.3071	13.3465
340	13.3858	13.4252	13.4646	13.5039	13.5433	13.5827	13.6220	13.6614	13.7008	13.7402
350	13.7795	13.8189	13.8583	13.8976	13.9370	13.9764	14.0157	14.0551	14.0945	14.1339
360	14.1732	14.2126	14.2520	14.2913	14.3307	14.3701	14.4094	14.4488	14.4882	14.5276
370	14.5669	14.6063	14.6457	14.6850	14.7244	14.7638	14.8031	14.8425	14.8819	14.9213
380	14.9606	15.0000	15.0394	15.0787	15.1181	15.1575	15.1969	15.2362	15.2756	15.3150
390	15.3543	15.3937	15.4331	15.4724	15.5118	15.5512	15.5906	15.6299	15.6693	15.7087
400	15.7480	15.7874	15.8268	15.8661	15.9055	15.9449	15.9843	16.0236	16.0630	16.1024
410	16.1417	16.1811	16.2205	16.2598	16.2992	16.3386	16.3780	16.4173	16.4567	16.4961
420	16.5354	16.5748	16.6142	16.6535	16.6929	16.7323	16.7717	16.8110	16.8504	16.8898
430	16.9291	16.9685	17.0079	17.0472	17.0866	17.1260	17.1654	17.2047	17.2441	17.2835
440	17.3228	17.3622	17.4016	17.4409	17.4803	17.5197	17.5591	17.5984	17.6378	17.6772
450	17.7165	17.7559	17.7953	17.8346	17.8740	17.9134	17.9528	17.9921	18.0315	18.0709
460	18.1102	18.1496	18.1890	18.2283	18.2677	18.3071	18.3465	18.3858	18.4252	18.4646
470	18.5039	18.5433	18.5827	18.6220	18.6614	18.7008	18.7402	18.7795	18.8189	18.8583
480	18.8976	18.9370	18.9764	19.0157	19.0551	19.0945	19.1339	19.1732	19.2126	19.2520
490	19.2913	19.3307	19.3701	19.4094	19.4488	19.4882	19.5276	19.5669	19.6063	19.6457

Table 7. Millimeters to Inches *(Continued)*
(Based on 1 inch = 25.4 millimeters, exactly)

Millimeters	0	1	2	3	4	5	6	7	8	9
	Inches									
500	19.6850	19.7244	19.7638	19.8031	19.8425	19.8819	19.9213	19.9606	20.0000	20.0394
510	20.0787	20.1181	20.1575	20.1969	20.2362	20.2756	20.3150	20.3543	20.3937	20.4331
520	20.4724	20.5118	20.5512	20.5906	20.6299	20.6693	20.7087	20.7480	20.7874	20.8268
530	20.8661	20.9055	20.9449	20.9843	21.0236	21.0630	21.1024	21.1417	21.1811	21.2205
540	21.2598	21.2992	21.3386	21.3780	21.4173	21.4567	21.4961	21.5354	21.5748	21.6142
550	21.6535	21.6929	21.7323	21.7717	21.8110	21.8504	21.8898	21.9291	21.9685	22.0079
560	22.0472	22.0866	22.1260	22.1654	22.2047	22.2441	22.2835	22.3228	22.3622	22.4016
570	22.4409	22.4803	22.5197	22.5591	22.5984	22.6378	22.6772	22.7165	22.7559	22.7953
580	22.8346	22.8740	22.9134	22.9528	22.9921	23.0315	23.0709	23.1102	23.1496	23.1890
590	23.2283	23.2677	23.3071	23.3465	23.3858	23.4252	23.4646	23.5039	23.5433	23.5827
600	23.6220	23.6614	23.7008	23.7402	23.7795	23.8189	23.8583	23.8976	23.9370	23.9764
610	24.0157	24.0551	24.0945	24.1339	24.1732	24.2126	24.2520	24.2913	24.3307	24.3701
620	24.4094	24.4488	24.4882	24.5276	24.5669	24.6063	24.6457	24.6850	24.7244	24.7638
630	24.8031	24.8425	24.8819	24.9213	24.9606	25.0000	25.0394	25.0787	25.1181	25.1575
640	25.1969	25.2362	25.2756	25.3150	25.3543	25.3937	25.4331	25.4724	25.5118	25.5512
650	25.5906	25.6299	25.6693	25.7087	25.7480	25.7874	25.8268	25.8661	25.9055	25.9449
660	25.9843	26.0236	26.0630	26.1024	26.1417	26.1811	26.2205	26.2598	26.2992	26.3386
670	26.3780	26.4173	26.4567	26.4961	26.5354	26.5748	26.6142	26.6535	26.6929	26.7323
680	26.7717	26.8110	26.8504	26.8898	26.9291	26.9685	27.0079	27.0472	27.0866	27.1260
690	27.1654	27.2047	27.2441	27.2835	27.3228	27.3622	27.4016	27.4409	27.4803	27.5197
700	27.5591	27.5984	27.6378	27.6772	27.7165	27.7559	27.7953	27.8346	27.8740	27.9134
710	27.9528	27.9921	28.0315	28.0709	28.1102	28.1496	28.1890	28.2283	28.2677	28.3071
720	28.3465	28.3858	28.4252	28.4646	28.5039	28.5433	28.5827	28.6220	28.6614	28.7008
730	28.7402	28.7795	28.8189	28.8583	28.8976	28.9370	28.9764	29.0157	29.0551	29.0945
740	29.1339	29.1732	29.2126	29.2520	29.2913	29.3307	29.3701	29.4094	29.4488	29.4882
750	29.5276	29.5669	29.6063	29.6457	29.6850	29.7244	29.7638	29.8031	29.8425	29.8819
760	29.9213	29.9606	30.0000	30.0394	30.0787	30.1181	30.1575	30.1969	30.2362	30.2756
770	30.3150	30.3543	30.3937	30.4331	30.4724	30.5118	30.5512	30.5906	30.6299	30.6693
780	30.7087	30.7480	30.7874	30.8268	30.8661	30.9055	30.949	30.9843	31.0236	31.0630
790	31.1024	31.1417	31.1811	31.2205	31.2598	31.2992	31.3386	31.3780	31.4173	31.4567
800	31.4961	31.5354	31.5748	31.6142	31.6535	31.6929	31.7323	31.7717	31.8110	31.8504
810	31.8898	31.9291	31.9685	32.0079	32.0472	32.0866	32.1260	32.1654	32.2047	32.2441
820	32.2835	32.3228	32.3622	32.4016	32.4409	32.4803	32.5197	32.5591	32.5984	32.6378
830	32.6772	32.7165	32.7559	32.7953	32.8346	32.8740	32.9134	32.9528	32.9921	33.0315
840	33.0709	33.1102	33.1496	33.1890	33.2283	33.2677	33.3071	33.3465	33.3858	33.4252
850	33.4646	33.5039	33.5433	33.5827	33.6220	33.6614	33.7008	33.7402	33.7795	33.8189
860	33.8583	33.8976	33.9370	33.9764	34.0157	34.0551	34.0945	34.1339	34.1732	34.2126
870	34.2520	34.2913	34.3307	34.3701	34.4094	34.4488	34.4882	34.5276	34.5669	34.6063
880	34.6457	34.6850	34.7244	34.7638	34.8031	34.8425	34.8819	34.9213	34.9606	35.0000
890	35.0394	35.0787	35.1181	35.1575	35.1969	35.2362	35.2756	35.3150	35.3543	35.3937
900	35.4331	35.4724	35.5118	35.5512	35.5906	35.6299	35.6693	35.7087	35.7480	35.7874
910	35.8268	35.8661	35.9055	35.9449	35.9843	36.0236	36.0630	36.1024	36.1417	36.1811
920	36.2205	36.2598	36.2992	36.3386	36.3780	36.4173	36.4567	36.4961	36.5354	36.5748
930	36.6142	36.6535	36.6929	36.7323	36.7717	36.8110	36.8504	36.8898	36.9291	36.9685
940	37.0079	37.0472	37.0866	37.1260	37.1654	37.2047	37.2441	37.2835	37.3228	37.3622
950	37.4016	37.409	37.4803	37.5197	37.5591	37.5984	37.6378	37.6772	37.7165	37.7559
960	37.7953	37.8346	37.8740	37.9134	37.9528	37.9921	38.0315	38.0709	38.1102	38.1496
970	38.1800	38.2283	38.2677	38.3071	38.3465	38.3858	38.4252	38.4646	38.5039	38.5433
980	38.5827	38.6220	38.6614	38.7008	38.7402	38.7795	38.8189	38.8583	38.8976	38.9370
990	38.9764	39.0157	39.0551	39.0945	39.1339	39.1732	39.2126	39.2520	39.2913	39.3307
1000	39.3701	…	…	…	…	…	…	…	…	…

Table 8. Microinches to Micrometers (microns)
(Based on 1 microinch = 0.0254 micrometers, exactly)

Micro-inches	0	1	2	3	4	5	6	7	8	9
	Micrometers (microns)									
0	…	0.025	0.051	0.076	0.102	0.127	0.152	0.178	0.203	0.229
10	0.254	0.279	0.305	0.330	0.356	0.381	0.406	0.432	0.457	0.483
20	0.508	0.533	0.559	0.584	0.610	0.635	0.660	0.686	0.711	0.737
30	0.762	0.787	0.813	0.838	0.864	0.889	0.914	0.940	0.965	0.991
40	1.016	1.041	1.067	1.092	1.118	1.143	1.168	1.194	1.219	1.245
50	1.270	1.295	1.321	1.346	1.372	1.397	1.422	1.448	1.473	1.499
60	1.524	1.549	1.575	1.600	1.626	1.651	1.676	1.702	1.727	1.753
70	1.778	1.803	1.829	1.854	1.880	1.905	1.930	1.956	1.981	2.007
80	2.032	2.057	2.083	2.108	2.134	2.159	2.184	2.210	2.235	2.261
90	2.286	2.311	2.337	2.362	2.388	2.413	2.438	2.464	2.489	2.515
100	2.540	2.565	2.591	2.616	2.642	2.667	2.692	2.718	2.743	2.769
110	2.794	2.819	2.845	2.870	2.896	2.921	2.946	2.972	2.997	3.023
120	3.048	3.073	3.099	3.124	3.150	3.175	3.200	3.226	3.251	3.277
130	3.302	3.327	3.353	3.378	3.404	3.429	3.454	3.480	3.505	3.531
140	3.556	3.581	3.607	3.632	3.658	3.683	3.708	3.734	3.759	3.785
150	3.810	3.835	3.861	3.886	3.912	3.937	3.962	3.988	4.013	4.039
160	4.064	4.089	4.115	4.140	4.166	4.191	4.216	4.242	4.267	4.293
170	4.318	4.343	4.369	4.394	4.420	4.445	4.470	4.496	4.521	4.547
180	4.572	4.597	4.623	4.648	4.674	4.699	4.724	4.750	4.775	4.801
190	4.826	4.851	4.877	4.902	4.928	4.953	4.978	5.004	5.029	5.055
200	5.080	5.105	5.131	5.156	5.182	5.207	5.232	5.258	5.283	5.309
210	5.334	5.359	5.385	5.410	5.436	5.461	5.486	5.512	5.537	5.563
220	5.588	5.613	5.639	5.664	5.690	5.715	5.740	5.766	5.791	5.817
230	5.842	5.867	5.893	5.918	5.944	5.969	5.994	6.020	6.045	6.071
240	6.096	6.121	6.147	6.172	6.198	6.223	6.248	6.274	6.299	6.325
250	6.350	6.375	6.401	6.426	6.452	6.477	6.502	6.528	6.553	6.579
260	6.604	6.629	6.655	6.680	6.706	6.731	6.756	6.782	6.807	6.833
270	6.858	6.883	6.909	6.934	6.960	6.985	7.010	7.036	7.061	7.087
280	7.112	7.137	7.163	7.188	7.214	7.239	7.264	7.290	7.315	7.341
290	7.366	7.391	7.417	7.442	7.468	7.493	7.518	7.544	7.569	7.595

The following short table permits conversion of microinches to micrometers for ranges higher than in the main table given above. Appropriate quantities chosen from both tables are simply added to obtain the higher converted value:

μin.	μm	μin.	μm	μin.	μm	μin.	μm	μin.	μm
300	7.620	900	22.860	1500	38.100	2100	53.340	2700	68.580
600	15.240	1200	30.480	1800	45.720	2400	60.960	3000	76.200

Example: Convert 1375 μin. to μm:

From above table:	1200 μin.	=	30.480 μm
From main table:	175 μin.	=	4.445 μm
	1375 μin.	=	34.925 μm

Table 9. Micrometers (microns) to Microinches
(Based on 1 microinch = 0.0254 micrometers, exactly)

Micrometers Microns	0	0.01	0.02	0.03	0.04	0.05	0.06	0.07	0.08	0.09
	Microinches									
0	...	0.4	0.8	1.2	1.6	2.0	2.4	2.8	3.1	3.5
0.10	3.9	4.3	4.7	5.1	5.5	5.9	6.3	6.7	7.1	7.5
0.20	7.9	8.3	8.7	9.1	9.4	9.8	10.2	10.6	11.0	11.4
0.30	11.8	12.2	12.6	13.0	13.4	13.8	14.2	14.6	15.0	15.4
0.40	15.7	16.1	16.5	16.9	17.3	17.7	18.1	18.5	18.9	19.3
0.50	19.7	20.1	20.5	20.9	21.3	21.7	22.0	22.4	22.8	23.2
0.60	23.6	24.0	24.4	24.8	25.2	25.6	26.0	26.4	26.8	27.2
0.70	27.6	28.0	28.3	28.7	29.1	29.5	29.9	30.3	30.7	31.1
0.80	31.5	31.9	32.3	32.7	33.1	33.5	33.9	34.3	34.6	35.0
0.90	35.4	35.8	36.2	36.6	37.0	37.4	37.8	38.2	38.6	39.0
1.00	39.4	39.8	40.2	40.6	40.9	41.3	41.7	42.1	42.5	42.9
1.10	43.3	43.7	44.1	44.5	44.9	45.3	45.7	46.1	46.5	46.9
1.20	47.2	47.6	48.0	48.4	48.8	49.2	49.6	50.0	50.4	50.8
1.30	51.2	51.6	52.0	52.4	52.8	53.1	53.5	53.9	54.3	54.7
1.40	55.1	55.5	55.9	56.3	56.7	57.1	57.5	57.9	58.3	58.7
1.50	59.1	59.4	59.8	60.2	60.6	61.0	61.4	61.8	62.2	62.6
1.60	63.0	63.4	63.8	64.2	64.6	65.0	65.4	65.7	66.1	66.5
1.70	66.9	67.3	67.7	68.1	68.5	68.9	69.3	69.7	70.1	70.5
1.80	70.9	71.3	71.7	72.0	72.4	72.8	73.2	73.6	74.0	74.4
1.90	74.8	75.2	75.6	76.0	76.4	76.8	75.0	77.6	78.0	78.3
2.00	78.7	79.1	79.5	79.9	80.3	80.7	81.1	81.5	81.9	82.3
2.10	82.7	83.1	83.5	83.9	84.3	84.6	85.0	85.4	85.8	86.2
2.20	86.6	87.0	87.4	87.8	88.2	88.6	89.0	89.4	89.8	90.2
2.30	90.6	90.9	91.3	91.7	92.1	92.5	92.9	93.3	93.7	94.1
2.40	94.5	94.9	95.3	95.7	96.1	96.5	96.9	97.2	97.6	98.0
2.50	98.4	98.8	99.2	99.6	100.0	100.4	100.8	101.2	101.6	102.0
2.60	102.4	102.8	103.1	103.5	103.9	104.3	104.7	105.1	105.5	105.9
2.70	106.3	106.7	107.1	107.5	107.9	108.3	108.7	109.1	109.4	109.8
2.80	110.2	110.6	111.0	111.4	111.8	112.2	112.6	113.0	113.4	113.8
2.90	114.2	114.6	115.0	115.4	115.7	116.1	116.5	116.9	117.3	117.7
3.00	118.1	118.5	118.9	119.3	119.7	120.1	120.5	120.9	121.3	121.7
3.10	122.0	122.4	122.8	123.2	123.6	124.0	124.4	124.8	125.2	125.6
3.20	126.0	126.4	126.8	127.2	127.6	128.0	128.3	128.7	129.1	129.5
3.30	129.9	130.3	130.7	131.1	131.5	131.9	132.3	132.7	133.1	133.5
3.40	133.9	134.3	134.6	135.0	135.4	135.8	136.2	136.6	137.0	137.4
3.50	137.8	138.2	138.6	139.0	139.4	139.8	140.2	140.6	140.9	141.3
3.60	141.7	142.1	142.5	142.9	143.3	143.7	144.1	144.5	144.9	145.3
3.70	145.7	146.1	146.5	146.9	147.2	147.6	148.0	148.4	148.8	149.2
3.80	149.6	150.0	150.4	150.8	151.2	151.6	152.0	152.4	152.8	153.1
3.90	153.5	153.9	154.3	154.7	155.1	155.5	155.9	156.3	156.7	157.1
4.00	157.5	157.9	158.3	158.7	159.1	159.4	159.8	160.2	160.6	161.0
4.10	161.4	161.8	162.2	162.6	163.0	163.4	163.8	164.2	164.6	165.0
4.20	165.4	165.7	166.1	166.5	166.9	167.3	167.7	168.1	168.5	168.9
4.30	169.3	169.7	170.1	170.5	170.9	171.3	171.7	172.0	172.4	172.8
4.40	173.2	173.6	174.0	174.4	174.8	175.2	175.6	176.0	176.4	176.8
4.50	177.2	177.6	178.0	178.3	178.7	179.1	179.5	179.9	180.3	180.7
4.60	181.1	181.5	181.9	182.3	182.7	183.1	183.5	183.9	184.3	184.6
4.70	185.0	185.4	185.8	186.2	186.6	187.0	187.4	187.8	188.2	188.6
4.80	189.0	189.4	189.8	190.2	190.6	190.9	191.3	191.7	192.1	192.5
4.90	192.9	193.3	193.7	194.1	194.5	194.9	195.3	195.7	196.1	196.5
5.00	196.9	197.2	197.6	198.0	198.4	198.8	199.2	199.6	200.0	200.4

The table given below can be used with the preceding main table to obtain higher converted values, simply by adding appropriate quantities chosen from each table:

μm	μin.	μm	μin.	μm	μin.	μm	μin.	μm	μin.
10	393.7	20	787.4	30	1,181.1	40	1,574.8	50	1,968.5
15	590.6	25	984.3	35	1,378.0	45	1,771.7	55	2,165.4

Example: Convert 23.55 μm to μin.:

From above table:	20.00 μm	=	787.4 μin.
From main table:	3.55 μm	=	139.8 μin.
	23.55 μm	=	927.2 μin.

Thermal Energy Or Heat

Thermometer Scales.—There are two thermometer scales in general use: the Fahrenheit (F), which is used in the United States and in other countries still using the English system of units, and the Celsius (C) or Centigrade used throughout the rest of the world.

In the Fahrenheit thermometer, the freezing point of water is marked at 32 degrees on the scale and the boiling point, at atmospheric pressure, at 212 degrees. The distance between these two points is divided into 180 degrees. On the Celsius scale, the freezing point of water is at 0 degrees and the boiling point at 100 degrees. The following formulas may be used for converting temperatures given on any one of the scales to the other scale:

$$\text{Degrees Fahrenheit} = \frac{9 \times \text{degrees C}}{5} + 32$$

$$\text{Degrees Celsius} = \frac{5 \times (\text{degrees F} - 32)}{9}$$

Tables on the pages that follow can be used to convert degrees Celsius into degrees Fahrenheit or vice versa. In the event that the conversions are not covered in the tables, use those applicable portions of the formulas given above for converting.

Absolute Temperature and Absolute Zero.—A point has been determined on the thermometer scale, by theoretical considerations, that is called the absolute zero and beyond which a further decrease in temperature is inconceivable. This point is located at −273.2 degrees Celsius or −459.7 degrees F. A temperature reckoned from this point, instead of from the zero on the ordinary thermometers, is called absolute temperature. Absolute temperature in degrees C is known as "degrees Kelvin" or the "Kelvin scale" (K) and absolute temperature in degrees F is known as "degrees Rankine" or the "Rankine scale" (R).

$$\text{Degrees Kelvin} = \text{degrees C} + 273.2$$

$$\text{Degrees Rankine} = \text{degrees F} + 459.7$$

Measures of the Quantity of Thermal Energy.—The unit of quantity of thermal energy used in the United States is the British thermal unit, which is the quantity of heat or thermal energy required to raise the temperature of one pound of pure water one degree F. (American National Standard abbreviation, Btu; conventional British symbol, B.Th.U.) The French thermal unit, or *kilogram calorie,* is the quantity of heat or thermal energy required to raise the temperature of one kilogram of pure water one degree C. One kilogram calorie = 3.968 British thermal units = 1000 gram calories. The number of foot-pounds of mechanical energy equivalent to one British thermal unit is called the *mechanical equivalent of heat,* and equals 778 foot-pounds.

In the modern metric or SI system of units, the unit for thermal energy is the *joule* (J); a commonly used multiple being the kilojoule (kJ), or 1000 joules. See page 2520 for an explanation of the SI System. One kilojoule = 0.9478 Btu. Also in the SI System, the *watt* (W), equal to joule per second (J/s), is used for power, where one watt = 3.412 Btu per hour.

Fahrenheit-Celsius (Centigrade) Conversion.—A simple way to convert a Fahrenheit temperature reading into a Celsius temperature reading or vice versa is to enter the accompanying table in the center or boldface column of figures. These figures refer to the temperature in either Fahrenheit or Celsius degrees. If it is desired to convert from Fahrenheit to Celsius degrees, consider the center column as a table of Fahrenheit temperatures and read the corresponding Celsius temperature in the column at the left. If it is desired to convert from Celsius to Fahrenheit degrees, consider the center column as a table of Celsius values, and read the corresponding Fahrenheit temperature on the right.

Interpolation Factors

deg C		deg F	deg C		deg F
0.56	**1**	1.8	3.33	**6**	10.8
1.11	**2**	3.6	3.89	**7**	12.6
1.67	**3**	5.4	4.44	**8**	14.4
2.22	**4**	7.2	5.00	**9**	16.2
2.78	**5**	9.0	5.56	**10**	18.0

Interpolation factors are given for use with that portion of the table in which the center column advances in increments of 10. To illustrate, suppose it is desired to find the Fahrenheit equivalent of 314 degrees C. The equivalent of 310 degrees C, found in the body of the main table, is seen to be 590.0 degrees F. The Fahrenheit equivalent of a 4-degree C difference is seen to be 7.2, as read in the table of interpolating factors. The answer is the sum or 597.2 degrees F.

Fahrenheit-Celsius (Centigrade) Conversion Table

deg C		deg F	deg C		deg F	deg C		deg F	deg C		deg F
−273	**−459.4**	...	−62	**−80**	−112	−1.1	**30**	86.0	20.6	**69**	156.2
−268	**−450**	...	−57	**−70**	−94	−0.6	**31**	87.8	21.1	**70**	158.0
−262	**−440**	...	−51	**−60**	−76	0−	**32**	89.6	21.7	**71**	159.8
−257	**−430**	...	−46	**−50**	−58	0.6	**33**	91.4	22.2	**72**	161.6
−251	**−420**	...	−40	**−40**	−40	1.1	**34**	93.2	22.8	**73**	163.4
−246	**−410**	...	−34	**−30**	−22	1.7	**35**	95.0	23.3	**74**	165.2
−240	**−400**	...	−29	**−20**	−4	2.2	**36**	96.8	23.9	**75**	167.0
−234	**−390**	...	−23	**−10**	14	2.7	**37**	98.6	24.4	**76**	168.8
−229	**−380**	...	−17.8	**0**	32−	3.3	**38**	100.4	25.0	**77**	170.6
−223	**−370**	...	−17.2	**1**	33.8	3.9	**39**	102.2	25.6	**78**	172.4
−218	**−360**	...	−16.7	**2**	35.6	4.4	**40**	104.0	26.1	**79**	174.2
−212	**−350**	...	−16.1	**3**	37.4	5.0	**41**	105.8	26.7	**80**	176.0
−207	**−340**	...	−15.6	**4**	39.2	5.6	**42**	107.6	27.2	**81**	177.8
−201	**−330**	...	−15.0	**5**	41.0	6.1	**43**	109.4	27.8	**82**	179.6
−196	**−320**	...	−14.4	**6**	42.8	6.7	**44**	111.2	28.3	**83**	181.4
−190	**−310**	...	−13.9	**7**	44.6	7.2	**45**	113.0	28.9	**84**	183.2
−184	**−300**	...	−13.3	**8**	46.4	7.8	**46**	114.8	29.4	**85**	185.0
−179	**−290**	...	−12.8	**9**	48.2	8.3	**47**	116.6	30.0	**86**	186.8
−173	**−280**	...	−12.2	**10**	50.0	8.9	**48**	118.4	30.6	**87**	188.6
−169	**−273**	−459.4	−11.7	**11**	51.8	9.4	**49**	120.2	31.1	**88**	190.4
−168	**−270**	−454	−11.1	**12**	53.6	10.0	**50**	122.0	31.7	**89**	192.2
−162	**−260**	−436	−10.6	**13**	55.4	10.6	**51**	123.8	32.2	**90**	194.0
−157	**−250**	−418	−10.0	**14**	57.2	11.1	**52**	125.6	32.8	**91**	195.8
−151	**−240**	−400	−9.4	**15**	59.0	11.7	**53**	127.4	33.3	**92**	197.6
−146	**−230**	−382	−8.9	**16**	60.8	12.2	**54**	129.2	33.9	**93**	199.4
−140	**−220**	−364	−8.3	**17**	62.6	12.8	**55**	131.0	34.4	**94**	201.2
−134	**−210**	−346	−7.8	**18**	64.4	13.3	**56**	132.8	35.0	**95**	203.0
−129	**−200**	−328	−7.2	**19**	66.2	13.9	**57**	134.6	35.6	**96**	204.8
−123	**−190**	−310	−6.7	**20**	68.0	14.4	**58**	136.4	36.1	**97**	206.6
−118	**−180**	−292	−6.1	**21**	69.8	15.0	**59**	138.2	36.7	**98**	208.4
−112	**−170**	−274	−5.6	**22**	71.6	15.6	**60**	140.0	37.2	**99**	210.2
−107	**−160**	−256	−5.0	**23**	73.4	16.1	**61**	141.8	37.8	**100**	212.0
−101	**−150**	−238	−4.4	**24**	75.2	16.7	**62**	143.6	38.3	**101**	213.8
−96	**−140**	−220	−3.9	**25**	77.0	17.2	**63**	145.4	38.9	**102**	215.6
−90	**−130**	−202	−3.3	**26**	78.8	17.8	**64**	147.2	39.4	**103**	217.4
−84	**−120**	−184	−2.8	**27**	80.6	18.3	**65**	149.0	40.0	**104**	219.2
−79	**−110**	−166	−2.2	**28**	82.4	18.9	**66**	150.8	40.6	**105**	221.0
−73	**−100**	−148	−1.7	**29**	84.2	19.4	**67**	152.6	41.1	**106**	222.8
−68	**−90**	−130	−8.3	**17**	62.6	20.0	**68**	154.4	41.7	**107**	224.6

Fahrenheit-Celsius (Centigrade) Conversion Table *(Continued)*

deg C		deg F	deg C		deg F	deg C		deg F	deg C		deg F
42.2	**108**	226.4	69.4	**157**	314.6	96.7	**206**	402.8	337.8	**640**	1184
42.8	**109**	228.2	70.0	**158**	316.4	97.2	**207**	404.6	343.3	**650**	1202
43.3	**110**	230.0	70.6	**159**	318.2	97.8	**208**	406.4	348.9	**660**	1220
43.9	**111**	231.8	71.1	**160**	320.0	98.3	**209**	408.2	354.4	**670**	1238
44.4	**112**	233.6	71.7	**161**	321.8	98.9	**210**	410.0	360.0	**680**	1256
45.0	**113**	235.4	72.2	**162**	323.6	99.4	**211**	411.8	365.6	**690**	1274
45.6	**114**	237.2	72.8	**163**	325.4	100.0	**212**	413.6	371.1	**700**	1292
46.1	**115**	239.0	73.3	**164**	327.2	104.4	**220**	428.0	376.7	**710**	1310
46.7	**116**	240.8	73.9	**165**	329.0	110.0	**230**	446.0	382.2	**720**	1328
47.2	**117**	242.6	74.4	**166**	330.8	115.6	**240**	464.0	387.8	**730**	1346
47.8	**118**	244.4	75.0	**167**	332.6	121.1	**250**	482.0	393.3	**740**	1364
48.3	**119**	246.2	75.6	**168**	334.4	126.7	**260**	500.0	398.9	**750**	1382
48.9	**120**	248.0	76.1	**169**	336.2	132.2	**270**	518.0	404.4	**760**	1400
49.4	**121**	249.8	76.7	**170**	338.0	137.8	**280**	536.0	410.0	**770**	1418
50.0	**122**	251.6	77.2	**171**	339.8	143.3	**290**	554.0	415.6	**780**	1436
50.6	**123**	253.4	77.8	**172**	341.6	148.9	**300**	572.0	421.1	**790**	1454
51.1	**124**	255.2	78.3	**173**	343.4	154.4	**310**	590.0	426.7	**800**	1472
51.7	**125**	257.0	78.9	**174**	345.2	160.0	**320**	608.0	432.2	**810**	1490
52.2	**126**	258.8	79.4	**175**	347.0	165.6	**330**	626.0	437.8	**820**	1508
52.8	**127**	260.6	80.0	**176**	348.8	171.1	**340**	644.0	443.3	**830**	1526
53.3	**128**	262.4	80.6	**177**	350.6	176.7	**350**	662.0	448.9	**840**	1544
53.9	**129**	264.2	81.1	**178**	352.4	182.2	**360**	680.0	454.4	**850**	1562
54.4	**130**	266.0	81.7	**179**	354.2	187.8	**370**	698.0	460.0	**860**	1580
55.0	**131**	267.8	82.2	**180**	356.0	193.3	**380**	716.0	465.6	**870**	1598
55.6	**132**	269.6	82.8	**181**	357.8	198.9	**390**	734.0	471.1	**880**	1616
56.1	**133**	271.4	83.3	**182**	359.6	204.4	**400**	752.0	476.7	**890**	1634
56.7	**134**	273.2	83.9	**183**	361.4	210.0	**410**	770.0	482.2	**900**	1652
57.2	**135**	275.0	84.4	**184**	363.2	215.6	**420**	788	487.8	**910**	1670
57.8	**136**	276.8	85.0	**185**	365.0	221.1	**430**	806	493.3	**920**	1688
58.3	**137**	278.6	85.6	**186**	366.8	226.7	**440**	824	498.9	**930**	1706
58.9	**138**	280.4	86.1	**187**	368.6	232.2	**450**	842	504.4	**940**	1724
59.4	**139**	282.2	86.7	**188**	370.4	237.8	**460**	860	510.0	**950**	1742
60.0	**140**	284.0	87.2	**189**	372.2	243.3	**470**	878	515.6	**960**	1760
60.6	**141**	285.8	87.8	**190**	374.0	248.9	**480**	896	521.1	**970**	1778
61.1	**142**	287.6	88.3	**191**	375.8	254.4	**490**	914	526.7	**980**	1796
61.7	**143**	289.4	88.9	**192**	377.6	260.0	**500**	932	532.2	**990**	1814
62.2	**144**	291.2	89.4	**193**	379.4	265.6	**510**	950	537.8	**1000**	1832
62.8	**145**	293.0	90.0	**194**	381.2	271.1	**520**	968	565.6	**1050**	1922
63.3	**146**	294.8	90.6	**195**	383.0	276.7	**530**	986	593.3	**1100**	2012
63.9	**147**	296.6	91.1	**196**	384.8	282.2	**540**	1004	621.1	**1150**	2102
64.4	**148**	298.4	91.7	**197**	386.6	287.8	**550**	1022	648.9	**1200**	2192
65.0	**149**	300.2	92.2	**198**	388.4	293.3	**560**	1040	676.7	**1250**	2282
65.6	**150**	302.0	92.8	**199**	390.2	298.9	**570**	1058	704.4	**1300**	2372
66.1	**151**	303.8	93.3	**200**	392.0	304.4	**580**	1076	732.2	**1350**	2462
66.7	**152**	305.6	93.9	**201**	393.8	310.0	**590**	1094	760.0	**1400**	2552
67.2	**153**	307.4	94.4	**202**	395.6	315.6	**600**	1112	787.8	**1450**	2642
67.8	**154**	309.2	95.0	**203**	397.4	321.1	**610**	1130	815.6	**1500**	2732
68.3	**155**	311.0	95.6	**204**	399.2	326.7	**620**	1148	1093.9	**2000**	3632
68.9	**156**	312.8	96.1	**205**	401.0	332.2	**630**	1166	1648.9	**3000**	5432

Above 1000 in the center column, the table increases in increments of 50. To convert 1462 degrees F to Celsius, for instance, add to the Celsius equivalent of 1400 degrees F ten times the interpolation factor for 6 and the interpolation factor for 2 or 760.0 + 33.3 + 1.11, which equals 794.4.

MATHEMATICS

Table 1. Mathematical Signs and Commonly Used Abbreviations

Sign	Meaning		Sign	Meaning
$+$	Plus (sign of addition)		π	Pi (3.1416)
$+$	Positive		Σ	Sigma (sign of summation)
$-$	Minus (sign of subtraction)		ω	Omega (angles measured in radians)
$-$	Negative		g	Acceleration due to gravity (32.16 ft. per sec. per sec.)
$\pm$ ($\mp$)	Plus or minus (minus or plus)		i (or j)	Imaginary quantity ($\sqrt{-1}$)
$\times$	Multiplied by (multiplication sign)		sin	Sine
$\cdot$	Multiplied by (multiplication sign)		cos	Cosine
$\div$	Divided by (division sign)		tan	Tangent
$/$	Divided by (division sign)		cot	Cotangent
$:$	Is to (in proportion)		sec	Secant
$=$	Equals		csc	Cosecant
$\neq$	Is not equal to		vers	Versed sine
$\equiv$	Is identical to		covers	Coversed sine
$::$	Equals (in proportion)		$\sin^{-1} a$ $\arcsin a$	Arc the sine of which is a
$\cong$	Approximately equals		$(\sin a)^{-1}$	Reciprocal of $\sin a$ ($1 \div \sin a$)
$\approx$			$\sin^n x$	nth power of $\sin x$
$>$	Greater than		$\sinh x$	Hyperbolic sine of x
$<$	Less than		$\cosh x$	Hyperbolic cosine of x
$\geq$	Greater than or equal to		Δ	Delta (increment of)
$\leq$	Less than or equal to		δ	Delta (variation of)
$\rightarrow$	Approaches as a limit		d	Differential (in calculus)
$\propto$	Varies directly as		∂	Partial differentiation (in calculus)
$\therefore$	Therefore		$\int$	Integral (in calculus)
$\sqrt{\ }$	Square root		$\int_b^a$	Integral between the limits a and b
$\sqrt[3]{\ }$	Cube root		!	$5! = 1 \times 2 \times 3 \times 4 \times 5$ (Factorial)
$\sqrt[4]{\ }$	4th root		$\angle$	Angle
$\sqrt[n]{\ }$	nth root		∟	Right angle
a^2	a squared (2nd power of a)		$\perp$	Perpendicular to
a^3	a cubed (3rd power of a)		$\triangle$	Triangle
a^4	4th power of a		○	Circle
a^n	nth power of a		▱	Parallelogram
a^{-n}	$1 \div a^n$		$^\circ$	Degree (circular arc or temperature)
$\frac{1}{n}$	Reciprocal value of n		$'$	Minutes or feet
log	Logarithm		$''$	Seconds or inches
$\log_e$	Natural or Napierian logarithm		a'	a prime
ln			a''	a double prime
e	Base of natural logarithms (2.71828)		a_1	a sub one
lim	Limit value (of an expression)		a_2	a sub two
∞	Infinity		a_n	a sub n
α	Alpha	commonly used to denote angles (Alpha–Phi)	( )	Parentheses
β	Beta		[ ]	Brackets
γ	Gamma		{ }	Braces
θ	Theta			
ϕ	Phi			
μ	Mu (coefficient of friction)			

Arithmetic*

Common Fractions.— Common fractions consist of two basic parts, a denominator, or bottom number, and a numerator, or top number. The denominator shows how many parts the whole unit has been divided into. The numerator indicates the number of parts of the whole that are being considered. A fraction having a value of $5/32$, means the whole unit has been divided into 32 equal parts and 5 of these parts are considered in the value of the fraction.

The following are the basic facts, rules, and definitions concerning common fractions.

1) A proper fraction is a common fraction having a numerator smaller than its denominator.

Example: $1/4$, $1/2$, $3/4$, $15/16$, $23/32$, and $47/64$.

2) An improper fraction is a common fraction having a numerator larger than its denominator

Example: $3/2$, $5/4$, $10/8$, $19/16$, $45/32$, and $84/64$.

3) A common fraction having the same numerator and denominator is equal to 1.

Example: $2/2$, $4/4$, $8/8$, $16/16$, $32/32$, and $64/64$ all equal 1.

4) A reducible fraction is a common fraction that can be reduced to lower terms.

Example: $2/4$, $6/8$, $12/16$, $28/32$, and $62/64$.

5) To reduce a common fraction to lower terms, divide both the numerator and the denominator by the same number.

Example: $24/32$: $24/32 \div 8/8 = 3/8$ or $6/8$: $6/8 \div 2/2 = 3/4$.

6) A fraction may also be converted to higher terms by multiplying the numerator and denominator by the same number.

Example: $1/4$ in 16ths = : $1/4 \times 4/4 = 4/16$ and $3/8$ in 32nds = $3/8 \times 4/4 = 12/32$.

7) A least common denominator is the smallest denominator value that is evenly divisible by the other denominator values in the problem.

Example: $1/2 + 1/4 + 3/8$: The least common denominator is 8.

8) A mixed number is a combination of a whole number and a common fraction.

Example: $2\,1/2$, $1\,7/8$, $3\,15/16$ and $1\,9/32$.

9) To convert mixed numbers to improper fractions, multiply the whole number by the denominator and add the numerator to obtain the new numerator. The denominator remains the same.

Example: $2\,1/2 : 2 \times 2 + 1 = 5/2$ or $3\,7/16 : 3 \times 16 + 7 = 55/16$.

10) To convert a whole number to an improper fractions place the whole number over 1.

Example: $4 = 4/1$, or $3 = 3/1$

11) To change a whole number to a common fraction with a specific denominator value, convert the whole number to a fraction and multiply the numerator and denominator by the desired denominator value.

Example: 4 in 16ths = $4/1 \times 16/16 = 64/16$ or 3 in 32nds = $3/1 \times 32/32 = 96/32$.

12) To convert an improper fraction to a mixed number, divide the numerator by the denominator and reduce the remaining fraction to its lowest terms.

Example: $17/8 = 17 \div 8 = 2\,1/8$ or $26/16 = 26 \div 16 = 1\,10/16 = 1\,5/8$.

* Portions of this section have been reprinted from *Print Reading for Industry*, by permission of South-Western Publishing Co., Cincinnati, OH.

Adding Fractions and Mixed Numbers.— *To add common fractions:*

1) Find and convert to the least common denominator.

2) Add the numerators.

3) Convert the answer to a mixed number, if necessary.

4) Reduce the fraction to its lowest terms.

Example:

$$\frac{1}{4} + \frac{3}{16} + \frac{7}{8} =$$

$$\frac{1}{4} = \frac{4}{16}$$

$$\frac{3}{16} = \frac{3}{16}$$

$$\frac{7}{8} = \frac{14}{16}$$

$$\frac{21}{16} = 1\frac{5}{16}$$

To add mixed numbers:

1) Find and convert to the least common denominator.

2) Add the numerators.

3) Add the whole numbers.

4) Reduce the answer to its lowest terms.

Example:

$$2\frac{1}{2} + 4\frac{1}{4} + 1\frac{15}{32} =$$

$$2\frac{1}{2} = 2\frac{16}{32}$$

$$4\frac{1}{4} = 4\frac{8}{32}$$

$$1\frac{15}{32} = 1\frac{15}{32}$$

$$7\frac{39}{32} = 8\frac{7}{32}$$

Subtracting Fractions and Mixed Numbers.— *To subtract common fractions:*

1) Find and convert to the least common denominator.

2) Subtract the numerators.

3) Reduce the answer to its lowest terms.

Example:

$$\frac{15}{16}-\frac{7}{32}=$$

$$\frac{15}{16}=\frac{30}{32}$$

$$\frac{-7}{32}=\frac{-7}{32}$$

$$\frac{23}{32}$$

To subtract mixed numbers:

1) Find and convert to the least common denominator.
2) Subtract the numerators.
3) Subtract the whole numbers.
4) Reduce the answer to its lowest terms.

Example:

$$2\frac{3}{8}-1\frac{1}{16}=$$

$$2\frac{3}{8}=2\frac{6}{16}$$

$$-1\frac{1}{16}=-1\frac{1}{16}$$

$$1\frac{5}{16}$$

Multiplying Fraction and Mixed Numbers.—*To multiply common fractions:*

1) Multiply the numerators.
2) Multiply the denominators.
3) Convert improper fractions to mixed numbers, if necessary.

Example

$$\frac{3}{4}\times\frac{7}{16}=\frac{3\times 7}{4\times 16}=\frac{21}{64}$$

To multiply mixed numbers:

1) Convert the mixed numbers to improper fractions.
2) Multiply the numerators.
3) Multiply the denominators.
4) Convert improper fractions to mixed numbers, if necessary.

Example

$$2\frac{1}{4}\times 3\frac{1}{2}=\frac{9\times 7}{4\times 2}=\frac{63}{8}=7\frac{7}{8}$$

Dividing Fractions and Mixed Numbers.—*To divide common fractions:*

1) Write the fractions to be divided.
2) Invert (switch) the numerator and denominator in the dividing fraction.
3) Multiply the numerators and denominators.
4) Convert improper fractions to mixed numbers, if necessary.

Example

$$\frac{3}{4} \div \frac{1}{2} = \frac{3 \times 2}{4 \times 1} = \frac{6}{4} = 1\frac{1}{2}$$

To divide mixed numbers:

1) Convert the mixed numbers to improper fractions.
2) Write the improper fraction to be divided.
3) Invert (switch) the numerator and denominator in the dividing fraction.
4) Multiplying numerators and denominators.
5) Convert improper fractions to mixed numbers, if necessary.

Example

$$2\frac{1}{2} \div 1\frac{7}{8} = \frac{5 \times 8}{2 \times 15} = \frac{40}{30} = 1\frac{1}{3}$$

Decimal Fractions.—Decimal fractions are fractional parts of a whole unit, which have implied denominators that are multiples of 10. A decimal fraction of 0.1 has a value of 1/10th, 0.01 has a value of 1/100th, and 0.001 has a value of 1/1000th. As the number of decimal place values increases, the value of the decimal number changes by a multiple of 10. A single number placed to the right of a decimal point has a value expressed in tenths; two numbers to the right of a decimal point have a value expressed in hundredths; three numbers to the right have a value expressed in thousandths; and four numbers are expressed in ten-thousandths. Since the denominator is implied, the number of decimal places in the numerator indicates the value of the decimal fraction. So a decimal fraction expressed as a 0.125 means the whole unit has been divided into 1000 parts and 125 of these parts are considered in the value of the decimal fraction.

In industry, most decimal fractions are expressed in terms of thousandths rather than tenths or hundredths. So a decimal fraction of 0.2 is expressed as 200 thousandths, not 2 tenths, and a value of 0.75 is expressed as 750 thousandths, rather than 75 hundredths. In the case of four place decimals, the values are expressed in terms of ten-thousandths. So a value of 0.1875 is expressed as 1 thousand 8 hundred and 75 ten-thousandths. When whole numbers and decimal fractions are used together, whole units are shown to the left of a decimal point, while fractional parts of a whole unit are shown to the right.

Example

10.125

Whole Units	Fraction Units

Adding Decimal Fractions: To add decimal fractions:

1) Write the problem with all decimal points aligned vertically.
2) Add the numbers as whole number values.
3) Insert the decimal point in the same vertical column in the answer.

Example

0.125		1.750
1.0625		0.875
2.50	or	0.125
0.1875		2.0005
3.8750		4.7505

Subtracting Decimal Fractions: To subtract decimal fractions:

1) Write the problem with all decimal points aligned vertically.

2) Subtract the numbers as whole number values.

3) Insert the decimal point in the same vertical column in the answer.

Example

$$\begin{array}{r} 1.750 \\ -0.250 \\ \hline 1.500 \end{array} \quad \text{or} \quad \begin{array}{r} 2.625 \\ -1.125 \\ \hline 1.500 \end{array}$$

Multiplying Decimal Fractions: To multiply decimal fractions:

1) Write the problem with the decimal points aligned.

2) Multiply the values as whole numbers.

3) Count the number of decimal places in both multiplied values.

4) Counting from right to left in the answer, insert the decimal point so the number of decimal places in the answer equals the total number of decimal places in the numbers multiplied.

Example

$$\begin{array}{r} 0.75 \\ 0.25 \\ \hline 375 \\ 150 \\ \hline 0.1875 \end{array} \quad \text{(four decimal places)} \qquad \begin{array}{r} 1.625 \\ 0.033 \\ \hline 4875 \\ 4875 \\ \hline 0.053625 \end{array} \quad \text{(six decimal places)}$$

Converting Common Fractions To Decimal Fractions.—*To convert common fractions to decimal fractions:*

1) Write the common fraction.

2) Divide the numerator by the denominator.

3) Continue the division until there is no remainder or until the desired number of decimal places in the answer is reached (normally, five places and round off to four).

Example

$$\frac{1}{2} = 2\overline{)1.0} = 0.5 \qquad \begin{array}{r} 10 \\ \hline 0 \end{array}$$

or

$$\frac{1}{8} = 8\overline{)1.000} = 0.125 \qquad \begin{array}{r} 8 \\ \hline 20 \\ 16 \\ \hline 40 \\ 40 \\ \hline 0 \end{array}$$

To convert decimal fractions to common fractions:

1) Write the decimal fraction as a common fraction. Remember to use the proper number of zeros in the denominator.

2) Reduce this common fraction to its lowest terms.

Example

$$0.5 = \frac{5}{10} = \frac{1}{2} \qquad \text{or} \qquad 0.125 = \frac{125}{1000} = \frac{1}{8}$$

Dividing Decimal Fractions: To divide decimal fractions:

1) Write the problem as whole number division by moving the decimal point of the dividing number to the extreme right (this will make the decimal fraction appear as a whole number). Note the number of places the decimal point was moved.

2) Move the decimal point in the divided number to the right the same number of places. Add zeros if necessary.

3) Divide as a whole number problem.

4) Locate the decimal point in the answer directly above the new position in the divided number.

5) Continue the division until the desired number of decimal places in the answer is reached. Add zeros if necessary.

Example

```
2.75 ÷ 0.625 =
            4.4
0625.)2750.0
      2500
       2500
       2500
          0
```

or

```
2.75 ÷ 0.625 =
           2.216738
0625)5668.200000
     5114
      5542
      5114
       4280
       2557
        17230
        15342
         18880
         17899
           9810
           7671
           21390
           20456
             934
```

Converting Common and Decimal Fractions.—Conversions between common fractions and decimal fractions may be accomplished by using a decimal equivalent chart (See page 10) or by calculating the equivalent values.

Principal Algebraic Expressions and Formulas

$$a \times a = aa = a^2$$

$$a \times a \times a = aaa = a^3$$

$$a \times b = ab$$

$$a^2b^2 = (ab)^2$$

$$a^2a^3 = a^{2+3} = a^5$$

$$a^4 \div a^3 = a^{4-3} = a$$

$$a^0 = 1$$

$$a^2 - b^2 = (a+b)(a-b)$$

$$(a+b)^2 = a^2 + 2ab + b^2$$

$$(a-b)^2 = a^2 - 2ab + b^2$$

$$\frac{a^3}{b^3} = \left(\frac{a}{b}\right)^3$$

$$\frac{1}{a^3} = \left(\frac{1}{a}\right)^3 = a^{-3}$$

$$(a^2)^3 = a^{2\times3} = (a^3)^2 = a^6$$

$$a^3 + b^3 = (a+b)(a^2 - ab + b^2)$$

$$a^3 - b^3 = (a-b)(a^2 + ab + b^2)$$

$$(a+b)^3 = a^3 + 3a^2b + 3ab^2 + b^3$$

$$(a-b)^3 = a^3 - 3a^2b + 3ab^2 - b^3$$

$$\sqrt{a} \times \sqrt{a} = a$$

$$\sqrt[3]{a} \times \sqrt[3]{a} \times \sqrt[3]{a} = a$$

$$(\sqrt[3]{a})^3 = a$$

$$\sqrt[3]{a^2} = (\sqrt[3]{a})^2 = a^{2/3}$$

$$\sqrt[4]{\sqrt[3]{a}} = \sqrt[4\times3]{a} = \sqrt[3]{\sqrt[4]{a}}$$

$$\sqrt{a} + \sqrt{b} = \sqrt{a + b + 2\sqrt{ab}}$$

$$\sqrt[3]{ab} = \sqrt[3]{a} \times \sqrt[3]{b}$$

$$\sqrt[3]{\frac{a}{b}} = \frac{\sqrt[3]{a}}{\sqrt[3]{b}}$$

$$\sqrt[3]{\frac{1}{a}} = \frac{1}{\sqrt[3]{a}} = a^{-1/3}$$

When	$a \times b = x$	then	$\log a + \log b = \log x$
	$a \div b = x$	then	$\log a - \log b = \log x$
	$a^3 = x$	then	$3\log a = \log x$
	$\sqrt[3]{a} = x$	then	$\frac{\log a}{3} = \log x$

Equations

An equation is a statement of equality between two expressions, as $5x = 105$. The unknown quantity in an equation is generally designated by the letter x. If there is more than one unknown quantity, the others are designated by letters also selected at the end of the alphabet, as y, z, u, t, etc..

An equation of the first degree is one which contains the unknown quantity only in the first power, as $3x = 9$. A quadratic equation is one which contains the unknown quantity in the second, but no higher, power, as $x^2 + 3x = 10$.

Solving Equations of the First Degree with One Unknown.—Transpose all the terms containing the unknown x to one side of the equals sign, and all the other terms to the other side. Combine and simplify the expressions as far as possible, and divide both sides by the coefficient of the unknown x. (See the rules given for transposition of formulas.)

Example:

$$22x - 11 = 15x + 10$$

$$22x - 15x = 10 + 11$$

$$7x = 21$$

$$x = 3$$

Solution of Equations of the First Degree with Two Unknowns.—The form of the simplified equations is

$$ax + by = c$$
$$a_1x + b_1y = c_1$$

Then,
$$x = \frac{cb_1 - c_1b}{ab_1 - a_1b} \qquad y = \frac{ac_1 - a_1c}{ab_1 - a_1b}$$

Example:
$$3x + 4y = 17$$
$$5x - 2y = 11$$
$$x = \frac{17 \times (-2) - 11 \times 4}{3 \times (-2) - 5 \times 4} = \frac{-34 - 44}{-6 - 20} = \frac{-78}{-26} = 3$$

The value of y can now be most easily found by inserting the value of x in one of the equations:

$$5 \times 3 - 2y = 11 \qquad 2y = 15 - 11 = 4 \qquad y = 2$$

Solution of Quadratic Equations with One Unknown.—If the form of the equation is $ax^2 + bx + c = 0$, then

$$x = \frac{-b \pm \sqrt{b^2 - 4ac}}{2a}$$

Example: Given the equation, $1x^2 + 6x + 5 = 0$, then $a = 1$, $b = 6$, and $c = 5$.

$$x = \frac{-6 \pm \sqrt{6^2 - 4 \times 1 \times 5}}{2 \times 1} = \frac{(-6) + 4}{2} = -1 \quad \text{or} \quad \frac{(-6) - 4}{2} = -5$$

If the form of the equation is $ax^2 + bx = c$, then

$$x = \frac{-b \pm \sqrt{b^2 + 4ac}}{2a}$$

Example: A right-angle triangle has a hypotenuse 5 inches long and one side which is one inch longer than the other; find the lengths of the two sides.

Let x = one side and $x + 1$ = other side; then $x^2 + (x + 1)^2 = 5^2$ or $x^2 + x^2 + 2x + 1 = 25$; or $2x^2 + 2x = 24$; or $x^2 + x = 12$. Now referring to the basic formula, $ax^2 + bx = c$, we find that $a = 1$, $b = 1$, and $c = 12$; hence,

$$x = \frac{-1 \pm \sqrt{1 + 4 \times 1 \times 12}}{2 \times 1} = \frac{(-1) + 7}{2} = 3 \quad \text{or} \quad x = \frac{(-1) - 7}{2} = -4$$

Since the positive value (3) would apply in this case, the lengths of the two sides are $x = 3$ inches and $x + 1 = 4$ inches.

Cubic Equations.—If the given equation has the form: $x^3 + ax + b = 0$ then

$$x = \left(-\frac{b}{2} + \sqrt{\frac{a^3}{27} + \frac{b^2}{4}}\right)^{1/3} + \left(-\frac{b}{2} - \sqrt{\frac{a^3}{27} + \frac{b^2}{4}}\right)^{1/3}$$

The equation $x^3 + px^2 + qx + r = 0$, may be reduced to the form $x_1{}^3 + ax_1 + b = 0$ by substituting $x_1 - \frac{p}{3}$ for x in the given equation.

Rearrangement and Transposition of Terms in Formulas

A formula is a rule for a calculation expressed by using letters and signs instead of writing out the rule in words; by this means, it is possible to condense, in a very small space, the essentials of long and cumbersome rules. The letters used in formulas simply stand in place of the figures that are to be substituted when solving a specific problem.

As an example, the formula for the horsepower transmitted by belting may be written

$$P = \frac{SVW}{33{,}000}$$

where P = horsepower transmitted; S = working stress of belt per inch of width in pounds; V = velocity of belt in feet per minute; and, W = width of belt in inches.

If the working stress S, the velocity V, and the width W are known, the horsepower can be found directly from this formula by inserting the given values. Assume $S = 33$; $V = 600$; and $W = 5$. Then

$$P = \frac{33 \times 600 \times 5}{33{,}000} = 3$$

Assume that the horsepower P, the stress S, and the velocity V are known, and that the width of belt, W, is to be found. The formula must then be rearranged so that the symbol W will be on one side of the equals sign and all the known quantities on the other. The rearranged formula is as follows:

$$\frac{P \times 33{,}000}{SV} = W$$

The quantities (S and V) that were in the numerator on the right side of the equals sign are moved to the denominator on the left side, and "33,000," which was in the denominator on the right side of the equals sign, is moved to the numerator on the other side. Symbols that are not part of a fraction, like "P" in the formula first given, are to be considered as being numerators (having the denominator 1).

Thus, any formula of the form $A = B/C$ can be rearranged as follows:

$$A \times C = B \qquad \text{and} \qquad C = \frac{B}{A}$$

Suppose a formula to be of the form

$$A = \frac{B \times C}{D}$$

Then

$$D = \frac{B \times C}{A} \qquad \frac{A \times D}{C} = B \qquad \frac{A \times D}{B} = C$$

The method given is only directly applicable when all the quantities in the numerator or denominator are standing independently or are *factors of a product*. If connected by + or − signs, the entire numerator or denominator must be moved as a unit, thus,

Given:

$$\frac{B + C}{A} = \frac{D + E}{F}$$

to solve for F

then

$$\frac{F}{A} = \frac{D + E}{B + C}$$

and

$$F = \frac{A(D + E)}{B + C}$$

A quantity preceded by a + or − sign can be transposed to the opposite side of the equals sign by changing its sign; if the sign is +, change it to − on the other side; if it is −, change it to +. This process is called *transposition* of terms.

Example:

$$B + C = A - D \quad \text{then} \quad A = B + C + D$$
$$B = A - D - C$$
$$C = A - D - B$$

Sequence of Performing Arithmetic Operations

When several numbers or quantities in a formula are connected by signs indicating that additions, subtractions, multiplications, and divisions are to be made, the multiplications and divisions should be carried out first, in the sequence in which they appear, before the additions or subtractions are performed.

Example:

$$10 + 26 \times 7 - 2 = 10 + 182 - 2 = 190$$
$$18 \div 6 + 15 \times 3 = 3 + 45 = 48$$
$$12 + 14 \div 2 - 4 = 12 + 7 - 4 = 15$$

When it is required that certain additions and subtractions should precede multiplications and divisions, use is made of parentheses ( ) and brackets [ ]. These signs indicate that the calculation inside the parentheses or brackets should be carried out completely by itself before the remaining calculations are commenced. If one bracket is placed inside another, the one inside is first calculated.

Example:

$$(6 - 2) \times 5 + 8 = 4 \times 5 + 8 = 20 + 8 = 28$$
$$6 \times (4 + 7) \div 22 = 6 \times 11 \div 22 = 66 \div 22 = 3$$
$$2 + [10 \times 6(8 + 2) - 4] \times 2 = 2 + [10 \times 6 \times 10 - 4] \times 2$$
$$= 2 + [600 - 4] \times 2 = 2 + 596 \times 2 = 2 + 1192 = 1194$$

The parentheses are considered as a sign of multiplication; for example, $6(8 + 2) = 6 \times (8 + 2)$.

The line or bar between the numerator and denominator in a fractional expression is to be considered as a division sign. For example,

$$\frac{12 + 16 + 22}{10} = (12 + 16 + 22) \div 10 = 50 \div 10 = 5$$

In formulas, the multiplication sign (×) is often left out between symbols or letters, the values of which are to be multiplied. Thus,

$$AB = A \times B \quad \text{and} \quad \frac{ABC}{D} = (A \times B \times C) \div D$$

Imaginary and Complex Numbers

Complex or Imaginary Numbers.—Complex or imaginary numbers represent a class of mathematical objects that are used to simplify certain problems, such as the solution of polynomial equations. The basis of the complex number system is the unit imaginary number i that satisfies the following relations:

$$i^2 = (-i)^2 = -1 \qquad i = \sqrt{-1} \qquad -i = -\sqrt{-1}$$

In electrical engineering and other fields, the unit imaginary number is often represented by j rather than i. However, the meaning of the two terms is identical.

Rectangular or Trigonometric Form: Every complex number, Z, can be written as the sum of a real number and an imaginary number. When expressed as a sum, $Z = a + bi$, the complex number is said to be in rectangular or trigonometric form. The real part of the number is a, and the imaginary portion is bi because it has the imaginary unit assigned to it.

Polar Form: A complex number $Z = a + bi$ can also be expressed in polar form, also known as phasor form. In polar form, the complex number Z is represented by a magnitude r and an angle θ as follows:

$$Z = r\angle\theta$$

where $\angle\theta$ = a direction, the angle whose tangent is $b \div a$, thus $\theta = \operatorname{atan}\frac{b}{a}$; and, r = a magnitude = $\sqrt{a^2 + b^2}$.

A complex number can be plotted on a real-imaginary coordinate system known as the complex plane. The figure below illustrates the relationship between the rectangular coordinates a and b, and the polar coordinates r and θ.

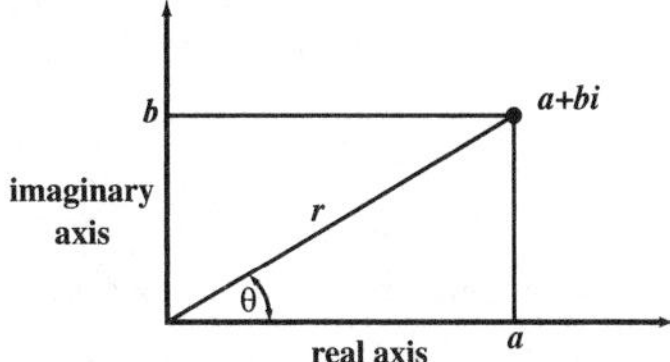

Complex Number in the Complex Plane

The rectangular form can be determined from r and θ as follows:

$$a = r\cos\theta \qquad b = r\sin\theta \qquad a + bi = r\cos\theta + ir\sin\theta = r(\cos\theta + i\sin\theta)$$

The rectangular form can also be written using Euler's Formula:

$$e^{\pm i\theta} = \cos\theta \pm i\sin\theta \qquad \sin\theta = \frac{e^{i\theta} - e^{-i\theta}}{2i} \qquad \cos\theta = \frac{e^{i\theta} + e^{-i\theta}}{2}$$

Complex Conjugate: Complex numbers commonly arise in finding the solution of polynomials. A polynomial of nth degree has n solutions, an even number of which are complex and the rest are real. The complex solutions always appear as complex conjugate pairs in the form $a + bi$ and $a - bi$. The product of these two conjugates, $(a + bi) \times (a - bi) = a^2 + b^2$, is the square of the magnitude r illustrated in the previous figure.

Operations on Complex Numbers

Example 3, Addition: When adding two complex numbers, the real parts and imaginary parts are added separately, the real parts added to real parts and the imaginary to imaginary parts. Thus,

$$(a_1 + ib_1) + (a_2 + ib_2) = (a_1 + a_2) + i(b_1 + b_2)$$
$$(a_1 + ib_1) - (a_2 + ib_2) = (a_1 - a_2) + i(b_1 - b_2)$$

$$(3 + 4i) + (2 + i) = (3 + 2) + (4 + 1)i = 5 + 5i$$

Example 4, Multiplication: Multiplication of two complex numbers requires the use of the imaginary unit, $i^2 = -1$ and the algebraic distributive law.

$$(a_1 + ib_1)(a_2 + ib_2) = a_1a_2 + ia_1b_2 + ia_2b_1 + i^2b_1b_2$$
$$= a_1a_2 + ia_1b_2 + ia_2b_1 - b_1b_2$$

$$
\begin{aligned}
(7+2i)\times(5-3i) &= (7)(5)-(7)(3i)+(2i)(5)-(2i)(3i)\\
&= 35-21i+10i-6i^2\\
&= 35-21i+10i-(6)(-1)\\
&= 41-11i
\end{aligned}
$$

Multiplication of two complex numbers, $Z_1 = r_1(\cos\theta_1 + i\sin\theta_1)$ and $Z_2 = r_2(\cos\theta_2 + i\sin\theta_2)$, results in the following:

$$Z_1 \times Z_2 = r_1(\cos\theta_1 + i\sin\theta_1)\times r_2(\cos\theta_2 + i\sin\theta_2) = r_1r_2[\cos(\theta_1+\theta_2)+i\sin(\theta_1+\theta_2)]$$

Example 5, Division: Divide the following two complex numbers, $2 + 3i$ and $4 - 5i$. Dividing complex numbers makes use of the complex conjugate.

$$\frac{2+3i}{4-5i} = \frac{(2+3i)(4+5i)}{(4-5i)(4+5i)} = \frac{8+12i+10i+15i^2}{16+20i-20i-25i^2} = \frac{-7+22i}{16+25} = \left(\frac{-7}{41}\right)+i\left(\frac{22}{41}\right)$$

Example 6: Convert the complex number $8+6i$ into phasor form.

First find the magnitude of the phasor vector and then the direction.

$$\text{magnitude} = \sqrt{8^2+6^2} = \sqrt{100} = 10 \qquad \text{direction} = \operatorname{atan}\frac{6}{8} = \operatorname{atan}0.75 = 36.87°$$

$$\text{phasor} = 10\angle 36.87°$$

Imaginary Quantities.—An imaginary quantity is an even root of a negative number, such as

$$\sqrt{-1},\ \sqrt{-a^2}$$

The square root of -1 is the simplest imaginary quantity. It is usually denoted by i. In electric work, where imaginary quantities appear frequently, the square root of -1 is denoted by j, since i is used to represent current flow.
The following examples illustrate the use of i in mathematical operations:

$$
\begin{aligned}
\sqrt{-9} &= \sqrt{(9)(-1)} = \sqrt{9}\sqrt{-1} = 3i\\
\sqrt{-25} &= \sqrt{(25)(-1)} = \sqrt{25}\sqrt{-1} = 5i\\
(6i)^2 &= 6^2\times i^2 = (6\times 6)\\
&= ((36(\sqrt{-1}\times\sqrt{-1})) = 36(-1)) = -36\\
\sqrt{-b^2} &= \sqrt{(b^2)(-1)} = bi\\
(bi)^2 &= b^2\times i^2 = (b^2)(-1) = -b^2
\end{aligned}
$$

Note that

$$i^2 = -1,\ i^3 = -\sqrt{-1} = -i,\ i^4 = i,\ i^5 = i$$

and so on, in a repeating pattern of i, -1, i ,etc.

Squares and Cubes of Numbers from 1/32 to 6 15/16

No.	Square	Cube	No.	Square	Cube	No.	Square	Cube
1/32	0.00098	0.00003	1 17/32	2.34473	3.59036	4	16.00000	64.00000
1/16	0.00391	0.00024	1 9/16	2.44141	3.81470	4 1/16	16.50391	67.04712
3/32	0.00879	0.00082	1 19/32	2.54004	4.04819	4 1/8	17.01563	70.18945
1/8	0.01563	0.00195	1 5/8	2.64063	4.29102	4 3/16	17.53516	73.42847
5/32	0.02441	0.00381	1 21/32	2.74316	4.54337	4 1/4	18.06250	76.76563
3/16	0.03516	0.00659	1 11/16	2.84766	4.80542	4 5/16	18.59766	80.20239
7/32	0.04785	0.01047	1 23/32	2.95410	5.07736	4 3/8	19.14063	83.74023
1/4	0.06250	0.01563	1 3/4	3.06250	5.35938	4 7/16	19.69141	87.38062
9/32	0.07910	0.02225	1 25/32	3.17285	5.65164	4 1/2	20.25000	91.12500
5/16	0.09766	0.03052	1 13/16	3.28516	5.95435	4 9/16	20.81641	94.97485
11/32	0.11816	0.04062	1 27/32	3.39941	6.26767	4 5/8	21.39063	98.93164
3/8	0.14063	0.05273	1 7/8	3.51563	6.59180	4 11/16	21.97266	102.99683
13/32	0.16504	0.06705	1 29/32	3.63379	6.92691	4 3/4	22.56250	107.17188
7/16	0.19141	0.08374	1 15/16	3.75391	7.27319	4 13/16	23.16016	111.45825
15/32	0.21973	0.10300	1 31/32	3.87598	7.63083	4 7/8	23.76563	115.85742
1/2	0.25000	0.12500	2	4.00000	8.00000	4 15/16	24.37891	120.37085
17/32	0.28223	0.14993	2 1/32	4.12598	8.38089	5	25.00000	125.00000
9/16	0.31641	0.17798	2 1/16	4.25391	8.77368	5 1/16	25.62891	129.74634
19/32	0.35254	0.20932	2 1/8	4.51563	9.59570	5 1/8	26.26563	134.61133
5/8	0.39063	0.24414	2 3/16	4.78516	10.46753	5 3/16	26.91016	139.59644
21/32	0.43066	0.28262	2 1/4	5.06250	11.39063	5 1/4	27.56250	144.70313
11/16	0.47266	0.32495	2 5/16	5.34766	12.36646	5 5/16	28.22266	149.93286
23/32	0.51660	0.37131	2 3/8	5.64063	13.39648	5 3/8	28.89063	155.28711
3/4	0.56250	0.42188	2 7/16	5.94141	14.48218	5 7/16	29.56641	160.76733
25/32	0.61035	0.47684	2 1/2	6.25000	15.62500	5 1/2	30.25000	166.37500
13/16	0.66016	0.53638	2 9/16	6.56641	16.82642	5 9/16	30.94141	172.11157
27/32	0.71191	0.60068	2 5/8	6.89063	18.08789	5 5/8	31.64063	177.97852
7/8	0.76563	0.66992	2 11/16	7.22266	19.41089	5 11/16	32.34766	183.97729
29/32	0.82129	0.74429	2 3/4	7.56250	20.79688	5 3/4	33.06250	190.10938
15/16	0.87891	0.82397	2 13/16	7.91016	22.24731	5 13/16	33.78516	196.37622
31/32	0.93848	0.90915	2 7/8	8.26563	23.76367	5 7/8	34.51563	202.77930
1	1.00000	1.00000	2 15/16	8.62891	25.34741	5 15/16	35.25391	209.32007
1 1/32	1.06348	1.09671	3	9.00000	27.00000	6	36.00000	216.00000
1 1/16	1.12891	1.19946	3 1/16	9.37891	28.72290	6 1/16	36.75391	222.82056
1 3/32	1.19629	1.30844	3 1/8	9.76563	30.51758	6 1/8	37.51563	229.78320
1 1/8	1.26563	1.42383	3 3/16	10.16016	32.38550	6 3/16	38.28516	236.88940
1 5/32	1.33691	1.54581	3 1/4	10.56250	34.32813	6 1/4	39.06250	244.14063
1 3/16	1.41016	1.67456	3 5/16	10.97266	36.34692	6 5/16	39.84766	251.53833
1 7/32	1.48535	1.81027	3 3/8	11.39063	38.44336	6 3/8	40.64063	259.08398
1 1/4	1.56250	1.95313	3 7/16	11.81641	40.61890	6 7/16	41.44141	266.77905
1 9/32	1.64160	2.10330	3 1/2	12.25000	42.87500	6 1/2	42.25000	274.62500
1 5/16	1.72266	2.26099	3 9/16	12.69141	45.21313	6 9/16	43.06641	282.62329
1 11/32	1.80566	2.42636	3 5/8	13.14063	47.63477	6 5/8	43.89063	290.77539
1 3/8	1.89063	2.59961	3 11/16	13.59766	50.14136	6 11/16	44.72266	299.08276
1 13/32	1.97754	2.78091	3 3/4	14.06250	52.73438	6 3/4	45.56250	307.54688
1 7/16	2.06641	2.97046	3 13/16	14.53516	55.41528	6 13/16	46.41016	316.16919
1 15/32	2.15723	3.16843	3 7/8	15.01563	58.18555	6 7/8	47.26563	324.95117
1 1/2	2.25000	3.37500	3 15/16	15.50391	61.04663	6 15/16	48.12891	333.89429

Squares and Cubes of Numbers from 7 to 21 7/8

No.	Square	Cube	No.	Square	Cube	No.	Square	Cube
7	49.00000	343.00000	10	100.00000	1000.00000	16	256.00000	4096.00000
1/16	49.87891	352.26978	1/8	102.51563	1037.97070	1/8	260.01563	4192.75195
1/8	50.76563	361.70508	1/4	105.06250	1076.89063	1/4	264.06250	4291.01563
3/16	51.66016	371.30737	3/8	107.64063	1116.77148	3/8	268.14063	4390.80273
1/4	52.56250	381.07813	1/2	110.25000	1157.62500	1/2	272.25000	4492.12500
5/16	53.47266	391.01880	5/8	112.89063	1199.46289	5/8	276.39063	4594.99414
3/8	54.39063	401.13086	3/4	115.56250	1242.29688	3/4	280.56250	4699.42188
7/16	55.31641	411.41577	7/8	118.26563	1286.13867	7/8	284.76563	4805.41992
1/2	56.25000	421.87500	11	121.00000	1331.00000	17	289.00000	4913.00000
9/16	57.19141	432.51001	1/8	123.76563	1376.89258	1/8	293.26563	5022.17383
5/8	58.14063	443.32227	1/4	126.56250	1423.82813	1/4	297.56250	5132.95313
11/16	59.09766	454.31323	3/8	129.39063	1471.81836	3/8	301.89063	5245.34961
3/4	60.06250	465.48438	1/2	132.25000	1520.87500	1/2	306.25000	5359.37500
13/16	61.03516	476.83716	5/8	135.14063	1571.00977	5/8	310.64063	5475.04102
7/8	62.01563	488.37305	3/4	138.06250	1622.23438	3/4	315.06250	5592.35938
15/16	63.00391	500.09351	7/8	141.01563	1674.56055	7/8	319.51563	5711.34180
8	64.00000	512.00000	12	144.00000	1728.00000	18	324.00000	5832.00000
1/16	65.00391	524.09399	1/8	147.01563	1782.56445	1/8	328.51563	5954.34570
1/8	66.01563	536.37695	1/4	150.06250	1838.26563	1/4	333.06250	6078.39063
3/16	67.03516	548.85034	3/8	153.14063	1895.11523	3/8	337.64063	6204.14648
1/4	68.06250	561.51563	1/2	156.25000	1953.12500	1/2	342.25000	6331.62500
5/16	69.09766	574.37427	5/8	159.39063	2012.30664	5/8	346.89063	6460.83789
3/8	70.14063	587.42773	3/4	162.56250	2072.67188	3/4	351.56250	6591.79688
7/16	71.19141	600.67749	7/8	165.76563	2134.23242	7/8	356.26563	6724.51367
1/2	72.25000	614.12500	13	169.00000	2197.00000	19	361.00000	6859.00000
9/16	73.31641	627.77173	1/8	172.26563	2260.98633	1/8	365.76563	6995.26758
5/8	74.39063	641.61914	1/4	175.56250	2326.20313	1/4	370.56250	7133.32813
11/16	75.47266	655.66870	3/8	178.89063	2392.66211	3/8	375.39063	7273.19336
3/4	76.56250	669.92188	1/2	182.25000	2460.37500	1/2	380.25000	7414.87500
13/16	77.66016	684.38013	5/8	185.64063	2529.35352	5/8	385.14063	7558.38477
7/8	78.76563	699.04492	3/4	189.06250	2599.60938	3/4	390.06250	7703.73438
15/16	79.87891	713.91772	7/8	192.51563	2671.15430	7/8	395.01563	7850.93555
9	81.00000	729.00000	14	196.00000	2744.00000	20	400.00000	8000.00000
1/16	82.12891	744.29321	1/8	199.51563	2818.15820	1/8	405.01563	8150.93945
1/8	83.26563	759.79883	1/4	203.06250	2893.64063	1/4	410.06250	8303.76563
3/16	84.41016	775.51831	3/8	206.64063	2970.45898	3/8	415.14063	8458.49023
1/4	85.56250	791.45313	1/2	210.25000	3048.62500	1/2	420.25000	8615.12500
5/16	86.72266	807.60474	5/8	213.89063	3128.15039	5/8	425.39063	8773.68164
3/8	87.89063	823.97461	3/4	217.56250	3209.04688	3/4	430.56250	8934.17188
7/16	89.06641	840.56421	7/8	221.26563	3291.32617	7/8	435.76563	9096.60742
1/2	90.25000	857.37500	15	225.00000	3375.00000	21	441.00000	9261.00000
9/16	91.44141	874.40845	1/8	228.76563	3460.08008	1/8	446.26563	9427.36133
5/8	92.64063	891.66602	1/4	232.56250	3546.57813	1/4	451.56250	9595.70313
11/16	93.84766	909.14917	3/8	236.39063	3634.50586	3/8	456.89063	9766.03711
3/4	95.06250	926.85938	1/2	240.25000	3723.87500	1/2	462.25000	9938.37500
13/16	96.28516	944.79810	5/8	244.14063	3814.69727	5/8	467.64063	10112.72852
7/8	97.51563	962.96680	3/4	248.06250	3906.98438	3/4	473.06250	10289.10938
15/16	98.75391	981.36694	7/8	252.01563	4000.74805	7/8	478.51563	10467.52930

Squares and Cubes of Numbers from 22 to 39 ⅞

No.	Square	Cube	No.	Square	Cube	No.	Square	Cube
22	484.00000	10648.00000	28	784.00000	21952.00000	34	1156.00000	39304.00000
⅛	489.51563	10830.53320	⅛	791.01563	22247.31445	⅛	1164.51563	39739.09570
¼	495.06250	11015.14063	¼	798.06250	22545.26563	¼	1173.06250	40177.39063
⅜	500.64063	11201.83398	⅜	805.14063	22845.86523	⅜	1181.64063	40618.89648
½	506.25000	11390.62500	½	812.25000	23149.12500	½	1190.25000	41063.62500
⅝	511.89063	11581.52539	⅝	819.39063	23455.05664	⅝	1198.89063	41511.58789
¾	517.56250	11774.54688	¾	826.56250	23763.67188	¾	1207.56250	41962.79688
⅞	523.26563	11969.70117	⅞	833.76563	24074.98242	⅞	1216.26563	42417.26367
23	529.00000	12167.00000	29	841.00000	24389.00000	35	1225.00000	42875.00000
⅛	534.76563	12366.45508	⅛	848.26563	24705.73633	⅛	1233.76563	43336.01758
¼	540.56250	12568.07813	¼	855.56250	25025.20313	¼	1242.56250	43800.32813
⅜	546.39063	12771.88086	⅜	862.89063	25347.41211	⅜	1251.39063	44267.94336
½	552.25000	12977.87500	½	870.25000	25672.37500	½	1260.25000	44738.87500
⅝	558.14063	13186.07227	⅝	877.64063	26000.10352	⅝	1269.14063	45213.13477
¾	564.06250	13396.48438	¾	885.06250	26330.60938	¾	1278.06250	45690.73438
⅞	570.01563	13609.12305	⅞	892.51563	26663.90430	⅞	1287.01563	46171.68555
24	576.00000	13824.00000	30	900.00000	27000.00000	36	1296.00000	46656.00000
⅛	582.01563	14041.12695	⅛	907.51563	27338.90820	⅛	1305.01563	47143.68945
¼	588.06250	14260.51563	¼	915.06250	27680.64063	¼	1314.06250	47634.76563
⅜	594.14063	14482.17773	⅜	922.64063	28025.20898	⅜	1323.14063	48129.24023
½	600.25000	14706.12500	½	930.25000	28372.62500	½	1332.25000	48627.12500
⅝	606.39063	14932.36914	⅝	937.89063	28722.90039	⅝	1341.39063	49128.43164
¾	612.56250	15160.92188	¾	945.56250	29076.04688	¾	1350.56250	49633.17188
⅞	618.76563	15391.79492	⅞	953.26563	29432.07617	⅞	1359.76563	50141.35742
25	625.00000	15625.00000	31	961.00000	29791.00000	37	1369.00000	50653.00000
⅛	631.26563	15860.54883	⅛	968.76563	30152.83008	⅛	1378.26563	51168.11133
¼	637.56250	16098.45313	¼	976.56250	30517.57813	¼	1387.56250	51686.70313
⅜	643.89063	16338.72461	⅜	984.39063	30885.25586	⅜	1396.89063	52208.78711
½	650.25000	16581.37500	½	992.25000	31255.87500	½	1406.25000	52734.37500
⅝	656.64063	16826.41602	⅝	1000.14063	31629.44727	⅝	1415.64063	53263.47852
¾	663.06250	17073.85938	¾	1008.06250	32005.98438	¾	1425.06250	53796.10938
⅞	669.51563	17323.71680	⅞	1016.01563	32385.49805	⅞	1434.51563	54332.27930
26	676.00000	17576.00000	32	1024.00000	32768.00000	38	1444.00000	54872.00000
⅛	682.51563	17830.72070	⅛	1032.01563	33153.50195	⅛	1453.51563	55415.28320
¼	689.06250	18087.89063	¼	1040.06250	33542.01563	¼	1463.06250	55962.14063
⅜	695.64063	18347.52148	⅜	1048.14063	33933.55273	⅜	1472.64063	56512.58398
½	702.25000	18609.62500	½	1056.25000	34328.12500	½	1482.25000	57066.62500
⅝	708.89063	18874.21289	⅝	1064.39063	34725.74414	⅝	1491.89063	57624.27539
¾	715.56250	19141.29688	¾	1072.56250	35126.42188	¾	1501.56250	58185.54688
⅞	722.26563	19410.88867	⅞	1080.76563	35530.16992	⅞	1511.26563	58750.45117
27	729.00000	19683.00000	33	1089.00000	35937.00000	39	1521.00000	59319.00000
⅛	735.76563	19957.64258	⅛	1097.26563	36346.92383	⅛	1530.76563	59891.20508
¼	742.56250	20234.82813	¼	1105.56250	36759.95313	¼	1540.56250	60467.07813
⅜	749.39063	20514.56836	⅜	1113.89063	37176.09961	⅜	1550.39063	61046.63086
½	756.25000	20796.87500	½	1122.25000	37595.37500	½	1560.25000	61629.87500
⅝	763.14063	21081.75977	⅝	1130.64063	38017.79102	⅝	1570.14063	62216.82227
¾	770.06250	21369.23438	¾	1139.06250	38443.35938	¾	1580.06250	62807.48438
⅞	777.01563	21659.31055	⅞	1147.51563	38872.09180	⅞	1590.01563	63401.87305

Squares and Cubes of Numbers from 40 to 57 ⅞

No.	Square	Cube	No.	Square	Cube	No.	Square	Cube
40	1600.00000	64000.00000	46	2116.00000	97336.00000	52	2704.00000	140608.00000
⅛	1610.01563	64601.87695	⅛	2127.51563	98131.65820	⅛	2717.01563	141624.43945
¼	1620.06250	65207.51563	¼	2139.06250	98931.64063	¼	2730.06250	142645.76563
⅜	1630.14063	65816.92773	⅜	2150.64063	99735.95898	⅜	2743.14063	143671.99023
½	1640.25000	66430.12500	½	2162.25000	100544.62500	½	2756.25000	144703.12500
⅝	1650.39063	67047.11914	⅝	2173.89063	101357.65039	⅝	2769.39063	145739.18164
¾	1660.56250	67667.92188	¾	2185.56250	102175.04688	¾	2782.56250	146780.17188
⅞	1670.76563	68292.54492	⅞	2197.26563	102996.82617	⅞	2795.76563	147826.10742
41	1681.00000	68921.00000	47	2209.00000	103823.00000	53	2809.00000	148877.00000
⅛	1691.26563	69553.29883	⅛	2220.76563	104653.58008	⅛	2822.26563	149932.86133
¼	1701.56250	70189.45313	¼	2232.56250	105488.57813	¼	2835.56250	150993.70313
⅜	1711.89063	70829.47461	⅜	2244.39063	106328.00586	⅜	2848.89063	152059.53711
½	1722.25000	71473.37500	½	2256.25000	107171.87500	½	2862.25000	153130.37500
⅝	1732.64063	72121.16602	⅝	2268.14063	108020.19727	⅝	2875.64063	154206.22852
¾	1743.06250	72772.85938	¾	2280.06250	108872.98438	¾	2889.06250	155287.10938
⅞	1753.51563	73428.46680	⅞	2292.01563	109730.24805	⅞	2902.51563	156373.02930
42	1764.00000	74088.00000	48	2304.00000	110592.00000	54	2916.00000	157464.00000
⅛	1774.51563	74751.47070	⅛	2316.01563	111458.25195	⅛	2929.51563	158560.03320
¼	1785.06250	75418.89063	¼	2328.06250	112329.01563	¼	2943.06250	159661.14063
⅜	1795.64063	76090.27148	⅜	2340.14063	113204.30273	⅜	2956.64063	160767.33398
½	1806.25000	76765.62500	½	2352.25000	114084.12500	½	2970.25000	161878.62500
⅝	1816.89063	77444.96289	⅝	2364.39063	114968.49414	⅝	2983.89063	162995.02539
¾	1827.56250	78128.29688	¾	2376.56250	115857.42188	¾	2997.56250	164116.54688
⅞	1838.26563	78815.63867	⅞	2388.76563	116750.91992	⅞	3011.26563	165243.20117
43	1849.00000	79507.00000	49	2401.00000	117649.00000	55	3025.00000	166375.00000
⅛	1859.76563	80202.39258	⅛	2413.26563	118551.67383	⅛	3038.76563	167511.95508
¼	1870.56250	80901.82813	¼	2425.56250	119458.95313	¼	3052.56250	168654.07813
⅜	1881.39063	81605.31836	⅜	2437.89063	120370.84961	⅜	3066.39063	169801.38086
½	1892.25000	82312.87500	½	2450.25000	121287.37500	½	3080.25000	170953.87500
⅝	1903.14063	83024.50977	⅝	2462.64063	122208.54102	⅝	3094.14063	172111.57227
¾	1914.06250	83740.23438	¾	2475.06250	123134.35938	¾	3108.06250	173274.48438
⅞	1925.01563	84460.06055	⅞	2487.51563	124064.84180	⅞	3122.01563	174442.62305
44	1936.00000	85184.00000	50	2500.00000	125000.00000	56	3136.00000	175616.00000
⅛	1947.01563	85912.06445	⅛	2512.51563	125939.84570	⅛	3150.01563	176794.62695
¼	1958.06250	86644.26563	¼	2525.06250	126884.39063	¼	3164.06250	177978.51563
⅜	1969.14063	87380.61523	⅜	2537.64063	127833.64648	⅜	3178.14063	179167.67773
½	1980.25000	88121.12500	½	2550.25000	128787.62500	½	3192.25000	180362.12500
⅝	1991.39063	88865.80664	⅝	2562.89063	129746.33789	⅝	3206.39063	181561.86914
¾	2002.56250	89614.67188	¾	2575.56250	130709.79688	¾	3220.56250	182766.92188
⅞	2013.76563	90367.73242	⅞	2588.26563	131678.01367	⅞	3234.76563	183977.29492
45	2025.00000	91125.00000	51	2601.00000	132651.00000	57	3249.00000	185193.00000
⅛	2036.26563	91886.48633	⅛	2613.76563	133628.76758	⅛	3263.26563	186414.04883
¼	2047.56250	92652.20313	¼	2626.56250	134611.32813	¼	3277.56250	187640.45313
⅜	2058.89063	93422.16211	⅜	2639.39063	135598.69336	⅜	3291.89063	188872.22461
½	2070.25000	94196.37500	½	2652.25000	136590.87500	½	3306.25000	190109.37500
⅝	2081.64063	94974.85352	⅝	2665.14063	137587.88477	⅝	3320.64063	191351.91602
¾	2093.06250	95757.60938	¾	2678.06250	138589.73438	¾	3335.06250	192599.85938
⅞	2104.51563	96544.65430	⅞	2691.01563	139596.43555	⅞	3349.51563	193853.21680

Squares and Cubes of Numbers from 58 to 75⅞

No.	Square	Cube	No.	Square	Cube	No.	Square	Cube
58	3364.00000	195112.00000	64	4096.00000	262144.00000	70	4900.00000	343000.00000
⅛	3378.51563	196376.22070	⅛	4112.01563	263683.00195	⅛	4917.51563	344840.78320
¼	3393.06250	197645.89063	¼	4128.06250	265228.01563	¼	4935.06250	346688.14063
⅜	3407.64063	198921.02148	⅜	4144.14063	266779.05273	⅜	4952.64063	348542.08398
½	3422.25000	200201.62500	½	4160.25000	268336.12500	½	4970.25000	350402.62500
⅝	3436.89063	201487.71289	⅝	4176.39063	269899.24414	⅝	4987.89063	352269.77539
¾	3451.56250	202779.29688	¾	4192.56250	271468.42188	¾	5005.56250	354143.54688
⅞	3466.26563	204076.38867	⅞	4208.76563	273043.66992	⅞	5023.26563	356023.95117
59	3481.00000	205379.00000	65	4225.00000	274625.00000	71	5041.00000	357911.00000
⅛	3495.76563	206687.14258	⅛	4241.26563	276212.42383	⅛	5058.76563	359804.70508
¼	3510.56250	208000.82813	¼	4257.56250	277805.95313	¼	5076.56250	361705.07813
⅜	3525.39063	209320.06836	⅜	4273.89063	279405.59961	⅜	5094.39063	363612.13086
½	3540.25000	210644.87500	½	4290.25000	281011.37500	½	5112.25000	365525.87500
⅝	3555.14063	211975.25977	⅝	4306.64063	282623.29102	⅝	5130.14063	367446.32227
¾	3570.06250	213311.23438	¾	4323.06250	284241.35938	¾	5148.06250	369373.48438
⅞	3585.01563	214652.81055	⅞	4339.51563	285865.59180	⅞	5166.01563	371307.37305
60	3600.00000	216000.00000	66	4356.00000	287496.00000	72	5184.00000	373248.00000
⅛	3615.01563	217352.81445	⅛	4372.51563	289132.59570	⅛	5202.01563	375195.37695
¼	3630.06250	218711.26563	¼	4389.06250	290775.39063	¼	5220.06250	377149.51563
⅜	3645.14063	220075.36523	⅜	4405.64063	292424.39648	⅜	5238.14063	379110.42773
½	3660.25000	221445.12500	½	4422.25000	294079.62500	½	5256.25000	381078.12500
⅝	3675.39063	222820.55664	⅝	4438.89063	295741.08789	⅝	5274.39063	383052.61914
¾	3690.56250	224201.67188	¾	4455.56250	297408.79688	¾	5292.56250	385033.92188
⅞	3705.76563	225588.48242	⅞	4472.26563	299082.76367	⅞	5310.76563	387022.04492
61	3721.00000	226981.00000	67	4489.00000	300763.00000	73	5329.00000	389017.00000
⅛	3736.26563	228379.23633	⅛	4505.76563	302449.51758	⅛	5347.26563	391018.79883
¼	3751.56250	229783.20313	¼	4522.56250	304142.32813	¼	5365.56250	393027.45313
⅜	3766.89063	231192.91211	⅜	4539.39063	305841.44336	⅜	5383.89063	395042.97461
½	3782.25000	232608.37500	½	4556.25000	307546.87500	½	5402.25000	397065.37500
⅝	3797.64063	234029.60352	⅝	4573.14063	309258.63477	⅝	5420.64063	399094.66602
¾	3813.06250	235456.60938	¾	4590.06250	310976.73438	¾	5439.06250	401130.85938
⅞	3828.51563	236889.40430	⅞	4607.01563	312701.18555	⅞	5457.51563	403173.96680
62	3844.00000	238328.00000	68	4624.00000	314432.00000	74	5476.00000	405224.00000
⅛	3859.51563	239772.40820	⅛	4641.01563	316169.18945	⅛	5494.51563	407280.97070
¼	3875.06250	241222.64063	¼	4658.06250	317912.76563	¼	5513.06250	409344.89063
⅜	3890.64063	242678.70898	⅜	4675.14063	319662.74023	⅜	5531.64063	411415.77148
½	3906.25000	244140.62500	½	4692.25000	321419.12500	½	5550.25000	413493.62500
⅝	3921.89063	245608.40039	⅝	4709.39063	323181.93164	⅝	5568.89063	415578.46289
¾	3937.56250	247082.04688	¾	4726.56250	324951.17188	¾	5587.56250	417670.29688
⅞	3953.26563	248561.57617	⅞	4743.76563	326726.85742	⅞	5606.26563	419769.13867
63	3969.00000	250047.00000	69	4761.00000	328509.00000	75	5625.00000	421875.00000
⅛	3984.76563	251538.33008	⅛	4778.26563	330297.61133	⅛	5643.76563	423987.89258
¼	4000.56250	253035.57813	¼	4795.56250	332092.70313	¼	5662.56250	426107.82813
⅜	4016.39063	254538.75586	⅜	4812.89063	333894.28711	⅜	5681.39063	428234.81836
½	4032.25000	256047.87500	½	4830.25000	335702.37500	½	5700.25000	430368.87500
⅝	4048.14063	257562.94727	⅝	4847.64063	337516.97852	⅝	5719.14063	432510.00977
¾	4064.06250	259083.98438	¾	4865.06250	339338.10938	¾	5738.06250	434658.23438
⅞	4080.01563	260610.99805	⅞	4882.51563	341165.77930	⅞	5757.01563	436813.56055

Squares and Cubes of Numbers from 76 to 93⅞

No.	Square	Cube	No.	Square	Cube	No.	Square	Cube
76	5776.00000	438976.00000	82	6724.00000	551368.00000	88	7744.00000	681472.00000
⅛	5795.01563	441145.56445	⅛	6744.51563	553893.34570	⅛	7766.01563	684380.12695
¼	5814.06250	443322.26563	¼	6765.06250	556426.39063	¼	7788.06250	687296.51563
⅜	5833.14063	445506.11523	⅜	6785.64063	558967.14648	⅜	7810.14063	690221.17773
½	5852.25000	447697.12500	½	6806.25000	561515.62500	½	7832.25000	693154.12500
⅝	5871.39063	449895.30664	⅝	6826.89063	564071.83789	⅝	7854.39063	696095.36914
¾	5890.56250	452100.67188	¾	6847.56250	566635.79688	¾	7876.56250	699044.92188
⅞	5909.76563	454313.23242	⅞	6868.26563	569207.51367	⅞	7898.76563	702002.79492
77	5929.00000	456533.00000	83	6889.00000	571787.00000	89	7921.00000	704969.00000
⅛	5948.26563	458759.98633	⅛	6909.76563	574374.26758	⅛	7943.26563	707943.54883
¼	5967.56250	460994.20313	¼	6930.56250	576969.32813	¼	7965.56250	710926.45313
⅜	5986.89063	463235.66211	⅜	6951.39063	579572.19336	⅜	7987.89063	713917.72461
½	6006.25000	465484.37500	½	6972.25000	582182.87500	½	8010.25000	716917.37500
⅝	6025.64063	467740.35352	⅝	6993.14063	584801.38477	⅝	8032.64063	719925.41602
¾	6045.06250	470003.60938	¾	7014.06250	587427.73438	¾	8055.06250	722941.85938
⅞	6064.51563	472274.15430	⅞	7035.01563	590061.93555	⅞	8077.51563	725966.71680
78	6084.00000	474552.00000	84	7056.00000	592704.00000	90	8100.00000	729000.00000
⅛	6103.51563	476837.15820	⅛	7077.01563	595353.93945	⅛	8122.51563	732041.72070
¼	6123.06250	479129.64063	¼	7098.06250	598011.76563	¼	8145.06250	735091.89063
⅜	6142.64063	481429.45898	⅜	7119.14063	600677.49023	⅜	8167.64063	738150.52148
½	6162.25000	483736.62500	½	7140.25000	603351.12500	½	8190.25000	741217.62500
⅝	6181.89063	486051.15039	⅝	7161.39063	606032.68164	⅝	8212.89063	744293.21289
¾	6201.56250	488373.04688	¾	7182.56250	608722.17188	¾	8235.56250	747377.29688
⅞	6221.26563	490702.32617	⅞	7203.76563	611419.60742	⅞	8258.26563	750469.88867
79	6241.00000	493039.00000	85	7225.00000	614125.00000	91	8281.00000	753571.00000
⅛	6260.76563	495383.08008	⅛	7246.26563	616838.36133	⅛	8303.76563	756680.64258
¼	6280.56250	497734.57813	¼	7267.56250	619559.70313	¼	8326.56250	759798.82813
⅜	6300.39063	500093.50586	⅜	7288.89063	622289.03711	⅜	8349.39063	762925.56836
½	6320.25000	502459.87500	½	7310.25000	625026.37500	½	8372.25000	766060.87500
⅝	6340.14063	504833.69727	⅝	7331.64063	627771.72852	⅝	8395.14063	769204.75977
¾	6360.06250	507214.98438	¾	7353.06250	630525.10938	¾	8418.06250	772357.23438
⅞	6380.01563	509603.74805	⅞	7374.51563	633286.52930	⅞	8441.01563	775518.31055
80	6400.00000	512000.00000	86	7396.00000	636056.00000	92	8464.00000	778688.00000
⅛	6420.01563	514403.75195	⅛	7417.51563	638833.53320	⅛	8487.01563	781866.31445
¼	6440.06250	516815.01563	¼	7439.06250	641619.14063	¼	8510.06250	785053.26563
⅜	6460.14063	519233.80273	⅜	7460.64063	644412.83398	⅜	8533.14063	788248.86523
½	6480.25000	521660.12500	½	7482.25000	647214.62500	½	8556.25000	791453.12500
⅝	6500.39063	524093.99414	⅝	7503.89063	650024.52539	⅝	8579.39063	794666.05664
¾	6520.56250	526535.42188	¾	7525.56250	652842.54688	¾	8602.56250	797887.67188
⅞	6540.76563	528984.41992	⅞	7547.26563	655668.70117	⅞	8625.76563	801117.98242
81	6561.00000	531441.00000	87	7569.00000	658503.00000	93	8649.00000	804357.00000
⅛	6581.26563	533905.17383	⅛	7590.76563	661345.45508	⅛	8672.26563	807604.73633
¼	6601.56250	536376.95313	¼	7612.56250	664196.07813	¼	8695.56250	810861.20313
⅜	6621.89063	538856.34961	⅜	7634.39063	667054.88086	⅜	8718.89063	814126.41211
½	6642.25000	541343.37500	½	7656.25000	669921.87500	½	8742.25000	817400.37500
⅝	6662.64063	543838.04102	⅝	7678.14063	672797.07227	⅝	8765.64063	820683.10352
¾	6683.06250	546340.35938	¾	7700.06250	675680.48438	¾	8789.06250	823974.60938
⅞	6703.51563	548850.34180	⅞	7722.01563	678572.12305	⅞	8812.51563	827274.90430

Squares and Cubes of Numbers from 94 to 100

No.	Square	Cube	No.	Square	Cube	No.	Square	Cube
94	8836.00000	830584.00000	96	9216.00000	884736.00000	98	9604.00000	941192.00000
⅛	8859.51563	833901.90820	⅛	9240.01563	888196.50195	⅛	9628.51563	944798.09570
¼	8883.06250	837228.64063	¼	9264.06250	891666.01563	¼	9653.06250	948413.39063
⅜	8906.64063	840564.20898	⅜	9288.14063	895144.55273	⅜	9677.64063	952037.89648
½	8930.25000	843908.62500	½	9312.25000	898632.12500	½	9702.25000	955671.62500
⅝	8953.89063	847261.90039	⅝	9336.39063	902128.74414	⅝	9726.89063	959314.58789
¾	8977.56250	850624.04688	¾	9360.56250	905634.42188	¾	9751.56250	962966.79688
⅞	9001.26563	853995.07617	⅞	9384.76563	909149.16992	⅞	9776.26563	966628.26367
95	9025.00000	857375.00000	97	9409.00000	912673.00000	99	9801.00000	970299.00000
⅛	9048.76563	860763.83008	⅛	9433.26563	916205.92383	⅛	9825.76563	973979.01758
¼	9072.56250	864161.57813	¼	9457.56250	919747.95313	¼	9850.56250	977668.32813
⅜	9096.39063	867568.25586	⅜	9481.89063	923299.09961	⅜	9875.39063	981366.94336
½	9120.25000	870983.87500	½	9506.25000	926859.37500	½	9900.25000	985074.87500
⅝	9144.14063	874408.44727	⅝	9530.64063	930428.79102	⅝	9925.14063	988792.13477
¾	9168.06250	877841.98438	¾	9555.06250	934007.35938	¾	9950.06250	992518.73438
⅞	9192.01563	881284.49805	⅞	9579.51563	937595.09180	⅞	9975.01563	996254.68555
						100	10,000.00	1,000,000

Fractions of Pi

Table of Fractions of $\pi = 3.14159265$

a	π/a	a	π/a	a	π/a	a	π/a	a	π/a
1	3.14159	21	0.14960	41	0.07662	61	0.05150	81	0.03879
2	1.57080	22	0.14280	42	0.07480	62	0.05067	82	0.03831
3	1.04720	23	0.13659	43	0.07306	63	0.04987	83	0.03785
4	0.78540	24	0.13090	44	0.07140	64	0.04909	84	0.03740
5	0.62832	25	0.12566	45	0.06981	65	0.04833	85	0.03696
6	0.52360	26	0.12083	46	0.06830	66	0.04760	86	0.03653
7	0.44880	27	0.11636	47	0.06684	67	0.04689	87	0.03611
8	0.39270	28	0.11220	48	0.06545	68	0.04620	88	0.03570
9	0.34907	29	0.10833	49	0.06411	69	0.04553	89	0.03530
10	0.31416	30	0.10472	50	0.06283	70	0.04488	90	0.03491
11	0.28560	31	0.10134	51	0.06160	71	0.04425	91	0.03452
12	0.26180	32	0.09817	52	0.06042	72	0.04363	92	0.03415
13	0.24166	33	0.09520	53	0.05928	73	0.04304	93	0.03378
14	0.22440	34	0.09240	54	0.05818	74	0.04245	94	0.03342
15	0.20944	35	0.08976	55	0.05712	75	0.04189	95	0.03307
16	0.19635	36	0.08727	56	0.05610	76	0.04134	96	0.03272
17	0.18480	37	0.08491	57	0.05512	77	0.04080	97	0.03239
18	0.17453	38	0.08267	58	0.05417	78	0.04028	98	0.03206
19	0.16535	39	0.08055	59	0.05325	79	0.03977	99	0.03173
20	0.15708	40	0.07854	60	0.05236	80	0.03927	100	0.03142

Functions of π

Constant	Numerical Value	Logarithm	Constant	Numerical Value	Logarithm
π	3.141593	0.49715	2π	6.283185	0.79818
3π	9.424778	0.97427	4π	12.566370	1.09921
$2\pi/3$	2.094395	0.32105	$4\pi/3$	4.188790	0.62209
$\pi \div 2$	1.570796	0.19611	$\pi \div 3$	1.047197	0.02003
$\pi \div 4$	0.785398	−0.10491	$\pi \div 6$	0.523598	−0.28100
$\pi\sqrt{2}$	4.442882	0.64766	$\pi\sqrt{3}$	5.441398	0.73571
$\pi/\sqrt{2}$	2.221441	0.34663	$\pi/\sqrt{3}$	1.813799	0.25859

Functions of π *(Continued)*

Constant	Numerical Value	Logarithm	Constant	Numerical Value	Logarithm
π^2	9.869604	0.99430	$1 \div \pi^2$	0.101321	–0.99430
$1 \div \pi$	0.318310	–0.49715	$1 \div 2\pi$	0.159155	–0.79818
$1 \div \pi^3$	0.032252	–1.49145	π^3	31.006277	1.49145
$\sqrt{\pi}$	1.772454	0.24858	$\sqrt[3]{\pi}$	1.464592	0.16572

Weights and Volumes

Constant	Numerical Value	Logarithm
Weight in pounds of:		
Water column, 1″ × 1″ × 1 ft.	0.4335	–0.36301
1 U.S. gallon of water, 39.1°F.	8.34	0.92117
1 cu. ft. of water, 39.1° F.	62.4245	1.79536
1 cu. in. of water, 39.1°F.	0.0361	–1.44249
1 cu. ft. of air, 32°F., atmospheric pressure	0.08073	–1.09297
Volume in cu. ft. of:		
1 pound of water, 39.1°F	0.01602	–1.79534
1 pound of air, 32°F., atmospheric pressure	12.387	1.09297
Volume in gallons of 1 pound of water, 39.1°F	0.1199	–0.92118
Volume in cu.in of 1 pound of water, 39.1 °F	27.70	1.44248
Gallons in one cubic feet	7.4805	0.87393
Atmospheric pressure in pounds per sq. inch	14.696	1.16720

Useful Constants Multiplied and Divided by 1 to 10

Constant	Multiplied by:							
	2	3	4	5	6	7	8	9
0.7854	1.5708	2.3562	3.1416	3.9270	4.7124	5.4978	6.2832	7.0686
3.1416	6.2832	9.4248	12.566	15.708	18.850	21.991	25.133	28.274
14.7	29.4	44.1	58.8	73.5	88.2	102.9	117.6	132.3
32.16	64.32	96.48	128.64	160.80	192.96	225.12	257.28	289.44
64.32	128.64	192.96	257.28	321.60	385.92	450.24	514.56	578.88
144	288	432	576	720	864	1,008	1,152	1,296
778	1,556	2,334	3,112	3,890	4,668	5,446	6,224	7,002
1,728	3,456	5,184	6,912	8,640	10,368	12,096	13,824	15,552
33,000	66,000	99,000	132,000	165,000	198,000	231,000	264,000	297,000

Constant	Divided by:							
	2	3	4	5	6	7	8	9
0.7854	0.3927	0.2618	0.1964	0.1571	0.1309	0.1122	0.0982	0.0873
3.1416	1.5708	1.0472	0.7854	0.6283	0.5236	0.4488	0.3927	0.3491
14.7	7.350	4.900	3.675	2.940	2.450	2.100	1.838	1.633
32.16	16.080	10.720	8.040	6.432	5.360	4.594	4.020	3.573
64.32	32.160	21.440	16.080	12.864	10.720	9.189	8.040	7.147
144	72	48	36	28.800	24	20.571	18	16
778	389	259.33	194.50	155.60	129.67	111.14	97.25	86.44
1,728	864	576	432	345.60	288	246.86	216	192
33,000	16,500	11,000	8250	6600	5500	4714.3	4125	3666.7

Circumferences and Areas of Circles

Circumferences and Areas of Circles From 1/64 to 9 7/8

Diameter	Circumfer-ence	Area	Diameter	Circumfer-ence	Area	Diameter	Circumfer-ence	Area
1/64	0.0491	0.0002	2	6.2832	3.1416	5	15.7080	19.635
1/32	0.0982	0.0008	2 1/16	6.4795	3.3410	5 1/16	15.9043	20.129
1/16	0.1963	0.0031	2 1/8	6.6759	3.5466	5 1/8	16.1007	20.629
3/32	0.2945	0.0069	2 3/16	6.8722	3.7583	5 3/16	16.2970	21.135
1/8	0.3927	0.0123	2 1/4	7.0686	3.9761	5 1/4	16.4934	21.648
5/32	0.4909	0.0192	2 5/16	7.2649	4.2000	5 5/16	16.6897	22.166
3/16	0.5890	0.0276	2 3/8	7.4613	4.4301	5 3/8	16.8861	22.691
7/32	0.6872	0.0376	2 7/16	7.6576	4.6664	5 7/16	17.0824	23.221
1/4	0.7854	0.0491	2 1/2	7.8540	4.9087	5 1/2	17.2788	23.758
9/32	0.8836	0.0621	2 9/16	8.0503	5.1572	5 9/16	17.4751	24.301
5/16	0.9817	0.0767	2 5/8	8.2467	5.4119	5 5/8	17.6715	24.850
11/32	1.0799	0.0928	2 11/16	8.4430	5.6727	5 11/16	17.8678	25.406
3/8	1.1781	0.1104	2 3/4	8.6394	5.9396	5 3/4	18.0642	25.967
13/32	1.2763	0.1296	2 13/16	8.8357	6.2126	5 13/16	18.2605	26.535
7/16	1.3744	0.1503	2 7/8	9.0321	6.4918	5 7/8	18.4569	27.109
15/32	1.4726	0.1726	2 15/16	9.2284	6.7771	5 15/16	18.6532	27.688
1/2	1.5708	0.1963	3	9.4248	7.0686	6	18.8496	28.274
17/32	1.6690	0.2217	3 1/16	9.6211	7.3662	6 1/8	19.2423	29.465
9/16	1.7671	0.2485	3 1/8	9.8175	7.6699	6 1/4	19.6350	30.680
19/32	1.8653	0.2769	3 3/16	10.0138	7.9798	6 3/8	20.0277	31.919
5/8	1.9635	0.3068	3 1/4	10.2102	8.2958	6 1/2	20.4204	33.183
21/32	2.0617	0.3382	3 5/16	10.4065	8.6179	6 5/8	20.8131	34.472
11/16	2.1598	0.3712	3 3/8	10.6029	8.9462	6 3/4	21.2058	35.785
23/32	2.2580	0.4057	3 7/16	10.7992	9.2806	6 7/8	21.5984	37.122
3/4	2.3562	0.4418	3 1/2	10.9956	9.6211	7	21.9911	38.485
25/32	2.4544	0.4794	3 9/16	11.1919	9.9678	7 1/8	22.3838	39.871
13/16	2.5525	0.5185	3 5/8	11.388	10.3206	7 1/4	22.7765	41.282
27/32	2.6507	0.5591	3 11/16	11.585	10.6796	7 3/8	23.1692	42.718
7/8	2.7489	0.6013	3 3/4	11.781	11.0447	7 1/2	23.5619	44.179
29/32	2.8471	0.6450	3 13/16	11.977	11.4159	7 5/8	23.9546	45.664
15/16	2.9452	0.6903	3 7/8	12.174	11.7932	7 3/4	24.3473	47.173
31/32	3.0434	0.7371	3 15/16	12.370	12.1767	7 7/8	24.7400	48.707
1	3.1416	0.7854	4	12.566	12.5664	8	25.1327	50.265
1 1/16	3.3379	0.8866	4 1/16	12.763	12.9621	8 1/8	25.5254	51.849
1 1/8	3.5343	0.9940	4 1/8	12.959	13.3640	8 1/4	25.9181	53.456
1 3/16	3.7306	1.1075	4 3/16	13.155	13.7721	8 3/8	26.3108	55.088
1 1/4	3.9270	1.2272	4 1/4	13.352	14.1863	8 1/2	26.7035	56.745
1 5/16	4.1233	1.3530	4 5/16	13.548	14.6066	8 5/8	27.0962	58.426
1 3/8	4.3197	1.4849	4 3/8	13.744	15.0330	8 3/4	27.4889	60.132
1 7/16	4.5160	1.6230	4 7/16	13.941	15.4656	8 7/8	27.8816	61.862
1 1/2	4.7124	1.7671	4 1/2	14.137	15.9043	9	28.2743	63.617
1 9/16	4.9087	1.9175	4 9/16	14.334	16.3492	9 1/8	28.6670	65.397
1 5/8	5.1051	2.0739	4 5/8	14.530	16.8002	9 1/4	29.0597	67.201
1 11/16	5.3014	2.2365	4 11/16	14.726	17.2573	9 3/8	29.4524	69.029
1 3/4	5.4978	2.4053	4 3/4	14.923	17.7205	9 1/2	29.8451	70.882
1 13/16	5.6941	2.5802	4 13/16	15.119	18.1899	9 5/8	30.2378	72.760
1 7/8	5.8905	2.7612	4 7/8	15.315	18.6655	9 3/4	30.6305	74.662
1 15/16	6.0868	2.9483	4 15/16	15.512	19.1471	9 7/8	31.0232	76.589

Circumferences and Areas of Circles From 10 to 28

Diameter	Circumfer-ence	Area	Diameter	Circumfer-ence	Area	Diameter	Circumfer-ence	Area
10	31.41593	78.53983	16	50.26549	201.06195	22	69.11505	380.13275
1/8	31.80863	80.51559	1/8	50.65819	204.21582	1/8	69.50775	384.46472
1/4	32.20133	82.51590	1/4	51.05089	207.39423	1/4	69.90044	388.82122
3/8	32.59403	84.54076	3/8	51.44359	210.59718	3/8	70.29314	393.20227
1/2	32.98673	86.59016	1/2	51.83628	213.82467	1/2	70.68584	397.60786
5/8	33.37943	88.66410	5/8	52.22898	217.07671	5/8	71.07854	402.03800
3/4	33.77212	90.76259	3/4	52.62168	220.35330	3/4	71.47124	406.49268
7/8	34.16482	92.88561	7/8	53.01438	223.65442	7/8	71.86394	410.97191
11	34.55752	95.03319	17	53.40708	226.98009	23	72.25664	415.47567
1/8	34.95022	97.20531	1/8	53.79978	230.33031	1/8	72.64934	420.00399
1/4	35.34292	99.40197	1/4	54.19248	233.70507	1/4	73.04204	424.55684
3/8	35.73562	101.62317	3/8	54.58518	237.10437	3/8	73.43474	429.13424
1/2	36.12832	103.86892	1/2	54.97788	240.52821	1/2	73.82744	433.73618
5/8	36.52102	106.13921	5/8	55.37058	243.97660	5/8	74.22013	438.36267
3/4	36.91372	108.43405	3/4	55.76328	247.44954	3/4	74.61283	443.01370
7/8	37.30642	110.75343	7/8	56.15597	250.94701	7/8	75.00553	447.68927
12	37.69912	113.09735	18	56.54867	254.46903	24	75.39823	452.38939
1/8	38.09182	115.46581	1/8	56.94137	258.01560	1/8	75.79093	457.11405
1/4	38.48451	117.85882	1/4	57.33407	261.58670	1/4	76.18363	461.86326
3/8	38.87721	120.27638	3/8	57.72677	265.18236	3/8	76.57633	466.63701
1/2	39.26991	122.71848	1/2	58.11947	268.80255	1/2	76.96903	471.43530
5/8	39.66261	125.18512	5/8	58.51217	272.44729	5/8	77.36173	476.25814
3/4	40.05531	127.67630	3/4	58.90487	276.11657	3/4	77.75443	481.10552
7/8	40.44801	130.19203	7/8	59.29757	279.81040	7/8	78.14713	485.97744
13	40.84071	132.73230	19	59.69027	283.52877	25	78.53983	490.87391
1/8	41.23341	135.29712	1/8	60.08297	287.27168	1/8	78.93252	495.79492
1/4	41.62611	137.88648	1/4	60.47567	291.03914	1/4	79.32522	500.74047
3/8	42.01881	140.50038	3/8	60.86836	294.83114	3/8	79.71792	505.71057
1/2	42.41151	143.13883	1/2	61.26106	298.64768	1/2	80.11062	510.70521
5/8	42.80420	145.80182	5/8	61.65376	302.48877	5/8	80.50332	515.72440
3/4	43.19690	148.48936	3/4	62.04646	306.35440	3/4	80.89602	520.76813
7/8	43.58960	151.20143	7/8	62.43916	310.24458	7/8	81.28872	525.83640
14	43.98230	153.93806	20	62.83186	314.15930	26	81.68142	530.92922
1/8	44.37500	156.69922	1/8	63.22456	318.09856	1/8	82.07412	536.04658
1/4	44.76770	159.48493	1/4	63.61726	322.06237	1/4	82.46682	541.18848
3/8	45.16040	162.29519	3/8	64.00996	326.05072	3/8	82.85952	546.35493
1/2	45.55310	165.12998	1/2	64.40266	330.06361	1/2	83.25221	551.54592
5/8	45.94580	167.98932	5/8	64.79536	334.10105	5/8	83.64491	556.76146
3/4	46.33850	170.87321	3/4	65.18805	338.16303	3/4	84.03761	562.00154
7/8	46.73120	173.78163	7/8	65.58075	342.24956	7/8	84.43031	567.26616
15	47.12390	176.71461	21	65.97345	346.36063	27	84.82301	572.55532
1/8	47.51659	179.67212	1/8	66.36615	350.49624	1/8	85.21571	577.86903
1/4	47.90929	182.65418	1/4	66.75885	354.65640	1/4	85.60841	583.20729
3/8	48.30199	185.66078	3/8	67.15155	358.84110	3/8	86.00111	588.57009
1/2	48.69469	188.69193	1/2	67.54425	363.05034	1/2	86.39381	593.95743
5/8	49.08739	191.74762	5/8	67.93695	367.28413	5/8	86.78651	599.36931
3/4	49.48009	194.82785	3/4	68.32965	371.54246	3/4	87.17921	604.80574
7/8	49.87279	197.93263	7/8	68.72235	375.82533	7/8	87.57190	610.26671
16	50.26549	201.06195	22	69.11505	380.13275	28	87.96460	615.75223

Circumferences and Areas of Circles From 28 to 46

Diameter	Circumference	Area	Diameter	Circumference	Area	Diameter	Circumference	Area
28	87.96460	615.75223	34	106.81416	907.92038	40	125.66372	1256.63720
1/8	88.35730	621.26229	1/8	107.20686	914.60853	1/8	126.05642	1264.50345
1/4	88.75000	626.79689	1/4	107.59956	921.32123	1/4	126.44912	1272.39425
3/8	89.14270	632.35604	3/8	107.99226	928.05848	3/8	126.84182	1280.30959
1/2	89.53540	637.93973	1/2	108.38496	934.82027	1/2	127.23452	1288.24948
5/8	89.92810	643.54796	5/8	108.77766	941.60660	5/8	127.62722	1296.21391
3/4	90.32080	649.18074	3/4	109.17036	948.41747	3/4	128.01991	1304.20288
7/8	90.71350	654.83806	7/8	109.56306	955.25289	7/8	128.41261	1312.21640
29	91.10620	660.51993	35	109.95576	962.11286	41	128.80531	1320.25446
1/8	91.49890	666.22634	1/8	110.34845	968.99736	1/8	129.19801	1328.31706
1/4	91.89160	671.95729	1/4	110.74115	975.90641	1/4	129.59071	1336.40421
3/8	92.28429	677.71279	3/8	111.13385	982.84001	3/8	129.98341	1344.51590
1/2	92.67699	683.49283	1/2	111.52655	989.79814	1/2	130.37611	1352.65214
5/8	93.06969	689.29741	5/8	111.91925	996.78083	5/8	130.76881	1360.81291
3/4	93.46239	695.12654	3/4	112.31195	1003.78805	3/4	131.16151	1368.99824
7/8	93.85509	700.98021	7/8	112.70465	1010.81982	7/8	131.55421	1377.20810
30	94.24779	706.85843	36	113.09735	1017.87613	42	131.94691	1385.44251
1/8	94.64049	712.76118	1/8	113.49005	1024.95699	1/8	132.33961	1393.70147
1/4	95.03319	718.68849	1/4	113.88275	1032.06239	1/4	132.73230	1401.98496
3/8	95.42589	724.64033	3/8	114.27545	1039.19233	3/8	133.12500	1410.29300
1/2	95.81859	730.61672	1/2	114.66814	1046.34682	1/2	133.51770	1418.62559
5/8	96.21129	736.61766	5/8	115.06084	1053.52585	5/8	133.91040	1426.98272
3/4	96.60398	742.64313	3/4	115.45354	1060.72942	3/4	134.30310	1435.36439
7/8	96.99668	748.69315	7/8	115.84624	1067.95754	7/8	134.69580	1443.77060
31	97.38938	754.76772	37	116.23894	1075.21020	43	135.08850	1452.20136
1/8	97.78208	760.86683	1/8	116.63164	1082.48741	1/8	135.48120	1460.65667
1/4	98.17478	766.99048	1/4	117.02434	1089.78916	1/4	135.87390	1469.13651
3/8	98.56748	773.13867	3/8	117.41704	1097.11545	3/8	136.26660	1477.64090
1/2	98.96018	779.31141	1/2	117.80974	1104.46629	1/2	136.65930	1486.16984
5/8	99.35288	785.50870	5/8	118.20244	1111.84167	5/8	137.05199	1494.72332
3/4	99.74558	791.73052	3/4	118.59514	1119.24159	3/4	137.44469	1503.30134
7/8	100.13828	797.97689	7/8	118.98783	1126.66606	7/8	137.83739	1511.90390
32	100.53098	804.24781	38	119.38053	1134.11507	44	138.23009	1520.53101
1/8	100.92368	810.54327	1/8	119.77323	1141.58863	1/8	138.62279	1529.18266
1/4	101.31637	816.86327	1/4	120.16593	1149.08673	1/4	139.01549	1537.85886
3/8	101.70907	823.20781	3/8	120.55863	1156.60937	3/8	139.40819	1546.55960
1/2	102.10177	829.57690	1/2	120.95133	1164.15656	1/2	139.80089	1555.28488
5/8	102.49447	835.97053	5/8	121.34403	1171.72829	5/8	140.19359	1564.03471
3/4	102.88717	842.38871	3/4	121.73673	1179.32456	3/4	140.58629	1572.80908
7/8	103.27987	848.83143	7/8	122.12943	1186.94538	7/8	140.97899	1581.60800
33	103.67257	855.29869	39	122.52213	1194.59074	45	141.37169	1590.43146
1/8	104.06527	861.79050	1/8	122.91483	1202.26064	1/8	141.76438	1599.27946
1/4	104.45797	868.30685	1/4	123.30753	1209.95509	1/4	142.15708	1608.15200
3/8	104.85067	874.84775	3/8	123.70022	1217.67408	3/8	142.54978	1617.04909
1/2	105.24337	881.41319	1/2	124.09292	1225.41762	1/2	142.94248	1625.97073
5/8	105.63606	888.00317	5/8	124.48562	1233.18570	5/8	143.33518	1634.91690
3/4	106.02876	894.61769	3/4	124.87832	1240.97832	3/4	143.72788	1643.88762
7/8	106.42146	901.25676	7/8	125.27102	1248.79549	7/8	144.12058	1652.88289
34	106.81416	907.92038	40	125.66372	1256.63720	46	144.51328	1661.90270

Circumferences and Areas of Circles From 46 to 64

Diameter	Circumference	Area	Diameter	Circumference	Area	Diameter	Circumference	Area
46	144.51328	1661.90270	52	163.36284	2123.71687	58	182.21239	2642.07971
1/8	144.90598	1670.94705	1/8	163.75554	2133.93932	1/8	182.60509	2653.48026
1/4	145.29868	1680.01594	1/4	164.14823	2144.18631	1/4	182.99779	2664.90535
3/8	145.69138	1689.10938	3/8	164.54093	2154.45785	3/8	183.39049	2676.35498
1/2	146.08407	1698.22737	1/2	164.93363	2164.75393	1/2	183.78319	2687.82916
5/8	146.47677	1707.36989	5/8	165.32633	2175.07455	5/8	184.17589	2699.32788
3/4	146.86947	1716.53696	3/4	165.71903	2185.41972	3/4	184.56859	2710.85115
7/8	147.26217	1725.72858	7/8	166.11173	2195.78943	7/8	184.96129	2722.39896
47	147.65487	1734.94473	53	166.50443	2206.18368	59	185.35399	2733.97131
1/8	148.04757	1744.18544	1/8	166.89713	2216.60248	1/8	185.74669	2745.56820
1/4	148.44027	1753.45068	1/4	167.28983	2227.04583	1/4	186.13939	2757.18964
3/8	148.83297	1762.74047	3/8	167.68253	2237.51371	3/8	186.53208	2768.83563
1/2	149.22567	1772.05480	1/2	168.07523	2248.00614	1/2	186.92478	2780.50615
5/8	149.61837	1781.39368	5/8	168.46792	2258.52311	5/8	187.31748	2792.20123
3/4	150.01107	1790.75710	3/4	168.86062	2269.06463	3/4	187.71018	2803.92084
7/8	150.40376	1800.14506	7/8	169.25332	2279.63069	7/8	188.10288	2815.66500
48	150.79646	1809.55757	54	169.64602	2290.22130	60	188.49558	2827.43370
1/8	151.18916	1818.99462	1/8	170.03872	2300.83645	1/8	188.88828	2839.22695
1/4	151.58186	1828.45621	1/4	170.43142	2311.47614	1/4	189.28098	2851.04473
3/8	151.97456	1837.94235	3/8	170.82412	2322.14037	3/8	189.67368	2862.88707
1/2	152.36726	1847.45303	1/2	171.21682	2332.82915	1/2	190.06638	2874.75394
5/8	152.75996	1856.98826	5/8	171.60952	2343.54248	5/8	190.45908	2886.64536
3/4	153.15266	1866.54803	3/4	172.00222	2354.28034	3/4	190.85177	2898.56133
7/8	153.54536	1876.13234	7/8	172.39492	2365.04275	7/8	191.24447	2910.50184
49	153.93806	1885.74120	55	172.78762	2375.82971	61	191.63717	2922.46689
1/8	154.33076	1895.37460	1/8	173.18031	2386.64120	1/8	192.02987	2934.45648
1/4	154.72346	1905.03254	1/4	173.57301	2397.47725	1/4	192.42257	2946.47062
3/8	155.11615	1914.71503	3/8	173.96571	2408.33783	3/8	192.81527	2958.50930
1/2	155.50885	1924.42206	1/2	174.35841	2419.22296	1/2	193.20797	2970.57253
5/8	155.90155	1934.15364	5/8	174.75111	2430.13263	5/8	193.60067	2982.66030
3/4	156.29425	1943.90976	3/4	175.14381	2441.06685	3/4	193.99337	2994.77261
7/8	156.68695	1953.69042	7/8	175.53651	2452.02561	7/8	194.38607	3006.90947
50	157.07965	1963.49563	56	175.92921	2463.00891	62	194.77877	3019.07087
1/8	157.47235	1973.32537	1/8	176.32191	2474.01676	1/8	195.17147	3031.25682
1/4	157.86505	1983.17967	1/4	176.71461	2485.04915	1/4	195.56416	3043.46731
3/8	158.25775	1993.05851	3/8	177.10731	2496.10609	3/8	195.95686	3055.70234
1/2	158.65045	2002.96189	1/2	177.50000	2507.18756	1/2	196.34956	3067.96191
5/8	159.04315	2012.88981	5/8	177.89270	2518.29359	5/8	196.74226	3080.24603
3/4	159.43584	2022.84228	3/4	178.28540	2529.42415	3/4	197.13496	3092.55470
7/8	159.82854	2032.81929	7/8	178.67810	2540.57926	7/8	197.52766	3104.88790
51	160.22124	2042.82085	57	179.07080	2551.75891	63	197.92036	3117.24565
1/8	160.61394	2052.84695	1/8	179.46350	2562.96311	1/8	198.31306	3129.62795
1/4	161.00664	2062.89759	1/4	179.85620	2574.19185	1/4	198.70576	3142.03479
3/8	161.39934	2072.97278	3/8	180.24890	2585.44514	3/8	199.09846	3154.46617
1/2	161.79204	2083.07251	1/2	180.64160	2596.72296	1/2	199.49116	3166.92209
5/8	162.18474	2093.19678	5/8	181.03430	2608.02534	5/8	199.88385	3179.40256
3/4	162.57744	2103.34560	3/4	181.42700	2619.35225	3/4	200.27655	3191.90758
7/8	162.97014	2113.51896	7/8	181.81969	2630.70371	7/8	200.66925	3204.43713
52	163.36284	2123.71687	58	182.21239	2642.07971	64	201.06195	3216.99123

Circumferences and Areas of Circles From 64 to 82

Diameter	Circumfer-ence	Area	Diameter	Circumfer-ence	Area	Diameter	Circumfer-ence	Area
64	201.06195	3216.99123	70	219.91151	3848.45143	76	238.76107	4536.46029
1/8	201.45465	3229.56988	1/8	220.30421	3862.20817	1/8	239.15377	4551.39513
1/4	201.84735	3242.17306	1/4	220.69691	3875.98945	1/4	239.54647	4566.35451
3/8	202.24005	3254.80079	3/8	221.08961	3889.79528	3/8	239.93917	4581.33844
1/2	202.63275	3267.45307	1/2	221.48231	3903.62565	1/2	240.33186	4596.34691
5/8	203.02545	3280.12989	5/8	221.87501	3917.48057	5/8	240.72456	4611.37992
3/4	203.41815	3292.83125	3/4	222.26770	3931.36003	3/4	241.11726	4626.43748
7/8	203.81085	3305.55716	7/8	222.66040	3945.26403	7/8	241.50996	4641.51958
65	204.20355	3318.30761	71	223.05310	3959.19258	77	241.90266	4656.62622
1/8	204.59624	3331.08260	1/8	223.44580	3973.14567	1/8	242.29536	4671.75741
1/4	204.98894	3343.88214	1/4	223.83850	3987.12330	1/4	242.68806	4686.91314
3/8	205.38164	3356.70622	3/8	224.23120	4001.12548	3/8	243.08076	4702.09342
1/2	205.77434	3369.55484	1/2	224.62390	4015.15220	1/2	243.47346	4717.29824
5/8	206.16704	3382.42801	5/8	225.01660	4029.20347	5/8	243.86616	4732.52760
3/4	206.55974	3395.32572	3/4	225.40930	4043.27928	3/4	244.25886	4747.78151
7/8	206.95244	3408.24798	7/8	225.80200	4057.37963	7/8	244.65155	4763.05996
66	207.34514	3421.19478	72	226.19470	4071.50453	78	245.04425	4778.36295
1/8	207.73784	3434.16612	1/8	226.58740	4085.65397	1/8	245.43695	4793.69049
1/4	208.13054	3447.16201	1/4	226.98009	4099.82795	1/4	245.82965	4809.04257
3/8	208.52324	3460.18244	3/8	227.37279	4114.02648	3/8	246.22235	4824.41920
1/2	208.91593	3473.22741	1/2	227.76549	4128.24955	1/2	246.61505	4839.82037
5/8	209.30863	3486.29693	5/8	228.15819	4142.49717	5/8	247.00775	4855.24608
3/4	209.70133	3499.39099	3/4	228.55089	4156.76933	3/4	247.40045	4870.69633
7/8	210.09403	3512.50960	7/8	228.94359	4171.06603	7/8	247.79315	4886.17113
67	210.48673	3525.65274	73	229.33629	4185.38727	79	248.18585	4901.67048
1/8	210.87943	3538.82044	1/8	229.72899	4199.73306	1/8	248.57855	4917.19437
1/4	211.27213	3552.01267	1/4	230.12169	4214.10340	1/4	248.97125	4932.74280
3/8	211.66483	3565.22945	3/8	230.51439	4228.49828	3/8	249.36394	4948.31577
1/2	212.05753	3578.47078	1/2	230.90709	4242.91770	1/2	249.75664	4963.91329
5/8	212.45023	3591.73664	5/8	231.29978	4257.36166	5/8	250.14934	4979.53535
3/4	212.84293	3605.02705	3/4	231.69248	4271.83017	3/4	250.54204	4995.18196
7/8	213.23562	3618.34201	7/8	232.08518	4286.32322	7/8	250.93474	5010.85311
68	213.62832	3631.68151	74	232.47788	4300.84082	80	251.32744	5026.54880
1/8	214.02102	3645.04555	1/8	232.87058	4315.38296	1/8	251.72014	5042.26904
1/4	214.41372	3658.43414	1/4	233.26328	4329.94964	1/4	252.11284	5058.01382
3/8	214.80642	3671.84727	3/8	233.65598	4344.54087	3/8	252.50554	5073.78314
1/2	215.19912	3685.28494	1/2	234.04868	4359.15664	1/2	252.89824	5089.57701
5/8	215.59182	3698.74716	5/8	234.44138	4373.79695	5/8	253.29094	5105.39542
3/4	215.98452	3712.23392	3/4	234.83408	4388.46181	3/4	253.68363	5121.23838
7/8	216.37722	3725.74522	7/8	235.22678	4403.15121	7/8	254.07633	5137.10588
69	216.76992	3739.28107	75	235.61948	4417.86516	81	254.46903	5152.99792
1/8	217.16262	3752.84146	1/8	236.01217	4432.60365	1/8	254.86173	5168.91450
1/4	217.55532	3766.42640	1/4	236.40487	4447.36668	1/4	255.25443	5184.85563
3/8	217.94801	3780.03587	3/8	236.79757	4462.15425	3/8	255.64713	5200.82131
1/2	218.34071	3793.66990	1/2	237.19027	4476.96637	1/2	256.03983	5216.81153
5/8	218.73341	3807.32846	5/8	237.58297	4491.80304	5/8	256.43253	5232.82629
3/4	219.12611	3821.01157	3/4	237.97567	4506.66425	3/4	256.82523	5248.86559
7/8	219.51881	3834.71923	7/8	238.36837	4521.55000	7/8	257.21793	5264.92944
70	219.91151	3848.45143	76	238.76107	4536.46029	82	257.61063	5281.01783

Circumferences and Areas of Circles From 82 to 100

Diameter	Circumference	Area	Diameter	Circumference	Area	Diameter	Circumference	Area
82	257.61063	5281.01783	88	276.46018	6082.12405	94	295.30974	6939.77894
1/8	258.00333	5297.13077	1/8	276.85288	6099.41508	1/8	295.70244	6958.24807
1/4	258.39602	5313.26825	1/4	277.24558	6116.73066	1/4	296.09514	6976.74174
3/8	258.78872	5329.43027	3/8	277.63828	6134.07078	3/8	296.48784	6995.25996
1/2	259.18142	5345.61684	1/2	278.03098	6151.43544	1/2	296.88054	7013.80272
5/8	259.57412	5361.82795	5/8	278.42368	6168.82465	5/8	297.27324	7032.37003
3/4	259.96682	5378.06360	3/4	278.81638	6186.23840	3/4	297.66594	7050.96188
7/8	260.35952	5394.32380	7/8	279.20908	6203.67670	7/8	298.05864	7069.57827
83	260.75222	5410.60854	89	279.60178	6221.13954	95	298.45134	7088.21921
1/8	261.14492	5426.91783	1/8	279.99448	6238.62692	1/8	298.84403	7106.88469
1/4	261.53762	5443.25166	1/4	280.38718	6256.13885	1/4	299.23673	7125.57471
3/8	261.93032	5459.61003	3/8	280.77987	6273.67532	3/8	299.62943	7144.28928
1/2	262.32302	5475.99295	1/2	281.17257	6291.23633	1/2	300.02213	7163.02839
5/8	262.71571	5492.40041	5/8	281.56527	6308.82189	5/8	300.41483	7181.79204
3/4	263.10841	5508.83241	3/4	281.95797	6326.43199	3/4	300.80753	7200.58024
7/8	263.50111	5525.28896	7/8	282.35067	6344.06664	7/8	301.20023	7219.39299
84	263.89381	5541.77005	90	282.74337	6361.72583	96	301.59293	7238.23027
1/8	264.28651	5558.27569	1/8	283.13607	6379.40956	1/8	301.98563	7257.09210
1/4	264.67921	5574.80587	1/4	283.52877	6397.11783	1/4	302.37833	7275.97848
3/8	265.07191	5591.36059	3/8	283.92147	6414.85065	3/8	302.77103	7294.88939
1/2	265.46461	5607.93985	1/2	284.31417	6432.60802	1/2	303.16372	7313.82485
5/8	265.85731	5624.54366	5/8	284.70687	6450.38992	5/8	303.55642	7332.78486
3/4	266.25001	5641.17202	3/4	285.09956	6468.19638	3/4	303.94912	7351.76941
7/8	266.64271	5657.82492	7/8	285.49226	6486.02737	7/8	304.34182	7370.77850
85	267.03541	5674.50236	91	285.88496	6503.88291	97	304.73452	7389.81213
1/8	267.42810	5691.20434	1/8	286.27766	6521.76299	1/8	305.12722	7408.87031
1/4	267.82080	5707.93087	1/4	286.67036	6539.66762	1/4	305.51992	7427.95304
3/8	268.21350	5724.68194	3/8	287.06306	6557.59679	3/8	305.91262	7447.06030
1/2	268.60620	5741.45756	1/2	287.45576	6575.55050	1/2	306.30532	7466.19211
5/8	268.99890	5758.25772	5/8	287.84846	6593.52876	5/8	306.69802	7485.34847
3/4	269.39160	5775.08242	3/4	288.24116	6611.53156	3/4	307.09072	7504.52937
7/8	269.78430	5791.93167	7/8	288.63386	6629.55890	7/8	307.48341	7523.73481
86	270.17700	5808.80546	92	289.02656	6647.61079	98	307.87611	7542.96479
1/8	270.56970	5825.70379	1/8	289.41926	6665.68722	1/8	308.26881	7562.21932
1/4	270.96240	5842.62667	1/4	289.81195	6683.78819	1/4	308.66151	7581.49839
3/8	271.35510	5859.57409	3/8	290.20465	6701.91371	3/8	309.05421	7600.80201
1/2	271.74779	5876.54606	1/2	290.59735	6720.06378	1/2	309.44691	7620.13017
5/8	272.14049	5893.54257	5/8	290.99005	6738.23838	5/8	309.83961	7639.48287
3/4	272.53319	5910.56362	3/4	291.38275	6756.43753	3/4	310.23231	7658.86012
7/8	272.92589	5927.60921	7/8	291.77545	6774.66123	7/8	310.62501	7678.26191
87	273.31859	5944.67935	93	292.16815	6792.90946	99	311.01771	7697.68825
1/8	273.71129	5961.77404	1/8	292.56085	6811.18225	1/8	311.41041	7717.13913
1/4	274.10399	5978.89327	1/4	292.95355	6829.47957	1/4	311.80311	7736.61455
3/8	274.49669	5996.03704	3/8	293.34625	6847.80144	3/8	312.19580	7756.11451
1/2	274.88939	6013.20535	1/2	293.73895	6866.14785	1/2	312.58850	7775.63902
5/8	275.28209	6030.39821	5/8	294.13164	6884.51881	5/8	312.98120	7795.18808
3/4	275.67479	6047.61561	3/4	294.52434	6902.91431	3/4	313.37390	7814.76167
7/8	276.06748	6064.85756	7/8	294.91704	6921.33435	7/8	313.76660	7834.35982
88	276.46018	6082.12405	94	295.30974	6939.77894	100	314.15930	7853.98250

Circumferences and Areas of Circles From 100 to 249

Diameter	Circumfer-ence	Area	Diameter	Circumfer-ence	Area	Diameter	Circumfer-ence	Area
100	314.15930	7853.98250	150	471.23895	17671.46063	200	628.31860	31415.93000
101	317.30089	8011.84755	151	474.38054	17907.86550	201	631.46019	31730.87470
102	320.44249	8171.28339	152	477.52214	18145.84117	202	634.60179	32047.39019
103	323.58408	8332.29003	153	480.66373	18385.38763	203	637.74338	32365.47648
104	326.72567	8494.86747	154	483.80532	18626.50490	204	640.88497	32685.13357
105	329.86727	8659.01571	155	486.94692	18869.19296	205	644.02657	33006.36146
106	333.00886	8824.73474	156	490.08851	19113.45181	206	647.16816	33329.16014
107	336.15045	8992.02456	157	493.23010	19359.28146	207	650.30975	33653.52961
108	339.29204	9160.88519	158	496.37169	19606.68191	208	653.45134	33979.46989
109	342.43364	9331.31661	159	499.51329	19855.65316	209	656.59294	34306.98096
110	345.57523	9503.31883	160	502.65488	20106.19520	210	659.73453	34636.06283
111	348.71682	9676.89184	161	505.79647	20358.30804	211	662.87612	34966.71549
112	351.85842	9852.03565	162	508.93807	20611.99167	212	666.01772	35298.93895
113	355.00001	10028.75025	163	512.07966	20867.24610	213	669.15931	35632.73320
114	358.14160	10207.03566	164	515.22125	21124.07133	214	672.30090	35968.09826
115	361.28320	10386.89186	165	518.36285	21382.46736	215	675.44250	36305.03411
116	364.42479	10568.31885	166	521.50444	21642.43418	216	678.58409	36643.54075
117	367.56638	10751.31664	167	524.64603	21903.97179	217	681.72568	36983.61819
118	370.70797	10935.88523	168	527.78762	22167.08021	218	684.86727	37325.26643
119	373.84957	11122.02462	169	530.92922	22431.75942	219	688.00887	37668.48547
120	376.99116	11309.73480	170	534.07081	22698.00943	220	691.15046	38013.27530
121	380.13275	11499.01578	171	537.21240	22965.83023	221	694.29205	38359.63593
122	383.27435	11689.86755	172	540.35400	23235.22183	222	697.43365	38707.56735
123	386.41594	11882.29012	173	543.49559	23506.18422	223	700.57524	39057.06957
124	389.55753	12076.28349	174	546.63718	23778.71742	224	703.71683	39408.14259
125	392.69913	12271.84766	175	549.77878	24052.82141	225	706.85843	39760.78641
126	395.84072	12468.98262	176	552.92037	24328.49619	226	710.00002	40115.00102
127	398.98231	12667.68837	177	556.06196	24605.74177	227	713.14161	40470.78642
128	402.12390	12867.96493	178	559.20355	24884.55815	228	716.28320	40828.14263
129	405.26550	13069.81228	179	562.34515	25164.94533	229	719.42480	41187.06963
130	408.40709	13273.23043	180	565.48674	25446.90330	230	722.56639	41547.56743
131	411.54868	13478.21937	181	568.62833	25730.43207	231	725.70798	41909.63602
132	414.69028	13684.77911	182	571.76993	26015.53163	232	728.84958	42273.27541
133	417.83187	13892.90964	183	574.91152	26302.20199	233	731.99117	42638.48559
134	420.97346	14102.61098	184	578.05311	26590.44315	234	735.13276	43005.26658
135	424.11506	14313.88311	185	581.19471	26880.25511	235	738.27436	43373.61836
136	427.25665	14526.72603	186	584.33630	27171.63786	236	741.41595	43743.54093
137	430.39824	14741.13975	187	587.47789	27464.59140	237	744.55754	44115.03430
138	433.53983	14957.12427	188	590.61948	27759.11575	238	747.69913	44488.09847
139	436.68143	15174.67959	189	593.76108	28055.21089	239	750.84073	44862.73344
140	439.82302	15393.80570	190	596.90267	28352.87683	240	753.98232	45238.93920
141	442.96461	15614.50261	191	600.04426	28652.11356	241	757.12391	45616.71576
142	446.10621	15836.77031	192	603.18586	28952.92109	242	760.26551	45996.06311
143	449.24780	16060.60881	193	606.32745	29255.29941	243	763.40710	46376.98126
144	452.38939	16286.01811	194	609.46904	29559.24854	244	766.54869	46759.47021
145	455.53099	16512.99821	195	612.61064	29864.76846	245	769.69029	47143.52996
146	458.67258	16741.54910	196	615.75223	30171.85917	246	772.83188	47529.16050
147	461.81417	16971.67078	197	618.89382	30480.52068	247	775.97347	47916.36183
148	464.95576	17203.36327	198	622.03541	30790.75299	248	779.11506	48305.13397
149	468.09736	17436.62655	199	625.17701	31102.55610	249	782.25666	48695.47690

Circumferences and Areas of Circles From 250 to 399

Diameter	Circumference	Area	Diameter	Circumference	Area	Diameter	Circumference	Area
250	785.39825	49087.39063	300	942.47790	70685.84250	350	1099.55755	96211.28563
251	788.53984	49480.87515	301	945.61949	71157.86685	351	1102.69914	96761.84980
252	791.68144	49875.93047	302	948.76109	71631.46199	352	1105.84074	97313.98477
253	794.82303	50272.55658	303	951.90268	72106.62793	353	1108.98233	97867.69053
254	797.96462	50670.75350	304	955.04427	72583.36467	354	1112.12392	98422.96710
255	801.10622	51070.52121	305	958.18587	73061.67221	355	1115.26552	98979.81446
256	804.24781	51471.85971	306	961.32746	73541.55054	356	1118.40711	99538.23261
257	807.38940	51874.76901	307	964.46905	74022.99966	357	1121.54870	100098.22156
258	810.53099	52279.24911	308	967.61064	74506.01959	358	1124.69029	100659.78131
259	813.67259	52685.30001	309	970.75224	74990.61031	359	1127.83189	101222.91186
260	816.81418	53092.92170	310	973.89383	75476.77183	360	1130.97348	101787.61320
261	819.95577	53502.11419	311	977.03542	75964.50414	361	1134.11507	102353.88534
262	823.09737	53912.87747	312	980.17702	76453.80725	362	1137.25667	102921.72827
263	826.23896	54325.21155	313	983.31861	76944.68115	363	1140.39826	103491.14200
264	829.38055	54739.11643	314	986.46020	77437.12586	364	1143.53985	104062.12653
265	832.52215	55154.59211	315	989.60180	77931.14136	365	1146.68145	104634.68186
266	835.66374	55571.63858	316	992.74339	78426.72765	366	1149.82304	105208.80798
267	838.80533	55990.25584	317	995.88498	78923.88474	367	1152.96463	105784.50489
268	841.94692	56410.44391	318	999.02657	79422.61263	368	1156.10622	106361.77261
269	845.08852	56832.20277	319	1002.16817	79922.91132	369	1159.24782	106940.61112
270	848.23011	57255.53243	320	1005.30976	80424.78080	370	1162.38941	107521.02043
271	851.37170	57680.43288	321	1008.45135	80928.22108	371	1165.53100	108103.00053
272	854.51330	58106.90413	322	1011.59295	81433.23215	372	1168.67260	108686.55143
273	857.65489	58534.94617	323	1014.73454	81939.81402	373	1171.81419	109271.67312
274	860.79648	58964.55902	324	1017.87613	82447.96669	374	1174.95578	109858.36562
275	863.93808	59395.74266	325	1021.01773	82957.69016	375	1178.09738	110446.62891
276	867.07967	59828.49709	326	1024.15932	83468.98442	376	1181.23897	111036.46299
277	870.22126	60262.82232	327	1027.30091	83981.84947	377	1184.38056	111627.86787
278	873.36285	60698.71835	328	1030.44250	84496.28533	378	1187.52215	112220.84355
279	876.50445	61136.18518	329	1033.58410	85012.29198	379	1190.66375	112815.39003
280	879.64604	61575.22280	330	1036.72569	85529.86943	380	1193.80534	113411.50730
281	882.78763	62015.83122	331	1039.86728	86049.01767	381	1196.94693	114009.19537
282	885.92923	62458.01043	332	1043.00888	86569.73671	382	1200.08853	114608.45423
283	889.07082	62901.76044	333	1046.15047	87092.02654	383	1203.23012	115209.28389
284	892.21241	63347.08125	334	1049.29206	87615.88718	384	1206.37171	115811.68435
285	895.35401	63793.97286	335	1052.43366	88141.31861	385	1209.51331	116415.65561
286	898.49560	64242.43526	336	1055.57525	88668.32083	386	1212.65490	117021.19766
287	901.63719	64692.46845	337	1058.71684	89196.89385	387	1215.79649	117628.31050
288	904.77878	65144.07245	338	1061.85843	89727.03767	388	1218.93808	118236.99415
289	907.92038	65597.24724	339	1065.00003	90258.75229	389	1222.07968	118847.24859
290	911.06197	66051.99283	340	1068.14162	90792.03770	390	1225.22127	119459.07383
291	914.20356	66508.30921	341	1071.28321	91326.89391	391	1228.36286	120072.46986
292	917.34516	66966.19639	342	1074.42481	91863.32091	392	1231.50446	120687.43669
293	920.48675	67425.65436	343	1077.56640	92401.31871	393	1234.64605	121303.97431
294	923.62834	67886.68314	344	1080.70799	92940.88731	394	1237.78764	121922.08274
295	926.76994	68349.28271	345	1083.84959	93482.02671	395	1240.92924	122541.76196
296	929.91153	68813.45307	346	1086.99118	94024.73690	396	1244.07083	123163.01197
297	933.05312	69279.19423	347	1090.13277	94569.01788	397	1247.21242	123785.83278
298	936.19471	69746.50619	348	1093.27436	95114.86967	398	1250.35401	124410.22439
299	939.33631	70215.38895	349	1096.41596	95662.29225	399	1253.49561	125036.18680

Circumferences and Areas of Circles From 400 to 549

Diameter	Circumference	Area	Diameter	Circumference	Area	Diameter	Circumference	Area
400	1256.63720	125663.72000	450	1413.71685	159043.14563	500	1570.79650	196349.56250
401	1259.77879	126292.82400	451	1416.85844	159750.78945	501	1573.93809	197135.74615
402	1262.92039	126923.49879	452	1420.00004	160460.00407	502	1577.07969	197923.50059
403	1266.06198	127555.74438	453	1423.14163	161170.78948	503	1580.22128	198712.82583
404	1269.20357	128189.56077	454	1426.28322	161883.14570	504	1583.36287	199503.72187
405	1272.34517	128824.94796	455	1429.42482	162597.07271	505	1586.50447	200296.18871
406	1275.48676	129461.90594	456	1432.56641	163312.57051	506	1589.64606	201090.22634
407	1278.62835	130100.43471	457	1435.70800	164029.63911	507	1592.78765	201885.83476
408	1281.76994	130740.53429	458	1438.84959	164748.27851	508	1595.92924	202683.01399
409	1284.91154	131382.20466	459	1441.99119	165468.48871	509	1599.07084	203481.76401
410	1288.05313	132025.44583	460	1445.13278	166190.26970	510	1602.21243	204282.08483
411	1291.19472	132670.25779	461	1448.27437	166913.62149	511	1605.35402	205083.97644
412	1294.33632	133316.64055	462	1451.41597	167638.54407	512	1608.49562	205887.43885
413	1297.47791	133964.59410	463	1454.55756	168365.03745	513	1611.63721	206692.47205
414	1300.61950	134614.11846	464	1457.69915	169093.10163	514	1614.77880	207499.07606
415	1303.76110	135265.21361	465	1460.84075	169822.73661	515	1617.92040	208307.25086
416	1306.90269	135917.87955	466	1463.98234	170553.94238	516	1621.06199	209116.99645
417	1310.04428	136572.11629	467	1467.12393	171286.71894	517	1624.20358	209928.31284
418	1313.18587	137227.92383	468	1470.26552	172021.06631	518	1627.34517	210741.20003
419	1316.32747	137885.30217	469	1473.40712	172756.98447	519	1630.48677	211555.65802
420	1319.46906	138544.25130	470	1476.54871	173494.47343	520	1633.62836	212371.68680
421	1322.61065	139204.77123	471	1479.69030	174233.53318	521	1636.76995	213189.28638
422	1325.75225	139866.86195	472	1482.83190	174974.16373	522	1639.91155	214008.45675
423	1328.89384	140530.52347	473	1485.97349	175716.36507	523	1643.05314	214829.19792
424	1332.03543	141195.75579	474	1489.11508	176460.13722	524	1646.19473	215651.50989
425	1335.17703	141862.55891	475	1492.25668	177205.48016	525	1649.33633	216475.39266
426	1338.31862	142530.93282	476	1495.39827	177952.39389	526	1652.47792	217300.84622
427	1341.46021	143200.87752	477	1498.53986	178700.87842	527	1655.61951	218127.87057
428	1344.60180	143872.39303	478	1501.68145	179450.93375	528	1658.76110	218956.46573
429	1347.74340	144545.47933	479	1504.82305	180202.55988	529	1661.90270	219786.63168
430	1350.88499	145220.13643	480	1507.96464	180955.75680	530	1665.04429	220618.36843
431	1354.02658	145896.36432	481	1511.10623	181710.52452	531	1668.18588	221451.67597
432	1357.16818	146574.16301	482	1514.24783	182466.86303	532	1671.32748	222286.55431
433	1360.30977	147253.53249	483	1517.38942	183224.77234	533	1674.46907	223123.00344
434	1363.45136	147934.47278	484	1520.53101	183984.25245	534	1677.61066	223961.02338
435	1366.59296	148616.98386	485	1523.67261	184745.30336	535	1680.75226	224800.61411
436	1369.73455	149301.06573	486	1526.81420	185507.92506	536	1683.89385	225641.77563
437	1372.87614	149986.71840	487	1529.95579	186272.11755	537	1687.03544	226484.50795
438	1376.01773	150673.94187	488	1533.09738	187037.88085	538	1690.17703	227328.81107
439	1379.15933	151362.73614	489	1536.23898	187805.21494	539	1693.31863	228174.68499
440	1382.30092	152053.10120	490	1539.38057	188574.11983	540	1696.46022	229022.12970
441	1385.44251	152745.03706	491	1542.52216	189344.59551	541	1699.60181	229871.14521
442	1388.58411	153438.54371	492	1545.66376	190116.64199	542	1702.74341	230721.73151
443	1391.72570	154133.62116	493	1548.80535	190890.25926	543	1705.88500	231573.88861
444	1394.86729	154830.26941	494	1551.94694	191665.44734	544	1709.02659	232427.61651
445	1398.00889	155528.48846	495	1555.08854	192442.20621	545	1712.16819	233282.91521
446	1401.15048	156228.27830	496	1558.23013	193220.53587	546	1715.30978	234139.78470
447	1404.29207	156929.63893	497	1561.37172	194000.43633	547	1718.45137	234998.22498
448	1407.43366	157632.57037	498	1564.51331	194781.90759	548	1721.59296	235858.23607
449	1410.57526	158337.07260	499	1567.65491	195564.94965	549	1724.73456	236719.81795

Circumferences and Areas of Circles From 550 to 699

Diameter	Circumference	Area	Diameter	Circumference	Area	Diameter	Circumference	Area
550	1727.87615	237582.97063	600	1884.95580	282743.37000	650	2042.03545	331830.76063
551	1731.01774	238447.69410	601	1888.09739	283686.63330	651	2045.17704	332852.56375
552	1734.15934	239313.98837	602	1891.23899	284631.46739	652	2048.31864	333875.93767
553	1737.30093	240181.85343	603	1894.38058	285577.87228	653	2051.46023	334900.88238
554	1740.44252	241051.28930	604	1897.52217	286525.84797	654	2054.60182	335927.39790
555	1743.58412	241922.29596	605	1900.66377	287475.39446	655	2057.74342	336955.48421
556	1746.72571	242794.87341	606	1903.80536	288426.51174	656	2060.88501	337985.14131
557	1749.86730	243669.02166	607	1906.94695	289379.19981	657	2064.02660	339016.36921
558	1753.00889	244544.74071	608	1910.08854	290333.45869	658	2067.16819	340049.16791
559	1756.15049	245422.03056	609	1913.23014	291289.28836	659	2070.30979	341083.53741
560	1759.29208	246300.89120	610	1916.37173	292246.68883	660	2073.45138	342119.47770
561	1762.43367	247181.32264	611	1919.51332	293205.66009	661	2076.59297	343156.98879
562	1765.57527	248063.32487	612	1922.65492	294166.20215	662	2079.73457	344196.07067
563	1768.71686	248946.89790	613	1925.79651	295128.31500	663	2082.87616	345236.72335
564	1771.85845	249832.04173	614	1928.93810	296091.99866	664	2086.01775	346278.94683
565	1775.00005	250718.75636	615	1932.07970	297057.25311	665	2089.15935	347322.74111
566	1778.14164	251607.04178	616	1935.22129	298024.07835	666	2092.30094	348368.10618
567	1781.28323	252496.89799	617	1938.36288	298992.47439	667	2095.44253	349415.04204
568	1784.42482	253388.32501	618	1941.50447	299962.44123	668	2098.58412	350463.54871
569	1787.56642	254281.32282	619	1944.64607	300933.97887	669	2101.72572	351513.62617
570	1790.70801	255175.89143	620	1947.78766	301907.08730	670	2104.86731	352565.27443
571	1793.84960	256072.03083	621	1950.92925	302881.76653	671	2108.00890	353618.49348
572	1796.99120	256969.74103	622	1954.07085	303858.01655	672	2111.15050	354673.28333
573	1800.13279	257869.02202	623	1957.21244	304835.83737	673	2114.29209	355729.64397
574	1803.27438	258769.87382	624	1960.35403	305815.22899	674	2117.43368	356787.57542
575	1806.41598	259672.29641	625	1963.49563	306796.19141	675	2120.57528	357847.07766
576	1809.55757	260576.28979	626	1966.63722	307778.72462	676	2123.71687	358908.15069
577	1812.69916	261481.85397	627	1969.77881	308762.82862	677	2126.85846	359970.79452
578	1815.84075	262388.98895	628	1972.92040	309748.50343	678	2130.00005	361035.00915
579	1818.98235	263297.69473	629	1976.06200	310735.74903	679	2133.14165	362100.79458
580	1822.12394	264207.97130	630	1979.20359	311724.56543	680	2136.28324	363168.15080
581	1825.26553	265119.81867	631	1982.34518	312714.95262	681	2139.42483	364237.07782
582	1828.40713	266033.23683	632	1985.48678	313706.91061	682	2142.56643	365307.57563
583	1831.54872	266948.22579	633	1988.62837	314700.43939	683	2145.70802	366379.64424
584	1834.69031	267864.78555	634	1991.76996	315695.53898	684	2148.84961	367453.28365
585	1837.83191	268782.91611	635	1994.91156	316692.20936	685	2151.99121	368528.49386
586	1840.97350	269702.61746	636	1998.05315	317690.45053	686	2155.13280	369605.27486
587	1844.11509	270623.88960	637	2001.19474	318690.26250	687	2158.27439	370683.62665
588	1847.25668	271546.73255	638	2004.33633	319691.64527	688	2161.41598	371763.54925
589	1850.39828	272471.14629	639	2007.47793	320694.59884	689	2164.55758	372845.04264
590	1853.53987	273397.13083	640	2010.61952	321699.12320	690	2167.69917	373928.10683
591	1856.68146	274324.68616	641	2013.76111	322705.21836	691	2170.84076	375012.74181
592	1859.82306	275253.81229	642	2016.90271	323712.88431	692	2173.98236	376098.94759
593	1862.96465	276184.50921	643	2020.04430	324722.12106	693	2177.12395	377186.72416
594	1866.10624	277116.77694	644	2023.18589	325732.92861	694	2180.26554	378276.07154
595	1869.24784	278050.61546	645	2026.32749	326745.30696	695	2183.40714	379366.98971
596	1872.38943	278986.02477	646	2029.46908	327759.25610	696	2186.54873	380459.47867
597	1875.53102	279923.00488	647	2032.61067	328774.77603	697	2189.69032	381553.53843
598	1878.67261	280861.55579	648	2035.75226	329791.86677	698	2192.83191	382649.16899
599	1881.81421	281801.67750	649	2038.89386	330810.52830	699	2195.97351	383746.37035

Circumferences and Areas of Circles From 700 to 849

Diameter	Circumference	Area	Diameter	Circumference	Area	Diameter	Circumference	Area
700	2199.11510	384845.14250	750	2356.19475	441786.51563	800	2513.27440	502654.88000
701	2202.25669	385945.48545	751	2359.33634	442965.39840	801	2516.41599	503912.30260
702	2205.39829	387047.39919	752	2362.47794	444145.85197	802	2519.55759	505171.29599
703	2208.53988	388150.88373	753	2365.61953	445327.87633	803	2522.69918	506431.86018
704	2211.68147	389255.93907	754	2368.76112	446511.47150	804	2525.84077	507693.99517
705	2214.82307	390362.56521	755	2371.90272	447696.63746	805	2528.98237	508957.70096
706	2217.96466	391470.76214	756	2375.04431	448883.37421	806	2532.12396	510222.97754
707	2221.10625	392580.52986	757	2378.18590	450071.68176	807	2535.26555	511489.82491
708	2224.24784	393691.86839	758	2381.32749	451261.56011	808	2538.40714	512758.24309
709	2227.38944	394804.77771	759	2384.46909	452453.00926	809	2541.54874	514028.23206
710	2230.53103	395919.25783	760	2387.61068	453646.02920	810	2544.69033	515299.79183
711	2233.67262	397035.30874	761	2390.75227	454840.61994	811	2547.83192	516572.92239
712	2236.81422	398152.93045	762	2393.89387	456036.78147	812	2550.97352	517847.62375
713	2239.95581	399272.12295	763	2397.03546	457234.51380	813	2554.11511	519123.89590
714	2243.09740	400392.88626	764	2400.17705	458433.81693	814	2557.25670	520401.73886
715	2246.23900	401515.22036	765	2403.31865	459634.69086	815	2560.39830	521681.15261
716	2249.38059	402639.12525	766	2406.46024	460837.13558	816	2563.53989	522962.13715
717	2252.52218	403764.60094	767	2409.60183	462041.15109	817	2566.68148	524244.69249
718	2255.66377	404891.64743	768	2412.74342	463246.73741	818	2569.82307	525528.81863
719	2258.80537	406020.26472	769	2415.88502	464453.89452	819	2572.96467	526814.51557
720	2261.94696	407150.45280	770	2419.02661	465662.62243	820	2576.10626	528101.78330
721	2265.08855	408282.21168	771	2422.16820	466872.92113	821	2579.24785	529390.62183
722	2268.23015	409415.54135	772	2425.30980	468084.79063	822	2582.38945	530681.03115
723	2271.37174	410550.44182	773	2428.45139	469298.23092	823	2585.53104	531973.01127
724	2274.51333	411686.91309	774	2431.59298	470513.24202	824	2588.67263	533266.56219
725	2277.65493	412824.95516	775	2434.73458	471729.82391	825	2591.81423	534561.68391
726	2280.79652	413964.56802	776	2437.87617	472947.97659	826	2594.95582	535858.37642
727	2283.93811	415105.75167	777	2441.01776	474167.70007	827	2598.09741	537156.63972
728	2287.07970	416248.50613	778	2444.15935	475388.99435	828	2601.23900	538456.47383
729	2290.22130	417392.83138	779	2447.30095	476611.85943	829	2604.38060	539757.87873
730	2293.36289	418538.72743	780	2450.44254	477836.29530	830	2607.52219	541060.85443
731	2296.50448	419686.19427	781	2453.58413	479062.30197	831	2610.66378	542365.40092
732	2299.64608	420835.23191	782	2456.72573	480289.87943	832	2613.80538	543671.51821
733	2302.78767	421985.84034	783	2459.86732	481519.02769	833	2616.94697	544979.20629
734	2305.92926	423138.01958	784	2463.00891	482749.74675	834	2620.08856	546288.46518
735	2309.07086	424291.76961	785	2466.15051	483982.03661	835	2623.23016	547599.29486
736	2312.21245	425447.09043	786	2469.29210	485215.89726	836	2626.37175	548911.69533
737	2315.35404	426603.98205	787	2472.43369	486451.32870	837	2629.51334	550225.66660
738	2318.49563	427762.44447	788	2475.57528	487688.33095	838	2632.65493	551541.20867
739	2321.63723	428922.47769	789	2478.71688	488926.90399	839	2635.79653	552858.32154
740	2324.77882	430084.08170	790	2481.85847	490167.04783	840	2638.93812	554177.00520
741	2327.92041	431247.25651	791	2485.00006	491408.76246	841	2642.07971	555497.25966
742	2331.06201	432412.00211	792	2488.14166	492652.04789	842	2645.22131	556819.08491
743	2334.20360	433578.31851	793	2491.28325	493896.90411	843	2648.36290	558142.48096
744	2337.34519	434746.20571	794	2494.42484	495143.33114	844	2651.50449	559467.44781
745	2340.48679	435915.66371	795	2497.56644	496391.32896	845	2654.64609	560793.98546
746	2343.62838	437086.69250	796	2500.70803	497640.89757	846	2657.78768	562122.09390
747	2346.76997	438259.29208	797	2503.84962	498892.03698	847	2660.92927	563451.77313
748	2349.91156	439433.46247	798	2506.99121	500144.74719	848	2664.07086	564783.02317
749	2353.05316	440609.20365	799	2510.13281	501399.02820	849	2667.21246	566115.84400

Circumferences and Areas of Circles From 850 to 999

Diameter	Circumference	Area	Diameter	Circumference	Area	Diameter	Circumference	Area
850	2670.35405	567450.23563	900	2827.43370	636172.58250	950	2984.51335	708821.92063
851	2673.49564	568786.19805	901	2830.57529	637587.08475	951	2987.65494	710314.96270
852	2676.63724	570123.73127	902	2833.71689	639003.15779	952	2990.79654	711809.57557
853	2679.77883	571462.83528	903	2836.85848	640420.80163	953	2993.93813	713305.75923
854	2682.92042	572803.51010	904	2840.00007	641840.01627	954	2997.07972	714803.51370
855	2686.06202	574145.75571	905	2843.14167	643260.80171	955	3000.22132	716302.83896
856	2689.20361	575489.57211	906	2846.28326	644683.15794	956	3003.36291	717803.73501
857	2692.34520	576834.95931	907	2849.42485	646107.08496	957	3006.50450	719306.20186
858	2695.48679	578181.91731	908	2852.56644	647532.58279	958	3009.64609	720810.23951
859	2698.62839	579530.44611	909	2855.70804	648959.65141	959	3012.78769	722315.84796
860	2701.76998	580880.54570	910	2858.84963	650388.29083	960	3015.92928	723823.02720
861	2704.91157	582232.21609	911	2861.99122	651818.50104	961	3019.07087	725331.77724
862	2708.05317	583585.45727	912	2865.13282	653250.28205	962	3022.21247	726842.09807
863	2711.19476	584940.26925	913	2868.27441	654683.63385	963	3025.35406	728353.98970
864	2714.33635	586296.65203	914	2871.41600	656118.55646	964	3028.49565	729867.45213
865	2717.47795	587654.60561	915	2874.55760	657555.04986	965	3031.63725	731382.48536
866	2720.61954	589014.12998	916	2877.69919	658993.11405	966	3034.77884	732899.08938
867	2723.76113	590375.22514	917	2880.84078	660432.74904	967	3037.92043	734417.26419
868	2726.90272	591737.89111	918	2883.98237	661873.95483	968	3041.06202	735937.00981
869	2730.04432	593102.12787	919	2887.12397	663316.73142	969	3044.20362	737458.32622
870	2733.18591	594467.93543	920	2890.26556	664761.07880	970	3047.34521	738981.21343
871	2736.32750	595835.31378	921	2893.40715	666206.99698	971	3050.48680	740505.67143
872	2739.46910	597204.26293	922	2896.54875	667654.48595	972	3053.62840	742031.70023
873	2742.61069	598574.78287	923	2899.69034	669103.54572	973	3056.76999	743559.29982
874	2745.75228	599946.87362	924	2902.83193	670554.17629	974	3059.91158	745088.47022
875	2748.89388	601320.53516	925	2905.97353	672006.37766	975	3063.05318	746619.21141
876	2752.03547	602695.76749	926	2909.11512	673460.14982	976	3066.19477	748151.52339
877	2755.17706	604072.57062	927	2912.25671	674915.49277	977	3069.33636	749685.40617
878	2758.31865	605450.94455	928	2915.39830	676372.40653	978	3072.47795	751220.85975
879	2761.46025	606830.88928	929	2918.53990	677830.89108	979	3075.61955	752757.88413
880	2764.60184	608212.40480	930	2921.68149	679290.94643	980	3078.76114	754296.47930
881	2767.74343	609595.49112	931	2924.82308	680752.57257	981	3081.90273	755836.64527
882	2770.88503	610980.14823	932	2927.96468	682215.76951	982	3085.04433	757378.38203
883	2774.02662	612366.37614	933	2931.10627	683680.53724	983	3088.18592	758921.68959
884	2777.16821	613754.17485	934	2934.24786	685146.87578	984	3091.32751	760466.56795
885	2780.30981	615143.54436	935	2937.38946	686614.78511	985	3094.46911	762013.01711
886	2783.45140	616534.48466	936	2940.53105	688084.26523	986	3097.61070	763561.03706
887	2786.59299	617926.99575	937	2943.67264	689555.31615	987	3100.75229	765110.62780
888	2789.73458	619321.07765	938	2946.81423	691027.93787	988	3103.89388	766661.78935
889	2792.87618	620716.73034	939	2949.95583	692502.13039	989	3107.03548	768214.52169
890	2796.01777	622113.95383	940	2953.09742	693977.89370	990	3110.17707	769768.82483
891	2799.15936	623512.74811	941	2956.23901	695455.22781	991	3113.31866	771324.69876
892	2802.30096	624913.11319	942	2959.38061	696934.13271	992	3116.46026	772882.14349
893	2805.44255	626315.04906	943	2962.52220	698414.60841	993	3119.60185	774441.15901
894	2808.58414	627718.55574	944	2965.66379	699896.65491	994	3122.74344	776001.74534
895	2811.72574	629123.63321	945	2968.80539	701380.27221	995	3125.88504	777563.90246
896	2814.86733	630530.28147	946	2971.94698	702865.46030	996	3129.02663	779127.63037
897	2818.00892	631938.50053	947	2975.08857	704352.21918	997	3132.16822	780692.92908
898	2821.15051	633348.29039	948	2978.23016	705840.54887	998	3135.30981	782259.79859
899	2824.29211	634759.65105	949	2981.37176	707330.44935	999	3138.45141	783828.23890

Lengths of Chords for Spacing Off the Circumferences of Circles

On the Table 1 are given lengths of chords for spacing off the circumferences of circles. The object of these tables is to make possible the division of the periphery into a number of equal parts without trials with the dividers. The first table is calculated for circles having a diameter equal to 1. For circles of other diameters, the length of chord given in the table should be multiplied by the diameter of the circle. This first table may be used by toolmakers when setting "buttons" in circular formation. Assume that it is required to divide the periphery of a circle of 20 inches diameter into thirty-two equal parts. From the table the length of the chord is found to be 0.098017 inch, if the diameter of the circle were 1 inch. With a diameter of 20 inches the length of the chord for one division would be 20 × 0.098017 = 1.9603 inches. Another example in metric units: For a 100 millimeter diameter requiring 5 equal divisions, the length of the chord for one division would be 100 × 0.587785 = 58.7785 millimeters.

The Table 2 and 3 give an additional table for the spacing off of circles which are being worked out for diameters from 1⁄16 inch to 14 inches. As an example, assume that it is required to divide a circle having a diameter of 6½ inches into seven equal parts. Find first, in the column headed "6" and in line with 7 divisions, the length of the chord for a 6-inch circle, which is 2.603 inches. Then find the length of the chord for a ½-inch diameter circle, 7 divisions, which is 0.217. The sum of these two values, 2.603 + 0.217 = 2.820 inches, is the length of the chord required for spacing off the circumference of a 6½-inch circle into seven equal divisions.

As another example, assume that it is required to divide a circle having a diameter of 9 23⁄32 inches into 15 equal divisions. First find the length of the chord for a 9-inch circle, which is 1.871 inch. The length of the chord for a 23⁄32-inch circle can easily be estimated from the table by taking the value that is exactly between those given for 11⁄16 and ¾ inch. The value for 11⁄16 inch is 0.143, and for ¾ inch, 0.156. For 23⁄32 the value would be 0.150. Then, 1.871 + 0.150 = 2.021 inches.

Table 1. Lengths of Chords for Spacing Off the Circumferences of Circles with a Diameter Equal to 1 (English or metric units)

No. of Spaces	Length of Chord	No. of Spaces	Length of Chord	No. of Spaces	Length of Chord	No. of Spaces	Length of Chord
3	0.866025	22	0.142315	41	0.076549	60	0.052336
4	0.707107	23	0.136167	42	0.074730	61	0.051479
5	0.587785	24	0.130526	43	0.072995	62	0.050649
6	0.500000	25	0.125333	44	0.071339	63	0.049846
7	0.433884	26	0.120537	45	0.069756	64	0.049068
8	0.382683	27	0.116093	46	0.068242	65	0.048313
9	0.342020	28	0.111964	47	0.066793	66	0.047582
10	0.309017	29	0.108119	48	0.065403	67	0.046872

Table 1. Lengths of Chords for Spacing Off the Circumferences of Circles with a Diameter Equal to 1 (English or metric units) *(Continued)*

No. of Spaces	Length of Chord	No. of Spaces	Length of Chord	No. of Spaces	Length of Chord	No. of Spaces	Length of Chord
11	0.281733	30	0.104528	49	0.064070	68	0.046183
12	0.258819	31	0.101168	50	0.062791	69	0.045515
13	0.239316	32	0.098017	51	0.061561	70	0.044865
14	0.222521	33	0.095056	52	0.060378	71	0.044233
15	0.207912	34	0.092268	53	0.059241	72	0.043619
16	0.195090	35	0.089639	54	0.058145	73	0.043022
17	0.183750	36	0.087156	55	0.057089	74	0.042441
18	0.173648	37	0.084806	56	0.056070	75	0.041876
19	0.164595	38	0.082579	57	0.055088	76	0.041325
20	0.156434	39	0.080467	58	0.054139	77	0.040789
21	0.149042	40	0.078459	59	0.053222	78	0.040266

For circles of other diameters, multiply length given in table by diameter of circle.

Hole Coordinate Dimension Factors for Jig Boring.—Tables of hole coordinate dimension factors for use in jig boring are given in Tables 4 through 7 starting on page 57. The coordinate axes shown in the figure accompanying each table are used to reference the tool path; the values listed in each table are for the end points of the tool path. In this machine coordinate system, a positive *Y* value indicates that the effective motion of the tool with reference to the work is toward the front of the jig borer (the actual motion of the jig borer table is toward the column). Similarly, a positive *X* value indicates that the effective motion of the tool with respect to the work is toward the right (the actual motion of the jig borer table is toward the left).

When entering data into most computer-controlled jig borers, current practice is to use the more familiar Cartesian coordinate axis system in which the positive *Y* direction is "up" (i.e., pointing toward the column of the jig borer). The computer will automatically change the signs of the entered *Y* values to the signs that they would have in the machine coordinate system. Therefore, before applying the coordinate dimension factors given in the tables, it is important to determine the coordinate system to be used. If a Cartesian coordinate system is to be used for the tool path, then the sign of the *Y* values in the tables must be changed, from positive to negative and from negative to positive. For example, when programming for a three-hole type *A* circle using Cartesian coordinates, the *Y* values from Table 6 would be $y1 = +0.50000$, $y2 = -0.25000$, and $y3 = -0.25000$.

Table 2. Table for Spacing Off the Circumferences of Circles

No. of Divisions	Degrees in Arc	Diameter of Circle to be Spaced Off														
		1/16	1/8	3/16	1/4	5/16	3/8	7/16	1/2	9/16	5/8	11/16	3/4	13/16	7/8	15/16
		Length of Chord														
3	120	0.054	0.108	0.162	0.217	0.271	0.325	0.379	0.433	0.487	0.541	0.595	0.650	0.704	0.758	0.812
4	90	0.044	0.088	0.133	0.177	0.221	0.265	0.309	0.354	0.398	0.442	0.486	0.530	0.575	0.619	0.663
5	72	0.037	0.073	0.110	0.147	0.184	0.220	0.257	0.294	0.331	0.367	0.404	0.441	0.478	0.514	0.551
6	60	0.031	0.063	0.094	0.125	0.156	0.188	0.219	0.250	0.281	0.313	0.344	0.375	0.406	0.438	0.469
7	51 3/7	0.027	0.054	0.081	0.108	0.136	0.163	0.190	0.217	0.244	0.271	0.298	0.325	0.353	0.380	0.407
8	45	0.024	0.048	0.072	0.096	0.120	0.144	0.167	0.191	0.215	0.239	0.263	0.287	0.311	0.335	0.359
9	40	0.021	0.043	0.064	0.086	0.107	0.128	0.150	0.171	0.192	0.214	0.235	0.257	0.278	0.299	0.321
10	36	0.019	0.039	0.058	0.077	0.097	0.116	0.135	0.155	0.174	0.193	0.212	0.232	0.251	0.270	0.290
11	32 8/11	0.018	0.035	0.053	0.070	0.088	0.106	0.123	0.141	0.158	0.176	0.194	0.211	0.229	0.247	0.264
12	30	0.016	0.032	0.049	0.065	0.081	0.097	0.113	0.129	0.146	0.162	0.178	0.194	0.210	0.226	0.243
13	27 9/13	0.015	0.030	0.045	0.060	0.075	0.090	0.105	0.120	0.135	0.150	0.165	0.179	0.194	0.209	0.224
14	25 5/7	0.014	0.028	0.042	0.056	0.069	0.083	0.097	0.111	0.125	0.139	0.153	0.167	0.181	0.195	0.209
15	24	0.013	0.026	0.039	0.052	0.065	0.078	0.091	0.104	0.117	0.130	0.143	0.156	0.169	0.182	0.195
16	22 1/2	0.012	0.024	0.037	0.049	0.061	0.073	0.085	0.098	0.110	0.122	0.134	0.146	0.159	0.171	0.183
17	21 3/17	0.011	0.023	0.034	0.046	0.057	0.069	0.080	0.092	0.103	0.115	0.126	0.138	0.149	0.161	0.172
18	20	0.011	0.022	0.033	0.043	0.054	0.065	0.076	0.087	0.098	0.109	0.119	0.130	0.141	0.152	0.163
19	18 18/19	0.010	0.021	0.031	0.041	0.051	0.062	0.072	0.082	0.093	0.103	0.113	0.123	0.134	0.144	0.154
20	18	0.010	0.020	0.029	0.039	0.049	0.059	0.068	0.078	0.088	0.098	0.108	0.117	0.127	0.137	0.147
21	17 1/7	0.009	0.019	0.028	0.037	0.047	0.056	0.065	0.075	0.084	0.093	0.102	0.112	0.121	0.130	0.140
22	16 4/11	0.009	0.018	0.027	0.036	0.044	0.053	0.062	0.071	0.080	0.089	0.098	0.107	0.116	0.125	0.133
23	15 15/23	0.009	0.017	0.026	0.034	0.043	0.051	0.060	0.068	0.077	0.085	0.094	0.102	0.111	0.119	0.128
24	15	0.008	0.016	0.024	0.033	0.041	0.049	0.057	0.065	0.073	0.082	0.090	0.098	0.106	0.114	0.122
25	14 2/5	0.008	0.016	0.023	0.031	0.039	0.047	0.055	0.063	0.070	0.078	0.086	0.094	0.102	0.110	0.117
26	13 11/13	0.008	0.015	0.023	0.030	0.038	0.045	0.053	0.060	0.068	0.075	0.083	0.090	0.098	0.105	0.113
28	12 6/7	0.007	0.014	0.021	0.028	0.035	0.042	0.049	0.056	0.063	0.070	0.077	0.084	0.091	0.098	0.105
30	12	0.007	0.013	0.020	0.026	0.033	0.039	0.046	0.052	0.059	0.065	0.072	0.078	0.085	0.091	0.098
32	11 1/4	0.006	0.012	0.018	0.025	0.031	0.037	0.043	0.049	0.055	0.061	0.067	0.074	0.080	0.086	0.092

See on page 53 for explanatory matter.

Table 3. Table for Spacing Off the Circumferences of Circles

No. of Divisions	Degrees in Arc	Diameter of Circle to be Spaced Off													
		1	2	3	4	5	6	7	8	9	10	11	12	13	14
		Length of Chord													
3	120	0.866	1.732	2.598	3.464	4.330	5.196	6.062	6.928	7.794	8.660	9.526	10.392	11.258	12.124
4	90	0.707	1.414	2.121	2.828	3.536	4.243	4.950	5.657	6.364	7.071	7.778	8.485	9.192	9.899
5	72	0.588	1.176	1.763	2.351	2.939	3.527	4.114	4.702	5.290	5.878	6.466	7.053	7.641	8.229
6	60	0.500	1.000	1.500	2.000	2.500	3.000	3.500	4.000	4.500	5.000	5.500	6.000	6.500	7.000
7	$51\frac{3}{7}$	0.434	0.868	1.302	1.736	2.169	2.603	3.037	3.471	3.905	4.339	4.773	5.207	5.640	6.074
8	45	0.383	0.765	1.148	1.531	1.913	2.296	2.679	3.061	3.444	3.827	4.210	4.592	4.975	5.358
9	40	0.342	0.684	1.026	1.368	1.710	2.052	2.394	2.736	3.078	3.420	3.762	4.104	4.446	4.788
10	36	0.309	0.618	0.927	1.236	1.545	1.854	2.163	2.472	2.781	3.090	3.399	3.708	4.017	4.326
11	$32\frac{8}{11}$	0.282	0.563	0.845	1.127	1.409	1.690	1.972	2.254	2.536	2.817	3.099	3.381	3.663	3.944
12	30	0.259	0.518	0.776	1.035	1.294	1.553	1.812	2.071	2.329	2.588	2.847	3.106	3.365	3.623
13	$27\frac{9}{13}$	0.239	0.479	0.718	0.957	1.197	1.436	1.675	1.915	2.154	2.393	2.632	2.872	3.111	3.350
14	$25\frac{5}{7}$	0.223	0.445	0.668	0.890	1.113	1.335	1.558	1.780	2.003	2.225	2.448	2.670	2.893	3.115
15	24	0.208	0.416	0.624	0.832	1.040	1.247	1.455	1.663	1.871	2.079	2.287	2.495	2.703	2.911
16	$22\frac{1}{2}$	0.195	0.390	0.585	0.780	0.975	1.171	1.366	1.561	1.756	1.951	2.146	2.341	2.536	2.731
17	$21\frac{3}{17}$	0.184	0.367	0.551	0.735	0.919	1.102	1.286	1.470	1.654	1.837	2.021	2.205	2.389	2.572
18	20	0.174	0.347	0.521	0.695	0.868	1.042	1.216	1.389	1.563	1.736	1.910	2.084	2.257	2.431
19	$18\frac{18}{19}$	0.165	0.329	0.494	0.658	0.823	0.988	1.152	1.317	1.481	1.646	1.811	1.975	2.140	2.304
20	18	0.156	0.313	0.469	0.626	0.782	0.939	1.095	1.251	1.408	1.564	1.721	1.877	2.034	2.190
21	$17\frac{1}{7}$	0.149	0.298	0.447	0.596	0.745	0.894	1.043	1.192	1.341	1.490	1.639	1.789	1.938	2.087
22	$16\frac{4}{11}$	0.142	0.285	0.427	0.569	0.712	0.854	0.996	1.139	1.281	1.423	1.565	1.708	1.850	1.992
23	$15\frac{15}{23}$	0.136	0.272	0.408	0.545	0.681	0.817	0.953	1.089	1.225	1.362	1.498	1.634	1.770	1.906
24	15	0.131	0.261	0.392	0.522	0.653	0.783	0.914	1.044	1.175	1.305	1.436	1.566	1.697	1.827
25	$14\frac{2}{5}$	0.125	0.251	0.376	0.501	0.627	0.752	0.877	1.003	1.128	1.253	1.379	1.504	1.629	1.755
26	$13\frac{11}{13}$	0.121	0.241	0.362	0.482	0.603	0.723	0.844	0.964	1.085	1.205	1.326	1.446	1.567	1.688
28	$12\frac{6}{7}$	0.112	0.224	0.336	0.448	0.560	0.672	0.784	0.896	1.008	1.120	1.232	1.344	1.456	1.568
30	12	0.105	0.209	0.314	0.418	0.523	0.627	0.732	0.836	0.941	1.045	1.150	1.254	1.359	1.463
32	$11\frac{1}{4}$	0.098	0.196	0.294	0.392	0.490	0.588	0.686	0.784	0.882	0.980	1.078	1.176	1.274	1.372

Table 4. Hole Coordinate Dimension Factors for Jig Boring — Type "A" Hole Circles (English or Metric Units)

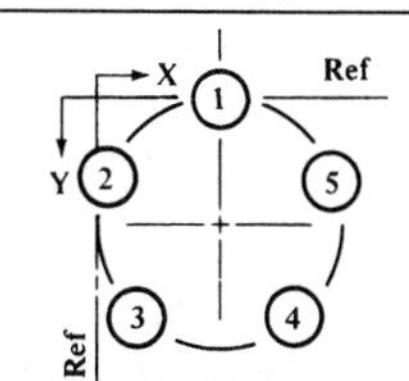

The diagram shows a type "A" circle for a 5-hole circle. Coordinates *x*, *y* are given in the table for hole circles of from 3 to 28 holes. Dimensions are for holes numbered in a counterclockwise direction (as shown). Dimensions given are based upon a hole circle of unit diameter. For a hole circle of, say, 3-inch or 3-centimeter diameter, multiply table values by 3.

3 Holes		4 Holes		5 Holes		6 Holes		7 Holes		8 Holes		9 Holes	
x1	0.50000	x1	0.50000	x1	0.50000	x1	0.50000	x1	0.50000	x1	0.50000	x1	0.50000
y1	0.00000	y1	0.00000	y1	0.00000	y1	0.00000	y1	0.00000	y1	0.00000	y1	0.00000
x2	0.06699	x2	0.00000	x2	0.02447	x2	0.06699	x2	0.10908	x2	0.14645	x2	0.17861
y2	0.75000	y2	0.50000	y2	0.34549	y2	0.25000	y2	0.18826	y2	0.14645	y2	0.11698
x3	0.93301	x3	0.50000	x3	0.20611	x3	0.06699	x3	0.01254	x3	0.00000	x3	0.00760
y3	0.75000	y3	1.00000	y3	0.90451	y3	0.75000	y3	0.61126	y3	0.50000	y3	0.41318
		x4	1.00000	x4	0.79389	x4	0.50000	x4	0.28306	x4	0.14645	x4	0.06699
		y4	0.50000	y4	0.90451	y4	1.00000	y4	0.95048	y4	0.85355	y4	0.75000
				x5	0.97553	x5	0.93301	x5	0.71694	x5	0.50000	x5	0.32899
				y5	0.34549	y5	0.75000	y5	0.95048	y5	1.00000	y5	0.96985
						x6	0.93301	x6	0.98746	x6	0.85355	x6	0.67101
						y6	0.25000	y6	0.61126	y6	0.85355	y6	0.96985
								x7	0.89092	x7	1.00000	x7	0.93301
								y7	0.18826	y7	0.50000	y7	0.75000
										x8	0.85355	x8	0.99240
										y8	0.14645	y8	0.41318
												x9	0.82139
												y9	0.11698

10 Holes		11 Holes		12 Holes		13 Holes		14 Holes		15 Holes		16 Holes	
x1	0.50000	x1	0.50000	x1	0.50000	x1	0.50000	x1	0.50000	x1	0.50000	x1	0.50000
y1	0.00000	y1	0.00000	y1	0.00000	y1	0.00000	y1	0.00000	y1	0.00000	y1	0.00000
x2	0.20611	x2	0.22968	x2	0.25000	x2	0.26764	x2	0.28306	x2	0.29663	x2	0.30866
y2	0.09549	y2	0.07937	y2	0.06699	y2	0.05727	y2	0.04952	y2	0.04323	y2	0.03806
x3	0.02447	x3	0.04518	x3	0.06699	x3	0.08851	x3	0.10908	x3	0.12843	x3	0.14645
y3	0.34549	y3	0.29229	y3	0.25000	y3	0.21597	y3	0.18826	y3	0.16543	y3	0.14645
x4	0.02447	x4	0.00509	x4	0.00000	x4	0.00365	x4	0.01254	x4	0.02447	x4	0.03806
y4	0.65451	y4	0.57116	y4	0.50000	y4	0.43973	y4	0.38874	y4	0.34549	y4	0.30866
x5	0.20611	x5	0.12213	x5	0.06699	x5	0.03249	x5	0.01254	x5	0.00274	x5	0.00000
y5	0.90451	y5	0.82743	y5	0.75000	y5	0.67730	y5	0.61126	y5	0.55226	y5	0.50000
x6	0.50000	x6	0.35913	x6	0.25000	x6	0.16844	x6	0.10908	x6	0.06699	x6	0.03806
y6	1.00000	y6	0.97975	y6	0.93301	y6	0.87426	y6	0.81174	y6	0.75000	y6	0.69134
x7	0.79389	x7	0.64087	x7	0.50000	x7	0.38034	x7	0.28306	x7	0.20611	x7	0.14645
y7	0.90451	y7	0.97975	y7	1.00000	y7	0.98547	y7	0.95048	y7	0.90451	y7	0.85355
x8	0.97553	x8	0.87787	x8	0.75000	x8	0.61966	x8	0.50000	x8	0.39604	x8	0.30866
y8	0.65451	y8	0.82743	y8	0.93301	y8	0.98547	y8	1.00000	y8	0.98907	y8	0.96194
x9	0.97553	x9	0.99491	x9	0.93301	x9	0.83156	x9	0.71694	x9	0.60396	x9	0.50000
y9	0.34549	y9	0.57116	y9	0.75000	y9	0.87426	y9	0.95048	y9	0.98907	y9	1.00000
x10	0.79389	x10	0.95482	x10	1.00000	x10	0.96751	x10	0.89092	x10	0.79389	x10	0.69134
y10	0.09549	y10	0.29229	y10	0.50000	y10	0.67730	y10	0.81174	y10	0.90451	y10	0.96194
		x11	0.77032	x11	0.93801	x11	0.99635	x11	0.98746	x11	0.93301	x11	0.85355
		y11	0.07937	y11	0.25000	y11	0.43973	y11	0.61126	y11	0.75000	y11	0.85355
				x12	0.75000	x12	0.91149	x12	0.98746	x12	0.99726	x12	0.96194
				y12	0.06699	y12	0.21597	y12	0.38874	y12	0.55226	y12	0.69134
						x13	0.73236	x13	0.89092	x13	0.97553	x13	1.00000
						y13	0.05727	y13	0.18826	y13	0.34549	y13	0.50000
								x14	0.71694	x14	0.87157	x14	0.96194
								y14	0.04952	y14	0.16543	y14	0.30866
										x15	0.70337	x15	0.85355
										y15	0.04323	y15	0.14645
												x16	0.69134
												y16	0.03806

Table 4. *(Continued)* **Hole Coordinate Dimension Factors for Jig Boring — Type "A" Hole Circles** (English or Metric Units)

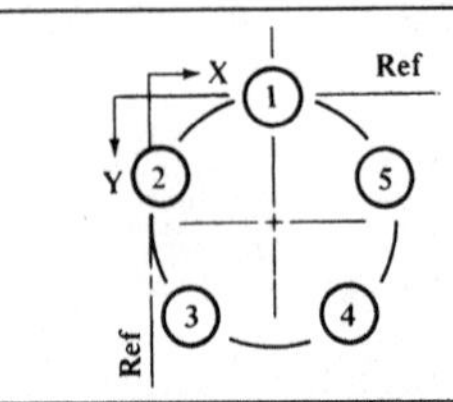

The diagram shows a type "A" circle for a 5-hole circle. Coordinates *x*, *y* are given in the table for hole circles of from 3 to 28 holes. Dimensions are for holes numbered in a counterclockwise direction (as shown). Dimensions given are based upon a hole circle of unit diameter. For a hole circle of, say, 3-inch or 3-centimeter diameter, multiply table values by 3.

	17 Holes	18 Holes	19 Holes	20 Holes	21 Holes	22 Holes	23 Holes
*x*1	0.50000	0.50000	0.50000	0.50000	0.50000	0.50000	0.50000
*y*1	0.00000	0.00000	0.00000	0.00000	0.00000	0.00000	0.00000
*x*2	0.31938	0.32899	0.33765	0.34549	0.35262	0.35913	0.36510
*y*2	0.03376	0.03015	0.02709	0.02447	0.02221	0.02025	0.01854
*x*3	0.16315	0.17861	0.19289	0.20611	0.21834	0.22968	0.24021
*y*3	0.13050	0.11698	0.10543	0.09549	0.08688	0.07937	0.07279
*x*4	0.05242	0.06699	0.08142	0.09549	0.10908	0.12213	0.13458
*y*4	0.27713	0.25000	0.22653	0.20611	0.18826	0.17257	0.15872
*x*5	0.00213	0.00760	0.01530	0.02447	0.03456	0.04518	0.05606
*y*5	0.45387	0.41318	0.37726	0.34549	0.31733	0.29229	0.26997
*x*6	0.01909	0.00760	0.00171	0.00000	0.00140	0.00509	0.01046
*y*6	0.63683	0.58682	0.54129	0.50000	0.46263	0.42884	0.39827
*x*7	0.10099	0.06699	0.04211	0.02447	0.01254	0.00509	0.00117
*y*7	0.80132	0.75000	0.70085	0.65451	0.61126	0.57116	0.53412
*x*8	0.23678	0.17861	0.13214	0.09549	0.06699	0.04518	0.02887
*y*8	0.92511	0.88302	0.83864	0.79389	0.75000	0.70771	0.66744
*x*9	0.40813	0.32899	0.26203	0.20611	0.15991	0.12213	0.09152
*y*9	0.99149	0.96985	0.93974	0.90451	0.86653	0.82743	0.78834
*x*10	0.59187	0.50000	0.41770	0.34549	0.28306	0.22968	0.18446
*y*10	0.99149	1.00000	0.99318	0.97553	0.95048	0.92063	0.88786
*x*11	0.76322	0.67101	0.58230	0.50000	0.42548	0.35913	0.30080
*y*11	0.92511	0.96985	0.99318	1.00000	0.99442	0.97975	0.95861
*x*12	0.89901	0.82139	0.73797	0.65451	0.57452	0.50000	0.43192
*y*12	0.80132	0.88302	0.93974	0.97553	0.99442	1.00000	0.99534
*x*13	0.98091	0.93301	0.86786	0.79389	0.71694	0.64087	0.56808
*y*13	0.63683	0.75000	0.83864	0.90451	0.95048	0.97975	0.99534
*x*14	0.99787	0.99240	0.95789	0.90451	0.84009	0.77032	0.69920
*y*14	0.45387	0.58682	0.70085	0.79389	0.86653	0.92063	0.95861
*x*15	0.94758	0.99240	0.99829	0.97553	0.93301	0.87787	0.81554
*y*15	0.27713	0.41318	0.54129	0.65451	0.75000	0.82743	0.88786
*x*16	0.83685	0.93301	0.98470	1.00000	0.98746	0.95482	0.90848
*y*16	0.13050	0.25000	0.37726	0.50000	0.61126	0.70771	0.78834
*x*17	0.68062	0.82139	0.91858	0.97553	0.99860	0.99491	0.97113
*y*17	0.03376	0.11698	0.22658	0.34549	0.46263	0.57116	0.66744
*x*18		0.67101	0.80711	0.90451	0.96544	0.99491	0.99883
*y*18		0.03015	0.10543	0.20611	0.31733	0.42884	0.53412
*x*19			0.66235	0.79389	0.89092	0.95482	0.98954
*y*19			0.02709	0.09549	0.18826	0.29229	0.39827
*x*20				0.65451	0.78166	0.87787	0.94394
*y*20				0.02447	0.08688	0.17257	0.26997
*x*21					0.64738	0.77032	0.86542
*y*21					0.02221	0.07937	0.15872
*x*22						0.64087	0.75979
*y*22						0.02025	0.07279
*x*23							0.63490
*y*23							0.01854

	24 Holes	25 Holes	26 Holes	27 Holes	28 Holes
*x*1	0.50000	0.50000	0.50000	0.50000	0.50000
*y*1	0.00000	0.00000	0.00000	0.00000	0.00000
*x*2	0.37059	0.37566	0.38034	0.38469	0.38874
*y*2	0.01704	0.01571	0.01453	0.01348	0.01254

Table 4. *(Continued)* **Hole Coordinate Dimension Factors for Jig Boring — Type "A" Hole Circles** (English or Metric Units)

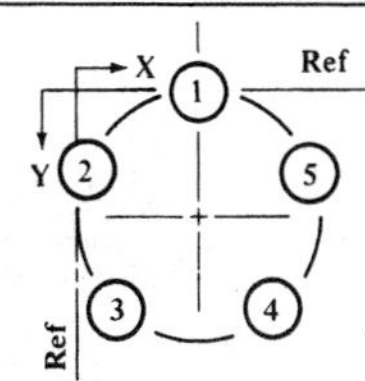

The diagram shows a type "A" circle for a 5-hole circle. Coordinates *x*, *y* are given in the table for hole circles of from 3 to 28 holes. Dimensions are for holes numbered in a counterclockwise direction (as shown). Dimensions given are based upon a hole circle of unit diameter. For a hole circle of, say, 3-inch or 3-centimeter diameter, multiply table values by 3.

24 Holes		25 Holes		26 Holes		27 Holes		28 Holes	
x3	0.25000	x3	0.25912	x3	0.26764	x3	0.27560	x3	0.28306
y3	0.06699	y3	0.06185	y3	0.05727	y3	0.05318	y3	0.04952
x4	0.14645	x4	0.15773	x4	0.16844	x4	0.17861	x4	0.18826
y4	0.14645	y4	0.13552	y4	0.12574	y4	0.11698	y4	0.10908
x5	0.06699	x5	0.07784	x5	0.08851	x5	0.09894	x5	0.10908
y5	0.25000	y5	0.23209	y5	0.21597	y5	0.20142	y5	0.18826
x6	0.01704	x6	0.02447	x6	0.03249	x6	0.04089	x6	0.04952
y6	0.37059	y6	0.34549	y6	0.32270	y6	0.30196	y6	0.28306
x7	0.00000	x7	0.00099	x7	0.00365	x7	0.00760	x7	0.01254
y7	0.50000	y7	0.46860	y7	0.43973	y7	0.41318	y7	0.38874
x8	0.01704	x8	0.00886	x8	0.00365	x8	0.00085	x8	0.00000
y8	0.62941	y8	0.59369	y8	0.56027	y8	0.52907	y8	0.50000
x9	0.06699	x9	0.04759	x9	0.03249	x9	0.02101	x9	0.01254
y9	0.75000	y9	0.71289	y9	0.67730	y9	0.64340	y9	0.61126
x10	0.14645	x10	0.11474	x10	0.08851	x10	0.06699	x10	0.04952
y10	0.85355	y10	0.81871	y10	0.78403	y10	0.75000	y10	0.71694
x11	0.25000	x11	0.20611	x11	0.16844	x11	0.13631	x11	0.10908
y11	0.93301	y11	0.90451	y11	0.87426	y11	0.84312	y11	0.81174
x12	0.37059	x12	0.31594	x12	0.26764	x12	0.22525	x12	0.18826
y12	0.98296	y12	0.96489	y12	0.94273	y12	0.91774	y12	0.89092
x13	0.50000	x13	0.43733	x13	0.38034	x13	0.32899	x13	0.28306
y13	1.00000	y13	0.99606	y13	0.98547	y13	0.96985	y13	0.95048
x14	0.62941	x14	0.56267	x14	0.50000	x14	0.44195	x14	0.38874
y14	0.98296	y14	0.99606	y14	1.00000	y14	0.99662	y14	0.98746
x15	0.75000	x15	0.68406	x15	0.61966	x15	0.55805	x15	0.50000
y15	0.93301	y15	0.96489	y15	0.98547	y15	0.99662	y15	1.00000
x16	0.85355	x16	0.79389	x16	0.73236	x16	0.67101	x16	0.61126
y16	0.85355	y16	0.90451	y16	0.94273	y16	0.96985	y16	0.98746
x17	0.93301	x17	0.88526	x17	0.83156	x17	0.77475	x17	0.71694
y17	0.75000	y17	0.81871	y17	0.87426	y17	0.91774	y17	0.95048
x18	0.98296	x18	0.95241	x18	0.91149	x18	0.86369	x18	0.81174
y18	0.62941	y18	0.71289	y18	0.78403	y18	0.84312	y18	0.89092
x19	1.00000	x19	0.99114	x19	0.96751	x19	0.93301	x19	0.89092
y19	0.50000	y19	0.59369	y19	0.67730	y19	0.75000	y19	0.81174
x20	0.98296	x20	0.99901	x20	0.99635	x20	0.97899	x20	0.95048
y20	0.37059	y20	0.46860	y20	0.56027	y20	0.64340	y20	0.71694
x21	0.93301	x21	0.97553	x21	0.99635	x21	0.99915	x21	0.98746
y21	0.25000	y21	0.34549	y21	0.43973	y21	0.52907	y21	0.61126
x22	0.85355	x22	0.92216	x22	0.96751	x22	0.99240	x22	1.00000
y22	0.14645	y22	0.23209	y22	0.32270	y22	0.41318	y22	0.50000
x23	0.75000	x23	0.84227	x23	0.91149	x23	0.95911	x23	0.98746
y23	0.6699	y23	0.13552	y23	0.21597	y23	0.30196	y23	0.38874
x24	0.62941	x24	0.74088	x24	0.83156	x24	0.90106	x24	0.95048
y24	0.01704	y24	0.06185	y24	0.12574	y24	0.20142	y24	0.28306
		x25	0.62434	x25	0.73236	x25	0.82139	x25	0.89092
		y25	0.01571	y25	0.05727	y25	0.11698	y25	0.18826
				x26	0.61966	x26	0.72440	x26	0.81174
				y26	0.01453	y26	0.05318	y26	0.10908
						x27	0.61531	x27	0.71694
						y27	0.01348	y27	0.04952
								x28	0.61126
								y28	0.01254

Table 5. Hole Coordinate Dimension Factors for Jig Boring — Type "B" Hole Circles (English or Metric Units)

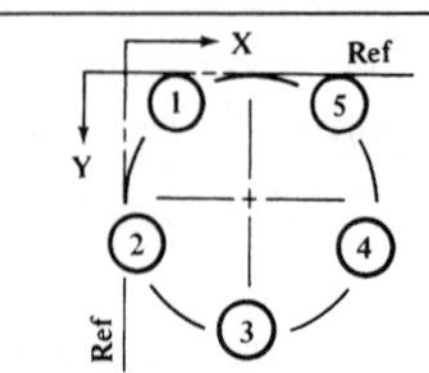

The diagram shows a type "B" circle for a 5-hole circle. Coordinates *x*, *y* are given in the table for hole circles of from 3 to 28 holes. Dimensions are for holes numbered in a counterclockwise direction (as shown). Dimensions given are based upon a hole circle of unit diameter. For a hole circle of, say, 3-inch or 3-centimeter diameter, multiply table values by 3.

3 Holes		4 Holes		5 Holes		6 Holes		7 Holes		8 Holes		9 Holes	
x1	0.06699	x1	0.14645	x1	0.20611	x1	0.25000	x1	0.28306	x1	0.30866	x1	0.32899
y1	0.25000	y1	0.14645	y1	0.09549	y1	0.06699	y1	0.04952	y1	0.03806	y1	0.03015
x2	0.50000	x2	0.14645	x2	0.02447	x2	0.00000	x2	0.01254	x2	0.03806	x2	0.06699
y2	1.00000	y2	0.85355	y2	0.65451	y2	0.50000	y2	0.38874	y2	0.30866	y2	0.25000
x3	0.93301	x3	0.85355	x3	0.50000	x3	0.25000	x3	0.10908	x3	0.03806	x3	0.00760
y3	0.25000	y3	0.85355	y3	1.00000	y3	0.93301	y3	0.81174	y3	0.69134	y3	0.58682
		x4	0.85355	x4	0.97553	x4	0.75000	x4	0.50000	x4	0.30866	x4	0.17861
		y4	0.14645	y4	0.65451	y4	0.93301	y4	1.00000	y4	0.96194	y4	0.88302
				x5	0.79389	x5	1.00000	x5	0.89092	x5	0.69134	x5	0.50000
				y5	0.09549	y5	0.50000	y5	0.81174	y5	0.96194	y5	1.00000
						x6	0.75000	x6	0.98746	x6	0.96194	x6	0.82139
						y6	0.06699	y6	0.38874	y6	0.69134	y6	0.88302
								x7	0.71694	x7	0.96194	x7	0.99240
								y7	0.04952	y7	0.30866	y7	0.58682
										x8	0.69134	x8	0.93301
										y8	0.03806	y8	0.25000
												x9	0.67101
												y9	0.03015

10 Holes		11 Holes		12 Holes		13 Holes		14 Holes		15 Holes		16 Holes	
x1	0.34549	x1	0.35913	x1	0.37059	x1	0.38034	x1	0.38874	x1	0.39604	x1	0.40245
y1	0.02447	y1	0.02025	y1	0.01704	y1	0.01453	y1	0.01254	y1	0.01093	y1	0.00961
x2	0.09549	x2	0.12213	x2	0.14645	x2	0.16844	x2	0.18826	x2	0.20611	x2	0.22221
y2	0.20611	y2	0.17257	y2	0.14645	y2	0.12574	y2	0.10908	y2	0.09549	y2	0.08427
x3	0.00000	x3	0.00509	x3	0.01704	x3	0.03249	x3	0.04952	x3	0.06699	x3	0.08427
y3	0.50000	y3	0.42884	y3	0.37059	y3	0.32270	y3	0.28306	y3	0.25000	y3	0.22221
x4	0.09549	x4	0.04518	x4	0.01704	x4	0.00365	x4	0.00000	x4	0.00274	x4	0.00961
y4	0.79389	y4	0.70771	y4	0.62941	y4	0.56027	y4	0.50000	y4	0.44774	y4	0.40245
x5	0.34549	x5	0.22968	x5	0.14645	x5	0.08851	x5	0.04952	x5	0.02447	x5	0.00961
y5	0.97553	y5	0.92063	y5	0.85355	y5	0.78403	y5	0.71694	y5	0.65451	y5	0.59755
x6	0.65451	x6	0.50000	x6	0.37059	x6	0.26764	x6	0.18826	x6	0.12843	x6	0.08427
y6	0.97553	y6	1.00000	y6	0.98296	y6	0.94273	y6	0.89092	y6	0.83457	y6	0.77779
x7	0.90451	x7	0.77032	x7	0.62941	x7	0.50000	x7	0.38874	x7	0.29663	x7	0.22221
y7	0.79389	y7	0.92063	y7	0.98296	y7	1.00000	y7	0.98746	y7	0.95677	y7	0.91573
x8	1.00000	x8	0.95482	x8	0.85355	x8	0.73236	x8	0.61126	x8	0.50000	x8	0.40245
y8	0.50000	y8	0.70771	y8	0.85355	y8	0.94273	y8	0.98746	y8	1.00000	y8	0.99039
x9	0.90451	x9	0.99491	x9	0.98296	x9	0.91149	x9	0.81174	x9	0.70337	x9	0.59755
y9	0.20611	y9	0.42884	y9	0.62941	y9	0.78403	y9	0.89092	y9	0.95677	y9	0.99039
x10	0.65451	x10	0.87787	x10	0.98296	x10	0.99635	x10	0.95048	x10	0.87157	x10	0.77779
y10	0.02447	y10	0.17257	y10	0.37059	y10	0.56027	y10	0.71694	y10	0.83457	y10	0.91573
		x11	0.64087	x11	0.85355	x11	0.96751	x11	1.00000	x11	0.97553	x11	0.91573
		y11	0.02025	y11	0.14645	y11	0.32270	y11	0.50000	y11	0.65451	y11	0.77779
				x12	0.62941	x12	0.83156	x12	0.95048	x12	0.99726	x12	0.99039
				y12	0.01704	y12	0.12574	y12	0.28306	y12	0.44774	y12	0.59755
						x13	0.61966	x13	0.81174	x13	0.93301	x13	0.99039
						y13	0.01453	y13	0.10908	y13	0.25000	y13	0.40245
								x14	0.61126	x14	0.79389	x14	0.91573
								y14	0.01254	y14	0.09549	y14	0.22221
										x15	0.60396	x15	0.77779
										y15	0.01093	y15	0.08427
												x16	0.59755
												y16	0.00961

Table 5. *(Continued)* **Hole Coordinate Dimension Factors for Jig Boring — Type "B" Hole Circles** (English or Metric Units)

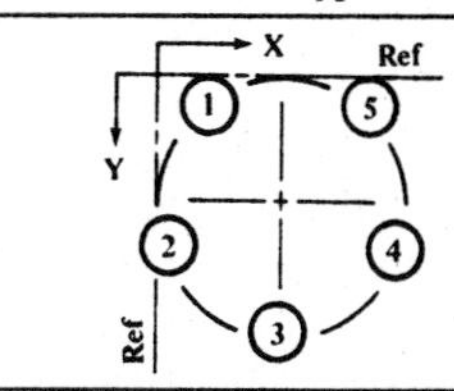

The diagram shows a type "B" circle for a 5-hole circle. Coordinates *x*, *y* are given in the table for hole circles of from 3 to 28 holes. Dimensions are for holes numbered in a counterclockwise direction (as shown). Dimensions given are based upon a hole circle of unit diameter. For a hole circle of, say, 3-inch or 3-centimeter diameter, multiply table values by 3.

17 Holes		18 Holes		19 Holes		20 Holes		21 Holes		22 Holes		23 Holes	
*x*1	0.40813	*x*1	0.41318	*x*1	0.41770	*x*1	0.42178	*x*1	0.42548	*x*1	0.42884	*x*1	0.43192
*y*1	0.00851	*y*1	0.00760	*y*1	0.00682	*y*1	0.00616	*y*1	0.00558	*y*1	0.00509	*y*1	0.00466
*x*2	0.23678	*x*2	0.25000	*x*2	0.26203	*x*2	0.27300	*x*2	0.28306	*x*2	0.29229	*x*2	0.30080
*y*2	0.07489	*y*2	0.06699	*y*2	0.06026	*y*2	0.05450	*y*2	0.04952	*y*2	0.04518	*y*2	0.04139
*x*3	0.10099	*x*3	0.11698	*x*3	0.13214	*x*3	0.14645	*x*3	0.15991	*x*3	0.17257	*x*3	0.18446
*y*3	0.19868	*y*3	0.17861	*y*3	0.16136	*y*3	0.14645	*y*3	0.13347	*y*3	0.12213	*y*3	0.11214
*x*4	0.01909	*x*4	0.03015	*x*4	0.04211	*x*4	0.05450	*x*4	0.06699	*x*4	0.07937	*x*4	0.09152
*y*4	0.36317	*y*4	0.32899	*y*4	0.29915	*y*4	0.27300	*y*4	0.25000	*y*4	0.22968	*y*4	0.21166
*x*5	0.00213	*x*5	0.00000	*x*5	0.00171	*x*5	0.00616	*x*5	0.01254	*x*5	0.02025	*x*5	0.02887
*y*5	0.54613	*y*5	0.50000	*y*5	0.45871	*y*5	0.42178	*y*5	0.38874	*y*5	0.35913	*y*5	0.33256
*x*6	0.05242	*x*6	0.03015	*x*6	0.01530	*x*6	0.00616	*x*6	0.00140	*x*6	0.00000	*x*6	0.00117
*y*6	0.72287	*y*6	0.67101	*y*6	0.62274	*y*6	0.57822	*y*6	0.53737	*y*6	0.50000	*y*6	0.46588
*x*7	0.16315	*x*7	0.11698	*x*7	0.08142	*x*7	0.05450	*x*7	0.03456	*x*7	0.02025	*x*7	0.01046
*y*7	0.86950	*y*7	0.82139	*y*7	0.77347	*y*7	0.72700	*y*7	0.68267	*y*7	0.64087	*y*7	0.60173
*x*8	0.31938	*x*8	0.25000	*x*8	0.19289	*x*8	0.14645	*x*8	0.10908	*x*8	0.07937	*x*8	0.05606
*y*8	0.96624	*y*8	0.93301	*y*8	0.89457	*y*8	0.85355	*y*8	0.81174	*y*8	0.77032	*y*8	0.73003
*x*9	0.50000	*x*9	0.41318	*x*9	0.33765	*x*9	0.27300	*x*9	0.21834	*x*9	0.17257	*x*9	0.13458
*y*9	1.00000	*y*9	0.99240	*y*9	0.97291	*y*9	0.94550	*y*9	0.91312	*y*9	0.87787	*y*9	0.84128
*x*10	0.68062	*x*10	0.58682	*x*10	0.50000	*x*10	0.42178	*x*10	0.35262	*x*10	0.29229	*x*10	0.24021
*y*10	0.96624	*y*10	0.99240	*y*10	1.00000	*y*10	0.99384	*y*10	0.97779	*y*10	0.95482	*y*10	0.92721
*x*11	0.83685	*x*11	0.75000	*x*11	0.66235	*x*11	0.57822	*x*11	0.50000	*x*11	0.42884	*x*11	0.36510
*y*11	0.86950	*y*11	0.93301	*y*11	0.97291	*y*11	0.99384	*y*11	1.00000	*y*11	0.99491	*y*11	0.98146
*x*12	0.94758	*x*12	0.88302	*x*12	0.80711	*x*12	0.72700	*x*12	0.64738	*x*12	0.57116	*x*12	0.50000
*y*12	0.72287	*y*12	0.82139	*y*12	0.89457	*y*12	0.94550	*y*12	0.97779	*y*12	0.99491	*y*12	1.00000
*x*13	0.99787	*x*13	0.96985	*x*13	0.91858	*x*13	0.85355	*x*13	0.78166	*x*13	0.70771	*x*13	0.63490
*y*13	0.54613	*y*13	0.67101	*y*13	0.77347	*y*13	0.85355	*y*13	0.91312	*y*13	0.95482	*y*13	0.98146
*x*14	0.98091	*x*14	1.00000	*x*14	0.98470	*x*14	0.94550	*x*14	0.89092	*x*14	0.82743	*x*14	0.75979
*y*14	0.36317	*y*14	0.50000	*y*14	0.62274	*y*14	0.72700	*y*14	0.81174	*y*14	0.87787	*y*14	0.92721
*x*15	0.89901	*x*15	0.96985	*x*15	0.99829	*x*15	0.99384	*x*15	0.96544	*x*15	0.92063	*x*15	0.86542
*y*15	0.19868	*y*15	0.32899	*y*15	0.45871	*y*15	0.57822	*y*15	0.68267	*y*15	0.77032	*y*15	0.84128
*x*16	0.76322	*x*16	0.88302	*x*16	0.95789	*x*16	0.99384	*x*16	0.99860	*x*16	0.97975	*x*16	0.94394
*y*16	0.07489	*y*16	0.17861	*y*16	0.29915	*y*16	0.42178	*y*16	0.53737	*y*16	0.64087	*y*16	0.73003
*x*17	0.59187	*x*17	0.75000	*x*17	0.86786	*x*17	0.94550	*x*17	0.98746	*x*17	1.00000	*x*17	0.98954
*y*17	0.00851	*y*17	0.06699	*y*17	0.16136	*y*17	0.27300	*y*17	0.38874	*y*17	0.50000	*y*17	0.60173
		*x*18	0.58682	*x*18	0.73797	*x*18	0.85355	*x*18	0.93301	*x*18	0.97975	*x*18	0.99883
		*y*18	0.00760	*y*18	0.06026	*y*18	0.14645	*y*18	0.25000	*y*18	0.35913	*y*18	0.46588
				*x*19	0.58230	*x*19	0.72700	*x*19	0.84009	*x*19	0.92063	*x*19	0.97113
				*y*19	0.00682	*y*19	0.05450	*y*19	0.13347	*y*19	0.22968	*y*19	0.33256
						*x*20	0.57822	*x*20	0.71694	*x*20	0.82743	*x*20	0.90848
						*y*20	0.00616	*y*20	0.04952	*y*20	0.12213	*y*20	0.21166
								*x*21	0.57452	*x*21	0.70771	*x*21	0.81554
								*y*21	0.00558	*y*21	0.04518	*y*21	0.11214
										*x*22	0.57116	*x*22	0.69920
										*y*22	0.00509	*y*22	0.04139
												*x*23	0.56808
												*y*23	0.00466

24 Holes		25 Holes		26 Holes		27 Holes		28 Holes	
*x*1	0.43474	*x*1	0.43733	*x*1	0.43973	*x*1	0.44195	*x*1	0.44402
*y*1	0.00428	*y*1	0.00394	*y*1	0.00365	*y*1	0.00338	*y*1	0.00314
*x*2	0.30866	*x*2	0.31594	*x*2	0.32270	*x*2	0.32899	*x*2	0.33486
*y*2	0.03806	*y*2	0.03511	*y*2	0.03249	*y*2	0.03015	*y*2	0.02806
*x*3	0.19562	*x*3	0.20611	*x*3	0.21597	*x*3	0.22525	*x*3	0.23398
*y*3	0.10332	*y*3	0.09549	*y*3	0.08851	*y*3	0.08226	*y*3	0.07664

Table 5. *(Continued)* **Hole Coordinate Dimension Factors for Jig Boring — Type "B" Hole Circles** (English or Metric Units)

The diagram shows a type "B" circle for a 5-hole circle. Coordinates *x*, *y* are given in the table for hole circles of from 3 to 28 holes. Dimensions are for holes numbered in a counterclockwise direction (as shown). Dimensions given are based upon a hole circle of unit diameter. For a hole circle of, say, 3-inch or 3-centimeter diameter, multiply table values by 3.

x4	0.10332	x4	0.11474	x4	0.12574	x4	0.13631	x4	0.14645
y4	0.19562	y4	0.18129	y4	0.16844	y4	0.15688	y4	0.14645
x5	0.03806	x5	0.04759	x5	0.05727	x5	0.06699	x5	0.07664
y5	0.30866	y5	0.28711	y5	0.26764	y5	0.25000	y5	0.23398
x6	0.00428	x6	0.00886	x6	0.01453	x6	0.02101	x6	0.02806
y6	0.43474	y6	0.40631	y6	0.38034	y6	0.35660	y6	0.33486
x7	0.00428	x7	0.00099	x7	0.00000	x7	0.00085	x7	0.00314
y7	0.56526	y7	0.53140	y7	0.50000	y7	0.47093	y7	0.44402
x8	0.03806	x8	0.02447	x8	0.01453	x8	0.00760	x8	0.00314
y8	0.69134	y8	0.65451	y8	0.61966	y8	0.58682	y8	0.55598
x9	0.10332	x9	0.07784	x9	0.05727	x9	0.04089	x9	0.02806
y9	0.80438	y9	0.76791	y9	0.73236	y9	0.69804	y9	0.66514
x10	0.19562	x10	0.15773	x10	0.12574	x10	0.09894	x10	0.07664
y10	0.89668	y10	0.86448	y10	0.83156	y10	0.79858	y10	0.76602
x11	0.30866	x11	0.25912	x11	0.21597	x11	0.17861	x11	0.14645
y11	0.96194	y11	0.93815	y11	0.91149	y11	0.88302	y11	0.85355
x12	0.43474	x12	0.37566	x12	0.32270	x12	0.27560	x12	0.23398
y12	0.99572	y12	0.98429	y12	0.96751	y12	0.94682	y12	0.92336
x13	0.56526	x13	0.50000	x13	0.43973	x13	0.38469	x13	0.33486
y13	0.99572	y13	1.00000	y13	0.99635	y13	0.98652	y13	0.97194
x14	0.69134	x14	0.62434	x14	0.56027	x14	0.50000	x14	0.44402
y14	0.96194	y14	0.98429	y14	0.99635	y14	1.00000	y14	0.99686
x15	0.80438	x15	0.74088	x15	0.67730	x15	0.61531	x15	0.55598
y15	0.89668	y15	0.93815	y15	0.96751	y15	0.98652	y15	0.99686
x16	0.89668	x16	0.84227	x16	0.78403	x16	0.72440	x16	0.66514
y16	0.80438	y16	0.86448	y16	0.91149	y16	0.94682	y16	0.97194
x17	0.96194	x17	0.92216	x17	0.87426	x17	0.82139	x17	0.76602
y17	0.69134	y17	0.76791	y17	0.83156	y17	0.88302	y17	0.92336
x18	0.99572	x18	0.97553	x18	0.94273	x18	0.90106	x18	0.85355
y18	0.56526	y18	0.65451	y18	0.73236	y18	0.79858	y18	0.85355
x19	0.99572	x19	0.99901	x19	0.98547	x19	0.95911	x19	0.92336
y19	0.43474	y19	0.53140	y19	0.61966	y19	0.69804	y19	0.76602
x20	0.96194	x20	0.99114	x20	1.00000	x20	0.99240	x20	0.97194
y20	0.30866	y20	0.40631	y20	0.50000	y20	0.58682	y20	0.66514
x21	0.89668	x21	0.95241	x21	0.98547	x21	0.99915	x21	0.99686
y21	0.19562	y21	0.28711	y21	0.38034	y21	0.47093	y21	0.55598
x22	0.80438	x22	0.88526	x22	0.94273	x22	0.97899	x22	0.99686
y22	0.10332	y22	0.18129	y22	0.26764	y22	0.35660	y22	0.44402
x23	0.69134	x23	0.79389	x23	0.87426	x23	0.93301	x23	0.97194
y23	0.03806	y23	0.09549	y23	0.16844	y23	0.25000	y23	0.33486
x24	0.56526	x24	0.68406	x24	0.78403	x24	0.86369	x24	0.92336
y24	0.00428	y24	0.03511	y24	0.08851	y24	0.15688	y24	0.23398
		x25	0.56267	x25	0.67730	x25	0.77475	x25	0.85355
		y25	0.00394	y25	0.03249	y25	0.08226	y25	0.14645
				x26	0.56027	x26	0.67101	x26	0.76602
				y26	0.00365	y26	0.03015	y26	0.07664
						x27	0.55805	x27	0.66514
						y27	0.00338	y27	0.02806
								x28	0.55598
								y28	0.00314

Table 6. Hole Coordinate Dimension Factors for Jig Boring— Type "A" Hole Circles, Central Coordinates (English or Metric Units)

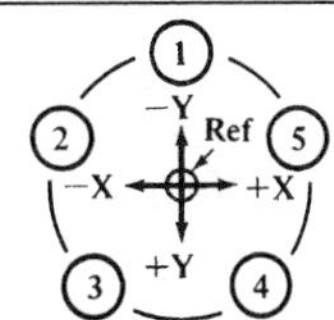

The diagram shows a type "A" circle for a 5-hole circle. Coordinates *x*, *y* are given in the table for hole circles of from 3 to 28 holes. Dimensions are for holes numbered in a counterclockwise direction (as shown). Dimensions given are based upon a hole circle of unit diameter. For a hole circle of, say, 3-inch or 3-centimeter diameter, multiply table values by 3.

3 Holes		4 Holes		5 Holes		6 Holes		7 Holes		8 Holes		9 Holes	
*x*1	0.00000	*x*1	0.00000	*x*1	0.00000	*x*1	0.00000	*x*1	0.00000	*x*1	0.00000	*x*1	0.00000
*y*1	−0.50000	*y*1	−0.50000	*y*1	−0.50000	*y*1	−0.50000	*y*1	−0.50000	*y*1	−0.50000	*y*1	−0.50000
*x*2	−0.43301	*x*2	−0.50000	*x*2	−0.47553	*x*2	−0.43301	*x*2	−0.39092	*x*2	−0.35355	*x*2	−0.32139
*y*2	+0.25000	*y*2	0.00000	*y*2	−0.15451	*y*2	−0.25000	*y*2	−0.31174	*y*2	−0.35355	*y*2	−0.38302
*x*3	+0.43301	*x*3	0.00000	*x*3	−0.29389	*x*3	−0.43301	*x*3	−0.48746	*x*3	−0.50000	*x*3	−0.49240
*y*3	+0.25000	*y*3	+0.50000	*y*3	+0.40451	*y*3	+0.25000	*y*3	+0.11126	*y*3	0.00000	*y*3	−0.08682
		*x*4	+0.50000	*x*4	+0.29389	*x*4	0.00000	*x*4	−0.21694	*x*4	−0.35355	*x*4	−0.43301
		*y*4	0.00000	*y*4	+0.40451	*y*4	+0.50000	*y*4	+0.45048	*y*4	+0.35355	*y*4	+0.25000
				*x*5	+0.47553	*x*5	+0.43301	*x*5	+0.21694	*x*5	0.00000	*x*5	−0.17101
				*y*5	−0.15451	*y*5	+0.25000	*y*5	+0.45048	*y*5	+0.50000	*y*5	+0.46985
						*x*6	+0.43301	*x*6	+0.48746	*x*6	+0.35355	*x*6	+0.17101
						*y*6	−0.25000	*y*6	+0.11126	*y*6	+0.35355	*y*6	+0.46985
								*x*7	+0.39092	*x*7	+0.50000	*x*7	+0.43301
								*y*7	−0.31174	*y*7	0.00000	*y*7	+0.25000
										*x*8	+0.35355	*x*8	+0.49240
										*y*8	−0.35355	*y*8	−0.08682
												*x*9	+0.32139
												*y*9	−0.38302

10 Holes		11 Holes		12 Holes		13 Holes		14 Holes		15 Holes		16 Holes	
*x*1	0.00000	*x*1	0.00000	*x*1	0.00000	*x*1	0.00000	*x*1	0.00000	*x*1	0.00000	*x*1	0.00000
*y*1	−0.50000	*y*1	−0.50000	*y*1	−0.50000	*y*1	−0.50000	*y*1	−0.50000	*y*1	−0.50000	*y*1	−0.50000
*x*2	−0.29389	*x*2	−0.27032	*x*2	−0.25000	*x*2	−0.23236	*x*2	−0.21694	*x*2	−0.20337	*x*2	−0.19134
*y*2	−0.40451	*y*2	−0.42063	*y*2	−0.43301	*y*2	−0.44273	*y*2	−0.45048	*y*2	−0.45677	*y*2	−0.46194
*x*3	−0.47553	*x*3	−0.45482	*x*3	−0.43301	*x*3	−0.41149	*x*3	−0.39092	*x*3	−0.37157	*x*3	−0.35355
*y*3	−0.15451	*y*3	−0.20771	*y*3	−0.25000	*y*3	−2.28403	*y*3	−0.31174	*y*3	−0.33457	*y*3	−0.35355
*x*4	−0.47553	*x*4	−0.49491	*x*4	−0.50000	*x*4	−0.49635	*x*4	−0.48746	*x*4	−0.47553	*x*4	−0.46194
*y*4	+0.15451	*y*4	+0.07116	*y*4	0.00000	*y*4	−0.06027	*y*4	−0.11126	*y*4	−0.15451	*y*4	−0.19134
*x*5	−0.29389	*x*5	−0.37787	*x*5	−0.43301	*x*5	−0.46751	*x*5	−0.48746	*x*5	−0.49726	*x*5	−0.50000
*y*5	+0.40451	*y*5	+0.32743	*y*5	+0.25000	*y*5	+0.17730	*y*5	+0.11126	*y*5	+0.05226	*y*5	0.00000
*x*6	0.00000	*x*6	−0.14087	*x*6	−0.25000	*x*6	− 0.33156	*x*6	−0.39092	*x*6	−0.43301	*x*6	−0.46194
*y*6	+0.50000	*y*6	+0.47975	*y*6	+0.43301	*y*6	+0.37426	*y*6	+0.31174	*y*6	+0.25000	*y*6	+0.19134
*x*7	+0.29389	*x*7	+0.14087	*x*7	0.00000	*x*7	−0.11966	*x*7	−0.21694	*x*7	−0.29389	*x*7	−0.35355
*y*7	+0.40451	*y*7	+0.47975	*y*7	+0.50000	*y*7	+0.48547	*y*7	+0.45048	*y*7	+0.40451	*y*7	+0.35355
*x*8	+0.47553	*x*8	+0.37787	*x*8	+0.25000	*x*8	+0.11966	*x*8	0.00000	*x*8	−0.10396	*x*8	−0.19134
*y*8	+0.15451	*y*8	+0.32743	*y*8	+0.43301	*y*8	+0.48547	*y*8	+0.50000	*y*8	+0.48907	*y*8	+0.46194
*x*9	+0.47553	*x*9	+0.49491	*x*9	+0.43301	*x*9	+0.33156	*x*9	+0.21694	*x*9	+0.10396	*x*9	0.00000
*y*9	−0.15451	*y*9	+0.07116	*y*9	+0.25000	*y*9	+0.37426	*y*9	+0.45048	*y*9	+0.48907	*y*9	+0.50000
*x*10	+0.29389	*x*10	+0.45482	*x*10	+0.50000	*x*10	+0.46751	*x*10	+0.39092	*x*10	+0.29389	*x*10	+0.19134
*y*10	−0.40451	*y*10	−0.20771	*y*10	0.00000	*y*10	+0.17730	*y*10	+0.31174	*y*10	+0.40451	*y*10	+0.46194
		*x*11	+0.27032	*x*11	+0.43301	*x*11	+0.49635	*x*11	+0.48746	*x*11	+0.43301	*x*11	+0.35355
		*y*11	−0.42063	*y*11	−0.25000	*y*11	−0.06027	*y*11	+0.11126	*y*11	+0.25000	*y*11	+0.35355
				*x*12	+0.25000	*x*12	+0.41149	*x*12	+0.48746	*x*12	+0.49726	*x*12	+0.46194
				*y*12	−0.43301	*y*12	−0.28403	*y*12	−0.11126	*y*12	+0.05226	*y*12	+0.19134
						*x*13	+0.23236	*x*13	+0.39092	*x*13	+0.47553	*x*13	+0.50000
						*y*13	−0.44273	*y*13	−0.31174	*y*13	−0.15451	*y*13	0.00000
								*x*14	+0.21694	*x*14	+0.37157	*x*14	+0.46194
								*y*14	−0.45048	*y*14	−0.33457	*y*14	−0.19134
										*x*15	+0.20337	*x*15	+0.35355
										*y*15	−0.45677	*y*15	−0.35355
												*x*16	+0.19134
												*y*16	−0.46194

Table 6. *(Continued)* **Hole Coordinate Dimension Factors for Jig Boring — Type "A" Hole Circles, Central Coordinates** (English or Metric Units)

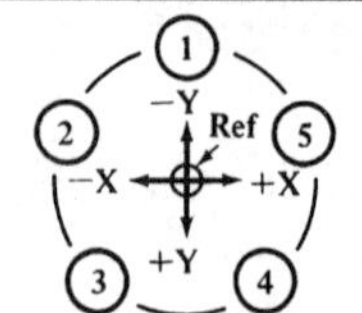

The diagram shows a type "A" circle for a 5-hole circle. Coordinates *x*, *y* are given in the table for hole circles of from 3 to 28 holes. Dimensions are for holes numbered in a counterclockwise direction (as shown). Dimensions given are based upon a hole circle of unit diameter. For a hole circle of, say, 3-inch or 3-centimeter diameter, multiply table values by 3.

17 Holes		18 Holes		19 Holes		20 Holes		21 Holes		22 Holes		23 Holes	
x1	0.00000	x1	0.00000	x1	0.00000	x1	0.000000	x1	0.00000	x1	0.00000	x1	0.00000
y1	–0.50000	y1	–0.50000	y1	–0.50000	y1	–0.50000	y1	–0.50000	y1	–0.50000	y1	–0.50000
x2	–0.18062	x2	–0.17101	x2	–0.16235	x2	–0.15451	x2	–0.14738	x2	–0.14087	x2	–0.13490
y2	–0.46624	y2	–0.46985	y2	–0.47291	y2	–0.47553	y2	–0.47779	y2	–0.47975	y2	–0.48146
x3	–0.33685	x3	+0.32139	x3	–0.30711	x3	–0.29389	x3	–0.28166	x3	–0.27032	x3	–0.25979
y3	–0.36950	y3	–0.38302	y3	–0.39457	y3	–0.40451	y3	–0.41312	y3	–0.42063	y3	–0.42721
x4	–0.44758	x4	–0.43301	x4	–0.41858	x4	–0.40451	x4	–0.39092	x4	–0.37787	x4	–0.36542
y4	–0.22287	y4	–0.25000	y4	–0.27347	y4	–0.29389	y4	–0.31174	y4	–0.32743	y4	–0.34128
x5	–0.49787	x5	–0.49240	x5	–0.48470	x5	–0.47553	x5	–.046544	x5	–0.45482	x5	–0.44394
y5	–0.04613	y5	–0.08682	y5	–0.12274	y5	–0.15451	y5	–0.18267	y5	–0.20771	y5	–0.23003
x6	–0.48091	x6	–0.49420	x6	–0.49829	x6	–0.50000	x6	–0.49860	x6	–0.49491	x6	–0.48954
y6	+0.13683	y6	+0.08682	y6	+0.04129	y6	0.00000	y6	–0.03737	y6	–0.07116	y6	–0.10173
x7	–0.39901	x7	–0.43301	x7	–0.45789	x7	–0.47553	x7	–0.48746	x7	–0.49491	x7	–0.49883
y7	+0.30132	y7	+0.25000	y7	+0.20085	y7	+0.15451	y7	+0.11126	y7	+0.07116	y7	+0.03412
x8	–0.26322	x8	–0.32139	x8	–0.36786	x8	–0.40451	x8	–0.43301	x8	–0.45482	x8	–0.47113
y8	+0.42511	y8	+0.38302	y8	+0.33864	y8	+0.29389	y8	+0.25000	y8	+0.20771	y8	+0.16744
x9	–0.09187	x9	–0.17101	x9	–0.23797	x9	–0.29389	x9	–0.34009	x9	–0.37787	x9	–0.40848
y9	+0.49149	y9	+0.46985	y9	+0.43974	y9	+0.40451	y9	+0.36653	y9	+0.32743	y9	+0.28834
x10	+0.09187	x10	0.00000	x10	–0.08230	x10	–0.15451	x10	–0.21694	x10	–0.27032	x10	–0.31554
y10	+0.49149	y10	+0.50000	y10	+0.49318	y10	+0.47553	y10	+0.45048	y10	+0.42063	y10	+0.38786
x11	+0.26322	x11	+0.17101	x11	+0.08230	x11	0.00000	x11	–0.07452	x11	–0.14087	x11	–0.19920
y11	+0.42511	y11	+0.46985	y11	+0.49318	y11	+0.50000	y11	+0.49442	y11	+0.47975	y11	+0.45861
x12	+0.39901	x12	+0.32139	x12	+0.23797	x12	+0.15451	x12	+0.07452	x12	0.00000	x12	–0.06808
y12	+0.30132	y12	+0.38302	y12	+0.43974	y12	+0.47553	y12	+0.49442	y12	+0.50000	y12	+0.49534
x13	+0.48091	x13	+0.43301	x13	+0.36786	x13	+0.29389	x13	+0.21694	x13	+0.14087	x13	+0.06808
y13	+0.13683	y13	+0.25000	y13	+0.33864	y13	+0.40451	y13	+0.45048	y13	+0.47975	y13	+0.49534
x14	+0.49787	x14	+0.49240	x14	+0.45789	x14	+0.40451	x14	+0.34009	x14	+0.27032	x14	+0.19920
y14	–0.04613	y14	+0.08682	y14	+0.20085	y14	+0.29389	y14	+0.36653	y14	+0.42063	y14	+0.45861
x15	+0.44758	x15	+0.49240	x15	+0.49829	x15	+0.47553	x15	+0.43301	x15	+0.37787	x15	+0.31554
y15	–0.22287	y15	–0.08682	y15	+0.04129	y15	+0.15451	y15	+0.25000	y15	+0.32743	y15	+0.38786
x16	+0.33685	x16	+0.43301	x16	+0.48470	x16	+0.50000	x16	+0.48746	x16	+0.45482	x16	+0.40848
y16	–0.36950	y16	–0.25000	y16	–0.12274	y16	0.00000	y16	+0.11126	y16	+0.20771	y16	+0.28834
x17	+0.18062	x17	+0.32139	x17	+0.41858	x17	+0.47553	x17	+0.49860	x17	+0.49491	x17	+0.47113
y17	–0.46624	y17	–0.38302	y17	–0.27347	y17	–0.15451	y17	–0.03737	y17	+0.07116	y17	+0.16744
		x18	+0.17101	x18	+0.30711	x18	+0.40451	x18	+0.46544	x18	+0.49491	x18	+0.49883
		y18	–0.46985	y18	–0.39457	y18	–0.29389	y18	–0.18267	y18	–0.07116	y18	+0.03412
				x19	+ 0.16235	x19	+0.29389	x19	+0.39092	x19	+0.45482	x19	+0.48954
				y19	–0.47291	y19	–0.40451	y19	–0.31174	y19	–0.20771	y19	–0.10173
						x20	+0.15451	x20	+0.28166	x20	+0.37787	x20	+0.44394
						y20	–0.47553	y20	–0.41312	y20	–0.32743	y20	–0.23003
								x21	+0.14738	x21	+0.27032	x21	+0.36542
								y21	–0.47779	y21	–0.42063	y21	–0.34128
										x22	+0.14087	x22	+0.25979
										y22	–0.47975	y22	–0.42721
												x23	+0.13490
												y23	–0.48146

24 Holes		25 Holes		26 Holes		27 Holes		28 Holes	
x1	0.00000	x1	0.00000	x1	0.00000	x1	0.00000	x1	0.00000
y1	–0.50000	y1	–0.50000	y1	–0.50000	y1	–0.50000	y1	–0.50000
x2	–0.12941	x2	–0.12434	x2	–0.11966	x2	–0.11531	x2	–0.11126
y2	–0.48296	y2	–0.48429	y2	–0.48547	y2	–0.48652	y2	–0.48746
x3	–0.25000	x3	–0.24088	x3	–0.23236	x3	–0.22440	x3	–0.21694
y3	– 0.43301	y3	–0.43815	y3	–0.44273	y3	–0.44682	y3	–0.45048

Table 6. *(Continued)* **Hole Coordinate Dimension Factors for Jig Boring — Type "A" Hole Circles, Central Coordinates** (English or Metric Units)

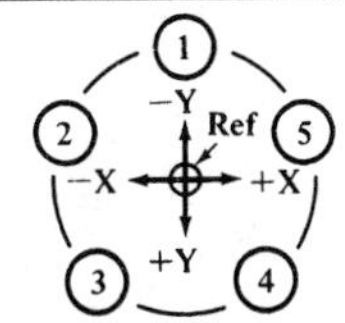

The diagram shows a type "A" circle for a 5-hole circle. Coordinates *x*, *y* are given in the table for hole circles of from 3 to 28 holes. Dimensions are for holes numbered in a counterclockwise direction (as shown). Dimensions given are based upon a hole circle of unit diameter. For a hole circle of, say, 3-inch or 3-centimeter diameter, multiply table values by 3.

24 Holes		25 Holes		26 Holes		27 Holes		28 Holes	
*x*4	−0.35355	*x*4	−0.34227	*x*4	−0.33156	*x*4	−0.32139	*x*4	−0.31174
*y*4	−0.35355	*y*4	−0.36448	*y*4	−0.37426	*y*4	−0.38302	*y*4	−0.39092
*x*5	−0.43301	*x*5	−0.42216	*x*5	−0.41149	*x*5	−0.40106	*x*5	−0.39092
*y*5	−0.25000	*y*5	−0.26791	*y*5	−0.28403	*y*5	−0.29858	*y*5	−0.31174
*x*6	−0.48296	*x*6	−0.47553	*x*6	−0.46751	*x*6	−0.45911	*x*6	−0.45048
*y*6	−0.12941	*y*6	−0.15451	*y*6	−0.17730	*y*6	−0.19804	*y*6	−0.21694
*x*7	−0.50000	*x*7	−0.49901	*x*7	−0.49635	*x*7	−0.49240	*x*7	−0.48746
*y*7	0.00000	*y*7	−0.03140	*y*7	−0.06027	*y*7	−0.08682	*y*7	−0.11126
*x*8	−0.48296	*x*8	−0.49114	*x*8	−0.49635	*x*8	−0.49915	*x*8	−0.50000
*y*8	+0.12941	*y*8	+0.09369	*y*8	+0.06027	*y*8	+0.02907	*y*8	0.00000
*x*9	−0.43301	*x*9	−0.45241	*x*9	−0.46751	*x*9	−0.47899	*x*9	−0.48746
*y*9	+0.25000	*y*9	+0.21289	*y*9	+0.17730	*y*9	+0.14340	*y*9	+0.11126
*x*10	−0.35355	*x*10	−0.38526	*x*10	−0.41149	*x*10	−0.43301	*x*10	−0.45048
*y*10	+0.35355	*y*10	+0.31871	*y*10	+0.28403	*y*10	+0.25000	*y*10	+0.21694
*x*11	−0.25000	*x*11	−0.29389	*x*11	−0.33156	*x*11	−0.36369	*x*11	−0.39092
*y*11	+0.43301	*y*11	+0.40451	*y*11	+0.37426	*y*11	+0.34312	*y*11	+0.31174
*x*12	−0.12941	*x*12	−0.18406	*x*12	−0.23236	*x*12	−0.27475	*x*12	−0.31174
*y*12	+0.48296	*y*12	+0.46489	*y*12	+0.44273	*y*12	+0.41774	*y*12	+0.39092
*x*13	0.00000	*x*13	−0.06267	*x*13	−0.11966	*x*13	−0.17101	*x*13	−0.21694
*y*13	+0.50000	*y*13	+0.49606	*y*13	+0.48547	*y*13	+0.46985	*y*13	+0.45048
*x*14	+0.12941	*x*14	+0.06267	*x*14	0.00000	*x*14	−0.05805	*x*14	−0.11126
*y*14	+0.48296	*y*14	+0.49606	*y*14	+0.50000	*y*14	+0.49662	*y*14	+0.48746
*x*15	+0.25000	*x*15	+0.18406	*x*15	+0.11966	*x*15	+0.05805	*x*15	0.00000
*y*15	+0.43301	*y*15	+0.46489	*y*15	+0.48547	*y*15	+0.49662	*y*15	+0.50000
*x*16	+0.35355	*x*16	+0.29389	*x*16	+0.23236	*x*16	+0.17101	*x*16	+0.11126
*y*16	+0.35355	*y*16	+0.40451	*y*16	+0.44273	*y*16	+0.46985	*y*16	+0.48746
*x*17	+0.43301	*x*17	+ 0.38526	*x*17	+0.33156	*x*17	+0.27475	*x*17	+0.21694
*y*17	+0.25000	*y*17	+0.31871	*y*17	+0.37426	*y*17	+0.41774	*y*17	+0.45048
*x*18	+0.48296	*x*18	+0.45241	*x*18	+0.41149	*x*18	+0.36369	*x*18	+0.31174
*y*18	+0.12941	*y*18	+0.21289	*y*18	+0.28403	*y*18	+0.34312	*y*18	+0.39092
*x*19	+0.50000	*x*19	+0.49114	*x*19	+0.46751	*x*19	+0.43301	*x*19	+0.39092
*y*19	0.00000	*y*19	+0.09369	*y*19	+0.17730	*y*19	+0.25000	*y*19	+0.31174
*x*20	+0.48296	*x*20	+0.49901	*x*20	+0.49635	*x*20	+0.47899	*x*20	+0.45048
*y*20	−0.12941	*y*20	−0.03140	*y*20	+0.06027	*y*20	+0.14340	*y*20	+0.21694
*x*21	+0.43301	*x*21	+0.47553	*x*21	+0.49635	*x*21	+0.49915	*x*21	+0.48746
*y*21	−0.25000	*y*21	−0.15451	*y*21	−0.06027	*y*21	+0.02907	*y*21	+0.11126
*x*22	+0.35355	*x*22	+0.42216	*x*22	+0.46751	*x*22	+0.49240	*x*22	+0.50000
*y*22	−0.35355	*y*22	−0.26791	*y*22	−0.17730	*y*22	−0.08682	*y*22	0.00000
*x*23	+0.25000	*x*23	+0.34227	*x*23	+0.41149	*x*23	+0.45911	*x*23	+0.48746
*y*23	−0.43301	*y*23	−0.36448	*y*23	−0.28403	*y*23	−0.19804	*y*23	−0.11126
*x*24	+0.12941	*x*24	+0.24088	*x*24	+0.33156	*x*24	+0.40106	*x*24	+0.45048
*y*24	−0.48296	*y*24	−0.43815	*y*24	−0.37426	*y*24	−0.29858	*y*24	−0.21694
		*x*25	+0.12434	*x*25	+0.23236	*x*25	+0.32139	*x*25	+0.39092
		*y*25	−0.48429	*y*25	−0.44273	*y*25	−0.38302	*y*25	−0.31174
				*x*26	+0.11966	*x*26	+0.22440	*x*26	+0.31174
				*y*26	−0.48547	*y*26	−0.44682	*y*26	−0.39092
						*x*27	+0.11531	*x*27	+0.21694
						*y*27	−0.48652	*y*27	−0.45048
								*x*28	+0.11126
								*y*28	−0.48746

Table 7. Hole Coordinate Dimension Factors for Jig Boring —Type "B" Hole Circles Central Coordinates (English or Metric units)

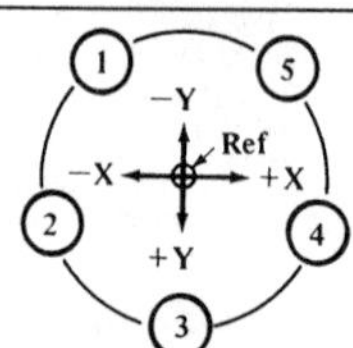

The diagram shows a type "B" circle for a 5-hole circle. Coordinates *x*, *y* are given in the table for hole circles of from 3 to 28 holes. Dimensions are for holes numbered in a counterclockwise direction (as shown). Dimensions given are based upon a hole circle of unit diameter. For a hole circle of, say, 3-inch or 3-centimeter diameter, multiply table values by 3.

	3 Holes	4 Holes	5 Holes	6 Holes	7 Holes	8 Holes	9 Holes
*x*1	−0.43301	−0.35355	−0.29389	−0.25000	−0.21694	−0.19134	−0.17101
*y*1	−0.25000	−0.35355	−0.40451	−0.43301	−0.45048	−0.46194	−0.46985
*x*2	0.00000	−0.35355	−0.47553	−0.50000	−0.48746	−0.46194	−0.43301
*y*2	+0.50000	+0.35355	+0.15451	0.00000	−0.11126	−0.19134	−0.25000
*x*3	+0.43301	+0.35355	0.00000	−0.25000	−0.39092	−0.46194	−0.49240
*y*3	−0.25000	+0.35355	+0.50000	+0.43301	+0.31174	+0.19134	+0.08682
*x*4		+0.35355	+0.47553	+0.25000	0.00000	−0.19134	−0.32139
*y*4		−0.35355	+0.15451	+0.43301	+0.50000	+0.46194	+0.38302
*x*5			+0.29389	+0.50000	+0.39092	+0.19134	0.00000
*y*5			−0.40451	0.00000	+0.31174	+0.46194	+0.50000
*x*6				+0.25000	+0.48746	+0.46194	+0.32139
*y*6				−0.43301	−0.11126	+0.19134	+0.38302
*x*7					+0.21694	+0.46194	+0.49240
*y*7					−0.45048	−0.19134	+0.08682
*x*8						+0.19134	+0.43301
*y*8						−0.46194	−0.25000
*x*9							+0.17101
*y*9							−0.46985

	10 Holes	11 Holes	12 Holes	13 Holes	14 Holes	15 Holes	16 Holes
*x*1	−0.15451	−0.14087	−0.12941	−0.11966	−0.11126	−0.10396	−0.09755
*y*1	−0.47553	−0.47975	−0.48296	−0.48547	−0.48746	−0.48907	−0.49039
*x*2	−0.40451	−0.37787	−0.35355	−0.33156	−0.31174	−0.29389	−0.27779
*y*2	−0.29389	−0.32743	−0.35355	−0.37426	−0.39092	−0.40451	−0.41573
*x*3	−0.50000	−0.49491	−0.48296	−0.46751	−0.45048	−0.43301	−0.41573
*y*3	0.00000	−0.07116	−0.12941	−0.17730	−0.21694	−0.25000	−0.27779
*x*4	−0.40451	−0.45482	−0.48296	−0.49635	−0.50000	−0.49726	−0.49039
*y*4	+0.29389	+0.20771	+0.12941	+0.06027	0.00000	−0.05226	−0.09755
*x*5	−0.15451	−0.27032	−0.35355	−0.41149	−0.45048	−0.47553	−0.49039
*y*5	+0.47553	+0.42063	+0.35355	+0.28403	+0.21694	+0.15451	+0.09755
*x*6	+0.15451	0.00000	−0.12941	−0.23236	−0.31174	−0.37157	−0.41573
*y*6	+0.47553	+0.50000	+0.48296	+0.44273	+0.39092	+0.33457	+0.27779
*x*7	+0.40451	+0.27032	+0.12941	0.00000	−0.11126	−0.20337	−0.27779
*y*7	+0.29389	+0.42063	+0.48296	+ 0.50000	+0.48746	+0.45677	+0.41573
*x*8	+0.50000	+0.45482	+0.35355	+0.23236	+0.11126	0.00000	−0.09755
*y*8	0.00000	+0.20771	+0.35355	+0.44273	+0.48746	+0.50000	+0.49039
*x*9	+0.40451	+0.49491	+0.48296	+0.41149	+0.31174	+0.20337	+0.09755
*y*9	−0.29389	−0.07116	+0.12941	+0.28403	+0.39092	+0.45677	+0.49039
*x*10	+0.15451	+0.37787	+0.48296	+0.49635	+0.45048	+0.37157	+0.27779
*y*10	−0.47553	−0.32743	−0.12941	+0.06027	+0.21694	+0.33457	+0.41573
*x*11		+0.14087	+0.35355	+0.46751	+0.50000	+0.47553	+0.41573
*y*11		−0.47975	−0.35355	−0.17730	0.00000	+0.15451	+0.27779
*x*12			+0.12941	+0.33156	+0.45048	+0.49726	+0.49039
*y*12			−0.48296	−0.37426	−0.21694	−0.05226	+0.09755
*x*13				+0.11966	+0.31174	+0.43301	+0.49039
*y*13				−0.48547	−0.39092	−0.25000	−0.09755
*x*14					+0.11126	+0.29389	+0.41573
*y*14					−0.48746	−0.40451	−0.27779
*x*15						+0.10396	+0.27779
*y*15						−0.48907	−0.41573
*x*16							+0.09755
*y*16							−0.49039

Table 7. *(Continued)* **Hole Coordinate Dimension Factors for Jig Boring —Type "B" Hole Circles Central Coordinates** (English or Metric units)

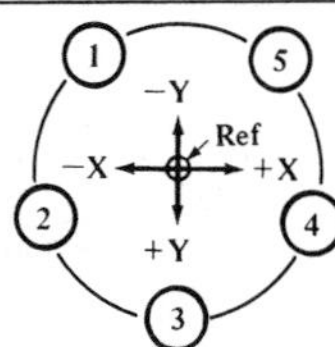

The diagram shows a type "B" circle for a 5-hole circle. Coordinates *x*, *y* are given in the table for hole circles of from 3 to 28 holes. Dimensions are for holes numbered in a counterclockwise direction (as shown). Dimensions given are based upon a hole circle of unit diameter. For a hole circle of, say, 3-inch or 3-centimeter diameter, multiply table values by 3.

17 Holes	18 Holes	19 Holes	20 Holes	21 Holes	22 Holes	23 Holes
x1 −0.09187	x1 −0.08682	x1 −0.08230	x1 −0.07822	x1 −0.07452	x1 −0.07116	x1 −0.06808
y1 −0.49149	y1 −0.49240	y1 −0.49318	y1 −0.49384	y1 −0.49442	y1 −0.49491	y1 −0.49534
x2 −0.26322	x2 −0.25000	x2 −0.23797	x2 −0.22700	x2 −0.21694	x2 −0.20771	x2 −0.19920
y2 −0.42511	y2 −0.43301	y2 −0.43974	y2 −0.44550	y2 −0.45048	y2 −0.45482	y2 −0.45861
x3 − 0.39901	x3 −0.38302	x3 −0.36786	x3 −0.35355	x3 −0.34009	x3 −0.32743	x3 −0.31554
y3 −0.30132	y3 −0.32139	y3 −0.33864	y3 −0.35355	y3 −0.36653	y3 −0.37787	y3 −0.38786
x4 −0.48091	x4 −0.46985	x4 −0.45789	x4 −0.44550	x4 −0.43301	x4 −0.42063	x4 −0.40848
y4 −0.13683	y4 −0.17101	y4 −0.20085	y4 −0.22700	y4 −0.25000	y4 −0.27032	y4 −0.28834
x5 −0.49787	x5 −0.50000	x5 −0.49829	x5 −0.49384	x5 −0.48746	x5 −0.47975	x5 −0.47113
y5 +0.04613	y5 0.00000	y5 −0.04129	y5 −0.07822	y5 −0.11126	y5 −0.14087	y5 −0.16744
x6 −0.44758	x6 −0.46985	x6 −0.48470	x6 −0.49384	x6 −0.49860	x6 −0.50000	x6 −0.49883
y6 +0.22287	y6 +0.17101	y6 +0.12274	y6 +0.07822	y6 +0.03737	y6 0.00000	y6 −0.03412
x7 −0.33685	x7 −0.38302	x7 −0.41858	x7 −0.44550	x7 −0.46544	x7 −0.47975	x7 −0.48954
y7 +0.36950	y7 +0.32139	y7 +0.27347	y7 +0.22700	y7 +0.18267	y7 +0.14087	y7 +0.10173
x8 −0.18062	x8 −0.25000	x8 −0.30711	x8 −0.35355	x8 −0.39092	x8 −0.42063	x8 −0.44394
y8 +0.46624	y8 +0.43301	y8 +0.39457	y8 +0.35355	y8 +0.31174	y8 +0.27032	y8 +0.23003
x9 0.00000	x9 −0.08682	x9 −0.16235	x9 −0.22700	x9 −0.28166	x9 −0.32743	x9 −0.36542
y9 +0.50000	y9 +0.49240	y9 +0.47291	y9 +0.44550	y9 +0.41312	y9 +0.37787	y9 +0.34128
x10 +0.18062	x10 +0.08682	x10 0.00000	x10 −0.07822	x10 −0.14738	x10 −0.20771	x10 −0.25979
y10 +0.46624	y10 +0.49240	y10 +0.50000	y10 +0.49384	y10 +0.47779	y10 +0.45482	y10 +0.42721
x11 +0.33685	x11 +0.25000	x11 +0.16235	x11 +0.07822	x11 0.00000	x11 −0.07116	x11 −0.13490
y11 +0.36950	y11 +0.43301	y11 +0.47291	y11 +0.49384	y11 +0.50000	y11 +0.49491	y11 +0.48146
x12 +0.44758	x12 +0.38302	x12 +0.30711	x12 +0.22700	x12 +0.14738	x12 + 0.07116	x12 0.00000
y12 +0.22287	y12 +0.32139	y12 +0.39457	y12 +0.44550	y12 +0.47779	y12 +0.49491	y12 +0.50000
x13 +0.49787	x13 +0.46985	x13 +0.41858	x13 +0.35355	x13 +0.28166	x13 +0.20771	x13 +0.13490
y13 +0.04613	y13 +0.17101	y13 +0.27347	y13 +0.35355	y13 +0.41312	y13 +0.45482	y13 +0.48146
x14 +0.48091	x14 +0.50000	x14 +0.48470	x14 +0.44550	x14 +0.39092	x14 +0.32743	x14 +0.25979
y14 −0.13683	y14 0.00000	y14 +0.12274	y14 +0.22700	y14 +0.31174	y14 +0.37787	y14 +0.42721
x15 +0.39901	x15 +0.46985	x15 +0.49829	x15 +0.49384	x15 +0.46544	x15 +0.42063	x15 +0.36542
y15 −0.30132	y15 −0.17101	y15 −0.04129	y15 +0.07822	y15 +0.18267	y15 +0.27032	y15 +0.34128
x16 +0.26322	x16 +0.38302	x16 +0.45789	x16 +0.49384	x16 +0.49860	x16 +0.47975	x16 +0.44394
y16 −0.42511	y16 −0.32139	y16 −0.20085	y16 −0.07822	y16 +0.03737	y16 +0.14087	y16 +0.23003
x17 +0.09187	x17 +0.25000	x17 +0.36786	x17 +0.44550	x17 +0.48746	x17 +0.50000	x17 +0.48954
y17 − 0.49149	y17 −0.43301	y17 −0.33864	y17 −0.22700	y17 −0.11126	y17 0.00000	y17 +0.10173
	x18 +0.08682	x18 +0.23797	x18 +0.35355	x18 +0.43301	x18 +0.47975	x18 +0.49883
	y18 −0.49240	y18 −0.43974	y18 −0.35355	y18 −0.25000	y18 −0.14087	y18 −0.03412
		x19 +0.08230	x19 +0.22700	x19 +0.34009	x19 +0.42063	x19 +0.47113
		y19 −0.49318	y19 −0.44550	y19 −0.36653	y19 −0.27032	y19 −0.16744
			x20 +0.07822	x20 +0.21694	x20 +0.32743	x20 +0.40848
			y20 −0.49384	y20 −0.45048	y20 −0.37787	y20 −0.28834
				x21 +0.07452	x21 +0.20771	x21 +0.31554
				y21 −0.49442	y21 −0.45482	y21 −0.38786
					x22 +0.07116	x22 +0.19920
					y22 −0.49491	y22 −0.45861
						x23 +0.06808
						y23 −0.49534

24 Holes	25 Holes	26 Holes	27 Holes	28 Holes		
x1 −0.06526	x1 −0.06267	x1 −0.06027	x1 −0.05805	x1 −0.05598		
y1 −0.49572	y1 −0.49606	y1 −0.49635	y1 −0.49662	y1 −0.49686		
x2 −0.19134	x2 −0.18406	x2 −0.17730	x2 −0.17101	x2 −0.16514		
y2 −0.46194	y2 −0.46489	y2 −0.46751	y2 −0.46985	y2 −0.47194		
x3 −0.30438	x3 −0.29389	x3 −0.28403	x3 −0.27475	x3 −0.26602		
y3 −0.39668	y3 −0.40451	y3 −0.41149	y3 −0.41774	y3 −0.42336		

Table 7. *(Continued)* **Hole Coordinate Dimension Factors for Jig Boring —Type "B" Hole Circles Central Coordinates** (English or Metric units)

The diagram shows a type "B" circle for a 5-hole circle. Coordinates *x*, *y* are given in the table for hole circles of from 3 to 28 holes. Dimensions are for holes numbered in a counterclockwise direction (as shown). Dimensions given are based upon a hole circle of unit diameter. For a hole circle of, say, 3-inch or 3-centimeter diameter, multiply table values by 3.

x4	−0.39668	x4	−0.38526	x4	−0.37426	x4	−0.36369	x4	−0.35355
y4	−0.30438	y4	−0.31871	y4	−0.33156	y4	−0.34312	y4	−0.35355
x5	−0.46194	x5	−0.45241	x5	−0.44273	x5	−0.43301	x5	−0.42336
y5	−0.19134	y5	−0.21289	y5	−0.23236	y5	−0.25000	y5	−0.26602
x6	−0.49572	x6	−0.49114	x6	−0.48547	x6	−0.47899	x6	−0.47194
y6	−0.06526	y6	−0.09369	y6	−0.11966	y6	−0.14340	y6	−0.16514
x7	−0.49572	x7	−0.49901	x7	−0.50000	x7	−0.49915	x7	−0.49686
y7	+0.06526	y7	+0.03140	y7	0.00000	y7	−0.02907	y7	−0.05598
x8	−0.46194	x8	−0.47553	x8	−0.48547	x8	−0.49240	x8	−0.49686
y8	+0.19134	y8	+0.15451	y8	+0.11966	y8	+0.08682	y8	+0.05598
x9	−0.39668	x9	−0.42216	x9	−0.44273	x9	−0.45911	x9	−0.47194
y9	+0.30438	y9	+0.26791	y9	+0.23236	y9	+0.19804	y9	+0.16514
x10	−0.30438	x10	−0.34227	x10	−0.37426	x10	−0.40106	x10	−0.42336
y10	+0.39668	y10	+0.36448	y10	+0.33156	y10	+0.29858	y10	+0.26602
x11	−0.19134	x11	−0.24088	x11	−0.28403	x11	−0.32139	x11	−0.35355
y11	+0.46194	y11	+0.43815	y11	+0.41149	y11	+0.38302	y11	+0.35355
x12	−0.06526	x12	−0.12434	x12	−0.17730	x12	−0.22440	x12	−0.26602
y12	+0.49572	y12	+0.48429	y12	+0.46751	y12	+0.44682	y12	+0.42336
x13	+0.06526	x13	0.00000	x13	−0.06027	x13	−0.11531	x13	−0.16514
y13	+0.49572	y13	+0.50000	y13	+0.49635	y13	+0.48652	y13	+0.47194
x14	+0.19134	x14	+0.12434	x14	+0.06027	x14	0.00000	x14	−0.05598
y14	+0.46194	y14	+0.48429	y14	+0.49635	y14	+0.50000	y14	+0.49686
x15	+0.30438	x15	+0.24088	x15	+0.17730	x15	+0.11531	x15	+0.05598
y15	+0.39668	y15	+0.43815	y15	+0.46751	y15	+0.48652	y15	+0.49686
x16	+0.39668	x16	+0.34227	x16	+0.28403	x16	+0.22440	x16	+0.16514
y16	+0.30438	y16	+0.36448	y16	+0.41149	y16	+0.44682	y16	+0.47194
x17	+0.46194	x17	+0.42216	x17	+0.37426	x17	+0.32139	x17	+0.26602
y17	+0.19134	y17	+0.26791	y17	+0.33156	y17	+0.38302	y17	+0.42336
x18	+0.49572	x18	+0.47553	x18	+0.44273	x18	+0.40106	x18	+0.35355
y18	+0.06526	y18	+0.15451	y18	+0.23236	y18	+0.29858	y18	+0.35355
x19	+0.49572	x19	+0.49901	x19	+0.48547	x19	+0.45911	x19	+0.42336
y19	−0.06526	y19	+0.03140	y19	+0.11966	y19	+0.19804	y19	+0.26602
x20	+0.46194	x20	+0.49114	x20	+0.50000	x20	+0.49240	x20	+0.47194
y20	−0.19134	y20	−0.09369	y20	0.00000	y20	+0.08682	y20	+0.16514
x21	+0.39668	x21	+0.45241	x21	+0.48547	x21	+0.49915	x21	+0.49686
y21	−0.30438	y21	−0.21289	y21	−0.11966	y21	−0.02907	y21	+0.05598
x22	+0.30438	x22	+0.38526	x22	+0.44273	x22	+0.47899	x22	+0.49686
y22	−0.39668	y22	−0.31871	y22	−0.23236	y22	−0.14340	y22	−0.05598
x23	+0.19134	x23	+0.29389	x23	+0.37426	x23	+0.43301	x23	+0.47194
y23	−0.46194	y23	−0.40451	y23	−0.33156	y23	−0.25000	y23	−0.16514
x24	+0.06526	x24	+0.18406	x24	+0.28403	x24	+0.36369	x24	+0.42336
y24	−0.49572	y24	−0.46489	y24	−0.41149	y24	−0.34312	y24	−0.26602
		x25	+0.06267	x25	+0.17730	x25	+0.27475	x25	+0.35355
		y25	−0.49606	y25	−0.46751	y25	−0.41774	y25	−0.35355
				x26	+0.06027	x26	+0.17101	x26	+0.26602
				y26	−0.49635	y26	−0.46985	y26	−0.42336
						x27	+0.05805	x27	+0.16514
						y27	−0.49662	y27	−0.47194
								x28	+0.05598
								y28	−0.49686

Tabulated Dimensions Of Geometric Shapes

Table 8. Diameters of Circles and Sides of Squares of Equal Area

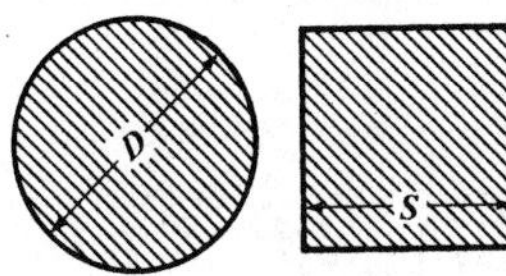

The table below will be found useful for determining the diameter of a circle of an area equal to that of a square, the side of which is known, or for determining the side of a square which has an area equal to that of a circle, the area or diameter of which is known. For example, if the diameter of a circle is 17½ inches, it is found from the table that the side of a square of the same area is 15.51 inches.

Diam. of Circle, *D*	Side of Square, *S*	Area of Circle or Square	Diam. of Circle, *D*	Side of Square, *S*	Area of Circle or Square	Diam. of Circle, *D*	Side of Square, *S*	Area of Circle or Square
½	0.44	0.196	20½	18.17	330.06	40½	35.89	1288.25
1	0.89	0.785	21	18.61	346.36	41	36.34	1320.25
1½	1.33	1.767	21½	19.05	363.05	41½	36.78	1352.65
2	1.77	3.142	22	19.50	380.13	42	37.22	1385.44
2½	2.22	4.909	22½	19.94	397.61	42½	37.66	1418.63
3	2.66	7.069	23	20.38	415.48	43	38.11	1452.20
3½	3.10	9.621	23½	20.83	433.74	43½	38.55	1486.17
4	3.54	12.566	24	21.27	452.39	44	38.99	1520.53
4½	3.99	15.904	24½	21.71	471.44	44½	39.44	1555.28
5	4.43	19.635	25	22.16	490.87	45	39.88	1590.43
5½	4.87	23.758	25½	22.60	510.71	45½	40.32	1625.97
6	5.32	28.274	26	23.04	530.93	46	40.77	1661.90
6½	5.76	33.183	26½	23.49	551.55	46½	41.21	1698.23
7	6.20	38.485	27	23.93	572.56	47	41.65	1734.94
7½	6.65	44.179	27½	24.37	593.96	47½	42.10	1772.05
8	7.09	50.265	28	24.81	615.75	48	42.54	1809.56
8½	7.53	56.745	28½	25.26	637.94	48½	42.98	1847.45
9	7.98	63.617	29	25.70	660.52	49	43.43	1885.74
9½	8.42	70.882	29½	26.14	683.49	49½	43.87	1924.42
10	8.86	78.540	30	26.59	706.86	50	44.31	1963.50
10½	9.31	86.590	30½	27.03	730.62	50½	44.75	2002.96
11	9.75	95.033	31	27.47	754.77	51	45.20	2042.82
11½	10.19	103.87	31½	27.92	779.31	51½	45.64	2083.07
12	10.63	113.10	32	28.36	804.25	52	46.08	2123.72
12½	11.08	122.72	32½	28.80	829.58	52½	46.53	2164.75
13	11.52	132.73	33	29.25	855.30	53	46.97	2206.18
13½	11.96	143.14	33½	29.69	881.41	53½	47.41	2248.01
14	12.41	153.94	34	30.13	907.92	54	47.86	2290.22
14½	12.85	165.13	34½	30.57	934.82	54½	48.30	2332.83
15	13.29	176.71	35	31.02	962.11	55	48.74	2375.83
15½	13.74	188.69	35½	31.46	989.80	55½	49.19	2419.22
16	14.18	201.06	36	31.90	1017.88	56	49.63	2463.01
16½	14.62	213.82	36½	32.35	1046.35	56½	50.07	2507.19
17	15.07	226.98	37	32.79	1075.21	57	50.51	2551.76
17½	15.51	240.53	37½	33.23	1104.47	57½	50.96	2596.72
18	15.95	254.47	38	33.68	1134.11	58	51.40	2642.08
18½	16.40	268.80	38½	34.12	1164.16	58½	51.84	2687.83
19	16.84	283.53	39	34.56	1194.59	59	52.29	2733.97
19½	17.28	298.65	39½	35.01	1225.42	59½	52.73	2780.51
20	17.72	314.16	40	35.45	1256.64	60	53.17	2827.43

Table 9. Distance Across Corners of Squares and Hexagons (English and metric units)

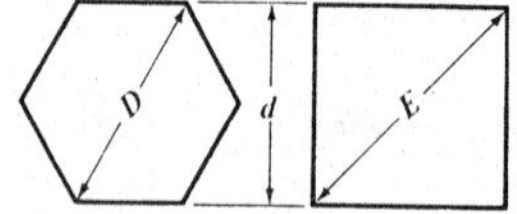

$D = 1.154701d$

$E = 1.414214d$

A desired value not given directly in the table can be obtained by the simple addition of two or more values taken directly from the table. Further values can be obtained by shifting the decimal point.

Example 1: Find D when $d = 2\ 5/16$ inches. From the table, 2 = 2.3094, and 5/16 = 0.3608. Therefore, D = 2.3094 + 0.3608 = 2.6702 inches.

Example 2: Find E when $d = 20.25$ millimeters. From the table, 20 = 28.2843; 0.2 = 0.2828; and 0.05 = 0.0707 (obtained by shifting the decimal point one place to the left at $d = 0.5$). Thus, $E = 28.2843 + 0.2828 + 0.0707 = 28.6378$ millimeters.

d	*D*	*E*	*d*	*D*	*E*	*d*	*D*	*E*	*d*	*D*	*E*
1/32	0.0361	0.0442	0.9	1.0392	1.2728	32	36.9504	45.2548	67	77.3650	94.7523
1/16	0.0722	0.0884	29/32	1.0464	1.2816	33	38.1051	46.6691	68	78.5197	96.1666
3/32	0.1083	0.1326	15/16	1.0825	1.3258	34	39.2598	48.0833	69	79.6744	97.5808
0.1	0.1155	0.1414	31/32	1.1186	1.3700	35	40.4145	49.4975	70	80.8291	98.9950
1/8	0.1443	0.1768	1.0	1.1547	1.4142	36	41.5692	50.9117	71	81.9838	100.409
5/32	0.1804	0.2210	2.0	2.3094	2.8284	37	42.7239	52.3259	72	83.1385	101.823
3/16	0.2165	0.2652	3.0	3.4641	4.2426	38	43.8786	53.7401	73	84.2932	103.238
0.2	0.2309	0.2828	4.0	4.6188	5.6569	39	45.0333	55.1543	74	85.4479	104.652
7/32	0.2526	0.3094	5.0	5.7735	7.0711	40	46.1880	56.5686	75	86.6026	106.066
1/4	0.2887	0.3536	6.0	6.9282	8.4853	41	47.3427	57.9828	76	87.7573	107.480
9/32	0.3248	0.3977	7.0	8.0829	9.8995	42	48.4974	59.3970	77	88.9120	108.894
0.3	0.3464	0.4243	8.0	9.2376	11.3137	43	49.6521	60.8112	78	90.0667	110.309
5/16	0.3608	0.4419	9.0	10.3923	12.7279	44	50.8068	62.2254	79	91.2214	111.723
11/32	0.3969	0.4861	10	11.5470	14.1421	45	51.9615	63.6396	80	92.3761	113.137
3/8	0.4330	0.5303	11	12.7017	15.5564	46	53.1162	65.0538	81	93.5308	114.551
0.4	0.4619	0.5657	12	13.8564	16.9706	47	54.2709	66.4681	82	94.6855	115.966
13/32	0.4691	0.5745	13	15.0111	18.3848	48	55.4256	67.8823	83	95.8402	117.380
7/16	0.5052	0.6187	14	16.1658	19.7990	49	56.5803	69.2965	84	96.9949	118.794
15/32	0.5413	0.6629	15	17.3205	21.2132	50	57.7351	70.7107	85	98.1496	120.208
0.5	0.5774	0.7071	16	18.4752	22.6274	51	58.8898	72.1249	86	99.3043	121.622
17/32	0.6134	0.7513	17	19.6299	24.0416	52	60.0445	73.5391	87	100.459	123.037
9/16	0.6495	0.7955	18	20.7846	25.4559	53	61.1992	74.9533	88	101.614	124.451
19/32	0.6856	0.8397	19	21.9393	26.8701	54	62.3539	76.3676	89	102.768	125.865
0.6	0.6928	0.8485	20	23.0940	28.2843	55	63.5086	77.7818	90	103.923	127.279
5/8	0.7217	0.8839	21	24.2487	29.6985	56	64.6633	79.1960	91	105.078	128.693
21/32	0.7578	0.9281	22	25.4034	31.1127	57	65.8180	80.6102	92	106.232	130.108
11/16	0.7939	0.9723	23	26.5581	32.5269	58	66.9727	82.0244	93	107.387	131.522
0.7	0.8083	0.9899	24	27.7128	33.9411	59	68.1274	83.4386	94	108.542	132.936
23/32	0.8299	1.0165	25	28.8675	35.3554	60	69.2821	84.8528	95	109.697	134.350
3/4	0.8660	1.0607	26	30.0222	36.7696	61	70.4368	86.2671	96	110.851	135.765
25/32	0.9021	1.1049	27	31.1769	38.1838	62	71.5915	87.6813	97	112.006	137.179
0.8	0.9238	1.1314	28	32.3316	39.5980	63	72.7462	89.0955	98	113.161	138.593
13/16	0.9382	1.1490	29	33.4863	41.0122	64	73.9009	90.5097	99	114.315	140.007
27/32	0.9743	1.1932	30	34.6410	42.4264	65	75.0556	91.9239	100	115.470	141.421
7/8	1.0104	1.2374	31	35.7957	43.8406	66	76.2103	93.3381	…	…	…

Formulas and Table for Regular Polygons.—The following formulas and table can be used to calculate the area, length of side, and radii of the inscribed and circumscribed circles of regular polygons (equal sided).

$$A = NS^2 \cot\alpha \div 4 = NR^2 \sin\alpha\cos\alpha = Nr^2 \tan\alpha$$

$$r = R\cos\alpha = (S\cot\alpha) \div 2 = \sqrt{(A \times \cot\alpha) \div N}$$

$$R = S \div (2\sin\alpha) = r \div \cos\alpha = \sqrt{A \div (N\sin\alpha\cos\alpha)}$$

$$S = 2R\sin\alpha = 2r\tan\alpha = 2\sqrt{(A \times \tan\alpha) \div N}$$

where N = number of sides

S = length of side

R = radius of circumscribed circle

r = radius of inscribed circle

A = area of polygon

$\alpha = 180° \div N$ = one-half center angle of one side (see *Regular Polygon* on page 74)

Table 10. Area, Length of Side, and Inscribed and Circumscribed Radii of Regular Polygons

No. of Sides	$\frac{A}{S^2}$	$\frac{A}{R^2}$	$\frac{A}{r^2}$	$\frac{R}{S}$	$\frac{R}{r}$	$\frac{S}{R}$	$\frac{S}{r}$	$\frac{r}{R}$	$\frac{r}{S}$
3	0.4330	1.2990	5.1962	0.5774	2.0000	1.7321	3.4641	0.5000	0.2887
4	1.0000	2.0000	4.0000	0.7071	1.4142	1.4142	2.0000	0.7071	0.5000
5	1.7205	2.3776	3.6327	0.8507	1.2361	1.1756	1.4531	0.8090	0.6882
6	2.5981	2.5981	3.4641	1.0000	1.1547	1.0000	1.1547	0.8660	0.8660
7	3.6339	2.7364	3.3710	1.1524	1.1099	0.8678	0.9631	0.9010	1.0383
8	4.8284	2.8284	3.3137	1.3066	1.0824	0.7654	0.8284	0.9239	1.2071
9	6.1818	2.8925	3.2757	1.4619	1.0642	0.6840	0.7279	0.9397	1.3737
10	7.6942	2.9389	3.2492	1.6180	1.0515	0.6180	0.6498	0.9511	1.5388
12	11.196	3.0000	3.2154	1.9319	1.0353	0.5176	0.5359	0.9659	1.8660
16	20.109	3.0615	3.1826	2.5629	1.0196	0.3902	0.3978	0.9808	2.5137
20	31.569	3.0902	3.1677	3.1962	1.0125	0.3129	0.3168	0.9877	3.1569
24	45.575	3.1058	3.1597	3.8306	1.0086	0.2611	0.2633	0.9914	3.7979
32	81.225	3.1214	3.1517	5.1011	1.0048	0.1960	0.1970	0.9952	5.0766
48	183.08	3.1326	3.1461	7.6449	1.0021	0.1308	0.1311	0.9979	7.6285
64	325.69	3.1365	3.1441	10.190	1.0012	0.0981	0.0983	0.9988	10.178

Example 1: A regular hexagon is inscribed in a circle of 6 inches diameter. Find the area and the radius of an inscribed circle. Here $R = 3$. From the table, area $A = 2.5981R^2 = 2.5981 \times 9 = 23.3829$ square inches. Radius of inscribed circle, $r = 0.866R = 0.866 \times 3 = 2.598$ inches.

Example 2: An octagon is inscribed in a circle of 100 millimeters diameter. Thus $R = 50$. Find the area and radius of an inscribed circle. $A = 2.8284R^2 = 2.8284 \times 2500 = 7071$ mm^2 = 70.7 cm^2. Radius of inscribed circle, $r = 0.9239R = 09239 \times 50 = 46.195$ mm.

Example 3: Thirty-two bolts are to be equally spaced on the periphery of a bolt-circle, 16 inches in diameter. Find the chordal distance between the bolts. Chordal distance equals the side S of a polygon with 32 sides. $R = 8$. Hence, $S = 0.196R = 0.196 \times 8 = 1.568$ inch.

Example 4: Sixteen bolts are to be equally spaced on the periphery of a bolt-circle, 250 millimeters diameter. Find the chordal distance between the bolts. Chordal distance equals the side S of a polygon with 16 sides. $R = 125$. Thus, $S = 0.3902R = 0.3902 \times 125 = 48.775$ millimeters.

Areas and Dimensions of Plane Figures

In the following tables are given formulas for the areas of plane figures, together with other formulas relating to their dimensions and properties; the surfaces of solids; and the volumes of solids. The notation used in the formulas is, as far as possible, given in the illustration accompanying them; where this has not been possible, it is given at the beginning of each set of formulas.

Examples are given with each entry, some in English and some in metric units, showing the use of the preceding formula.

Square:

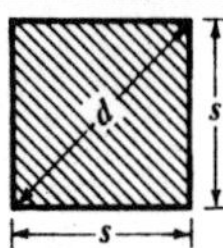

$$\text{Area} = A = s^2 = \tfrac{1}{2}d^2$$

$$s = 0.7071d = \sqrt{A}$$

$$d = 1.414s = 1.414\sqrt{A}$$

Example: Assume that the side s of a square is 15 inches. Find the area and the length of the diagonal.

$$\text{Area} = A = s^2 = 15^2 = 225 \text{ square inches}$$

$$\text{Diagonal} = d = 1.414s = 1.414 \times 15 = 21.21 \text{ inches}$$

Example: The area of a square is 625 square inches. Find the length of the side s and the diagonal d.

$$s = \sqrt{A} = \sqrt{625} = 25 \text{ inches}$$

$$d = 1.414\sqrt{A} = 1.414 \times 25 = 35.35 \text{ inches}$$

Rectangle:

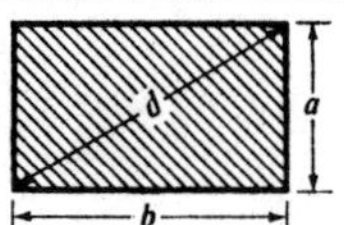

$$\text{Area} = A = ab = a\sqrt{d^2 - a^2} = b\sqrt{d^2 - b^2}$$

$$d = \sqrt{a^2 + b^2}$$

$$a = \sqrt{d^2 - b^2} = A \div b$$

$$a = \sqrt{d^2 - a^2} = A \div a$$

Example: The side a of a rectangle is 12 centimeters, and the area 70.5 square centimeters. Find the length of the side b, and the diagonal d.

$$b = A \div a = 70.5 \div 12 = 5.875 \text{ centimeters}$$

$$d = \sqrt{a^2 + b^2} = \sqrt{12^2 + 5.875^2} = \sqrt{178.516} = 13.361 \text{ centimeters}$$

Example: The sides of a rectangle are 30.5 and 11 centimeters long. Find the area.

$$\text{Area} = A = a \times b = 30.5 \times 11 = 335.5 \text{ square centimeters}$$

Parallelogram:

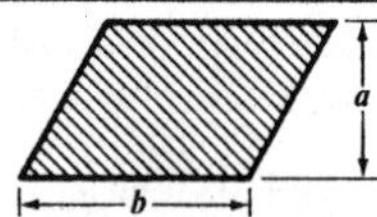

$$\text{Area} = A = ab$$

$$a = A \div b$$

$$b = A \div a$$

Note: The dimension a is measured at right angles to line b.

Example: The base b of a parallelogram is 16 feet. The height a is 5.5 feet. Find the area.

$$\text{Area} = A = a \times b = 5.5 \times 16 = 88 \text{ square feet}$$

Example: The area of a parallelogram is 12 square inches. The height is 1.5 inches. Find the length of the base b.

$$b = A \div a = 12 \div 1.5 = 8 \text{ inches}$$

Right-Angled Triangle:

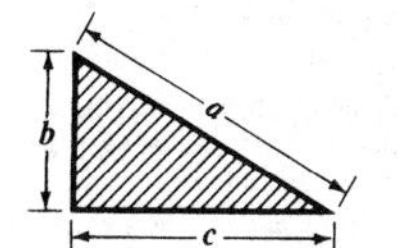

$$\text{Area} = A = \frac{bc}{2}$$

$$a = \sqrt{b^2 + c^2}$$

$$b = \sqrt{a^2 - c^2}$$

$$c = \sqrt{a^2 - b^2}$$

Example: The sides b and c in a right-angled triangle are 6 and 8 inches. Find side a and the area

$$a = \sqrt{b^2 + c^2} = \sqrt{6^2 + 8^2} = \sqrt{36 + 64} = \sqrt{100} = 10 \text{ inches}$$

$$A = \frac{b \times c}{2} = \frac{6 \times 8}{2} = \frac{48}{2} = 24 \text{ square inches}$$

Example: If $a = 10$ and $b = 6$ had been known, but not c, the latter would have been found as follows:

$$c = \sqrt{a^2 - b^2} = \sqrt{10^2 - 6^2} = \sqrt{100 - 36} = \sqrt{64} = 8 \text{ inches}$$

Acute-Angled Triangle:

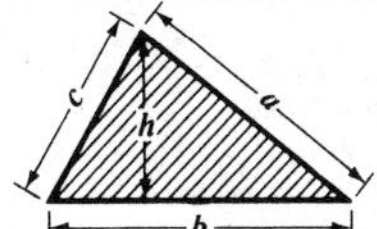

$$\text{Area} = A = \frac{bh}{2} = \frac{b}{2}\sqrt{a^2 - \left(\frac{a^2 + b^2 - c^2}{2b}\right)^2}$$

$$\text{If } S = \tfrac{1}{2}(a + b + c), \text{ then}$$

$$A = \sqrt{S(S-a)(S-b)(S-c)}$$

Example: If $a = 10$, $b = 9$, and $c = 8$ centimeters, what is the area of the triangle?

$$A = \frac{b}{2}\sqrt{a^2 - \left(\frac{a^2 + b^2 - c^2}{2b}\right)^2} = \frac{9}{2}\sqrt{10^2 - \left(\frac{10^2 + 9^2 - 8^2}{2 \times 9}\right)^2} = 4.5\sqrt{100 - \left(\frac{117}{18}\right)^2}$$

$$= 4.5\sqrt{100 - 42.25} = 4.5\sqrt{57.75} = 4.5 \times 7.60 = 34.20 \text{ square centimeters}$$

Obtuse-Angled Triangle:

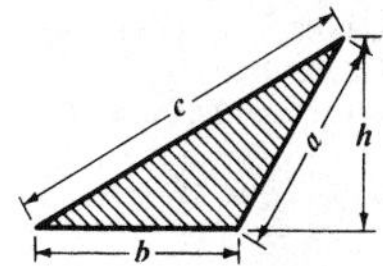

$$\text{Area} = A = \frac{bh}{2} = \frac{b}{2}\sqrt{a^2 - \left(\frac{c^2 - a^2 - b^2}{2b}\right)^2}$$

$$\text{If S} = \tfrac{1}{2}(a + b + c), \text{ then}$$

$$A = \sqrt{S(S-a)(S-b)(S-c)}$$

Example: The side $a = 5$, side $b = 4$, and side $c = 8$ inches. Find the area.

$$S = \tfrac{1}{2}(a + b + c) = \tfrac{1}{2}(5 + 4 + 8) = \tfrac{1}{2} \times 17 = 8.5$$

$$A = \sqrt{S(S-a)(S-b)(S-c)} = \sqrt{8.5(8.5-5)(8.5-4)(8.5-8)}$$

$$= \sqrt{8.5 \times 3.5 \times 4.5 \times 0.5} = \sqrt{66.937} = 8.18 \text{ square inches}$$

Trapezoid:

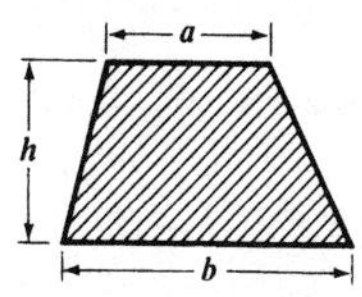

$$\text{Area} = A = \frac{(a + b)h}{2}$$

Note: In Britain, this figure is called a *trapezium* and the one below it is known as a *trapezoid*, the terms being reversed.

Example: Side $a = 23$ meters, side $b = 32$ meters, and height $h = 12$ meters. Find the area.

$$A = \frac{(a + b)h}{2} = \frac{(23 + 32)12}{2} = \frac{55 \times 12}{2} = 330 \text{ square meters}$$

Trapezium:

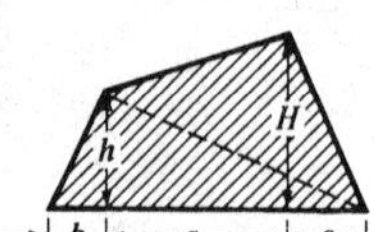

$$\text{Area} = A = \frac{(H+h)a + bh + cH}{2}$$

A trapezium can also be divided into two triangles as indicated by the dashed line. The area of each of these triangles is computed, and the results added to find the area of the trapezium.

Example: Let $a = 10$, $b = 2$, $c = 3$, $h = 8$, and $H = 12$ inches. Find the area.

$$A = \frac{(H+h)a + bh + cH}{2} = \frac{(12+8)10 + 2\times 8 + 3\times 12}{2}$$

$$= \frac{20\times 10 + 16 + 36}{2} = \frac{252}{2} = 126 \text{ square inches}$$

Regular Hexagon:

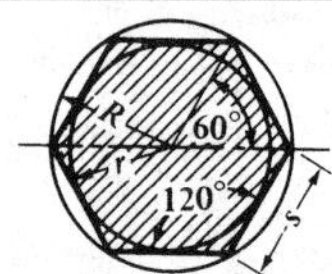

$A = 2.598s^2 = 2.598R^2 = 3.464r^2$

$R = s$ = radius of circumscribed circle = $1.155r$

r = radius of inscribed circle = $0.866s = 0.866R$

$s = R = 1.155r$

Example: The side s of a regular hexagon is 40 millimeters. Find the area and the radius r of the inscribed circle.

$$A = 2.598s^2 = 2.598\times 40^2 = 2.598\times 1600 = 4156.8 \text{ square millimeters}$$

$$r = 0.866s = 0.866\times 40 = 34.64 \text{ millimeters}$$

Example: What is the length of the side of a hexagon that is drawn around a circle of 50 millimeters radius? — Here $r = 50$. Hence, $s = 1.155r = 1.155\times 50 = 57.75$ millimeters

Regular Octagon:

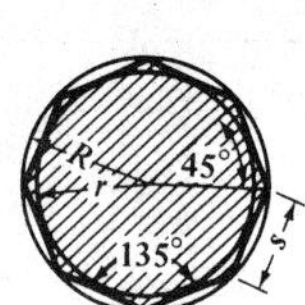

A = area = $4.828s^2 = 2.828R^2 = 3.314r^2$

R = radius of circumscribed circle = $1.307s = 1.082r$

r = radius of inscribed circle = $1.207s = 0.924R$

$s = 0.765R = 0.828r$

Example: Find the area and the length of the side of an octagon that is inscribed in a circle of 12 inches diameter.

Diameter of circumscribed circle = 12 inches; hence, $R = 6$ inches.

$$A = 2.828R^2 = 2.828\times 6^2 = 2.828\times 36 = 101.81 \text{ squre inches}$$

$$s = 0.765R = 0.765\times 6 = 4.590 \text{ inches}$$

Regular Polygon:

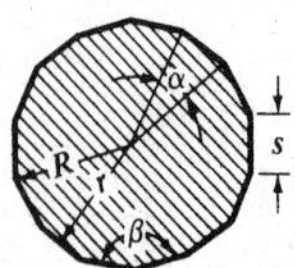

A = area $\quad n$ = number of sides

$\alpha = 360° \div n \quad \beta = 180° - \alpha$

$$A = \frac{nsr}{2} = \frac{ns}{2}\sqrt{R^2 - \frac{s^2}{4}}$$

$$R = \sqrt{r^2 + \frac{s^2}{4}} \qquad r = \sqrt{R^2 - \frac{s^2}{4}} \qquad s = 2\sqrt{R^2 - r^2}$$

Example: Find the area of a polygon having 12 sides, inscribed in a circle of 8 centimeters radius. The length of the side s is 4.141 centimeters.

$$A = \frac{ns}{2}\sqrt{R^2 - \frac{s^2}{4}} = \frac{12\times 4.141}{2}\sqrt{8^2 - \frac{4.141^2}{4}} = 24.846\sqrt{59.713}$$

$$= 24.846\times 7.727 = 191.98 \text{ square centimeters}$$

Circle:

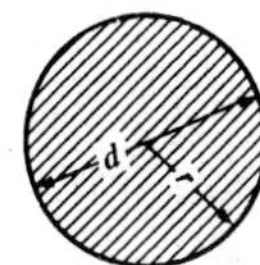

$$\text{Area} = A = \pi r^2 = 3.1416r^2 = 0.7854d^2$$
$$\text{Circumference} = C = 2\pi r = 6.832r = 3.1416d$$
$$r = C \div 6.2832 = \sqrt{A \div 3.1416} = 0.564\sqrt{A}$$
$$d = C \div 3.1416 = \sqrt{A \div 0.7854} = 1.128\sqrt{A}$$

Length of arc for center angle of 1° = 0.008727*d*

Length of arc for center angle of *n*° = 0.008727*nd*

Example: Find the area *A* and circumference *C* of a circle with a diameter of 2¾ inches.

$$A = 0.7854d^2 = 0.7854 \times 2.75^2 = 0.7854 \times 2.75 \times 2.75 = 5.9396 \text{ square inches}$$
$$C = 3.1416d = 3.1416 \times 2.75 = 8.6394 \text{ inches}$$

Example: The area of a circle is 16.8 square inches. Find its diameter.

$$d = 1.128\sqrt{A} = 1.128\sqrt{16.8} = 1.128 \times 4.099 = 4.624 \text{ inches}$$

Circular Sector:

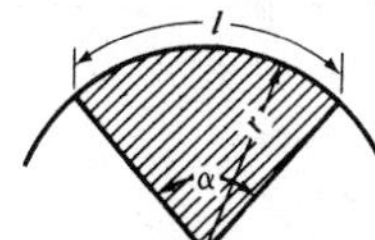

$$\text{Length of arc} = l = \frac{r \times \alpha \times 3.1416}{180} = 0.01745r\alpha = \frac{2A}{r}$$
$$\text{Area} = A = \tfrac{1}{2}rl = 0.008727\alpha r^2$$
$$\text{Angle, in degrees} = \alpha = \frac{57.296\ l}{r} \qquad r = \frac{2A}{l} = \frac{57.296\ l}{\alpha}$$

Example: The radius of a circle is 35 millimeters, and angle α of a sector of the circle is 60 degrees. Find the area of the sector and the length of arc *l*.

$$A = 0.008727\alpha r^2 = 0.008727 \times 60 \times 35^2 = 641.41\,\text{mm}^2 = 6.41\,\text{cm}^2$$
$$l = 0.01745r\alpha = 0.01745 \times 35 \times 60 = 36.645 \text{ millimeters}$$

Circular Segment:

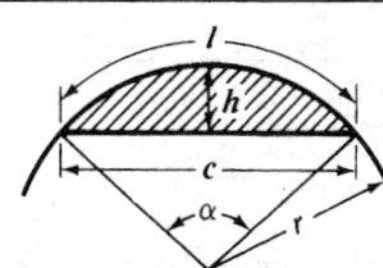

A = area $\quad l$ = length of arc $\quad \alpha$ = angle, in degrees

$$c = 2\sqrt{h(2r-h)} \qquad A = \tfrac{1}{2}[rl - c(r-h)]$$
$$r = \frac{c^2 + 4h^2}{8h} \qquad l = 0.01745r\alpha$$
$$h = r - \tfrac{1}{2}\sqrt{4r^2 - c^2} = r[1 - \cos(\alpha/2)] \qquad \alpha = \frac{57.296\ l}{r}$$

Example: The radius *r* is 60 inches and the height *h* is 8 inches. Find the length of the chord *c*.

$$c = 2\sqrt{h(2r-h)} = 2\sqrt{8 \times (2 \times 60 - 8)} = 2\sqrt{896} = 2 \times 29.93 = 59.86 \text{ inches}$$

Example: If *c* = 16, and *h* = 6 inches, what is the radius of the circle of which the segment is a part?

$$r = \frac{c^2 + 4h^2}{8h} = \frac{16^2 + 4 \times 6^2}{8 \times 6} = \frac{256 + 144}{48} = \frac{400}{48} = 8\tfrac{1}{3} \text{ inches}$$

Cycloid:

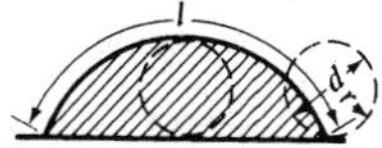

$$\text{Area} = A = 3\pi r^2 = 9.4248r^2 = 2.3562d^2$$
$$= 3 \times \text{area of generating circle}$$
$$\text{Length of cycloid} = l = 8r = 4d$$

Example: The diameter of the generating circle of a cycloid is 6 inches. Find the length *l* of the cycloidal curve, and the area enclosed between the curve and the base line.

$$l = 4d = 4 \times 6 = 24 \text{ inches}$$
$$A = 2.3562d^2 = 2.3562 \times 6^2 = 84.82 \text{ square inches}$$

Circular Ring:

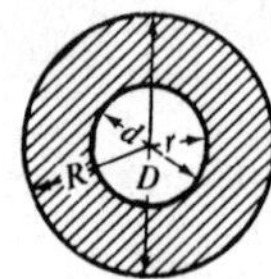

$$\text{Area} = A = \pi(R^2 - r^2) = 3.1416(R^2 - r^2)$$
$$= 3.1416(R + r)(R - r)$$
$$= 0.7845(D^2 - d^2) = 0.7854(D + d)(D - d)$$

Example: Let the outside diameter $D = 12$ centimeters and the inside diameter $d = 8$ centimeters. Find the area of the ring.

$$A = 0.7854(D^2 - d^2) = 0.7854(12^2 - 8^2) = 0.7854(144 - 64) = 0.7854 \times 80$$
$$= 62.83 \text{ square centimeters}$$

By the alternative formula:

$$A = 0.7854(D + d)(D - d) = 0.7854(12 + 8)(12 - 8) = 0.7854 \times 20 \times 4$$
$$= 62.83 \text{ square centimeters}$$

Circular Ring Sector:

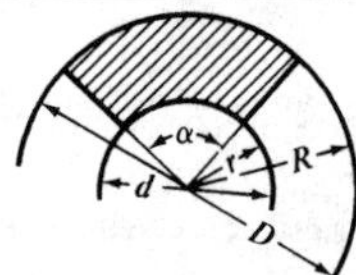

A = area α = angle, in degrees

$$A = \frac{\alpha\pi}{360}(R^2 - r^2) = 0.00873\alpha(R^2 - r^2)$$
$$= \frac{\alpha\pi}{4 \times 360}(D^2 - d^2) = 0.00218\alpha(D^2 - d^2)$$

Example: Find the area, if the outside radius $R = 5$ inches, the inside radius $r = 2$ inches, and $\alpha = 72$ degrees.

$$A = 0.00873\alpha(R^2 - r^2) = 0.00873 \times 72(5^2 - 2^2)$$
$$= 0.6286(25 - 4) = 0.6286 \times 21 = 13.2 \text{ square inches}$$

Spandrel or Fillet:

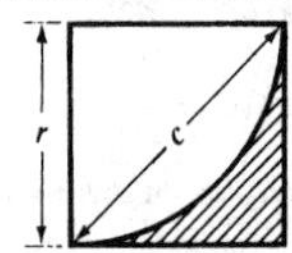

$$\text{Area} = A = r^2 - \frac{\pi r^2}{4} = 0.215r^2 = 0.1075c^2$$

Example: Find the area of a spandrel, the radius of which is 0.7 inch.

$$A = 0.215r^2 = 0.215 \times 0.7^2 = 0.105 \text{ square inch}$$

Example: If chord c were given as 2.2 inches, what would be the area?

$$A = 0.1075c^2 = 0.1075 \times 2.2^2 = 0.520 \text{ square inch}$$

Parabola:

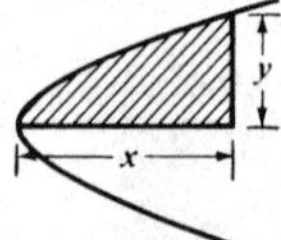

$$\text{Area} = A = \tfrac{2}{3}xy$$

(The area is equal to two-thirds of a rectangle which has x for its base and y for its height.)

Example: Let x in the illustration be 15 centimeters, and y, 9 centimeters. Find the area of the shaded portion of the parabola.

$$A = \tfrac{2}{3} \times xy = \tfrac{2}{3} \times 15 \times 9 = 10 \times 9 = 90 \text{ square centimeters}$$

Parabola:

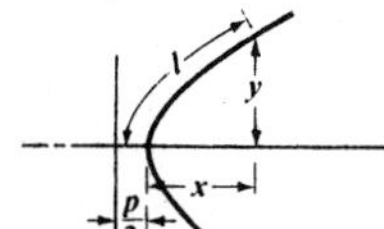

$$l = \text{length of arc} = \frac{p}{2}\left[\sqrt{\frac{2x}{p}\left(1+\frac{2x}{p}\right)} + \ln\left(\sqrt{\frac{2x}{p}} + \sqrt{1+\frac{2x}{p}}\right)\right]$$

When x is small in proportion to y, the following is a close approximation:

$$l = y\left[1 + \frac{2}{3}\left(\frac{x}{y}\right)^2 - \frac{2}{5}\left(\frac{x}{y}\right)^4\right] \text{ or } l = \sqrt{y^2 + \frac{4}{3}x^2}$$

Example: If $x = 2$ and $y = 24$ feet, what is the approximate length l of the parabolic curve?

$$l = y\left[1 + \frac{2}{3}\left(\frac{x}{y}\right)^2 - \frac{2}{5}\left(\frac{x}{y}\right)^4\right] = 24\left[1 + \frac{2}{3}\left(\frac{2}{24}\right)^2 - \frac{2}{5}\left(\frac{2}{24}\right)^4\right]$$

$$= 24\left[1 + \frac{2}{3} \times \frac{1}{144} - \frac{2}{5} \times \frac{1}{20{,}736}\right] = 24 \times 1.0046 = 24.11 \text{ feet}$$

Segment of Parabola:

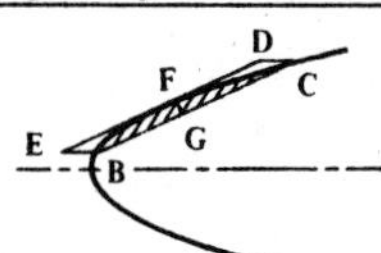

Area BFC $= A = \frac{2}{3}$ area of parallelogram BCDE

If FG is the height of the segment, measured at right angles to BC, then:

Area of segment BFC $= \frac{2}{3}\text{BC} \times \text{FG}$

Example: The length of the chord $BC = 19.5$ inches. The distance between lines BC and DE, measured at right angles to BC, is 2.25 inches. This is the height of the segment. Find the area.

$$\text{Area} = A = \tfrac{2}{3}\text{BC} \times \text{FG} = \tfrac{2}{3} \times 19.5 \times 2.25 = 29.25 \text{ square inches}$$

Hyperbola:

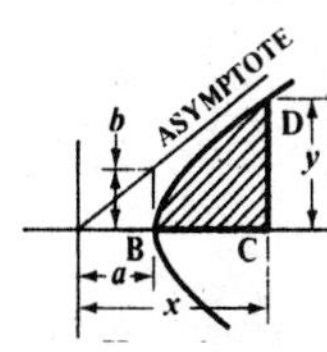

$$\text{Area BCD} = A = \frac{xy}{2} - \frac{ab}{2}\ln\left(\frac{x}{a} + \frac{y}{b}\right)$$

Example: The half-axes a and b are 3 and 2 inches, respectively. Find the area shown shaded in the illustration for $x = 8$ and $y = 5$.

Inserting the known values in the formula:

$$A = \frac{8 \times 5}{2} - \frac{3 \times 2}{2} \times \ln\left(\frac{8}{3} + \frac{5}{2}\right) = 20 - 3 \times \ln 5.167$$

$$= 20 - 3 \times 1.6423 = 20 - 4.927 = 15.073 \text{ square inches}$$

Ellipse:

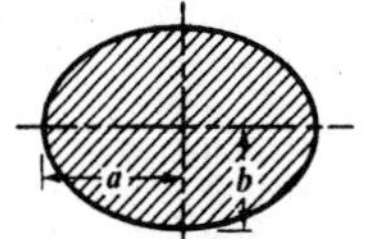

$$\text{Area} = A = \pi ab = 3.1416ab$$

An approximate formula for the perimeter is

$$\text{Perimeter} = P = 3.1416\sqrt{2(a^2 + b^2)}$$

A closer approximation is $P = 3.1416\sqrt{2(a^2 + b^2) - \frac{(a-b)^2}{2.2}}$

Example: The larger or major axis is 200 millimeters. The smaller or minor axis is 150 millimeters. Find the area and the approximate circumference. Here, then, $a = 100$, and $b = 75$.

$$A = 3.1416ab = 3.1416 \times 100 \times 75 = 23{,}562 \text{ square millimeters} = 235.62 \text{ square centimeters}$$

$$P = 3.1416\sqrt{2(a^2 + b^2)} = 3.1416\sqrt{2(100^2 + 75^2)} = 3.1416\sqrt{2 \times 15{,}625}$$

$$= 3.1416\sqrt{31{,}250} = 3.1416 \times 176.78 = 555.37 \text{ millimeters} = (55.537 \text{ centimeters})$$

Table 11. Segments of Circles for Radius = 1 (English or metric units)

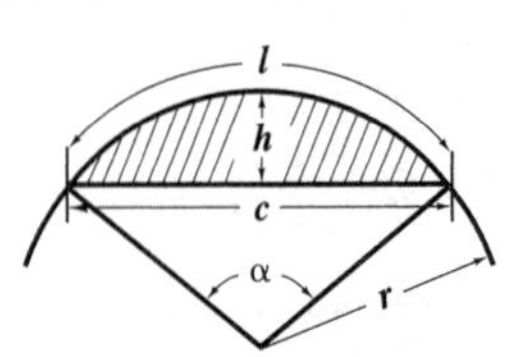

Formulas for segments of circles are given on page 75. When the central angle α and radius *r* are known, the tables on these pages can be used to find the length of arc *l*, height of segment *h*, chord length *c*, and segment area *A*. When angle α and radius *r* are not known, but segment height *h* and chord length *c* are known or can be measured, the ratio *h*/*c* can be used to enter the table and find α, *l*, and *A* by linear interpolation. Radius *r* is found by the formula on page 75. The value of *l* is then multiplied by the radius *r* and the area *A* by r^2, the square of the radius.

Angle α can be found thus with an accuracy of about 0.001 degree; arc length *l* with an error of about 0.02 per cent; and area *A* with an error ranging from about 0.02 per cent for the highest entry value of *h*/*c* to about 1 per cent for values of *h*/*c* of about 0.050. For lower values of *h*/*c*, and where greater accuracy is required, area *A* should be found by the formula on page 75.

θ, Deg.	*l*	*h*	*c*	*Area A*	*h*/*c*	θ, Deg.	*l*	*h*	*c*	*Area A*	*h*/*c*
1	0.01745	0.00004	0.01745	0.0000	0.00218	41	0.71558	0.06333	0.70041	0.0298	0.09041
2	0.03491	0.00015	0.03490	0.0000	0.00436	42	0.73304	0.06642	0.71674	0.0320	0.09267
3	0.05236	0.00034	0.05235	0.0000	0.00655	43	0.75049	0.06958	0.73300	0.0342	0.09493
4	0.06981	0.00061	0.06980	0.0000	0.00873	44	0.76794	0.07282	0.74921	0.0366	0.09719
5	0.08727	0.00095	0.08724	0.0001	0.01091	45	0.78540	0.07612	0.76537	0.0391	0.09946
6	0.10472	0.00137	0.10467	0.0001	0.01309	46	0.80285	0.07950	0.78146	0.0418	0.10173
7	0.12217	0.00187	0.12210	0.0002	0.01528	47	0.82030	0.08294	0.79750	0.0445	0.10400
8	0.13963	0.00244	0.13951	0.0002	0.01746	48	0.83776	0.08645	0.81347	0.0473	0.10628
9	0.15708	0.00308	0.15692	0.0003	0.01965	49	0.85521	0.09004	0.82939	0.0503	0.10856
10	0.17453	0.00381	0.17431	0.0004	0.02183	50	0.87266	0.09369	0.84524	0.0533	0.11085
11	0.19199	0.00460	0.19169	0.0006	0.02402	51	0.89012	0.09741	0.86102	0.0565	0.11314
12	0.20944	0.00548	0.20906	0.0008	0.02620	52	0.90757	0.10121	0.87674	0.0598	0.11543
13	0.22689	0.00643	0.22641	0.0010	0.02839	53	0.92502	0.10507	0.89240	0.0632	0.11773
14	0.24435	0.00745	0.24374	0.0012	0.03058	54	0.94248	0.10899	0.90798	0.0667	0.12004
15	0.26180	0.00856	0.26105	0.0015	0.03277	55	0.95993	0.11299	0.92350	0.0704	0.12235
16	0.27925	0.00973	0.27835	0.0018	0.03496	56	0.97738	0.11705	0.93894	0.0742	0.12466
17	0.29671	0.01098	0.29562	0.0022	0.03716	57	0.99484	0.12118	0.95432	0.0781	0.12698
18	0.31416	0.01231	0.31287	0.0026	0.03935	58	1.01229	0.12538	0.96962	0.0821	0.12931
19	0.33161	0.01371	0.33010	0.0030	0.04155	59	1.02974	0.12964	0.98485	0.0863	0.13164
20	0.34907	0.01519	0.34730	0.0035	0.04374	60	1.04720	0.13397	1.00000	0.0906	0.13397
21	0.36652	0.01675	0.36447	0.0041	0.04594	61	1.06465	0.13837	1.01508	0.0950	0.13632
22	0.38397	0.01837	0.38162	0.0047	0.04814	62	1.08210	0.14283	1.03008	0.0996	0.13866
23	0.40143	0.02008	0.39874	0.0053	0.05035	63	1.09956	0.14736	1.04500	0.1043	0.14101
24	0.41888	0.02185	0.41582	0.0061	0.05255	64	1.11701	0.15195	1.05984	0.1091	0.14337
25	0.43633	0.02370	0.43288	0.0069	0.05476	65	1.13446	0.15661	1.07460	0.1141	0.14574
26	0.45379	0.02563	0.44990	0.0077	0.05697	66	1.15192	0.16133	1.08928	0.1192	0.14811
27	0.47124	0.02763	0.46689	0.0086	0.05918	67	1.16937	0.16611	1.10387	0.1244	0.15048
28	0.48869	0.02970	0.48384	0.0096	0.06139	68	1.18682	0.17096	1.11839	0.1298	0.15287
29	0.50615	0.03185	0.50076	0.0107	0.06361	69	1.20428	0.17587	1.13281	0.1353	0.15525
30	0.52360	0.03407	0.51764	0.0118	0.06583	70	1.22173	0.18085	1.14715	0.1410	0.15765
31	0.54105	0.03637	0.53448	0.0130	0.06805	71	1.23918	0.18588	1.16141	0.1468	0.16005
32	0.55851	0.03874	0.55127	0.0143	0.07027	72	1.25664	0.19098	1.17557	0.1528	0.16246
33	0.57596	0.04118	0.56803	0.0157	0.07250	73	1.27409	0.19614	1.18965	0.1589	0.16488
34	0.59341	0.04370	0.58474	0.0171	0.07473	74	1.29154	0.20136	1.20363	0.1651	0.16730
35	0.61087	0.04628	0.60141	0.0186	0.07696	75	1.30900	0.20665	1.21752	0.1715	0.16973
36	0.62832	0.04894	0.61803	0.0203	0.07919	76	1.32645	0.21199	1.23132	0.1781	0.17216
37	0.64577	0.05168	0.63461	0.0220	0.08143	77	1.34390	0.21739	1.24503	0.1848	0.17461
38	0.66323	0.05448	0.65114	0.0238	0.08367	78	1.36136	0.22285	1.25864	0.1916	0.17706
39	0.68068	0.05736	0.66761	0.0257	0.08592	79	1.37881	0.22838	1.27216	0.1986	0.17952
40	0.69813	0.06031	0.68404	0.0277	0.08816	80	1.39626	0.23396	1.28558	0.2057	0.18199

Table 12. Segments of Circles for Radius = 1 (English or metric units)

θ, Deg.	*l*	*h*	*c*	*Area A*	*h/c*	θ, Deg.	*l*	*h*	*c*	*Area A*	*h/c*
81	1.41372	0.23959	1.29890	0.2130	0.18446	131	2.28638	0.58531	1.81992	0.7658	0.32161
82	1.43117	0.24529	1.31212	0.2205	0.18694	132	2.30383	0.59326	1.82709	0.7803	0.32470
83	1.44862	0.25104	1.32524	0.2280	0.18943	133	2.32129	0.60125	1.83412	0.7950	0.32781
84	1.46608	0.25686	1.33826	0.2358	0.19193	134	2.33874	0.60927	1.84101	0.8097	0.33094
85	1.48353	0.26272	1.35118	0.2437	0.19444	135	2.35619	0.61732	1.84776	0.8245	0.33409
86	1.50098	0.26865	1.36400	0.2517	0.19696	136	2.37365	0.62539	1.85437	0.8395	0.33725
87	1.51844	0.27463	1.37671	0.2599	0.19948	137	2.39110	0.63350	1.86084	0.8546	0.34044
88	1.53589	0.28066	1.38932	0.2682	0.20201	138	2.40855	0.64163	1.86716	0.8697	0.34364
89	1.55334	0.28675	1.40182	0.2767	0.20456	139	2.42601	0.64979	1.87334	0.8850	0.34686
90	1.57080	0.29289	1.41421	0.2854	0.20711	140	2.44346	0.65798	1.87939	0.9003	0.35010
91	1.58825	0.29909	1.42650	0.2942	0.20967	141	2.46091	0.66619	1.88528	0.9158	0.35337
92	1.60570	0.30534	1.43868	0.3032	0.21224	142	2.47837	0.67443	1.89104	0.9314	0.35665
93	1.62316	0.31165	1.45075	0.3123	0.21482	143	2.49582	0.68270	1.89665	0.9470	0.35995
94	1.64061	0.31800	1.46271	0.3215	0.21741	144	2.51327	0.69098	1.90211	0.9627	0.36327
95	1.65806	0.32441	1.47455	0.3309	0.22001	145	2.53073	0.69929	1.90743	0.9786	0.36662
96	1.67552	0.33087	1.48629	0.3405	0.22261	146	2.54818	0.70763	1.91261	0.9945	0.36998
97	1.69297	0.33738	1.49791	0.3502	0.22523	147	2.56563	0.71598	1.91764	1.0105	0.37337
98	1.71042	0.34394	1.50942	0.3601	0.22786	148	2.58309	0.72436	1.92252	1.0266	0.37678
99	1.72788	0.35055	1.52081	0.3701	0.23050	149	2.60054	0.73276	1.92726	1.0428	0.38021
100	1.74533	0.35721	1.53209	0.3803	0.23315	150	2.61799	0.74118	1.93185	1.0590	0.38366
101	1.76278	0.36392	1.54325	0.3906	0.23582	151	2.63545	0.74962	1.93630	1.0753	0.38714
102	1.78024	0.37068	1.55429	0.4010	0.23849	152	2.65290	0.75808	1.94059	1.0917	0.39064
103	1.79769	0.37749	1.56522	0.4117	0.24117	153	2.67035	0.76655	1.94474	1.1082	0.39417
104	1.81514	0.38434	1.57602	0.4224	0.24387	154	2.68781	0.77505	1.94874	1.1247	0.39772
105	1.83260	0.39124	1.58671	0.4333	0.24657	155	2.70526	0.78356	1.95259	1.1413	0.40129
106	1.85005	0.39818	1.59727	0.4444	0.24929	156	2.72271	0.79209	1.95630	1.1580	0.40489
107	1.86750	0.40518	1.60771	0.4556	0.25202	157	2.74017	0.80063	1.95985	1.1747	0.40852
108	1.88496	0.41221	1.61803	0.4669	0.25476	158	2.75762	0.80919	1.96325	1.1915	0.41217
109	1.90241	0.41930	1.62823	0.4784	0.25752	159	2.77507	0.81776	1.96651	1.2084	0.41585
110	1.91986	0.42642	1.63830	0.4901	0.26028	160	2.79253	0.82635	1.96962	1.2253	0.41955
111	1.93732	0.43359	1.64825	0.5019	0.26306	161	2.80998	0.83495	1.97257	1.2422	0.42328
112	1.95477	0.44081	1.65808	0.5138	0.26585	162	2.82743	0.84357	1.97538	1.2592	0.42704
113	1.97222	0.44806	1.66777	0.5259	0.26866	163	2.84489	0.85219	1.97803	1.2763	0.43083
114	1.98968	0.45536	1.67734	0.5381	0.27148	164	2.86234	0.86083	1.98054	1.2934	0.43464
115	2.00713	0.46270	1.68678	0.5504	0.27431	165	2.87979	0.86947	1.98289	1.3105	0.43849
116	2.02458	0.47008	1.69610	0.5629	0.27715	166	2.89725	0.87813	1.98509	1.3277	0.44236
117	2.04204	0.47750	1.70528	0.5755	0.28001	167	2.91470	0.88680	1.98714	1.3449	0.44627
118	2.05949	0.48496	1.71433	0.5883	0.28289	168	2.93215	0.89547	1.98904	1.3621	0.45020
119	2.07694	0.49246	1.72326	0.6012	0.28577	169	2.94961	0.90415	1.99079	1.3794	0.45417
120	2.09440	0.50000	1.73205	0.6142	0.28868	170	2.96706	0.91284	1.99239	1.3967	0.45817
121	2.11185	0.50758	1.74071	0.6273	0.29159	171	2.98451	0.92154	1.99383	1.4140	0.46220
122	2.12930	0.51519	1.74924	0.6406	0.29452	172	3.00197	0.93024	1.99513	1.4314	0.46626
123	2.14675	0.52284	1.75763	0.6540	0.29747	173	3.01942	0.93895	1.99627	1.4488	0.47035
124	2.16421	0.53053	1.76590	0.6676	0.30043	174	3.03687	0.94766	1.99726	1.4662	0.47448
125	2.18166	0.53825	1.77402	0.6813	0.30341	175	3.05433	0.95638	1.99810	1.4836	0.47865
126	2.19911	0.54601	1.78201	0.6950	0.30640	176	3.07178	0.96510	1.99878	1.5010	0.48284
127	2.21657	0.55380	1.78987	0.7090	0.30941	177	3.08923	0.97382	1.99931	1.5184	0.48708
128	2.23402	0.56163	1.79759	0.7230	0.31243	178	3.10669	0.98255	1.99970	1.5359	0.49135
129	2.25147	0.56949	1.80517	0.7372	0.31548	179	3.12414	0.99127	1.99992	1.5533	0.49566
130	2.26893	0.57738	1.81262	0.7514	0.31854	180	3.14159	1.00000	2.00000	1.5708	0.50000

Volumes of Solids

Cube:

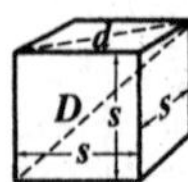

Diagonal of cube face $= d = s\sqrt{2}$

Diagonal of cube $= D = \sqrt{\frac{3d^2}{2}} = s\sqrt{3} = 1.732s$

Volume $= V = s^3$

$s = \sqrt[3]{V}$

Example: The side of a cube equals 9.5 centimeters. Find its volume.

Volume $= V = s^3 = 9.5^3 = 9.5 \times 9.5 \times 9.5 = 857.375$ cubic centimeters

Example: The volume of a cube is 231 cubic centimeters. What is the length of the side?

$s = \sqrt[3]{V} = \sqrt[3]{231} = 6.136$ centimeters

Square Prism:

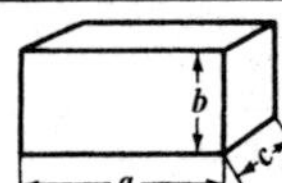

Volume $= V = abc$

$a = \frac{V}{bc} \qquad b = \frac{V}{ac} \qquad c = \frac{V}{ab}$

Example: In a square prism, $a = 6$, $b = 5$, $c = 4$. Find the volume.

$V = a \times b \times c = 6 \times 5 \times 4 = 120$ cubic inches

Example: How high should a box be made to contain 25 cubic feet, if it is 4 feet long and 2½ feet wide? Here, $a = 4$, $c = 2.5$, and $V = 25$. Then,

$$b = \text{depth} = \frac{V}{ac} = \frac{25}{4 \times 2.5} = \frac{25}{10} = 2.5 \text{ feet}$$

Prism:

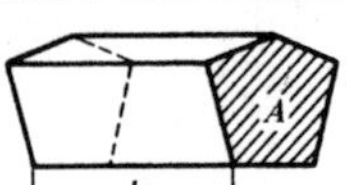

V = volume

A = area of end surface

$V = h \times A$

The area A of the end surface is found by the formulas for areas of plane figures on the preceding pages. Height h must be measured perpendicular to the end surface.

Example: A prism, having for its base a regular hexagon with a side s of 7.5 centimeters, is 25 centimeters high. Find the volume.

Area of hexagon $= A = 2.598s^2 = 2.598 \times 56.25 = 146.14$ square centimeters

Volume of prism $= h \times A = 25 \times 146.14 = 3653.5$ cubic centimeters

Pyramid:

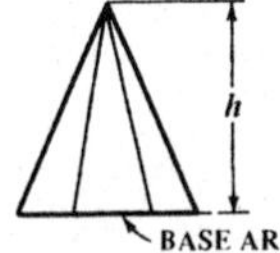

Volume $= V = \frac{1}{3}h \times$ area of base

If the base is a regular polygon with n sides, and s = length of side, r = radius of inscribed circle, and R = radius of circumscribed circle, then:

$$V = \frac{nsrh}{6} = \frac{nsh}{6}\sqrt{R^2 - \frac{s^2}{4}}$$

Example: A pyramid, having a height of 9 feet, has a base formed by a rectangle, the sides of which are 2 and 3 feet, respectively. Find the volume.

Area of base $= 2 \times 3 = 6$ square feet; $h = 9$ feet

Volume $= V = \frac{1}{3}h \times$ area of base $= \frac{1}{3} \times 9 \times 6 = 18$ cubic feet

Frustum of Pyramid:

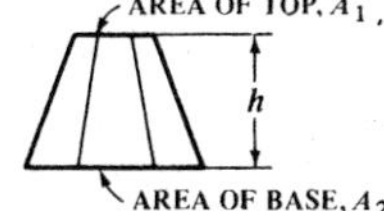

$$\text{Volume} = V = \frac{h}{3}(A_1 + A_2 + \sqrt{A_1 \times A_2})$$

Example: The pyramid in the previous example is cut off 4½ feet from the base, the upper part being removed. The sides of the rectangle forming the top surface of the frustum are, then, 1 and 1½ feet long, respectively. Find the volume of the frustum.

$$\text{Area of top} = A_1 = 1 \times 1\tfrac{1}{2} = 1\tfrac{1}{2} \text{ sq. ft.} \qquad \text{Area of base} = A_2 = 2 \times 3 = 6 \text{ sq. ft.}$$

$$V = \frac{4 \cdot 5}{3}(1.5 + 6 + \sqrt{1.5 \times 6}) = 1.5(7.5 + \sqrt{9}) = 1.5 \times 10.5 = 15.75 \text{ cubic feet}$$

Wedge:

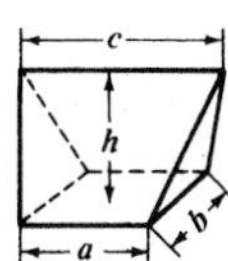

$$\text{Volume} = V = \frac{(2a + c)bh}{6}$$

Example: Let $a = 4$ inches, $b = 3$ inches, and $c = 5$ inches. The height $h = 4.5$ inches. Find the volume.

$$V = \frac{(2a + c)bh}{6} = \frac{(2 \times 4 + 5) \times 3 \times 4.5}{6} = \frac{(8 + 5) \times 13.5}{6}$$

$$= \frac{175.5}{6} = 29.25 \text{ cubic inches}$$

Cylinder:

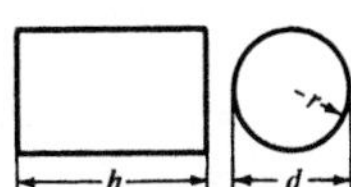

$$\text{Volume} = V = 3.1416r^2h = 0.7854d^2h$$

$$\text{Area of cylindrical surface} = S = 6.2832rh = 3.1416dh$$

Total area A of cylindrical surface and end surfaces:

$$A = 6.2832r(r + h) = 3.1416d(\tfrac{1}{2}d + h)$$

Example: The diameter of a cylinder is 2.5 inches. The length or height is 20 inches. Find the volume and the area of the cylindrical surface S.

$$V = 0.7854d^2h = 0.7854 \times 2.5^2 \times 20 = 0.7854 \times 6.25 \times 20 = 98.17 \text{ cubic inches}$$

$$S = 3.1416dh = 3.1416 \times 2.5 \times 20 = 157.08 \text{ square inches}$$

Portion of Cylinder:

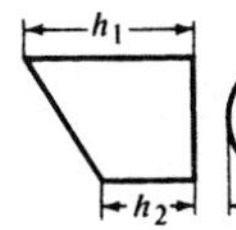

$$\text{Volume} = V = 1.5708r^2(h_1 + h_2)$$

$$= 0.3927d^2(h_1 + h_2)$$

$$\text{Cylindrical surface area} = S = 3.1416r(h_1 + h_2)$$

$$= 1.5708d(h_1 + h_2)$$

Example: A cylinder 125 millimeters in diameter is cut off at an angle, as shown in the illustration. Dimension $h_1 = 150$, and $h_2 = 100$ mm. Find the volume and the area S of the cylindrical surface.

$$V = 0.3927d^2(h_1 + h_2) = 0.3927 \times 125^2 \times (150 + 100)$$

$$= 0.3927 \times 15{,}625 \times 250 = 1{,}533{,}984 \text{ cubic millimeters} = 1534 \text{ cm}^3$$

$$S = 1.5708d(h_1 + h_2) = 1.5708 \times 125 \times 250$$

$$= 49{,}087.5 \text{ square millimeters} = 490.9 \text{ square centimeters}$$

Portion of Cylinder:

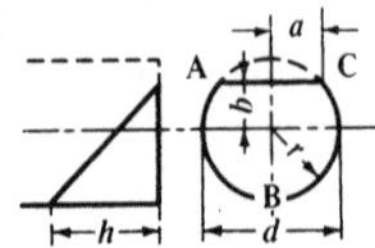

$$\text{Volume} = V = (\tfrac{2}{3}a^3 \pm b \times \text{area ABC})\frac{h}{r \pm b}$$

$$\text{Cylindrical surface area} = S = (ad \pm b \times \text{length of arc ABC})\frac{h}{r \pm b}$$

Use + when base area is larger, and – when base area is less than one-half the base circle.

Example: Find the volume of a cylinder so cut off that line AC passes through the center of the base circle — that is, the base area is a half-circle. The diameter of the cylinder = 5 inches, and the height h = 2 inches.

In this case, $a = 2.5$; $b = 0$; area $ABC = 0.5 \times 0.7854 \times 5^2 = 9.82$; $r = 2.5$.

$$V = \left(\frac{2}{3} \times 2.5^3 + 0 \times 9.82\right)\frac{2}{2.5 + 0} = \frac{2}{3} \times 15.625 \times 0.8 = 8.33 \text{ cubic inches}$$

Hollow Cylinder:

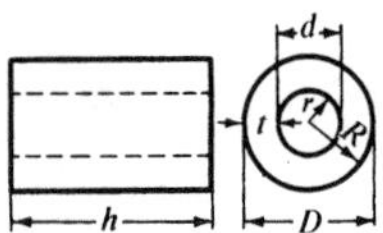

$$\begin{aligned}\text{Volume} = V &= 3.1416h(R^2 - r^2) = 0.7854h(D^2 - d^2)\\ &= 3.1416ht(2R - t) = 3.1416ht(D - t)\\ &= 3.1416ht(2r + t) = 3.1416ht(d + t)\\ &= 3.1416ht(R + r) = 1.5708ht(D + d)\end{aligned}$$

Example: A cylindrical shell, 28 centimeters high, is 36 centimeters in outside diameter, and 4 centimeters thick. Find its volume.

$$\begin{aligned}V &= 3.1416ht(D - t) = 3.1416 \times 28 \times 4(36 - 4) = 3.1416 \times 28 \times 4 \times 32\\ &= 11{,}259.5 \text{ cubic centimeters}\end{aligned}$$

Cone:

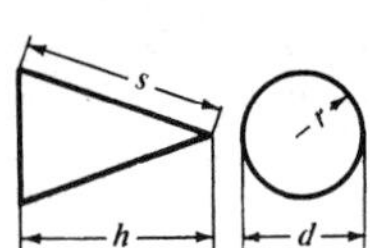

$$\text{Volume} = V = \frac{3.1416r^2h}{3} = 1.0472r^2h = 0.2618d^2h$$

$$\begin{aligned}\text{Conical surface area} = A &= 3.1416r\sqrt{r^2 + h^2} = 3.1416rs\\ &= 1.5708ds\end{aligned}$$

$$s = \sqrt{r^2 + h^2} = \sqrt{\frac{d^2}{4} + h^2}$$

Example: Find the volume and area of the conical surface of a cone, the base of which is a circle of 6 inches diameter, and the height of which is 4 inches.

$$V = 0.2618d^2h = 0.2618 \times 6^2 \times 4 = 0.2618 \times 36 \times 4 = 37.7 \text{ cubic inches}$$

$$\begin{aligned}A &= 3.1416r\sqrt{r^2 + h^2} = 3.1416 \times 3 \times \sqrt{3^2 + 4^2} = 9.4248 \times \sqrt{25}\\ &= 47.124 \text{ square inches}\end{aligned}$$

Frustum of Cone:

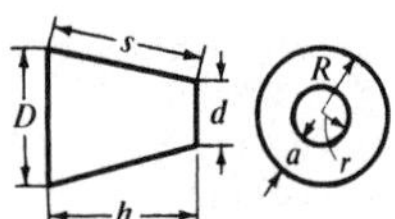

V = volume $\quad A$ = area of conical surface

$$V = 1.0472h(R^2 + Rr + r^2) = 0.2618h(D^2 + Dd + d^2)$$

$$A = 3.1416s(R + r) = 1.5708s(D + d)$$

$$a = R - r \qquad s = \sqrt{a^2 + h^2} = \sqrt{(R - r)^2 + h^2}$$

Example: Find the volume of a frustum of a cone of the following dimensions: $D = 8$ centimeters; $d = 4$ centimeters; $h = 5$ centimeters.

$$\begin{aligned}V &= 0.2618 \times 5(8^2 + 8 \times 4 + 4^2) = 0.2618 \times 5(64 + 32 + 16)\\ &= 0.2618 \times 5 \times 112 = 146.61 \text{ cubic centimeters}\end{aligned}$$

Sphere:

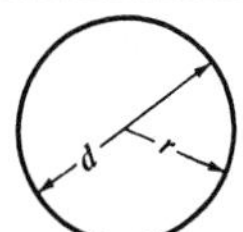

$$\text{Volume} = V = \frac{4\pi r^3}{3} = \frac{\pi d^3}{6} = 4.1888r^3 = 0.5236d^3$$

$$\text{Surface area} = A = 4\pi r^2 = \pi d^2 = 12.5664r^2 = 3.1416d^2$$

$$r = \sqrt[3]{\frac{3V}{4\pi}} = 0.6024\sqrt[3]{V}$$

Example: Find the volume and the surface of a sphere 6.5 centimeters diameter.

$$V = 0.5236d^3 = 0.5236 \times 6.5^3 = 0.5236 \times 6.5 \times 6.5 \times 6.5 = 143.79 \text{ cm}^3$$

$$A = 3.1416d^2 = 3.1416 \times 6.5^2 = 3.1416 \times 6.5 \times 6.5 = 132.73 \text{ cm}^2$$

Example: The volume of a sphere is 64 cubic centimeters. Find its radius.

$$r = 0.6204\sqrt[3]{64} = 0.6204 \times 4 = 2.4816 \text{ centimeters}$$

Spherical Sector:

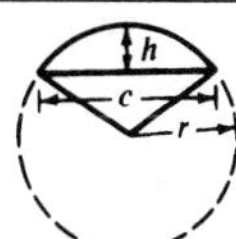

$$V = \frac{2\pi r^2 h}{3} = 2.0944r^2h = \text{Volume}$$

$$A = 3.1416r(2h + \tfrac{1}{2}c)$$

$$= \text{total area of conical and spherical surface}$$

$$c = 2\sqrt{h(2r-h)}$$

Example: Find the volume of a sector of a sphere 6 inches in diameter, the height h of the sector being 1.5 inch. Also find the length of chord c. Here $r = 3$ and $h = 1.5$.

$$V = 2.0944r^2h = 2.0944 \times 3^2 \times 1.5 = 2.0944 \times 9 \times 1.5 = 28.27 \text{ cubic inches}$$

$$c = 2\sqrt{h(2r-h)} = 2\sqrt{1.5(2\times 3 - 1.5)} = 2\sqrt{6.75} = 2 \times 2.598 = 5.196 \text{ inches}$$

Spherical Segment:

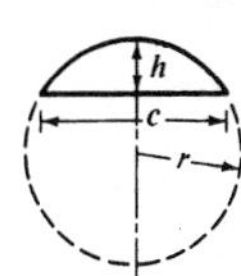

V = volume $\qquad$ A = area of spherical surface

$$V = 3.1416h^2\left(r - \frac{h}{3}\right) = 3.1416h\left(\frac{c^2}{8} + \frac{h^2}{6}\right)$$

$$A = 2\pi rh = 6.2832rh = 3.1416\left(\frac{c^2}{4} + h^2\right)$$

$$c = 2\sqrt{h(2r-h)}; \qquad r = \frac{c^2 + 4h^2}{8h}$$

Example: A segment of a sphere has the following dimensions: $h = 50$ millimeters; $c = 125$ millimeters. Find the volume V and the radius of the sphere of which the segment is a part.

$$V = 3.1416 \times 50 \times \left(\frac{125^2}{8} + \frac{50^2}{6}\right) = 157.08 \times \left(\frac{15{,}625}{8} + \frac{2500}{6}\right) = 372{,}247 \text{ mm}^3 = 372 \text{ cm}^3$$

$$r = \frac{125^2 + 4 \times 50^2}{8 \times 50} = \frac{15{,}625 + 10{,}000}{400} = \frac{25{,}625}{400} = 64 \text{ millimeters}$$

Ellipsoid:

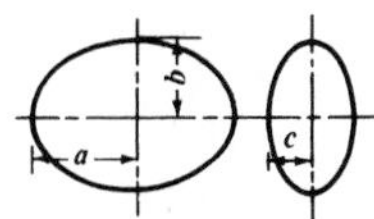

$$\text{Volume} = V = \frac{4\pi}{3}abc = 4.1888abc$$

In an ellipsoid of revolution, or spheroid, where $c = b$:

$$V = 4.1888ab^2$$

Example: Find the volume of a spheroid in which $a = 5$, and $b = c = 1.5$ inches.

$$V = 4.1888 \times 5 \times 1.5^2 = 47.124 \text{ cubic inches}$$

Spherical Zone:

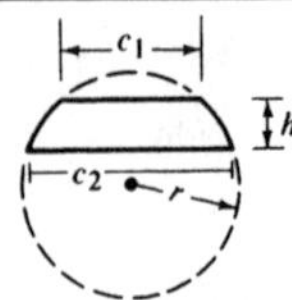

$$\text{Volume} = V = 0.5236h\left(\frac{3c_1^2}{4} + \frac{3c_2^2}{4} + h^2\right)$$

$$A = 2\pi rh = 6.2832rh = \text{area of spherical surface}$$

$$r = \sqrt{\frac{c_2^2}{4} + \left(\frac{c_2^2 - c_1^2 - 4h^2}{8h}\right)^2}$$

Example: In a spherical zone, let $c_1 = 3$; $c_2 = 4$; and $h = 1.5$ inch. Find the volume.

$$V = 0.5236 \times 1.5 \times \left(\frac{3 \times 3^2}{4} + \frac{3 \times 4^2}{4} + 1.5^2\right) = 0.5236 \times 1.5 \times \left(\frac{27}{4} + \frac{48}{4} + 2.25\right) = 16.493 \text{ in}^3$$

Spherical Wedge:

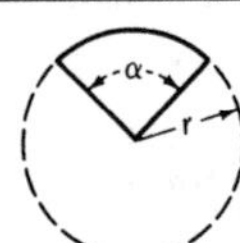

V = volume A = area of spherical surface

α = center angle in degrees

$$V = \frac{\alpha}{360} \times \frac{4\pi r^3}{3} = 0.0116\alpha r^3$$

$$A = \frac{\alpha}{360} \times 4\pi r^2 = 0.0349\alpha r^2$$

Example: Find the area of the spherical surface and the volume of a wedge of a sphere. The diameter of the sphere is 100 millimeters, and the center angle α is 45 degrees.

$$V = 0.0116 \times 45 \times 50^3 = 0.0116 \times 45 \times 125{,}000 = 65{,}250 \text{ mm}^3 = 65.25 \text{ cm}^3$$

$$A = 0.0349 \times 45 \times 50^2 = 3926.25 \text{ square millimeters} = 39.26 \text{ cm}^2$$

Hollow Sphere:

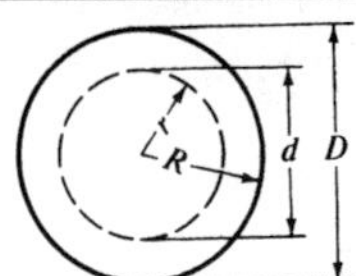

V = volume of material used to make a hollow sphere

$$V = \frac{4\pi}{3}(R^3 - r^3) = 4.1888(R^3 - r^3)$$

$$= \frac{\pi}{6}(D^3 - d^3) = 0.5236(D^3 - d^3)$$

Example: Find the volume of a hollow sphere, 8 inches in outside diameter, with a thickness of material of 1.5 inch.

Here $R = 4$; $r = 4 - 1.5 = 2.5$.

$$V = 4.1888(4^3 - 2.5^3) = 4.1888(64 - 15.625) = 4.1888 \times 48.375 = 202.63 \text{ cubic inches}$$

Paraboloid:

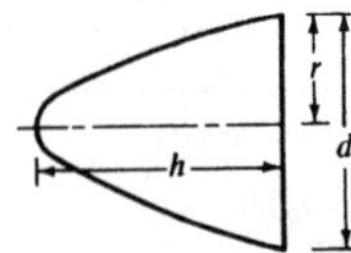

$$\text{Volume} = V = \tfrac{1}{2}\pi r^2 h = 0.3927d^2h$$

$$\text{Area} = A = \frac{2\pi}{3p}\left[\sqrt{\left(\frac{d^2}{4} + p^2\right)^3} - p^3\right]$$

in which $p = \dfrac{d^2}{8h}$

Example: Find the volume of a paraboloid in which $h = 300$ millimeters and $d = 125$ millimeters.

$$V = 0.3927d^2h = 0.3927 \times 125^2 \times 300 = 1{,}840{,}781 \text{ mm}^3 = 1{,}840.8 \text{ cm}^3$$

Paraboloidal Segment:

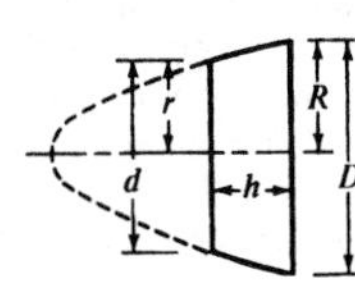

$$\text{Volume} = V = \frac{\pi}{2}h(R^2 + r^2) = 1.5708h(R^2 + r^2)$$

$$= \frac{\pi}{8}h(D^2 + d^2) = 0.3927h(D^2 + d^2)$$

Example: Find the volume of a segment of a paraboloid in which $D = 5$ inches, $d = 3$ inches, and $h = 6$ inches.

$$V = 0.3927h(D^2 + d^2) = 0.3927 \times 6 \times (5^2 + 3^2)$$

$$= 0.3927 \times 6 \times 34 = 80.11 \text{ cubic inches}$$

Torus:

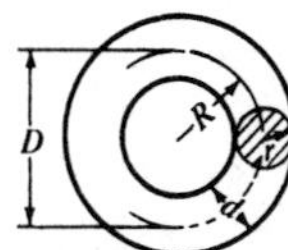

$$\text{Volume} = V = 2\pi^2 Rr^2 = 19.739Rr^2$$

$$= \frac{\pi^2}{4}Dd^2 = 2.4674Dd^2$$

$$\text{Area of surface} = A = 4\pi^2 Rr = 39.478Rr$$

$$= \pi^2 Dd = 9.8696Dd$$

Example: Find the volume and area of surface of a torus in which $d = 1.5$ and $D = 5$ inches.

$$V = 2.4674 \times 5 \times 1.5^2 = 2.4674 \times 5 \times 2.25 = 27.76 \text{ cubic inches}$$

$$A = 9.8696 \times 5 \times 1.5 = 74.022 \text{ square inches}$$

Barrel:

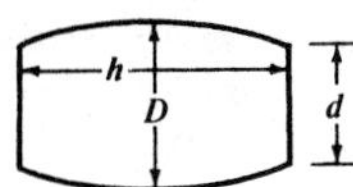

V = approximate volume.

If the sides are bent to the arc of a circle:

$$V = \frac{1}{12}\pi h(2D^2 + d^2) = 0.262h(2D^2 + d^2)$$

If the sides are bent to the arc of a parabola:

$$V = 0.209h(2D^2 + Dd + \tfrac{3}{4}d^2)$$

Example: Find the approximate contents of a barrel, the inside dimensions of which are $D = 60$ centimeters, $d = 50$ centimeters; $h = 120$ centimeters.

$$V = 0.262h(2D^2 + d^2) = 0.262 \times 120 \times (2 \times 60^2 + 50^2)$$

$$= 0.262 \times 120 \times (7200 + 2500) = 0.262 \times 120 \times 9700$$

$$= 304{,}968 \text{ cubic centimeters} = 0.305 \text{ cubic meter}$$

Ratio of Volumes:

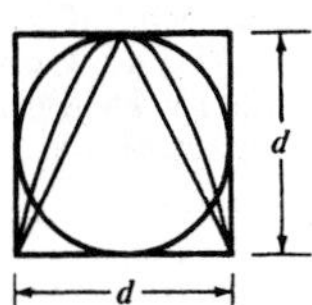

If d = base diameter and height of a cone, a paraboloid and a cylinder, and the diameter of a sphere, then the volumes of these bodies are to each other as follows:

$$\text{Cone} : \text{paraboloid} : \text{sphere} : \text{cylinder} = \tfrac{1}{3} : \tfrac{1}{2} : \tfrac{2}{3} : 1$$

Example: Assume, as an example, that the diameter of the base of a cone, paraboloid, and cylinder is 2 inches, that the height is 2 inches, and that the diameter of a sphere is 2 inches. Then the volumes, written in formula form, are as follows:

Cone Paraboloid Sphere Cylinder

$$\frac{3.1416 \times 2^2 \times 2}{12} : \frac{3.1416 \times (2p)^2 \times 2}{8} : \frac{3.1416 \times 2^3}{6} : \frac{3.1416 \times 2^2 \times 2}{4} = \tfrac{1}{3} : \tfrac{1}{2} : \tfrac{2}{3} : 1$$

Ratio and Proportion

The *ratio* between two quantities is the quotient obtained by dividing the first quantity by the second. For example, the ratio between 3 and 12 is ¼, and the ratio between 12 and 3 is 4. Ratio is generally indicated by the sign (:); thus, 12 : 3 indicates the ratio of 12 to 3.

A *reciprocal*, or *inverse* ratio, is the opposite of the original ratio. Thus, the inverse ratio of 5 : 7 is 7 : 5.

In a *compound* ratio, each term is the product of the corresponding terms in two or more simple ratios. Thus, when

$$8:2 = 4 \qquad 9:3 = 3 \qquad 10:5 = 2$$

then the compound ratio is

$$8 \times 9 \times 10 : 2 \times 3 \times 5 = 4 \times 3 \times 2$$

$$720:30 = 24$$

Proportion is the equality of ratios. Thus,

$$6:3 = 10:5 \qquad \text{or} \qquad 6:3::10:5$$

The first and last terms in a proportion are called the *extremes;* the second and third, the *means*. The product of the extremes is equal to the product of the means. Thus,

$$25:2 = 100:8 \qquad \text{and} \qquad 25 \times 8 = 2 \times 100$$

If three terms in a proportion are known, the remaining term may be found by the following rules:

The first term is equal to the product of the second and third terms, divided by the fourth.

The second term is equal to the product of the first and fourth terms, divided by the third.

The third term is equal to the product of the first and fourth terms, divided by the second.

The fourth term is equal to the product of the second and third terms, divided by the first.

Example: Let x be the term to be found, then,

$$x : 12 = 3.5 : 21 \qquad x = \frac{12 \times 3.5}{21} = \frac{42}{21} = 2$$

$$\tfrac{1}{4} : x = 14 : 42 \qquad x = \frac{\tfrac{1}{4} \times 42}{14} = \frac{1}{4} \times 3 = \frac{3}{4}$$

$$5 : 9 = x : 63 \qquad x = \frac{5 \times 63}{9} = \frac{315}{9} = 35$$

$$\tfrac{1}{4} : \tfrac{7}{8} = 4 : x \qquad x = \frac{\tfrac{7}{8} \times 4}{\tfrac{1}{4}} = \frac{3\tfrac{1}{2}}{\tfrac{1}{4}} = 14$$

If the second and third terms are the same, that number is the *mean proportional* between the other two. Thus, 8 : 4 = 4 : 2, and 4 is the mean proportional between 8 and 2. The mean proportional between two numbers may be found by multiplying the numbers together and extracting the square root of the product. Thus, the mean proportional between 3 and 12 is found as follows:

$$3 \times 12 = 36 \qquad \text{and} \qquad \sqrt{36} = 6$$

which is the mean proportional.

Practical Examples Involving Simple Proportion.—If it takes 18 days to assemble 4 lathes, how long would it take to assemble 14 lathes?

Let the number of days to be found be x. Then write out the proportion as follows:

$$4:18 = 14:x$$

$$(\text{lathes} : \text{days} = \text{lathes} : \text{days})$$

Now find the fourth term by the rule given:

$$x = \frac{18 \times 14}{4} = 63 \text{ days}$$

Thirty-four linear feet of bar stock are required for the blanks for 100 clamping bolts. How many feet of stock would be required for 912 bolts?

Let x = total length of stock required for 912 bolts.

$$34:100 = x:912$$

$$(\text{feet} : \text{bolts} = \text{feet} : \text{bolts})$$

Then, the third term $x = (34 \times 912)/100 = 310$ feet, approximately.

Inverse Proportion.—In an inverse proportion, as one of the items involved *increases*, the corresponding item in the proportion *decreases*, or vice versa. For example, a factory employing 270 men completes a given number of typewriters weekly, the number of working hours being 44 per week. How many men would be required for the same production if the working hours were reduced to 40 per week?

The time per week is in an inverse proportion to the number of men employed; the shorter the time, the more men. The inverse proportion is written:

$$270 : x = 40 : 44$$

(men, 44-hour basis: men, 40-hour basis = time, 40-hour basis: time, 44-hour basis) Thus

$$\frac{270}{x} = \frac{40}{44} \quad \text{and} \quad x = \frac{270 \times 44}{40} = 297 \text{ men}$$

Problems Involving Both Simple and Inverse Proportions.—If two groups of data are related both by direct (simple) and inverse proportions among the various quantities, then a simple mathematical relation that may be used in solving problems is as follows:

$$\frac{\text{Product of all directly proportional items in first group}}{\text{Product of all inversely proportional items in first group}}$$

$$= \frac{\text{Product of all directly proportional items in second group}}{\text{Product of all inversely proportional items in second group}}$$

Example: If a man capable of turning 65 studs in a day of 10 hours is paid $6.50 per hour, how much per hour ought a man be paid who turns 72 studs in a 9-hour day, if compensated in the same proportion?

The first group of data in this problem consists of the number of hours worked by the first man, his hourly wage, and the number of studs which he produces per day; the second group contains similar data for the second man except for his unknown hourly wage, which may be indicated by x.

The labor cost per stud, as may be seen, is directly proportional to the number of hours worked and the hourly wage. These quantities, therefore, are used in the numerators of the fractions in the formula. The labor cost per stud is inversely proportional to the number of studs produced per day. (The greater the number of studs produced in a given time the less the cost per stud.) The numbers of studs per day, therefore, are placed in the denominators of the fractions in the formula. Thus,

$$\frac{10 \times 6.50}{65} = \frac{9 \times x}{72}$$

$$x = \frac{10 \times 6.50 \times 72}{65 \times 9} = \$8.00 \text{ per hour}$$

Powers, Roots, and Reciprocals

The *square* of a number (or quantity) is the product of that number multiplied by itself. Thus, the square of 9 is $9 \times 9 = 81$. The square of a number is indicated by the *exponent* (^{2}), thus: $9^2 = 9 \times 9 = 81$.

The *cube* or *third power* of a number is the product obtained by using that number as a factor three times. Thus, the cube of 4 is $4 \times 4 \times 4 = 64$, and is written 4^3.

If a number is used as a factor four or five times, respectively, the product is the fourth or fifth power. Thus, $3^4 = 3 \times 3 \times 3 \times 3 = 81$, and $2^5 = 2 \times 2 \times 2 \times 2 \times 2 = 32$. A number can be raised to any power by using it as a factor the required number of times.

The *square root* of a given number is that number which, when multiplied by itself, will give a product equal to the given number. The square root of 16 (written $\sqrt{16}$) equals 4, because $4 \times 4 = 16$.

The *cube root* of a given number is that number which, when used as a factor three times, will give a product equal to the given number. Thus, the cube root of 64 (written $\sqrt[3]{64}$) equals 4, because $4 \times 4 \times 4 = 64$.

The fourth, fifth, etc., roots of a given number are those numbers which when used as factors four, five, etc., times, will give as a product the given number. Thus, $\sqrt[4]{16} = 2$, because $2 \times 2 \times 2 \times 2 = 16$.

In some formulas, there may be such expressions as $(a^2)^3$ and $a^{3/2}$. The first of these, $(a^2)^3$, means that the number a is first to be squared, a^2, and the result then cubed to give a^6. Thus, $(a^2)^3$ is equivalent to a^6 which is obtained by *multiplying* the exponents 2 and 3. Similarly, $a^{3/2}$ may be interpreted as the cube of the square root of a, $(\sqrt{a})^3$, or $(a^{1/2})^3$, so that, for example, $16^{3/2} = (\sqrt{16})^3 = 64$.

The multiplications required for raising numbers to powers and the extracting of roots are greatly facilitated by the use of logarithms. Extracting the square root and cube root by the regular arithmetical methods is a slow and cumbersome operation, and any roots can be more rapidly found by using logarithms.

When the power to which a number is to be raised is not an integer, say 1.62, the use of either logarithms or a scientific calculator becomes the only practical means of solution.

The *reciprocal* R of a number N is obtained by dividing 1 by the number; $R = 1/N$. Reciprocals are useful in some calculations because they avoid the use of negative characteristics as in calculations with logarithms and in trigonometry. In trigonometry, the values *cosecant, secant*, and *cotangent* are often used for convenience and are the reciprocals of the *sine, cosine*, and *tangent*, respectively (see page 103). The reciprocal of a fraction, for instance $3/4$, is the fraction inverted, since $1 \div 3/4 = 1 \times 4/3 = 4/3$.

Powers of Ten Notation

Powers of ten notation is used to simplify calculations and ensure accuracy, particularly with respect to the position of decimal points, and also simplifies the expression of numbers which are so large or so small as to be unwieldy. For example, the metric (SI) pressure unit pascal is equivalent to 0.00000986923 atmosphere or 0.0001450377 pound/inch^2. In powers of ten notation, these figures are 9.86923×10^{-6} atmosphere and 1.450377×10^{-4} pound/inch^2. The notation also facilitates adaptation of numbers for electronic data processing and computer readout.

Expressing Numbers in Powers of Ten Notation.—In this system of notation, every number is expressed by two factors, one of which is some integer from 1 to 9 followed by a decimal and the other is some power of 10.

Thus, 10,000 is expressed as 1.0000×10^4 and 10,463 as 1.0463×10^4. The number 43 is expressed as 4.3×10 and 568 is expressed. as 5.68×10^2.

In the case of decimals, the number 0.0001, which as a fraction is $\frac{1}{10{,}000}$, is expressed as 1×10^{-4} and 0.0001463 is expressed as 1.463×10^{-4}. The decimal 0.498 is expressed as 4.98×10^{-1} and 0.03146 is expressed as 3.146×10^{-2}.

Rules for Converting Any Number to Powers of Ten Notation.—Any number can be converted to the powers of ten notation by means of one of two rules.

Rule 1: If the number is a whole number or a whole number and a decimal so that it has digits to the left of the decimal point, the decimal point is moved a sufficient number of places to the *left* to bring it to the immediate right of the first digit. With the decimal point shifted to this position, the number so written comprises the *first* factor when written in powers of ten notation.

The number of places that the decimal point is moved to the left to bring it immediately to the right of the first digit is the *positive* index or power of 10 that comprises the *second* factor when written in powers of ten notation.

Thus, to write 4639 in this notation, the decimal point is moved three places to the left giving the two factors: 4.639×10^3. Similarly,

$$431.412 = 4.31412 \times 10^2$$

$$986,388 = 9.86388 \times 10^5$$

Rule 2: If the number is a decimal, i.e., it has digits entirely to the right of the decimal point, then the decimal point is moved a sufficient number of places to the *right* to bring it immediately to the right of the first digit. With the decimal point shifted to this position, the number so written comprises the *first* factor when written in powers of ten notation.

The number of places that the decimal point is moved to the *right* to bring it immediately to the right of the first digit is the *negative* index or power of 10 that follows the number when written in powers of ten notation.

Thus, to bring the decimal point in 0.005721 to the immediate right of the first digit, which is 5, it must be moved *three* places to the right, giving the two factors: 5.721×10^{-3}. Similarly,

$$0.469 = 4.69 \times 10^{-1}$$

$$0.0000516 = 5.16 \times 10^{-5}$$

Multiplying Numbers Written in Powers of Ten Notation.—When multiplying two numbers written in the powers of ten notation together, the procedure is as follows:

1) Multiply the first factor of one number by the first factor of the other to obtain the first factor of the product.

2) Add the index of the second factor (which is some power of 10) of one number to the index of the second factor of the other number to obtain the index of the second factor (which is some power of 10) in the product. Thus:

$$(4.31 \times 10^{-2}) \times (9.0125 \times 10) = (4.31 \times 9.0125) \times 10^{-2+1} = 38.844 \times 10^{-1}$$

$$(5.986 \times 10^4) \times (4.375 \times 10^3) = (5.986 \times 4.375) \times 10^{4+3} = 26.189 \times 10^7$$

In the preceeding calculations, neither of the results shown are in the conventional powers of ten form since the first factor in each has two digits. In the conventional powers of ten notation, the results would be in each case rounding off the first factor to three decimal places.

$$38.844 \times 10^{-1} = 3.884 \times 10^{0} = 3.884$$

since $10^0 = 1$, and

$$26.189 \times 10^{7} = 2.619 \times 10^{8}$$

When multiplying several numbers written in this notation together, the procedure is the same. All of the first factors are multiplied together to get the first factor of the product and all of the indices of the respective powers of ten are added together, taking into account their respective signs, to get the index of the second factor of the product. Thus, $(4.02 \times 10^{-3}) \times (3.987 \times 10) \times (4.863 \times 10^{5}) = (4.02 \times 3.987 \times 4.863) \times (10^{-3+1+5}) = 77.94 \times 10^{3} = 7.79 \times 10^{4}$ rounding off the first factor to two decimal places.

Dividing Numbers Written in Powers of Ten Notation.—When dividing one number by another when both are written in this notation, the procedure is as follows:

1) Divide the first factor of the dividend by the first factor of the divisor to get the first factor of the quotient.

2) Subtract the index of the second factor of the divisor from the index of the second factor of the dividend, taking into account their respective signs, to get the index of the second factor of the quotient. Thus:

$$\begin{aligned}(4.31 \times 10^{-2}) \div (9.0125 \times 10) &= \\ (4.31 \div 9.0125) \times (10^{-2-1}) &= 0.4782 \times 10^{-3} = 4.782 \times 10^{-4}\end{aligned}$$

It can be seen that this system of notation is helpful where several numbers of different magnitudes are to be multiplied and divided.

Example: Find the quotient of

$$\frac{250 \times 4698 \times 0.00039}{43678 \times 0.002 \times 0.0147}$$

Solution: Changing all these numbers to powers of ten notation and performing the operations indicated:

$$\begin{aligned}\frac{(2.5 \times 10^{2}) \times (4.698 \times 10^{3}) \times (3.9 \times 10^{-4})}{(4.3678 \times 10^{4}) \times (2 \times 10^{-3}) \times (1.47 \times 10^{-2})} &= \\ = \frac{(2.5 \times 4.698 \times 3.9)(10^{2+3-4})}{(4.3678 \times 2 \times 1.47)(10^{4-3-2})} &= \frac{45.8055 \times 10}{12.8413 \times 10^{-1}} \\ &= 3.5670 \times 10^{1-(-1)} \\ &= 3.5670 \times 10^{2} \\ &= 356.70\end{aligned}$$

Geometrical Propositions

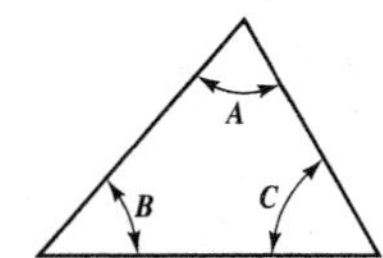	The sum of the three angles in a triangle always equals 180 degrees. Hence, if two angles are known, the third angle can always be found. $A + B + C = 180°$ $A = 180° - (B + C)$ $B = 180° - (A + C)$ $C = 180° - (A + B)$
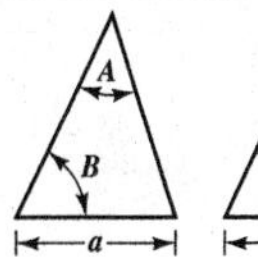 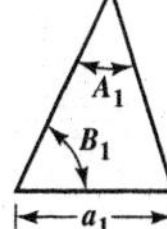	If one side and two angles in one triangle are equal to one side and similarly located angles in another triangle, then the remaining two sides and angle also are equal. If $a = a_1$, $A = A_1$, and $B = B_1$, then the two other sides and the remaining angle also are equal.
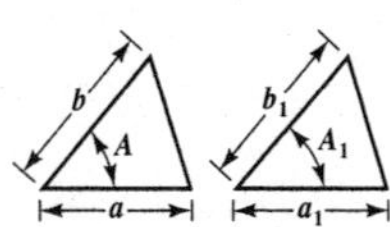	If two sides and the angle between them in one triangle are equal to two sides and a similarly located angle in another triangle, then the remaining side and angles also are equal. If $a = a_1$, $b = b_1$, and $A = A_1$, then the remaining side and angles also are equal.
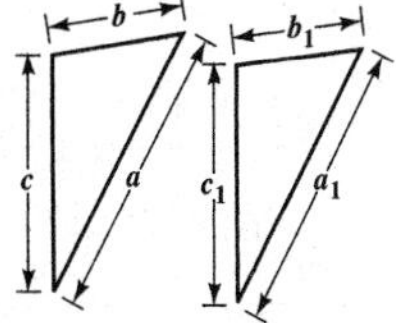	If the three sides in one triangle are equal to the three sides of another triangle, then the angles in the two triangles also are equal. If $a = a_1$, $b = b_1$, and $c = c_1$, then the angles between the respective sides also are equal.
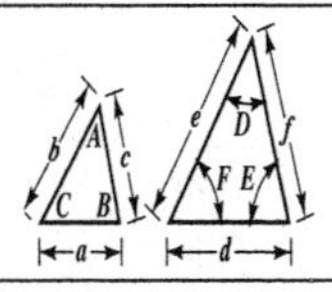	If the three sides of one triangle are proportional to corresponding sides in another triangle, then the triangles are called *similar*, and the angles in the one are equal to the angles in the other. If $a : b : c = d : e : f$, then $A = D$, $B = E$, and $C = F$.
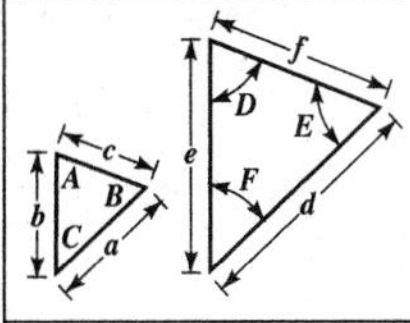	If the angles in one triangle are equal to the angles in another triangle, then the triangles are similar and their corresponding sides are proportional. If $A = D, B = E$, and $C = F$, then $a : b : c = d : e : f$.
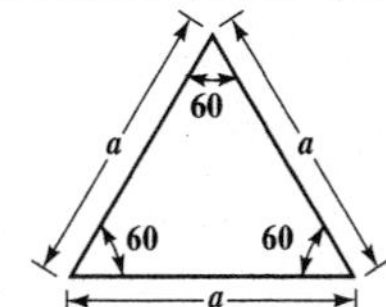	If the three sides in a triangle are equal—that is, if the triangle is *equilateral*—then the three angles also are equal. Each of the three equal angles in an equilateral triangle is 60 degrees. If the three angles in a triangle are equal, then the three sides also are equal.

Geometrical Propositions

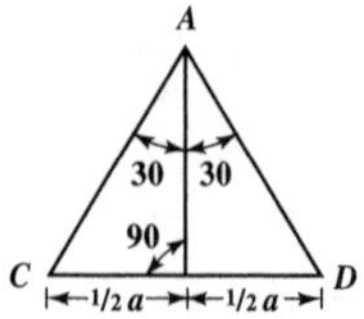

A line in an equilateral triangle that bisects or divides any of the angles into two equal parts also bisects the side opposite the angle and is at right angles to it.

If line *AB* divides angle *CAD* into two equal parts, it also divides line *CD* into two equal parts and is at right angles to it.

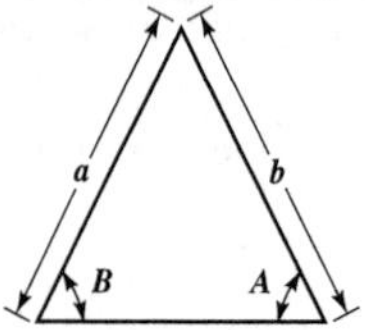

If two sides in a triangle are equal—that is, if the triangle is an *isosceles* triangle—then the angles opposite these sides also are equal.

If side *a* equals side *b*, then angle *A* equals angle *B*.

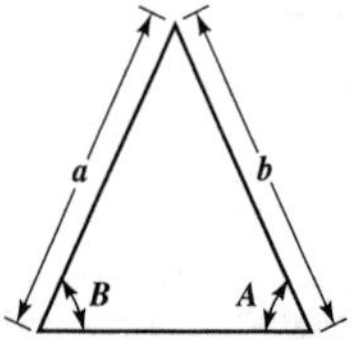

If two angles in a triangle are equal, the sides opposite these angles also are equal.

If angles *A* and *B* are equal, then side *a* equals side *b*.

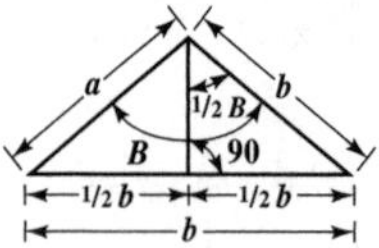

In an isosceles triangle, if a straight line is drawn from the point where the two equal sides meet, so that it bisects the third side or base of the triangle, then it also bisects the angle between the equal sides and is perpendicular to the base.

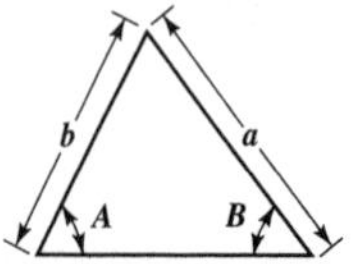

In every triangle, that angle is greater that is opposite a longer side. In every triangle, that side is greater which is opposite a greater angle.

If *a* is longer than *b*, then angle *A* is greater than *B*. If angle *A* is greater than *B*, then side *a* is longer than *b*.

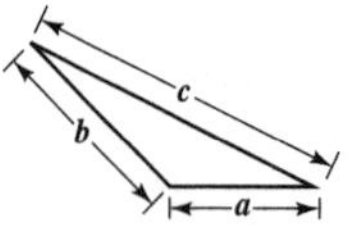

In every triangle, the sum of the lengths of two sides is always greater than the length of the third.

Side *a* + side *b* is always greater than side *c*.

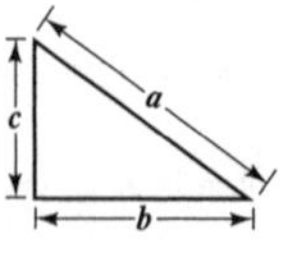

In a right-angle triangle, the square of the hypotenuse or the side opposite the right angle is equal to the sum of the squares on the two sides that form the right angle.

$$a^2 = b^2 + c^2$$

Geometrical Propositions

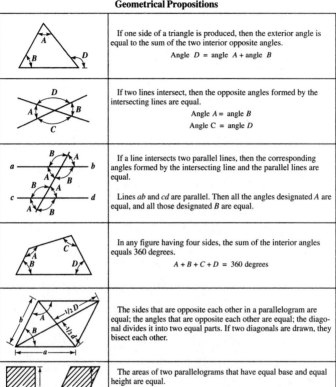

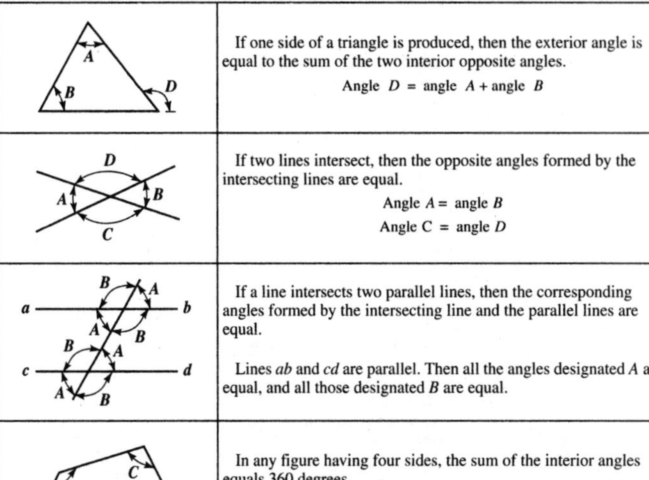	If one side of a triangle is produced, then the exterior angle is equal to the sum of the two interior opposite angles. Angle D = angle A + angle B
	If two lines intersect, then the opposite angles formed by the intersecting lines are equal. Angle A = angle B Angle C = angle D
	If a line intersects two parallel lines, then the corresponding angles formed by the intersecting line and the parallel lines are equal. Lines *ab* and *cd* are parallel. Then all the angles designated A are equal, and all those designated B are equal.
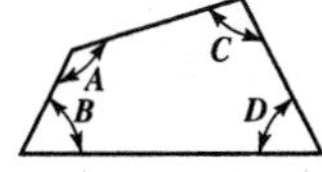	In any figure having four sides, the sum of the interior angles equals 360 degrees. $A + B + C + D$ = 360 degrees
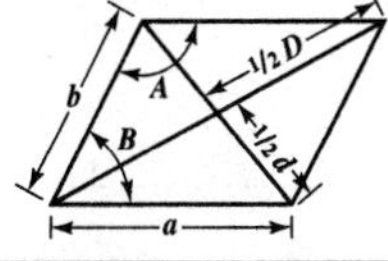	The sides that are opposite each other in a parallelogram are equal; the angles that are opposite each other are equal; the diagonal divides it into two equal parts. If two diagonals are drawn, they bisect each other.
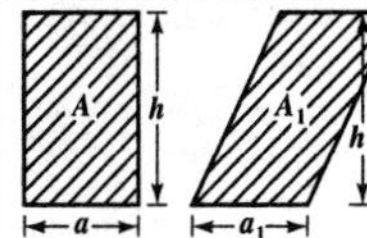	The areas of two parallelograms that have equal base and equal height are equal. If $a = a_1$ and $h = h_1$, then Area A = area A_1
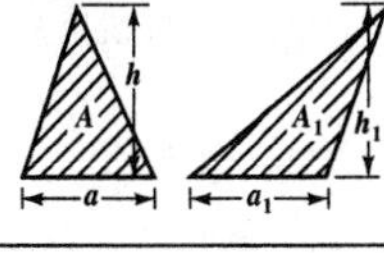	The areas of triangles having equal base and equal height are equal. If $a = a_1$ and $h = h_1$, then Area A = area A_1
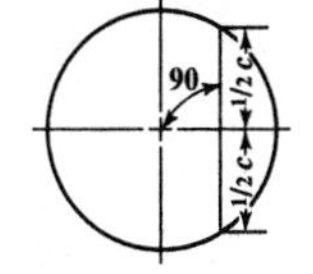	If a diameter of a circle is at right angles to a chord, then it bisects or divides the chord into two equal parts.

Geometrical Propositions

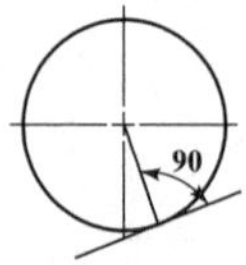	If a line is tangent to a circle, then it is also at right angles to a line drawn from the center of the circle to the point of tangency—that is, to a radial line through the point of tangency.
Point of Tangency 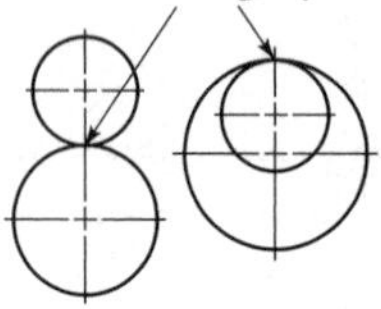	If two circles are tangent to each other, then the straight line that passes through the centers of the two circles must also pass through the point of tangency.
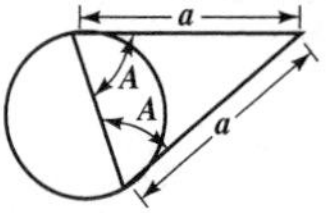	If from a point outside a circle, tangents are drawn to a circle, the two tangents are equal and make equal angles with the chord joining the points of tangency.
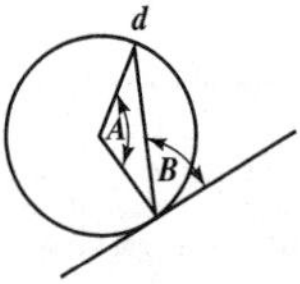	The angle between a tangent and a chord drawn from the point of tangency equals one-half the angle at the center subtended by the chord. Angle $B = \frac{1}{2}$ angle A
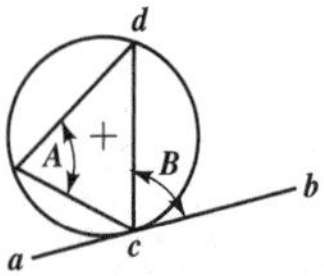	The angle between a tangent and a chord drawn from the point of tangency equals the angle at the periphery subtended by the chord. Angle *B*, between tangent *ab* and chord *cd*, equals angle *A* subtended at the periphery by chord *cd*.
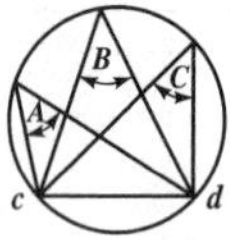	All angles having their vertex at the periphery of a circle and subtended by the same chord are equal. Angles *A*, *B*, and *C*, all subtended by chord *cd*, are equal.
	If an angle at the circumference of a circle, between two chords, is subtended by the same arc as the angle at the center, between two radii, then the angle at the circumference is equal to one-half of the angle at the center. Angle $A = \frac{1}{2}$ angle B

Geometrical Propositions

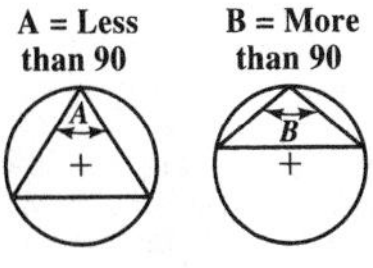

A = Less than 90; B = More than 90	An angle subtended by a chord in a circular segment larger than one-half the circle is an acute angle—an angle less than 90 degrees. An angle subtended by a chord in a circular segment less than one-half the circle is an obtuse angle—an angle greater than 90 degrees.
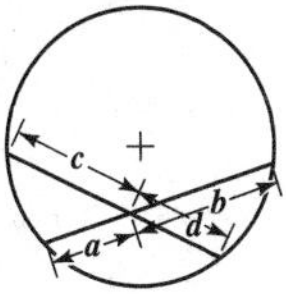	If two chords intersect each other in a circle, then the rectangle of the segments of the one equals the rectangle of the segments of the other. $a \times b = c \times d$
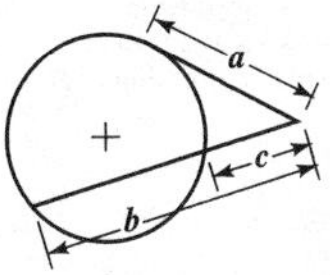	If from a point outside a circle two lines are drawn, one of which intersects the circle and the other is tangent to it, then the rectangle contained by the total length of the intersecting line, and that part of it that is between the outside point and the periphery, equals the square of the tangent. $a^2 = b \times c$
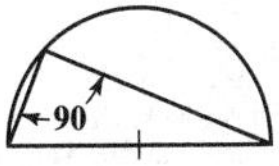 90	If a triangle is inscribed in a semicircle, the angle opposite the diameter is a right (90-degree) angle. All angles at the periphery of a circle, subtended by the diameter, are right (90-degree) angles.
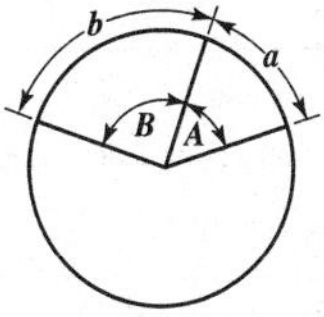	The lengths of circular arcs of the same circle are proportional to the corresponding angles at the center. $A : B = a : b$
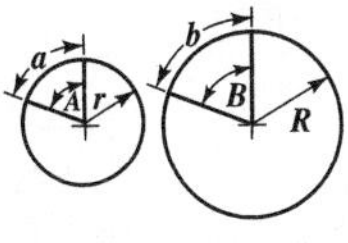	The lengths of circular arcs having the same center angle are proportional to the lengths of the radii. If $A = B$, then $a : b = r : R$.
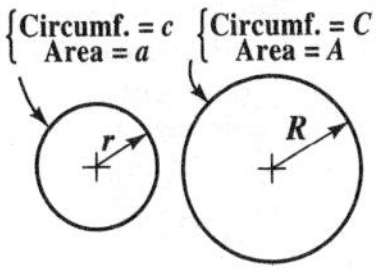 {Circumf. = c, Area = a}; {Circumf. = C, Area = A}	The circumferences of two circles are proportional to their radii. The areas of two circles are proportional to the squares of their radii. $c : C = r : R$; $a : A = r^2 : R^2$

Geometrical Constructions

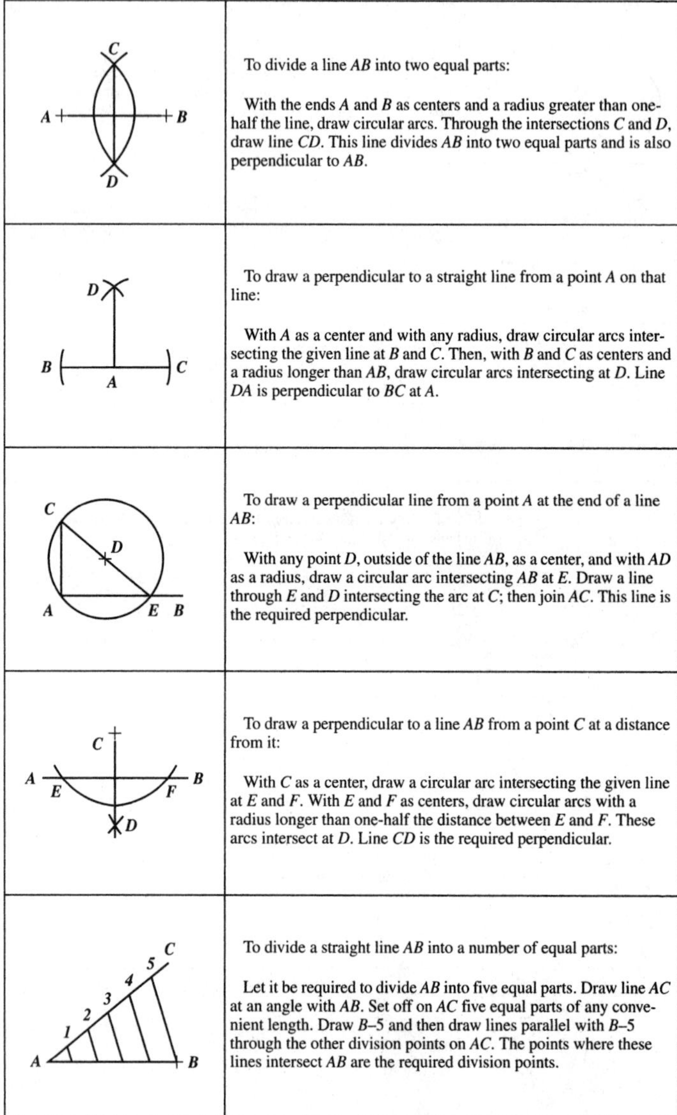

	To divide a line *AB* into two equal parts: With the ends *A* and *B* as centers and a radius greater than one-half the line, draw circular arcs. Through the intersections *C* and *D*, draw line *CD*. This line divides *AB* into two equal parts and is also perpendicular to *AB*.
	To draw a perpendicular to a straight line from a point *A* on that line: With *A* as a center and with any radius, draw circular arcs intersecting the given line at *B* and *C*. Then, with *B* and *C* as centers and a radius longer than *AB*, draw circular arcs intersecting at *D*. Line *DA* is perpendicular to *BC* at *A*.
	To draw a perpendicular line from a point *A* at the end of a line *AB*: With any point *D*, outside of the line *AB*, as a center, and with *AD* as a radius, draw a circular arc intersecting *AB* at *E*. Draw a line through *E* and *D* intersecting the arc at *C*; then join *AC*. This line is the required perpendicular.
	To draw a perpendicular to a line *AB* from a point *C* at a distance from it: With *C* as a center, draw a circular arc intersecting the given line at *E* and *F*. With *E* and *F* as centers, draw circular arcs with a radius longer than one-half the distance between *E* and *F*. These arcs intersect at *D*. Line *CD* is the required perpendicular.
	To divide a straight line *AB* into a number of equal parts: Let it be required to divide *AB* into five equal parts. Draw line *AC* at an angle with *AB*. Set off on *AC* five equal parts of any convenient length. Draw *B*–5 and then draw lines parallel with *B*–5 through the other division points on *AC*. The points where these lines intersect *AB* are the required division points.

Geometrical Constructions

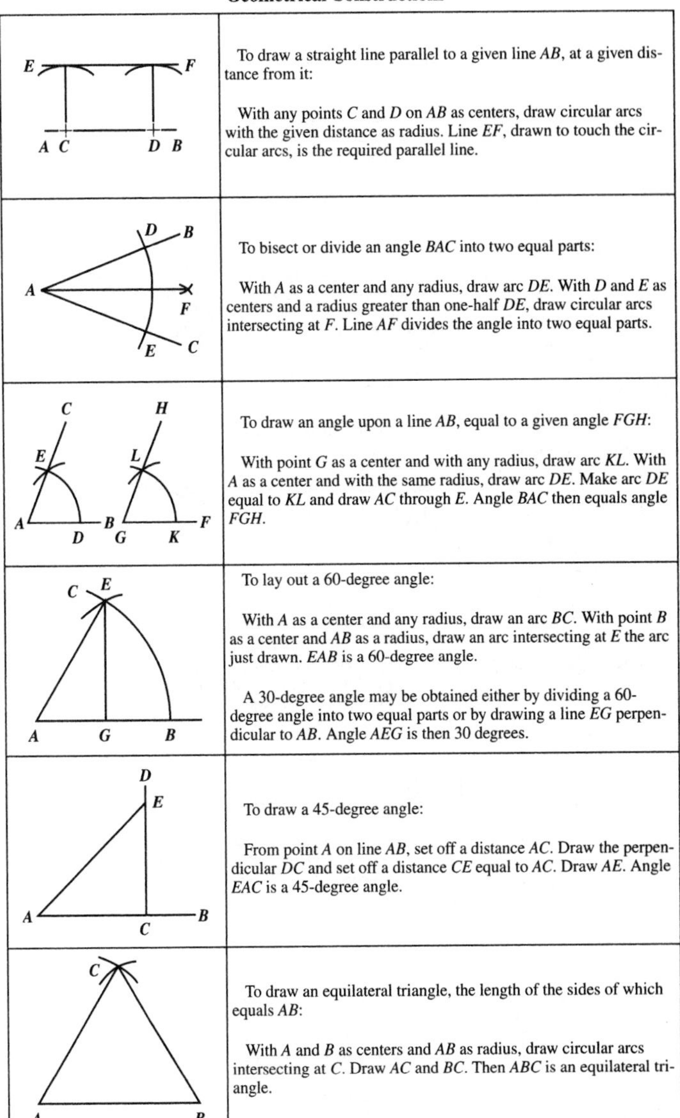

To draw a straight line parallel to a given line *AB*, at a given distance from it:

With any points *C* and *D* on *AB* as centers, draw circular arcs with the given distance as radius. Line *EF*, drawn to touch the circular arcs, is the required parallel line.

To bisect or divide an angle *BAC* into two equal parts:

With *A* as a center and any radius, draw arc *DE*. With *D* and *E* as centers and a radius greater than one-half *DE*, draw circular arcs intersecting at *F*. Line *AF* divides the angle into two equal parts.

To draw an angle upon a line *AB*, equal to a given angle *FGH*:

With point *G* as a center and with any radius, draw arc *KL*. With *A* as a center and with the same radius, draw arc *DE*. Make arc *DE* equal to *KL* and draw *AC* through *E*. Angle *BAC* then equals angle *FGH*.

To lay out a 60-degree angle:

With *A* as a center and any radius, draw an arc *BC*. With point *B* as a center and *AB* as a radius, draw an arc intersecting at *E* the arc just drawn. *EAB* is a 60-degree angle.

A 30-degree angle may be obtained either by dividing a 60-degree angle into two equal parts or by drawing a line *EG* perpendicular to *AB*. Angle *AEG* is then 30 degrees.

To draw a 45-degree angle:

From point *A* on line *AB*, set off a distance *AC*. Draw the perpendicular *DC* and set off a distance *CE* equal to *AC*. Draw *AE*. Angle *EAC* is a 45-degree angle.

To draw an equilateral triangle, the length of the sides of which equals *AB*:

With *A* and *B* as centers and *AB* as radius, draw circular arcs intersecting at *C*. Draw *AC* and *BC*. Then *ABC* is an equilateral triangle.

Geometrical Constructions

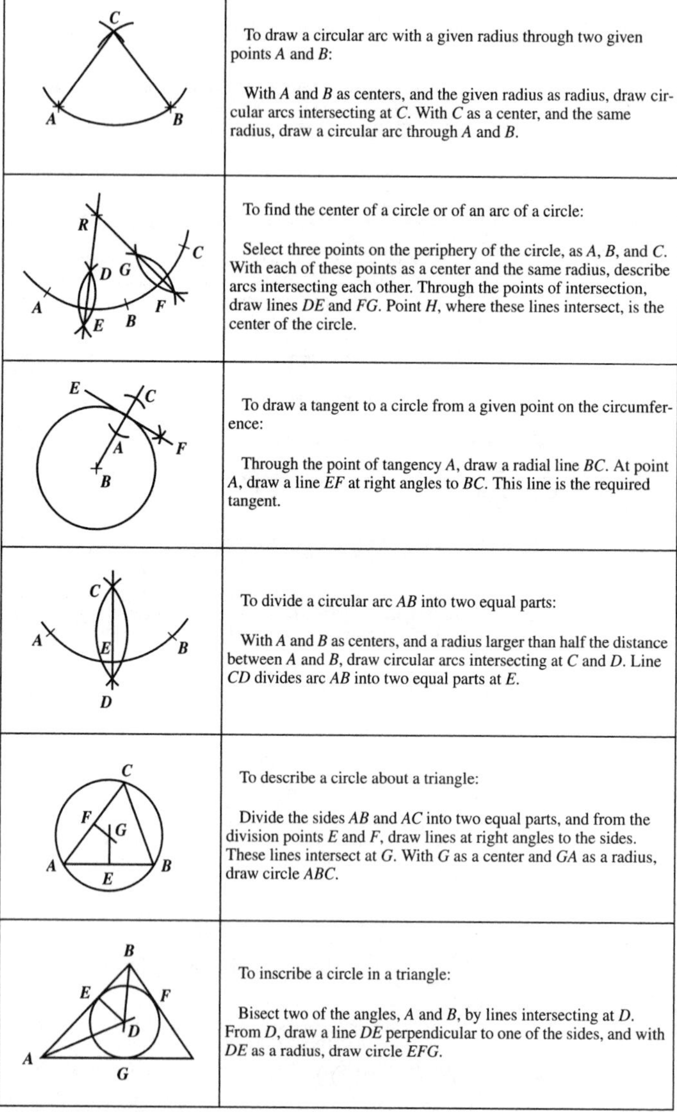

To draw a circular arc with a given radius through two given points *A* and *B*:

With *A* and *B* as centers, and the given radius as radius, draw circular arcs intersecting at *C*. With *C* as a center, and the same radius, draw a circular arc through *A* and *B*.

To find the center of a circle or of an arc of a circle:

Select three points on the periphery of the circle, as *A*, *B*, and *C*. With each of these points as a center and the same radius, describe arcs intersecting each other. Through the points of intersection, draw lines *DE* and *FG*. Point *H*, where these lines intersect, is the center of the circle.

To draw a tangent to a circle from a given point on the circumference:

Through the point of tangency *A*, draw a radial line *BC*. At point *A*, draw a line *EF* at right angles to *BC*. This line is the required tangent.

To divide a circular arc *AB* into two equal parts:

With *A* and *B* as centers, and a radius larger than half the distance between *A* and *B*, draw circular arcs intersecting at *C* and *D*. Line *CD* divides arc *AB* into two equal parts at *E*.

To describe a circle about a triangle:

Divide the sides *AB* and *AC* into two equal parts, and from the division points *E* and *F*, draw lines at right angles to the sides. These lines intersect at *G*. With *G* as a center and *GA* as a radius, draw circle *ABC*.

To inscribe a circle in a triangle:

Bisect two of the angles, *A* and *B*, by lines intersecting at *D*. From *D*, draw a line *DE* perpendicular to one of the sides, and with *DE* as a radius, draw circle *EFG*.

Geometrical Constructions

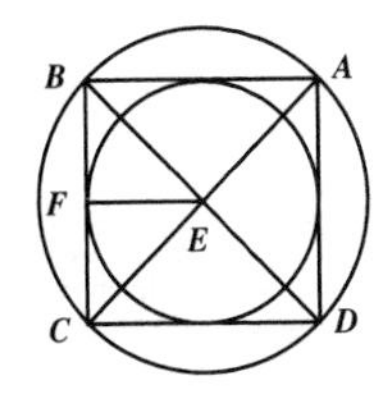

To describe a circle about a square and to inscribe a circle in a square:

The centers of both the circumscribed and inscribed circles are located at the point *E*, where the two diagonals of the square intersect. The radius of the circumscribed circle is *AE*, and of the inscribed circle, *EF*.

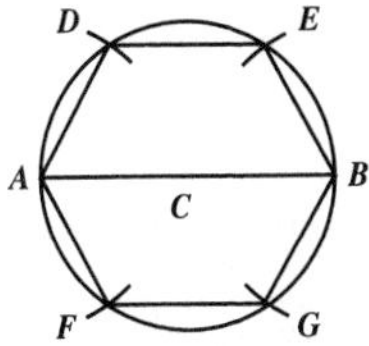

To inscribe a hexagon in a circle:

Draw a diameter *AB*. With *A* and *B* as centers and with the radius of the circle as radius, describe circular arcs intersecting the given circle at *D*, *E*, *F*, and *G*. Draw lines *AD*, *DE*, etc., forming the required hexagon.

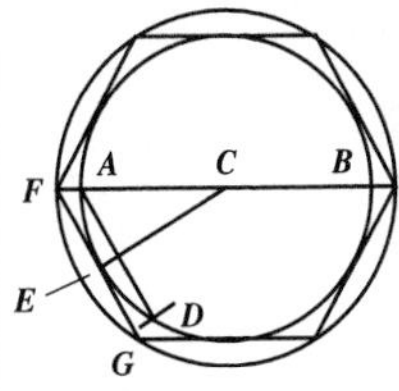

To describe a hexagon about a circle:

Draw a diameter *AB*, and with *A* as a center and the radius of the circle as radius, cut the circumference of the given circle at *D*. Join *AD* and bisect it with radius *CE*. Through *E*, draw *FG* parallel to *AD* and intersecting line *AB* at *F*. With *C* as a center and *CF* as radius, draw a circle. Within this circle, inscribe the hexagon as in the preceding problem.

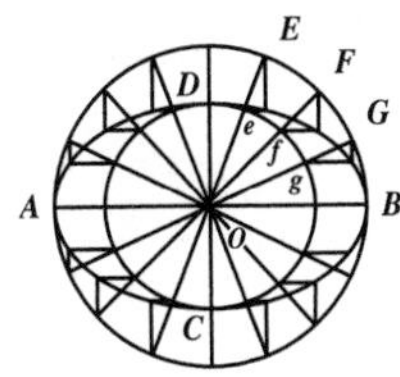

To describe an ellipse with the given axes *AB* and *CD*:

Describe circles with *O* as a center and *AB* and *CD* as diameters. From a number of points, *E*, *F*, *G*, etc., on the outer circle, draw radii intersecting the inner circle at *e*, *f*, and *g*. From *E*, *F*, and *G*, draw lines perpendicular to *AB*, and from *e*, *f*, and *g*, draw lines parallel to *AB*. The intersections of these perpendicular and parallel lines are points on the curve of the ellipse.

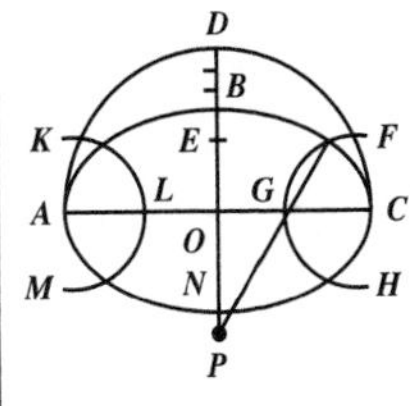

To construct an approximate ellipse by circular arcs:

Let *AC* be the major axis and *BN* the minor. Draw half circle *ADC* with *O* as a center. Divide *BD* into three equal parts and set off *BE* equal to one of these parts. With *A* and *C* as centers and *OE* as radius, describe circular arcs *KLM* and *FGH*; with *G* and *L* as centers, and the same radius, describe arcs *FCH* and *KAM*. Through *F* and *G*, drawn line *FP*, and with *P* as a center, draw the arc *FBK*. Arc *HNM* is drawn in the same manner.

Geometrical Constructions

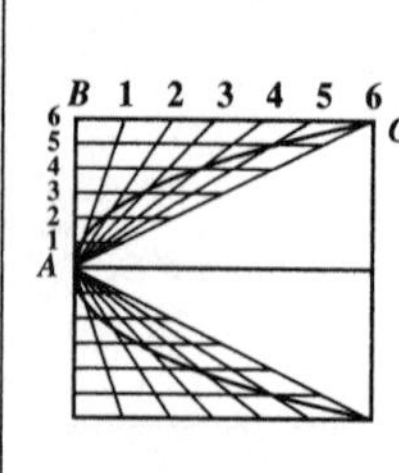

To construct a parabola:

Divide line *AB* into a number of equal parts and divide *BC* into the same number of parts. From the division points on *AB*, draw horizontal lines. From the division points on *BC*, draw lines to point *A*. The points of intersection between lines drawn from points numbered alike are points on the parabola.

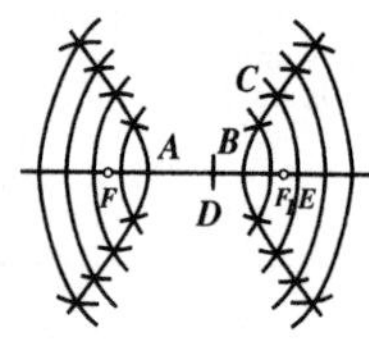

To construct a hyperbola:

From focus *F*, lay off a distance *FD* equal to the transverse axis, or the distance *AB* between the two branches of the curve. With *F* as a center and any distance *FE* greater than *FB* as a radius, describe a circular arc. Then with F_1 as a center and *DE* as a radius, describe arcs intersecting at *C* and *G* the arc just described. *C* and *G* are points on the hyperbola. Any number of points can be found in a similar manner.

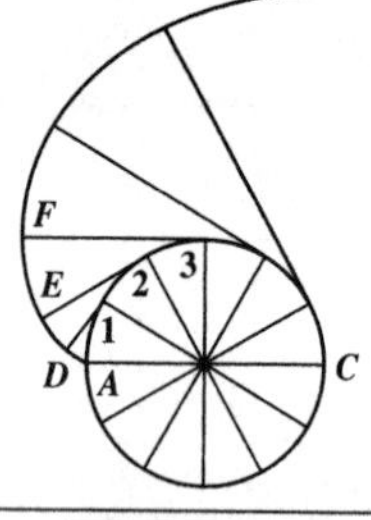

To construct an involute:

Divide the circumference of the base circle *ABC* into a number of equal parts. Through the division points 1, 2, 3, etc., draw tangents to the circle and make the lengths *D*–1, *E*–2, *F*–3, etc., of these tangents equal to the actual length of the arcs *A*–1, *A*–2, *A*–3, etc.

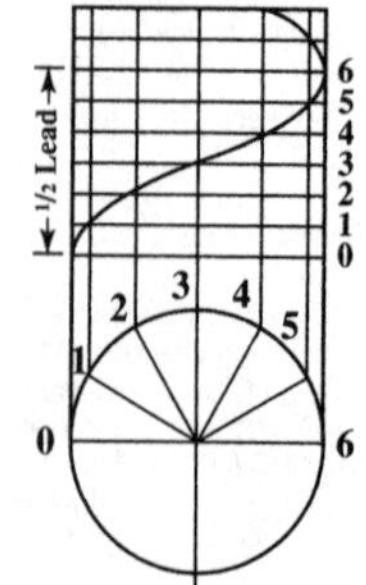

To construct a helix:

Divide half the circumference of the cylinder, on the surface of which the helix is to be described, into a number of equal parts. Divide half the lead of the helix into the same number of equal parts. From the division points on the circle representing the cylinder, draw vertical lines, and from the division points on the lead, draw horizontal lines as shown. The intersections between lines numbered alike are points on the helix.

Solution of Triangles

Any figure bounded by three straight lines is called a triangle. Any one of the three lines may be called the base, and the line drawn perpendicular to the base from the angle opposite it is called the height or altitude of the triangle.

A triangle has three angles, as its name indicates. The sum of the angle measures in any triangle is 180°.

If all three sides of a triangle are of equal length, the triangle is called equilateral. An equilateral triangle is also called equiangular, as all its angles have the same measure. Since the sum of the angle measures in any triangle is 180°, each of the three angles in an equilateral triangle measures 180°/3 = 60°.

If at least two sides of a triangle are of equal length, the triangle is an isosceles triangle. An isosceles triangle has at least two angles of equal measure.

If one angle of a triangle is a right angle, that is, a 90° angle, the triangle is called a right triangle.

Right Triangles.—The right triangle is an important geometric figure in shop calculations. From setting sine bars to spacing holes on a circle, the right triangle is used virtually everywhere in shop work. Depending on the specific application, right triangle calculations be simple or quite detailed.

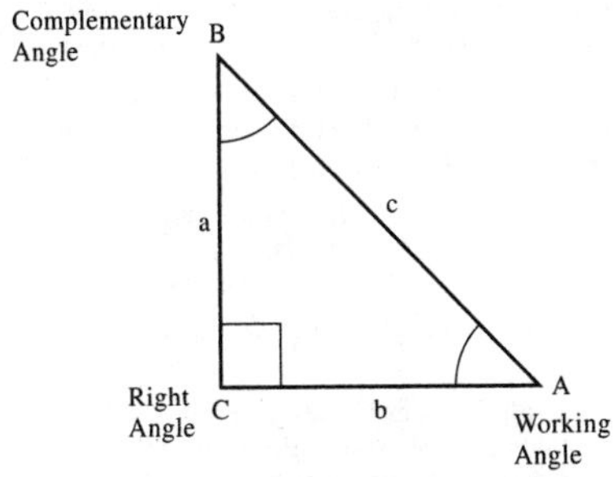

Fig. 1.

Right Angle: The right angle in any right triangle is the only angle that is 90°, regardless of the size of shape of the triangle. The right angle is easy to identify because it forms a square corner. A small square is usually drawn in the right angle to identify it. The right angle is labeled *C* in the Fig. 1.

Working Angle: The working angle of a right triangle is the dimensioned angle, that is the angle used to calculate whatever is to be calculated. When only one angle, say angle $A\angle A$ is called the dimensioned angle, then this is the angle intended on the drawing to be used in all calculations. When both angles are dimensioned, the working angle is the angle used for the calculation. The working angle is labeled *A* in the Fig. 1.

Complementary Angle: The complementary angle of a right triangle is the angle that is not dimensioned. The complementary angle is labeled *B* in the Fig. 1. When both angles are dimensioned, the complementary angle is not normally used for the calculation.

Basic Angle Calculation: When working with the angles of a right triangle, unknown values are relatively easy to calculate. Since the right angle measures 90°, and the sum of all three angles is 180°, the sum of the working angle and complementary angle is 180° − 90° = 90°. For example, a working angle of 30° gives a complementary angle of 60°. (90° − 60° = 30°). If the complementary angle is 44°, then the working angle must be 46° (90° − 44°= 46°).

Just as the angles of a right triangle have different names, so, too, do the sides of the triangle. The names associated with the different sides of the triangle are as follows. (Notice how the sides are labeled with lower case letters corresponding to the angle letters directly opposite them.)

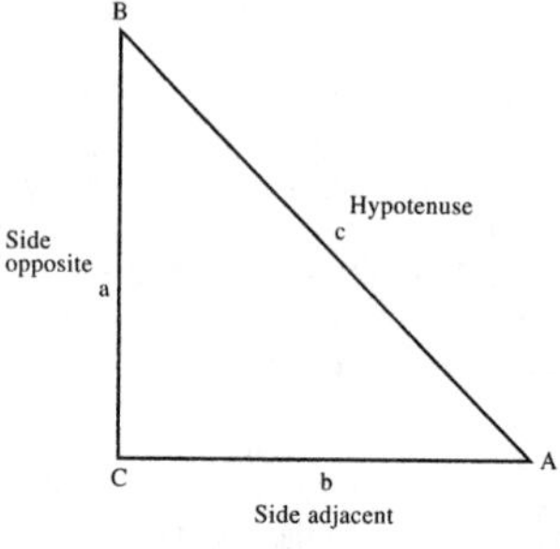

Fig. 2.

Hypotenuse: The hypotenuse is the longest side of any right triangle. It is the side directly opposite the right angle of the triangle. The hypotenuse is labeled c in the Fig. 2.

Side Adjacent: The side adjacent is the side of the triangle that forms the working angle when intersected with the hypotenuse. The term "adjacent" refers to the fact that this side is adjacent to the working angle. The side adjacent is labeled b in the Fig. 2.

Side Opposite: The side opposite is the side of the triangle that forms the complementary angle when intersected with the hypotenuse. The term "opposite" refers to the fact that this side is opposite to the working angle. The side opposite is labeled a in the Fig. 2.

If the designation of the working angle changes, the location of the side adjacent and side opposite also changes.

Trigonometric Relationships of Right Triangle Parts.—The calculations necessary to determine the length or angular value of the various parts of a right triangle are based on the trigonometric relationships of these parts, which exist in all right triangles. There are relationships between the angular parts, the linear parts, and both the angular and linear parts. They are named by their function names, sine, cosine, tangent, cosecant, secant, and cotangent.

Angular Relationships: As we have seen the angle measure of one angle $\angle B$ other than the right angle is related to the angle measure of the other angle $90° - m\angle A = m\angle B$ by the formula or $90° - m\angle A = m\angle B$.

Linear Relationships: If the lengths of two sides of a right triangle are known, the length of the third side can be determined by the Pythagorean Theorem. In the figure, the relationship among the sides is: $a^2 + b^2 = c^2$

Example: If $a = 6$ inches and $b = 8$ inches, then c is found by $6^2 + 8^2 = c^2$ or $c = \sqrt{6^2 + 8^2} = 10$ inch

Angular/Linear Relationships.—If the lengths of any two sides of the right triangle are known, the measure of either the working or complementary angle can be found by using the trigonometric relationships and the Table of Trigonometric Functions. Likewise, if the measure of either the working or complementary angle and the length of one side are known, then the length of either of the other two sides can be found also by the methods of trigonometry.

Trigonometric Functions.—The trigonometric functions give the relationship of the working angle of a right triangle to the sides of the triangle. The relationships described by the functions are as follows:

sine: The ratio of the length of the side opposite the working angle A to the length of the hypotenuse. It is written as "sin A."

cosine: The ratio of the length of the side adjacent to the working angle to the length of the hypotenuse. It is written as "cos A."

tangent: The ratio of the length of the side opposite the working angle to the length of the side adjacent to it. It is written as "tan A."

These three relationships are summed up as follows:

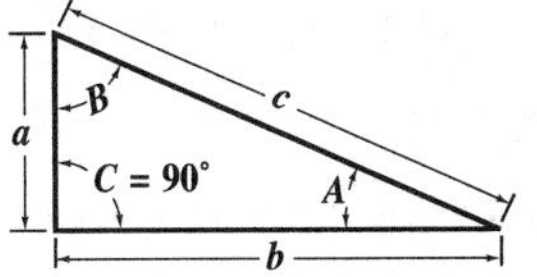

$$\sin A = \frac{\text{Length of side opposite}}{\text{Length of hypotenuse}} = \frac{a}{c}$$

$$\cos A = \frac{\text{Length of side adjacent}}{\text{Length of hypotenuse}} = \frac{b}{c}$$

$$\tan A = \frac{\text{Length of side opposite}}{\text{Length of side adjacent}} = \frac{a}{b}$$

where a, b, c are the sides as shown in the diagram.

The next three trigonometric functions are simply the reciprocal ratios of the first three functions:

cotangent: The ratio of the length of the side adjacent the working angle to the length of the side opposite it. (The reciprocal of the tangent.)

secant: The ratio of the length of the hypotenuse to the side adjacent the working angle. (The reciprocal of the cosine.)

cosecant: The ratio of the length of the hypotenuse to the side opposite the working angle. (The reciprocal of the sine.)

Although there are six different functions, the sine, cosine, and tangent functions are the most frequently used.

The first letter of each word in the simple three-part equations or formulas of the trigonometric functions can be used to make memorization of the formulas easy. First, just the words describing the functions are written:

sine	opposite/hypotenuse
cosine	adjacent/hypotenuse
tangent	opposite/adjacent

Then the names of the functions and the sides are abbreviated to their first letters, reducing the formulas to:

s	o/h
c	a/h
t	o/a

or "sohcahtoa" (pronounced "so-ca-to-a."). Be careful not to confuse these abbreviations with the mathematical variables representing the lengths of the sides, namely a, b, and c.

Examples of how to apply trigonometric functions to solving for unknown parts of a right triangle problem follows.

Example: The measure of working angle A is sought. The lengths of the side opposite and the hypotenuse are known: $a = 8$ inches and $c = 10$ inches. The measure of $\angle A$ is found using the sine relationship as follows:

$\sin A = 8$ in./10 in. = 0.800. Notice that the units cancel. This is important since ratios of right triangle sides are unitless.

Trigonometrical Functions of Angles: The *sine* of an angle equals the opposite side divided by the hypotenuse. Hence, $\sin B = b/c$, and $\sin A = a/c$.

The *cosine* of an angle equals the adjacent side divided by the hypotenuse. Hence, $\cos B = a/c$, and $\sin B = b/c$.

The *tangent* of an angle equals the opposite side divided by the adjacent side. Hence, $\tan B = b/a$, and $\tan A = a/b$.

The *cotangent* of an angle equals the adjacent side divided by the opposite side. Hence, $\cot B = a/b$, and $\cot A = b/a$.

The *secant* of an angle equals the hypotenuse side divided by the adjacent side. Hence, $\sec B = c/a$, and $\sec A = c/b$.

The *cosecant* of an angle equals the hypotenuse side divided by the opposite side. Hence, $\operatorname{cosec} B = c/b$, and $\operatorname{cosec} A = c/a$.

It should be noted that the functions of the angles can be found in this manner only when the triangle is right-angled.

It is now known that $\sin A = 0.800$. To find the value of $\angle A$, refer to the Table of *Trigonometric Functions* in the following manner.

First note that the table lists every angle from 0° to 180° in 1-minute increments. Each page of the table is devoted to one degree and lists values for the degree and all its 60-minute subdivisions. The top and bottom rows of the table have the names of the functions. The values in the columns are the ratios of the lengths of the sides that is, the values of the trig functions for the angles being considered. The most common method of reading these tables is to read the values of the trig functions of any angle from 0° to 90° from left to right under the names of the functions at the top of the page. Values of the trig functions of angles from 90° to 180° are read from right to left above the names of the functions at the bottom of the page.

To find the angle measurement for A, the value 0.8000 or one very close to it is located in the table by looking under the sine column. The sine column does not reach this value when read top to bottom, so the sine column is then followed from bottom to top. The closest value is 0.8003, which corresponds to angle measure of 52 degrees, 8 minutes. Hence $\sin A = 0.800$ means $\angle A = 52° 8'$.

Example: Given the angle 27°, find sin A. In the Table of *Trigonometric Functions* the sin27° is found by reading across from 27° in the sine column. It is thus found that $\sin 27° = 0.45399$. This means that in any right triangle with the working angle measuring 27°, the ratio of the side opposite to the hypotenuse is 0.45399.

If the values of the angle measure and the length of one side is known, the length of either of the other sides can be found by solving the appropriate trigonometric function for the missing length.

Example: If $\angle A = 30°$ and the length of the side opposite measures 6 units, the length of the hypotenuse c is found by the calculation: $\sin 30° = 6/c$, which is solved for c by rearranging the equation to read $c = 6 / \sin 30° = 6 / 0.500 = 12$ units

From this result we can find the length of the adjacent side a by using the Pythagorean theorm $a^2 + b^2 = c^2$, $a^2 + 6^2\text{units}^2 = c^2\text{units}^2$ or

$$a = \sqrt{12^2 - 6^2}\text{units} = \sqrt{108}\text{ units} = 10.4\text{units}$$

Alternately, the length of adjacent side is found by using the cosine relationship: $\cos 30° = a/12$ units or $a = 12$ units $\times \cos 30° = 12$ units $\times 0.86603 = 10.4$ units

This method applies to all the trigonometric functions. More examples of right angle solutions are given in *Examples of the Solution of Right-Angled Triangles (English and metric units)* starting on page 107.

Chart For The Rapid Solution of Right-Angle and Oblique-Angle Triangles

$C = \sqrt{A^2 - B^2}$	$\sin d = \frac{B}{A}$	$e = 90° - d$	$B = \sqrt{A^2 - C^2}$
$\sin e = \frac{C}{A}$	$d = 90° - e$	$A = \sqrt{B^2 + C^2}$	$\tan d = \frac{B}{C}$
$B = A \times \sin d$	$C = A \times \cos d$	$B = A \times \cos e$	$C = A \times \sin e$
$A = \frac{B}{\sin d}$	$C = B \times \cot d$	$A = \frac{B}{\cot e}$	$C = B \times \tan e$
$A = \frac{C}{\cos d}$	$A = \frac{C}{\cos d}$	$A = \frac{C}{\sin e}$	$B = C \times \cot e$
$B = \frac{A \times \sin f}{\sin d}$	$e = 180° - (d + f)$	$C = \frac{A \times \sin e}{\sin d}$	$\tan d = \frac{A \times \sin e}{B - A \times \cos e}$
$f = 180° - (d + e)$	$C = \frac{A \times \sin e}{\sin d}$	$\sin f = \frac{B \times \sin d}{A}$	$e = 180° - (d + f)$
$\text{Area} = \frac{A \times B \times \sin e}{2}$	$\cos d = \frac{B^2 + C^2 - A^2}{2 \times B \times C}$	$\sin f = \frac{B \times \sin d}{A}$	$e = 180° - (d + f)$

Solution of Right-Angled Triangles

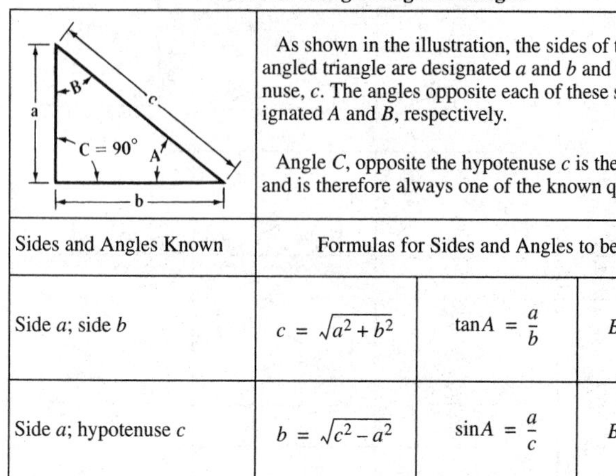

As shown in the illustration, the sides of the right-angled triangle are designated a and b and the hypotenuse, c. The angles opposite each of these sides are designated A and B, respectively.

Angle C, opposite the hypotenuse c is the right angle, and is therefore always one of the known quantities.

Sides and Angles Known	Formulas for Sides and Angles to be Found		
Side a; side b	$c = \sqrt{a^2 + b^2}$	$\tan A = \frac{a}{b}$	$B = 90° - A$
Side a; hypotenuse c	$b = \sqrt{c^2 - a^2}$	$\sin A = \frac{a}{c}$	$B = 90° - A$
Side b; hypotenuse c	$a = \sqrt{c^2 - b^2}$	$\sin B = \frac{b}{c}$	$A = 90° - B$
Hypotenuse c; angle B	$b = c \times \sin B$	$a = c \times \cos B$	$A = 90° - B$
Hypotenuse c; angle A	$b = c \times \cos A$	$a = c \times \sin A$	$B = 90° - A$
Side b; angle B	$c = \frac{b}{\sin B}$	$a = b \times \cot B$	$A = 90° - B$
Side b; angle A	$c = \frac{b}{\cos A}$	$a = b \times \tan A$	$B = 90° - A$
Side a; angle B	$c = \frac{a}{\cos B}$	$b = a \times \tan B$	$A = 90° - B$
Side a; angle A	$c = \frac{a}{\sin A}$	$b = a \times \cot A$	$B = 90° - A$

Examples of the Solution of Right-Angled Triangles (English and metric units)

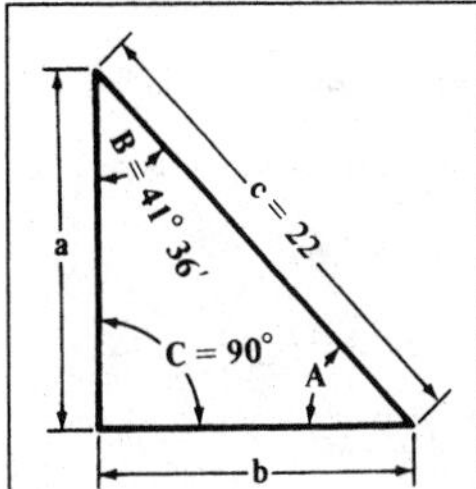

Hypotenuse and one angle known

$c = 22$ inches; $B = 41°\ 36'$.

$$a = c \times \cos B = 22 \times \cos 41^\circ 36' = 22 \times 0.74780$$
$$= 16.4516 \text{ inches}$$
$$b = c \times \sin B = 22 \times \sin 41^\circ 36' = 22 \times 0.66393$$
$$= 14.6065 \text{ inches}$$
$$A = 90^\circ - B = 90^\circ - 41^\circ 36' = 48^\circ 24'$$

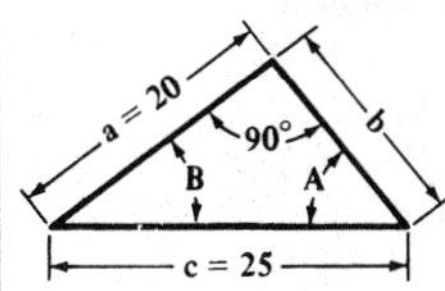

Hypotenuse and one side known

$c = 25$ centimeters; $a = 20$ centimeters.

$$b = \sqrt{c^2 - a^2} = \sqrt{25^2 - 20^2} = \sqrt{625 - 400}$$
$$= \sqrt{225} = 15 \text{ centimeters}$$
$$\sin A = \frac{a}{c} = \frac{20}{25} = 0.8$$

Hence,
$$A = 53°8'$$
$$B = 90° - A = 90° - 53°8' = 36°52'$$

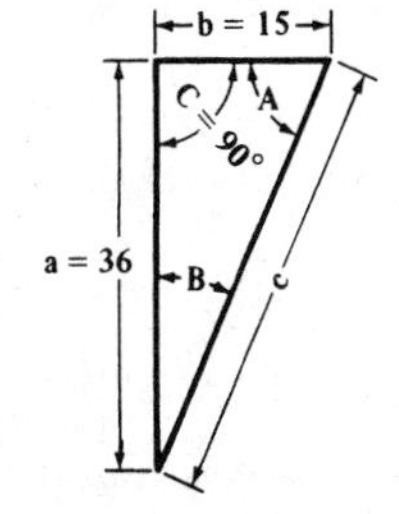

Two sides known

$a = 36$ inches; $b = 15$ inches.

$$c = \sqrt{a^2 + b^2} = \sqrt{36^2 + 15^2} = \sqrt{1296 + 225}$$
$$= \sqrt{1521} = 39 \text{ inches}$$
$$\tan A = \frac{a}{b} = \frac{36}{15} = 2.4$$

Hence,
$$A = 67^\circ 23'$$
$$B = 90^\circ - A = 90^\circ - 67^\circ 23' = 22^\circ 37'$$

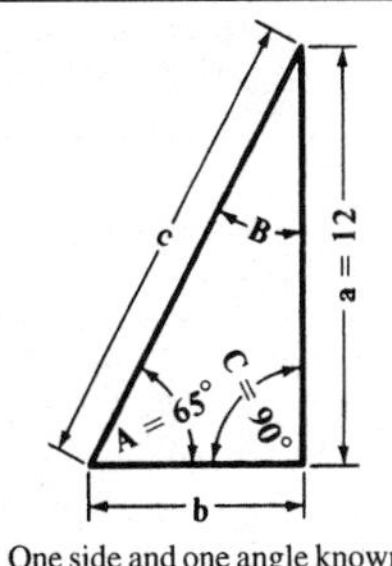

One side and one angle known

$a = 12$ meters; $A = 65°$.

$$c = \frac{a}{\sin A} = \frac{12}{\sin 65^\circ} = \frac{12}{0.90631} = 13.2405 \text{ meters}$$
$$b = a \times \cot A = 12 \times \cot 65^\circ = 12 \times 0.46631$$
$$= 5.5957 \text{ meters}$$
$$B = 90^\circ - A = 90^\circ - 65^\circ = 25^\circ$$

Solution of Oblique-Angled Triangles

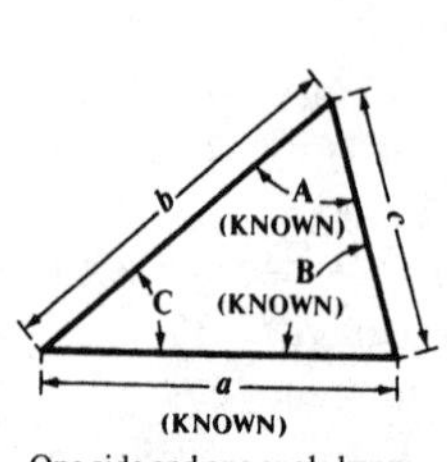

One side and one angle known

Call the known side a, the angle opposite it A, and the other known angle B. Then:

$C = 180° - (A + B)$; or if angles B and C are given, but not A, then $A = 180° - (B + C)$.

$$C = 180^\circ - (A + B)$$

$$b = \frac{a \times \sin B}{\sin A} \qquad c = \frac{a \times \sin C}{\sin A}$$

$$\text{Area} = \frac{a \times b \times \sin C}{2}$$

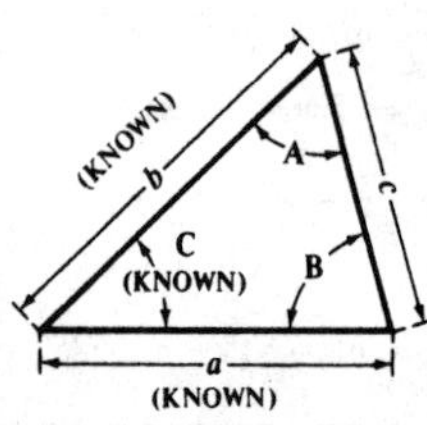

Two sides and the angle between them known

Call the known sides a and b, and the known angle between them C. Then:

$$\tan A = \frac{a \times \sin C}{b - (a \times \cos C)}$$

$$B = 180^\circ - (A + C) \qquad c = \frac{a \times \sin C}{\sin A}$$

Side c may also be found directly as below:

$$c = \sqrt{a^2 + b^2 - (2ab \times \cos C)}$$

$$\text{Area} = \frac{a \times b \times \sin C}{2}$$

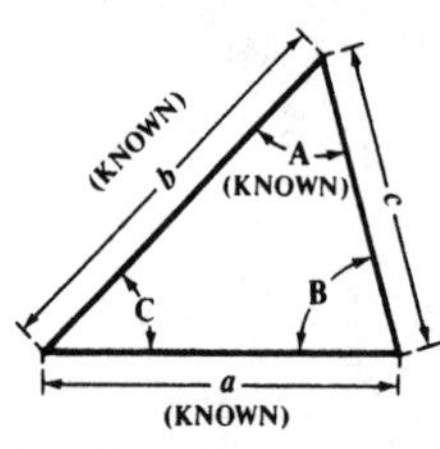

Two sides and the angle opposite one of the sides known

Call the known angle A, the side opposite it a, and the other known side b. Then:

$$\sin B = \frac{b \times \sin A}{a} \qquad C = 180° - (A + B)$$

$$c = \frac{a \times \sin C}{\sin A} \qquad \text{Area} = \frac{a \times b \times \sin C}{2}$$

If, in the above, angle $B >$ angle A but $<90°$, then a second solution B_2, C_2, c_2 exists for which: $B_2 = 180° - B$; $C_2 = 180° - (A + B_2)$; $c_2 = (a \times \sin C_2) \div \sin A$; area $= (a \times b \times \sin C_2) \div 2$. If $a \geq b$, then the first solution only exists. If $a < b \times \sin A$, then no solution exists.

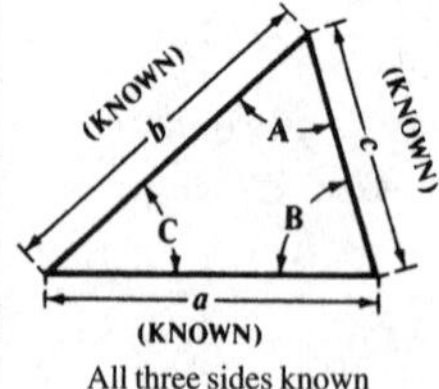

All three sides known

Call the sides a, b, and c, and the angles opposite them, A, B, and C. Then:

$$\cos A = \frac{b^2 + c^2 - a^2}{2bc} \qquad \sin B = \frac{b \times \sin A}{a}$$

$$C = 180° - (A + B) \qquad \text{Area} = \frac{a \times b \times \sin C}{2}$$

Examples of the Solution of Oblique-Angled Triangles (English and metric units)

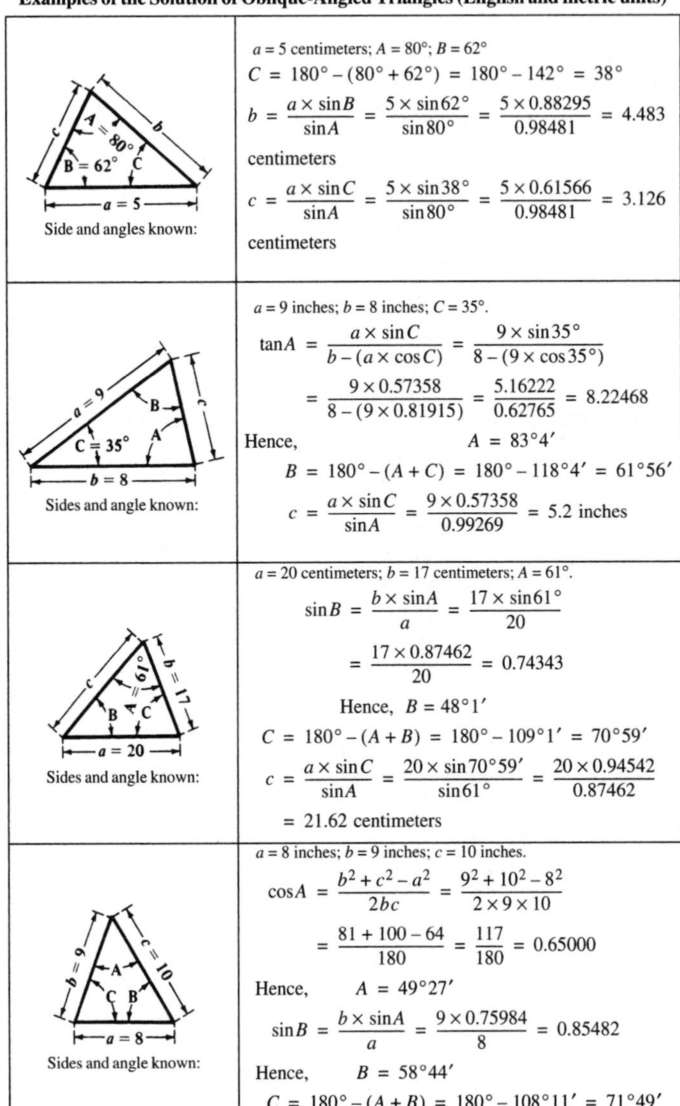

Side and angles known:	$a = 5$ centimeters; $A = 80°$; $B = 62°$ $C = 180° - (80° + 62°) = 180° - 142° = 38°$ $b = \frac{a \times \sin B}{\sin A} = \frac{5 \times \sin 62°}{\sin 80°} = \frac{5 \times 0.88295}{0.98481} = 4.483$ centimeters $c = \frac{a \times \sin C}{\sin A} = \frac{5 \times \sin 38°}{\sin 80°} = \frac{5 \times 0.61566}{0.98481} = 3.126$ centimeters
Sides and angle known:	$a = 9$ inches; $b = 8$ inches; $C = 35°$. $\tan A = \frac{a \times \sin C}{b - (a \times \cos C)} = \frac{9 \times \sin 35°}{8 - (9 \times \cos 35°)}$ $= \frac{9 \times 0.57358}{8 - (9 \times 0.81915)} = \frac{5.16222}{0.62765} = 8.22468$ Hence, $A = 83°4'$ $B = 180° - (A + C) = 180° - 118°4' = 61°56'$ $c = \frac{a \times \sin C}{\sin A} = \frac{9 \times 0.57358}{0.99269} = 5.2$ inches
Sides and angle known:	$a = 20$ centimeters; $b = 17$ centimeters; $A = 61°$. $\sin B = \frac{b \times \sin A}{a} = \frac{17 \times \sin 61°}{20}$ $= \frac{17 \times 0.87462}{20} = 0.74343$ Hence, $B = 48°1'$ $C = 180° - (A + B) = 180° - 109°1' = 70°59'$ $c = \frac{a \times \sin C}{\sin A} = \frac{20 \times \sin 70°59'}{\sin 61°} = \frac{20 \times 0.94542}{0.87462}$ $= 21.62$ centimeters
Sides and angle known:	$a = 8$ inches; $b = 9$ inches; $c = 10$ inches. $\cos A = \frac{b^2 + c^2 - a^2}{2bc} = \frac{9^2 + 10^2 - 8^2}{2 \times 9 \times 10}$ $= \frac{81 + 100 - 64}{180} = \frac{117}{180} = 0.65000$ Hence, $A = 49°27'$ $\sin B = \frac{b \times \sin A}{a} = \frac{9 \times 0.75984}{8} = 0.85482$ Hence, $B = 58°44'$ $C = 180° - (A + B) = 180° - 108°11' = 71°49'$

Conversion Tables of Angular Measure.—The accompanying tables of degrees , minutes, and seconds into radians; radians into degrees, minutes, and seconds; radians into degrees and decimals of a degree; and minutes and seconds into decimals of a degree and vice versa facilitate the conversion of measurements.

Example: The Degrees, Minutes, and Seconds into Radians table is used to find the number of radians in 324 degrees, 25 minutes, 13 seconds as follows:

300 degrees	= 5.235988 radians
20 degrees	= 0.349066 radian
4 degrees	= 0.069813 radian
25 minutes	= 0.007272 radian
13 seconds	= 0.000063 radian
324°25′13″	= 5.662202 radians

Example: The Radians into Degrees and Decimals of a Degree, and Radians into Degrees, Minutes and Seconds tables are used to find the number of decimal degrees or degrees, minutes and seconds in 0.734 radian as follows:

0.7 radian =	40.1070 degrees	0.7 radian =	40° 6′25″
0.03 radian =	1.7189 degrees	0.03 radian =	1°43′8″
0.004 radian =	0.2292 degree	0.004 radian =	0°13′45″
0.734 radian =	42.0551 degrees	0.734 radian =	41°62′78″ or 42°3′18″

Degrees, Minutes, and Seconds into Radians (Based on 180 degrees = π radians)

Degrees into Radians											
Deg.	Rad.	Deg.	Rad.	Deg.	Rad.	Deg.	Rad.	Deg.	Rad.	Deg.	Rad.
1000	17.453293	100	1.745329	10	0.174533	1	0.017453	0.1	0.001745	0.01	0.000175
2000	34.906585	200	3.490659	20	0.349066	2	0.034907	0.2	0.003491	0.02	0.000349
3000	52.359878	300	5.235988	30	0.523599	3	0.052360	0.3	0.005236	0.03	0.000524
4000	69.813170	400	6.981317	40	0.698132	4	0.069813	0.4	0.006981	0.04	0.000698
5000	87.266463	500	8.726646	50	0.872665	5	0.087266	0.5	0.008727	0.05	0.000873
6000	104.719755	600	10.471976	60	1.047198	6	0.104720	0.6	0.010472	0.06	0.001047
7000	122.173048	700	12.217305	70	1.221730	7	0.122173	0.7	0.012217	0.07	0.001222
8000	139.626340	800	13.962634	80	1.396263	8	0.139626	0.8	0.013963	0.08	0.001396
9000	157.079633	900	15.707963	90	1.570796	9	0.157080	0.9	0.015708	0.09	0.001571
10000	174.532925	1000	17.453293	100	1.745329	10	0.174533	1.0	0.017453	0.10	0.001745
Minutes into Radians											
Min.	Rad.	Min.	Rad.	Min.	Rad.	Min.	Rad.	Min.	Rad.	Min.	Rad.
1	0.000291	11	0.003200	21	0.006109	31	0.009018	41	0.011926	51	0.014835
2	0.000582	12	0.003491	22	0.006400	32	0.009308	42	0.012217	52	0.015126
3	0.000873	13	0.003782	23	0.006690	33	0.009599	43	0.012508	53	0.015417
4	0.001164	14	0.004072	24	0.006981	34	0.009890	44	0.012799	54	0.015708
5	0.001454	15	0.004363	25	0.007272	35	0.010181	45	0.013090	55	0.015999
6	0.001745	16	0.004654	26	0.007563	36	0.010472	46	0.013381	56	0.016290
7	0.002036	17	0.004945	27	0.007854	37	0.010763	47	0.013672	57	0.016581
8	0.002327	18	0.005236	28	0.008145	38	0.011054	48	0.013963	58	0.016872
9	0.002618	19	0.005527	29	0.008436	39	0.011345	49	0.014254	59	0.017162
10	0.002909	20	0.005818	30	0.008727	40	0.011636	50	0.014544	60	0.017453
Seconds into Radians											
Sec.	Rad.	Sec.	Rad.	Sec.	Rad.	Sec.	Rad.	Sec.	Rad.	Sec.	Rad.
1	0.000005	11	0.000053	21	0.000102	31	0.000150	41	0.000199	51	0.000247
2	0.000010	12	0.000058	22	0.000107	32	0.000155	42	0.000204	52	0.000252
3	0.000015	13	0.000063	23	0.000112	33	0.000160	43	0.000208	53	0.000257
4	0.000019	14	0.000068	24	0.000116	34	0.000165	44	0.000213	54	0.000262
5	0.000024	15	0.000073	25	0.000121	35	0.000170	45	0.000218	55	0.000267
6	0.000029	16	0.000078	26	0.000126	36	0.000175	46	0.000223	56	0.000271
7	0.000034	17	0.000082	27	0.000131	37	0.000179	47	0.000228	57	0.000276
8	0.000039	18	0.000087	28	0.000136	38	0.000184	48	0.000233	58	0.000281
9	0.000044	19	0.000092	29	0.000141	39	0.000189	49	0.000238	59	0.000286
10	0.000048	20	0.000097	30	0.000145	40	0.000194	50	0.000242	60	0.000291

Radians into Degrees and Decimals of a Degree
(Based on π radians = 180 degrees)

Rad.	Deg.	Rad.	Deg.	Rad.	Deg.	Rad.	Deg.	Rad.	Deg.	Rad.	Deg.
10	572.9578	1	57.2958	0.1	5.7296	0.01	0.5730	0.001	0.0573	0.0001	0.0057
20	1145.9156	2	114.5916	0.2	11.4592	0.02	1.1459	0.002	0.1146	0.0002	0.0115
30	1718.8734	3	171.8873	0.3	17.1887	0.03	1.7189	0.003	0.1719	0.0003	0.0172
40	2291.8312	4	229.1831	0.4	22.9183	0.04	2.2918	0.004	0.2292	0.0004	0.0229
50	2864.7890	5	286.4789	0.5	28.6479	0.05	2.8648	0.005	0.2865	0.0005	0.0286
60	3437.7468	6	343.7747	0.6	34.3775	0.06	3.4377	0.006	0.3438	0.0006	0.0344
70	4010.7046	7	401.0705	0.7	40.1070	0.07	4.0107	0.007	0.4011	0.0007	0.0401
80	4583.6624	8	458.3662	0.8	45.8366	0.08	4.5837	0.008	0.4584	0.0008	0.0458
90	5156.6202	9	515.6620	0.9	51.5662	0.09	5.1566	0.009	0.5157	0.0009	0.0516
100	5729.5780	10	572.9578	1.0	57.2958	0.10	5.7296	0.010	0.5730	0.0010	0.0573

Radians into Degrees, Minutes, and Seconds
(Based on π radians = 180 degrees)

Rad.	Angle	Rad.	Angle	Rad.	Angle	Rad.	Angle	Rad.	Angle	Rad.	Angle
10	572°57′28″	1	57°17′45″	0.1	5°43′46″	0.01	0°34′23″	0.001	0°3′26″	0.0001	0°0′21″
20	1145°54′56″	2	114°35′30″	0.2	11°27′33″	0.02	1°8′45″	0.002	0°6′53″	0.0002	0°0′41″
30	1718°52′24″	3	171°53′14″	0.3	17°11′19″	0.03	1°43′8″	0.003	0°10′19″	0.0003	0°1′ 2″
40	2291°49′52″	4	229°10′59″	0.4	22°55′6″	0.04	2°17′31″	0.004	0°13′45″	0.0004	0°1′23″
50	2864°47′20″	5	286°28′44″	0.5	28°38′52″	0.05	2°51′53″	0.005	0°17′11″	0.0005	0°1′43″
60	3437°44′48″	6	343°46′29″	0.6	34°22′39″	0.06	3°26′16″	0.006	0°20′38″	0.0006	0°2′ 4″
70	4010°42′16″	7	401°4′14″	0.7	40°6′25″	0.07	4°0′39″	0.007	0°24′4″	0.0007	0°2′24″
80	4583°39′44″	8	458°21′58″	0.8	45°50′12″	0.08	4°35′1″	0.008	0°27′30″	0.0008	0°2′45″
90	5156°37′13″	9	515°39′43″	0.9	51°33′58″	0.09	5°9′24″	0.009	0°30′56″	0.0009	0°3′6″
100	5729°34′41″	10	572°57′28″	1.0	57°17′45″	0.10	5°43′46″	0.010	0°34′23″	0.0010	0°3′26″

Minutes and Seconds into Decimal of a Degree and Vice Versa
(Based on 1 second = 0.00027778 degree)

Minutes into Decimals of a Degree						Seconds into Decimals of a Degree					
Min.	Deg.	Min.	Deg.	Min.	Deg.	Sec.	Deg.	Sec.	Deg.	Sec.	Deg.
1	0.0167	21	0.3500	41	0.6833	1	0.0003	21	0.0058	41	0.0114
2	0.0333	22	0.3667	42	0.7000	2	0.0006	22	0.0061	42	0.0117
3	0.0500	23	0.3833	43	0.7167	3	0.0008	23	0.0064	43	0.0119
4	0.0667	24	0.4000	44	0.7333	4	0.0011	24	0.0067	44	0.0122
5	0.0833	25	0.4167	45	0.7500	5	0.0014	25	0.0069	45	0.0125
6	0.1000	26	0.4333	46	0.7667	6	0.0017	26	0.0072	46	0.0128
7	0.1167	27	0.4500	47	0.7833	7	0.0019	27	0.0075	47	0.0131
8	0.1333	28	0.4667	48	0.8000	8	0.0022	28	0.0078	48	0.0133
9	0.1500	29	0.4833	49	0.8167	9	0.0025	29	0.0081	49	0.0136
10	0.1667	30	0.5000	50	0.8333	10	0.0028	30	0.0083	50	0.0139
11	0.1833	31	0.5167	51	0.8500	11	0.0031	31	0.0086	51	0.0142
12	0.2000	32	0.5333	52	0.8667	12	0.0033	32	0.0089	52	0.0144
13	0.2167	33	0.5500	53	0.8833	13	0.0036	33	0.0092	53	0.0147
14	0.2333	34	0.5667	54	0.9000	14	0.0039	34	0.0094	54	0.0150
15	0.2500	35	0.5833	55	0.9167	15	0.0042	35	0.0097	55	0.0153
16	0.2667	36	0.6000	56	0.9333	16	0.0044	36	0.0100	56	0.0156
17	0.2833	37	0.6167	57	0.9500	17	0.0047	37	0.0103	57	0.0158
18	0.3000	38	0.6333	58	0.9667	18	0.0050	38	0.0106	58	0.0161
19	0.3167	39	0.6500	59	0.9833	19	0.0053	39	0.0108	59	0.0164
20	0.3333	40	0.6667	60	1.0000	20	0.0056	40	0.0111	60	0.0167

Example 1: Convert 11′37″ to decimals of a degree. From the left table, 11′ = 0.1833 degree. From the right table, 37″ = 0.0103 degree. Adding, 11′37″ = 0.1833 + 0.0103 = 0.1936 degree.

Example 2: Convert 0.1234 degree to minutes and seconds. From the left table, 0.1167 degree = 7′. Subtracting 0.1167 from 0.1234 gives 0.0067. From the right table, 0.0067 = 24″ so that 0.1234 = 7′24″.

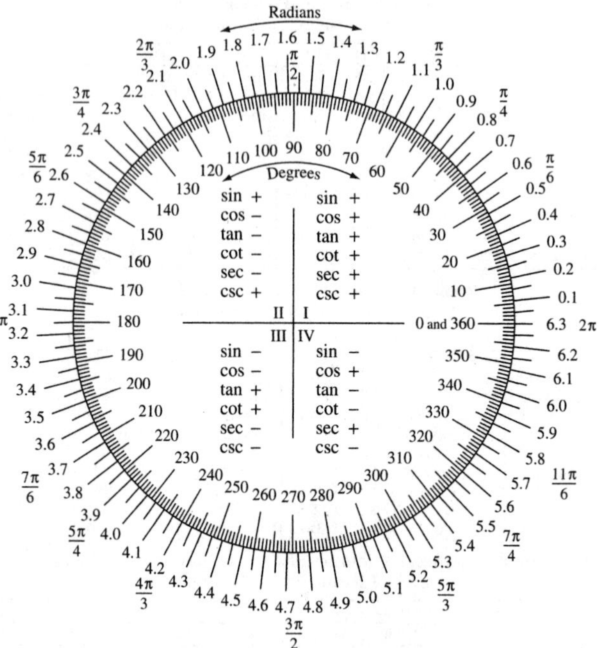

Signs of Trigonometric Functions, Fractions of π, and Degree–Radian Conversion

Tables of Trigonometric Functions.—The trigonometric (trig) tables on the following pages give numerical values for sine, cosine, tangent, and cotangent functions of angles from 0 to 90 degrees. Function values for all other angles can be obtained from the tables by applying the rules for signs of trigonometric functions and the useful relationships among angles given in the following. Secant and cosecant functions can be found from sec $A = 1/\cos A$ and csc $A = 1/\sin A$.

The trig tables are divided in half by a double line. The body of each half table consists of four labeled columns of data between columns listing angles. The angles listed to the left of the data increase, moving down the table, and angles listed to the right of the data increase, moving up the table. Labels above the data identify the trig functions corresponding to angles listed in the left column of each half table. Labels below the data correspond to angles listed in the right column of each half table. To find the value of a function for a particular angle, first locate the angle in the table, then find the appropriate function label across the top or bottom row of the table, and find the function value at the intersection of the angle row and label column. Angles opposite each other are complementary angles (i.e., their sum equals 90°) and are related. For example, sin 10° = cos 80° and cos 10° = sin 80°.

All the trig functions of angles between 0° and 90° have positive values. For other angles, consult the chart below to find the sign of the function in the quadrant where the angle is located. To determine trig functions of angles greater than 90° subtract 90, 180, 270, or 360 from the angle to get an angle less than 90° and use Table to find the equivalent first-quadrant function and angle to look up in the trig tables.

Useful Relationships Among Angles

Angle Function	θ	−θ	90° ± θ	180° ± θ	270° ± θ	360° ± θ
sin	sin θ	−sin θ	+cos θ	∓sin θ	−cos θ	±sin θ
cos	cos θ	+cos θ	∓sin θ	−cos θ	±sin θ	+cos θ
tan	tan θ	−tan θ	∓cot θ	±tan θ	∓cot θ	±tan θ
cot	cot θ	−cot θ	∓tan θ	±cot θ	∓tan θ	±cot θ
sec	sec θ	+sec θ	∓csc θ	−sec θ	±csc θ	+sec θ
csc	csc θ	−csc θ	+sec θ	∓csc θ	−sec θ	±csc θ

Examples: cos (270° − θ) = −sin θ; tan (90° + θ) = −cot θ.

Example: Find the cosine of 336°40′. The diagram in *Signs of Trigonometric Functions, Fractions of p, and Degree–Radian Conversion* shows that the cosine of every angle in Quadrant IV (270° to 360°) is positive. To find the angle and trig function to use when entering the trig table, subtract 270 from 336 to get cos 336°40′ = cos (270° + 66°40′) and then find the intersection of the cos row and the 270 ± θ column in Table . Because cos (270 ± θ) in the fourth quadrant is equal to ± sin θ in the first quadrant, find sin 66°40′ in the trig table. Therefore, cos 336°40′ = sin 66°40′ = 0.918216.

The Law of Sines.—In any triangle, any side is to the sine of the angle opposite that side as any other side is to the sine of the angle opposite that side. If *a*, *b*, and *c* are the sides, and *A*, *B*, and *C* their opposite angles, respectively, then:

$$\frac{a}{\sin A} = \frac{b}{\sin B} = \frac{c}{\sin C}, \quad \text{so that:}$$

$$a = \frac{b\sin A}{\sin B} \quad \text{or} \quad a = \frac{c\sin A}{\sin C}$$

$$b = \frac{a\sin B}{\sin A} \quad \text{or} \quad b = \frac{c\sin B}{\sin C}$$

$$c = \frac{a\sin C}{\sin A} \quad \text{or} \quad c = \frac{b\sin C}{\sin B}$$

The Law of Cosines.—In any triangle, the square of any side is equal to the sum of the squares of the other two sides minus twice their product times the cosine of the included angle; or if *a*, *b* and *c* are the sides and *A*, *B*, and *C* are the opposite angles, respectively, then:

$$a^2 = b^2 + c^2 - 2bc\cos A$$

$$b^2 = a^2 + c^2 - 2ac\cos B$$

$$c^2 = a^2 + b^2 - 2ab\cos C$$

These two laws, together with the proposition that the sum of the three angles equals 180 degrees, are the basis of all formulas relating to the solution of triangles.

Formulas for the solution of right-angled and oblique-angled triangles, arranged in tabular form, are given on the following pages.

Signs of Trigonometric Functions.—The diagram, *Signs of Trigonometric Functions, Fractions of p, and Degree–Radian Conversion* on page 112, shows the proper sign (+ or −) for the trigonometric functions of angles in each of the four quadrants, 0 to 90, 90 to 180, 180 to 270, and 270 to 360 degrees. Thus, the cosine of an angle between 90 and 180 degrees is negative; the sine of the same angle is positive.

Trigonometric Identities.—Trigonometric identities are formulas that show the relationship between different trigonometric functions. They may be used to change the form of some trigonometric expressions to simplify calculations. For example, if a formula has a term, 2sin*A*cos*A*, the equivalent but simpler in terms sin2*A* may be substituted. The identities that follow may themselves be combined or rearranged in various ways to form new identities.

Basic

$$\tan A = \frac{\sin A}{\cos A} = \frac{1}{\cot A} \qquad \sec A = \frac{1}{\cos A} \qquad \csc A = \frac{1}{\sin A}$$

Negative Angle

$$\sin(-A) = -\sin A \qquad \cos(-A) = \cos A \qquad \tan(-A) = -\tan A$$

Pythagorean

$$\sin^2 A + \cos^2 A = 1 \qquad 1 + \tan^2 A = \sec^2 A \qquad 1 + \cot^2 A = \csc^2 A$$

Sum and Difference of Angles

$$\tan(A+B) = \frac{\tan A + \tan B}{1 - \tan A \tan B} \qquad \tan(A-B) = \frac{\tan A - \tan B}{1 + \tan A \tan B}$$

$$\cot(A+B) = \frac{\cot A \cot B - 1}{\cot B + \cot A} \qquad \cot(A-B) = \frac{\cot A \cot B + 1}{\cot B - \cot A}$$

$$\sin(A+B) = \sin A \cos B + \cos A \sin B \qquad \sin(A-B) = \sin A \cos B - \cos A \sin B$$

$$\cos(A+B) = \cos A \cos B - \sin A \sin B \qquad \cos(A-B) = \cos A \cos B - \sin A \sin B$$

Double-Angle

$$\cos 2A = \cos^2 A - \sin^2 A = 2\cos^2 A - 1 = 1 - 2\sin^2 A$$

$$\sin 2A = 2\sin A \cos A \qquad \tan 2A = \frac{2\tan A}{1 - \tan^2 A} = \frac{2}{\cot A - \tan A}$$

Half-Angle

$$\sin \tfrac{1}{2}A = \sqrt{\tfrac{1}{2}(1 - \cos A)} \qquad \cos \tfrac{1}{2}A = \sqrt{\tfrac{1}{2}(1 + \cos A)}$$

$$\tan \tfrac{1}{2}A = \sqrt{\frac{1 - \cos A}{1 + \cos A}} = \frac{1 - \cos A}{\sin A} = \frac{\sin A}{1 + \cos A}$$

Product-to-Sum

$$\sin A \cos B = \tfrac{1}{2}[\sin(A+B) + \sin(A-B)]$$

$$\cos A \cos B = \tfrac{1}{2}[\cos(A+B) + \cos(A-B)]$$

$$\sin A \sin B = \tfrac{1}{2}[\cos(A-B) - \cos(A+B)]$$

$$\tan A \tan B = \frac{\tan A + \tan B}{\cot A + \cot B}$$

Sum and Difference of Functions

$$\sin A + \sin B = 2[\sin \tfrac{1}{2}(A+B) \cos \tfrac{1}{2}(A-B)]$$

$$\sin A - \sin B = 2[\sin \tfrac{1}{2}(A-B) \cos \tfrac{1}{2}(A+B)]$$

$$\cos A + \cos B = 2[\cos \tfrac{1}{2}(A+B) \cos \tfrac{1}{2}(A-B)]$$

$$\cos A - \cos B = -2[\cos \tfrac{1}{2}(A+B) \cos \tfrac{1}{2}(A-B)]$$

$$\tan A + \tan B = \frac{\sin(A+B)}{\cos A \cos B} \qquad \tan A - \tan B = \frac{\sin(A-B)}{\cos A \cos B}$$

$$\cot A + \cot B = \frac{\sin(B+A)}{\sin A \sin B} \qquad \cot A - \cot B = \frac{\sin(B-A)}{\sin A \sin B}$$

Trigonometric Functions of Angles from 0° to 15° and 75° to 90°

Angle	sin	cos	tan	cot		Angle	sin	cos	tan	cot	
0° 0′	0.000000	1.000000	0.000000	—	**90° 0′**	**7° 30′**	0.130526	0.991445	0.131652	7.595754	**82° 30′**
10	0.002909	0.999996	0.002909	343.7737	**50**	**40**	0.133410	0.991061	0.134613	7.428706	**20**
20	0.005818	0.999983	0.005818	171.8854	**40**	**50**	0.136292	0.990669	0.137576	7.268725	**10**
30	0.008727	0.999962	0.008727	114.5887	**30**	**8° 0′**	0.139173	0.990268	0.140541	7.115370	**82° 0′**
40	0.011635	0.999932	0.011636	85.93979	**20**	**10**	0.142053	0.989859	0.143508	6.968234	**50**
50	0.014544	0.999894	0.014545	68.75009	**10**	**20**	0.144932	0.989442	0.146478	6.826944	**40**
1° 0′	0.017452	0.999848	0.017455	57.28996	**89° 0′**	**30**	0.147809	0.989016	0.149451	6.691156	**30**
10	0.020361	0.999793	0.020365	49.10388	**50**	**40**	0.150686	0.988582	0.152426	6.560554	**20**
20	0.023269	0.999729	0.023275	42.96408	**40**	**50**	0.153561	0.988139	0.155404	6.434843	**10**
30	0.026177	0.999657	0.026186	38.18846	**30**	**9° 0′**	0.156434	0.987688	0.158384	6.313752	**81° 0′**
40	0.029085	0.999577	0.029097	34.36777	**20**	**10**	0.159307	0.987229	0.161368	6.197028	**50**
50	0.031992	0.999488	0.032009	31.24158	**10**	**20**	0.162178	0.986762	0.164354	6.084438	**40**
2° 0′	0.034899	0.999391	0.034921	28.63625	**88° 0′**	**30**	0.165048	0.986286	0.167343	5.975764	**30**
10	0.037806	0.999285	0.037834	26.43160	**50**	**40**	0.167916	0.985801	0.170334	5.870804	**20**
20	0.040713	0.999171	0.040747	24.54176	**40**	**50**	0.170783	0.985309	0.173329	5.769369	**10**
30	0.043619	0.999048	0.043661	22.90377	**30**	**10° 0′**	0.173648	0.984808	0.176327	5.671282	**80° 0′**
40	0.046525	0.998917	0.046576	21.47040	**20**	**10**	0.176512	0.984298	0.179328	5.576379	**50**
50	0.049431	0.998778	0.049491	20.20555	**10**	**20**	0.179375	0.983781	0.182332	5.484505	**40**
3° 0′	0.052336	0.998630	0.052408	19.08114	**87° 0′**	**30**	0.182236	0.983255	0.185339	5.395517	**30**
10	0.055241	0.998473	0.055325	18.07498	**50**	**40**	0.185095	0.982721	0.188349	5.309279	**20**
20	0.058145	0.998308	0.058243	17.16934	**40**	**50**	0.187953	0.982178	0.191363	5.225665	**10**
30	0.061049	0.998135	0.061163	16.34986	**30**	**11° 0′**	0.190809	0.981627	0.194380	5.144554	**79° 0′**
40	0.063952	0.997953	0.064083	15.60478	**20**	**10**	0.193664	0.981068	0.197401	5.065835	**50**
50	0.066854	0.997763	0.067004	14.92442	**10**	**20**	0.196517	0.980500	0.200425	4.989403	**40**
4° 0′	0.069756	0.997564	0.069927	14.30067	**86° 0′**	**30**	0.199368	0.979925	0.203452	4.915157	**30**
10	0.072658	0.997357	0.072851	13.72674	**50**	**40**	0.202218	0.979341	0.206483	4.843005	**20**
20	0.075559	0.997141	0.075775	13.19688	**40**	**50**	0.205065	0.978748	0.209518	4.772857	**10**
30	0.078459	0.996917	0.078702	12.70621	**30**	**12° 0′**	0.207912	0.978148	0.212557	4.704630	**78° 0′**
40	0.081359	0.996685	0.081629	12.25051	**20**	**10**	0.210756	0.977539	0.215599	4.638246	**50**
50	0.084258	0.996444	0.084558	11.82617	**10**	**20**	0.213599	0.976921	0.218645	4.573629	**40**
5° 0′	0.087156	0.996195	0.087489	11.43005	**85° 0′**	**30**	0.216440	0.976296	0.221695	4.510709	**30**
10	0.090053	0.995937	0.090421	11.05943	**50**	**40**	0.219279	0.975662	0.224748	4.449418	**20**
20	0.092950	0.995671	0.093354	10.71191	**40**	**50**	0.222116	0.975020	0.227806	4.389694	**10**
30	0.095846	0.995396	0.096289	10.38540	**30**	**13° 0′**	0.224951	0.974370	0.230868	4.331476	**77° 0′**
40	0.098741	0.995113	0.099226	10.07803	**20**	**10**	0.227784	0.973712	0.233934	4.274707	**50**
50	0.101635	0.994822	0.102164	9.788173	**10**	**20**	0.230616	0.973045	0.237004	4.219332	**40**
6° 0′	0.104528	0.994522	0.105104	9.514364	**84° 0′**	**30**	0.233445	0.972370	0.240079	4.165300	**30**
10	0.107421	0.994214	0.108046	9.255304	**50**	**40**	0.236273	0.971687	0.243157	4.112561	**20**
20	0.110313	0.993897	0.110990	9.009826	**40**	**50**	0.239098	0.970995	0.246241	4.061070	**10**
30	0.113203	0.993572	0.113936	8.776887	**30**	**14° 0′**	0.241922	0.970296	0.249328	4.010781	**76° 0′**
40	0.116093	0.993238	0.116883	8.555547	**20**	**10**	0.244743	0.969588	0.252420	3.961652	**50**
50	0.118982	0.992896	0.119833	8.344956	**10**	**20**	0.247563	0.968872	0.255516	3.913642	**40**
7° 0′	0.121869	0.992546	0.122785	8.144346	**83° 0′**	**30**	0.250380	0.968148	0.258618	3.866713	**30**
10	0.124756	0.992187	0.125738	7.953022	**50**	**40**	0.253195	0.967415	0.261723	3.820828	**20**
20	0.127642	0.991820	0.128694	7.770351	**40**	**50**	0.256008	0.966675	0.264834	3.775952	**10**
7° 30′	0.130526	0.991445	0.131652	7.595754	**82° 30**	**15° 0′**	0.258819	0.965926	0.267949	3.732051	**75° 0′**
	cos	sin	cot	tan	Angle		cos	sin	cot	tan	Angle

For angles 0° to 15° 0′ (angles found in a column to the left of the data), use the column labels at the top of the table; for angles 75° to 90° 0′ (angles found in a column to the right of the data), use the column labels at the bottom of the table.

Trigonometric Functions of Angles from 15° to 30° and 60° to 75°

Angle	sin	cos	tan	cot		Angle	sin	cos	tan	cot	
15° 0′	0.258819	0.965926	0.267949	3.732051	75° 0′	22° 30′	0.382683	0.923880	0.414214	2.414214	67° 30
10	0.261628	0.965169	0.271069	3.689093	50	40	0.385369	0.922762	0.417626	2.394489	20
20	0.264434	0.964404	0.274194	3.647047	40	50	0.388052	0.921638	0.421046	2.375037	10
30	0.267238	0.963630	0.277325	3.605884	30	23° 0′	0.390731	0.920505	0.424475	2.355852	67° 0′
40	0.270040	0.962849	0.280460	3.565575	20	10	0.393407	0.919364	0.427912	2.336929	50
50	0.272840	0.962059	0.283600	3.526094	10	20	0.396080	0.918216	0.431358	2.318261	40
16° 0′	0.275637	0.961262	0.286745	3.487414	74° 0′	30	0.398749	0.917060	0.434812	2.299843	30
10	0.278432	0.960456	0.289896	3.449512	50	40	0.401415	0.915896	0.438276	2.281669	20
20	0.281225	0.959642	0.293052	3.412363	40	50	0.404078	0.914725	0.441748	2.263736	10
30	0.284015	0.958820	0.296213	3.375943	30	24° 0′	0.406737	0.913545	0.445229	2.246037	66° 0′
40	0.286803	0.957990	0.299380	3.340233	20	10	0.409392	0.912358	0.448719	2.228568	50
50	0.289589	0.957151	0.302553	3.305209	10	20	0.412045	0.911164	0.452218	2.211323	40
17° 0′	0.292372	0.956305	0.305731	3.270853	73° 0′	30	0.414693	0.909961	0.455726	2.194300	30
10	0.295152	0.955450	0.308914	3.237144	50	40	0.417338	0.908751	0.459244	2.177492	20
20	0.297930	0.954588	0.312104	3.204064	40	50	0.419980	0.907533	0.462771	2.160896	10
30	0.300706	0.953717	0.315299	3.171595	30	25° 0′	0.422618	0.906308	0.466308	2.144507	65° 0′
40	0.303479	0.952838	0.318500	3.139719	20	10	0.425253	0.905075	0.469854	2.128321	50
50	0.306249	0.951951	0.321707	3.108421	10	20	0.427884	0.903834	0.473410	2.112335	40
18° 0′	0.309017	0.951057	0.324920	3.077684	72° 0′	30	0.430511	0.902585	0.476976	2.096544	30
10	0.311782	0.950154	0.328139	3.047492	50	40	0.433135	0.901329	0.480551	2.080944	20
20	0.314545	0.949243	0.331364	3.017830	40	50	0.435755	0.900065	0.484137	2.065532	10
30	0.317305	0.948324	0.334595	2.988685	30	26° 0′	0.438371	0.898794	0.487733	2.050304	64° 0′
40	0.320062	0.947397	0.337833	2.960042	20	10	0.440984	0.897515	0.491339	2.035256	50
50	0.322816	0.946462	0.341077	2.931888	10	20	0.443593	0.896229	0.494955	2.020386	40
19° 0′	0.325568	0.945519	0.344328	2.904211	71° 0′	30	0.446198	0.894934	0.498582	2.005690	30
10	.0.328317	0.944568	0.347585	2.876997	50	40	0.448799	0.893633	0.502219	1.991164	20
20	0.331063	0.943609	0.350848	2.850235	40	50	0.451397	0.892323	0.505867	1.976805	10
30	0.333807	0.942641	0.354119	2.823913	30	27° 0′	0.453990	0.891007	0.509525	1.962611	63° 0′
40	0.336547	0.941666	0.357396	2.798020	20	10	0.456580	0.889682	0.513195	1.948577	50
50	0.339285	0.940684	0.360679	2.772545	10	20	0.459166	0.888350	0.516875	1.934702	40
20° 0′	0.342020	0.939693	0.363970	2.747477	70° 0′	30	0.461749	0.887011	0.520567	1.920982	30
10	0.344752	0.938694	0.367268	2.722808	50	40	0.464327	0.885664	0.524270	1.907415	20
20	0.347481	0.937687	0.370573	2.698525	40	50	0.466901	0.884309	0.527984	1.893997	10
30	0.350207	0.936672	0.373885	2.674621	30	28° 0′	0.469472	0.882948	0.531709	1.880726	62° 0′
40	0.352931	0.935650	0.377204	2.651087	20	10	0.472038	0.881578	0.535446	1.867600	50
50	0.355651	0.934619	0.380530	2.627912	10	20	0.474600	0.880201	0.539195	1.854616	40
21° 0′	0.358368	0.933580	0.383864	2.605089	69° 0′	30	0.477159	0.878817	0.542956	1.841771	30
10	0.361082	0.932534	0.387205	2.582609	50	40	0.479713	0.877425	0.546728	1.829063	20
20	0.363793	0.931480	0.390554	2.560465	40	50	0.482263	0.876026	0.550513	1.816489	10
30	0.366501	0.930418	0.393910	2.538648	30	29° 0′	0.484810	0.874620	0.554309	1.804048	61° 0′
40	0.369206	0.929348	0.397275	2.517151	20	10	0.487352	0.873206	0.558118	1.791736	50
50	0.371908	0.928270	0.400646	2.495966	10	20	0.489890	0.871784	0.561939	1.779552	40
22° 0′	0.374607	0.927184	0.404026	2.475087	68° 0′	30	0.492424	0.870356	0.565773	1.767494	30
10	0.377302	0.926090	0.407414	2.454506	50	40	0.494953	0.868920	0.569619	1.755559	20
20	0.379994	0.924989	0.410810	2.434217	40	50	0.497479	0.867476	0.573478	1.743745	10
22° 30	0.382683	0.923880	0.414214	2.414214	67° 30	30° 0′	0.500000	0.866025	0.577350	1.732051	60° 0′
	cos	sin	cot	tan	Angle		cos	sin	cot	tan	Angle

For angles 15° to 30° 0′ (angles found in a column to the left of the data), use the column labels at the top of the table; for angles 60° to 75° 0′ (angles found in a column to the right of the data), use the column labels at the bottom of the table.

Trigonometric Functions of Angles from 30° to 60°

Angle	sin	cos	tan	cot		Angle	sin	cos	tan	cot	
30° 0′	0.500000	0.866025	0.577350	1.732051	**60° 0′**	**37° 30′**	0.608761	0.793353	0.767327	1.303225	**52° 30′**
10	0.502517	0.864567	0.581235	1.720474	**50**	**40**	0.611067	0.791579	0.771959	1.295406	**20**
20	0.505030	0.863102	0.585134	1.709012	**40**	**50**	0.613367	0.789798	0.776612	1.287645	**10**
30	0.507538	0.861629	0.589045	1.697663	**30**	**38° 0′**	0.615661	0.788011	0.781286	1.279942	**52° 0′**
40	0.510043	0.860149	0.592970	1.686426	**20**	**10**	0.617951	0.786217	0.785981	1.272296	**50**
50	0.512543	0.858662	0.596908	1.675299	**10**	**20**	0.620235	0.784416	0.790697	1.264706	**40**
31° 0′	0.515038	0.857167	0.600861	1.664279	**59° 0′**	**30**	0.622515	0.782608	0.795436	1.257172	**30**
10	0.517529	0.855665	0.604827	1.653366	**50**	**40**	0.624789	0.780794	0.800196	1.249693	**20**
20	0.520016	0.854156	0.608807	1.642558	**40**	**50**	0.627057	0.778973	0.804979	1.242268	**10**
30	0.522499	0.852640	0.612801	1.631852	**30**	**39° 0′**	0.629320	0.777146	0.809784	1.234897	**51° 0′**
40	0.524977	0.851117	0.616809	1.621247	**20**	**10**	0.631578	0.775312	0.814612	1.227579	**50**
50	0.527450	0.849586	0.620832	1.610742	**10**	**20**	0.633831	0.773472	0.819463	1.220312	**40**
32° 0′	0.529919	0.848048	0.624869	1.600335	**58° 0′**	**30**	0.636078	0.771625	0.824336	1.213097	**30**
10	0.532384	0.846503	0.628921	1.590024	**50**	**40**	0.638320	0.769771	0.829234	1.205933	**20**
20	0.534844	0.844951	0.632988	1.579808	**40**	**50**	0.640557	0.767911	0.834155	1.198818	**10**
30	0.537300	0.843391	0.637070	1.569686	**30**	**40° 0′**	0.642788	0.766044	0.839100	1.191754	**50° 0′**
40	0.539751	0.841825	0.641167	1.559655	**20**	**10**	0.645013	0.764171	0.844069	1.184738	**50**
50	0.542197	0.840251	0.645280	1.549715	**10**	**20**	0.647233	0.762292	0.849062	1.177770	**40**
33° 0′	0.544639	0.838671	0.649408	1.539865	**57° 0′**	**30**	0.649448	0.760406	0.854081	1.170850	**30**
10	0.547076	0.837083	0.653551	1.530102	**50**	**40**	0.651657	0.758514	0.859124	1.163976	**20**
20	0.549509	0.835488	0.657710	1.520426	**40**	**50**	0.653861	0.756615	0.864193	1.157149	**10**
30	0.551937	0.833886	0.661886	1.510835	**30**	**41° 0′**	0.656059	0.754710	0.869287	1.150368	**49° 0′**
40	0.554360	0.832277	0.666077	1.501328	**20**	**10**	0.658252	0.752798	0.874407	1.143633	**50**
50	0.556779	0.830661	0.670284	1.491904	**10**	**20**	0.660439	0.750880	0.879553	1.136941	**40**
34° 0′	0.559193	0.829038	0.674509	1.482561	**56° 0′**	**30**	0.662620	0.748956	0.884725	1.130294	**30**
10	0.561602	0.827407	0.678749	1.473298	**50**	**40**	0.664796	0.747025	0.889924	1.123691	**20**
20	0.564007	0.825770	0.683007	1.464115	**40**	**50**	0.666966	0.745088	0.895151	1.117130	**10**
30	0.566406	0.824126	0.687281	1.455009	**30**	**42° 0′**	0.669131	0.743145	0.900404	1.110613	**48° 0′**
40	0.568801	0.822475	0.691572	1.445980	**20**	**10**	0.671289	0.741195	0.905685	1.104137	**50**
50	0.571191	0.820817	0.695881	1.437027	**10**	**20**	0.673443	0.739239	0.910994	1.097702	**40**
35° 0′	0.573576	0.819152	0.700208	1.428148	**55° 0′**	**30**	0.675590	0.737277	0.916331	1.091309	**30**
10	0.575957	0.817480	0.704551	1.419343	**50**	**40**	0.677732	0.735309	0.921697	1.084955	**20**
20	0.578332	0.815801	0.708913	1.410610	**40**	**50**	0.679868	0.733334	0.927091	1.078642	**10**
30	0.580703	0.814116	0.713293	1.401948	**30**	**43° 0′**	0.681998	0.731354	0.932515	1.072369	**47° 0′**
40	0.583069	0.812423	0.717691	1.393357	**20**	**10**	0.684123	0.729367	0.937968	1.066134	**50**
50	0.585429	0.810723	0.722108	1.384835	**10**	**20**	0.686242	0.727374	0.943451	1.059938	**40**
36° 0′	0.587785	0.809017	0.726543	1.376382	**54° 0′**	**30**	0.688355	0.725374	0.948965	1.053780	**30**
10	0.590136	0.807304	0.730996	1.367996	**50**	**40**	0.690462	0.723369	0.954508	1.047660	**20**
20	0.592482	0.805584	0.735469	1.359676	**40**	**50**	0.692563	0.721357	0.960083	1.041577	**10**
30	0.594823	0.803857	0.739961	1.351422	**30**	**44° 0′**	0.694658	0.719340	0.965689	1.035530	**46° 0′**
40	0.597159	0.802123	0.744472	1.343233	**20**	**10**	0.696748	0.717316	0.971326	1.029520	**50**
50	0.599489	0.800383	0.749003	1.335108	**10**	**20**	0.698832	0.715286	0.976996	1.023546	**40**
37° 0′	0.601815	0.798636	0.753554	1.327045	**53° 0′**	**30**	0.700909	0.713250	0.982697	1.017607	**30**
10	0.604136	0.796882	0.758125	1.319044	**50**	**40**	0.702981	0.711209	0.988432	1.011704	**20**
20	0.606451	0.795121	0.762716	1.311105	**40**	**50**	0.705047	0.709161	0.994199	1.005835	**10**
37° 30	0.608761	0.793353	0.767327	1.303225	**52° 30**	**45° 0′**	0.707107	0.707107	1.000000	1.000000	**45° 0′**
	cos	sin	cot	tan	Angle		cos	sin	cot	tan	Angle

For angles 30° to 45° 0′ (angles found in a column to the left of the data), use the column labels at the top of the table; for angles 45° to 60° 0′ (angles found in a column to the right of the data), use the column labels at the bottom of the table.

Using a Calculator to Find Trig Functions.—A scientific calculator is quicker and more accurate than tables for finding trig functions and angles corresponding to trig functions. On scientific calculators, the keys labeled **sin**, **cos**, and **tan** are used to find the common trig functions. The other functions can be found by using the same keys and the **1/x** key, noting that $\csc A = 1/\sin A$, $\sec A = 1/\cos A$, and $\cot A = 1/\tan A$. The specific keystrokes used will vary slightly from one calculator to another. To find the angle corresponding to a given trig function use the keys labeled **sin^{-1}**, **cos^{-1}**, and **tan^{-1}**. On some other calculators, the **sin**, **cos**, and **tan** are used in combination with the **INV**, or inverse, key to find the number corresponding to a given trig function.

If a scientific calculator or computer is not available, tables are the easiest way to find trig values. However, trig function values can be calculated very accurately without a scientific calculator by using the following formulas:

$$\sin A = A - \frac{A^3}{3!} + \frac{A^5}{5!} - \frac{A^7}{7!} \pm \cdots \qquad \cos A = 1 - \frac{A^2}{2!} + \frac{A^4}{4!} - \frac{A^6}{6!} \pm \cdots$$

$$\sin^{-1} A = \frac{1}{2} \times \frac{A^3}{3} + \frac{1}{2} \times \frac{3}{4} \times \frac{A^5}{5} + \cdots \qquad \tan^{-1} A = A - \frac{A^3}{3} + \frac{A^5}{5} - \frac{A^7}{7} \pm \cdots$$

where the angle A is expressed in radians (convert degrees to radians by multiplying degrees by $\pi/180 = 0.0174533$). The three dots at the ends of the formulas indicate that the expression continues with more terms following the sequence established by the first few terms. Generally, calculating just three or four terms of the expression is sufficient for accuracy. In these formulas, a number followed by the symbol **!** is called a factorial (for example, 3! is three factorial). Except for 0!, which is defined as 1, a factorial is found by multiplying together all the integers greater than zero and less than or equal to the factorial number wanted. For example: $3! = 1 \times 2 \times 3 = 6$; $4! = 1 \times 2 \times 3 \times 4 = 24$; $7! = 1 \times 2 \times 3 \times 4 \times 5 \times 6 \times 7 = 5040$; etc.

Versed Sine and Versed Cosine.—These functions are sometimes used in formulas for segments of a circle and may be obtained using the relationships:

$$\text{versed } \sin\theta = 1 - \cos\theta; \text{ versed } \cos\theta = 1 - \sin\theta.$$

Sevolute Functions.—Sevolute functions are used in calculating the form diameter of involute splines. They are computed by subtracting the involute function of an angle from the secant of the angle (1/cosine = secant). Thus, sevolute of 20 degrees = secant of 20 degrees − involute function of 20 degrees = 1.064178 − 0.014904 = 1.049274.

Involute Functions.—Involute functions are used in certain formulas relating to the design and measurement of gear teeth as well as measurement of threads over wires.

The tables on the following pages provide values of involute functions for angles from 14 to 51 degrees in increments of 1 minute. These involute functions were calculated from the following formulas: Involute of $\theta = \tan\theta - \theta$, for θ in radians, and involute of $\theta = \tan\theta - \pi \times \theta/180$, for θ in degrees.

Example: For an angle of 14 degrees and 10 minutes, the involute function is found as follows: 10 minutes = 10/60 = 0.166666 degrees, 14 + 0.166666 = 14.166666 degree, so that the involute of 14.166666 degrees = tan 14.166666 − $\pi \times 14.166666/180$ = 0.252420 − 0.247255 = 0.005165. This value is the same as that in Table for 14 degrees and 10 minutes. The same result would be obtained from using the conversion tables beginning on page 110 to convert 14 degrees and 10 minutes to radians and then applying the first of the formulas given above.

Involute Functions for Angles from 14 to 23 Degrees

	Degrees								
	14	15	16	17	18	19	20	21	22
Minutes	Involute Functions								
0	0.004982	0.006150	0.007493	0.009025	0.010760	0.012715	0.014904	0.017345	0.020054
1	0.005000	0.006171	0.007517	0.009052	0.010791	0.012750	0.014943	0.017388	0.020101
2	0.005018	0.006192	0.007541	0.009079	0.010822	0.012784	0.014982	0.017431	0.020149
3	0.005036	0.006213	0.007565	0.009107	0.010853	0.012819	0.015020	0.017474	0.020197
4	0.005055	0.006234	0.007589	0.009134	0.010884	0.012854	0.015059	0.017517	0.020244
5	0.005073	0.006255	0.007613	0.009161	0.010915	0.012888	0.015098	0.017560	0.020292
6	0.005091	0.006276	0.007637	0.009189	0.010946	0.012923	0.015137	0.017603	0.020340
7	0.005110	0.006297	0.007661	0.009216	0.010977	0.012958	0.015176	0.017647	0.020388
8	0.005128	0.006318	0.007686	0.009244	0.011008	0.012993	0.015215	0.017690	0.020436
9	0.005146	0.006340	0.007710	0.009272	0.011039	0.013028	0.015254	0.017734	0.020484
10	0.005165	0.006361	0.007735	0.009299	0.011071	0.013063	0.015293	0.017777	0.020533
11	0.005184	0.006382	0.007759	0.009327	0.011102	0.013098	0.015333	0.017821	0.020581
12	0.005202	0.006404	0.007784	0.009355	0.011133	0.013134	0.015372	0.017865	0.020629
13	0.005221	0.006425	0.007808	0.009383	0.011165	0.013169	0.015411	0.017908	0.020678
14	0.005239	0.006447	0.007833	0.009411	0.011196	0.013204	0.015451	0.017952	0.020726
15	0.005258	0.006469	0.007857	0.009439	0.011228	0.013240	0.015490	0.017996	0.020775
16	0.005277	0.006490	0.007882	0.009467	0.011260	0.013275	0.015530	0.018040	0.020824
17	0.005296	0.006512	0.007907	0.009495	0.011291	0.013311	0.015570	0.018084	0.020873
18	0.005315	0.006534	0.007932	0.009523	0.011323	0.013346	0.015609	0.018129	0.020921
19	0.005334	0.006555	0.007957	0.009552	0.011355	0.013382	0.015649	0.018173	0.020970
20	0.005353	0.006577	0.007982	0.009580	0.011387	0.013418	0.015689	0.018217	0.021019
21	0.005372	0.006599	0.008007	0.009608	0.011419	0.013454	0.015729	0.018262	0.021069
22	0.005391	0.006621	0.008032	0.009637	0.011451	0.013490	0.015769	0.018306	0.021118
23	0.005410	0.006643	0.008057	0.009665	0.011483	0.013526	0.015809	0.018351	0.021167
24	0.005429	0.006665	0.008082	0.009694	0.011515	0.013562	0.015850	0.018395	0.021217
25	0.005448	0.006687	0.008107	0.009722	0.011547	0.013598	0.015890	0.018440	0.021266
26	0.005467	0.006709	0.008133	0.009751	0.011580	0.013634	0.015930	0.018485	0.021316
27	0.005487	0.006732	0.008158	0.009780	0.011612	0.013670	0.015971	0.018530	0.021365
28	0.005506	0.006754	0.008183	0.009808	0.011644	0.013707	0.016011	0.018575	0.021415
29	0.005525	0.006776	0.008209	0.009837	0.011677	0.013743	0.016052	0.018620	0.021465
30	0.005545	0.006799	0.008234	0.009866	0.011709	0.013779	0.016092	0.018665	0.021514
31	0.005564	0.006821	0.008260	0.009895	0.011742	0.013816	0.016133	0.018710	0.021564
32	0.005584	0.006843	0.008285	0.009924	0.011775	0.013852	0.016174	0.018755	0.021614
33	0.005603	0.006866	0.008311	0.009953	0.011807	0.013889	0.016215	0.018800	0.021665
34	0.005623	0.006888	0.008337	0.009982	0.011840	0.013926	0.016255	0.018846	0.021715
35	0.005643	0.006911	0.008362	0.010011	0.011873	0.013963	0.016296	0.018891	0.021765
36	0.005662	0.006934	0.008388	0.010041	0.011906	0.013999	0.016337	0.018937	0.021815
37	0.005682	0.006956	0.008414	0.010070	0.011939	0.014036	0.016379	0.018983	0.021866
38	0.005702	0.006979	0.008440	0.010099	0.011972	0.014073	0.016420	0.019028	0.021916
39	0.005722	0.007002	0.008466	0.010129	0.012005	0.014110	0.016461	0.019074	0.021967
40	0.005742	0.007025	0.008492	0.010158	0.012038	0.014148	0.016502	0.019120	0.022018
41	0.005762	0.007048	0.008518	0.010188	0.012071	0.014185	0.016544	0.019166	0.022068
42	0.005782	0.007071	0.008544	0.010217	0.012105	0.014222	0.016585	0.019212	0.022119
43	0.005802	0.007094	0.008571	0.010247	0.012138	0.014259	0.016627	0.019258	0.022170
44	0.005822	0.007117	0.008597	0.010277	0.012172	0.014297	0.016669	0.019304	0.022221
45	0.005842	0.007140	0.008623	0.010307	0.012205	0.014334	0.016710	0.019350	0.022272
46	0.005862	0.007163	0.008650	0.010336	0.012239	0.014372	0.016752	0.019397	0.022324
47	0.005882	0.007186	0.008676	0.010366	0.012272	0.014409	0.016794	0.019443	0.022375
48	0.005903	0.007209	0.008702	0.010396	0.012306	0.014447	0.016836	0.019490	0.022426
49	0.005923	0.007233	0.008729	0.010426	0.012340	0.014485	0.016878	0.019536	0.022478
50	0.005943	0.007256	0.008756	0.010456	0.012373	0.014523	0.016920	0.019583	0.022529
51	0.005964	0.007280	0.008782	0.010486	0.012407	0.014560	0.016962	0.019630	0.022581
52	0.005984	0.007303	0.008809	0.010517	0.012441	0.014598	0.017004	0.019676	0.022633
53	0.006005	0.007327	0.008836	0.010547	0.012475	0.014636	0.017047	0.019723	0.022684
54	0.006025	0.007350	0.008863	0.010577	0.012509	0.014674	0.017089	0.019770	0.022736
55	0.006046	0.007374	0.008889	0.010608	0.012543	0.014713	0.017132	0.019817	0.022788
56	0.006067	0.007397	0.008916	0.010638	0.012578	0.014751	0.017174	0.019864	0.022840
57	0.006087	0.007421	0.008943	0.010669	0.012612	0.014789	0.017217	0.019912	0.022892
58	0.006108	0.007445	0.008970	0.010699	0.012646	0.014827	0.017259	0.019959	0.022944
59	0.006129	0.007469	0.008998	0.010730	0.012681	0.014866	0.017302	0.020006	0.022997
60	0.006150	0.007493	0.009025	0.010760	0.012715	0.014904	0.017345	0.020054	0.023049

Involute Functions for Angles from 23 to 32 Degrees

Minutes	Degrees								
	23	24	25	26	27	28	29	30	31
	Involute Functions								
0	0.023049	0.026350	0.029975	0.033947	0.038287	0.043017	0.048164	0.053752	0.059809
1	0.023102	0.026407	0.030039	0.034016	0.038362	0.043100	0.048253	0.053849	0.059914
2	0.023154	0.026465	0.030102	0.034086	0.038438	0.043182	0.048343	0.053946	0.060019
3	0.023207	0.026523	0.030166	0.034155	0.038514	0.043264	0.048432	0.054043	0.060124
4	0.023259	0.026581	0.030229	0.034225	0.038590	0.043347	0.048522	0.054140	0.060230
5	0.023312	0.026639	0.030293	0.034294	0.038666	0.043430	0.048612	0.054238	0.060335
6	0.023365	0.026697	0.030357	0.034364	0.038742	0.043513	0.048702	0.054336	0.060441
7	0.023418	0.026756	0.030420	0.034434	0.038818	0.043596	0.048792	0.054433	0.060547
8	0.023471	0.026814	0.030484	0.034504	0.038894	0.043679	0.048883	0.054531	0.060653
9	0.023524	0.026872	0.030549	0.034574	0.038971	0.043762	0.048973	0.054629	0.060759
10	0.023577	0.026931	0.030613	0.034644	0.039047	0.043845	0.049064	0.054728	0.060866
11	0.023631	0.026989	0.030677	0.034714	0.039124	0.043929	0.049154	0.054826	0.060972
12	0.023684	0.027048	0.030741	0.034785	0.039201	0.044012	0.049245	0.054924	0.061079
13	0.023738	0.027107	0.030806	0.034855	0.039278	0.044096	0.049336	0.055023	0.061186
14	0.023791	0.027166	0.030870	0.034926	0.039355	0.044180	0.049427	0.055122	0.061292
15	0.023845	0.027225	0.030935	0.034997	0.039432	0.044264	0.049518	0.055221	0.061400
16	0.023899	0.027284	0.031000	0.035067	0.039509	0.044348	0.049609	0.055320	0.061507
17	0.023952	0.027343	0.031065	0.035138	0.039586	0.044432	0.049701	0.055419	0.061614
18	0.024006	0.027402	0.031130	0.035209	0.039664	0.044516	0.049792	0.055518	0.061721
19	0.024060	0.027462	0.031195	0.035280	0.039741	0.044601	0.049884	0.055617	0.061829
20	0.024114	0.027521	0.031260	0.035352	0.039819	0.044685	0.049976	0.055717	0.061937
21	0.024169	0.027581	0.031325	0.035423	0.039897	0.044770	0.050068	0.055817	0.062045
22	0.024223	0.027640	0.031390	0.035494	0.039974	0.044855	0.050160	0.055916	0.062153
23	0.024277	0.027700	0.031456	0.035566	0.040052	0.044940	0.050252	0.056016	0.062261
24	0.024332	0.027760	0.031521	0.035637	0.040131	0.045024	0.050344	0.056116	0.062369
25	0.024386	0.027820	0.031587	0.035709	0.040209	0.045110	0.050437	0.056217	0.062478
26	0.024441	0.027880	0.031653	0.035781	0.040287	0.045195	0.050529	0.056317	0.062586
27	0.024495	0.027940	0.031718	0.035853	0.040366	0.045280	0.050622	0.056417	0.062695
28	0.024550	0.028000	0.031784	0.035925	0.040444	0.045366	0.050715	0.056518	0.062804
29	0.024605	0.028060	0.031850	0.035997	0.040523	0.045451	0.050808	0.056619	0.062913
30	0.024660	0.028121	0.031917	0.036069	0.040602	0.045537	0.050901	0.056720	0.063022
31	0.024715	0.028181	0.031983	0.036142	0.040680	0.045623	0.050994	0.056821	0.063131
32	0.024770	0.028242	0.032049	0.036214	0.040759	0.045709	0.051087	0.056922	0.063241
33	0.024825	0.028302	0.032116	0.036287	0.040839	0.045795	0.051181	0.057023	0.063350
34	0.024881	0.028363	0.032182	0.036359	0.040918	0.045881	0.051274	0.057124	0.063460
35	0.024936	0.028424	0.032249	0.036432	0.040997	0.045967	0.051368	0.057226	0.063570
36	0.024992	0.028485	0.032315	0.036505	0.041077	0.046054	0.051462	0.057328	0.063680
37	0.025047	0.028546	0.032382	0.036578	0.041156	0.046140	0.051556	0.057429	0.063790
38	0.025103	0.028607	0.032449	0.036651	0.041236	0.046227	0.051650	0.057531	0.063901
39	0.025159	0.028668	0.032516	0.036724	0.041316	0.046313	0.051744	0.057633	0.064011
40	0.025214	0.028729	0.032583	0.036798	0.041395	0.046400	0.051838	0.057736	0.064122
41	0.025270	0.028791	0.032651	0.036871	0.041475	0.046487	0.051933	0.057838	0.064232
42	0.025326	0.028852	0.032718	0.036945	0.041556	0.046575	0.052027	0.057940	0.064343
43	0.025382	0.028914	0.032785	0.037018	0.041636	0.046662	0.052122	0.058043	0.064454
44	0.025439	0.028976	0.032853	0.037092	0.041716	0.046749	0.052217	0.058146	0.064565
45	0.025495	0.029037	0.032920	0.037166	0.041797	0.046837	0.052312	0.058249	0.064677
46	0.025551	0.029099	0.032988	0.037240	0.041877	0.046924	0.052407	0.058352	0.064788
47	0.025608	0.029161	0.033056	0.037314	0.041958	0.047012	0.052502	0.058455	0.064900
48	0.025664	0.029223	0.033124	0.037388	0.042039	0.047100	0.052597	0.058558	0.065012
49	0.025721	0.029285	0.033192	0.037462	0.042120	0.047188	0.052693	0.058662	0.065123
50	0.025778	0.029348	0.033260	0.037537	0.042201	0.047276	0.052788	0.058765	0.065236
51	0.025834	0.029410	0.033328	0.037611	0.042282	0.047364	0.052884	0.058869	0.065348
52	0.025891	0.029472	0.033397	0.037686	0.042363	0.047452	0.052980	.058973	0.065460
53	0.025948	0.029535	0.033465	0.037761	0.042444	0.047541	0.053076	0.059077	0.065573
54	0.026005	0.029598	0.033534	0.037835	0.042526	0.047630	0.053172	0.059181	0.065685
55	0.026062	0.029660	0.033602	0.037910	0.042608	0.047718	0.053268	0.059285	0.065798
56	0.026120	0.029723	0.033671	0.037985	0.042689	0.047807	0.053365	0.059390	0.065911
57	0.026177	0.029786	0.033740	0.038060	0.042771	0.047896	0.053461	0.059494	0.066024
58	0.026235	0.029849	0.033809	0.038136	0.042853	0.047985	0.053558	0.059599	0.066137
59	0.026292	0.029912	0.033878	0.038211	0.042935	0.048074	0.053655	0.059704	0.066251
60	0.026350	0.029975	0.033947	0.038287	0.043017	0.048164	0.053752	0.059809	0.066364

Involute Functions for Angles from 32 to 41 Degrees

	Degrees								
	32	33	34	35	36	37	38	39	40
Minutes	Involute Functions								
0	0.066364	0.073449	0.081097	0.089342	0.098224	0.107782	0.118061	0.129106	0.140968
1	0.066478	0.073572	0.081229	0.089485	0.098378	0.107948	0.118238	0.129297	0.141173
2	0.066591	0.073695	0.081362	0.089628	0.098532	0.108113	0.118416	0.129488	0.141378
3	0.066705	0.073818	0.081494	0.089771	0.098686	0.108279	0.118594	0.129679	0.141584
4	0.066820	0.073941	0.081627	0.089914	0.098840	0.108445	0.118773	0.129870	0.141789
5	0.066934	0.074064	0.081760	0.090058	0.098994	0.108611	0.118951	0.130062	0.141995
6	0.067048	0.074188	0.081894	0.090201	0.099149	0.108777	0.119130	0.130254	0.142201
7	0.067163	0.074312	0.082027	0.090345	0.099303	0.108943	0.119309	0.130446	0.142408
8	0.067277	0.074435	0.082161	0.090489	0.099458	0.109110	0.119488	0.130639	0.142614
9	0.067392	0.074559	0.082294	0.090633	0.099614	0.109277	0.119667	0.130832	0.142821
10	0.067507	0.074684	0.082428	0.090777	0.099769	0.109444	0.119847	0.131025	0.143028
11	0.067622	0.074808	0.082562	0.090922	0.099924	0.109611	0.120027	0.131218	0.143236
12	0.067738	0.074932	0.082697	0.091067	0.100080	0.109779	0.120207	0.131411	0.143443
13	0.067853	0.075057	0.082831	0.091211	0.100236	0.109947	0.120387	0.131605	0.143651
14	0.067969	0.075182	0.082966	0.091356	0.100392	0.110114	0.120567	0.131799	0.143859
15	0.068084	0.075307	0.083101	0.091502	0.100549	0.110283	0.120748	0.131993	0.144068
16	0.068200	0.075432	0.083235	0.091647	0.100705	0.110451	0.120929	0.132187	0.144276
17	0.068316	0.075557	0.083371	0.091793	0.100862	0.110619	0.121110	0.132381	0.144485
18	0.068432	0.075683	0.083506	0.091938	0.101019	0.110788	0.121291	0.132576	0.144694
19	0.068549	0.075808	0.083641	0.092084	0.101176	0.110957	0.121473	0.132771	0.144903
20	0.068665	0.075934	0.083777	0.092230	0.101333	0.111126	0.121655	0.132966	0.145113
21	0.068782	0.076060	0.083913	0.092377	0.101490	0.111295	0.121837	0.133162	0.145323
22	0.068899	0.076186	0.084049	0.092523	0.101648	0.111465	0.122019	0.133358	0.145533
23	0.069016	0.076312	0.084185	0.092670	0.101806	0.111635	0.122201	0.133553	0.145743
24	0.069133	0.076439	0.084321	0.092816	0.101964	0.111805	0.122384	0.133750	0.145954
25	0.069250	0.076565	0.084458	0.092963	0.102122	0.111975	0.122567	0.133946	0.146165
26	0.069367	0.076692	0.084594	0.093111	0.102280	0.112145	0.122750	0.134143	0.146376
27	0.069485	0.076819	0.084731	0.093258	0.102439	0.112316	0.122933	0.134339	0.146587
28	0.069602	0.076946	0.084868	0.093406	0.102598	0.112486	0.123117	0.134537	0.146799
29	0.069720	0.077073	0.085005	0.093553	0.102757	0.112657	0.123300	0.134734	0.147010
30	0.069838	0.077200	0.085142	0.093701	0.102916	0.112829	0.123484	0.134931	0.147222
31	0.069956	0.077328	0.085280	0.093849	0.103075	0.113000	0.123668	0.135129	0.147435
32	0.070075	0.077455	0.085418	0.093998	0.103235	0.113172	0.123853	0.135327	0.147647
33	0.070193	0.077583	0.085555	0.094146	0.103395	0.113343	0.124037	0.135525	0.147860
34	0.070312	0.077711	0.085693	0.094295	0.103555	0.113515	0.124222	0.135724	0.148073
35	0.070430	0.077839	0.085832	0.094443	0.103715	0.113688	0.124407	0.135923	0.148286
36	0.070549	0.077968	0.085970	0.094593	0.103875	0.113860	0.124592	0.136122	0.148500
37	0.070668	0.078096	0.086108	0.094742	0.104036	0.114033	0.124778	0.136321	0.148714
38	0.070788	0.078225	0.086247	0.094891	0.104196	0.114205	0.124964	0.136520	0.148928
39	0.070907	0.078354	0.086386	0.095041	0.104357	0.114378	0.125150	0.136720	0.149142
40	0.071026	0.078483	0.086525	0.095190	0.104518	0.114552	0.125336	0.136920	0.149357
41	0.071146	0.078612	0.086664	0.095340	0.104680	0.114725	0.125522	0.137120	0.149572
42	0.071266	0.078741	0.086804	0.095490	0.104841	0.114899	0.125709	0.137320	0.149787
43	0.071386	0.078871	0.086943	0.095641	0.105003	0.115073	0.125896	0.137521	0.150002
44	0.071506	0.079000	0.087083	0.095791	0.105165	0.115247	0.126083	0.137722	0.150218
45	0.071626	0.079130	0.087223	0.095942	0.105327	0.115421	0.126270	0.137923	0.150434
46	0.071747	0.079260	0.087363	0.096093	0.105489	0.115595	0.126457	0.138124	0.150650
47	0.071867	0.079390	0.087503	0.096244	0.105652	0.115770	0.126645	0.138326	0.150866
48	0.071988	0.079520	0.087644	0.096395	0.105814	0.115945	0.126833	0.138528	0.151083
49	0.072109	0.079651	0.087784	0.096546	0.105977	0.116120	0.127021	0.138730	0.151299
50	0.072230	0.079781	0.087925	0.096698	0.106140	0.116296	0.127209	0.138932	0.151517
51	0.072351	0.079912	0.088066	0.096850	0.106304	0.116471	0.127398	0.139134	0.151734
52	0.072473	0.080043	0.088207	0.097002	0.106467	0.116647	0.127587	0.139337	0.151952
53	0.072594	0.080174	0.088348	0.097154	0.106631	0.116823	0.127776	0.139540	0.152169
54	0.072716	0.080306	0.088490	0.097306	0.106795	0.116999	0.127965	0.139743	0.152388
55	0.072838	0.080437	0.088631	0.097459	0.106959	0.117175	0.128155	0.139947	0.152606
56	0.072960	0.080569	0.088773	0.097611	0.107123	0.117352	0.128344	0.140151	0.152825
57	0.073082	0.080700	0.088915	0.097764	0.107288	0.117529	0.128534	0.140355	0.153044
58	0.073204	0.080832	0.089057	0.097917	0.107452	0.117706	0.128725	0.140559	0.153263
59	0.073326	0.080964	0.089200	0.098071	0.107617	0.117883	0.128915	0.140763	0.153482
60	0.073449	0.081097	0.089342	0.098224	0.107782	0.118061	0.129106	0.140968	0.153702

Involute Functions for Angles from 41 to 50 Degrees

	Degrees								
	41	42	43	44	45	46	47	48	49
Minutes	Involute Functions								
0	0.153702	0.167366	0.182024	0.197744	0.214602	0.232679	0.252064	0.272855	0.295157
1	0.153922	0.167602	0.182277	0.198015	0.214893	0.232991	0.252399	0.273214	0.295542
2	0.154142	0.167838	0.182530	0.198287	0.215184	0.233304	0.252734	0.273573	0.295928
3	0.154362	0.168075	0.182784	0.198559	0.215476	0.233616	0.253069	0.273933	0.296314
4	0.154583	0.168311	0.183038	0.198832	0.215768	0.233930	0.253405	0.274293	0.296701
5	0.154804	0.168548	0.183292	0.199104	0.216061	0.234243	0.253742	0.274654	0.297088
6	0.155025	0.168786	0.183547	0.199377	0.216353	0.234557	0.254078	0.275015	0.297475
7	0.155247	0.169023	0.183801	0.199651	0.216646	0.234871	0.254415	0.275376	0.297863
8	0.155469	0.169261	0.184057	0.199924	0.216940	0.235186	0.254753	0.275738	0.298251
9	0.155691	0.169500	0.184312	0.200198	0.217234	0.235501	0.255091	0.276101	0.298640
10	0.155913	0.169738	0.184568	0.200473	0.217528	0.235816	0.255429	0.276464	0.299029
11	0.156135	0.169977	0.184824	0.200747	0.217822	0.236132	0.255767	0.276827	0.299419
12	0.156358	0.170216	0.185080	0.201022	0.218117	0.236448	0.256106	0.277191	0.299809
13	0.156581	0.170455	0.185337	0.201297	0.218412	0.236765	0.256446	0.277555	0.300200
14	0.156805	0.170695	0.185594	0.201573	0.218708	0.237082	0.256786	0.277919	0.300591
15	0.157028	0.170935	0.185851	0.201849	0.219004	0.237399	0.257126	0.278284	0.300983
16	0.157252	0.171175	0.186109	0.202125	0.219300	0.237717	0.257467	0.278649	0.301375
17	0.157476	0.171415	0.186367	0.202401	0.219596	0.238035	0.257808	0.279015	0.301767
18	0.157701	0.171656	0.186625	0.202678	0.219893	0.238353	0.258149	0.279381	0.302160
19	0.157925	0.171897	0.186883	0.202956	0.220190	0.238672	0.258491	0.279748	0.302553
20	0.158150	0.172138	0.187142	0.203233	0.220488	0.238991	0.258833	0.280115	0.302947
21	0.158375	0.172380	0.187401	0.203511	0.220786	0.239310	0.259176	0.280483	0.303342
22	0.158601	0.172621	0.187661	0.203789	0.221084	0.239630	0.259519	0.280851	0.303736
23	0.158826	0.172864	0.187920	0.204067	0.221383	0.239950	0.259862	0.281219	0.304132
24	0.159052	0.173106	0.188180	0.204346	0.221682	0.240271	0.260206	0.281588	0.304527
25	0.159279	0.173349	0.188440	0.204625	0.221981	0.240592	0.260550	0.281957	0.304924
26	0.159505	0.173592	0.188701	0.204905	0.222281	0.240913	0.260895	0.282327	0.305320
27	0.159732	0.173835	0.188962	0.205185	0.222581	0.241235	0.261240	0.282697	0.305718
28	0.159959	0.174078	0.189223	0.205465	0.222881	0.241557	0.261585	0.283067	0.306115
29	0.160186	0.174322	0.189485	0.205745	0.223182	0.241879	0.261931	0.283438	0.306513
30	0.160414	0.174566	0.189746	0.206026	0.223483	0.242202	0.262277	0.283810	0.306912
31	0.160642	0.174811	0.190009	0.206307	0.223784	0.242525	0.262624	0.284182	0.307311
32	0.160870	0.175055	0.190271	0.206588	0.224086	0.242849	0.262971	0.284554	0.307710
33	0.161098	0.175300	0.190534	0.206870	0.224388	0.243173	0.263318	0.284927	0.308110
34	0.161327	0.175546	0.190797	0.207152	0.224690	0.243497	0.263666	0.285300	0.308511
35	0.161555	0.175791	0.191060	0.207434	0.224993	0.243822	0.264014	0.285673	0.308911
36	0.161785	0.176037	0.191324	0.207717	0.225296	0.244147	0.264363	0.286047	0.309313
37	0.162014	0.176283	0.191588	0.208000	0.225600	0.244472	0.264712	0.286422	0.309715
38	0.162244	0.176529	0.191852	0.208284	0.225904	0.244798	0.265062	0.286797	0.310117
39	0.162474	0.176776	0.192116	0.208567	0.226208	0.245125	0.265412	0.287172	0.310520
40	0.162704	0.177023	0.192381	0.208851	0.226512	0.245451	0.265762	0.287548	0.310923
41	0.162934	0.177270	0.192646	0.209136	0.226817	0.245778	0.266113	0.287924	0.311327
42	0.163165	0.177518	0.192912	0.209420	0.227123	0.246106	0.266464	0.288301	0.311731
43	0.163396	0.177766	0.193178	0.209705	0.227428	0.246433	0.266815	0.288678	0.312136
44	0.163628	0.178014	0.193444	0.209991	0.227734	0.246761	0.267167	0.289056	0.312541
45	0.163859	0.178262	0.193710	0.210276	0.228041	0.247090	0.267520	0.289434	0.312947
46	0.164091	0.178511	0.193977	0.210562	0.228347	0.247419	0.267872	0.289812	0.313353
47	0.164323	0.178760	0.194244	0.210849	0.228654	0.247748	0.268225	0.290191	0.313759
48	0.164556	0.179009	0.194511	0.211136	0.228962	0.248078	0.268579	0.290570	0.314166
49	0.164788	0.179259	0.194779	0.211423	0.229270	0.248408	0.268933	0.290950	0.314574
50	0.165021	0.179509	0.195047	0.211710	0.229578	0.248738	0.269287	0.291330	0.314982
51	0.165254	0.179759	0.195315	0.211998	0.229886	0.249069	0.269642	0.291711	0.315391
52	0.165488	0.180009	0.195584	0.212286	0.230195	0.249400	0.269998	0.292092	0.315800
53	0.165722	0.180260	0.195853	0.212574	0.230504	0.249732	0.270353	0.292474	0.316209
54	0.165956	0.180511	0.196122	0.212863	0.230814	0.250064	0.270709	0.292856	0.316619
55	0.166190	0.180763	0.196392	0.213152	0.231124	0.250396	0.271066	0.293238	0.317029
56	0.166425	0.181014	0.196661	0.213441	0.231434	0.250729	0.271423	0.293621	0.317440
57	0.166660	0.181266	0.196932	0.213731	0.231745	0.251062	0.271780	0.294004	0.317852
58	0.166895	0.181518	0.197202	0.214021	0.232056	0.251396	0.272138	0.294388	0.318264
59	0.167130	0.181771	0.197473	0.214311	0.232367	0.251730	0.272496	0.294772	0.318676
60	0.167366	0.182024	0.197744	0.214602	0.232679	0.252064	0.272855	0.295157	0.319089

Compound Angles

Table 1. Formulas for Compound Angles

<table>
<tr>
<td>
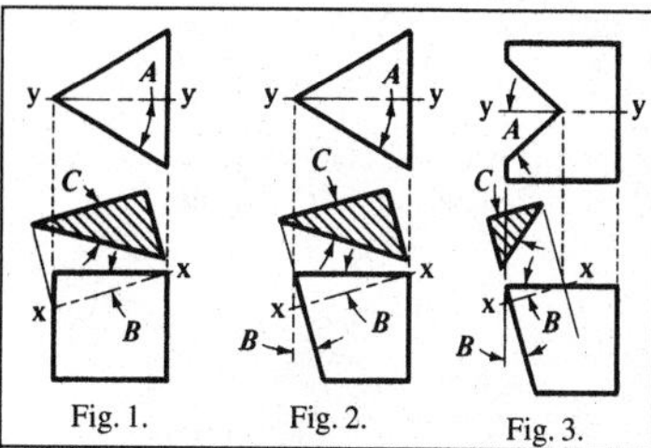

Fig. 1. Fig. 2. Fig. 3.
</td>
<td>
For given angles A and B, find the resultant angle C in plane x–x. Angle B is measured in vertical plane y–y of midsection.

(Fig. 1) $\tan C = \tan A \times \cos B$

(Fig. 2) $\tan C = \frac{\tan A}{\cos B}$

(Fig. 3) (Same formula as for Fig. 2)
</td>
</tr>
<tr>
<td>
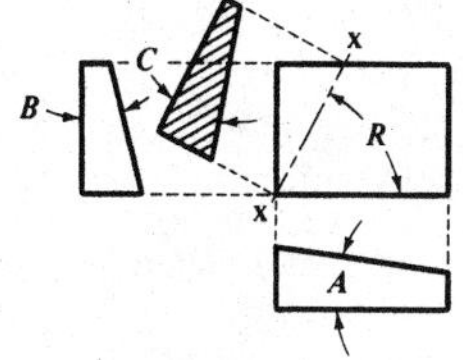

Fig. 4.
</td>
<td>
Fig. 4. In machining plate to angles A and B, it is held at angle C in plane x–x. Angle of rotation R in plane parallel to base (or complement of R) is for locating plate so that plane x–x is perpendicular to axis of pivot on angle-plate or work-holding vise.

$\tan R = \frac{\tan B}{\tan A}$; $\tan C = \frac{\tan A}{\cos R}$
</td>
</tr>
<tr>
<td>
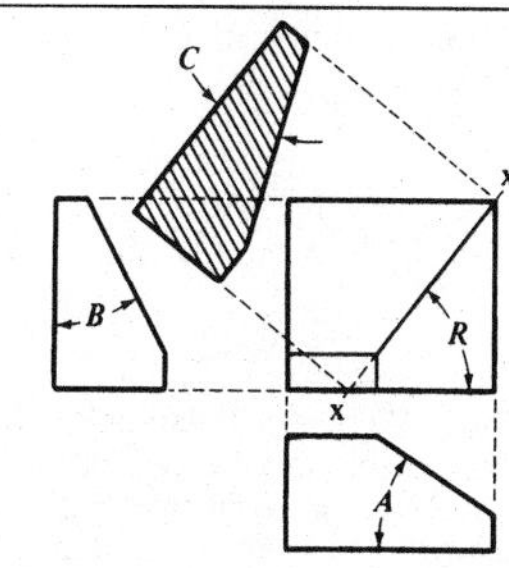

Fig. 5.
</td>
<td>
Fig. 5. Angle R in horizontal plane parallel to base is angle from plane x–x to side having angle A.

$\tan R = \frac{\tan A}{\tan B}$

$\tan C = \tan A \cos R = \tan B \sin R$
Compound angle C is angle in plane x–x from base to corner formed by intersection of planes inclined to angles A and B. This formula for C may be used to find cot of complement of C_1, Fig. 6.
</td>
</tr>
<tr>
<td>
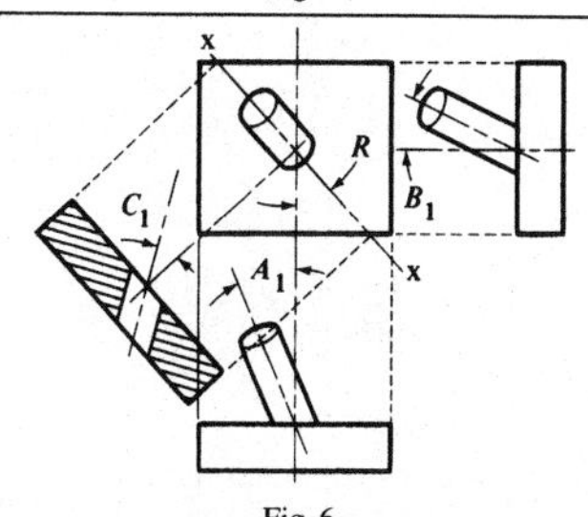

Fig. 6.
</td>
<td>
Fig. 6. Angles A_1 and B_1 are measured in vertical planes of front and side elevations. Plane x–x is located by angle R from center-line or from plane of angle B_1.

$\tan R = \frac{\tan A_1}{\tan B_1}$

$\tan C_1 = \frac{\tan A_1}{\sin R} = \frac{\tan B_1}{\cos R}$

The resultant angle C_1 would be required in drilling hole for pin.
</td>
</tr>
</table>

C = compound angle in plane x–x and is the resultant of angles A and B

Three types of compound angles are illustrated by Figs. 1 through 6. The first type is shown in Figs. 1, 2, and 3; the second in Fig. 4; and the third in Figs. 5 and 6.

In Fig. 1 is shown what might be considered as a thread-cutting tool without front clearance. A is a known angle in plane y–y of the top surface. C is the corresponding angle in plane x–x that is at some given angle B with plane y–y. Thus, angles A and B are components of the compound angle C.

Example Problem Referring to Fig. 1: Angle $2A$ in plane y–y is known, as is also angle B between planes x–x and y–y. It is required to find compound angle $2C$ in plane x–x.

Solution:

$$\text{Let } 2A = 60 \text{ and } B = 15$$

Then

$$\tan C = \tan A \cos B$$
$$\tan C = \tan 30 \cos 15$$
$$\tan C = 0.57735 \times 0.96592$$
$$\tan C = 0.55767$$
$$C = 29\ 8.8' \qquad 2C = 58\ 17.6'$$

Fig. 2 shows a thread-cutting tool with front clearance angle B. Angle A equals one-half the angle between the cutting edges in plane y–y of the top surface and compound angle C is one-half the angle between the cutting edges in a plane x–x at right angles to the inclined front edge of the tool. The angle between planes y–y and x–x is, therefore, equal to clearance angle B.

Example Problem Referring to Fig. 2: Find the angle $2C$ between the front faces of a thread-cutting tool having a known clearance angle B, which will permit the grinding of these faces so that their top edges will form the desired angle $2A$ for cutting the thread.

Solution:

$$\text{Let } 2A = 60 \text{ and } B = 15$$

Then

$$\tan C = \frac{\tan A}{\cos B} = \frac{\tan 30°}{\cos 15°} = \frac{0.57735}{0.96592}$$
$$\tan C = 0.59772$$
$$C = 30\ 52' \qquad 2C = 61\ 44'$$

In Fig. 3 is shown a form-cutting tool in which the angle A is one-half the angle between the cutting edges in plane y–y of the top surface; B is the front clearance angle; and C is one-half the angle between the cutting edges in plane x–x at right angles to the front edges of the tool. The formula for finding angle C when angles A and B are known is the same as that for Fig. 2.

Example Problem Referring to Fig. 3: Find the angle $2C$ between the front faces of a form-cutting tool having a known clearance angle B that will permit the grinding of these faces so that their top edges will form the desired angle $2A$ for form cutting.

Solution:

$$\text{Let } 2A = 46 \text{ and } B = 12$$

Then

$$\tan C = \frac{\tan A}{\cos B} = \frac{\tan 23°}{\cos 12°} = \frac{0.42447}{0.97815}$$
$$\tan C = 0.43395$$
$$C = 23\ 27.5' \qquad 2C = 46\ 55'$$

In Fig. 4 is shown a wedge-shaped block, the top surface of which is inclined at compound angle C with the base in a plane at right angles with the base and at angle R with the front edge. Angle A in the vertical plane of the front of the plate and angle B in the vertical plane of one side that is at right angles to the front are components of angle C.

Example Problem Referring to Fig. 4: Find the compound angle C of a wedge-shaped block having known component angles A and B in sides at right angles to each other.

Solution:

Let $A = 47\ 14'$ and $B = 38\ 10'$

$$\tan R = \frac{\tan B}{\tan A} = \frac{\tan 38°10'}{\tan 47°14'} \qquad \tan C = \frac{\tan A}{\cos R} = \frac{\tan 47°14'}{\cos 36°0.9'}$$

$$\tan R = \frac{0.78598}{1.0812} = 0.72695 \qquad \tan C = \frac{1.0812}{0.80887} = 1.3367$$

$$R = 36°09' \qquad C = 53°12'$$

In Fig. 5 is shown a four-sided block, two sides of which are at right angles to each other and to the base of the block. The other two sides are inclined at an oblique angle with the base. Angle C is a compound angle formed by the intersection of these two inclined sides and the intersection of a vertical plane passing through x–x, and the base of the block. The components of angle C are angles A and B and angle R is the angle in the base plane of the block between the plane of angle C and the plane of angle A.

Example Problem Referring to Fig. 5**:** Find the angles C and R in the block shown in Fig. 5 when angles A and B are known.

Solution:

Let angle $A = 27°$ and $B = 36°$

$$\tan R = \frac{\cot B}{\cot A} = \frac{\cot 36°}{\cot 27°} = \frac{1.3764}{1.9626} \qquad \cot C = \sqrt{\cot^2 A + \cot^2 B}$$

$$\tan R = 0.70131 \qquad R = 35°2.5' \qquad = \sqrt{1.9626^2 + 1.3764^2}$$

$$= \sqrt{5.74627572} = 2.3971$$

$$C = 22°38.6'$$

Example Problem Referring to Fig. 6: A rod or pipe is inserted into a rectangular block at an angle. Angle C_1 is the compound angle of inclination (measured from the vertical) in a plane passing through the center line of the rod or pipe and at right angles to the top surface of the block. Angles A_1 and B_1 are the angles of inclination of the rod or pipe when viewed respectively in the front and side planes of the block. Angle R is the angle between the plane of angle C_1 and the plane of angle B_1. Find angles C_1 and R when a rod or pipe is inclined at known angles A_1 and B_1.

Solution:

Let $A_1 = 39$ and $B_1 = 34$

Then

$$\tan C_1 = \sqrt{\tan^2 A_1 + \tan^2 B_1} = \sqrt{0.80978^2 + 0.67451^2}$$

$$\tan C_1 = \sqrt{1.1107074} = 1.0539$$

$$C_1 = 46\ 30.2'$$

$$\tan R = \frac{\tan A_1}{\tan B_1} = \frac{0.80978}{0.67451}$$

$$\tan R = 1.2005 \qquad R = 50\ 12.4'$$

ENGINEERING DRAWINGS

The following information represents the standard methods and procedures used to prepare engineering drawings. [This section is taken from ANSI Y14.5M–1994, Dimensioning and Tolerancing, by permission of the American Society of Mechanical Engineers.]

Fundamental Rules

Dimensioning and tolerancing shall clearly define engineering intent and shall conform to the following:

A) Each dimension shall have a tolerance, except for those dimensions specifically identified as reference, maximum, minimum, or stock (commercial stock size). The tolerance may be applied directly to the dimension (or indirectly in the case of basic dimensions), indicated by a general note, or located in a supplementary block or the drawing format ANSI Y14.1.

B) Dimensioning and tolerancing shall be complete so there is full understanding of the characteristics of each feature. Neither scaling (measuring the size of a feature directly from an engineering drawing) nor assumption of a distance or size is permitted, except as follows: Undimensioned drawings, such as loft, printed wiring, templates, and master layouts prepared on stable material, are excluded, provided the necessary control dimensions are specified.

C) Each necessary dimension of an end product shall be shown. No more dimensions than those necessary for complete definition shall be given. The use of reference dimensions on a drawing should be minimized.

D) Dimensions shall be selected and arranged to suit the function and mating relationship of a part and shall not be subject to more than one interpretation.

E) The drawing should define a part without specifying manufacturing methods. Thus, for example, the diameter of a hole is given without indicating whether it is to be drilled, (reamed, punched, or made by any other operation. However, in those instances where manufacturing processing, quality assurance, or environmental information is essential to the definition of engineering requirements, it shall be specified on the drawing or in a document referenced on the drawing.

F) It is permissible to identify as nonmandatory certain processing dimensions that provide for finish allowance, shrink allowance, and other requirements, provided the final dimensions are given on the drawing. Nonmandatory processing dimensions shall be identified by an appropriate note, such as *nonmandatory (mfg. data).*

G) Dimensions should be arranged to provide required information for optimum readability. Dimensions should be shown in true profile views and refer to visible outlines.

H) Wires, cables, sheets, rods, and other materials manufactured to gage or code numbers shall be specified by linear dimensions indicating the diameter or thickness. Gage or code numbers may be shown in parentheses following the dimension

I) A 90° angle applies where center lines and lines depicting features on a drawing are shown at right angles and no angle is specified.

J) A 90° basic angle applies where center lines of features in a pattern or surfaces shown at right angles on the drawing are located or defined by basic dimensions with no angle is specified

K) Unless otherwise specified, all dimensions are applicable at 20°C (68°F). Compensation may be made for measurements made at other temperatures.

L) All dimensions and tolerances apply in a free state condition. This principle does not apply for full depth, length, and width of a feature.

M) Unless otherwise specified, all geometric tolerances apply for full depth, length, and width of the feature.

N) Dimensions and tolerances apply only at the drawing level where they are specified. A dimension specified for a given feature on one level of drawing (for example, a detail drawing), is not mandatory for feature at any other level (for example, an assembly drawing).

Units of Measurement.—Note that the unit of measurement selected (SI or U.S. customary linear units) should be in accordance with the dimensioning policy of the user.

SI (Metric) Linear Units: The commonly used SI linear unit used on engineering drawings is the millimeter.

U.S. Customary Linear Units: The commonly used U.S. customary linear unit used on engineering drawings is the decimal inch.

Identification of Linear Units: On drawings where all dimensions are either in millimeters or inches, individual identification of linear units is not required. However, the drawing shall contain a note stating **unless otherwise specified, all dimensions are in millimeters** (or **in inches**, as applicable)

Combination SI (Metric) and U.S. Customary Units: Where some inch dimensions are shown on a millimeter-dimensioned drawing, the abbreviation **in.** shall follow the inch values. Where some millimeter dimensions are shown on an inch-dimensioned drawing, the symbol mm shall follow the millimeter values.

Angular Units: Angular dimensions are expressed in either degrees and decimal parts of a degree or in degrees, minutes, and seconds. These latter dimensions are expressed by symbols: for degrees, °, for minutes, ′, for seconds, ″. Where degrees are indicated alone, the numerical value shall be followed by the symbol. Where only minutes or seconds are specified, the number of minutes or seconds are specified, the number of minutes or seconds shall be preceded by 0° or 0°0′, as applicable. See Fig. 1.

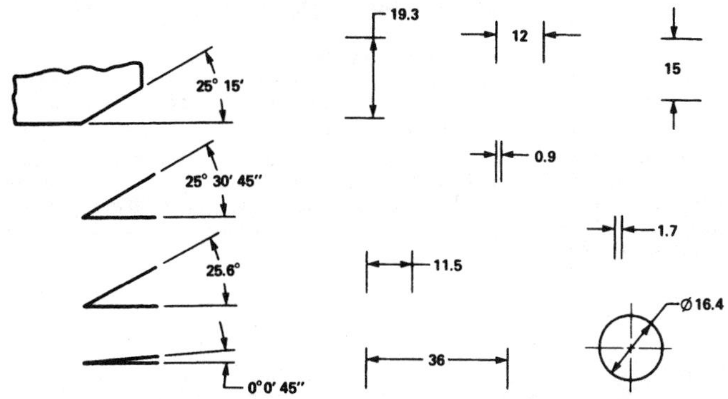

Fig. 1. Angular units

Fig. 2. Millimeter dimensions

Types of Dimensioning.—Decimal dimensioning shall be used on drawings except where certain commercial commodities are identified by standardized nominal designations, such as pipe and lumber sizes.

Millimeter Dimensioning: The following practices shall be observed when millimeter dimensions are specified on drawings:

A) Where the dimension is less than one millimeter, a zero precedes the decimal point. See Fig. 2.

B) Where the dimension is a whole number, neither the decimal point nor a zero is shown. See Fig. 2.

C) Where the dimension exceeds a whole number by a decimal fraction of one millimeter, the last digit to the right of the decimal point is not followed by a zero. See Fig. 2.

D) Neither commas nor spaces shall be used to separate digits into groups in specifying millimeter dimensions on drawings.

Decimal Inch Dimensioning: The following shall be observed where specifying decimal inch dimensions on drawings:

A) A zero is not used to the left of the decimal point for values less than one inch.

B) A dimension is expressed to the same number of decimal places as its tolerance. Zeros are added to the right of the decimal point where necessary. See Fig. 3.

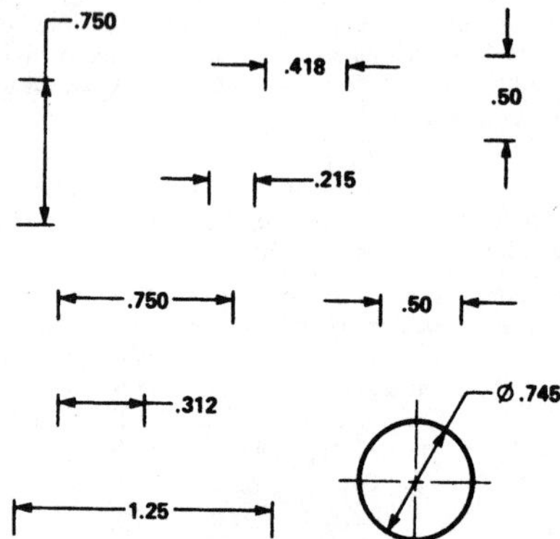

Fig. 3. Decimal inch dimensions

Decimal Points: Decimal points must be uniform, dense, and large enough to be clearly visible and meet the reproduction requirements of ASME Y14.2M. Decimal points are placed in line with the bottom of the associated digits.

Application of Dimensions.—Dimensions are applied by means of dimension lines, extension lines, chain lines, or a leader from a dimension, note, or specification directed to the appropriate feature. See Fig. 4. General notes are used to convey additional information. For further information on dimension lines, extension lines, chain lines, and leaders, see ASME Y14.2M.

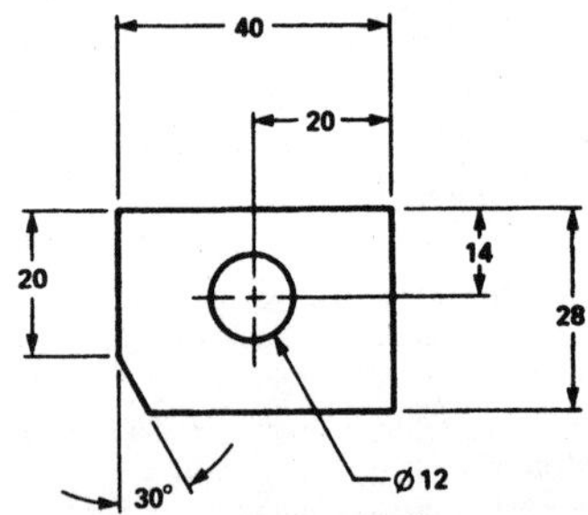

Fig. 4. Application of dimensions

Dimension Lines: A dimension line, with its arrowheads, shows the direction and extent of a dimension. Numerals indicate the number of units of a measurement. Preferably, dimension lines should be broken for insertion of numerals as shown in Fig. 4. Where horizontal dimension lines are not broken, numerals are placed above and parallel to the dimension lines.

Note: The following shall not be used as a dimension line: a center line, an extension line, a phantom line, a line that is part of the outline of the object, or a continuation of any of these lines. A dimension line is not used as an extension line, except where a simplified method of coordinate dimensioning is used to define curved outlines.

Alignment: Dimension lines shall be aligned if practicable and grouped for a uniform appearance. See Fig. 5.

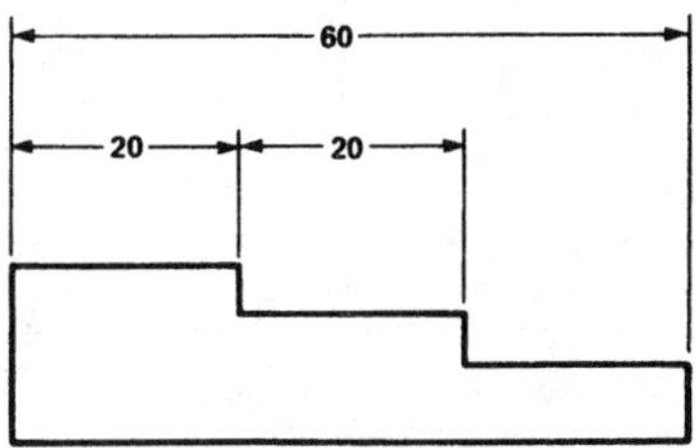

Fig. 5. Grouping of dimensions

Spacing: Dimension lines are drawn parallel to the direction of measurement. The space between the first dimension line and the part outline should be not less than 10 mm; the space between succeeding dimension lines should be not less than 6 mm. See Fig. 6.

Note: These spacings are intended as guides only. If the drawing meets the reproduction requirements of the accepted industry or military reproduction specification, nonconformance to these spacing requirements is not a basis for rejection of the drawing.

Where there are several parallel dimension lines, the numerals should be staggered for easier reading. See Fig. 7.

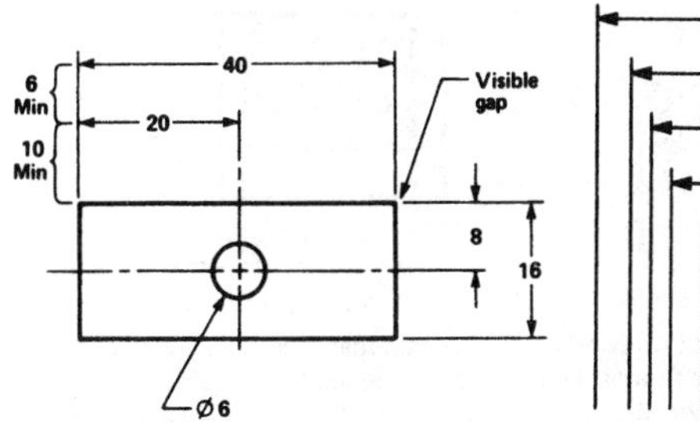

Fig. 6. Spacing of dimension lines

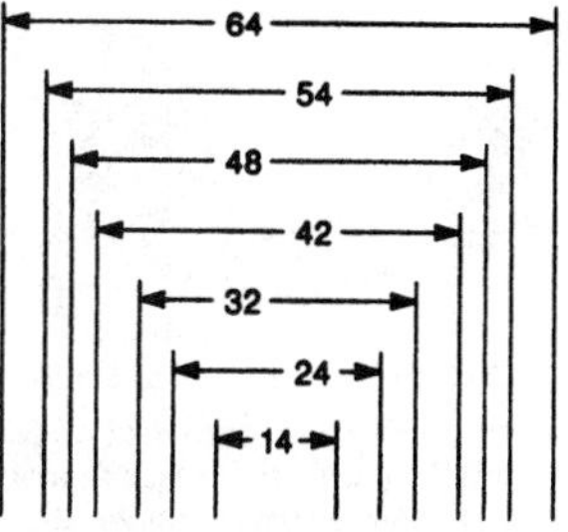

Fig. 7. Staggered Dimensions

Angle Dimensions: The dimension line of an angle is drawn with its center at the apex of the angle. The arrowheads terminate at the extensions of the two sides. See Fig. 1 and 4.

Crossing Dimension Lines: Crossing dimension lines should be avoided. Where unavoidable, the dimension lines are unbroken when crossed.

Extension (Projection) Lines: Extension lines are used to indicate the extension of a surface or point to a location preferably outside the part outline. Extension lines start with a short visible gap from the outline of the part and extend beyond the outermost related dimension line. See Fig. 6. Extension lines are drawn perpendicular to dimension lines.

Where space is limited, extension lines may be drawn in the direction in which they apply. See Fig. 8.

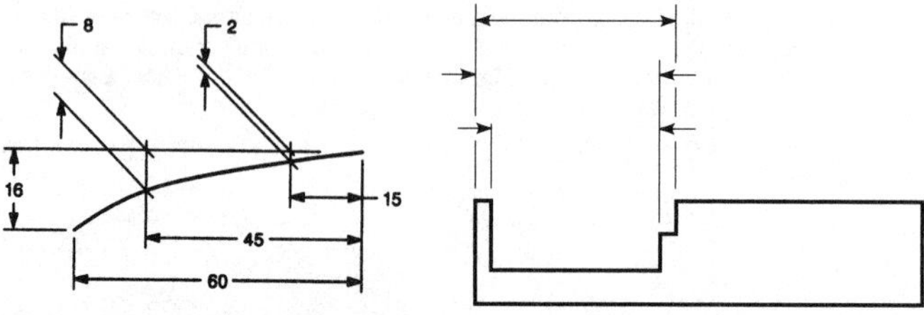

Fig. 8. Oblique Extension Lines

Fig. 9. Breaks in extension lines

Limited Length or Area Indication: Where it is desired to indicate that a limited length or area of a surface is to receive additional treatment or consideration within limits specified on the drawing, the extent of these limits may be indicated by use of a chain line.

Locating Points: Where a point is located by extension lines only, the extension lines from surfaces should pass through the point. See Fig. 10.

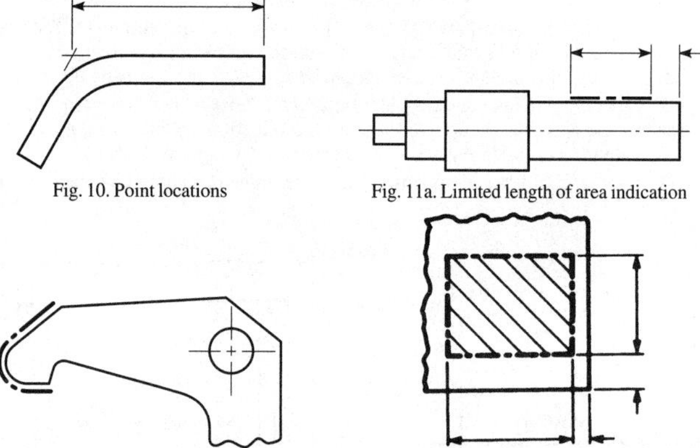

Fig. 10. Point locations

Fig. 11a. Limited length of area indication

Fig. 11b. Limited length of area indication

Fig. 11c. Limited length of area indication

Chain Lines: In an appropriate view or section, a chain line is drawn parallel to the surface profile at a short distance from it. Dimensions are added for length and location. If applied to a surface of revolution, the indication may be shown on one side only. See Fig. 11a.

Crossing Extension Lines: Wherever practicable, extension lines should neither cross one another nor cross dimension lines. To minimize such crossings, the shortest dimension line is shown nearest the outline of the object. See Fig. 7. Where extension lines must cross other extension lines, dimension lines, or lines depicting features, they are not broken. Where extension lines cross arrowheads, a break in the extension line is permissible. See Fig. 9.

Omitting Chain Line Dimensions: If the chain line clearly indicates the location and extent of the surface area, dimensions may be omitted. See Fig. 11b.

Area Indication Identification: Where the desired area is shown on a direct view of the surface, the area is section-lined within the chain line boundary and appropriately dimensioned. See Fig. 11c.

Leaders (Leader Lines): A leader is used to direct a dimension, note, or symbol to the intended place on the drawing. Normally a leader terminates in an arrowhead. However, where it is intended for a leader to refer to a surface by ending within the outline of that surface, the leader should terminate in a dot. A leader should be an inclined straight line except for a short horizontal portion extending to the mid-height of the first or last letter or digit of the note or dimension. Two or more leaders to adjacent areas on the drawing should be drawn parallel to each other. See Fig. 12.

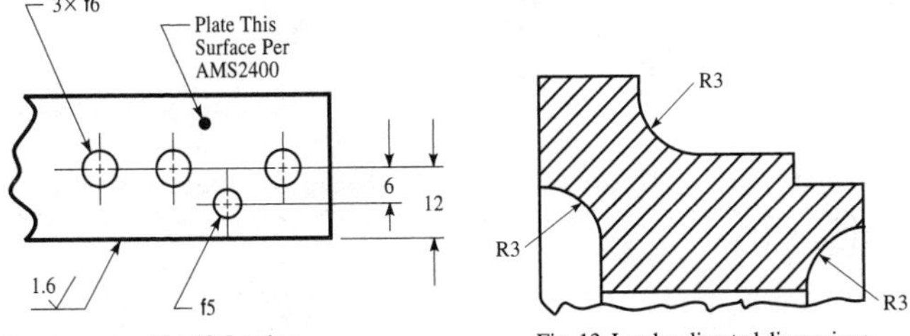

Fig. 12. Leaders

Fig. 13. Leader-directed dimensions

Leader–Directed Dimensions: Leader-directed dimensions are specified individually to avoid complicated leaders. See Fig. 13. If too many leaders would impair the legibility of the drawing, letters or symbols should be used to identify features. See Fig. 14.

Circle and Arc: Where a leader is directed to a circle or an arc, its direction should be radial. See Fig. 15.

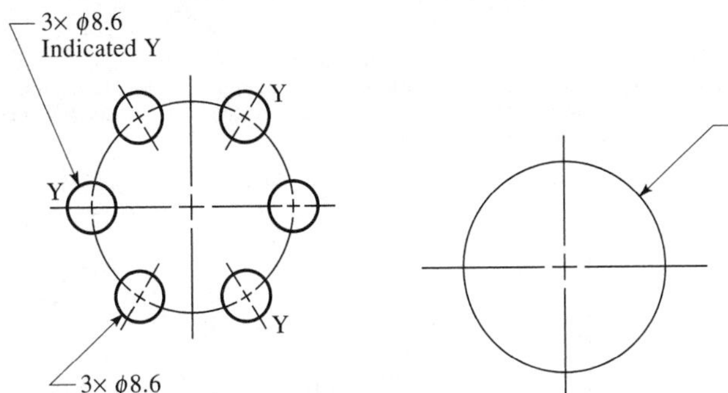

Fig. 14. Minimizing leaders

Fig. 15. Leader directions

Reading Direction: Reading direction for the following specifications apply:

Notes: Notes should be placed to be read from the bottom of the drawing with regard to the orientation of the drawing format.

Dimensions: Dimensions shown with dimension lines and arrowheads should be placed to read from the bottom of the drawing. See Fig. 16.

Baseline Dimensioning: Baseline dimensions are shown aligned to their extension lines and read from the bottom or right side of the drawing.

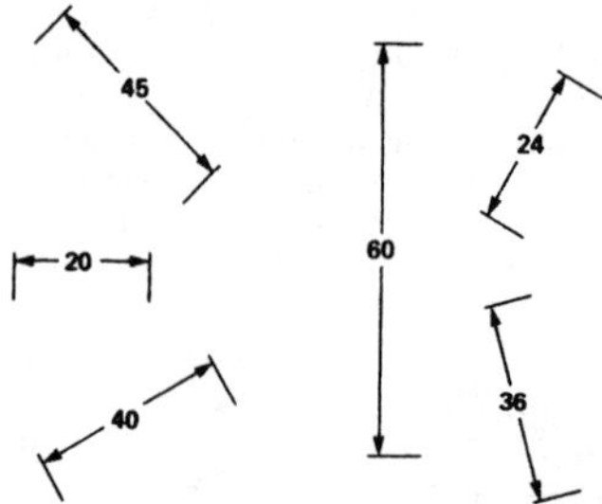

Fig. 16. Reading direction

Reference Dimensions: The method used for identifying a reference dimension (or reference data) on drawings is to enclose the dimension (or data) within parentheses. See Fig. 17 and Fig. 18.

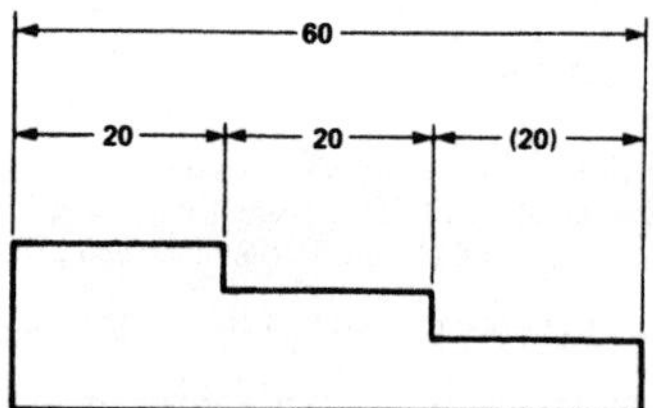

Fig. 17. Intermediate reference dimension

Overall Dimensions: Where an overall dimension is specified, one intermediate dimension is omitted or identified as a reference dimension. See Fig. 17. Where the intermediate dimensions are more important than the overall dimension, the overall dimension, if used, is identified as a reference dimension. See Fig. 18.

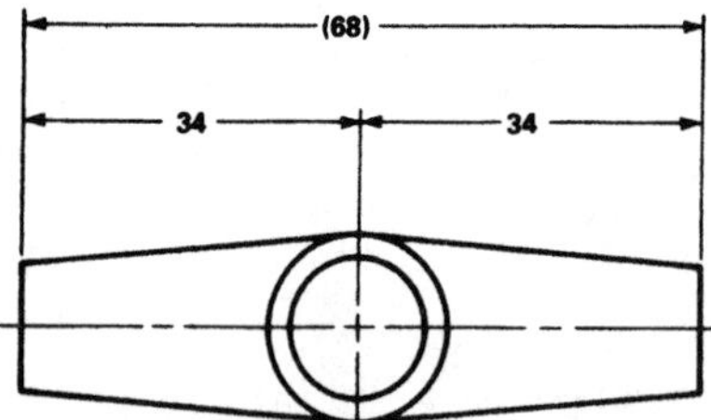

Fig. 18. Overall reference dimension

Dimensioning Within the Outline of a View: Dimensions are usually placed outside the outline of a view. Where directness of application makes it desirable, or where extension lines or leader lines would be excessively long, dimensions may be placed within the outline of a view.

Dimensions Not to Scale: Agreement should exist between the pictorial presentation of a feature and its defining dimension. Where a change to a feature is made, the following, as applicable, must be observed:

A) Where the sole authority for the product definition is a hard copy original drawing prepared either manually or on an interactive computer graphics system, and it is not feasible to update the pictorial view of the feature, the defining dimension is to be underlined with a straight thick line.

B) Where the sole authority for the product definition is a data set prepared on a computer graphics system, agreement shall be maintained between the defining dimension and the graphics presentation of the feature, in all views. The defining dimension and the true size, location, and direction of the feature shall always be in complete agreement.

Dimensioning Features.—Various characteristics and features of parts require unique methods of dimensioning.

Diameters: The diameter symbol precedes all diametral values. See Fig. 19. Where the diameter of a spherical feature is specified, the diametral value is preceded by the spherical diameter symbol. Where the diameters of a number of concentric cylindrical features are specified, such diameters should be dimensioned in a longitudinal view, if practicable.

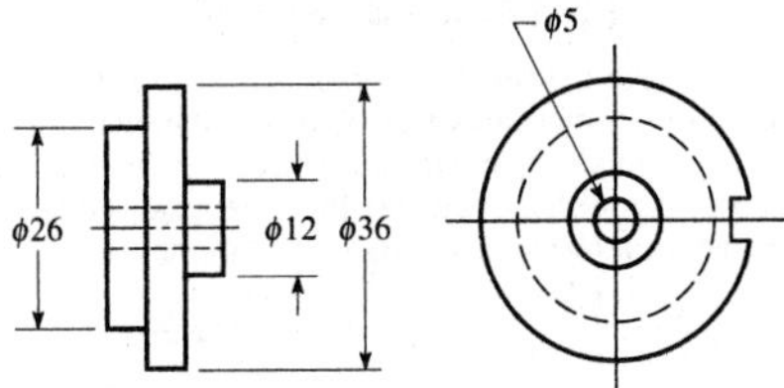

Fig. 19. Diameters

Radii: Each radius value is preceded by the appropriate radius symbol. See Fig. 20. A radius dimension line uses one arrowhead, located at the arc end. An arrowhead is never used at the radius center. Where location of the center is important and space permits, a dimension line is drawn from the radius center with the arrowhead touching the arc, and the dimension is placed between the arrowhead and the center. Where space is limited, the dimension line is extended through the radius center. Where it is inconvenient to place the arrowhead between the radius center and the arc, the arrowhead may be placed outside the arc with a leader. Where the center of a radius is not dimensionally located, the center shall not be indicated. See Fig. 20.

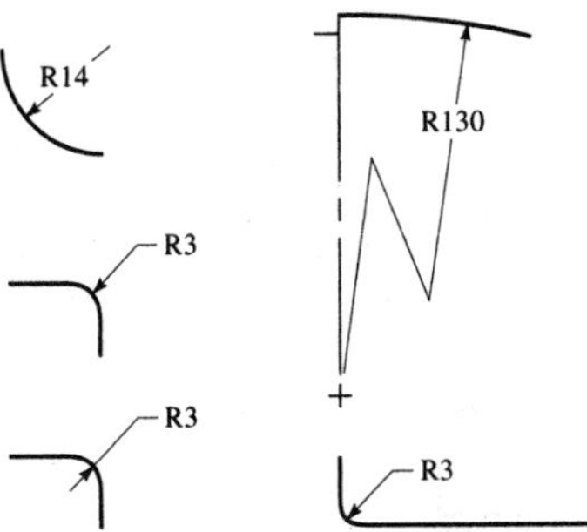

Fig. 20. Radii

Center of Radius: Where a dimension is given to the center of a radius, a small cross is drawn at the center. Extension lines and dimension lines are used to locate the center. See Fig. 21. Where location of the center is unimportant, the drawing must clearly show that the arc location is controlled by other dimension features, such as tangent surfaces. See Fig. 22.

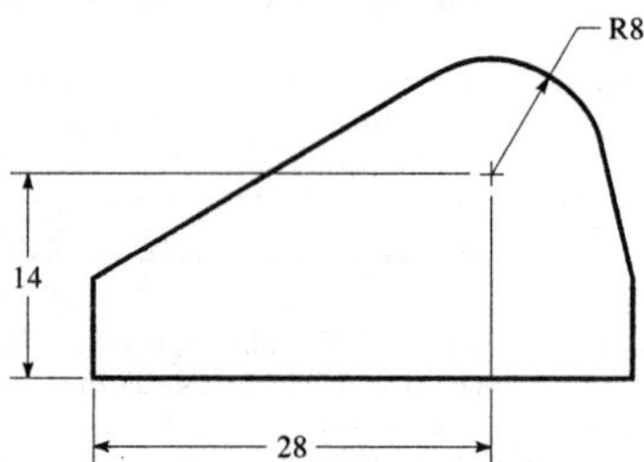

Fig. 21. Radius with located center

Foreshortened Radii: Where the center of a radius is outside the drawing or interferes with another view, the radius dimension line may be foreshortened. See Fig. 23. That portion of the dimension line extending from the arrowhead is radial relative to the arc. Where the radius dimension line is foreshortened and the center is located by coordinate dimensions, the dimension line locating the center is also foreshortened.

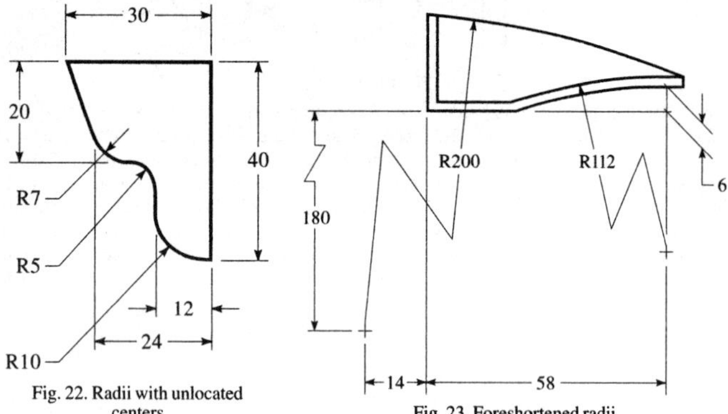

Fig. 22. Radii with unlocated centers

Fig. 23. Foreshortened radii

True Radius: Where a radius is dimensioned in a view that does not show the true shape of the radius, TRUE R is placed before the radius dimension. See Fig. 24a and 24b.

Fig. 24a. True radius

Fig. 24b. True radius

Multiple Radii: Where a part has a number of radii of the same dimension, a note may be used instead of dimensioning each radius dimension.

Spherical Radii: Where a spherical surface is dimensioned by a radius, the radius dimension is preceded by the symbol **SR**. See Fig. 25.

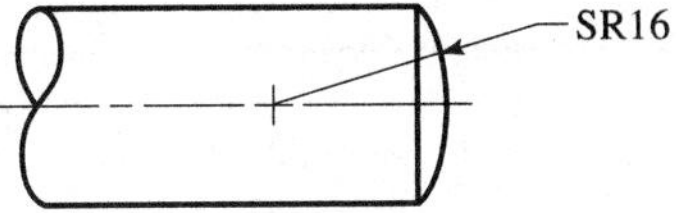

Fig. 25. Spherical radius

Chords, Arcs, and Angles: The dimensioning of chords, arcs, and angles shall be as shown in Fig. 26a, 26b and 26c.

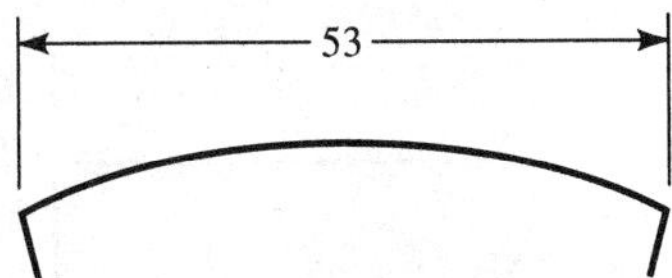

Fig. 26a. Dimensioning chords

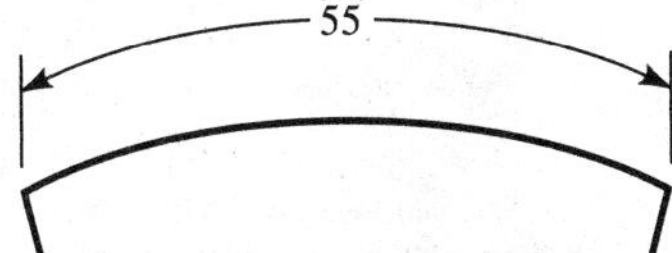

Fig. 26b. Dimensioning arcs

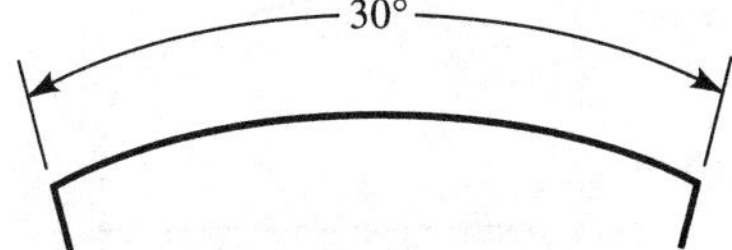

Fig. 26c. Dimensioning angles

Rounded Ends: Overall dimensions are used for features having rounded ends. For fully rounded ends, the radii are indicated, but not dimensioned. See Fig. 27. For features with partially rounded ends, the radii are dimensioned. See Fig. 28.

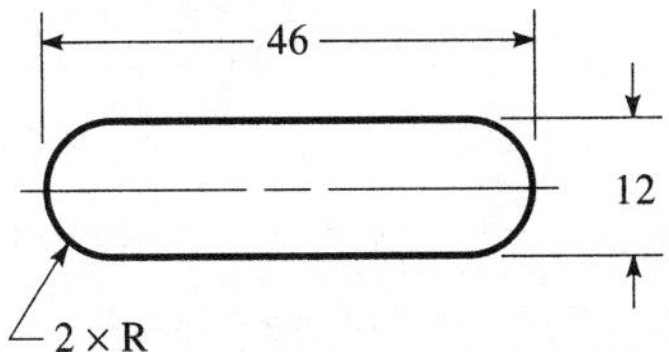

Fig. 27. Fully rounded ends

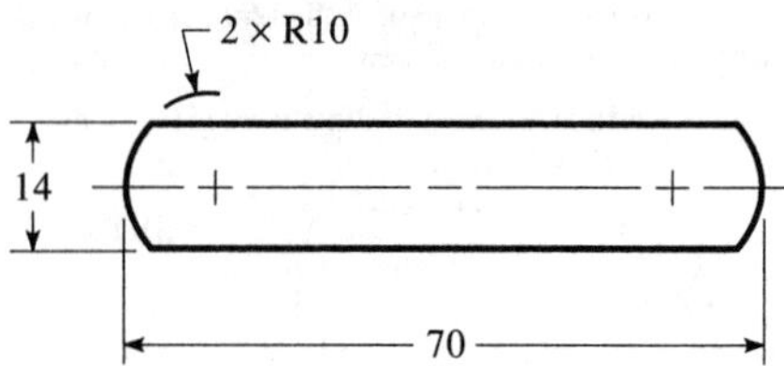

Fig. 28. Partially rounded ends

Rounded Corners: Where corners are rounded, dimensions define the edges, and the arcs are tangent. See Fig. 29.

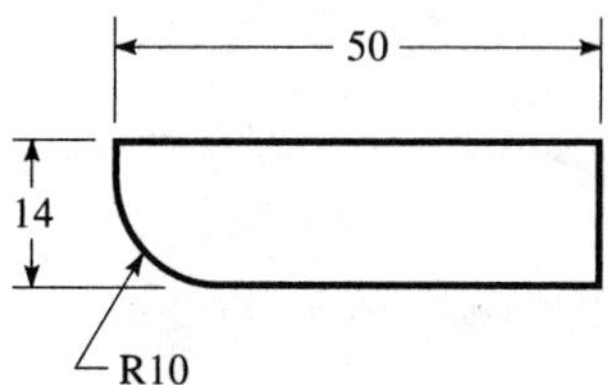

Fig. 29. Rounded corners

Outlines Consisting of Arcs: A curved outline consisting of two or more arcs is dimensioned by giving the radii of all arcs and locating the necessary centers with coordinate dimensions. Other radii are located on the basis of their points of tangency. See Fig. 30.

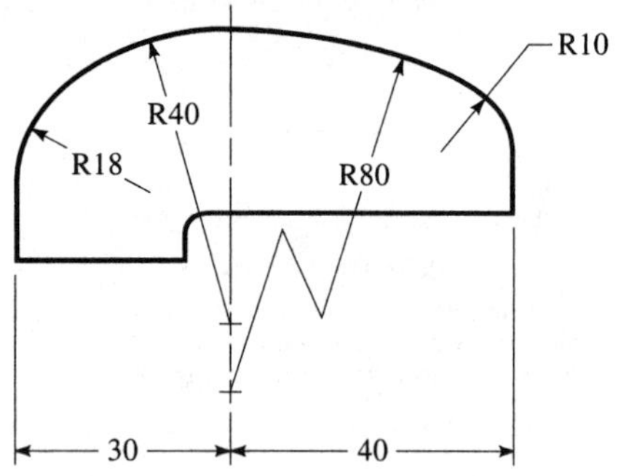

Fig. 30. Circular arc outline

Irregular Outlines: Irregular outlines may be dimensioned as shown in Fig. 31 and 32. Circular and noncircular outlines may be dimensioned by the rectangular coordinate or offset method. See Fig. 31. Coordinates are dimensioned from base lines. Where many coordinates are required to define and outline, the vertical and horizontal coordinate dimensions may be tabulated, as in Fig. 32.

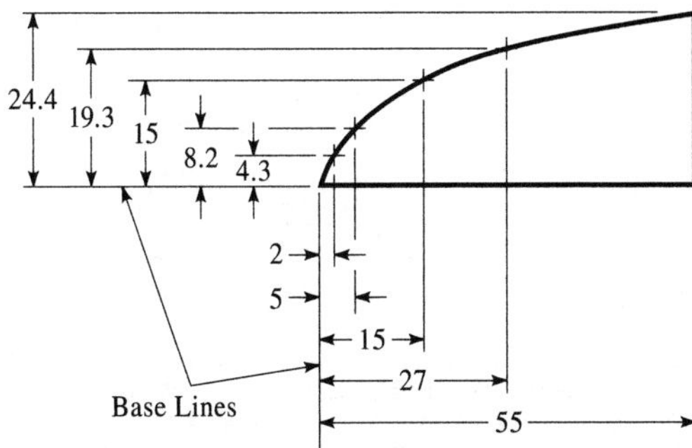

Fig. 31. Coordinate or offset outline

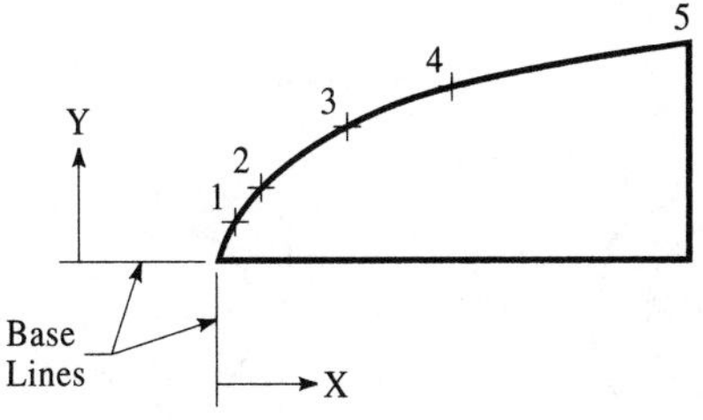

Fig. 32. Tabulated outline

Station	1	2	3	4	5
X	2	5	15	27	55
Y	4.3	8.2	15	19.3	24.4

Grid System: Curved pieces that represent patterns may be defined by a grid system with numbered grid lines.

Symmetrical Outlines: Symmetrical outlines may be dimensioned on one side of the center line of symmetry. Such is the case where, due to the size of the part or space limitations, only part of the outline can be conveniently shown. See Fig. 33. One-half the outline of the symmetrical shape is shown and symmetry is indicated by applying symbols for part symmetry to the center line. See ASME Y14.2M.

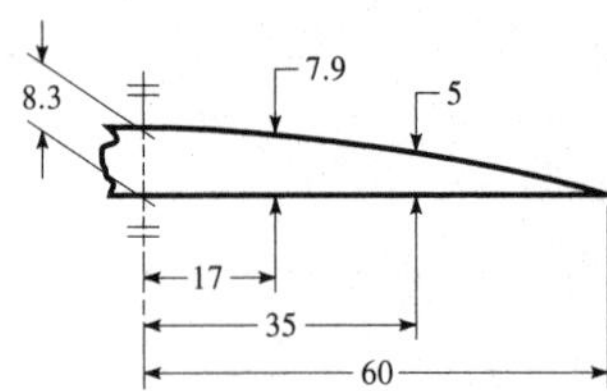

Fig. 33. Symmetrical outline

Round Holes: Round holes are dimensioned as shown in Fig. 34a, 34b, 34c, 34d, 34e, 34f and 34g. Where it is not clear that a hole goes through, the abbreviation **THRU** follows a dimension. The depth dimension of a blind hole is the depth of the full diameter from the outer surface of the part. Where the depth dimension is not clear, as from a curved surface, the depth should be dimensioned. For methods of specifying blind holes, see Fig. 34a, 34b, 34c, 36d, 34e, 34f and 34f.

Round Holes

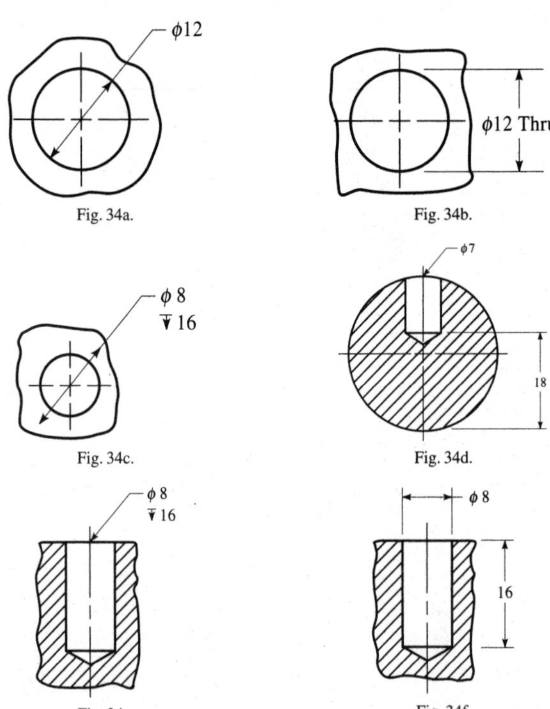

Fig. 34a.

Fig. 34b.

Fig. 34c.

Fig. 34d.

Fig. 34e.

Fig. 34f.

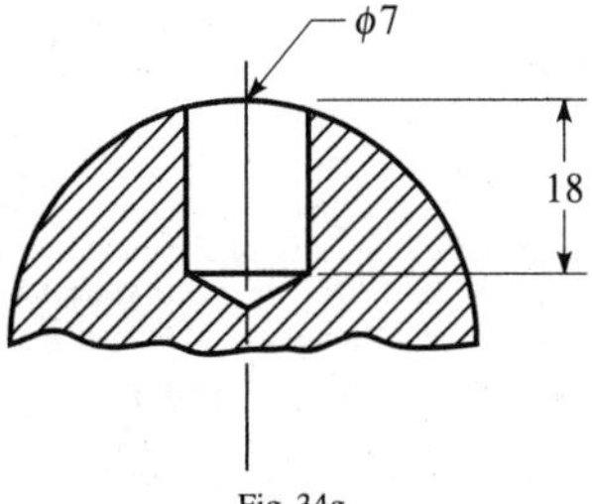

Fig. 34g.

Slotted Holes: Slotted holes are dimensioned as shown in Fig. 35a, 35b, and 35c. The end radii are indicated but not dimensioned.

Slotted Holes

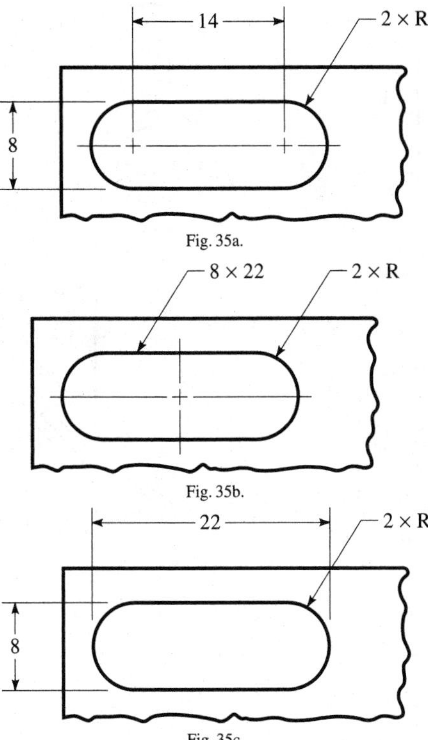

Fig. 35a.

Fig. 35b.

Fig. 35c.

Counterbored Holes: Counterbored holes may be specified as shown in Fig. 36a, 36b, 36c and 36d. Where the thickness of the remaining material has significance, this thickness (rather than the depth) is dimensioned. For holes having more than one counterbore, see Fig. 37a, 37b, 37c and 37d.

Countersunk and Counterdrilled Holes: For countersunk holes, the diameter and included angle of the countersink are specified. For counterdrilled holes, the diameter and depth of the counterdrill are specified. Specifying the included angle of the counterdrill is optional. See Fig. 38a, 38b, 38c and 38d. The depth dimension is the depth of the full diameter of the counterdrill from the outer surface of the part.

Counterbored Holes

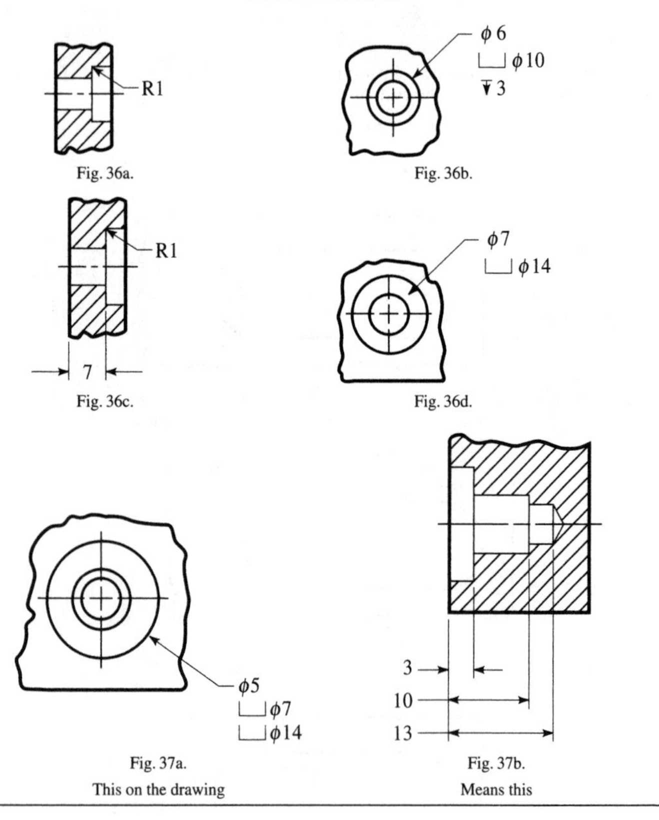

Fig. 36a. Fig. 36b. Fig. 36c. Fig. 36d. Fig. 37a. Fig. 37b.

This on the drawing — Means this

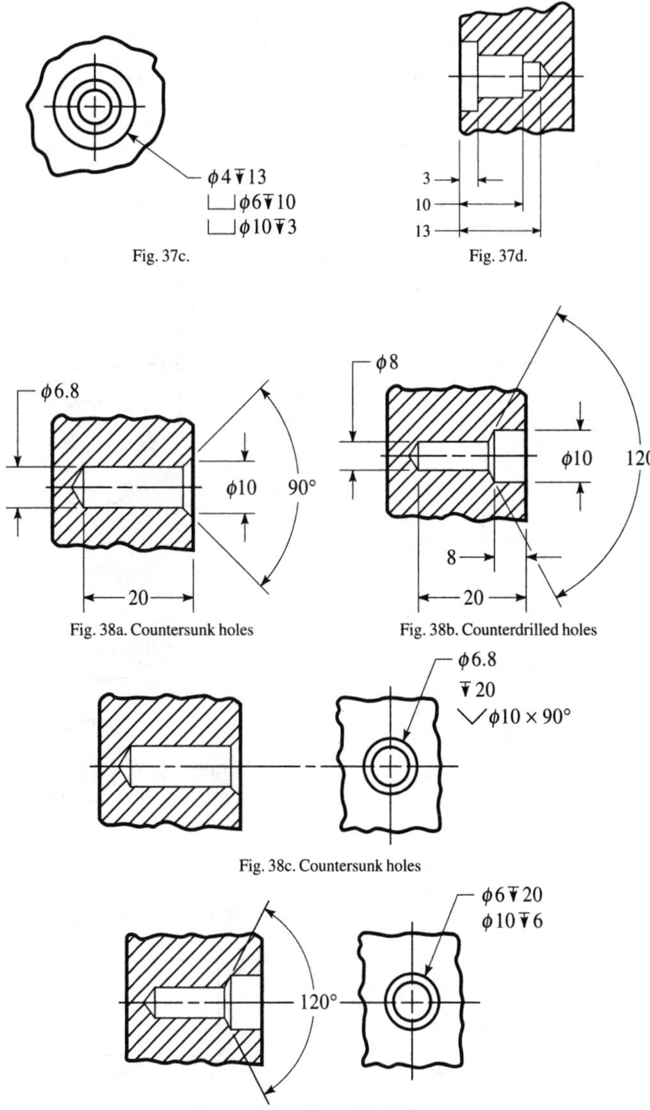

Fig. 37c.

Fig. 37d.

Fig. 38a. Countersunk holes

Fig. 38b. Counterdrilled holes

Fig. 38c. Countersunk holes

Fig. 38d. Counterdrilled holes

Chamfered and Countersunk Holes on Curved Surfaces: Where a hole is chamfered or countersunk on a curved surface, the diameter specified on the drawing applies at the minor diameter of the chamfer or countersink. See Fig. 39.

Spotfaces: The diameter of the spotfaced area is specified. Either the depth or the remaining thickness of material may be specified. See Fig. 40. A spotface may be specified by note only and need not be delineated on the drawing. If no depth or remaining thickness of material is specified, the spotface is the minimum depth necessary to clean up the surface to the specified diameter.

Machining Centers: Where machining centers are to remain on the finished part, they are indicated by a note or dimensioned on the drawing.

Chamfers: Chamfers are dimensioned by a linear dimension and an angle, or by two linear dimensions. See Fig. 41 through 44b, Where an angle and a linear dimension are specified, the linear dimension is the distance from the indicated surface of the part to the start of the chamfer. See Fig. 41.

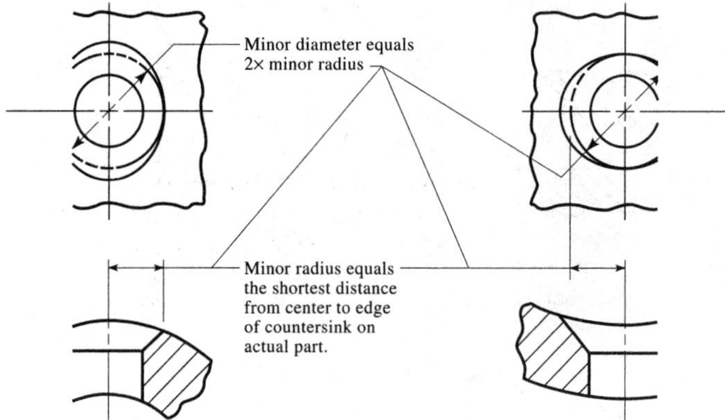

Fig. 39. Countersunk on a curved surface

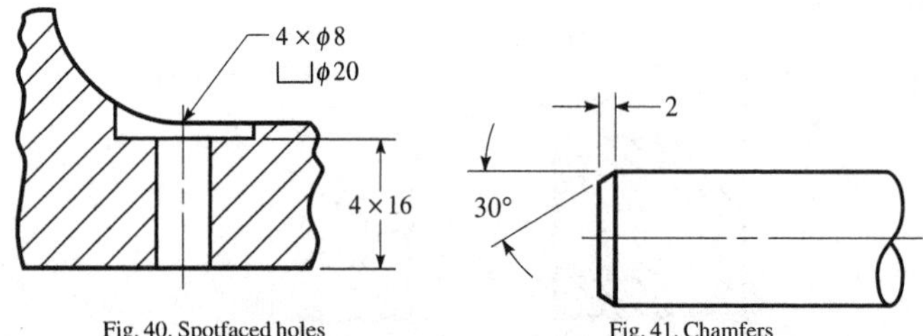

Fig. 40. Spotfaced holes

Fig. 41. Chamfers

Chamfers Specified by Note: A note may used to specify 45° chamfers, as in Fig. 42. This method is used only with 45° chamfers, as the linear value applies in either direction.

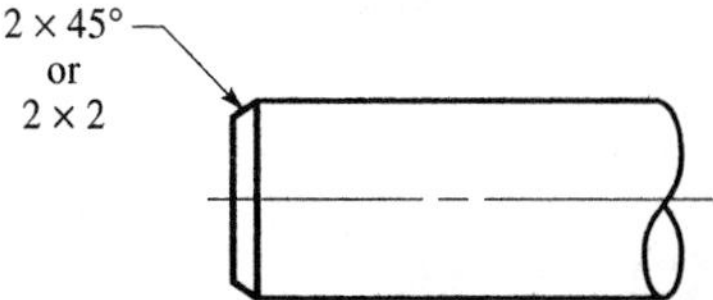

Fig. 42. 45° chamfers

Round Holes: Where the edge of a round hole is chamfered, the practice of using a note is followed, except where the chamfer diameter requires dimensional control. See Fig. 43a and 43a.This type of control may also be applied to the chamfer diameter on a shaft.

Intersecting Surfaces: Where chamfers are required for surfaces intersecting at other than right angles, the methods shown in Fig. 44a and 44b are used.

Keyseats: Keyseats are dimensioned by width, depth, location, and if required, length. The depth is dimensioned from the opposite side of the shaft or hole. See Fig. 45a and 45b.

Internal Chamfers

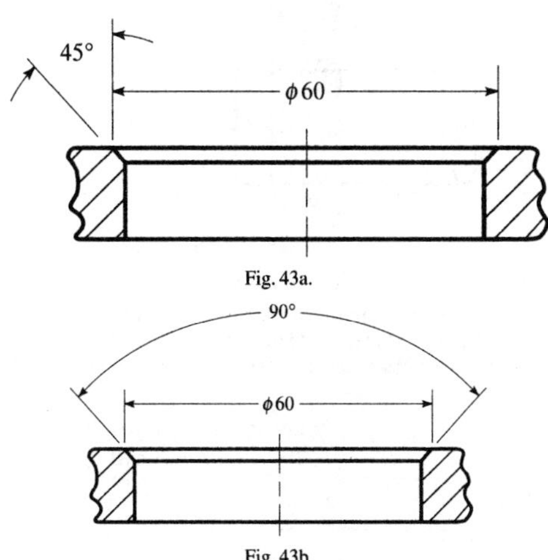

Fig. 43b.

Chamfers Between Surfaces at other than 90 Degrees

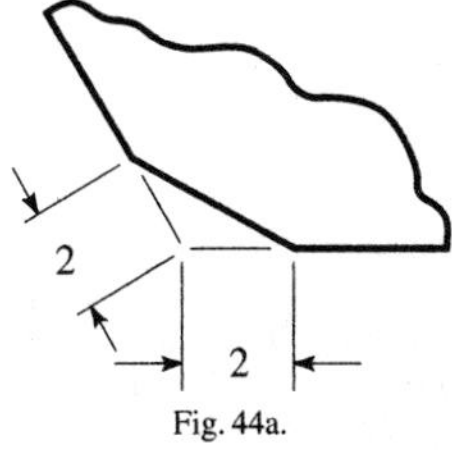

Fig. 44a.

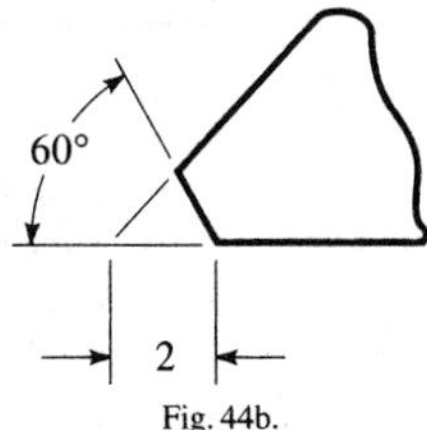

Fig. 44b.

Keyseats

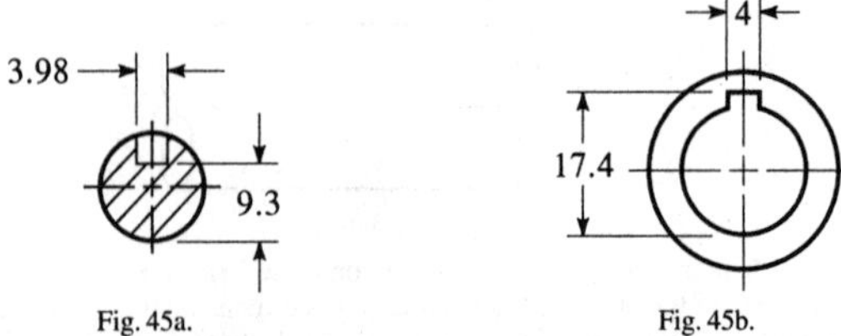

Fig. 45a. Fig. 45b.

Knurling: Knurling is specified in terms of type, pitch, and diameter before and after knurling. Where control is not required, the diameter after knurling is omitted. Where only a portion of a feature requires knurling, axial dimensioning is provided. See Fig. 46.

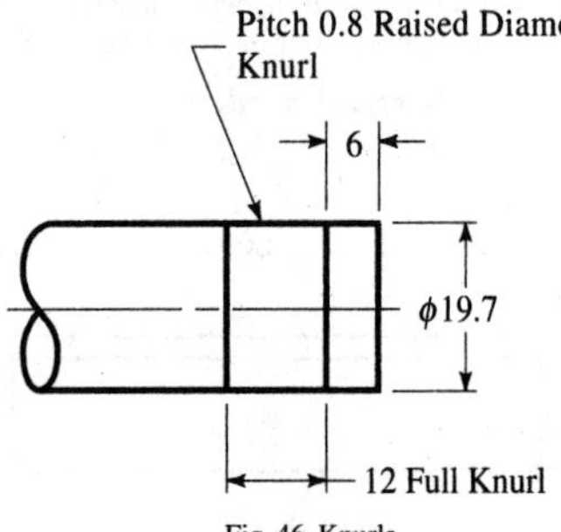

Fig. 46. Knurls

Knurling for Press Fit: Where required to provide a press fit between parts, knurling is specified by a note that includes the type of knurl required, its pitch, the toleranced diameter of the feature before knurling, and the minimum acceptable diameter after knurling. See Fig. 47.

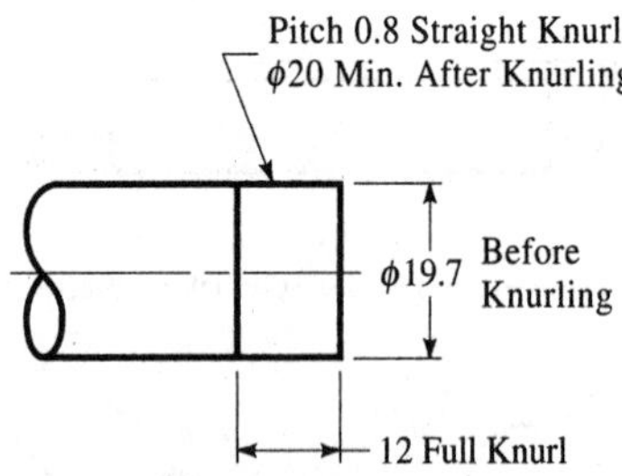

Fig. 47. Knurls for press fits

Knurling Standard: For information on inch knurling, see ANSI/ASME B94.6.

Rods and Tubing Details: Rods and tubing are dimensioned in three coordinate directions and toleranced using geometric principles or by specifying the straight lengths, bend radii, angles of bend, and angles of twist for all portions of the item. This may be done by means of auxiliary views, tabulation, or supplementary data.

Screw Threads: Methods of specifying and dimensioning screw threads are covered in ANSI Y14.36 and ANSI Y14.6aM.

Surface Texture: Methods of specifying surface texture requirements are covered in ANSI Y14.36. For additional information, see ANSI/ASME B46.1.

Gears and Involute Splines: Methods of specifying gear requirements are covered in the ASME Y14.7 series of standards. Methods of specifying involute spline requirements are covered in the ANSI B92 series of standards.

Castings and Forgings: Methods of specifying requirements peculiar to castings and forgings are covered in the ASME Y14.8.

Direct Tolerancing Methods.—Limits and directly applied tolerance values are specified as follows.

Limit Dimensioning: The high limit (maximum value) is placed above the low limit (minimum value). When expressed in a single line, the low limit precedes the high limit and a dash separates the two values. See Fig. 48.

Plus and Minus Tolerancing: The dimension is given first and is followed by a plus and minus expression of tolerance. See Fig. 49.

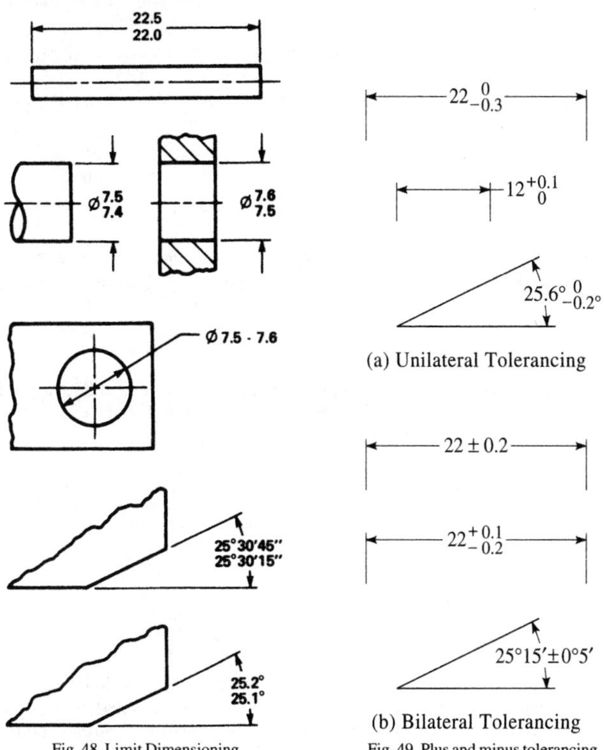

Fig. 48. Limit Dimensioning

Fig. 49. Plus and minus tolerancing

Metric Limits and Fits: For metric application of limits and fits, the tolerance may be indicated by a basic size and tolerance symbol as in Fig. 50a, 50b and 50c. See ANSI B4.2 for complete information on this system.

Limits and Tolerance Symbols: The method shown in Fig. 50a is recommended when the system is introduced by an organization. In this case, limit dimensions are specified, and the basic size and tolerance symbol are identified as reference.

Tolerance Symbol and Limits: As experience is gained, the method shown in Fig. 50b may be used. When the system is established and standard tools, gages, and stock materials are available with size and symbol identification, the method shown in Fig. 50c may be used.

Symbols for Metric Limits and Fits

$\begin{matrix}29.980\\29.959\end{matrix}$ (30 f7)	30 f7$\begin{pmatrix}29.980\\29.959\end{pmatrix}$	30 f7
Fig. 50a.	Fig. 50b.	Fig. 50c.

Tolerance Expression.— The conventions shown in the following paragraphs shall be observed pertaining to the number of decimal places carried in the tolerance.

Millimeter Tolerances: Where millimeter dimensions are used on the drawings, the following apply.

(a) Where unilateral tolerancing is used and either the plus or minus value is nil, a single zero is shown without a plus or minus sign.

Example:

$$32\begin{matrix}+0\\-0.02\end{matrix} \quad \text{or} \quad 32\begin{matrix}+0.02\\-0\end{matrix}$$

(b) Where bilateral tolerancing is used, both the plus and minus values have the same number of decimal places, using zeros where necessary.

Example:

$$32\begin{matrix}+0.25\\-0.10\end{matrix} \quad \text{not} \quad 32\begin{matrix}+0.25\\-0.1\end{matrix}$$

(c) Where limit dimensioning is used and either the maximum or minimum value has digits following a decimal point, the other value has zeros added for uniformity.

Example:

$$\begin{matrix}25.45\\25.00\end{matrix} \quad \text{not} \quad \begin{matrix}25.45\\25\end{matrix}$$

(d) Where basic dimensions are used, associated tolerances contain the number of decimal places necessary for control.

Example:

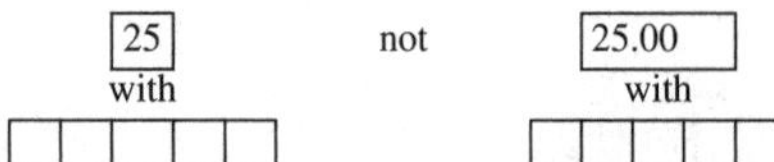

Inch Tolerances: Where inch dimensions are used on the drawing, the following apply:

(a) Where unilateral tolerancing is used and either the plus or minus value is nil, its dimension shall be expressed with the same number of decimal places, and the appropriate plus or minus sign.

Example:

$0.500\ {}^{+0.005}_{-0.000}$ not $0.500\ {}^{+0.005}_{0}$

(b) Where bilateral tolerancing is used, both the plus and minus values and the dimension have the same number of decimal places.

Example:

0.500 ± .005 not .50 ± .005

(c) Where limit dimensioning is used and either the maximum or minimum value has digits following a decimal point, the other value has zeros added for uniformity.

Example:

.750
.748 not .75
.748

(d) Where basic dimensions are used, associated tolerances contain the number of decimal places necessary for control. The basic dimension value is expressed with the same number of decimal places as the tolerance.

Example:

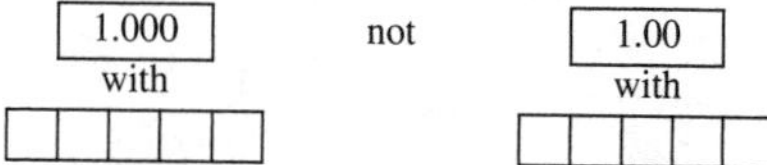

Angle Tolerances: Where angle dimensions are used, both the plus and minus values and the angle have the same number of decimal places.

Example:

25.0° ± .2° not 25° ± .2°

Interpretation of Limits.—All limits are absolute. Dimensional limits, regardless of the number of decimal places, are used as if they were continued with zeros.

Example:

12.2	means	12.20....0
12.0	means	12.00....0
12.01	means	12.010....0

To determine conformance within limits, the measured value is compared directly with the specified value and any deviation outside the specified limiting value signifies nonconformance with the limits.

Plated or Coated Parts: Where a part is to be coated or plated, the drawing or referenced document shall specify whether the dimensions are before or after plating. Typical examples of notes are the following:

1) Dimensional limits apply after plating.

2) Dimensional limits apply before plating. (For processes other than plating, substitute the appropriate term.)

Single Limits.—Min. or **Max.** is placed after a dimension where other elements of the design definitely determine the other unspecified limit. Features, such as depths of holes, lengths of threads, corner radii, chamfers, etc., may be limited in this way. Single limits are used where the intent will be clear, and the unspecified limit can be zero or approach infinity and will not result in a condition detrimental to the design.

Tolerance Accumulation.—Figure 51a, 51b and 51c compares the tolerance values resulting from three methods of dimensioning.

Chain Dimensioning: The maximum variation between two features is equal to the sum of the tolerances on the intermediate distances; this results in the greatest tolerance accumulation. In Fig. 51a the tolerance accumulation between surfaces X and Y is ±0.15.

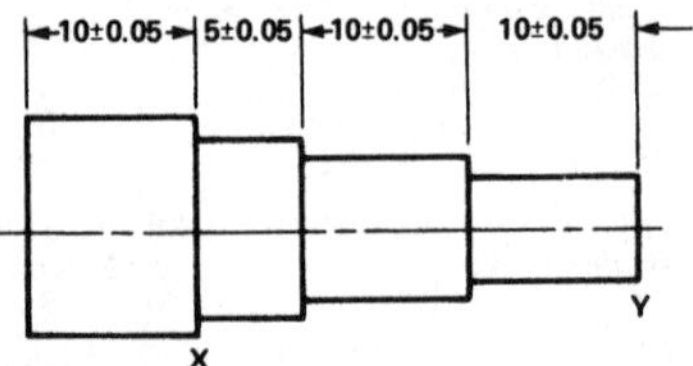

Fig. 51a. Chain dimensioning–greatest tolerance accumulation between X and Y.

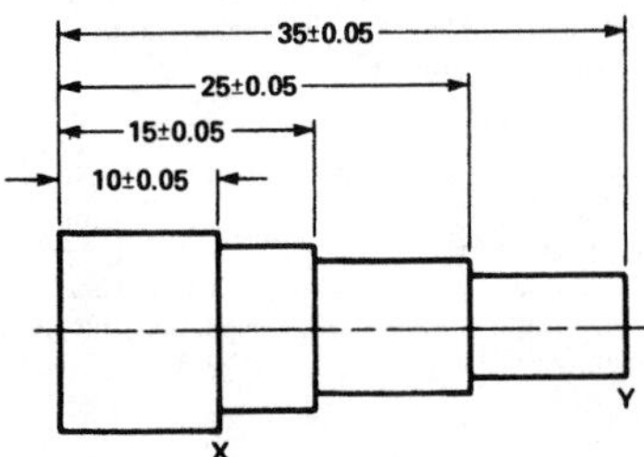

Fig. 51b. Base line dimensioning–lesser tolerance accumulation between X and Y.

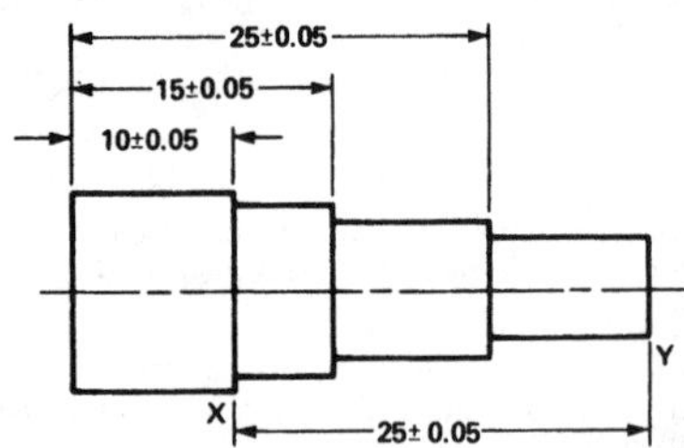

Fig. 51c. Direct dimensioning–least tolerance between X and Y.

Base Line Dimensioning: The maximum variation between two features is equal to the sum of the tolerances on the two dimensions from their origin to the features; this results in a reduction of the tolerance accumulation. In Fig. 51b, the tolerance accumulation between surfaces X and Y is ±0.1.

Direct Dimensioning: The maximum variation between two features is controlled by the tolerance on the dimension between the features; this results in the least tolerance. In Fig. 51c, the tolerance between surfaces X and Y is ±0.05.

Table 2. Relating Dimensional Limits to an Origin

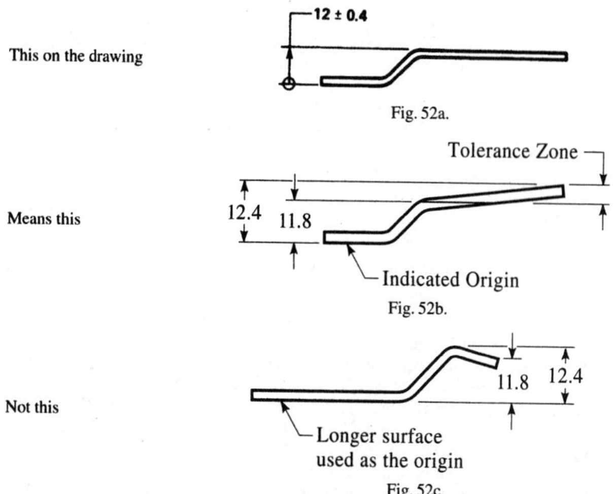

Fig. 52a.

Fig. 52b.

Fig. 52c.

Table 3. Application of Geometric Control Symbols

<table>
<tr><th>Type</th><th colspan="2">Geometric Characteristics</th><th>Pertains To</th><th>Basic Dimensions</th><th>Feature Modifier</th><th>Datum Modifier</th></tr>
<tr><td rowspan="4">Form</td><td>⏤</td><td>Straightness</td><td rowspan="4">ONLY individual feature</td><td rowspan="4"></td><td rowspan="6">Modifier not applicable</td><td rowspan="4">No datum</td></tr>
<tr><td>○</td><td>Circularity</td></tr>
<tr><td>▱</td><td>Flatness</td></tr>
<tr><td>⌭</td><td>Cylindricity</td></tr>
<tr><td rowspan="2">Profile</td><td>⌒</td><td>Profile (Line)</td><td rowspan="2">Individual or related</td><td rowspan="2">Yes if related</td><td rowspan="6">RFS implied unless MMC or LMC is stated</td></tr>
<tr><td>⌓</td><td>Profile (Surface)</td></tr>
<tr><td rowspan="3">Orientation</td><td>∠</td><td>Angularity</td><td rowspan="6">ALWAYS related fea-ture(s)</td><td>Yes</td><td rowspan="4">RFS implied unless MMC or LMC is stated</td></tr>
<tr><td>⊥</td><td>Perpendicularity</td><td rowspan="2"></td></tr>
<tr><td>//</td><td>Parallelism</td></tr>
<tr><td rowspan="3">Location</td><td>⌖</td><td>Position</td><td>Yes</td></tr>
<tr><td>◎</td><td>Concentricity</td><td rowspan="4"></td><td rowspan="4">Only RFS</td><td rowspan="4">Only RFS</td></tr>
<tr><td>⌯</td><td>Symmetry</td></tr>
<tr><td rowspan="2">Runout</td><td>↗</td><td>Circular Runout</td><td rowspan="2"></td></tr>
<tr><td>⌰</td><td>Total Runout</td></tr>
</table>

Five types of geometric control, when datums are indicated, when basic dimensions are required, and when MMC and LMC modifiers may be used.

Dimensional Limits Related to an Origin: In certain cases, it is necessary to indicate that a dimension between two features shall originate from one of these features and not the other. The high points of the surface indicated as the origin define a plane for measurement. The dimensions related to the origin are taken from the plane or axis and define a zone within which the other features must lie. This concept does not establish a datum reference frame. Such a case is illustrated in Fig. 52a, 52b, and 51c, where a part having two parallel surfaces of unequal length is to be mounted on the shorter surface. In this example, the dimension origin symbol signifies that the dimension originates from the plane established by the shorter surface and dimensional limits apply to the other surface. Without such indication, the longer surface could have been selected as the origin, thus permitting a greater angular variation between surfaces.

Geometric Characteristic Symbols: The symbolic means of indicating geometric characteristics are shown in Table 3.

Datum Feature Symbol: The symbolic means of indicating a datum feature consists of a capital letter enclosed in a square frame and a leader line extending from the frame to the concerned feature, terminating with a triangle. The triangle may be filled or not filled. See Fig. 53. Letters of the alphabet (except I, O, and Q) are used as datum identifying letters. Each datum feature of a part requiring identification shall be assigned a different letter. When datum features requiring identification on a drawing are so numerous as to exhaust the single alpha series, the double alpha series (AA through AZ, BA through BZ, etc.) shall be used and enclosed in a rectangular frame. Where the same datum feature symbol is repeated to identify the same feature in other locations of a drawing, it need not be identified as a reference. The datum feature is applied to the concerned feature surface outline, extension line, dimension line, or feature control frame as follows:

1) placed on the outline of a feature surface, or on an extension line of the feature outline, clearly separated from the dimension line, when the datum feature is the surface itself. See Fig. 54.

2) placed on an extension of the dimension line of a feature of size when the datum is the axis or center plane. If there is sufficient space for the two arrows, one of them may be replaced by the datum feature triangle. See Fig. 55a – 55f.

3) placed on the outline of a cylindrical feature surface or an extension line of the feature outline, separated from the size dimension, when the datum is the axis. For CAD systems, the triangle may be tangent to the feature. See Fig. 55d and 55f.

4) placed on dimension leader line to the feature size dimension where no geometrical tolerance and feature control frame are used. See Fig. 55e.

5) placed on the planes established by datum targets on complex or irregular datum features, or to reidentify previously established datum axis or planes on repeated or multistage drawing requirements.

6) placed above or below and attached to the feature control frame when the feature (or group of features) controlled is the datum axis or datum center plane. See Fig. 56.

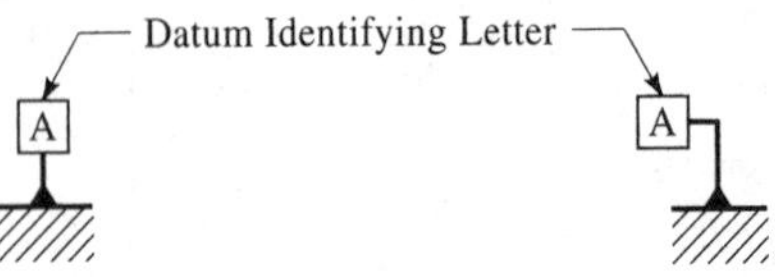

Fig. 53. Datum Feature Symbol

Datum feature triangle may be filled or not filled. Leader may be appropriately directed to a feature.

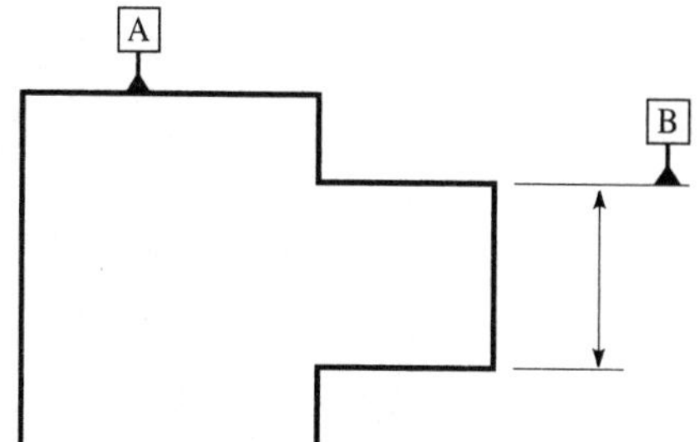

Fig. 54. Datum feature symbols on a feature surface and an extension line

Datum Target Symbol: The symbolic means of indicating a datum target is a circle divided horizontally into halves. See Fig. 57. The lower half contains a letter identifying the associated datum, followed by the target number assigned sequentially starting with 1 for each datum. A radial line attached to the symbol is directed to a target point, target line, or target area, as applicable. Where the datum target is an area, the area size is entered in the upper half of the symbol; otherwise, the upper half is left blank. If there is not sufficient space within the compartment, the area size may be placed outside and connected to the compartment by a leader line.

Placement of Datum Feature Symbols on Features of Size

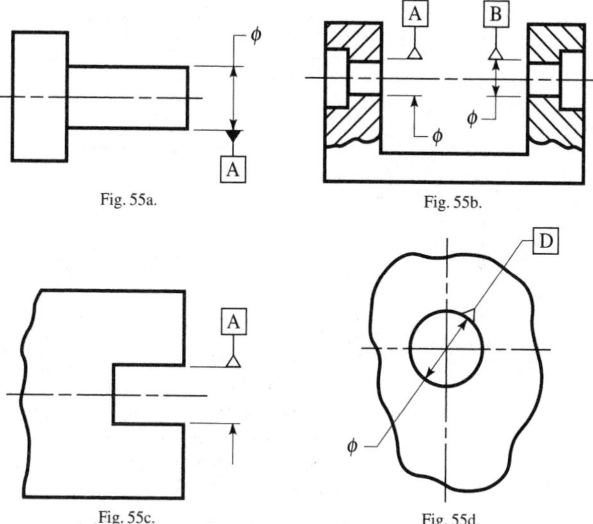

Fig. 55a.

Fig. 55b.

Fig. 55c.

Fig. 55d.

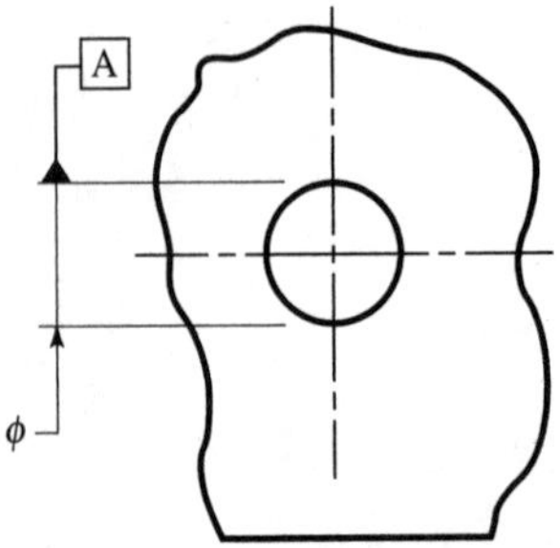

Fig. 55e.

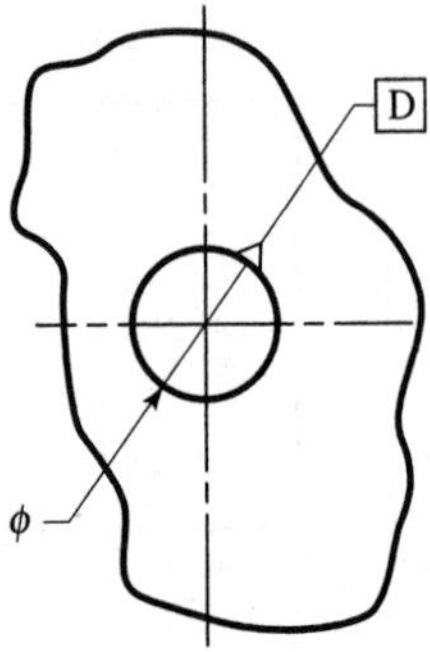

Fig. 55f.

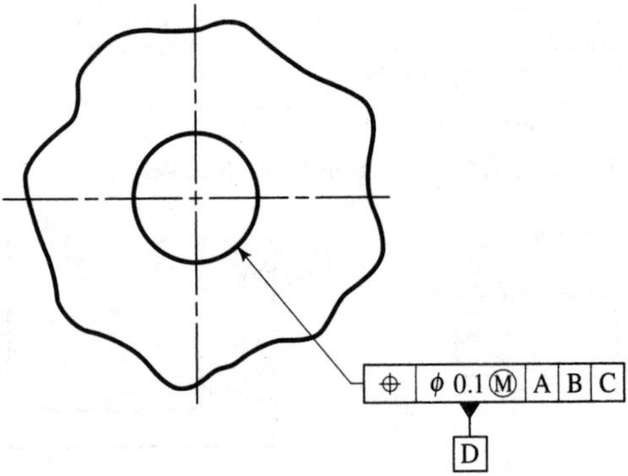

Fig. 56. Placement of datum feature symbols on features of size

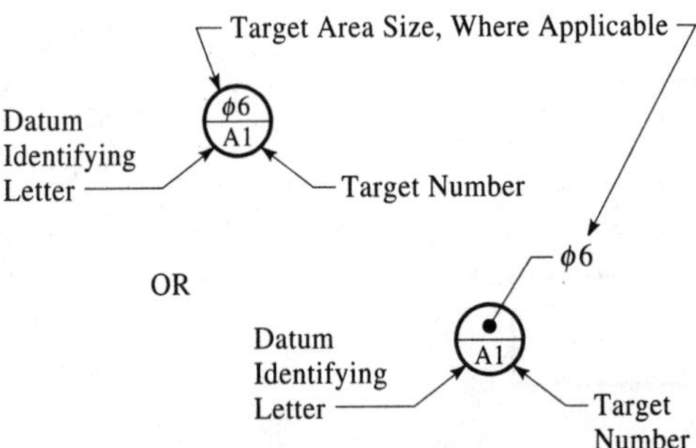

Fig. 57. Datum target symbol

Basic Dimension Symbol: The symbolic means of indicating a basic dimension is shown in Fig. 58.

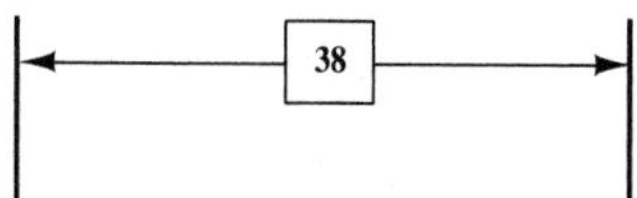

Fig. 58. Basic Dimensions

Material Condition Symbols: The symbolic means of indicating "at maximum material condition" and "at least material condition" are shown in Table 4. The use of these symbols in local and general notes is prohibited.

Diameter and Radius Symbols: The symbols used to indicate diameter, spherical diameter, radius, spherical radius, and controlled radius are shown in Table 4. These symbols precede the value of a dimension or tolerance given as a diameter or radius, as applicable. The symbol and the value are not separated by a space.

Reference Symbols: The symbolic means of indicating a dimension or other dimensional data as reference is by enclosing the dimension (or dimensional data) within parentheses. See Fig. 4. In written notes, parentheses retain their grammatical interpretation unless otherwise specified.

Arc Length: The symbolic means of indicating that a linear dimension is an arc length measured on a curved outline is shown in Table 4. The symbol is placed above the dimension.

Table 4. Modifying Symbols

Term	Symbol	Term	Symbol
At Maximum Material Condition	Ⓜ	Arc Length	$\frown$ **105**
At Least Material Condition	Ⓛ	Spherical Diameter	**SØ**
Projected Tolerance Zone	Ⓟ	Reference Dimension	**(50)**
Diameter	Ø	Radius	**R**
Spherical Radius	**SR**	Controlled radius	**CR**

Ⓕ	Ⓜ	Ⓛ	Ⓣ	Ⓟ	⟨ST⟩
Free State	**MMC**	**LMC**	**Tangent Plane**	**Projected Tolerance Zone**	**Statistical Tolerance**

Fig. 59. Tolerance modifiers

Projected Tolerance Zone Symbol: The symbolic means of indicating a projected tolerance zone is s hown in Table 4. The use of the symbol in local and general notes is prohibited.

Statistical Tolerancing Symbol: The symbolic means of indicating that a tolerance is based on statistical tolerancing is shown in Table 4. If the tolerance is a statistical geometric tolerance, the symbol is placed in the feature control frame following the stated tolerance and any modifier. See Fig. 60. If the tolerance is a statistical size tolerance, the symbol is placed adjacent to the size dimension. See Fig. 60 and 61.

Between Symbol: The symbolic means of indicating that a tolerance applies to a limited segment of a surface between designated extremities is shown in Fig. 59 and 62. In Fig. 62, for example, the tolerance applies only between point G and point H.

⌖	ϕ 0.1Ⓜ ⟨ST⟩	A	B	C

Fig. 60. Symbol indicating the specified tolerance is a statistical geometric tolerance

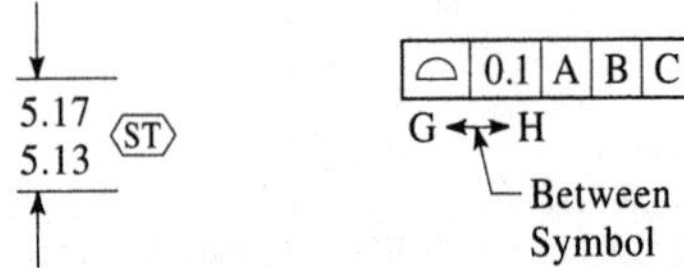

Fig. 61. Statistical tolerance symbol

Fig. 62. Between symbol

Counterbore or Spotface Symbol: The symbolic means of indicating a counterbore or a spotface is shown in Fig. 63. The symbol precedes, with no space, the dimension of the counterbore or spotface.

Countersink Symbol: The symbolic means of indicating a countersink is shown in Fig. 64. The symbol precedes, with no space, the dimensions of the countersink.

Depth Symbol: The symbolic means of indicating that a dimension applies to the depth of a feature is to precede that dimension with the depth symbol, as shown in Fig. 65. The symbol and the value are not separated by a space.

Square Symbol: The symbolic means of indicating that a single dimension applies to a square shape is to precede that dimension with the square symbol, as shown in Fig. 66. The symbol and the value are not separated by a space.

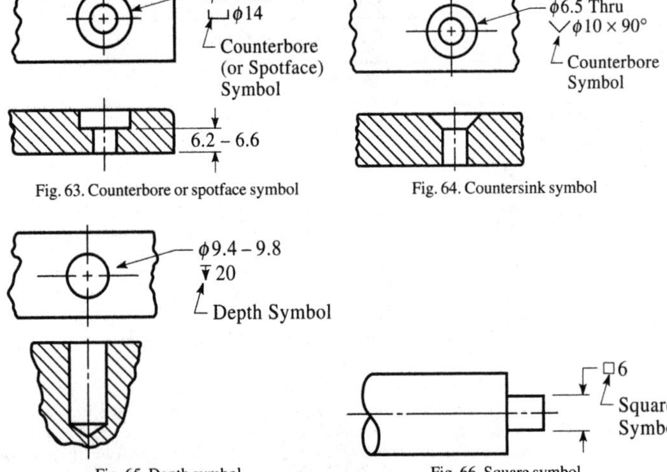

Fig. 63. Counterbore or spotface symbol

Fig. 64. Countersink symbol

Fig. 65. Depth symbol

Fig. 66. Square symbol

Dimension Origin Symbol: The symbolic means of indicating that a toleranced dimension between two features originates from one of these features and not the other is shown in Fig. 67.

Taper and Slope Symbols: The symbolic means of indicating taper and slope for conical and flat tapers are shown in Fig. 68a, 69, and 69a. These symbols are always shown with the vertical leg to the left.

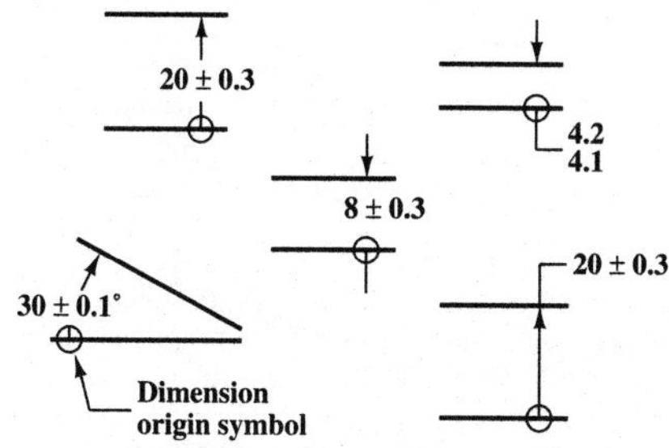

Fig. 67. Dimension Origin Symbol

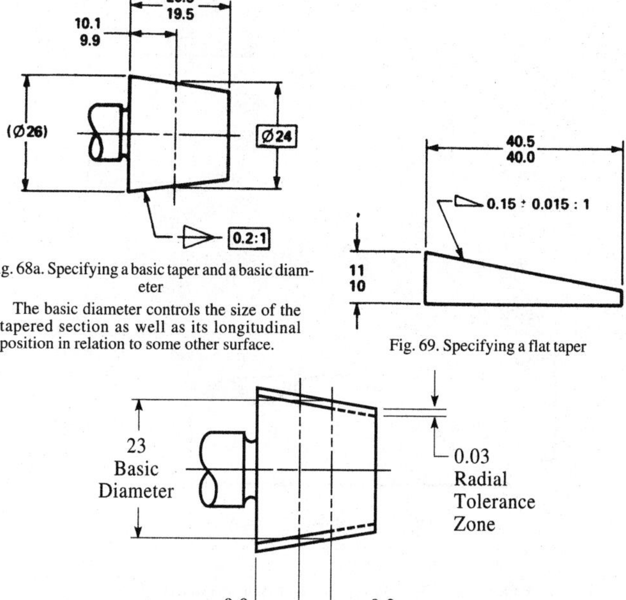

Fig. 68a. Specifying a basic taper and a basic diameter

The basic diameter controls the size of the tapered section as well as its longitudinal position in relation to some other surface.

Fig. 69. Specifying a flat taper

Fig. 69a. Specifying a basic taper and a basic diameter

The taper must fall within the zone created by the basic taper and the locating dimension of the basic diameter.

Fig. 70. Symbol for all around

Fig. 71. Feature control frame with free state symbol

All Around Symbol: The symbolic means of indicating that a tolerance is all around the part is a circle located at the junction of the leader from the feature control frame. See Fig. 70.

Free State Symbol: The symbolic means of indicating features subject to *free state variation* (the term used to describe the distortion of a part after removal of forces applied during manufacture), is shown in Table 4. The symbol is placed in the feature control frame following the stated tolerance and any modifier. See Fig. 71.

Tangent Plane Symbol: The symbolic means of indicating a tangent plane is shown in Table 4. The symbol is placed in the feature control frame following the stated tolerance as shown in Fig. 72a and 72b.

This on the drawing

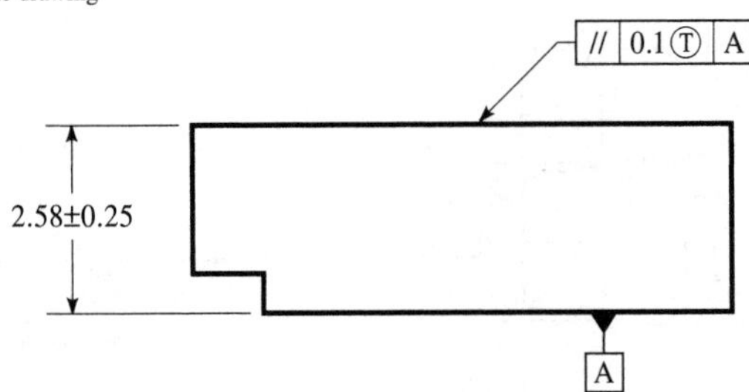

Fig. 72a. Specifying a tangent plane

Means this

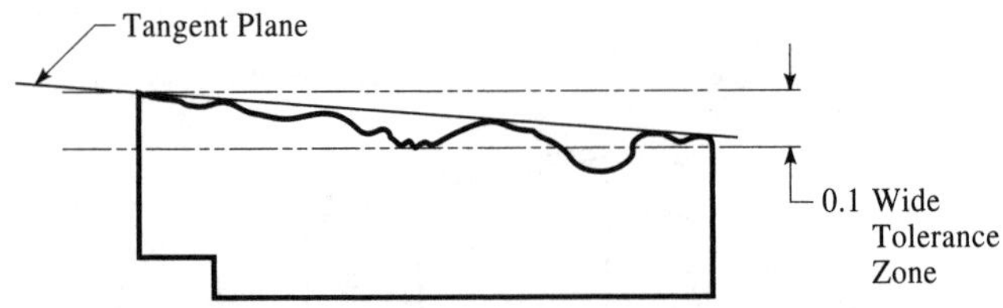

Fig. 72b. Specifying a tangent plane

A plane contacting the high points of the surface shall lie within two parallel planes 0.1 apart. The surface must be within the specified limits of size.

Geometric Tolerance Symbols.—Geometric characteristic symbols, the tolerance value, and datum reference letters, where applicable, are combined in a feature control frame to express a geometric tolerance.

Feature Control Frame: A geometric tolerance for an individual feature is specified by means of a feature control frame divided into compartments containing the geometric characteristic symbol followed by the tolerance. See Fig. 73. Where applicable, the tolerance is preceded by the diameter symbol and followed by a material condition symbol.

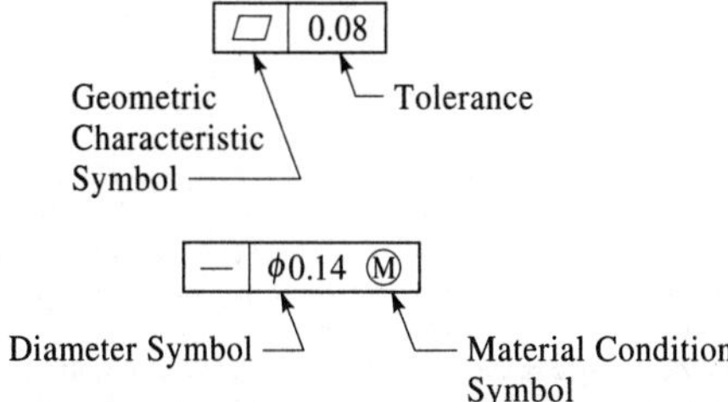

Fig. 73. Feature control frame

Feature Control Frame Incorporating One Datum Reference: Where a geometric tolerance is related to a datum, this relationship is indicated by entering the datum reference letter in a compartment following the tolerance. Where applicable, the datum reference letter is followed by a material condition symbol. See Fig. 74. Where a datum is established by two datum features—for example, an axis established by two datum diameters—both datum reference letters, separated by a dash, are entered in a single compartment. See Fig. 75a. Where applicable, each datum reference letter is followed by a material condition symbol.

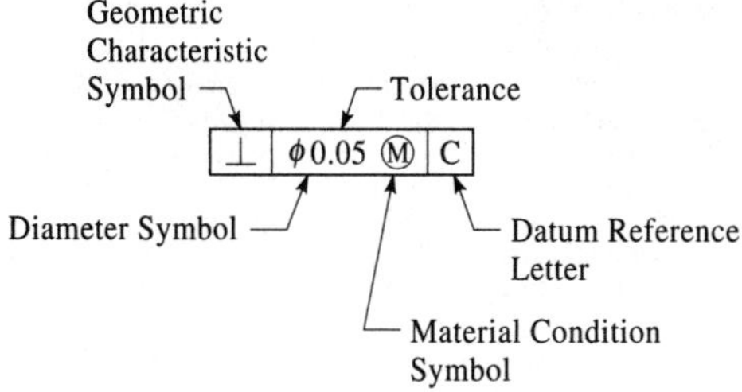

Fig. 74. Specifying a flat taper

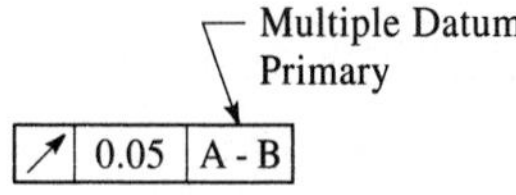

Fig. 75a. One datum reference

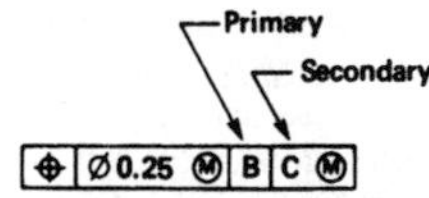

Fig. 75b. Two datum reference

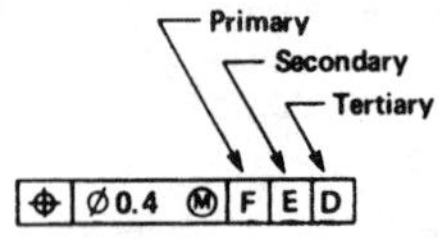

Fig. 75c. Three datum reference

Feature Control Frame Incorporating Two or Three Datum References: Where more than one datum is required, the datum reference letters (each followed by a material condition symbol, where applicable) are entered in separate compartments in the desired order of precedence, from left to right. See Fig. 75b and 75c. Datum reference letters need not be in alphabetical order in the feature control frame.

Composite Feature Control Frame: The composite feature control frame contains a single entry of the geometric characteristic symbol followed by each tolerance and datum requirement, one above the other. See Fig. 76a.

⌖	φ0.6 Ⓜ	D	E	F
⌖	φ0.25 Ⓜ	D		

Fig. 76a. Composite

⌖	φ0.6 Ⓜ	D	E	F
⌖	φ0.25 Ⓜ	D		

Fig. 76b. Two single segments

Two Single-Segment Feature Control Frames: The symbolic means of representing two singe-segment feature control frames is shown in Fig. 76b.

Combined Feature Control Frame and Datum Feature Symbol: Where a feature or pattern of features controlled by a geometric tolerance also serves as a datum feature, the feature control frame and datum feature symbol are combined. See Fig. 77. Wherever a feature control frame and datum feature symbol are combined, datums referenced on the feature control frame are not considered part of the datum feature symbol. In the positional tolerance example, Fig. 77, a feature is controlled for position in relation to datums A and B, and identified as datum feature C. Whenever datum C is referenced elsewhere on the drawing, the reference applies to datum C, not to datums A and B.

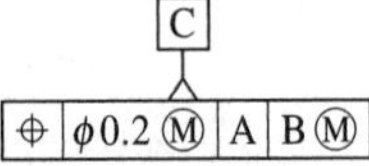

Fig. 77. Combined feature control frame and datum feature symbol

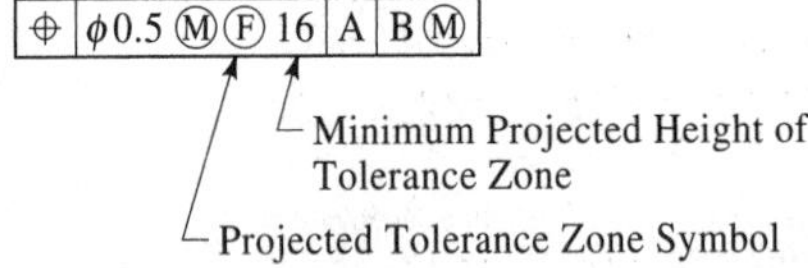

Fig. 78. Feature control frame with a projected tolerance zone symbol

Feature Control Frame With a Projected Tolerance Zone: Where a positional or an orientation tolerance is specified as a projected tolerance zone, the projected tolerance zone symbol is placed in the feature control frame, along with the dimension indicating the minimum height of the tolerance zone. This is to follow the stated tolerance and any modifier. See Fig. 78. Where necessary for clarification, the projected tolerance zone is indicated with a chain line and the minimum height of the tolerance zone is specified in a drawing view. The height dimension may then be omitted from the feature control frame. See Fig. 79.

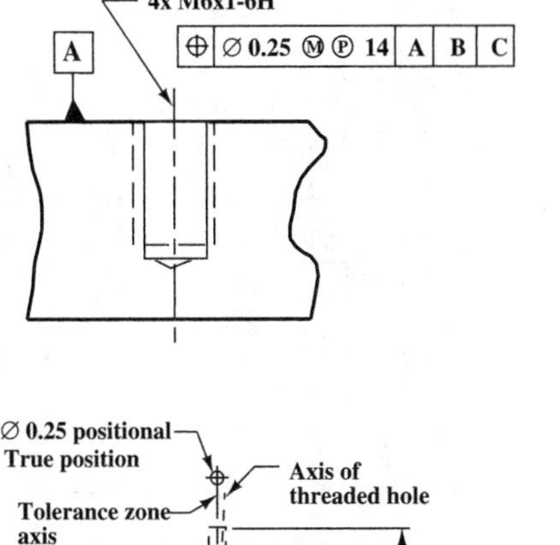

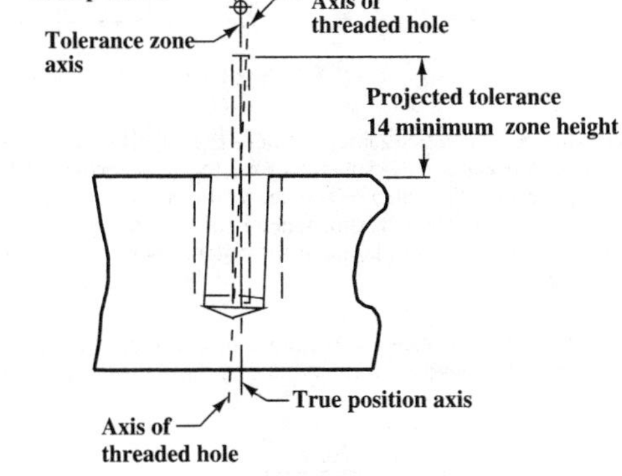

Fig. 79. Projected tolerance zone application

Feature Control Frame Placement.—The feature control frame is related to the considered feature by one of the following methods and as depicted in Fig. 80.

1) locating the frame below or attached to a leader-directed callout or dimension pertaining to the feature;

2) running a leader from the frame to the feature;

3) attaching a side or an end of the frame to an extension line from the feature, provided it is a plain surface;

4) attaching a side or an end of the frame to an extension of the dimension line pertaining to the feature of size.

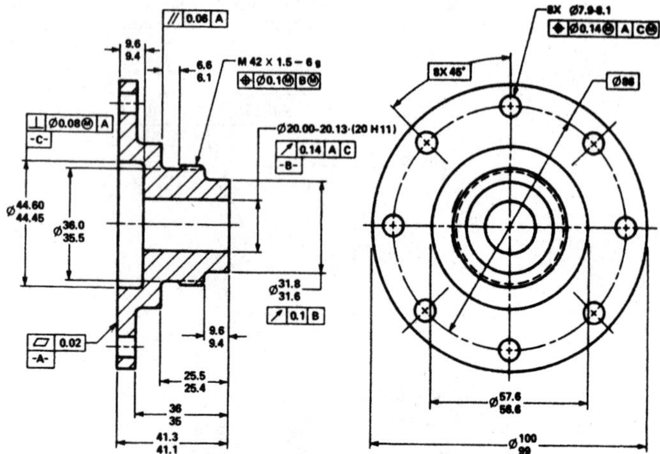

Fig. 80. Feature control frame placement

Definition of Tolerance Zone.—Where the specified tolerance value represents the diameter of a cylindrical or spherical zone, the diameter or spherical diameter symbol shall precede the tolerance value. Where the tolerance zone is other than a diameter, identification is unnecessary, and the specified tolerance value represents the distance between two parallel straight lines or planes, or the distance between two uniform boundaries, as the specific case may be.

Tabulated Tolerances.—Where a tolerance in a feature control frame is tabulated, a letter representing the tolerance, preceded by the abbreviation **TOL**, is entered as shown in Fig. 81.

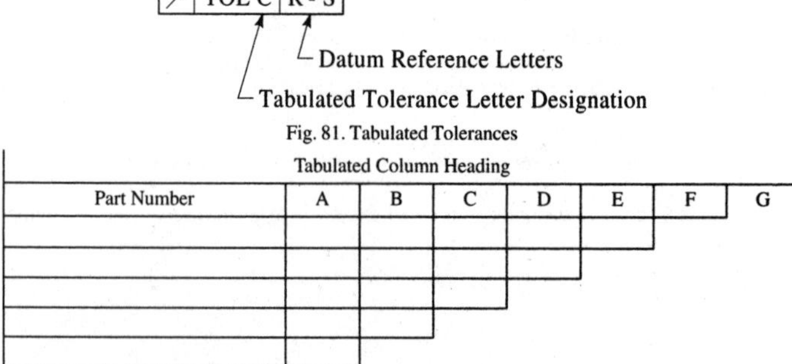

Fig. 81. Tabulated Tolerances

Tabulated Column Heading							
Part Number	A	B	C	D	E	F	G

American National Standard Drafting Practices

Several American National Standards for use in preparing engineering drawings and related documents are referred to for use.

Sizes of Drawing Sheets.—Recommended trimmed sheet sizes, based on ANSI Y14.1-1980 (R1987), are shown in the following table.

Size, inches				Metric Size, mm			
A	$8\frac{1}{2} \times 11$	*D*	22×34	*A0*	841×1189	*A3*	297×420
B	11×17	*E*	34×44	*A1*	594×841	*A4*	210×297
C	17×22	*F*	28×40	*A2*	420×594		

The standard sizes shown by the left-hand section of the table are based on the dimensions of the commercial letter head, $8\frac{1}{2} \times 11$ inches, in general use in the United States. The use of the basic sheet size $8\frac{1}{2} \times 11$ inches and its multiples permits filing of small tracings and folded blueprints in commercial standard letter files with or without correspondence. These sheet sizes also cut without unnecessary waste from the present 36-inch rolls of paper and cloth.

For drawings made in the metric system of units or for foreign correspondence, it is recommended that the metric standard trimmed sheet sizes be used. (Right-hand section of table.) These sizes are based on the width-to-length ratio of 1 to $\sqrt{2}$.

Line Conventions and Drawings.—American National Standard Y14.2M-1979 (R1987) establishes line and lettering practices for engineering drawings. The line conventions and the symbols for section lining are as shown on pages 162 and 163.

Approximate width of THICK lines for metric drawings are 0.6 mm, and for inch drawings, 0.032 inch. Approximate width of THIN lines for metric drawings are 0.3 mm, and for inch drawings, 0.016 inch. These approximate line widths are intended to differentiate between THICK and THIN lines and are not values for control of acceptance or rejection of the drawings.

Control and Production of Surface Texture.—Surface characteristics should not be controlled on a drawing or specification unless such control is essential to functional performance or appearance of the product. Imposition of such restrictions when unnecessary may increase production costs and in any event will serve to lessen the emphasis on the control specified for important surfaces.

Smoothness and roughness are relative, i.e., surfaces may be either smooth or rough for the purpose intended; what is smooth for one purpose may be rough for another purpose. In the mechanical field comparatively few surfaces require any control of surface texture beyond that afforded by the processes required to obtain the necessary dimensional characteristics.

Working surfaces such as on bearings, pistons, and gears are typical of surfaces for which optimum performance may require control of the surface characteristics. Non-working surfaces such as on the walls of transmission cases, crankcases, or housings seldom require any surface control. Experimentation or experience with surfaces performing similar functions is the best criterion on which to base selection of optimum surface characteristics.

Determination of required characteristics for working surfaces may involve consideration of such conditions as the area of contact, the load, speed, direction of motion, type and amount of lubricant, temperature, material and physical characteristics of component parts. Variations in any one of the conditions also may require a change in the specified surface characteristics.

Production: Surface texture is a result of the processing method. The surface obtained from casting, forging, or burnishing is the result of plastic deformation; if machined or ground, lapped or honed, the surface obtained is the result of action of cutting tools, abrasives or other forces. It is important to understand that surfaces with similar roughness average ratings may not have the same performance, due to tempering, sub-surface effects, different profile waveforms, etc.

Table 1. American National Standard for Engineering Drawings
ANSI/ASME Y14.2M-1992

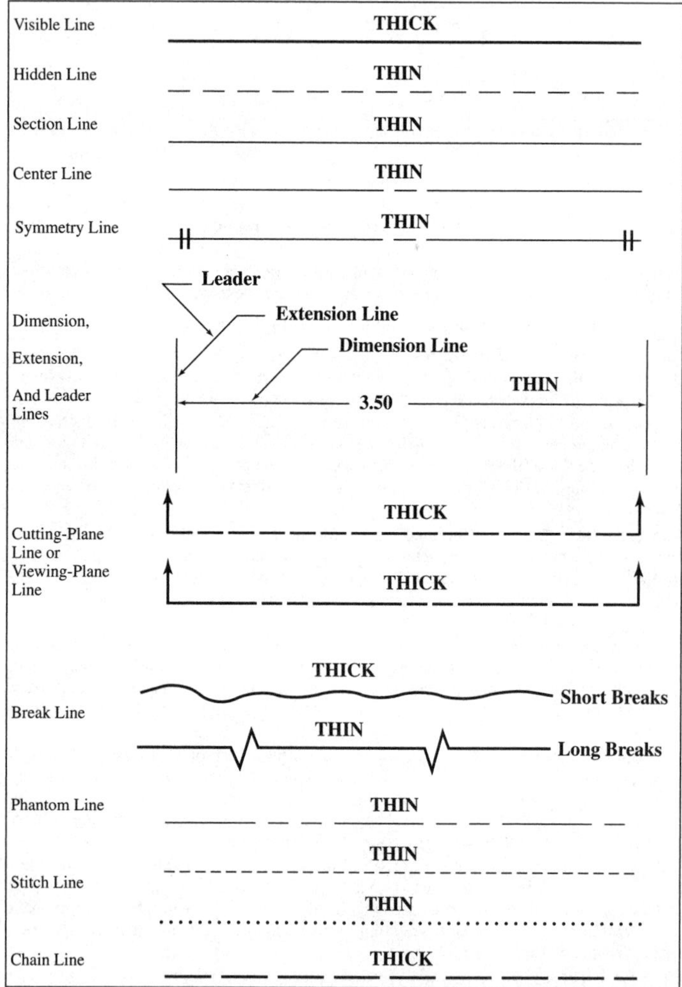

Surface-Texture Symbols.—A detailed explanation of the use of surface-texture symbols from American National Standard Y14.36-1978 (R1987) begins on Table .

Table 2. American National Standard Symbols for Section Lining
ANSI Y14.2M-1979 (R1987)

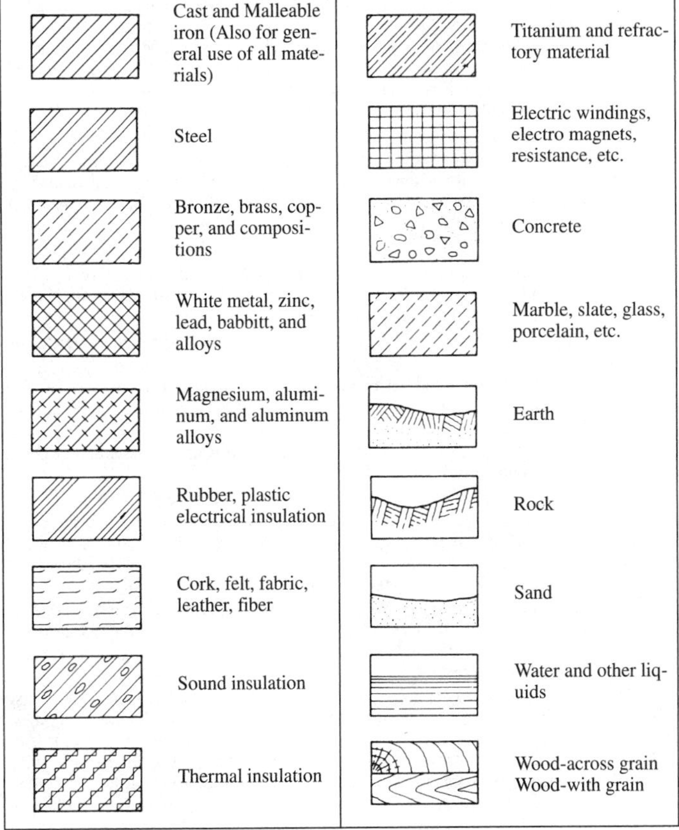

Symbol	Material	Symbol	Material
	Cast and Malleable iron (Also for general use of all materials)		Titanium and refractory material
	Steel		Electric windings, electro magnets, resistance, etc.
	Bronze, brass, copper, and compositions		Concrete
	White metal, zinc, lead, babbitt, and alloys		Marble, slate, glass, porcelain, etc.
	Magnesium, aluminum, and aluminum alloys		Earth
	Rubber, plastic electrical insulation		Rock
	Cork, felt, fabric, leather, fiber		Sand
	Sound insulation		Water and other liquids
	Thermal insulation		Wood-across grain Wood-with grain

American National Standard Surface Texture (Surface Roughness, Waviness, and Lay).—American National Standard ANSI/ASME B46.1-1995 is concerned with the geometric irregularities of surfaces of solid materials, physical specimens for gaging roughness, and the characteristics of stylus instrumentation for measuring roughness. The standard defines surface texture and its constituents: roughness, waviness, lay, and flaws. A set of symbols for drawings, specifications, and reports is established. To ensure a uniform basis for measurements the standard also provides specifications for Precision Reference Specimens, and Roughness Comparison Specimens, and establishes requirements for stylus-type instruments. The standard is not concerned with luster, appearance, color, corrosion resistance, wear resistance, hardness, subsurface microstructure, surface integrity, and many other characteristics that may be governing considerations in specific applications.

The standard is expressed in SI metric units but U.S. customary units may be used without prejudice. The standard does not define the degrees of surface roughness and waviness or type of lay suitable for specific purposes, nor does it specify the means by which any degree of such irregularities may be obtained or produced. However, criteria for selection of surface qualities and information on instrument techniques and methods of producing, controlling and inspecting surfaces are included in Appendixes attached to the standard. The Appendix sections are not considered a part of the standard: they are included for clarification or information purposes only.

Surfaces, in general, are very complex in character. The standard deals only with the height, width, and direction of surface irregularities because these characteristics are of practical importance in specific applications. Surface texture designations as delineated in this standard may not be a sufficient index to performance. Other part characteristics such as dimensional and geometrical relationships, material, metallurgy, and stress must also be controlled.

Definitions of Terms Relating to the Surfaces of Solid Materials.—The terms and ratings in the standard relate to surfaces produced by such means as abrading, casting, coating, cutting, etching, plastic deformation, sintering, wear, and erosion.

Surface is the boundary of an object that separates that object from another object, substance or space.

Surface, nominal is the intended surface contour (exclusive of any intended surface roughness), the shape and extent of which is usually shown and dimensioned on a drawing or descriptive specification.

Surface, real is the actual boundary of the object. Manufacturing processes determine its deviation from the nominal surface.

Surface, measured is the real surface obtained by instrumental or other means.

Surface texture is repetitive or random deviations from the real surface that forms the three-dimensional topography of the surface. Surface texture includes roughness, waviness, lay and flaws. Fig. 1 is an example of a unidirectional lay surface. Roughness and waviness parallel to the lay are not represented in the expanded views.

Roughness consists of the finer irregularities of the surface texture, usually including those irregularities that result from the inherent action of the production process. These irregularities are considered to include traverse feed marks and other irregularities within the limits of the roughness sampling length.

Waviness is the more widely spaced component of surface texture. Unless otherwise noted, waviness includes all irregularities whose spacing is greater than the roughness sampling length and less than the waviness sampling length. Waviness may result from such factors as machine or work deflections, vibration, chatter, heat-treatment or warping strains. Roughness may be considered as being superposed on a 'wavy' surface.

Lay is the direction of the predominant surface pattern, ordinarily determined by the production method used.

Flaws are unintentional, unexpected, and unwanted interruptions in the topography typical of a part surface and are defined as such only when agreed upon by buyer and seller. If flaws are defined, the surface should be inspected specifically to determine whether flaws are present, and rejected or accepted prior to performing final surface roughness measurements. If defined flaws are not present, or if flaws are not defined, then interruptions in the part surface may be included in roughness measurements.

Error of form is considered to be that deviation from the nominal surface caused by errors in machine tool ways, guides, insecure clamping or incorrect alignment of the workpiece or wear, all of which are not included in surface texture.

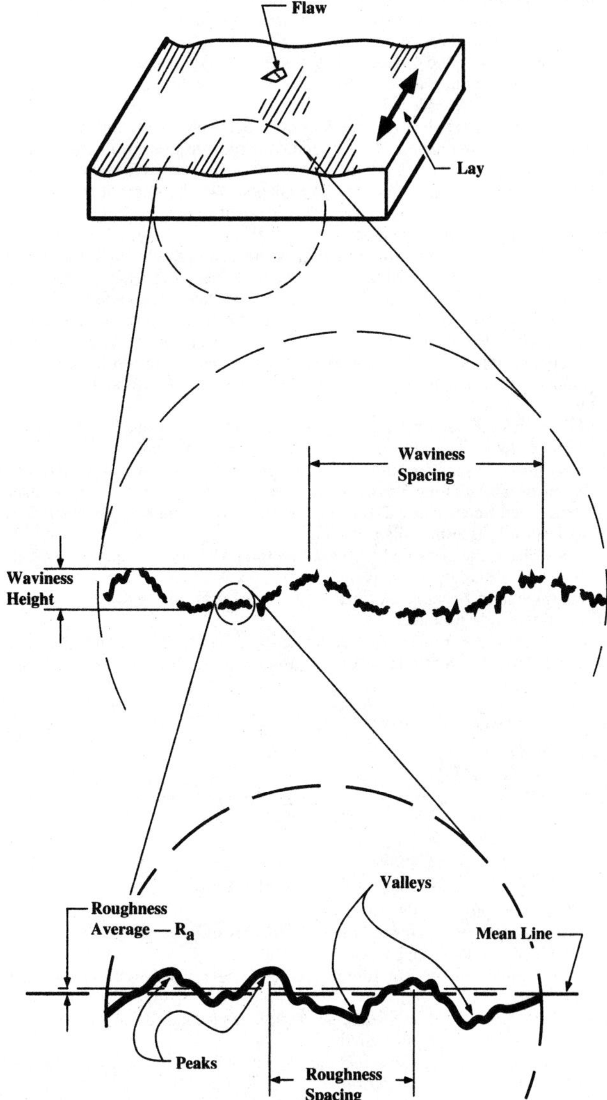

Fig. 1. Pictorial Display of Surface Characteristics

Definitions of Terms Relating to the Measurement of Surface Texture.—Terms regarding surface texture pertain to the geometric irregularities of surfaces and include roughness, waviness and lay.

Profile is the contour of the surface in a plane measured normal, or perpendicular, to the surface, unless another other angle is specified.

Graphical centerline. See Mean Line.

Height (z) is considered to be those measurements of the profile in a direction normal, or perpendicular, to the nominal profile. For digital instruments, the profile Z(x) is approximated by a set of digitized values. Height parameters are expressed in micrometers (μm).

Height range (z) is the maximum peak-to-valley surface height that can be detected accurately with the instrument. It is measurement normal, or perpendicular, to the nominal profile and is another key specification.

Mean line (M) is the line about which deviations are measured and is a line parallel to the general direction of the profile within the limits of the sampling length. See Fig. 2. The mean line may be determined in one of two ways. The filtered mean line is the centerline established by the selected cutoff and its associated circuitry in an electronic roughness average measuring instrument. The least squares mean line is formed by the nominal profile but by dividing into selected lengths the sum of the squares of the deviations minimizes the deviation from the nominal form. The form of the nominal profile could be a curve or a straight line.

Peak is the point of maximum height on that portion of a profile that lies above the mean line and between two intersections of the profile with the mean line.

Profile measured is a representation of the real profile obtained by instrumental or other means. When the measured profile is a graphical representation, it will usually be distorted through the use of different vertical and horizontal magnifications but shall otherwise be as faithful to the profile as technically possible.

Profile, modified is the measured profile where filter mechanisms (including the instrument datum) are used to minimize certain surface texture characteristics and emphasize others. Instrument users apply profile modifications typically to differentiate surface roughness from surface waviness.

Profile, nominal is the profile of the nominal surface; it is the intended profile (exclusive of any intended roughness profile). Profile is usually drawn in an x-z coordinate system. See Fig. 2.

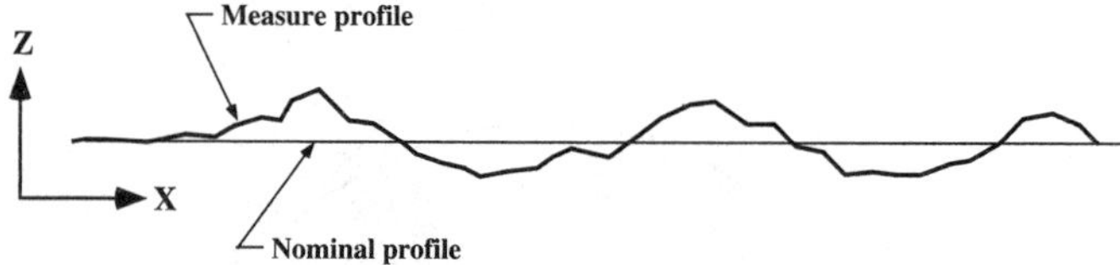

Fig. 2. Nominal and Measured Profiles

Profile, real is the profile of the real surface.

Profile, total is the measured profile where the heights and spacing may be amplified differently but otherwise no filtering takes place.

Roughness profile is obtained by filtering out the longer wavelengths characteristic of waviness.

Roughness spacing is the average spacing between adjacent peaks of the measured profile within the roughness sampling length.

Roughness topography is the modified topography obtained by filtering out the longer wavelengths of waviness and form error.

Sampling length is the nominal spacing within which a surface characteristic is determined. The range of sampling lengths is a key specification of a measuring instrument.

Spacing is the distance between specified points on the profile measured parallel to the nominal profile.

Spatial (x) resolution is the smallest wavelength which can be resolved to 50% of the actual amplitude. This also is a key specification of a measuring instrument.

System height resolution is the minimum height that can be distinguished from background noise of the measurement instrument. Background noise values can be determined by measuring approximate rms roughness of a sample surface where actual roughness is significantly less than the background noise of the measuring instrument. It is a key instrumentation specification.

Topography is the three-dimensional representation of geometric surface irregularities.

Topography, measured is the three-dimensional representation of geometric surface irregularities obtained by measurement.

Topography, modified is the three-dimensional representation of geometric surface irregularities obtained by measurement but filtered to minimize certain surface characteristics and accentuate others.

Valley is the point of maximum depth on that portion of a profile that lies below the mean line and between two intersections of the profile with the mean line.

Waviness, evaluation length (L), is the length within which waviness parameters are determined.

Waviness, long-wavelength cutoff (lcw) the spatial wavelength above which the undulations of waviness profile are removed to identify form parameters. A digital Gaussian filter can be used to separate form error from waviness but its use must be specified.

Waviness profile is obtained by filtering out the shorter roughness wavelengths characteristic of roughness and the longer wavelengths associated with the part form parameters.

Waviness sampling length is a concept no longer used. See waviness long-wavelength cutoff and waviness evaluation length.

Waviness short-wavelength cutoff (lsw) is the spatial wavelength below which roughness parameters are removed by electrical or digital filters.

Waviness topography is the modified topography obtained by filtering out the shorter wavelengths of roughness and the longer wavelengths associated with form error.

Waviness spacing is the average spacing between adjacent peaks of the measured profile within the waviness sampling length.

Sampling Lengths.—Sampling length is the normal interval for a single value of a surface parameter. Generally it is the longest spatial wavelength to be included in the profile measurement. Range of sampling lengths is an important specification for a measuring instrument.

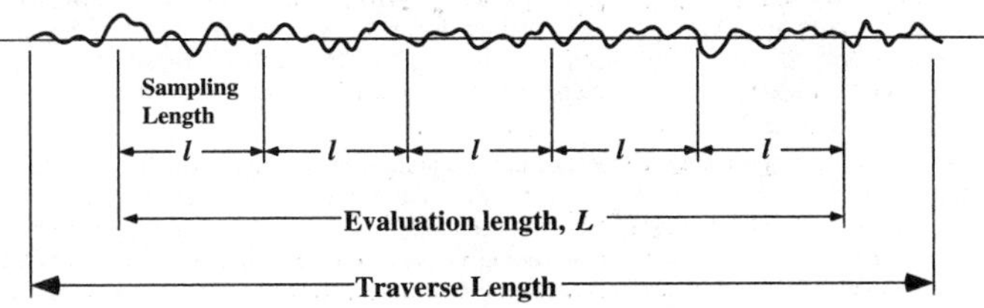

Fig. 3. Traverse Length

Roughness sampling length (l) is the sampling length within which the roughness average is determined. This length is chosen to separate the profile irregularities which are designated as roughness from those irregularities designated as waviness. It is different from evaluation length (*L*) and the traversing length. See Fig. 3.

Evaluation length (L) is the length the surface characteristics are evaluated. The evaluation length is a key specification of a measuring instrument.

Traversing length is profile length traversed to establish a representative evaluation length. It is always longer than the evaluation length. See Section 4.4.4 of ANSI/ASME B46.1-1995 for values which should be used for different type measurements.

Cutoff is the electrical response characteristic of the measuring instrument which is selected to limit the spacing of the surface irregularities to be included in the assessment of surface texture. Cutoff is rated in millimeters. In most electrical averaging instruments, the cutoff can be user selected and is a characteristic of the instrument rather than of the surface being measured. In specifying the cutoff, care must be taken to choose a value which will include all the surface irregularities to be assessed.

Waviness sampling length (l) is a concept no longer used. See waviness long-wavelength cutoff and waviness evaluation length.

Roughness Parameters.—Roughness is the fine irregularities of the surface texture resulting from the production process or material condition.

Roughness average (Ra), also known as arithmetic average (AA) is the arithmetic average of the absolute values of the measured profile height deviations divided by the evaluation length, L. This is shown as the shaded area of Fig. 4 and generally includes sampling lengths or cutoffs. For graphical determinations of roughness average, the height deviations are measured normal, or perpendicular, to the chart center line.

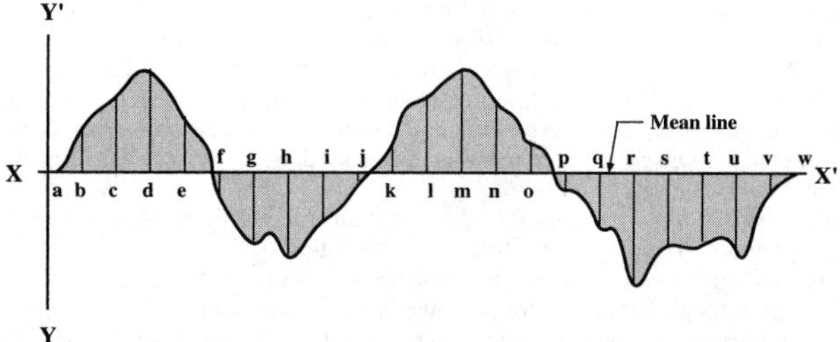

Fig. 4.

Roughness average is expressed in micrometers (μm). A micrometer is one millionth of a meter (0.000001 meter). A microinch (μin) is one millionth of an inch (0.000001 inch). One microinch equals 0.0254 micrometer (1 μin. = 0.0254 μm).

Roughness Average Value (Ra) From Continuously Averaging Meter Reading: So that uniform interpretation may be made of readings from stylus-type instruments of the continuously averaging type, it should be understood that the reading that is considered significant is the mean reading around which the needle tends to dwell or fluctuate with a small amplitude.

Roughness is also indicated by the root-mean-square (rms) average, which is the square root of the average value squared, within the evaluation length and measured from the mean line shown in Fig. 4, expressed in micrometers. A roughness-measuring instrument calibrated for rms average usually reads about 11 per cent higher than an instrument calibrated for arithmetical average. Such instruments usually can be recalibrated to read arithmetical average. Some manufacturers consider the difference between rms and AA to be small enough that rms on a drawing may be read as AA for many purposes.

Roughness evaluation length (L), for statistical purposes should, whenever possible, consist of five sampling lengths (*l*). Use of other than five sampling lengths must be clearly indicated.

Waviness Parameters.—Waviness is the more widely spaced component of surface texture. Roughness may be thought of as superimposed on waviness.

Waviness height (Wt) is the peak-to-valley height of the modified profile with roughness and part form errors removed by filtering, smoothing or other means. This value is typically three or more times the roughness average. The measurement is taken normal, or perpendicular, to the nominal profile within the limits of the waviness sampling length.

Waviness evaluation length (Lw) is the evaluation length required to determine waviness parameters. For waviness, the sampling length concept is no longer used. Rather, only waviness evaluation length (*Lw*) and waviness long-wavelength cutoff (*lew*) are defined. For better statistics, the waviness evaluation length should be several times the waviness long-wavelength cutoff.

Relation of Surface Roughness to Tolerances.—Because the measurement of surface roughness involves the determination of the average linear deviation of the measured surface from the nominal surface, there is a direct relationship between the dimensional tolerance on a part and the permissible surface roughness. It is evident that a requirement for the accurate measurement of a dimension is that the variations introduced by surface roughness should not exceed the dimensional tolerances. If this is not the case, the measurement of the dimension will be subject to an uncertainty greater than the required tolerance, as illustrated in Fig. 5.

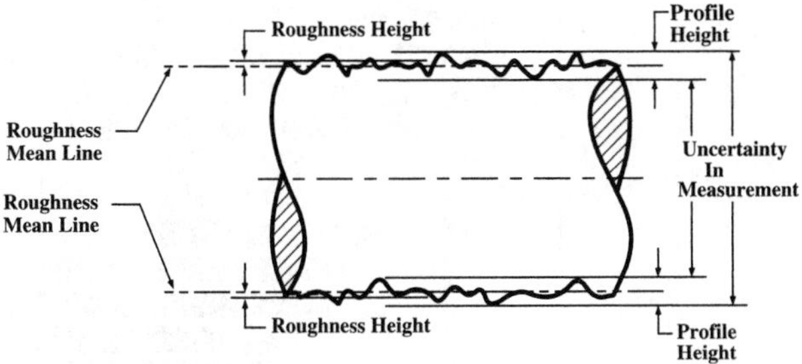

Fig. 5.

The standard method of measuring surface roughness involves the determination of the average deviation from the mean surface. On most surfaces the total profile height of the surface roughness (peak-to-valley height) will be approximately four times (4×) the measured average surface roughness. This factor will vary somewhat with the character of the surface under consideration, but the value of four may be used to establish approximate profile heights.

From these considerations it follows that if the arithmetical average value of surface roughness specified on a part exceeds one eighth of the dimensional tolerance, the whole tolerance will be taken up by the roughness height. In most cases, a smaller roughness specification than this will be found; but on parts where very small dimensional tolerances are given, it is necessary to specify a suitably small surface roughness so useful dimensional measurements can be made.

Values for surface roughness produced by common processing methods are shown in Table 3. The ability of a processing operation to produce a specific surface roughness depends on many factors. For example, in surface grinding, the final surface depends on the peripheral speed of the wheel, the speed of the traverse, the rate of feed, the grit size, bonding material and state of dress of the wheel, the amount and type of lubrication at the point of cutting, and the mechanical properties of the piece being ground. A small change in any of the above factors can have a marked effect on the surface produced.

Table 3. Surface Roughness Produced by Common Production Methods

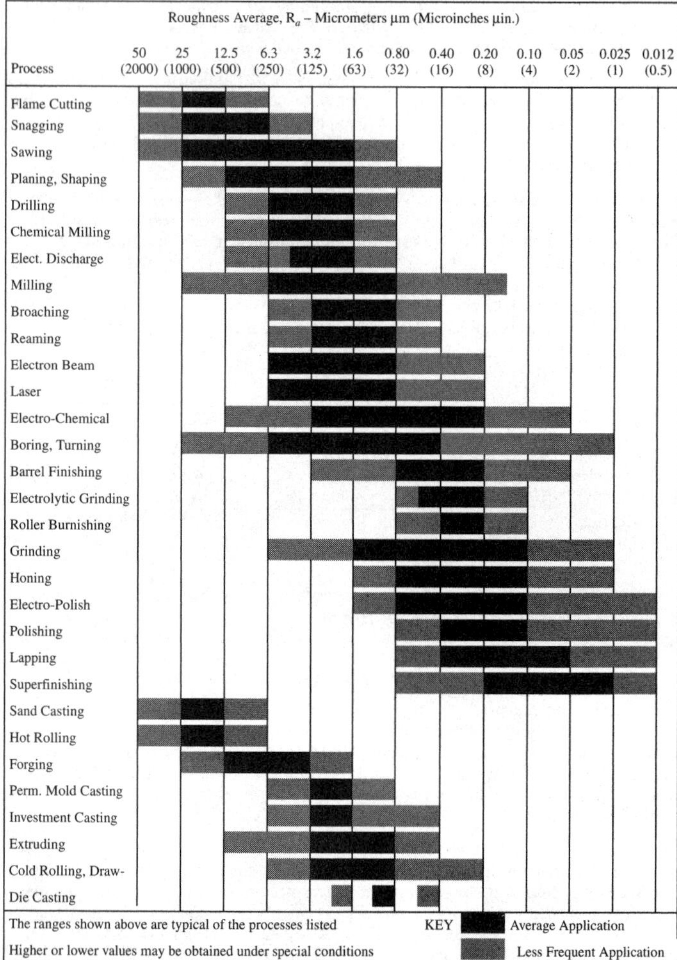

Selecting Cutoff for Roughness Measurements.—In general, surfaces will contain irregularities with a large range of widths. Surface texture instruments are designed to respond only to irregularity spacings less than a given value, called cutoff. In some cases, such as surfaces in which actual contact area with a mating surface is important, the largest convenient cutoff will be used. In other cases, such as surfaces subject to fatigue failure only the irregularities of small width will be important, and more significant values will be obtained when a short cutoff is used. In still other cases, such as identifying chatter marks on machined surfaces, information is needed on only the widely space irregularities. For such measurements, a large cutoff value and a larger radius stylus should be used.

The effect of variation in cutoff can be understood better by reference to Fig. 6. The profile at the top is the true movement of a stylus on a surface having a roughness spacing of about 1 mm and the profiles below are interpretations of the same surface with cutoff value settings of 0.8 mm, 0.25 mm and 0.08 mm, respectively. It can be seen that the trace based on 0.8 mm cutoff includes most of the coarse irregularities and all of the fine irregularities of the surface. The trace based on 0.25 mm excludes the coarser irregularities but includes the fine and medium fine. The trace based on 0.08 mm cutoff includes only the very fine irregularities. In this example the effect of reducing the cutoff has been to reduce the roughness average indication. However, had the surface been made up only of irregularities as fine as those of the bottom trace, the roughness average values would have been the same for all three cutoff settings.

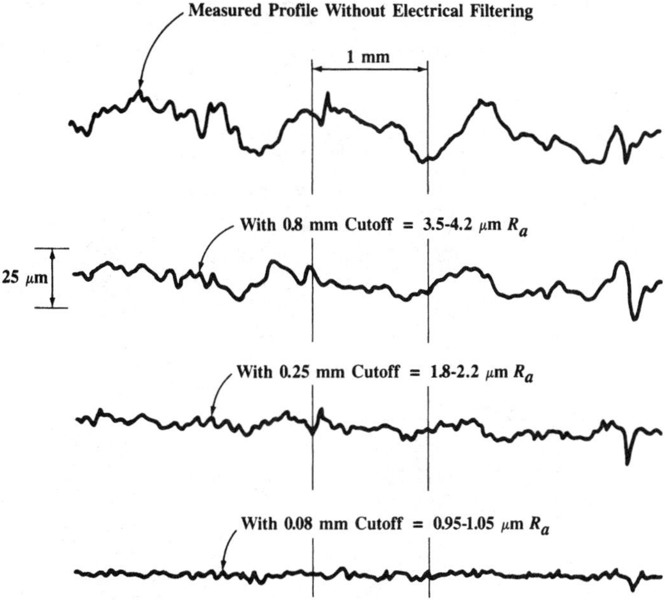

Fig. 6. Effects of Various Cutoff Values

In other words, all irregularities having a spacing less than the value of the cutoff used are included in a measurement. Obviously, if the cutoff value is too small to include coarser irregularities of a surface, the measurements will not agree with those taken with a larger cutoff. For this reason, care must be taken to choose a cutoff value which will include all of the surface irregularities it is desired to assess.

To become proficient in the use of continuously averaging stylus-type instruments the inspector or machine operator must realize that for uniform interpretation, the reading which is considered significant is the mean reading around which the needle tends to dwell or fluctuate under small amplitude.

Drawing Practices for Surface Texture Symbols.—American National Standard ANSI/ASME Y14.36M-1996 establishes the method to designate symbolic controls for surface texture of solid materials. It includes methods for controlling roughness, waviness, and lay, and provides a set of symbols for use on drawings, specifications, or other documents. The standard is expressed in SI metric units but U.S. customary units may be used without prejudice. Units used (metric or non-metric) should be consistent with the other units used on the drawing or documents. Approximate non-metric equivalents are shown for reference.

Surface Texture Symbol.—The symbol used to designate control of surface irregularities is shown in Fig. 7b and Fig. 7d. Where surface texture values other than roughness average are specified, the symbol must be drawn with the horizontal extension as shown in Fig. 7f.

Surface Texture Symbols and Construction

Symbol	Meaning
Fig. 7a.	Basic Surface Texture Symbol. Surface may be produced by any method except when the bar or circle (Fig. 7b or 7d) is specified.
Fig. 7b.	Material Removal By Machining Is Required. The horizontal bar indicates that material removal by machining is required to produce the surface and that material must be provided for that purpose.
3.5 Fig. 7c.	Material Removal Allowance. The number indicates the amount of stock to be removed by machining in millimeters (or inches). Tolerances may be added to the basic value shown or in general note.
Fig. 7d.	Material Removal Prohibited. The circle in the vee indicates that the surface must be produced by processes such as casting, forging, hot finishing, cold finishing, die casting, powder metallurgy or injection molding without subsequent removal of material.
 Fig. 7e.	Surface Texture Symbol. To be used when any surface characteristics are specified above the horizontal line or the right of the symbol. Surface may be produced by any method except when the bar or circle (Fig. 7b and 7d) is specified.

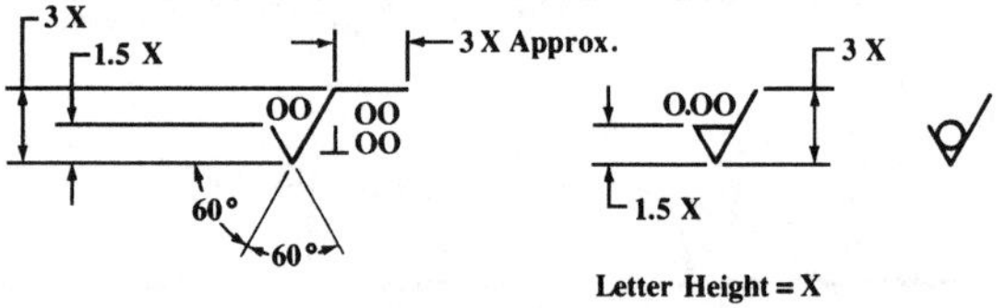

Fig. 7f.

Use of Surface Texture Symbols: When required from a functional standpoint, the desired surface characteristics should be specified. Where no surface texture control is specified, the surface produced by normal manufacturing methods is satisfactory provided it is within the limits of size (and form) specified in accordance with ANSI/ASME Y14.5M-1994, Dimensioning and Tolerancing. It is considered good practice to always specify some maximum value, either specifically or by default (for example, in the manner of the note shown in Fig. 2).

Material Removal Required or Prohibited: The surface texture symbol is modified when necessary to require or prohibit removal of material. When it is necessary to indicate that a surface must be produced by removal of material by machining, specify the symbol shown in Fig. 7b. When required, the amount of material to be removed is specified as shown in Fig. 7c, in millimeters for metric drawings and in inches for non-metric drawings. Tolerance for material removal may be added to the basic value shown or specified in a general note. When it is necessary to indicate that a surface must be produced without material removal, specify the machining prohibited symbol as shown in Fig. 7d.

Proportions of Surface Texture Symbols: The recommended proportions for drawing the surface texture symbol are shown in Fig. 7f. The letter height and line width should be the same as that for dimensions and dimension lines.

Applying Surface Texture Symbols.—The point of the symbol should be on a line representing the surface, an extension line of the surface, or a leader line directed to the surface, or to an extension line. The symbol may be specified following a diameter dimension. Although ANSI/ASME Y14.5M-1994, "Dimensioning and Tolerancing" specifies that normally all textual dimensions and notes should be read from the bottom of the drawing, the surface texture symbol itself with its textual values may be rotated as required. Regardless, the long leg (and extension) must be to the right as the symbol is read. For parts requiring extensive and uniform surface roughness control, a general note may be added to the drawing which applies to each surface texture symbol specified without values as shown in Fig. 8.

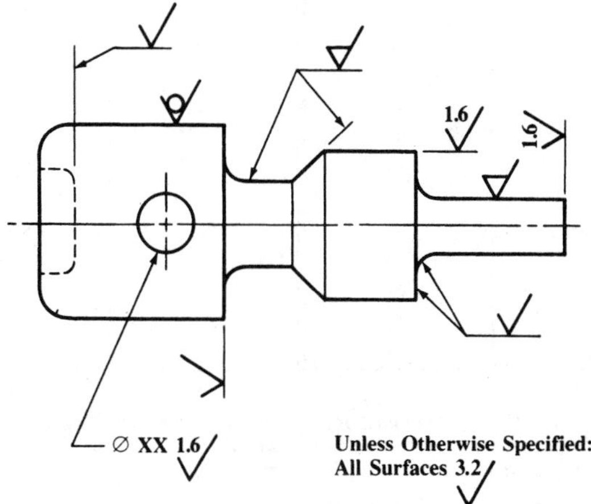

Fig. 8. Application of Surface Texture Symbols

When the symbol is used with a dimension, it affects the entire surface defined by the dimension. Areas of transition, such as chamfers and fillets, shall conform with the roughest adjacent finished area unless otherwise indicated.

Surface texture values, unless otherwise specified, apply to the complete surface. Drawings or specifications for plated or coated parts shall indicate whether the surface texture values apply before plating, after plating, or both before and after plating.

Only those values required to specify and verify the required texture characteristics should be included in the symbol. Values should be in metric units for metric drawing and non-metric units for non-metric drawings. Minority units on dual dimensioned drawings are enclosed in brackets.

Roughness and waviness measurements, unless otherwise specified, apply in a direction which gives the maximum reading; generally across the lay.

Cutoff or Roughness Sampling Length, (l): Standard values are listed in Table 4. When no value is specified, the value 0.8 mm (0.030 in.) applies.

Table 4. Standard Roughness Sampling Length (Cutoff) Values

mm	in.	mm	in.
0.08	0.003	2.5	0.1
0.25	0.010	8.0	0.3
0.80	0.030	25.0	1.0

Roughness Average (R_a): The preferred series of specified roughness average values is given in Table 5.

Table 5. Preferred Series Roughness Average Values (R_a)

μm	μin	μm	μin
0.012	0.5	1.25	50
0.025[a]	1[a]	1.60[a]	63[a]
0.050[a]	2[a]	2.0	80
0.075[a]	3	2.5	100
0.10[a]	4[a]	3.2[a]	125[a]
0.125	5	4.0	160
0.15	6	5.0	200
0.20[a]	8[a]	6.3[a]	250[a]
0.25	10	8.0	320
0.32	13	10.0	400
0.40[a]	16[a]	12.5[a]	500[a]
0.50	20	15	600
0.63	25	20	800
0.80[a]	32[a]	25[a]	1000[a]
1.00	40	…	…

[a] Recommended

Waviness Height (Wt): The preferred series of maximum waviness height values is listed in Table 6. Waviness height is not currently shown in U.S. or ISO Standards. It is included here to follow present industry practice in the United States.

Table 6. Preferred Series Maximum Waviness Height Values

mm	in.	mm	in.	mm	in.
0.0005	0.00002	0.008	0.0003	0.12	0.005
0.0008	0.00003	0.012	0.0005	0.20	0.008
0.0012	0.00005	0.020	0.0008	0.25	0.010
0.0020	0.00008	0.025	0.001	0.38	0.015
0.0025	0.0001	0.05	0.002	0.50	0.020
0.005	0.0002	0.08	0.003	0.80	0.030

Lay: Symbols for designating the direction of lay are shown and interpreted in Table 7.

Example Designations.—Table 8 illustrates examples of designations of roughness, waviness, and lay by insertion of values in appropriate positions relative to the symbol.

Table 7. Lay Symbols

Lay Symbol	Meaning	Example Showing Direction of Tool Marks
=	Lay approximately parallel to the line representing the surface to which the symbol is applied.	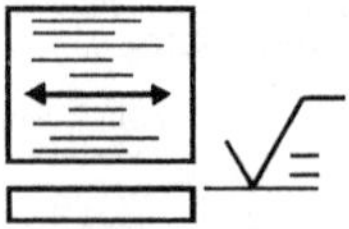
⊥	Lay approximately perpendicular to the line representing the surface to which the symbol is applied.	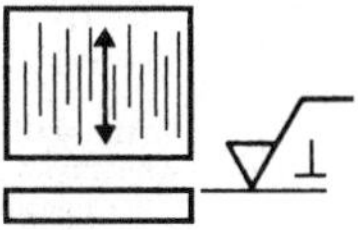
X	Lay angular in both directions to line representing the surface to which the symbol is applied.	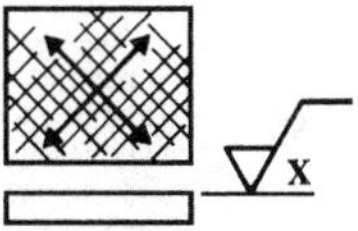
M	Lay multidirectional.	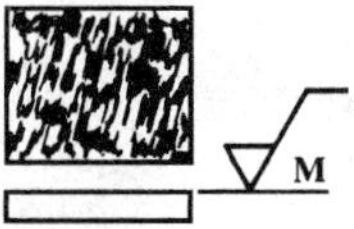
C	Lay approximately circular relative to the center of the surface to which the symbol is applied.	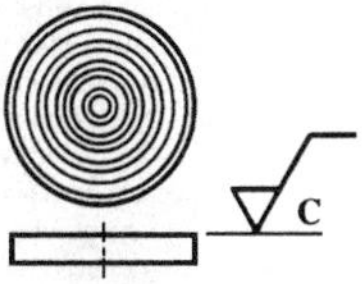
R	Lay approximately radial relative to the center of the surface to which the symbol is applied.	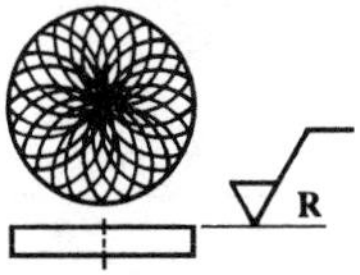
P	Lay particulate, non-directional, or protuberant.	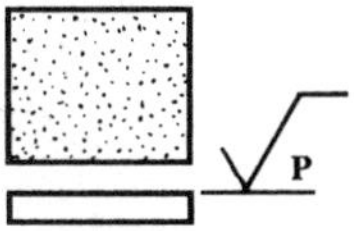

Where surface roughness control of several operations is required within a given area, or on a given surface, surface qualities may be designated, as in Fig. 9a. If a surface must be produced by one particular process or a series of processes, they should be specified as shown in Fig. 9b. Where special requirements are needed on a designated surface, a note should be added at the symbol giving the requirements and the area involved. An example is illustrated in Fig. 9c.

Table 8. Application of Surface Texture Values to Symbol

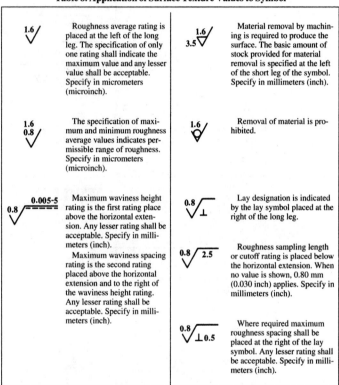

Symbol	Description	Symbol	Description
1.6	Roughness average rating is placed at the left of the long leg. The specification of only one rating shall indicate the maximum value and any lesser value shall be acceptable. Specify in micrometers (microinch).	1.6 3.5	Material removal by machining is required to produce the surface. The basic amount of stock provided for material removal is specified at the left of the short leg of the symbol. Specify in millimeters (inch).
1.6 0.8	The specification of maximum and minimum roughness average values indicates permissible range of roughness. Specify in micrometers (microinch).	1.6	Removal of material is prohibited.
0.8 0.005-5	Maximum waviness height rating is the first rating place above the horizontal extension. Any lesser rating shall be acceptable. Specify in millimeters (inch). Maximum waviness spacing rating is the second rating placed above the horizontal extension and to the right of the waviness height rating. Any lesser rating shall be acceptable. Specify in millimeters (inch).	0.8 ⊥	Lay designation is indicated by the lay symbol placed at the right of the long leg.
		0.8 2.5	Roughness sampling length or cutoff rating is placed below the horizontal extension. When no value is shown, 0.80 mm (0.030 inch) applies. Specify in millimeters (inch).
		0.8 ⊥0.5	Where required maximum roughness spacing shall be placed at the right of the lay symbol. Any lesser rating shall be acceptable. Specify in millimeters (inch).

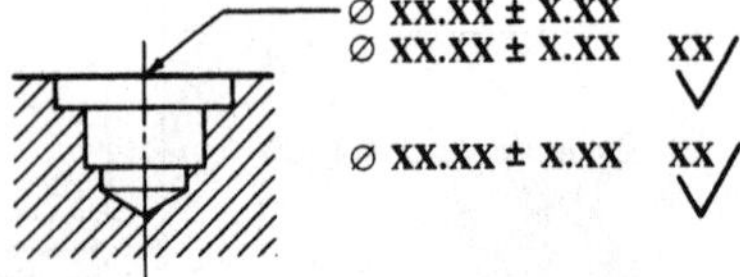

Fig. 9a.

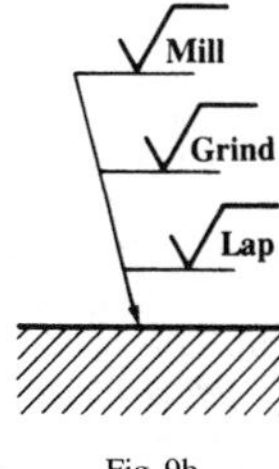

Fig. 9b.

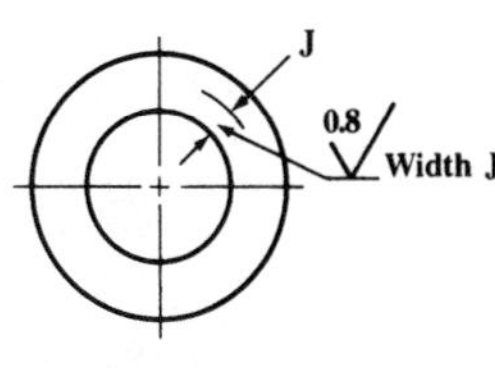

Fig. 9c.

Surface Texture of Castings.—Surface characteristics should not be controlled on a drawing or specification unless such control is essential to functional performance or appearance of the product. Imposition of such restrictions when unnecessary may increase production costs and in any event will serve to lessen the emphasis on the control specified for important surfaces. Surface characteristics of castings should never be considered on the same basis as machined surfaces. Castings are characterized by random distribution of non-directional deviations from the nominal surface.

Surfaces of castings rarely need control beyond that provided by the production method necessary to meet dimensional requirements. Comparison specimens are frequently used for evaluating surfaces having specific functional requirements. Surface texture control should not be specified unless required for appearance or function of the surface. Specification of such requirements may increase cost to the user.

Engineers should recognize that different areas of the same castings may have different surface textures. It is recommended that specifications of the surface be limited to defined areas of the casting. Practicality of and methods of determining that a casting's surface texture meets the specification shall be coordinated with the producer. The Society of Automotive Engineers standard J435 "Automotive Steel Castings" describes methods of evaluating steel casting surface texture used in the automotive and related industries.

Metric Dimensions on Drawings.—The length units of the metric system that are most generally used in connection with any work relating to mechanical engineering are the meter (39.37 inches) and the millimeter (0.03937 inch). One meter equals 1000 millimeters. On mechanical drawings, all dimensions are generally given in millimeters, no matter how large the dimensions may be. In fact, dimensions of such machines as locomotives and large electrical apparatus are given exclusively in millimeters. This practice is adopted to avoid mistakes due to misplacing decimal points, or misreading dimensions as when other units are used as well. When dimensions are given in millimeters, many of them can be given without resorting to decimal points, as a millimeter is only a little more than $\frac{1}{32}$ inch. Only dimensions of precision need be given in decimals of a millimeter; such dimensions are generally given in hundredths of a millimeter—for example, 0.02 millimeter, which is equal to 0.0008 inch. As 0.01 millimeter is equal to 0.0004 inch, dimensions are seldom given with greater accuracy than to hundredths of a millimeter.

Scales of Metric Drawings: Drawings made to the metric system are not made to scales of $\frac{1}{2}$, $\frac{1}{4}$, $\frac{1}{8}$, etc., as with drawings made to the English system. If the object cannot be drawn full size, it may be drawn $\frac{1}{2}$, $\frac{1}{5}$, $\frac{1}{10}$, $\frac{1}{20}$, $\frac{1}{50}$, $\frac{1}{100}$, $\frac{1}{200}$, $\frac{1}{500}$, or $\frac{1}{1000}$ size. If the object is too small and has to be drawn larger, it is drawn 2, 5, or 10 times its actual size.

Checking Drawings.—In order that the drawings may have a high standard of excellence, a set of instructions, as given in the following, has been issued to the checkers, and also to the draftsmen and tracers in the engineering department of a well-known machine-building company.

Inspecting a New Design: When a new design is involved, first inspect the layouts carefully to see that the parts function correctly under all conditions, that they have the proper relative proportions, that the general design is correct in the matters of strength, rigidity, bearing areas, appearance, convenience of assembly, and direction of motion of the parts, and that there are no interferences. Consider the design as a whole to see if any improvements can be made. If the design appears to be unsatisfactory in any particular, or improvements appear to be possible, call the matter to the attention of the chief engineer.

Checking for Strength: Inspect the design of the part being checked for strength, rigidity, and appearance by comparing it with other parts for similar service whenever possible, giving preference to the later designs in such comparison, unless the later designs are known to be unsatisfactory. If there is any question regarding the matter, compute the stresses and deformations or find out whether the chief engineer has approved the stresses or deformations that will result from the forces applied to the part in service. In checking parts that are to go on a machine of increased size, be sure that standard parts used in similar machines and proposed for use on the larger machine, have ample strength and rigidity under the new and more severe service to which they will be put.

Materials Specified: Consider the kind of material required for the part and the various possibilities of molding, forging, welding, or otherwise forming the rough part from this material. Then consider the machining operations to see whether changes in form or design will reduce the number of operations or the cost of machining.

See that parts are designed with reference to the economical use of material, and whenever possible, utilize standard sizes of stock and material readily obtainable from local dealers. In the case of alloy steel, special bronze, and similar materials, be sure that the material can be obtained in the size required.

Method of Making Drawing: Inspect the drawing to see that the projections and sections are made in such a way as to show most clearly the form of the piece and the work to be done on it. Make sure that any worker looking at the drawing will understand what the shape of the piece is and how it is to be molded or machined. Make sure that the delineation is correct in every particular, and that the information conveyed by the drawing as to the form of the piece is complete.

Checking Dimensions: Check all dimensions to see that they are correct. Scale all dimensions and see that the drawing is to scale. See that the dimensions on the drawing agree with the dimensions scaled from the lay-out. Wherever any dimension is out of scale, see that the dimension is so marked. Investigate any case where the dimension, the scale of the drawing, and the scale of the lay-out do not agree. All dimensions not to scale must be underlined on the tracing. In checking dimensions, note particularly the following points:

See that all figures are correctly formed and that they will print clearly, so that the workers can easily read them correctly.

See that the overall dimensions are given.

See that all witness lines go to the correct part of the drawing.

See that all arrow points go to the correct witness lines.

See that proper allowance is made for all fits.

See that the tolerances are correctly given where necessary.

See that all dimensions given agree with the corresponding dimensions of adjacent parts.

Be sure that the dimensions given on a drawing are those that the machinist will use, and that the worker will not be obliged to do addition or subtraction to obtain the necessary measurements for machining or checking his work.

Avoid strings of dimensions where errors can accumulate. It is generally better to give a number of dimensions from the same reference surface or center line.

When holes are to be located by boring on a horizontal spindle boring machine or other similar machine, give dimensions to centers of bored holes in rectangular coordinates and from the center lines of the first hole to be bored, so that the operator will not be obliged to add measurements or transfer gages.

Checking Assembly: See that the part can readily be assembled with the adjacent parts. If necessary, provide tapped holes for eyebolts and cored holes for tongs, lugs, or other methods of handling.

Make sure that, in being assembled, the piece will not interfere with other pieces already in place and that the assembly can be taken apart without difficulty.

Check the sum of a number of tolerances; this sum must not be great enough to permit two pieces that should not be in contact to come together.

Checking Castings: In checking castings, study the form of the pattern, the methods of molding, the method of supporting and venting the cores, and the effect of draft and rough molding on clearances.

Avoid undue metal thickness, and especially avoid thick and thin sections in the same casting.

Indicate all metal thicknesses, so that the molder will know what chaplets to use for supporting the cores.

See that ample fillets are provided, and that they are properly dimensioned.

See that the cores can be assembled in the mold without crushing or interference.

See that swelling, shrinkage, or misalignment of cores will not make trouble in machining.

See that the amount of extra material allowed for finishing is indicated.

See that there is sufficient extra material for finishing on large castings to permit them to be "cleaned up," even though they warp. In such castings, make sure that the metal thickness will be sufficient after finishing, even though the castings do warp.

Make sure that sufficient sections are shown so that the pattern makers and molders will not be compelled to make assumptions about the form of any part of the casting. These details are particularly important when a number of sections of the casting are similar in form, while others differ slightly.

Checking Machined Parts: Study the sequences of operations in machining and see that all finish marks are indicated.

See that the finish marks are placed on the lines to which dimensions are given.

See that methods of machining are indicated where necessary.

Give all drill, reamer, tap, and rose bit sizes.

See that jig and gage numbers are indicated at the proper places.

See that all necessary bosses, lugs, and openings are provided for lifting, handling, clamping, and machining the piece.

See that adequate wrench room is provided for all nuts and bolt heads.

Avoid special tools, such as taps, drills, reamers, etc., unless such tools are specifically authorized.

Where parts are right- and left-hand, be sure that the hand is correctly designated. When possible, mark parts as symmetrical, so as to avoid having them right- and left-hand, but do not sacrifice correct design or satisfactory operation on this account.

When heat-treatment is required, the heat-treatment should be specified.

Check the title, size of machine, the scale, and the drawing number on both the drawing and the drawing record card.

INSPECTION*

Measuring Instruments

Each dimension on a shop print represents a specific measurement required for the workpiece. Correctly reading and interpreting these dimensions requires an understanding and working knowledge of the tools and instruments used to make measurements in the shop.

Most shop measurements are classified as either *linear* measurements or *angular* measurements. Linear measurements are measurements of length and are typically expressed in units of inches or millimeters. Angular measurements are measurements of angles and are expressed in degrees and minutes.

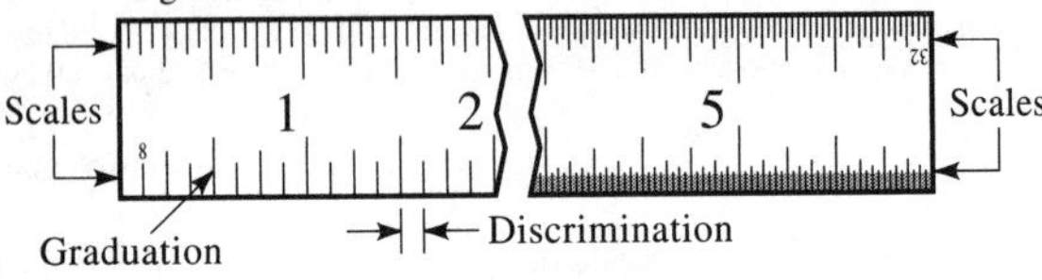

Fig. 1.

Steel Rules.—The *steel rule* is the basic measuring tool used for linear measurements. This tool is a flat, thin strip of steel with evenly spaced marks along each edge. These marks are called *graduations*. Each series of marks, or lines, along the edges of the rule are called *scales*. The spacing of the graduations determines the *discrimination* of the scale, meaning the finest or smallest division of a scale that can be read reliably. For example, a steel rule with a fractional scale having eight divisions per inch has a discrimination of $1/8$ inch. A scale with 64 divisions per inch has a discrimination of $1/64$ inch. The most common scales used on steel rules have 8, 16, 32, and 64 graduations per inch. Other variations of the basic steel rule are the *decimal inch rule* and the *millimeter rule*. Decimal inch rules have 10, 50, and 100 graduations per inch, and millimeter rules are graduated in millimeters and half millimeters.

The scale used for a measurement is determined by the stated dimensional size. For example, a $3/4$-inch dimension may be read with any common scale, but the $1/8$ scale is the easiest to read. If the dimension were $23/64$ inch, only the $1/64$ scale could be used to accurately make this measurement.

When reading a steel rule, the scale that is easiest to read should always be used to reduce errors. As a general rule, the scale used is both that which best suits the dimension to be measured, and is also the easiest to read.

Reading a Fractional Steel Rule: The rule is positioned on the workpiece with the *reference point* aligned with the starting point of the measurement. Next, the graduation line that is aligned with the *measured point*, or end of the measurement, is read. When the end of the rule is used as the reference point, the value of this graduation must be subtracted from the measured point to determine the measured value.

To determine the measurements obtained with a fractional steel rule, the value is written as a common fraction with the number of graduations between the reference point and the measured point as the numerator and the scale discrimination as the denominator. For example, a measurement of 7 inches on the $1/16$ scale is expressed as $7/16$ inch. Whenever possible, measurements expressed as common fractions should be reduced to lowest terms. Thus, a measurement of 6 lines on the $1/8$ scale should be expressed as $3/4$ inch, not $6/8$ inch. When making measurements larger than 1 inch, always add the value of the full inch units to the total measurement.

* Portions of the chapter have been taken from *Print Reading for Industry*, by permission of the South-Western Publishing Co., Cincinnati, OH.

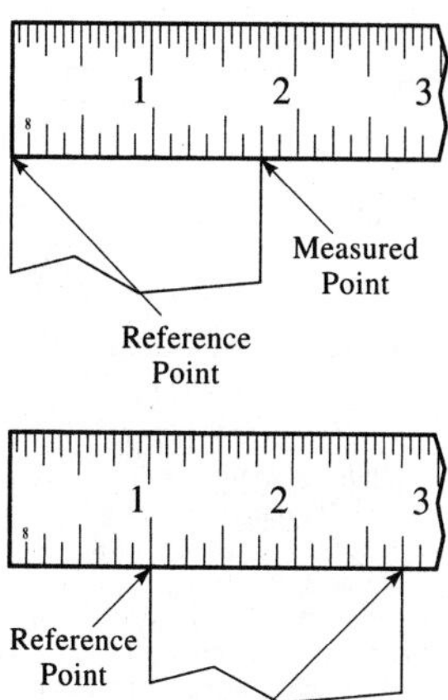

Fig. 2.

Fractional vs. Decimal Steel Rules: Although fractional rules are the most common steel rule in general shop work, decimal rules are also used for a variety of applications. The two most common styles of decimal rules are the decimal inch and millimeter rules. Unlike fractional rules, which are graduated with base 8 numbers, both decimal and millimeter rules are graduated with base 10 values. In use, these rules yield readings that are most often expressed as decimals rather than as fractions.

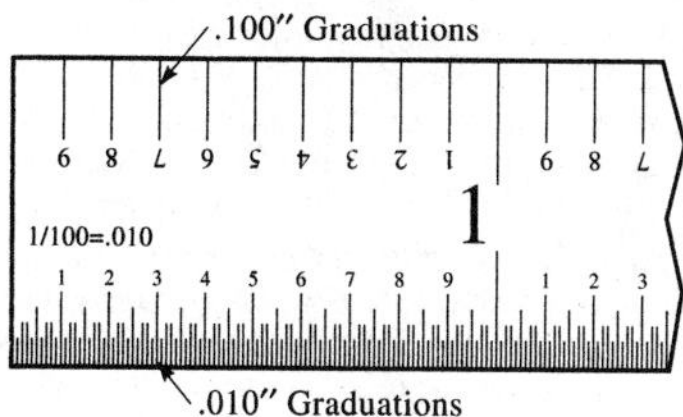

Fig. 3a.

For example, when reading a decimal rule, a value of $^{50}/_{100}$, while expressed as a half inch, would be written as 0.50 inch, not as the common fraction $^{1}/_{2}$ inch. More often, the values read from a decimal rule are expressed in terms of thousandths of an inch, instead of hundredths. For example, a .50-inch reading is usually expressed as $^{500}/_{1000}$ inch and written as .500 inch. For the sake of uniformity, decimal values used in the machine shop are typically expressed in thousandths. This reduces any confusion that might result from expressing sizes as tenths or hundredths.

The term "tenths," as used in most shops, refers to $^{1}/_{10}$ of $^{1}/_{1000}$ inch, that is, $^{1}/_{10,000}$ inch or 0.0001 inch. Rather than saying a dimension is accurate to $^{2}/_{10,000}$ inch, most machinists shorten this expression and describe this value as "$^{2}/_{10}$."

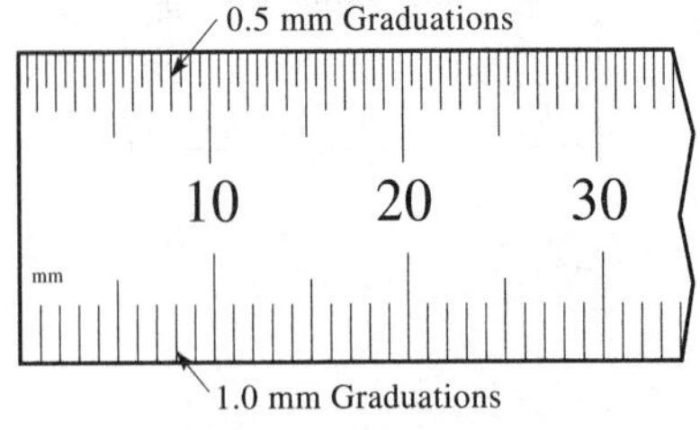

Fig. 3b.

Reading a Decimal Inch Steel Rule: As before, when reading a decimal steel rule, the scale that is easiest to read is selected to reduce the chance of error. The two most common scales found on decimal inch rules are tenths and hundredths. Measurements made with this rule are done in the same fashion as for the fractional rule. First, the rule is positioned with either the end or full inch graduation mark aligned with the reference point of the measurement. Next, the graduation line that is aligned with the measured point of the measurement is read. When the end of the rule is used as the reference point, the measured point should be read directly. However, if another graduation, such as one of the full inch lines, is used as the reference point, the value of this graduation must be subtracted from the measured point to determine the measured value.

To read the measured value, the number of graduations between the reference point and the measured point is counted. For example, if there are 25 lines on the hundredth scale, the value would be expressed as $^{25}/_{100}$ inch. If the tenth scale is used, and the measurement is made to a point 7 lines beyond the 1-inch graduation, the value is expressed as either 1 and $^{7}/_{10}$ inches. When measurements are larger than 1 inch, the value of the full inch units must be added to the measurement. The *quick reading numbers* above the graduations assist in counting the number of graduations on the scale.

Reading a Millimeter Steel Rule: Again, the proper scale is selected, that which is easiest to read so that the chance of error is reduced. Most millimeter rules have scales with one millimeter (1 mm) and one-half millimeter (1/2 mm) graduations. Once the proper scale is selected, measurements are made in the same way as for other rules. Measured values are expressed as either full millimeters or in units of millimeters and half millimeters.

In addition to using the *Sine Bar Constants* Table on page 191, a standard Table of *Trigonometric Functions* on page 103 may be used when working with a 10-inch sine bar. For example, to find the gage block height necessary to set the 10-inch sine bar to an angle of 29°30″, the following procedure may be used. First, the known values in the problem and those that must be calculated are identified. In this example, the hypotenuse, working angle, and right angle are known. The hypotenuse is represented by the sine bar itself and is 10.000 inches. The working angle is 29°30″ and the right angle is 90°. Here the side opposite, side adjacent, and complementary angle are to be calculated. As shown, using standard trigonometric formulas, these values are found to be:

side opposite = 4.9242 inches

side adjacent = 8.7036 inches

complementary angle = 60°30″

The 4.9242-inch measurement represents the gage block stack height required to set the 29° 30″ angle.

Micrometers.—Micrometers are linear measuring instruments used to make measurements to accuracies of 0.001 and 0.0001 of an inch and 0.01 and 0.002 of a millimeter. The most common type of micrometer is the micrometer caliper, See Fig. 4. This instrument consists of several parts: The frame is the main unit of the micrometer caliper, which maintains the alignment of the contact points on the ends of the spindle and anvil. The sleeve is also attached to the frame and, on inch-type micrometers, has graduations that divide the travel of the spindle into 40 equal parts, Fig. 5a. The thimble, which rotates around the sleeve, is also graduated. These graduations divide the rotation of the thimble into 25 equal parts. So together, the sleeve and thimble graduations of the micrometer caliper divide an inch into 1000 ($40 \times 25 = 1000$) parts. The ratchet stop, located at the end of the thimble, is used to limit the pressure between the contact points. Using the ratchet stop when making measurements with a micrometer will ensure the correct measuring pressure. Some micrometer calipers also have a graduated scale on the top side of the sleeve. This scale is called the vernier scale and is used to make measurements to an accuracy of .0001 inch.

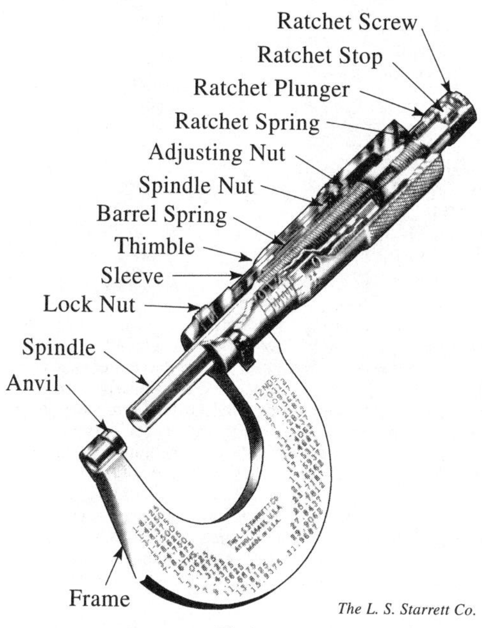

The L. S. Starrett Co.

Fig. 4.

Millimeter micrometers, Fig. 5b, have the same basic construction as the inch type micrometer, the only difference being in the graduations. The sleeve on a millimeter micrometer is graduated in half millimeter units. The thimble is divided into 50 equal parts each equal to 0.01 mm. So together, the sleeve and thimble graduations divide a millimeter into 100 parts. The millimeter micrometer may also have a vernier scale on the top side of the sleeve. This vernier scale further divides the micrometer reading and permits readings to an accuracy of 0.002 mm.

Reading An Inch Micrometer: Each graduation on the sleeve of a micrometer is equal to 1/40″. Since micrometers are read in decimal inches, this value is expressed as 0.025″ ($1 \div 40$). As an aid to reading the micrometer, every fourth line is numbered. Each of these numbered lines is equal to 0.100″. To determine the value of the sleeve graduations, simply count the number of graduations visible to the left of the thimble and multiply by 25. Do not count any lines that are partially covered by the thimble. Next, note the value of the thimble graduation aligned with the reading line on the sleeve and add this amount to the value of the sleeve graduations. The micrometer reading shown in Fig. 6a has a sleeve value of 0.350″ (14 lines). The thimble graduation aligned with the reading line is 15 that is, 0.015″. So adding these together the total reading is expressed as 0.365″.

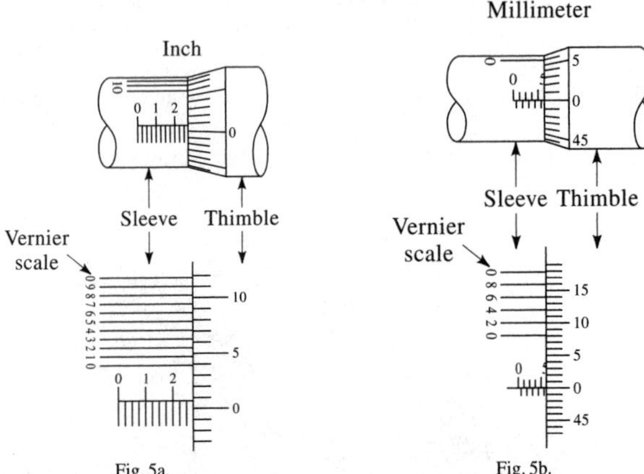

Fig. 5a. Fig. 5b.

To read a vernier micrometer, simply read the micrometer as outlined above and add the value of the vernier scale. To read the vernier scale, first turn the micrometer so the complete vernier scale is visible. Now, find the one line on the vernier scale that is aligned with one of the thimble lines. There will only be one line perfectly aligned at any one time. Remember, use the value of the vernier scale, not the thimble graduation. The purpose of a vernier scale on a micrometer is to divide the space between the thimble graduations into 10 equal parts. So, when reading this micrometer, the thimble graduations may not be exactly aligned with the reading line on the sleeve. In these cases, always use the lesser value. The micrometer reading shown in Fig. 6b shows how this type of micrometer should be read. The sleeve graduations visible to the left of the thimble equal 0.425″ (17 lines). The thimble location is between the thimble numbers 3 and 4 and the vernier is aligned on the vernier number 3. Adding all this together, the reading is expressed as 0.4283″. This value should be read as 4 thousand 2 hundred and 83 ten-thousandths of an inch.

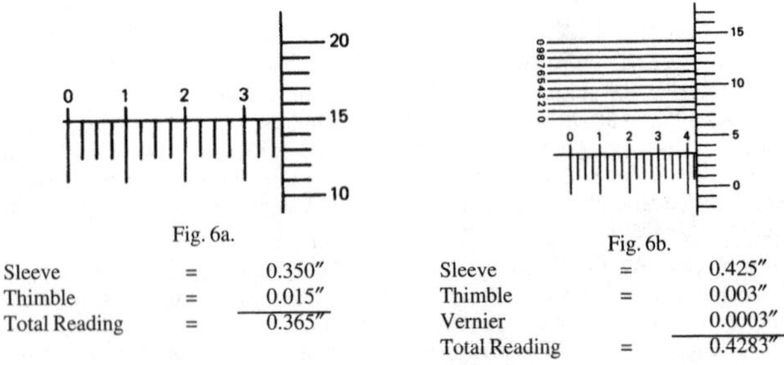

Fig. 6a.

Sleeve	=	0.350″
Thimble	=	0.015″
Total Reading	=	0.365″

Fig. 6b.

Sleeve	=	0.425″
Thimble	=	0.003″
Vernier		0.0003″
Total Reading	=	0.4283″

Reading A Millimeter Micrometer: When reading a millimeter micrometer, the process is identical to that used for the inch-type micrometer. First, find the value of the sleeve graduations visible to the left of the thimble. On millimeter micrometers the graduations are each equal to 0.5 mm. Most millimeter micrometers only number every 10 mm on the sleeve so you will have to pay close attention when counting these lines. Once you know the value of the sleeve graduations, note the value of the thimble graduation aligned with the reading line and add the two values. In the example shown in Fig. 6a, the sleeve graduations equal 14.5 mm and the thimble is aligned on the number 12. Together, these values equal 14.62 mm.

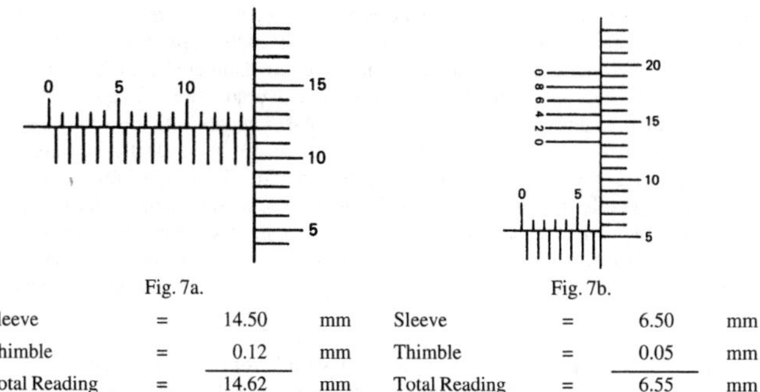

Fig. 7a. Fig. 7b.

Sleeve	= 14.50	mm		Sleeve	= 6.50	mm
Thimble	= 0.12	mm		Thimble	= 0.05	mm
Total Reading	= 14.62	mm		Total Reading	= 6.55	mm

To read a millimeter vernier micrometer, first determine the value of the measurement to an accuracy of .01 mm. Then find the one line on the vernier scale that is aligned with a thimble graduation and add this amount to the first reading. As shown in Fig. 7b, the sleeve graduations equal 6.5 mm. The thimble is between the numbers and 6 and the vernier is aligned on the number 8. Added together this value is mm. This value should be read as 6 and hundred and 58 thousandths millimeter.

Verniers.—Verniers are another form of linear measuring tool frequently used for making measurements in the shop. The most popular type of vernier instrument is the *vernier caliper*, Fig. 8.

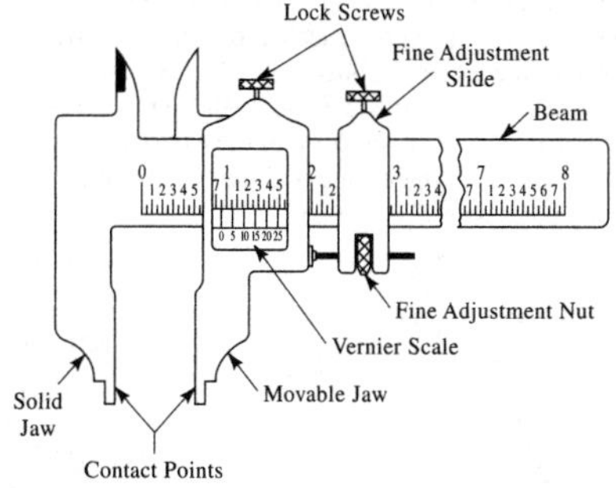

Fig. 8. Vernier Caliper

The vernier caliper consists of a *beam*, which contains the *solid jaw* and the *main scale* graduations. The *movable jaw* is mounted on the beam and can be moved and clamped in any desired position along the length of the beam with the *clamp screws*. The *vernier scale* is contained within the movable jaw and permits the vernier caliper to measure to within 0.001 or 0.02 mm. Some types of vernier calipers also have a *fine adjustment slide* to aid in making accurate measurements. The major variations of the vernier caliper are the 25-division vernier, the 50-division vernier, the millimeter vernier, and the dial caliper.

Reading A 25-division Vernier: The beam of the 25-division vernier has a main scale that is graduated in increments of 1/40″. Each graduation on the main scale is equal to 0.025″, just like the scale on the sleeve of a micrometer. The vernier plate on this vernier, like the micrometer thimble, has 25 divisions and is used to further divide the main scale graduations to 0.001″. To read this vernier, first locate the left zero on the vernier plate with reference to the main scale. Include all graduations to the left of the zero in your reading. Like the micrometer sleeve, the main scale graduations are also numbered at each 0.100″ value. The larger numbers on the main scale represent full inch graduations. You should always remember to include any full inch values in your measurements with a vernier. Once the zero is located and the value noted, find the one line on the vernier scale that is aligned with a line on the main scale and add this value to the total reading. Here again, remember to use the value of the vernier graduation and not the main scale number when determining this value. To read the vernier shown in Fig. 9a, first find the main scale value. In this case this value is 4.625″. Now find the vernier scale line that is aligned with a line on the main scale. In this example the number 16 is aligned. So, adding these values the total reading becomes 4.641″.

Reading A 50-division Vernier: The 50-division vernier is read in much the same way as the 25-division vernier. The only difference is in the main scale and vernier scale graduations. The graduations on the main scale of this vernier are equal to 0.050 and the vernier scale has 50 rather than 25, divisions. To read this vernier, first find the value of the main scale graduations to the left of the zero. Then, find the one vernier scale graduation that is aligned with one of the lines on the main scale and add the values together. To read the vernier shown in Fig. 9b, first find the main scale value. In this case the value is 1.000″. Now find the vernier scale value. On this vernier the number 25 is the only number that is aligned. So, adding the values together, the total reading is 1.025″.

Reading A Millimeter Vernier: Millimeter verniers are again read in much the same way as the inch-type verniers. The main scale graduations on the millimeter vernier is graduated in full millimeter units, and the vernier scale is divided into 50 parts, each equal to 0.02 mm. To read this vernier, first locate the position of the zero and note the value of the reading to the left of the zero. Now, find the one line on the vernier scale that is aligned with a main scale graduation and add the two values. In the example shown in Fig. 9c, the main scale reading is 32 mm and the vernier scale is aligned at the number 62. So, adding these values the total reading is expressed as 32.62 mm.

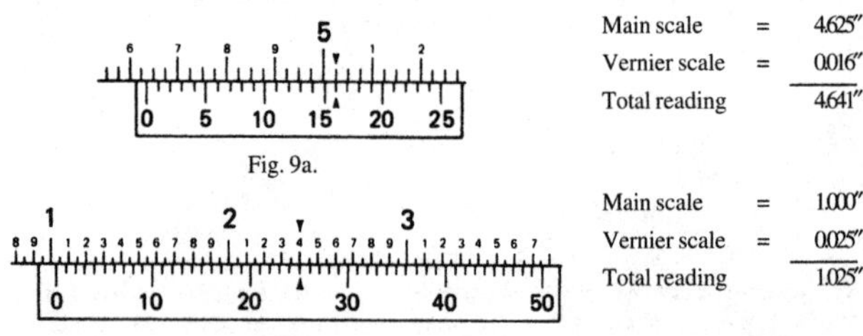

Main scale = 4.625″
Vernier scale = 0.016″
Total reading 4.641″
Main scale = 1.000″
Vernier scale = 0.025″
Total reading 1.025″

Fig. 9a.

Fig. 9b.

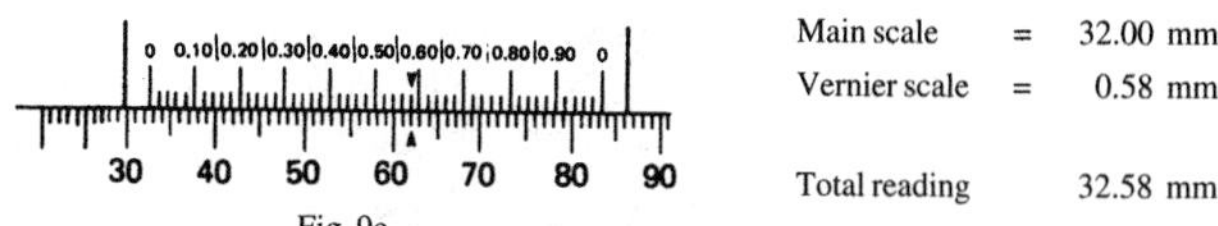

Fig. 9c.

Main scale	=	32.00 mm
Vernier scale	=	0.58 mm
Total reading		32.58 mm

Dial Calipers.—Another variation of the vernier caliper is the *dial caliper*. These calipers are read by finding the value of the main scale graduations visible to the left of the reference edge and adding the dial reading, Fig. 10a and 10b. With these calipers the dial replaces the vernier scale, generally make reading the instrument easier and faster than the standard vernier caliper. Inch-type dial calipers are accurate to 0.001″ and millimeter dial calipers are accurate to 0.02 mm. As shown, to read the inch-type dial caliper, Fig. 10a, the main scale value is first determined by reading the value shown to the left of the reference edge. In this reading it is 4.300″. The dial value is then read and added to the main scale value. In this case the measurement value is 4.307″. The millimeter dial vernier is read in the same way. As shown in Fig. 10b, the main scale value is 56.00 mm and the dial reading is .40 mm. Together these show a measurement value of 56.40 mm.

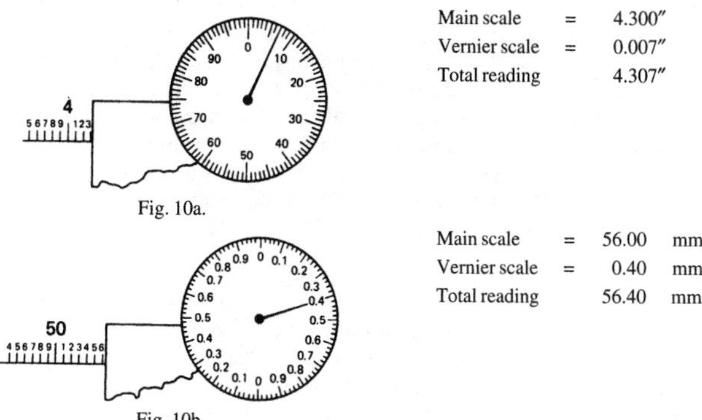

Fig. 10a.

Main scale	=	4.300″
Vernier scale	=	0.007″
Total reading		4.307″

Fig. 10b.

Main scale	=	56.00	mm
Vernier scale	=	0.40	mm
Total reading		56.40	mm

Protractors.—Protractors are measuring tools used to measure angles. The two basic types of protractors used in the shop are the *steel protractor* and the *vernier bevel protractor*, Fig. 11a and 11b. The basic difference between these two forms of protractors is accuracy. The steel protractor, and its variations, such as the bevel protractor head on a combination square, are accurate to 1° while the vernier bevel protractor is accurate to 5′.

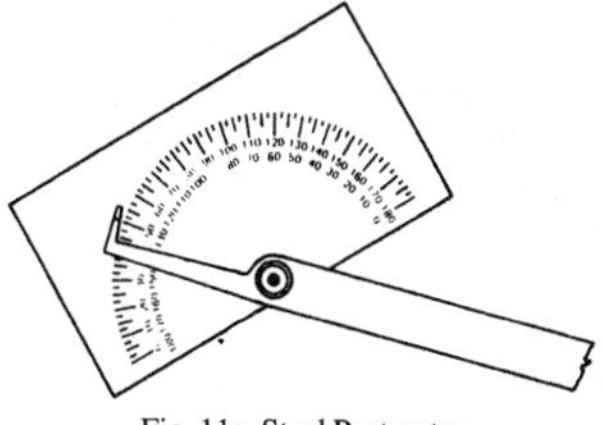
Fig. 11a. Steel Protractor

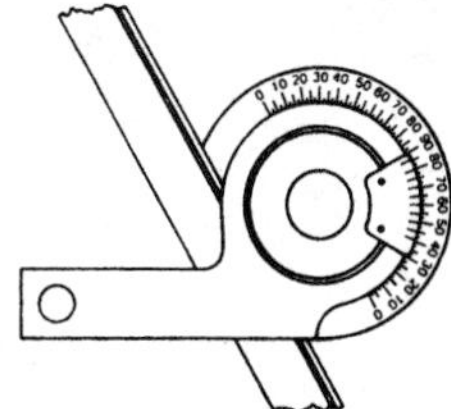
Fig. 11b. Vernier Bevel Protractor

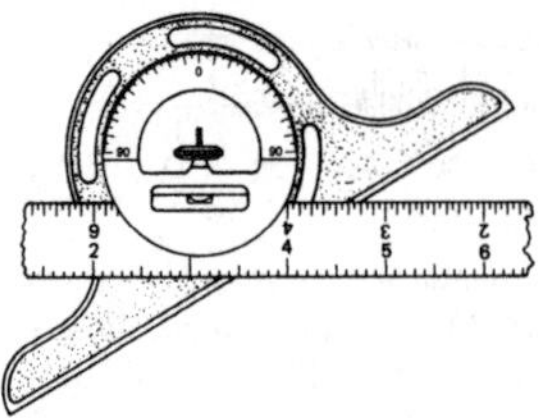

Fig. 11c. Bevel Protractor

The first step in measuring angles is understanding how angles are expressed. When an angle is specified on a print, it is normally noted in units of *degrees* and *minutes*. Every circle contains 360°. Each degree contains 60′. So, if an angle were specified as 15° 30′, it would be the same as 15 and ½ degrees. Minutes are also divided into *seconds* (″), but for all practical purposes, minutes are as close as you will be able to measure in the shop.

When reading a steel protractor, align the two sides of the protractor with the sides of the angle to be measured. Then, remove the protractor and note the position of the reference line against the protractor scale, Fig. 12a. Steel protractors are read directly, and the value shown on the protractor scale is the total value of the angle within 1° Since this type of protractor is designed to be used for angles to 180° the main scale on the protractor can be read from 0° to 180° in both directions. As shown, any measured value can be expressed in terms relative to 90° or 180°, depending the way the protractor is set. With this protractor setting, one side of the bar is set at 40° while the other side is at 140°.

When angles must be measured to an accuracy closer than 1°, a vernier bevel protractor may be used. To read a vernier bevel protractor, note the position of the zero reference line on the vernier scale. The vernier scale on this tool may be read in either direction, depending on the direction of rotation of the main scale. If the main scale is rotated to the right, or clockwise, the vernier scale to the right of the zero is used for the measurement. If, however, the main scale is rotated to the left, or counterclockwise, then the scale to the left of the zero is read. In the example shown in Fig. 12b, the main scale is rotated to the left and the zero is positioned between the 20° and 21° graduations. This means the main scale graduation is 20°. The one vernier scale graduation that is aligned with a line on the main scale is the number 15. So, adding these values together, the total reading is expressed as 20°15′.

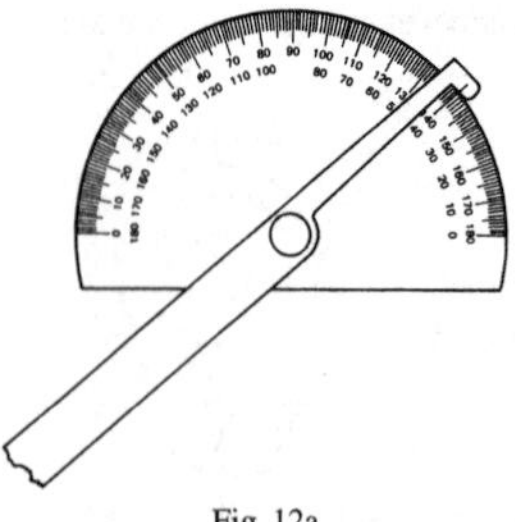

Fig. 12a.

Fig. 12b.

Main scale	=	20°
Vernier scale	=	15′
Total reading	=	20°15′

Sine-bar

The sine-bar is used either for very accurate angular measurements or for locating work at a given angle as, for example, in surface grinding templets, gages, etc. The sine-bar is especially useful in measuring or checking angles when the limit of accuracy is 5 minutes or less. Some bevel protractors are equipped with verniers which read to 5 minutes but the setting depends upon the alignment of graduations whereas a sine-bar usually is located by positive contact with precision gage-blocks selected for whatever dimension is required for obtaining a given angle.

Types of Sine-bars.—A sine-bar consists of a hardened, ground and lapped steel bar with very accurate cylindrical plugs of equal diameter attached to or near each end. The form illustrated by Fig. 15 has notched ends for receiving the cylindrical plugs so that they are held firmly against both faces of the notch. The standard center-to-center distance C between the plugs is either 5 or 10 inches. The upper and lower sides of sine-bars are parallel to the center line of the plugs within very close limits. The body of the sine-bar ordinarily has several through holes to reduce the weight. In the making of the sine-bar shown in Fig. 16, if too much material is removed from one locating notch, regrinding the shoulder at the opposite end would make it possible to obtain the correct center distance. That is the reason for this change in form. The type of sine-bar illustrated by Fig. 15 has the cylindrical disks or plugs attached to one side. These differences in form or arrangement do not, of course, affect the principle governing the use of the sine-bar. An accurate surface plate or master flat is always used in conjunction with a sine-bar in order to form the base from which the vertical measurements are made.

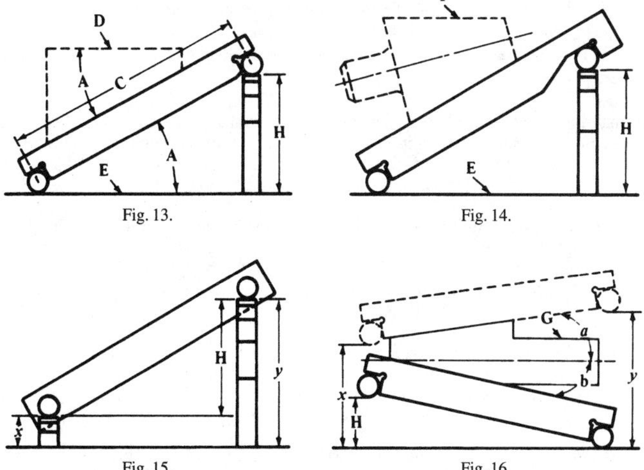

Fig. 13. Fig. 14.

Fig. 15. Fig. 16.

Setting a Sine Bar to a Given Angle.—To find the vertical distance *H*, for setting a sine bar to the required angle, convert the angle to decimal form on a pocket calculator, take the sine of that angle, and multiply by the distance between the cylinders. For example, if an angle of 31 degrees, 30 minutes is required, the equivalent angle is 31 degrees plus $^{30}/_{60}$ = 31 + 0.5, or 31.5 degrees. (For conversions from minutes and seconds to decimals of degrees and vice versa, see page 90). The sine of 31.5 degrees is 0.5225 and multiplying this value by the sine bar length gives 2.613 in. for the height *H*, Fig. 13 and 15, of the gage blocks.

Finding Angle when Height *H* of Sine Bar is Known.—To find the angle equivalent to a given height *H*, reverse the above procedure. Thus, if the height *H* is 1.4061 in., dividing by 5 gives a sine of 0.28122, which corresponds to an angle of 16.333 degrees, or 16 degrees 20 minutes.

Checking Angle of Templet or Gage by Using Sine Bar.—Place templet or gage on sine bar as indicated by dotted lines, Fig. 13. Clamps may be used to hold work in place. Place upper end of sine bar on gage blocks having total height *H* corresponding to the required angle. If upper edge *D* of work is parallel with surface plate *E*, then angle *A* of work equals angle *A* to which sine bar is set. Parallelism between edge *D* and surface plate may be tested by checking the height at each end with a dial gage or some type of indicating comparator.

Measuring Angle of Templet or Gage with Sine Bar.—To measure such an angle, adjust height of gage blocks and sine bar until edge *D*, Fig. 13, is parallel with surface plate *E*; then find angle corresponding to height *H*, of gage blocks. For example, if height *H* is 2.5939 inches when *D* and *E* are parallel, the calculator will show that the angle *A* of the work is 31 degrees, 15 minutes.

Checking Taper per Foot with Sine Bar.—As an example, assume that the plug gage in Fig. 14 is supposed to have a taper of 6⅛ inches per foot and taper is to be checked by using a 5-inch sine bar. The table of *Tapers per Foot and Corresponding Angles* on page 200 shows that the included angle for a taper of 6⅛ inches per foot is 28 degrees 38 minutes 1 second, or 28.6336 degrees from the calculator. For a 5-inch sine bar, the calculator gives a value of 2.396 in. for the height *H* of the gage blocks. Using this height, if the upper surface *F* of the plug gage is parallel to the surface plate the angle corresponds to a taper of 6 ⅛ inches per foot.

Setting Sine Bar having Plugs Attached to Side.—If the lower plug does not rest directly on the surface plate, as in Fig. 15, the height *H* for the sine bar is the difference between heights *x* and *y*, or the difference between the heights of the plugs; otherwise, the procedure in setting the sine bar and checking angles is the same as previously described.

Checking Templets Having Two Angles.—Assume that angle *a* of templet, Fig. 16, is 9 degrees, angle *b* 12 degrees, and that edge *G* is parallel to the surface plate. For an angle *b* of 12 degrees, the calculator shows that the height *H* is 1.03956 inches. For an angle *a* of 9 degrees, the difference between measurements *x* and *y* when the sine bar is in contact with the upper edge of the templet is 0.78217 inch.

Setting 10-inch Sine Bar to Given Angle.—A 10-inch sine bar may sometimes be preferred because of its longer working surface or because the longer center distance is conducive to greater precision. To obtain the vertical distances *H* for setting a 10-inch sine bar, multiply the sine of the angle by 10, by shifting the decimal point one place to the right.

For example, the sine of 39 degrees is 0.62932, hence the vertical height *H* for setting a 10-inch sine bar is 6.2932 inches.

Constants for Setting a 5-inch Sine-Bar for 1° to 7°

Min.	0°	1°	2°	3°	4°	5°	6°	7°
0	0.00000	0.08726	0.17450	0.26168	0.34878	0.43578	0.52264	0.60935
1	0.00145	0.08872	0.17595	0.26313	0.35023	0.43723	0.52409	0.61079
2	0.00291	0.09017	0.17740	0.26458	0.35168	0.43868	0.52554	0.61223
3	0.00436	0.09162	0.17886	0.26604	0.35313	0.44013	0.52698	0.61368
4	0.00582	0.09308	0.18031	0.26749	0.35459	0.44157	0.52843	0.61512
5	0.00727	0.09453	0.18177	0.26894	0.35604	0.44302	0.52987	0.61656
6	0.00873	0.09599	0.18322	0.27039	0.35749	0.44447	0.53132	0.61801
7	0.01018	0.09744	0.18467	0.27185	0.35894	0.44592	0.53277	0.61945
8	0.01164	0.09890	0.18613	0.27330	0.36039	0.44737	0.53421	0.62089
9	0.01309	0.10035	0.18758	0.27475	0.36184	0.44882	0.53566	0.62234
10	0.01454	0.10180	0.18903	0.27620	0.36329	0.45027	0.53710	0.62378
11	0.01600	0.10326	0.19049	0.27766	0.36474	0.45171	0.53855	0.62522
12	0.01745	0.10471	0.19194	0.27911	0.36619	0.45316	0.54000	0.62667
13	0.01891	0.10617	0.19339	0.28056	0.36764	0.45461	0.54144	0.62811
14	0.02036	0.10762	0.19485	0.28201	0.36909	0.45606	0.54289	0.62955
15	0.02182	0.10907	0.19630	0.28346	0.37054	0.45751	0.54433	0.63099
16	0.02327	0.11053	0.19775	0.28492	0.37199	0.45896	0.54578	0.63244
17	0.02473	0.11198	0.19921	0.28637	0.37344	0.46040	0.54723	0.63388
18	0.02618	0.11344	0.20066	0.28782	0.37489	0.46185	0.54867	0.63532
19	0.02763	0.11489	0.20211	0.28927	0.37634	0.46330	0.55012	0.63677
20	0.02909	0.11634	0.20357	0.29072	0.37779	0.46475	0.55156	0.63821
21	0.03054	0.11780	0.20502	0.29218	0.37924	0.46620	0.55301	0.63965
22	0.03200	0.11925	0.20647	0.29363	0.38069	0.46765	0.55445	0.64109
23	0.03345	0.12071	0.20793	0.29508	0.38214	0.46909	0.55590	0.64254
24	0.03491	0.12216	0.20938	0.29653	0.38360	0.47054	0.55734	0.64398
25	0.03636	0.12361	0.21083	0.29798	0.38505	0.47199	0.55879	0.64542
26	0.03782	0.12507	0.21228	0.29944	0.38650	0.47344	0.56024	0.64686
27	0.03927	0.12652	0.21374	0.30089	0.38795	0.47489	0.56168	0.64830
28	0.04072	0.12798	0.21519	0.30234	0.38940	0.47633	0.56313	0.64975
29	0.04218	0.12943	0.21664	0.30379	0.39085	0.47778	0.56457	0.65119
30	0.04363	0.13088	0.21810	0.30524	0.39230	0.47923	0.56602	0.65263
31	0.04509	0.13234	0.21955	0.30669	0.39375	0.48068	0.56746	0.65407
32	0.04654	0.13379	0.22100	0.30815	0.39520	0.48212	0.56891	0.65551
33	0.04800	0.13525	0.22246	0.30960	0.39665	0.48357	0.57035	0.65696
34	0.04945	0.13670	0.22391	0.31105	0.39810	0.48502	0.57180	0.65840
35	0.05090	0.13815	0.22536	0.31250	0.39954	0.48647	0.57324	0.65984
36	0.05236	0.13961	0.22681	0.31395	0.40099	0.48791	0.57469	0.66128
37	0.05381	0.14106	0.22827	0.31540	0.40244	0.48936	0.57613	0.66272
38	0.05527	0.14252	0.22972	0.31686	0.40389	0.49081	0.57758	0.66417
39	0.05672	0.14397	0.23117	0.31831	0.40534	0.49226	0.57902	0.66561
40	0.05818	0.14542	0.23263	0.31976	0.40679	0.49370	0.58046	0.66705
41	0.05963	0.14688	0.23408	0.32121	0.40824	0.49515	0.58191	0.66849
42	0.06109	0.14833	0.23553	0.32266	0.40969	0.49660	0.58335	0.66993
43	0.06254	0.14979	0.23699	0.32411	0.41114	0.49805	0.58480	0.67137
44	0.06399	0.15124	0.23844	0.32556	0.41259	0.49949	0.58624	0.67281
45	0.06545	0.15269	0.23989	0.32702	0.41404	0.50094	0.58769	0.67425
46	0.06690	0.15415	0.24134	0.32847	0.41549	0.50239	0.58913	0.67570
47	0.06836	0.15560	0.24280	0.32992	0.41694	0.50383	0.59058	0.67714
48	0.06981	0.15705	0.24425	0.33137	0.41839	0.50528	0.59202	0.67858
49	0.07127	0.15851	0.24570	0.33282	0.41984	0.50673	0.59346	0.68002
50	0.07272	0.15996	0.24715	0.33427	0.42129	0.50818	0.59491	0.68146
51	0.07417	0.16141	0.24861	0.33572	0.42274	0.50962	0.59635	0.68290
52	0.07563	0.16287	0.25006	0.33717	0.42419	0.51107	0.59780	0.68434
53	0.07708	0.16432	0.25151	0.33863	0.42564	0.51252	0.59924	0.68578
54	0.07854	0.16578	0.25296	0.34008	0.42708	0.51396	0.60068	0.68722
55	0.07999	0.16723	0.25442	0.34153	0.42853	0.51541	0.60213	0.68866
56	0.08145	0.16868	0.25587	0.34298	0.42998	0.51686	0.60357	0.69010
57	0.08290	0.17014	0.25732	0.34443	0.43143	0.51830	0.60502	0.69154
58	0.08435	0.17159	0.25877	0.34588	0.43288	0.51975	0.60646	0.69298
59	0.08581	0.17304	0.26023	0.34733	0.43433	0.52120	0.60790	0.69443
60	0.08726	0.17450	0.26168	0.34878	0.43578	0.52264	0.60935	0.69587

Constants for Setting a 5-inch Sine-Bar for 8° to 15°

Min.	8°	9°	10°	11°	12°	13°	14°	15°
0	0.69587	0.78217	0.86824	0.95404	1.03956	1.12476	1.20961	1.29410
1	0.69731	0.78361	0.86967	0.95547	1.04098	1.12617	1.21102	1.29550
2	0.69875	0.78505	0.87111	0.95690	1.04240	1.12759	1.21243	1.29690
3	0.70019	0.78648	0.87254	0.95833	1.04383	1.12901	1.21384	1.29831
4	0.70163	0.78792	0.87397	0.95976	1.04525	1.13042	1.21525	1.29971
5	0.70307	0.78935	0.87540	0.96118	1.04667	1.13184	1.21666	1.30112
6	0.70451	0.79079	0.87683	0.96261	1.04809	1.13326	1.21808	1.30252
7	0.70595	0.79223	0.87827	0.96404	1.04951	1.13467	1.21949	1.30393
8	0.70739	0.79366	0.87970	0.96546	1.05094	1.13609	1.22090	1.30533
9	0.70883	0.79510	0.88113	0.96689	1.05236	1.13751	1.22231	1.30673
10	0.71027	0.79653	0.88256	0.96832	1.05378	1.13892	1.22372	1.30814
11	0.71171	0.79797	0.88399	0.96974	1.05520	1.14034	1.22513	1.30954
12	0.71314	0.79941	0.88542	0.97117	1.05662	1.14175	1.22654	1.31095
13	0.71458	0.80084	0.88686	0.97260	1.05805	1.14317	1.22795	1.31235
14	0.71602	0.80228	0.88829	0.97403	1.05947	1.14459	1.22936	1.31375
15	0.71746	0.80371	0.88972	0.97545	1.06089	1.14600	1.23077	1.31516
16	0.71890	0.80515	0.89115	0.97688	1.06231	1.14742	1.23218	1.31656
17	0.72034	0.80658	0.89258	0.97830	1.06373	1.14883	1.23359	1.31796
18	0.72178	0.80802	0.89401	0.97973	1.06515	1.15025	1.23500	1.31937
19	0.72322	0.80945	0.89544	0.98116	1.06657	1.15166	1.23640	1.32077
20	0.72466	0.81089	0.89687	0.98258	1.06799	1.15308	1.23781	1.32217
21	0.72610	0.81232	0.89830	0.98401	1.06941	1.15449	1.23922	1.32357
22	0.72754	0.81376	0.89973	0.98544	1.07084	1.15591	1.24063	1.32498
23	0.72898	0.81519	0.90117	0.98686	1.07226	1.15732	1.24204	1.32638
24	0.73042	0.81663	0.90260	0.98829	1.07368	1.15874	1.24345	1.32778
25	0.73185	0.81806	0.90403	0.98971	1.07510	1.16015	1.24486	1.32918
26	0.73329	0.81950	0.90546	0.99114	1.07652	1.16157	1.24627	1.33058
27	0.73473	0.82093	0.90689	0.99256	1.07794	1.16298	1.24768	1.33199
28	0.73617	0.82237	0.90832	0.99399	1.07936	1.16440	1.24908	1.33339
29	0.73761	0.82380	0.90975	0.99541	1.08078	1.16581	1.25049	1.33479
30	0.73905	0.82524	0.91118	0.99684	1.08220	1.16723	1.25190	1.33619
31	0.74049	0.82667	0.91261	0.99826	1.08362	1.16864	1.25331	1.33759
32	0.74192	0.82811	0.91404	0.99969	1.08504	1.17006	1.25472	1.33899
33	0.74336	0.82954	0.91547	1.00112	1.08646	1.17147	1.25612	1.34040
34	0.74480	0.83098	0.91690	1.00254	1.08788	1.17288	1.25753	1.34180
35	0.74624	0.83241	0.91833	1.00396	1.08930	1.17430	1.25894	1.34320
36	0.74768	0.83384	0.91976	1.00539	1.09072	1.17571	1.26035	1.34460
37	0.74911	0.83528	0.92119	1.00681	1.09214	1.17712	1.26175	1.34600
38	0.75055	0.83671	0.92262	1.00824	1.09355	1.17854	1.26316	1.34740
39	0.75199	0.83815	0.92405	1.00966	1.09497	1.17995	1.26457	1.34880
40	0.75343	0.83958	0.92547	1.01109	1.09639	1.18136	1.26598	1.35020
41	0.75487	0.84101	0.92690	1.01251	1.09781	1.18278	1.26738	1.35160
42	0.75630	0.84245	0.92833	1.01394	1.09923	1.18419	1.26879	1.35300
43	0.75774	0.84388	0.92976	1.01536	1.10065	1.18560	1.27020	1.35440
44	0.75918	0.84531	0.93119	1.01678	1.10207	1.18702	1.27160	1.35580
45	0.76062	0.84675	0.93262	1.01821	1.10349	1.18843	1.27301	1.35720
46	0.76205	0.84818	0.93405	1.01963	1.10491	1.18984	1.27442	1.35860
47	0.76349	0.84961	0.93548	1.02106	1.10632	1.19125	1.27582	1.36000
48	0.76493	0.85105	0.93691	1.02248	1.10774	1.19267	1.27723	1.36140
49	0.76637	0.85248	0.93834	1.02390	1.10916	1.19408	1.27863	1.36280
50	0.76780	0.85391	0.93976	1.02533	1.11058	1.19549	1.28004	1.36420
51	0.76924	0.85535	0.94119	1.02675	1.11200	1.19690	1.28145	1.36560
52	0.77068	0.85678	0.94262	1.02817	1.11342	1.19832	1.28285	1.36700
53	0.77211	0.85821	0.94405	1.02960	1.11483	1.19973	1.28426	1.36840
54	0.77355	0.85965	0.94548	1.03102	1.11625	1.20114	1.28566	1.36980
55	0.77499	0.86108	0.94691	1.03244	1.11767	1.20255	1.28707	1.37119
56	0.77643	0.86251	0.94833	1.03387	1.11909	1.20396	1.28847	1.37259
57	0.77786	0.86394	0.94976	1.03529	1.12050	1.20538	1.28988	1.37399
58	0.77930	0.86538	0.95119	1.03671	1.12192	1.20679	1.29129	1.37539
59	0.78074	0.86681	0.95262	1.03814	1.12334	1.20820	1.29269	1.37679
60	0.78217	0.86824	0.95404	1.03956	1.12476	1.20961	1.29410	1.37819

Constants for Setting a 5-inch Sine-Bar for 16° to 23°

Min.	16°	17°	18°	19°	20°	21°	22°	23°
0	1.37819	1.46186	1.54509	1.62784	1.71010	1.79184	1.87303	1.95366
1	1.37958	1.46325	1.54647	1.62922	1.71147	1.79320	1.87438	1.95499
2	1.38098	1.46464	1.54785	1.63059	1.71283	1.79456	1.87573	1.95633
3	1.38238	1.46603	1.54923	1.63197	1.71420	1.79591	1.87708	1.95767
4	1.38378	1.46742	1.55062	1.63334	1.71557	1.79727	1.87843	1.95901
5	1.38518	1.46881	1.55200	1.63472	1.71693	1.79863	1.87977	1.96035
6	1.38657	1.47020	1.55338	1.63609	1.71830	1.79998	1.88112	1.96169
7	1.38797	1.47159	1.55476	1.63746	1.71966	1.80134	1.88247	1.96302
8	1.38937	1.47298	1.55615	1.63884	1.72103	1.80270	1.88382	1.96436
9	1.39076	1.47437	1.55753	1.64021	1.72240	1.80405	1.88516	1.96570
10	1.39216	1.47576	1.55891	1.64159	1.72376	1.80541	1.88651	1.96704
11	1.39356	1.47715	1.56029	1.64296	1.72513	1.80677	1.88786	1.96837
12	1.39496	1.47854	1.56167	1.64433	1.72649	1.80812	1.88920	1.96971
13	1.39635	1.47993	1.56306	1.64571	1.72786	1.80948	1.89055	1.97105
14	1.39775	1.48132	1.56444	1.64708	1.72922	1.81083	1.89190	1.97238
15	1.39915	1.48271	1.56582	1.64845	1.73059	1.81219	1.89324	1.97372
16	1.40054	1.48410	1.56720	1.64983	1.73195	1.81355	1.89459	1.97506
17	1.40194	1.48549	1.56858	1.65120	1.73331	1.81490	1.89594	1.97639
18	1.40333	1.48687	1.56996	1.65257	1.73468	1.81626	1.89728	1.97773
19	1.40473	1.48826	1.57134	1.65394	1.73604	1.81761	1.89863	1.97906
20	1.40613	1.48965	1.57272	1.65532	1.73741	1.81897	1.89997	1.98040
21	1.40752	1.49104	1.57410	1.65669	1.73877	1.82032	1.90132	1.98173
22	1.40892	1.49243	1.57548	1.65806	1.74013	1.82168	1.90266	1.98307
23	1.41031	1.49382	1.57687	1.65943	1.74150	1.82303	1.90401	1.98440
24	1.41171	1.49520	1.57825	1.66081	1.74286	1.82438	1.90535	1.98574
25	1.41310	1.49659	1.57963	1.66218	1.74422	1.82574	1.90670	1.98707
26	1.41450	1.49798	1.58101	1.66355	1.74559	1.82709	1.90804	1.98841
27	1.41589	1.49937	1.58238	1.66492	1.74695	1.82845	1.90939	1.98974
28	1.41729	1.50075	1.58376	1.66629	1.74831	1.82980	1.91073	1.99108
29	1.41868	1.50214	1.58514	1.66766	1.74967	1.83115	1.91207	1.99241
30	1.42008	1.50353	1.58652	1.66903	1.75104	1.83251	1.91342	1.99375
31	1.42147	1.50492	1.58790	1.67041	1.75240	1.83386	1.91476	1.99508
32	1.42287	1.50630	1.58928	1.67178	1.75376	1.83521	1.91610	1.99641
33	1.42426	1.50769	1.59066	1.67315	1.75512	1.83657	1.91745	1.99775
34	1.42565	1.50908	1.59204	1.67452	1.75649	1.83792	1.91879	1.99908
35	1.42705	1.51046	1.59342	1.67589	1.75785	1.83927	1.92013	2.00041
36	1.42844	1.51185	1.59480	1.67726	1.75921	1.84062	1.92148	2.00175
37	1.42984	1.51324	1.59617	1.67863	1.76057	1.84198	1.92282	2.00308
38	1.43123	1.51462	1.59755	1.68000	1.76193	1.84333	1.92416	2.00441
39	1.43262	1.51601	1.59893	1.68137	1.76329	1.84468	1.92550	2.00574
40	1.43402	1.51739	1.60031	1.68274	1.76465	1.84603	1.92685	2.00708
41	1.43541	1.51878	1.60169	1.68411	1.76601	1.84738	1.92819	2.00841
42	1.43680	1.52017	1.60307	1.68548	1.76737	1.84873	1.92953	2.00974
43	1.43820	1.52155	1.60444	1.68685	1.76873	1.85009	1.93087	2.01107
44	1.43959	1.52294	1.60582	1.68821	1.77010	1.85144	1.93221	2.01240
45	1.44098	1.52432	1.60720	1.68958	1.77146	1.85279	1.93355	2.01373
46	1.44237	1.52571	1.60857	1.69095	1.77282	1.85414	1.93490	2.01506
47	1.44377	1.52709	1.60995	1.69232	1.77418	1.85549	1.93624	2.01640
48	1.44516	1.52848	1.61133	1.69369	1.77553	1.85684	1.93758	2.01773
49	1.44655	1.52986	1.61271	1.69506	1.77689	1.85819	1.93892	2.01906
50	1.44794	1.53125	1.61408	1.69643	1.77825	1.85954	1.94026	2.02039
51	1.44934	1.53263	1.61546	1.69779	1.77961	1.86089	1.94160	2.02172
52	1.45073	1.53401	1.61683	1.69916	1.78097	1.86224	1.94294	2.02305
53	1.45212	1.53540	1.61821	1.70053	1.78233	1.86359	1.94428	2.02438
54	1.45351	1.53678	1.61959	1.70190	1.78369	1.86494	1.94562	2.02571
55	1.45490	1.53817	1.62096	1.70327	1.78505	1.86629	1.94696	2.02704
56	1.45629	1.53955	1.62234	1.70463	1.78641	1.86764	1.94830	2.02837
57	1.45769	1.54093	1.62371	1.70600	1.78777	1.86899	1.94964	2.02970
58	1.45908	1.54232	1.62509	1.70737	1.78912	1.87034	1.95098	2.03103
59	1.46047	1.54370	1.62647	1.70873	1.79048	1.87168	1.95232	2.03235
60	1.46186	1.54509	1.62784	1.71010	1.79184	1.87303	1.95366	2.03368

Constants for Setting a 5-inch Sine-Bar for 24° to 31°

Min.	24°	25°	26°	27°	28°	29°	30°	31°
0	2.03368	2.11309	2.19186	2.26995	2.34736	2.42405	2.50000	2.57519
1	2.03501	2.11441	2.19316	2.27125	2.34864	2.42532	2.50126	2.57644
2	2.03634	2.11573	2.19447	2.27254	2.34993	2.42659	2.50252	2.57768
3	2.03767	2.11704	2.19578	2.27384	2.35121	2.42786	2.50378	2.57893
4	2.03900	2.11836	2.19708	2.27513	2.35249	2.42913	2.50504	2.58018
5	2.04032	2.11968	2.19839	2.27643	2.35378	2.43041	2.50630	2.58142
6	2.04165	2.12100	2.19970	2.27772	2.35506	2.43168	2.50755	2.58267
7	2.04298	2.12231	2.20100	2.27902	2.35634	2.43295	2.50881	2.58391
8	2.04431	2.12363	2.20231	2.28031	2.35763	2.43422	2.51007	2.58516
9	2.04563	2.12495	2.20361	2.28161	2.35891	2.43549	2.51133	2.58640
10	2.04696	2.12626	2.20492	2.28290	2.36019	2.43676	2.51259	2.58765
11	2.04829	2.12758	2.20622	2.28420	2.36147	2.43803	2.51384	2.58889
12	2.04962	2.12890	2.20753	2.28549	2.36275	2.43930	2.51510	2.59014
13	2.05094	2.13021	2.20883	2.28678	2.36404	2.44057	2.51636	2.59138
14	2.05227	2.13153	2.21014	2.28808	2.36532	2.44184	2.51761	2.59262
15	2.05359	2.13284	2.21144	2.28937	2.36660	2.44311	2.51887	2.59387
16	2.05492	2.13416	2.21275	2.29066	2.36788	2.44438	2.52013	2.59511
17	2.05625	2.13547	2.21405	2.29196	2.36916	2.44564	2.52138	2.59635
18	2.05757	2.13679	2.21536	2.29325	2.37044	2.44691	2.52264	2.59760
19	2.05890	2.13810	2.21666	2.29454	2.37172	2.44818	2.52389	2.59884
20	2.06022	2.13942	2.21796	2.29583	2.37300	2.44945	2.52515	2.60008
21	2.06155	2.14073	2.21927	2.29712	2.37428	2.45072	2.52640	2.60132
22	2.06287	2.14205	2.22057	2.29842	2.37556	2.45198	2.52766	2.60256
23	2.06420	2.14336	2.22187	2.29971	2.37684	2.45325	2.52891	2.60381
24	2.06552	2.14468	2.22318	2.30100	2.37812	2.45452	2.53017	2.60505
25	2.06685	2.14599	2.22448	2.30229	2.37940	2.45579	2.53142	2.60629
26	2.06817	2.14730	2.22578	2.30358	2.38068	2.45705	2.53268	2.60753
27	2.06950	2.14862	2.22708	2.30487	2.38196	2.45832	2.53393	2.60877
28	2.07082	2.14993	2.22839	2.30616	2.38324	2.45959	2.53519	2.61001
29	2.07214	2.15124	2.22969	2.30745	2.38452	2.46085	2.53644	2.61125
30	2.07347	2.15256	2.23099	2.30874	2.38579	2.46212	2.53769	2.61249
31	2.07479	2.15387	2.23229	2.31003	2.38707	2.46338	2.53894	2.61373
32	2.07611	2.15518	2.23359	2.31132	2.38835	2.46465	2.54020	2.61497
33	2.07744	2.15649	2.23489	2.31261	2.38963	2.46591	2.54145	2.61621
34	2.07876	2.15781	2.23619	2.31390	2.39091	2.46718	2.54270	2.61745
35	2.08008	2.15912	2.23749	2.31519	2.39218	2.46844	2.54396	2.61869
36	2.08140	2.16043	2.23880	2.31648	2.39346	2.46971	2.54521	2.61993
37	2.08273	2.16174	2.24010	2.31777	2.39474	2.47097	2.54646	2.62117
38	2.08405	2.16305	2.24140	2.31906	2.39601	2.47224	2.54771	2.62241
39	2.08537	2.16436	2.24270	2.32035	2.39729	2.47350	2.54896	2.62364
40	2.08669	2.16567	2.24400	2.32163	2.39857	2.47477	2.55021	2.62488
41	2.08801	2.16698	2.24530	2.32292	2.39984	2.47603	2.55146	2.62612
42	2.08934	2.16830	2.24660	2.32421	2.40112	2.47729	2.55271	2.62736
43	2.09066	2.16961	2.24789	2.32550	2.40239	2.47856	2.55397	2.62860
44	2.09198	2.17092	2.24919	2.32679	2.40367	2.47982	2.55522	2.62983
45	2.09330	2.17223	2.25049	2.32807	2.40494	2.48108	2.55647	2.63107
46	2.09462	2.17354	2.25179	2.32936	2.40622	2.48235	2.55772	2.63231
47	2.09594	2.17485	2.25309	2.33065	2.40749	2.48361	2.55896	2.63354
48	2.09726	2.17616	2.25439	2.33193	2.40877	2.48487	2.56021	2.63478
49	2.09858	2.17746	2.25569	2.33322	2.41004	2.48613	2.56146	2.63602
50	2.09990	2.17877	2.25698	2.33451	2.41132	2.48739	2.56271	2.63725
51	2.10122	2.18008	2.25828	2.33579	2.41259	2.48866	2.56396	2.63849
52	2.10254	2.18139	2.25958	2.33708	2.41386	2.48992	2.56521	2.63972
53	2.10386	2.18270	2.26088	2.33836	2.41514	2.49118	2.56646	2.64096
54	2.10518	2.18401	2.26217	2.33965	2.41641	2.49244	2.56771	2.64219
55	2.10650	2.18532	2.26347	2.34093	2.41769	2.49370	2.56895	2.64343
56	2.10782	2.18663	2.26477	2.34222	2.41896	2.49496	2.57020	2.64466
57	2.10914	2.18793	2.26606	2.34350	2.42023	2.49622	2.57145	2.64590
58	2.11045	2.18924	2.26736	2.34479	2.42150	2.49748	2.57270	2.64713
59	2.11177	2.19055	2.26866	2.34607	2.42278	2.49874	2.57394	2.64836
60	2.11309	2.19186	2.26995	2.34736	2.42405	2.50000	2.57519	2.64960

Constants for Setting a 5-inch Sine-Bar for 32° to 39°

Min.	32°	33°	34°	35°	36°	37°	38°	39°
0	2.64960	2.72320	2.79596	2.86788	2.93893	3.00908	3.07831	3.14660
1	2.65083	2.72441	2.79717	2.86907	2.94010	3.01024	3.07945	3.14773
2	2.65206	2.72563	2.79838	2.87026	2.94128	3.01140	3.08060	3.14886
3	2.65330	2.72685	2.79958	2.87146	2.94246	3.01256	3.08174	3.14999
4	2.65453	2.72807	2.80079	2.87265	2.94363	3.01372	3.08289	3.15112
5	2.65576	2.72929	2.80199	2.87384	2.94481	3.01488	3.08403	3.15225
6	2.65699	2.73051	2.80319	2.87503	2.94598	3.01604	3.08518	3.15338
7	2.65822	2.73173	2.80440	2.87622	2.94716	3.01720	3.08632	3.15451
8	2.65946	2.73295	2.80560	2.87741	2.94833	3.01836	3.08747	3.15564
9	2.66069	2.73416	2.80681	2.87860	2.94951	3.01952	3.08861	3.15676
10	2.66192	2.73538	2.80801	2.87978	2.95068	3.02068	3.08976	3.15789
11	2.66315	2.73660	2.80921	2.88097	2.95185	3.02184	3.09090	3.15902
12	2.66438	2.73782	2.81042	2.88216	2.95303	3.02300	3.09204	3.16015
13	2.66561	2.73903	2.81162	2.88335	2.95420	3.02415	3.09318	3.16127
14	2.66684	2.74025	2.81282	2.88454	2.95538	3.02531	3.09433	3.16240
15	2.66807	2.74147	2.81402	2.88573	2.95655	3.02647	3.09547	3.16353
16	2.66930	2.74268	2.81523	2.88691	2.95772	3.02763	3.09661	3.16465
17	2.67053	2.74390	2.81643	2.88810	2.95889	3.02878	3.09775	3.16578
18	2.67176	2.74511	2.81763	2.88929	2.96007	3.02994	3.09890	3.16690
19	2.67299	2.74633	2.81883	2.89048	2.96124	3.03110	3.10004	3.16803
20	2.67422	2.74754	2.82003	2.89166	2.96241	3.03226	3.10118	3.16915
21	2.67545	2.74876	2.82123	2.89285	2.96358	3.03341	3.10232	3.17028
22	2.67668	2.74997	2.82243	2.89403	2.96475	3.03457	3.10346	3.17140
23	2.67791	2.75119	2.82364	2.89522	2.96592	3.03572	3.10460	3.17253
24	2.67913	2.75240	2.82484	2.89641	2.96709	3.03688	3.10574	3.17365
25	2.68036	2.75362	2.82604	2.89759	2.96827	3.03803	3.10688	3.17478
26	2.68159	2.75483	2.82723	2.89878	2.96944	3.03919	3.10802	3.17590
27	2.68282	2.75605	2.82843	2.89996	2.97061	3.04034	3.10916	3.17702
28	2.68404	2.75726	2.82963	2.90115	2.97178	3.04150	3.11030	3.17815
29	2.68527	2.75847	2.83083	2.90233	2.97294	3.04265	3.11143	3.17927
30	2.68650	2.75969	2.83203	2.90351	2.97411	3.04381	3.11257	3.18039
31	2.68772	2.76090	2.83323	2.90470	2.97528	3.04496	3.11371	3.18151
32	2.68895	2.76211	2.83443	2.90588	2.97645	3.04611	3.11485	3.18264
33	2.69018	2.76332	2.83563	2.90707	2.97762	3.04727	3.11599	3.18376
34	2.69140	2.76453	2.83682	2.90825	2.97879	3.04842	3.11712	3.18488
35	2.69263	2.76575	2.83802	2.90943	2.97996	3.04957	3.11826	3.18600
36	2.69385	2.76696	2.83922	2.91061	2.98112	3.05073	3.11940	3.18712
37	2.69508	2.76817	2.84042	2.91180	2.98229	3.05188	3.12053	3.18824
38	2.69630	2.76938	2.84161	2.91298	2.98346	3.05303	3.12167	3.18936
39	2.69753	2.77059	2.84281	2.91416	2.98463	3.05418	3.12281	3.19048
40	2.69875	2.77180	2.84401	2.91534	2.98579	3.05533	3.12394	3.19160
41	2.69998	2.77301	2.84520	2.91652	2.98696	3.05648	3.12508	3.19272
42	2.70120	2.77422	2.84640	2.91771	2.98813	3.05764	3.12621	3.19384
43	2.70243	2.77543	2.84759	2.91889	2.98929	3.05879	3.12735	3.19496
44	2.70365	2.77664	2.84879	2.92007	2.99046	3.05994	3.12848	3.19608
45	2.70487	2.77785	2.84998	2.92125	2.99162	3.06109	3.12962	3.19720
46	2.70610	2.77906	2.85118	2.92243	2.99279	3.06224	3.13075	3.19831
47	2.70732	2.78027	2.85237	2.92361	2.99395	3.06339	3.13189	3.19943
48	2.70854	2.78148	2.85357	2.92479	2.99512	3.06454	3.13302	3.20055
49	2.70976	2.78269	2.85476	2.92597	2.99628	3.06568	3.13415	3.20167
50	2.71099	2.78389	2.85596	2.92715	2.99745	3.06683	3.13529	3.20278
51	2.71221	2.78510	2.85715	2.92833	2.99861	3.06798	3.13642	3.20390
52	2.71343	2.78631	2.85834	2.92950	2.99977	3.06913	3.13755	3.20502
53	2.71465	2.78752	2.85954	2.93068	3.00094	3.07028	3.13868	3.20613
54	2.71587	2.78873	2.86073	2.93186	3.00210	3.07143	3.13982	3.20725
55	2.71709	2.78993	2.86192	2.93304	3.00326	3.07257	3.14095	3.20836
56	2.71831	2.79114	2.86311	2.93422	3.00443	3.07372	3.14208	3.20948
57	2.71953	2.79235	2.86431	2.93540	3.00559	3.07487	3.14321	3.21059
58	2.72076	2.79355	2.86550	2.93657	3.00675	3.07601	3.14434	3.21171
59	2.72198	2.79476	2.86669	2.93775	3.00791	3.07716	3.14547	3.21282
60	2.72320	2.79596	2.86788	2.93893	3.00908	3.07831	3.14660	3.21394

Constants for Setting a 5-inch Sine-Bar for 40° to 47°

Min.	40°	41°	42°	43°	44°	45°	46°	47°
0	3.21394	3.28030	3.34565	3.40999	3.47329	3.53553	3.59670	3.65677
1	3.21505	3.28139	3.34673	3.41106	3.47434	3.53656	3.59771	3.65776
2	3.21617	3.28249	3.34781	3.41212	3.47538	3.53759	3.59872	3.65875
3	3.21728	3.28359	3.34889	3.41318	3.47643	3.53862	3.59973	3.65974
4	3.21839	3.28468	3.34997	3.41424	3.47747	3.53965	3.60074	3.66073
5	3.21951	3.28578	3.35105	3.41531	3.47852	3.54067	3.60175	3.66172
6	3.22062	3.28688	3.35213	3.41637	3.47956	3.54170	3.60276	3.66271
7	3.22173	3.28797	3.35321	3.41743	3.48061	3.54273	3.60376	3.66370
8	3.22284	3.28907	3.35429	3.41849	3.48165	3.54375	3.60477	3.66469
9	3.22395	3.29016	3.35537	3.41955	3.48270	3.54478	3.60578	3.66568
10	3.22507	3.29126	3.35645	3.42061	3.48374	3.54580	3.60679	3.66667
11	3.22618	3.29235	3.35753	3.42168	3.48478	3.54683	3.60779	3.66766
12	3.22729	3.29345	3.35860	3.42274	3.48583	3.54785	3.60880	3.66865
13	3.22840	3.29454	3.35968	3.42380	3.48687	3.54888	3.60981	3.66964
14	3.22951	3.29564	3.36076	3.42486	3.48791	3.54990	3.61081	3.67063
15	3.23062	3.29673	3.36183	3.42592	3.48895	3.55093	3.61182	3.67161
16	3.23173	3.29782	3.36291	3.42697	3.48999	3.55195	3.61283	3.67260
17	3.23284	3.29892	3.36399	3.42803	3.49104	3.55297	3.61383	3.67359
18	3.23395	3.30001	3.36506	3.42909	3.49208	3.55400	3.61484	3.67457
19	3.23506	3.30110	3.36614	3.43015	3.49312	3.55502	3.61584	3.67556
20	3.23617	3.30219	3.36721	3.43121	3.49416	3.55604	3.61684	3.67655
21	3.23728	3.30329	3.36829	3.43227	3.49520	3.55707	3.61785	3.67753
22	3.23838	3.30438	3.36936	3.43332	3.49624	3.55809	3.61885	3.67852
23	3.23949	3.30547	3.37044	3.43438	3.49728	3.55911	3.61986	3.67950
24	3.24060	3.30656	3.37151	3.43544	3.49832	3.56013	3.62086	3.68049
25	3.24171	3.30765	3.37259	3.43649	3.49936	3.56115	3.62186	3.68147
26	3.24281	3.30874	3.37366	3.43755	3.50039	3.56217	3.62286	3.68245
27	3.24392	3.30983	3.37473	3.43861	3.50143	3.56319	3.62387	3.68344
28	3.24503	3.31092	3.37581	3.43966	3.50247	3.56421	3.62487	3.68442
29	3.24613	3.31201	3.37688	3.44072	3.50351	3.56523	3.62587	3.68540
30	3.24724	3.31310	3.37795	3.44177	3.50455	3.56625	3.62687	3.68639
31	3.24835	3.31419	3.37902	3.44283	3.50558	3.56727	3.62787	3.68737
32	3.24945	3.31528	3.38010	3.44388	3.50662	3.56829	3.62887	3.68835
33	3.25056	3.31637	3.38117	3.44494	3.50766	3.56931	3.62987	3.68933
34	3.25166	3.31746	3.38224	3.44599	3.50869	3.57033	3.63087	3.69031
35	3.25277	3.31854	3.38331	3.44704	3.50973	3.57135	3.63187	3.69130
36	3.25387	3.31963	3.38438	3.44810	3.51077	3.57236	3.63287	3.69228
37	3.25498	3.32072	3.38545	3.44915	3.51180	3.57338	3.63387	3.69326
38	3.25608	3.32181	3.38652	3.45020	3.51284	3.57440	3.63487	3.69424
39	3.25718	3.32289	3.38759	3.45126	3.51387	3.57542	3.63587	3.69522
40	3.25829	3.32398	3.38866	3.45231	3.51491	3.57643	3.63687	3.69620
41	3.25939	3.32507	3.38973	3.45336	3.51594	3.57745	3.63787	3.69718
42	3.26049	3.32615	3.39080	3.45441	3.51697	3.57846	3.63886	3.69816
43	3.26159	3.32724	3.39187	3.45546	3.51801	3.57948	3.63986	3.69913
44	3.26270	3.32832	3.39294	3.45651	3.51904	3.58049	3.64086	3.70011
45	3.26380	3.32941	3.39400	3.45757	3.52007	3.58151	3.64186	3.70109
46	3.26490	3.33049	3.39507	3.45862	3.52111	3.58252	3.64285	3.70207
47	3.26600	3.33158	3.39614	3.45967	3.52214	3.58354	3.64385	3.70305
48	3.26710	3.33266	3.39721	3.46072	3.52317	3.58455	3.64484	3.70402
49	3.26820	3.33375	3.39827	3.46177	3.52420	3.58557	3.64584	3.70500
50	3.26930	3.33483	3.39934	3.46281	3.52523	3.58658	3.64683	3.70598
51	3.27040	3.33591	3.40041	3.46386	3.52627	3.58759	3.64783	3.70695
52	3.27150	3.33700	3.40147	3.46491	3.52730	3.58861	3.64882	3.70793
53	3.27260	3.33808	3.40254	3.46596	3.52833	3.58962	3.64982	3.70890
54	3.27370	3.33916	3.40360	3.46701	3.52936	3.59063	3.65081	3.70988
55	3.27480	3.34025	3.40467	3.46806	3.53039	3.59164	3.65181	3.71085
56	3.27590	3.34133	3.40573	3.46910	3.53142	3.59266	3.65280	3.71183
57	3.27700	3.34241	3.40680	3.47015	3.53245	3.59367	3.65379	3.71280
58	3.27810	3.34349	3.40786	3.47120	3.53348	3.59468	3.65478	3.71378
59	3.27920	3.34457	3.40893	3.47225	3.53451	3.59569	3.65578	3.71475
60	3.28030	3.34565	3.40999	3.47329	3.53553	3.59670	3.65677	3.71572

Constants for Setting a 5-inch Sine-Bar for 48° to 55°

Min.	48°	49°	50°	51°	52°	53°	54°	55°
0	3.71572	3.77355	3.83022	3.88573	3.94005	3.99318	4.04508	4.09576
1	3.71670	3.77450	3.83116	3.88665	3.94095	3.99405	4.04594	4.09659
2	3.71767	3.77546	3.83209	3.88756	3.94184	3.99493	4.04679	4.09743
3	3.71864	3.77641	3.83303	3.88847	3.94274	3.99580	4.04765	4.09826
4	3.71961	3.77736	3.83396	3.88939	3.94363	3.99668	4.04850	4.09909
5	3.72059	3.77831	3.83489	3.89030	3.94453	3.99755	4.04936	4.09993
6	3.72156	3.77927	3.83583	3.89122	3.94542	3.99842	4.05021	4.10076
7	3.72253	3.78022	3.83676	3.89213	3.94631	3.99930	4.05106	4.10159
8	3.72350	3.78117	3.83769	3.89304	3.94721	4.00017	4.05191	4.10242
9	3.72447	3.78212	3.83862	3.89395	3.94810	4.00104	4.05277	4.10325
10	3.72544	3.78307	3.83956	3.89487	3.94899	4.00191	4.05362	4.10409
11	3.72641	3.78402	3.84049	3.89578	3.94988	4.00279	4.05447	4.10492
12	3.72738	3.78498	3.84142	3.89669	3.95078	4.00366	4.05532	4.10575
13	3.72835	3.78593	3.84235	3.89760	3.95167	4.00453	4.05617	4.10658
14	3.72932	3.78688	3.84328	3.89851	3.95256	4.00540	4.05702	4.10741
15	3.73029	3.78783	3.84421	3.89942	3.95345	4.00627	4.05787	4.10823
16	3.73126	3.78877	3.84514	3.90033	3.95434	4.00714	4.05872	4.10906
17	3.73222	3.78972	3.84607	3.90124	3.95523	4.00801	4.05957	4.10989
18	3.73319	3.79067	3.84700	3.90215	3.95612	4.00888	4.06042	4.11072
19	3.73416	3.79162	3.84793	3.90306	3.95701	4.00975	4.06127	4.11155
20	3.73513	3.79257	3.84886	3.90397	3.95790	4.01062	4.06211	4.11238
21	3.73609	3.79352	3.84978	3.90488	3.95878	4.01148	4.06296	4.11320
22	3.73706	3.79446	3.85071	3.90579	3.95967	4.01235	4.06381	4.11403
23	3.73802	3.79541	3.85164	3.90669	3.96056	4.01322	4.06466	4.11486
24	3.73899	3.79636	3.85257	3.90760	3.96145	4.01409	4.06550	4.11568
25	3.73996	3.79730	3.85349	3.90851	3.96234	4.01495	4.06635	4.11651
26	3.74092	3.79825	3.85442	3.90942	3.96322	4.01582	4.06720	4.11733
27	3.74189	3.79919	3.85535	3.91032	3.96411	4.01669	4.06804	4.11816
28	3.74285	3.80014	3.85627	3.91123	3.96500	4.01755	4.06889	4.11898
29	3.74381	3.80109	3.85720	3.91214	3.96588	4.01842	4.06973	4.11981
30	3.74478	3.80203	3.85812	3.91304	3.96677	4.01928	4.07058	4.12063
31	3.74574	3.80297	3.85905	3.91395	3.96765	4.02015	4.07142	4.12145
32	3.74671	3.80392	3.85997	3.91485	3.96854	4.02101	4.07227	4.12228
33	3.74767	3.80486	3.86090	3.91576	3.96942	4.02188	4.07311	4.12310
34	3.74863	3.80581	3.86182	3.91666	3.97031	4.02274	4.07395	4.12392
35	3.74959	3.80675	3.86274	3.91756	3.97119	4.02361	4.07480	4.12475
36	3.75056	3.80769	3.86367	3.91847	3.97207	4.02447	4.07564	4.12557
37	3.75152	3.80863	3.86459	3.91937	3.97296	4.02533	4.07648	4.12639
38	3.75248	3.80958	3.86551	3.92027	3.97384	4.02619	4.07732	4.12721
39	3.75344	3.81052	3.86644	3.92118	3.97472	4.02706	4.07817	4.12803
40	3.75440	3.81146	3.86736	3.92208	3.97560	4.02792	4.07901	4.12885
41	3.75536	3.81240	3.86828	3.92298	3.97649	4.02878	4.07985	4.12967
42	3.75632	3.81334	3.86920	3.92388	3.97737	4.02964	4.08069	4.13049
43	3.75728	3.81428	3.87012	3.92478	3.97825	4.03050	4.08153	4.13131
44	3.75824	3.81522	3.87104	3.92568	3.97913	4.03136	4.08237	4.13213
45	3.75920	3.81616	3.87196	3.92658	3.98001	4.03222	4.08321	4.13295
46	3.76016	3.81710	3.87288	3.92748	3.98089	4.03308	4.08405	4.13377
47	3.76112	3.81804	3.87380	3.92839	3.98177	4.03394	4.08489	4.13459
48	3.76207	3.81898	3.87472	3.92928	3.98265	4.03480	4.08572	4.13540
49	3.76303	3.81992	3.87564	3.93018	3.98353	4.03566	4.08656	4.13622
50	3.76399	3.82086	3.87656	3.93108	3.98441	4.03652	4.08740	4.13704
51	3.76495	3.82179	3.87748	3.93198	3.98529	4.03738	4.08824	4.13785
52	3.76590	3.82273	3.87840	3.93288	3.98616	4.03823	4.08908	4.13867
53	3.76686	3.82367	3.87931	3.93378	3.98704	4.03909	4.08991	4.13949
54	3.76782	3.82461	3.88023	3.93468	3.98792	4.03995	4.09075	4.14030
55	3.76877	3.82554	3.88115	3.93557	3.98880	4.04081	4.09158	4.14112
56	3.76973	3.82648	3.88207	3.93647	3.98967	4.04166	4.09242	4.14193
57	3.77068	3.82742	3.88298	3.93737	3.99055	4.04252	4.09326	4.14275
58	3.77164	3.82835	3.88390	3.93826	3.99143	4.04337	4.09409	4.14356
59	3.77259	3.82929	3.88481	3.93916	3.99230	4.04423	4.09493	4.14437
60	3.77355	3.83022	3.88573	3.94005	3.99318	4.04508	4.09576	4.14519

Measuring Tapers with Vee-Block and Sine-Bar.—The taper on a conical part may be checked or found by placing the part in a vee-block which rests on the surface of a sine-plate or sine-bar as shown in the accompanying diagram. The advantage of this method is that the axis of the vee-block may be aligned with the sides of the sine-bar. Thus when the tapered part is placed in the vee-block it will be aligned perpendicular to the transverse axis of the sine-bar.

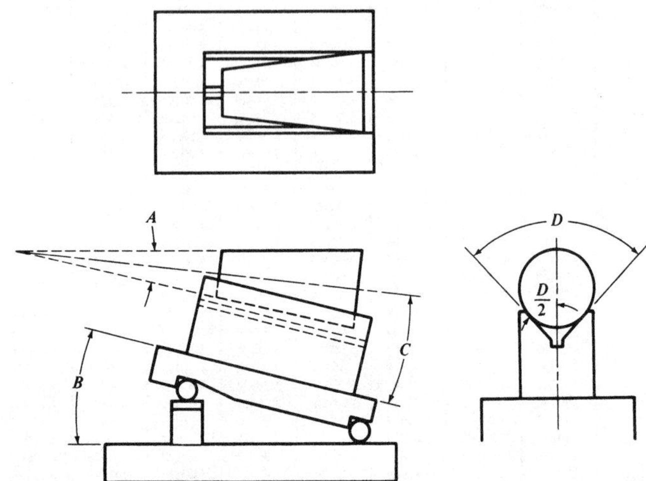

The sine-bar is set to angle $B = (C + A/2)$ where $A/2$ is one-half the included angle of the tapered part. If D is the included angle of the precision vee-block, the angle C is calculated from the formula:

$$\sin C = \frac{\sin(A/2)}{\sin(D/2)}$$

If dial indicator readings show no change across all points along the top of the taper surface, then this checks that the angle A of the taper is correct.

If the indicator readings vary, proceed as follows to find the actual angle of taper:

1) Adjust the angle of the sine-bar until the indicator reading is constant. Then find the new angle B' as explained in the paragraph *Measuring Angle of Templet or Gage with Sine Bar* on page 190; and 2) Using the angle B' calculate the actual half-angle $A'/2$ of the taper from the formula.

$$\tan\frac{A'}{2} = \frac{\sin B'}{\csc\frac{D}{2} + \cos B'}$$

The taper per foot corresponding to certain half-angles of taper may be found in the table on page 200.

Measuring Dovetail Slides.—Dovetail slides that must be machined accurately to a given width are commonly gaged by using pieces of cylindrical rod or wire and measuring as indicated by the dimensions x and y of the accompanying illustrations.

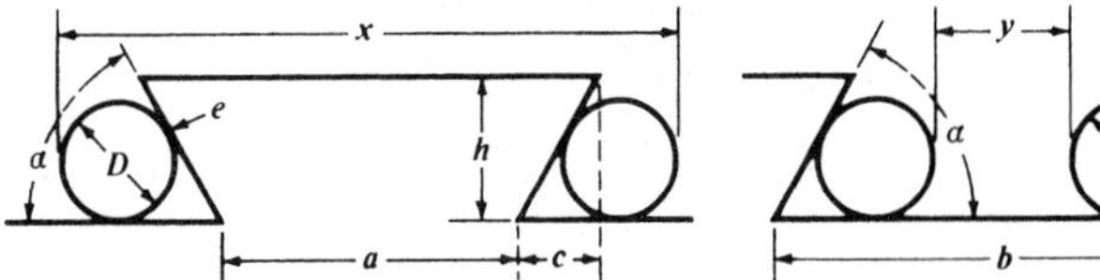

To obtain dimension x for measuring male dovetails, add I to the cotangent of one-half the dovetail angle α, multiply by diameter D of the rods used, and add the product to dimension α. To obtain dimension y for measuring a female dovetail, add 1 to the cotangent of one-half the dovetail angle α, multiply by diameter D of the rod used, and subtract the result from dimension b. Expressing these rules as formulas:

$$x = D(1 + \cot\tfrac{1}{2}\alpha) + a$$

$$y = b - D(1 + \cot\tfrac{1}{2}\alpha)$$

$$c = h \times \cot\alpha$$

The rod or wire used should be small enough so that the point of contact e is somewhat below the corner or edge of the dovetail.

Accurate Measurement of Angles and Tapers

When great accuracy is required in the measurement of angles, or when originating tapers, disks are commonly used. The principle of the disk method of taper measurement is that if two disks of unequal diameters are placed either in contact or a certain distance apart, lines tangent to their peripheries will represent an angle or taper, the degree of which depends upon the diameters of the two disks and the distance between them.

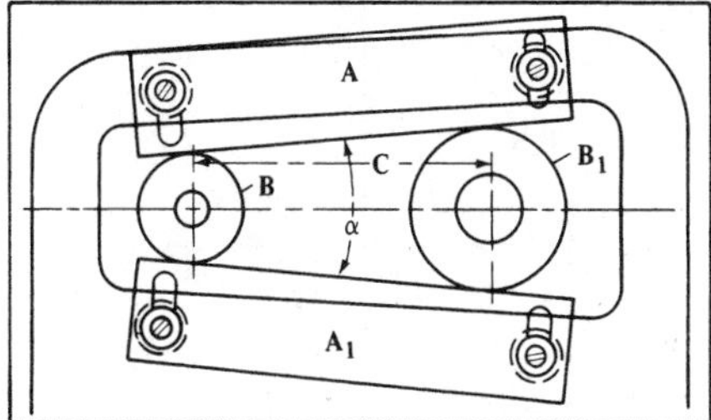

The gage shown in the accompanying illustration, which is a form commonly used for originating tapers or measuring angles accurately, is set by means of disks. This gage consists of two adjustable straight edges A and A_1, which are in contact with disks B and B_1. The angle α or the taper between the straight edges depends, of course, upon the diameters of the disks and the center distance C, and as these three dimensions can be measured accurately, it is possible to set the gage to a given angle within very close limits. Moreover, if a record of the three dimensions is kept, the exact setting of the gage can be reproduced quickly at any time. The following rules may be used for adjusting a gage of this type, and cover all problems likely to arise in practice. Disks are also occasionally used for the setting of parts in angular positions when they are to be machined accurately to a given angle: the rules are applicable to these conditions also.

Tapers per Foot and Corresponding Angles

Taper per Foot	Included Angle	Angle with Center Line	Taper per Foot	Included Angle	Angle with Center Line
1/64	0° 4′ 29″	0° 2′ 14″	1 7/8	8° 56′ 4″	4° 28′ 2″
1/32	0 8 57	0 4 29	1 15/16	9 13 51	4 36 56
1/16	0 17 54	0 8 57	2	9 31 38	4 45 49
3/32	0 26 51	0 13 26	2 1/8	10 7 11	5 3 36
1/8	0 35 49	0 17 54	2 1/4	10 42 42	5 21 21
5/32	0 44 46	0 22 23	2 3/8	11 18 11	5 39 5
3/16	0 53 43	0 26 51	2 1/2	11 53 37	5 56 49
7/32	1 2 40	0 31 20	2 5/8	12 29 2	6 14 31
1/4	1 11 37	0 35 49	2 3/4	13 4 24	6 32 12
9/32	1 20 34	0 40 17	2 7/8	13 39 43	6 49 52
5/16	1 29 31	0 44 46	3	14 15 0	7 7 30
11/32	1 38 28	0 49 14	3 1/8	14 50 14	7 25 7
3/8	1 47 25	0 53 43	3 1/4	15 25 26	7 42 43
13/32	1 56 22	0 58 11	3 3/8	16 0 34	8 0 17
7/16	2 5 19	1 2 40	3 1/2	16 35 39	8 17 50
15/32	2 14 16	1 7 8	3 5/8	17 10 42	8 35 21
1/2	2 23 13	1 11 37	3 3/4	17 45 41	8 52 50
17/32	2 32 10	1 16 5	3 7/8	18 20 36	9 10 18
9/16	2 41 7	1 20 33	4	18 55 29	9 27 44
19/32	2 50 4	1 25 2	4 1/8	19 30 17	9 45 9
5/8	2 59 1	1 29 30	4 1/4	20 5 3	10 2 31
21/32	3 7 57	1 33 59	4 3/8	20 39 44	10 19 52
11/16	3 16 54	1 38 27	4 1/2	21 14 22	10 37 11
23/32	3 25 51	1 42 55	4 5/8	21 48 55	10 54 28
3/4	3 34 47	1 47 24	4 3/4	22 23 25	11 11 42
25/32	3 43 44	1 51 52	4 7/8	22 57 50	11 28 55
13/16	3 52 41	1 56 20	5	23 32 12	11 46 6
27/32	4 1 37	2 0 49	5 1/8	24 6 29	12 3 14
7/8	4 10 33	2 5 17	5 1/4	24 40 41	12 20 21
29/32	4 19 30	2 9 45	5 3/8	25 14 50	12 37 25
15/16	4 28 26	2 14 13	5 1/2	25 48 53	12 54 27
31/32	4 37 23	2 18 41	5 5/8	26 22 52	13 11 26
1	4 46 19	2 23 9	5 3/4	26 56 47	13 28 23
1 1/16	5 4 11	2 32 6	5 7/8	27 30 36	13 45 18
1 1/8	5 22 3	2 41 2	6	28 4 21	14 2 10
1 3/16	5 39 55	2 49 57	6 1/8	28 38 1	14 19 0
1 1/4	5 57 47	2 58 53	6 1/4	29 11 35	14 35 48
1 5/16	6 15 38	3 7 49	6 3/8	29 45 5	14 52 32
1 3/8	6 33 29	3 16 44	6 1/2	30 18 29	15 9 15
1 7/16	6 51 19	3 25 40	6 5/8	30 51 48	15 25 54
1 1/2	7 9 10	3 34 35	6 3/4	31 25 2	15 42 31
1 9/16	7 27 0	3 43 30	6 7/8	31 58 11	15 59 5
1 5/8	7 44 49	3 52 25	7	32 31 13	16 15 37
1 11/16	8 2 38	4 1 19	7 1/8	33 4 11	16 32 5
1 3/4	8 20 27	4 10 14	7 1/4	33 37 3	16 48 31
1 13/16	8 38 16	4 19 8	7 3/8	34 9 49	17 4 54

For conversions into decimal degrees and radians see *Conversion Tables of Angular Measure* on page 90.

Rules for Figuring Tapers

Given	To Find	Rule
The taper per foot.	The taper per inch.	Divide the taper per foot by 12.
The taper per inch.	The taper per foot.	Multiply the taper per inch by 12.
End diameters and length of taper in inches.	The taper per foot.	Subtract small diameter from large; divide by length of taper; and multiply quotient by 12.
Large diameter and length of taper in inches, and taper per foot.	Diameter at small end in inches	Divide taper per foot by 12; multiply by length of taper; and subtract result from large diameter.
Small diameter and length of taper in inches, and taper per foot.	Diameter at large end in inches.	Divide taper per foot by 12; multiply by length of taper; and add result to small diameter.
The taper per foot and two diameters in inches.	Distance between two given diameters in inches.	Subtract small diameter from large; divide remainder by taper per foot; and multiply quotient by 12.
The taper per foot.	Amount of taper in a certain length in inches.	Divide taper per foot by 12; multiply by given length of tapered part.

Simplified Taper Calculations.—Tapers are devices widely used throughout the shop for a variety of purposes. The term "taper" as most commonly applied to shop work describes a machine element that has a uniform change of diameter between two points on, or in, a cylindrical workpiece. Tapers may be machined on either the outside surface of a shaft or the inside of a hole. The two primary styles of tapers used today are described as self-holding tapers and self-releasing tapers. The self-holding style of taper typically has an angle between 2° and 3° and is commonly used to centrally locate and securely hold tools such as tapered shank drills, reamers, and lathe centers. The self-releasing style of taper has an angle greater than 10° and is used mainly for centrally locating machine elements, such as milling machine arbors, lathe chucks, and collets. Due to the steepness of the taper, the self-releasing style must be held in place with a secondary clamping device, such as a drawbolt or drawnut.

Taper Per Foot and Taper Per Inch: The term taper per foot expresses the size difference between large and small end diameters per foot of taper length. The other term frequently applied to tapers is taper per inch, which expresses the size difference between large and small end diameters per inch of taper length.

Taper Formulas: Several standardized formulas, as shown in the following table of Taper Formulas, may be used to determine a part's taper per foot, taper per inch, and the other specific dimensions required for a particular taper. The relevant formula is selected based on the values needed, using the values known.

Taper Formulas

TPI	TPF	$\text{TPI} = \frac{\text{TPF}}{12}$
TPI	D, d, L	$\text{TPI} = \frac{D-d}{l}$
TPF	TPI	$\text{TPF} = \text{TPI} \times 12$
TPF	D, d, L	$\text{TPF} = \left(\frac{D-d}{L}\right) \times 12$
L	D, d, TPI	$L = \frac{D-d}{\text{TPI}}$

Taper Formulas *(Continued)*

L	D, d, TPF	$L = \left(\frac{D-d}{\text{TPF}}\right) \times 12$
D	d, L, TPI	$D = (\text{TPI} \times L) + d$
D	d, L, TPF	$D = \left(\frac{\text{TPF} \times L}{12}\right) + d$
d	D, L, TPI	$d = D - (L \times \text{TPI})$
d	D, L, TPF	$d = D - \left(\frac{L \times \text{TPF}}{12}\right)$

TPI = Taper per Inch,
TPF = Taper per Foot,
L = Length of taper,
D = Large diameter,
d = small diameter,

To find angle α for given taper *T* in inches per foot.—

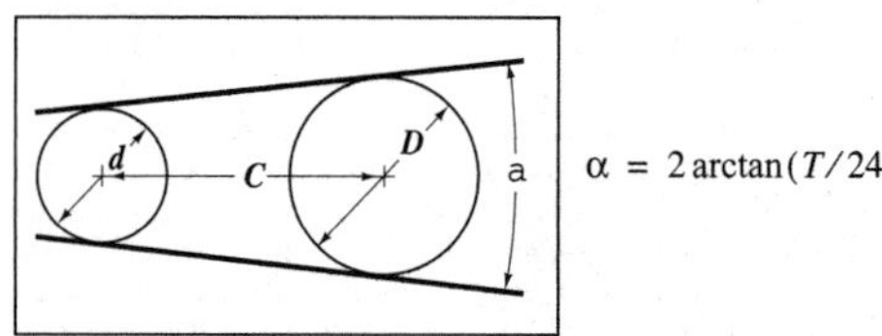

$$\alpha = 2\arctan(T/24).$$

Example: What angle α is equivalent to a taper of 1.5 inches per foot?

$$\alpha = 2 \times \arctan(1.5/24) = 7.153°$$

To find taper per foot *T* given angle α in degrees.—

$$T = 24 \ \tan(\alpha/2) \text{ inches per foot}$$

Example: What taper *T* is equivalent to an angle of 7.153°?

$$T = 24\tan(7.153/2) = 1.5 \text{ inches per foot}$$

To find angle α given dimensions *D*, *d*, and *C*.— Let *K* be the difference in the disk diameters divided by twice the center distance. $K = (D - d)/(2C)$, then $\alpha = 2\arcsin K$

Example: If the disk diameters *d* and *D* are 1 and 1.5 inches, respectively, and the center distance *C* is 5 inches, find the included angle α.

$$K = (1.5 - 1)/(2 \times 5) = 0.05 \qquad \alpha = 2 \times \arcsin 0.05 = 5.732°$$

To find taper *T* measured at right angles to a line through the disk centers given dimensions *D*, *d*, and distance *C*.— Find *K* using the formula in the previous example, then $T = 24K/\sqrt{1 - K^2}$ inches per foot

Example: If disk diameters *d* and *D* are 1 and 1.5 inches, respectively, and the center distance *C* is 5 inches, find the taper per foot.

$$K = (1.5 - 1)/(2 \times 5) = 0.05 \qquad T = \frac{24 \times 0.05}{\sqrt{1 - (0.05)^2}} = 1.2015 \text{ inches per foot}$$

To find center distance C for a given taper T in inches per foot.—

$$C = \frac{D-d}{2} \times \frac{\sqrt{1+(T/24)^2}}{T/24} \text{ inches}$$

Example: Gage is to be set to ¾ inch per foot, and disk diameters are 1.25 and 1.5 inches, respectively. Find the required center distance for the disks.

$$C = \frac{1.5-1.25}{2} \times \frac{\sqrt{1+(0.75/24)^2}}{0.75/24} = 4.002 \text{ inches}$$

To find center distance C for a given angle α and dimensions D and d.—

$$C = (D-d)/2\sin(\alpha/2) \text{ inches}$$

Example: If an angle α of 20° is required, and the disks are 1 and 3 inches in diameter, respectively, find the required center distance C.

$$C = (3-1)/(2 \times \sin 10°) = 5.759 \text{ inches}$$

To find taper T measured at right angles to one side .—When one side is taken as a base line and the taper is measured at right angles to that side, calculate K as explained above and use the following formula for determining the taper T:

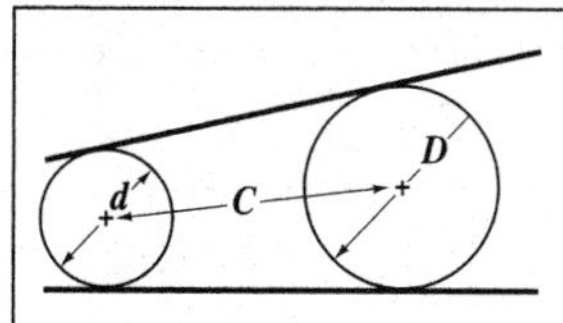

$$T = 24K\frac{\sqrt{1-K^2}}{1-2K^2} \text{ inches per foot}$$

Example: If the disk diameters are 2 and 3 inches, respectively, and the center 1 distance is 5 inches, what is the taper per foot measured at right angles to one side?

$$K = \frac{3-2}{2 \times 5} = 0.1 \qquad T = 24 \times 0.1 \times \frac{\sqrt{1-(0.1)^2}}{1-[2 \times (0.1)^2]} = 2.4367 \text{ in. per ft.}$$

To find center distance C when taper T is measured from one side.—

$$C = \frac{D-d}{\sqrt{2-2/\sqrt{1+(T/12)^2}}} \text{ inches}$$

Example: If the taper measured at right angles to one side is 6.9 inches per foot, and the disks are 2 and 5 inches in diameter, respectively, what is center distance C?

$$C = \frac{5-2}{\sqrt{2-2/\sqrt{1+(6.9/12)^2}}} = 5.815 \text{ inches.}$$

To find diameter D of a large disk in contact with a small disk of diameter d given angle α.—

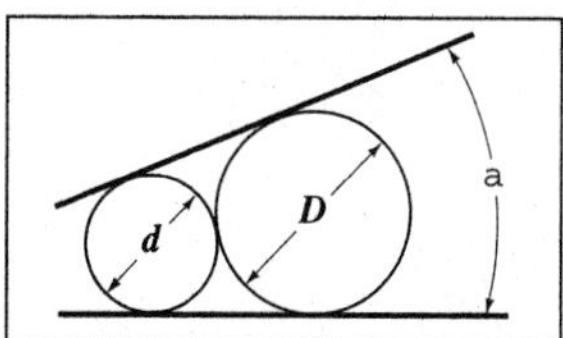

$$D = d \times \frac{1+\sin(\alpha/2)}{1-\sin(\alpha/2)} \text{ inches}$$

Example: The required angle α is 15°. Find diameter *D* of a large disk that is in contact with a standard 1-inch reference disk.

$$D = 1 \times \frac{1 + \sin 7.5°}{1 - \sin 7.5°} = 1.3002 \text{ inches}$$

Measurement over Pins.— When the distance across a bolt circle is too large to measure using ordinary measuring tools, then the required distance may be found from the distance across adacent or alternate holes using one of the methods that follow:

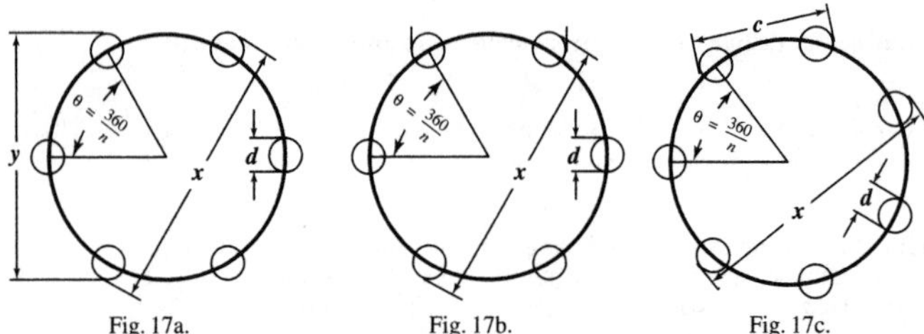

Fig. 17a. Fig. 17b. Fig. 17c.

Even Number of Holes in Circle: To measure the unknown distance *x* over opposite plugs in a bolt circle of *n* holes (*n* is even and greater than 4), as shown in Fig. 17a, where *y* is the distance over alternate plugs, *d* is the diameter of the holes, and θ = 360°/*n* is the angle between adjacent holes, use the following general equation for obtaining *x*:

$$x = \frac{y - d}{\sin\theta} + d$$

Example: In a die that has six 3/4-inch diameter holes equally spaced on a circle, where the distance *y* over alternate holes is 4½ inches, and the angle θ between adjacent holes is 60°, then

$$x = \frac{4.500 - 0.7500}{\sin 60°} + 0.7500 = 5.0801$$

In a similar problem, the distance *c* over adjacent plugs is given, as shown in Fig. 17b. If the number of holes is even and greater than 4, the distance *x* over opposite plugs is given in the following formula:

$$x = 2(c - d)\left(\frac{\sin\left(\frac{180 - \theta}{2}\right)}{\sin\theta}\right) + d$$

where *d* and θ are as defined above.

Odd Number of Holes in Circle: In a circle as shown in Fig. 17c, where the number of holes *n* is odd and greater than 3, and the distance *c* over adjacent holes is given, then θ equals 360/*n* and the distance *x* across the most widely spaced holes is given by:

$$x = \frac{\frac{c - d}{2}}{\sin\frac{\theta}{4}} + d$$

ALLOWANCES AND TOLERANCES FOR FITS

Limits and Fits.—Fits between cylindrical parts, i.e., cylindrical fits, govern the proper assembly and performance of many mechanisms. Clearance fits permit relative freedom of motion between a shaft and a hole—axially, radially, or both. Interference fits secure a certain amount of tightness between parts, whether these are meant to remain permanently assembled or to be taken apart from time to time. Or again, two parts may be required to fit together snugly—without apparent tightness or looseness. The designer's problem is to specify these different types of fits in such a way that the shop can produce them. Establishing the specifications requires the adoption of two manufacturing limits for the hole and two for the shaft, and, hence, the adoption of a manufacturing tolerance on each part.

In selecting and specifying limits and fits for various applications, it is essential in the interests of interchangeable manufacturing that 1) standard definitions of terms relating to limits and fits be used; 2) preferred basic sizes be selected wherever possible to reduce material and tooling costs; 3) limits be based upon a series of preferred tolerances and allowances; and 4) a uniform system of applying tolerances (preferably unilateral) be used. These principles have been incorporated in both the American and British standards for limits and fits. Information about these standards is given beginning on page 207.

Basic Dimensions.—The basic size of a screw thread or machine part is the theoretical or nominal standard size from which variations are made. For example, a shaft may have a *basic* diameter of 2 inches, but a maximum variation of minus 0.010 inch may be permitted. The minimum hole should be of basic size wherever the use of standard tools represents the greatest economy. The maximum shaft should be of basic size wherever the use of standard purchased material, without further machining, represents the greatest economy, even though special tools are required to machine the mating part.

Tolerances.—Tolerance is the amount of variation permitted on dimensions or surfaces of machine parts. The tolerance is equal to the difference between the maximum and minimum limits of any specified dimension. For example, if the maximum limit for the diameter of a shaft is 2.000 inches and its minimum limit 1.990 inches, the tolerance for this diameter is 0.010 inch. The extent of these tolerances is established by determining the maximum and minimum clearances required on operating surfaces. As applied to the fitting of machine parts, the word tolerance means the amount that duplicate parts are allowed to vary in size in connection with manufacturing operations, owing to unavoidable imperfections of workmanship. Tolerance may also be defined as the amount that duplicate parts are permitted to vary in size to secure sufficient accuracy without unnecessary refinement. The terms "tolerance" and "allowance" are often used interchangeably, but, according to common usage, *allowance* is a difference in dimensions prescribed to secure various classes of fits between different parts.

Unilateral and Bilateral Tolerances.—The term "unilateral tolerance" means that the total tolerance, as related to a basic dimension, is in *one* direction only. For example, if the basic dimension were 1 inch and the tolerance were expressed as 1.000 – 0.002, or as 1.000 + 0.002, these would be unilateral tolerances because the total tolerance in each is in one direction. On the contrary, if the tolerance were divided, so as to be partly plus and partly minus, it would be classed as "bilateral."

Thus,

$$1.000 \begin{array}{l} +0.001 \\ -0.001 \end{array}$$

is an example of bilateral tolerance, because the total tolerance of 0.002 is given in two directions—plus and minus.

When unilateral tolerances are used, one of the three following methods should be used to express them:

1) Specify, limiting dimensions only as

Diameter of hole: 2.250, 2.252

Diameter of shaft: 2.249, 2.247

2) One limiting size may be specified with its tolerances as

Diameter of hole: 2.250 + 0.002, −0.000

Diameter of shaft: 2.249 + 0.000, −0.002

3) The nominal size may be specified for both parts, with a notation showing both allowance and tolerance, as

Diameter of hole: $2\frac{1}{4}$ + 0.002, −0.000

Diameter of shaft: $2\frac{1}{4}$ − 0.001, −0.003

Bilateral tolerances should be specified as such, usually with plus and minus tolerances of equal amount. An example of the expression of bilateral tolerances is

$$2 \pm 0.001 \quad \text{or} \quad 2\begin{array}{l}+0.001\\-0.001\end{array}$$

Application of Tolerances.—According to common practice, tolerances are applied in such a way as to show the permissible amount of dimensional variation in the direction that is less dangerous. When a variation in either direction is equally dangerous, a bilateral tolerance should be given. When a variation in one direction is more dangerous than a variation in another, a unilateral tolerance should be given in the less dangerous direction.

For nonmating surfaces, or atmospheric fits, the tolerances may be bilateral, or unilateral, depending entirely upon the nature of the variations that develop in manufacture. On mating surfaces, with few exceptions, the tolerances should be unilateral.

Where tolerances are required on the distances between holes, usually they should be bilateral, as variation in either direction is normally equally dangerous. The variation in the distance between shafts carrying gears, however, should always be unilateral and plus; otherwise, the gears might run too tight. A slight increase in the backlash between gears is seldom of much importance.

One exception to the use of unilateral tolerances on mating surfaces occurs when tapers are involved; either bilateral or unilateral tolerances may then prove advisable, depending upon conditions. These tolerances should be determined in the same manner as the tolerances on the distances between holes. When a variation either in or out of the position of the mating taper surfaces is equally dangerous, the tolerances should be bilateral. When a variation in one direction is of less danger than a variation in the opposite direction, the tolerance should be unilateral and in the less dangerous direction.

Locating Tolerance Dimensions.—Only one dimension in the same straight line can be controlled within fixed limits. That dimension is the distance between the cutting surface of the tool and the locating or registering surface of the part being machined. Therefore, it is incorrect to locate any point or surface with tolerances from more than one point in the same straight line.

Every part of a mechanism must be located in each plane. Every operating part must be located with proper operating allowances. After such requirements of location are met, all other surfaces should have liberal clearances. Dimensions should be given between those points or surfaces that it is essential to hold in a specific relation to each other. This restriction applies particularly to those surfaces in each plane that control the location of other component parts. Many dimensions are relatively unimportant in this respect. It is good practice to establish a common locating point in each plane and give, as far as possible, all such dimensions from these common locating points. The locating points on the drawing, the locating or registering points used for machining the surfaces and the locating points for measuring should all be identical.

The initial dimensions placed on component drawings should be the exact dimensions that would be used if it were possible to work without tolerances. Tolerances should be given in that direction in which variations will cause the least harm or danger. When a variation in either direction is equally dangerous, the tolerances should be of equal amount in both directions, or bilateral. The initial clearance, or allowance, between operating parts should be as small as the operation of the mechanism will permit. The maximum clearance should be as great as the proper functioning of the mechanism will permit.

ANSI Standard Limits and Fits (ANSI B4.1-1967 , R1994).—This American National Standard for Preferred Limits and Fits for Cylindrical Parts presents definitions of terms applying to fits between plain (non threaded) cylindrical parts and makes recommendations on preferred sizes, allowances, tolerances, and fits for use wherever they are applicable. This standard is in accord with the recommendations of American-British-Canadian (ABC) conferences up to a diameter of 20 inches. Experimental work is being carried on with the objective of reaching agreement in the range above 20 inches. The recommendations in the standard are presented for guidance and for use where they might serve to improve and simplify products, practices, and facilities. They should have application for a wide range of products.

As revised in 1967, and reaffirmed in 1979, the definitions in ANSI B4.1 have been expanded and some of the limits in certain classes have been changed.

Factors Affecting Selection of Fits.—Many factors, such as length of engagement, bearing load, speed, lubrication, temperature, humidity, and materials must be taken into consideration in the selection of fits for a particular application, and modifications in the ANSI recommendations may be required to satisfy extreme conditions. Subsequent adjustments may also be found desirable as a result of experience in a particular application to suit critical functional requirements or to permit optimum manufacturing economy.

Definitions.—The following terms are defined in this standard:

Nominal Size: The nominal size is the designation used for the purpose of general identification.

Dimension: A dimension is a geometrical characteristic such as diameter, length, angle, or center distance.

Size: Size is a designation of magnitude. When a value is assigned to a dimension, it is referred to as the size of that dimension. (It is recognized that the words "dimension" and "size" are both used at times to convey the meaning of magnitude.)

Allowance: An allowance is a prescribed difference between the maximum material limits of mating parts. (See definition of *Fit*). It is a minimum clearance (positive allowance) or maximum interference (negative allowance) between such parts.

Tolerance: A tolerance is the total permissible variation of a size. The tolerance is the difference between the limits of size.

Basic Size: The basic size is that size from which the limits of size are derived by the application of allowances and tolerances.

Design Size: The design size is the basic size with allowance applied, from which the limits of size are derived by the application of tolerances. Where there is no allowance, the design size is the same as the basic size.

Actual Size: An actual size is a measured size.

Limits of Size: The limits of size are the applicable maximum and minimum sizes.

Maximum Material Limit: A maximum material limit is that limit of size that provides the maximum amount of material for the part. Normally it is the maximum limit of size of an external dimension or the minimum limit of size of an internal dimension.*

Minimum Material Limit: A minimum material limit is that limit of size that provides the minimum amount of material for the part. Normally it is the minimum limit of size of an external dimension or the maximum limit of size of an internal dimension.*

Tolerance Limit: A tolerance limit is the variation, positive or negative, by which a size is permitted to depart from the design size.

Unilateral Tolerance: A unilateral tolerance is a tolerance in which variation is permitted in only one direction from the design size.

Bilateral Tolerance: A bilateral tolerance is a tolerance in which variation is permitted in both directions from the design size.

Unilateral Tolerance System: A design plan that uses only unilateral tolerances is known as a Unilateral Tolerance System.

Bilateral Tolerance System: A design plan that uses only bilateral tolerances is known as a Bilateral Tolerance System.

Fit: Fit is the general term used to signify the range of tightness that may result from the application of a specific combination of allowances and tolerances in the design of mating parts.

Actual Fit: The actual fit between two mating parts is the relation existing between them with respect to the amount of clearance or interference that is present when they are assembled. (Fits are of three general types: clearance, transition, and interference.)

Clearance Fit: A clearance fit is one having limits of size so specified that a clearance always results when mating parts are assembled.

Interference Fit: An interference fit is one having limits of size so specified that an interference always results when mating parts are assembled.

Transition Fit: A transition fit is one having limits of size so specified that either a clearance or an interference may result when mating parts are assembled.

Basic Hole System: A basic hole system is a system of fits in which the design size of the hole is the basic size and the allowance, if any, is applied to the shaft.

Basic Shaft System: A basic shaft system is a system of fits in which the design size of the shaft is the basic size and the allowance, if any, is applied to the hole.

Preferred Basic Sizes.—In specifying fits, the basic size of mating parts may be chosen from the decimal series or the fractional series in the following table.

Table 1. Preferred Basic Sizes

Decimal			Fractional					
0.010	2.00	8.50	1/64	0.015625	2¼	2.2500	9½	9.5000
0.012	2.20	9.00	1/32	0.03125	2½	2.5000	10	10.0000
0.016	2.40	9.50	1/16	0.0625	2¾	2.7500	10½	10.5000
0.020	2.60	10.00	3/32	0.09375	3	3.0000	11	11.0000
0.025	2.80	10.50	1/8	0.1250	3¼	3.2500	11½	11.5000
0.032	3.00	11.00	5/32	0.15625	3½	3.5000	12	12.0000
0.040	3.20	11.50	3/16	0.1875	3¾	3.7500	12½	12.5000
0.05	3.40	12.00	1/4	0.2500	4	4.0000	13	13.0000

* An example of exceptions: an exterior corner radius where the maximum radius is the minimum material limit and the minimum radius is the maximum material limit.

Table 1. *(Continued)* **Preferred Basic Sizes**

Decimal			Fractional					
0.06	3.60	12.50	5/16	0.3125	4 1/4	4.2500	13 1/2	13.5000
0.08	3.80	13.00	3/8	0.3750	4 1/2	4.5000	14	14.0000
0.10	4.00	13.50	7/16	0.4375	4 3/4	4.7500	14 1/2	14.5000
0.12	4.20	14.00	1/2	0.5000	5	5.0000	15	15.0000
0.16	4.40	14.50	9/16	0.5625	5 1/4	5.2500	15 1/2	15.5000
0.20	4.60	15.00	5/8	0.6250	5 1/2	5.5000	16	16.0000
0.24	4.80	15.50	11/16	0.6875	5 3/4	5.7500	16 1/2	16.5000
0.30	5.00	16.00	3/4	0.7500	6	6.0000	17	17.0000
0.40	5.20	16.50	7/8	0.8750	6 1/2	6.5000	17 1/2	17.5000
0.50	5.40	17.00	1	1.0000	7	7.0000	18	18.0000
0.60	5.60	17.50	1 1/4	1.2500	7 1/2	7.5000	18 1/2	18.5000
0.80	5.80	18.00	1 1/2	1.5000	8	8.0000	19	19.0000
1.00	6.00	18.50	1 3/4	1.7500	8 1/2	8.5000	19 1/2	19.5000
1.20	6.50	19.00	2	2.0000	9	9.0000	20	20.0000
1.40	7.00	19.50	…	…	…	…	…	…
1.60	7.50	20.00	…	…	…	…	…	…
1.80	8.00	…	…	…	…	…	…	…

All dimensions are in inches.

Preferred Series of Tolerances and Allowances (In thousandths of an Inch)

0.1	1	10	100	0.3	3	30	…
…	1.2	12	125	…	3.5	35	…
0.15	1.4	14	…	0.4	4	40	…
…	1.6	16	160	…	4.5	45	…
…	1.8	18	…	0.5	5	50	…
0.2	2	20	200	0.6	6	60	…
…	2.2	22	…	0.7	7	70	…
0.25	2.5	25	250	0.8	8	80	…
…	2.8	28	…	0.9	9	…	…

Standard Tolerances.—The series of standard tolerances shown in Table 2 are so arranged that for any one grade they represent approximately similar production difficulties throughout the range of sizes. This table provides a suitable range from which appropriate tolerances for holes and shafts can be selected and enables standard gages to be used. The tolerances shown in Table 2 have been used in the succeeding tables for different classes of fits.

Table 2. ANSI Standard Tolerances *ANSI B4.1-1967 (R1987)*

Nominal Size, Inches		Grade									
		4	5	6	7	8	9	10	11	12	13
Over	To	Tolerances in thousandths of an inch[a]									
0	0.12	0.12	0.15	0.25	0.4	0.6	1.0	1.6	2.5	4	6
0.12	0.24	0.15	0.20	0.3	0.5	0.7	1.2	1.8	3.0	5	7
0.24	0.40	0.15	0.25	0.4	0.6	0.9	1.4	2.2	3.5	6	9
0.40	0.71	0.2	0.3	0.4	0.7	1.0	1.6	2.8	4.0	7	10
0.71	1.19	0.25	0.4	0.5	0.8	1.2	2.0	3.5	5.0	8	12
1.19	1.97	0.3	0.4	0.6	1.0	1.6	2.5	4.0	6	10	16
1.97	3.15	0.3	0.5	0.7	1.2	1.8	3.0	4.5	7	12	18
3.15	4.73	0.4	0.6	0.9	1.4	2.2	3.5	5	9	14	22
4.73	7.09	0.5	0.7	1.0	1.6	2.5	4.0	6	10	16	25
7.09	9.85	0.6	0.8	1.2	1.8	2.8	4.5	7	12	18	28
9.85	12.41	0.6	0.9	1.2	2.0	3.0	5.0	8	12	20	30
12.41	15.75	0.7	1.0	1.4	2.2	3.5	6	9	14	22	35
15.75	19.69	0.8	1.0	1.6	2.5	4	6	10	16	25	40
19.69	30.09	0.9	1.2	2.0	3	5	8	12	20	30	50
30.09	41.49	1.0	1.6	2.5	4	6	10	16	25	40	60
41.49	56.19	1.2	2.0	3	5	8	12	20	30	50	80
56.19	76.39	1.6	2.5	4	6	10	16	25	40	60	100
76.39	100.9	2.0	3	5	8	12	20	30	50	80	125
100.9	131.9	2.5	4	6	10	16	25	40	60	100	160
131.9	171.9	3	5	8	12	20	30	50	80	125	200
171.9	200	4	6	10	16	25	40	60	100	160	250

[a] All tolerances above heavy line are in accordance with American-British-Canadian (ABC) agreements.

Table 3. Relation of Machining Processes to Tolerance Grades

This chart may be used as a general guide to determine the machining processes that will under normal conditions, produce work within the tolerance grades indicated.

(See also *Relation of Surface Roughness to Tolerances* starting on page 169.)

MACHINING OPERATION	TOLERANCE GRADES									
	4	5	6	7	8	9	10	11	12	13
Lapping & Honing	■	■								
Cylindrical Grinding		■	■	■						
Surface Grinding		■	■	■	■					
Diamond Turning		■	■	■						
Diamond Boring		■	■	■						
Broaching		■	■	■	■					
Reaming			■	■	■	■	■			
Turning				■	■	■	■	■	■	■
Boring					■	■	■	■	■	■
Milling							■	■	■	■
Planing & Shaping							■	■	■	■
Drilling							■	■	■	■

ANSI Standard Fits.—Tables 4 through 10 inclusive show a series of standard types and classes of fits on a unilateral hole basis, such that the fit produced by mating parts in any one class will produce approximately similar performance throughout the range of sizes. These tables prescribe the fit for any given size, or type of fit; they also prescribe the standard limits for the mating parts that will produce the fit. The fits listed in these tables contain all those that appear in the approved American-British-Canadian proposal.

Selection of Fits: In selecting limits of size for any application, the type of fit is determined first, based on the use or service required from the equipment being designed; then the limits of size of the mating parts are established, to insure that the desired fit will be produced.

Theoretically, an infinite number of fits could be chosen, but the number of standard fits shown in the accompanying tables should cover most applications.

Designation of Standard Fits: Standard fits are designated by means of the following symbols which, facilitate reference to classes of fit for educational purposes. The symbols are not intended to be shown on manufacturing drawings; instead, sizes should be specified on drawings.

The letter symbols used are as follows:

RC = Running or Sliding Clearance Fit

LC = Locational Clearance Fit

LT = Transition Clearance or Interference Fit

LN = Locational Interference Fit

FN = Force or Shrink Fit

These letter symbols are used in conjunction with numbers representing the class of fit; thus FN 4 represents a Class 4, force fit.

Each of these symbols (two letters and a number) represents a complete fit for which the minimum and maximum clearance or interference and the limits of size for the mating parts are given directly in the tables.

Description of Fits.—The classes of fits are arranged in three general groups: running and sliding fits, locational fits, and force fits.

Running and Sliding Fits (RC): Running and sliding fits, for which limits of clearance are given in Table 3, are intended to provide a similar running performance, with suitable lubrication allowance, throughout the range of sizes. The clearances for the first two classes, used chiefly as slide fits, increase more slowly with the diameter than for the other classes, so that accurate location is maintained even at the expense of free relative motion.

These fits may be described as follows:

RC 1 *Close sliding fits* are intended for the accurate location of parts that must assemble without perceptible play.

RC 2 *Sliding fits* are intended for accurate location, but with greater maximum clearance than class RC 1. Parts made to this fit move and turn easily but are not intended to run freely, and in the larger sizes may seize with small temperature changes.

RC 3 *Precision running fits* are about the closest fits that can be expected to run freely, and are intended for precision work at slow speeds and light journal pressures, but are not suitable where appreciable temperature differences are likely to be encountered.

RC 4 *Close running fits* are intended chiefly for running fits on accurate machinery with moderate surface speeds and journal pressures, where accurate location and minimum play are desired.

RC 5 and RC 6 *Medium running fits* are intended for higher running speeds, or heavy journal pressures, or both.

RC 7 *Free running fits* are intended for use where accuracy is not essential, or where large temperature variations are likely to be encountered, or under both these conditions.

RC 8 and RC 9 *Loose running fits* are intended for use where wide commercial tolerances may be necessary, together with an allowance, on the external member.

Locational Fits (LC, LT, and LN): Locational fits are fits intended to determine only the location of the mating parts; they may provide rigid or accurate location, as with interference fits, or provide some freedom of location, as with clearance fits. Accordingly, they are divided into three groups: clearance fits (LC), transition fits (LT), and interference fits (LN).

These are described as follows:

LC *Locational clearance fits* are intended for parts which are normally stationary, but that can be freely assembled or disassembled. They range from snug fits for parts requiring accuracy of location, through the medium clearance fits for parts such as spigots, to the looser fastener fits where freedom of assembly is of prime importance.

LT *Locational transition fits* are a compromise between clearance and interference fits, for applications where accuracy of location is important, but either a small amount of clearance or interference is permissible.

LN *Locational interference fits* are used where accuracy of location is of prime importance, and for parts requiring rigidity and alignment with no special requirements for bore pressure. Such fits are not intended for parts designed to transmit frictional loads from one part to another by virtue of the tightness of fit. These conditions are covered by force fits.

Force Fits: (FN): Force or shrink fits constitute a special type of interference fit, normally characterized by maintenance of constant bore pressures throughout the range of sizes. The interference therefore varies almost directly with diameter, and the difference between its minimum and maximum value is small, to maintain the resulting pressures within reasonable limits.

These fits are described as follows:

FN 1 *Light drive fits* are those requiring light assembly pressures, and produce more or less permanent assemblies. They are suitable for thin sections or long fits, or in cast-iron external members.

FN 2 *Medium drive fits* are suitable for ordinary steel parts, or for shrink fits on light sections. They are about the tightest fits that can be used with high-grade cast-iron external members.

FN 3 *Heavy drive fits* are suitable for heavier steel parts or for shrink fits in medium sections.

FN 4 and FN 5 *Force fits* are suitable for parts that can be highly stressed, or for shrink fits where the heavy pressing forces required are impractical.

Graphical Representation of Limits and Fits.—A visual comparison of the hole and shaft tolerances and the clearances or interferences provided by the various types and classes of fits can be obtained from the diagrams on page 213. These diagrams have been drawn to scale for a nominal diameter of 1 inch.

Use of Standard Fit Tables.—*Example 1:* A Class RC 1 fit is to be used in assembling a mating hole and shaft of 2-inch nominal diameter. This class of fit was selected because the application required accurate location of the parts with no perceptible play (see *Description of Fits*, RC 1 close sliding fits). From the data in Table 3, establish the limits of size and clearance of the hole and shaft.

Maximum hole = 2 + 0.005 = 2.00005; Minimum hole = 2 inches

Maximum shaft = 2 − 0.0004 = 1.9996; Minimum shaft = 2 − 0.0007 = 1.9993 inches

Minimum clearance = 0.0004; Maximum clearance = 0.0012 inch

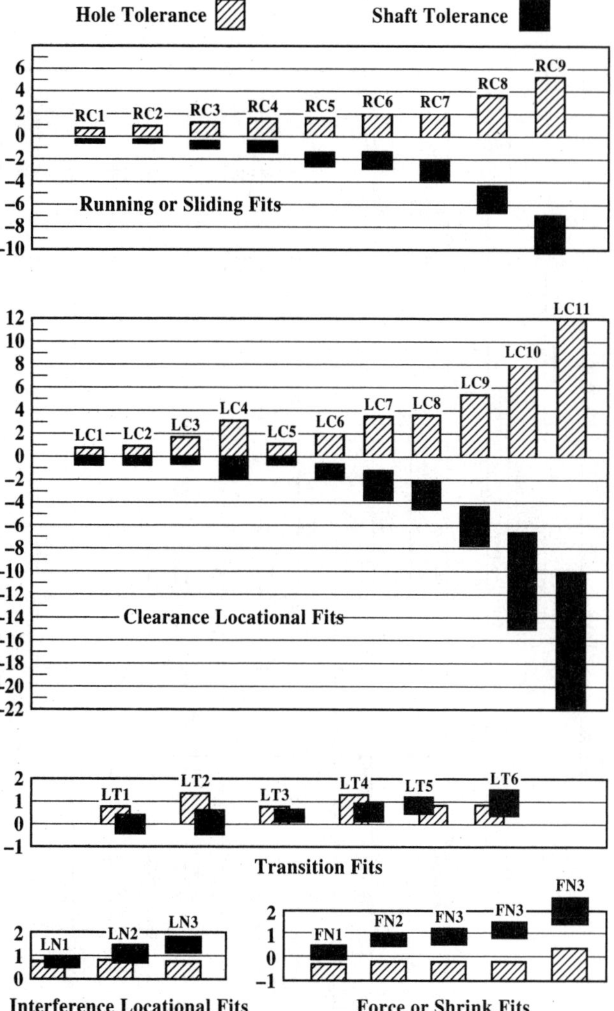

Diagrams show disposition of hole and shaft tolerances (in thousandths of an inch) with respect to basic size (0) for a diameter of 1 inch.

Table 4. American National Standard Running and Sliding Fits *ANSI B4.1-1967 (R1987)*

Nominal Size Range, Inches		Class RC 1			Class RC 2			Class RC 3			Class RC 4		
			Standard Tolerance Limits			Standard Tolerance Limits			Standard Tolerance Limits			Standard Tolerance Limits	
		Clear-ance[a]	Hole H5	Shaft g4	Clearance[a]	Hole H6	Shaft g5	Clearance[a]	Hole H7	Shaft f6	Clearance[a]	Hole H8	Shaft f7
Over	To	Values shown below are in thousandths of an inch											
0 –	0.12	0.1 0.45	+0.2 0	−0.1 −0.25	0.1 0.55	+0.25 0	−0.1 −0.3	0.3 0.95	+0.4 0	−0.3 −0.55	0.3 1.3	+0.6 0	−0.3 −0.7
0.12 –	0.24	0.15 0.5	+0.2 0	−0.15 −0.3	0.15 0.65	+0.3 0	−0.15 −0.35	0.4 1.12	+0.5 0	−0.4 −0.7	0.4 1.6	+0.7 0	−0.4 −0.9
0.24 –	0.40	0.2 0.6	+0.25 0	−0.2 −0.35	0.2 0.85	+0.4 0	−0.2 −0.45	0.5 1.5	+0.6 0	−0.5 −0.9	0.5 2.0	+0.9 0	−0.5 −1.1
0.40 –	0.71	0.25 0.75	+0.3 0	−0.25 −0.45	0.25 0.95	+0.4 0	−0.25 −0.55	0.6 1.7	+0.7 0	−0.6 −1.0	0.6 2.3	+1.0 0	−0.6 −1.3
0.71 –	1.19	0.3 0.95	+0.4 0	−0.3 −0.55	0.3 1.2	+0.5 0	−0.3 −0.7	0.8 2.1	+0.8 0	−0.8 −1.3	0.8 2.8	+1.2 0	−0.8 −1.6
1.19 –	1.97	0.4 1.1	+0.4 0	−0.4 −0.7	0.4 1.4	+0.6 0	−0.4 −0.8	1.0 2.6	+1.0 0	−1.0 −1.6	1.0 3.6	+1.6 0	−1.0 −2.0
1.97 –	3.15	0.4 1.2	+0.5 0	−0.4 −0.7	0.4 1.6	+0.7 0	−0.4 −0.9	1.2 3.1	+1.2 0	−1.2 −1.9	1.2 4.2	+1.8 0	−1.2 −2.4
3.15 –	4.73	0.5 1.5	+0.6 0	−0.5 −0.9	0.5 2.0	+0.9 0	−0.5 −1.1	1.4 3.7	+1.4 0	−1.4 −2.3	1.4 5.0	+2.2 0	−1.4 −2.8
4.73 –	7.09	0.6 1.8	+0.7 0	−0.6 −1.1	0.6 2.3	+1.0 0	−0.6 −1.3	1.6 4.2	+1.6 0	−1.6 −2.6	1.6 5.7	+2.5 0	−1.6 −3.2
7.09 –	9.85	0.6 2.0	+0.8 0	−0.6 −1.2	0.6 2.6	+1.2 0	−0.6 −1.4	2.0 5.0	+1.8 0	−2.0 −3.2	2.0 6.6	+2.8 0	−2.0 −3.8
9.85 –	12.41	0.8 2.3	+0.9 0	−0.8 −1.4	0.8 2.9	+1.2 0	−0.8 −1.7	2.5 5.7	+2.0 0	−2.5 −3.7	2.5 7.5	+3.0 0	−2.5 −4.5
12.41 –	15.75	1.0 2.7	+1.0 0	−1.0 −1.7	1.0 3.4	+1.4 0	−1.0 −2.0	3.0 6.6	+2.2 0	−3.0 −4.4	3.0 8.7	+3.5 0	−3.0 −5.2
15.75 –	19.69	1.2 3.0	+1.0 0	−1.2 −2.0	1.2 3.8	+1.6 0	−1.2 −2.2	4.0 8.1	+2.5 0	−4.0 −5.6	4.0 10.5	+4.0 0	−4.0 −6.5

[a] Pairs of values shown represent minimum and maximum amounts of clearance resulting from application of standard tolerance limits.

Table 5. American National Standard Running and Sliding Fits *ANSI B4.1-1967 (R1987)*

Nominal Size Range, Inches Over – To	Class RC 5			Class RC 6			Class RC 7			Class RC 8			Class RC 9		
	Clearance[a]	Standard Tolerance Limits		Clearance[a]	Standard Tolerance Limits		Clearance[a]	Standard Tolerance Limits		Clearance[a]	Standard Tolerance Limits		Clearance[a]	Standard Tolerance Limits	
		Hole H8	Shaft e7		Hole H9	Shaft e8		Hole H9	Shaft d8		Hole H10	Shaft c9		Hole H11	Shaft
	Values shown below are in thousandths of an inch														
0 – 0.12	0.6	+0.6	− 0.6	0.6	+1.0	− 0.6	1.0	+1.0	− 1.0	2.5	+1.6	− 2.5	4.0	+2.5	− 4.0
	1.6	0	− 1.0	2.2	0	− 1.2	2.6	0	− 1.6	5.1	0	− 3.5	8.1	0	− 5.6
0.12 – 0.24	0.8	+0.7	− 0.8	0.8	+1.2	− 0.8	1.2	+1.2	− 1.2	2.8	+1.8	− 2.8	4.5	+3.0	− 4.5
	2.0	0	− 1.3	2.7	0	− 1.5	3.1	0	− 1.9	5.8	0	− 4.0	9.0	0	− 6.0
0.24 – 0.40	1.0	+0.9	− 1.0	1.0	+1.4	− 1.0	1.6	+1.4	− 1.6	3.0	+2.2	− 3.0	5.0	+3.5	− 5.0
	2.5	0	− 1.6	3.3	0	− 1.9	3.9	0	− 2.5	6.6	0	− 4.4	10.7	0	− 7.2
0.40 – 0.71	1.2	+1.0	− 1.2	1.2	+1.6	− 1.2	2.0	+1.6	− 2.0	3.5	+2.8	− 3.5	6.0	+4.0	− 6.0
	2.9	0	− 1.9	3.8	0	− 2.2	4.6	0	− 3.0	7.9	0	− 5.1	12.8	0	− 8.8
0.71 – 1.19	1.6	+1.2	− 1.6	1.6	+2.0	− 1.6	2.5	+2.0	− 2.5	4.5	+3.5	− 4.5	7.0	+5.0	− 7.0
	3.6	0	− 2.4	4.8	0	− 2.8	5.7	0	− 3.7	10.0	0	− 6.5	15.5	0	−10.5
1.19 – 1.97	2.0	+1.6	− 2.0	2.0	+2.5	− 2.0	3.0	+2.5	− 3.0	5.0	+4.0	− 5.0	8.0	+6.0	− 8.0
	4.6	0	− 3.0	6.1	0	− 3.6	7.1	0	− 4.6	11.5	0	− 7.5	18.0	0	−12.0
1.97 – 3.15	2.5	+1.8	− 2.5	2.5	+3.0	− 2.5	4.0	+3.0	− 4.0	6.0	+4.5	− 6.0	9.0	+7.0	− 9.0
	5.5	0	− 3.7	7.3	0	− 4.3	8.8	0	− 5.8	13.5	0	− 9.0	20.5	0	−13.5
3.15 – 4.73	3.0	+2.2	− 3.0	3.0	+3.5	− 3.0	5.0	+3.5	− 5.0	7.0	+5.0	− 7.0	10.0	+9.0	−10.0
	6.6	0	− 4.4	8.7	0	− 5.2	10.7	0	− 7.2	15.5	0	−10.5	24.0	0	−15.0
4.73 – 7.09	3.5	+2.5	− 3.5	3.5	+4.0	− 3.5	6.0	+4.0	− 6.0	8.0	+6.0	− 8.0	12.0	+10.0	−12.0
	7.6	0	− 5.1	10.0	0	− 6.0	12.5	0	− 8.5	18.0	0	−12.0	28.0	0	−18.0
7.09 – 9.85	4.0	+2.8	− 4.0	4.0	+4.5	− 4.0	7.0	+4.5	− 7.0	10.0	+7.0	−10.0	15.0	+12.0	−15.0
	8.6	0	− 5.8	11.3	0	− 6.8	14.3	0	− 9.8	21.5	0	−14.5	34.0	0	−22.0
9.85 – 12.41	5.0	+3.0	− 5.0	5.0	+5.0	− 5.0	8.0	+5.0	− 8.0	12.0	+8.0	−12.0	18.0	+12.0	−18.0
	10.0	0	− 7.0	13.0	0	− 8.0	16.0	0	−11.0	25.0	0	−17.0	38.0	0	−26.0
12.41 – 15.75	6.0	+3.5	− 6.0	6.0	+6.0	− 6.0	10.0	+6.0	−10.0	14.0	+9.0	−14.0	22.0	+14.0	−22.0
	11.7	0	− 8.2	15.5	0	− 9.5	19.5	0	−13.5	29.0	0	−20.0	45.0	0	−31.0
15.75 – 19.69	8.0	+4.0	− 8.0	8.0	+6.0	− 8.0	12.0	+6.0	−12.0	16.0	+10.0	−16.0	25.0	+16.0	−25.0
	14.5	0	−10.5	18.0	0	−12.0	22.0	0	−16.0	32.0	0	−22.0	51.0	0	−35.0

Tolerance limits given in body of table are added to or subtracted from basic size (as indicated by + or − sign) to obtain maximum and minimum sizes of mating parts.

All data above heavy lines are in accord with ABC agreements. Symbols H5, g4, etc. are hole and shaft designations in ABC system. Limits for sizes above 19.69 inches are also given in the ANSI Standard.

Table 6. American National Standard Clearance Locational Fits *ANSI B4.1-1967 (R1987)*

Nominal Size Range, Inches	Class LC 1			Class LC 2			Class LC 3			Class LC 4			Class LC 5		
		Standard Tolerance Limits			Standard Tolerance Limits			Standard Tolerance Limits			Standard Tolerance Limits			Standard Tolerance Limits	
	Clearance[a]	Hole H6	Shaft h5	Clearance[a]	Hole H7	Shaft h6	Clearance[a]	Hole H8	Shaft h7	Clearance[a]	Hole H10	Shaft h9	Clearance[a]	Hole H7	Shaft g6
Over To	Values shown below are in thousandths of an inch														
0 – 0.12	0	+0.25	0	0	+0.4	0	0	+0.6	0	0	+1.6	0	0.1	+0.4	−0.1
	0.45	0	−0.2	0.65	0	−0.25	1	0	−0.4	2.6	0	−1.0	0.75	0	−0.35
0.12 – 0.24	0	+0.3	0	0	+0.5	0	0	+0.7	0	0	+1.8	0	0.15	+0.5	−0.15
	0.5	0	−0.2	0.8	0	−0.3	1.2	0	−0.5	3.0	0	−1.2	0.95	0	−0.45
0.24 – 0.40	0	+0.4	0	0	+0.6	0	0	+0.9	0	0	+2.2	0	0.2	+0.6	−0.2
	0.65	0	−0.25	1.0	0	−0.4	1.5	0	−0.6	3.6	0	−1.4	1.2	0	−0.6
0.40 – 0.71	0	+0.4	0	0	+0.7	0	0	+1.0	0	0	+2.8	0	0.25	+0.7	−0.25
	0.7	0	−0.3	1.1	0	−0.4	1.7	0	−0.7	4.4	0	−1.6	1.35	0	−0.65
0.71 – 1.19	0	+0.5	0	0	+0.8	0	0	+1.2	0	0	+3.5	0	0.3	+0.8	−0.3
	0.9	0	−0.4	1.3	0	−0.5	2	0	−0.8	5.5	0	−2.0	1.6	0	−0.8
1.19 – 1.97	0	+0.6	0	0	+1.0	0	0	+1.6	0	0	+4.0	0	0.4	+1.0	−0.4
	1.0	0	−0.4	1.6	0	−0.6	2.6	0	−1	6.5	0	−2.5	2.0	0	−1.0
1.97 – 3.15	0	+0.7	0	0	+1.2	0	0	+1.8	0	0	+4.5	0	0.4	+1.2	−0.4
	1.2	0	−0.5	1.9	0	−0.7	3	0	−1.2	7.5	0	−3	2.3	0	−1.1
3.15 – 4.73	0	+0.9	0	0	+1.4	0	0	+2.2	0	0	+5.0	0	0.5	+1.4	−0.5
	1.5	0	−0.6	2.3	0	−0.9	3.6	0	−1.4	8.5	0	−3.5	2.8	0	−1.4
4.73 – 7.09	0	+1.0	0	0	+1.6	0	0	+2.5	0	0	+6.0	0	0.6	+1.6	−0.6
	1.7	0	−0.7	2.6	0	−1.0	4.1	0	−1.6	10.0	0	−4	3.2	0	−1.6
7.09 – 9.85	0	+1.2	0	0	+1.8	0	0	+2.8	0	0	+7.0	0	0.6	+1.8	−0.6
	2.0	0	−0.8	3.0	0	−1.2	4.6	0	−1.8	11.5	0	−4.5	3.6	0	−1.8
9.85 – 12.41	0	+1.2	0	0	+2.0	0	0	+3.0	0	0	+8.0	0	0.7	+2.0	−0.7
	2.1	0	−0.9	3.2	0	−1.2	5	0	−2.0	13.0	0	−5	3.9	0	−1.9
12.41 – 15.75	0	+1.4	0	0	+2.2	0	0	+3.5	0	0	+9.0	0	0.7	+2.2	−0.7
	2.4	0	−1.0	3.6	0	−1.4	5.7	0	−2.2	15.0	0	−6	4.3	0	−2.1
15.75 – 19.69	0	+1.6	0	0	+2.5	0	0	+4	0	0	+10.0	0	0.8	+2.5	−0.8
	2.6	0	−1.0	4.1	0	−1.6	6.5	0	−2.5	16.0	0	−6	4.9	0	−2.4

[a] Pairs of values shown represent minimum and maximum amounts of interference resulting from application of standard tolerance limits.

Table 7. American National Standard Clearance Locational Fits *ANSI B4.1-1967 (R1987)*

Nominal Size Range, Inches	Class LC 6			Class LC 7			Class LC 8			Class LC 9			Class LC 10			Class LC 11		
		Std. Tolerance Limits			Std. Tolerance Limits			Std. Tolerance Limits			Std. Tolerance Limits			Std. Tolerance Limits			Std. Tolerance Limits	
Over To	Clearance[a]	Hole H9	Shaft f8	Clearance[a]	Hole H10	Shaft e9	Clearance[a]	Hole H10	Shaft d9	Clearance[a]	Hole H11	Shaft c10	Clearance[a]	Hole H12	Shaft	Clearance[a]	Hole H13	Shaft
	Values shown below are in thousandths of an inch																	
0 – 0.12	0.3	+1.0	−0.3	0.6	+1.6	− 0.6	1.0	+1.6	− 1.0	2.5	+2.5	− 2.5	4	+4	− 4	5	+6	− 5
	1.9	0	−0.9	3.2	0	− 1.6	2.0	0	− 2.0	6.6	0	− 4.1	12	0	− 8	17	0	− 11
0.12 – 0.24	0.4	+1.2	−0.4	0.8	+1.8	− 0.8	1.2	+1.8	− 1.2	2.8	+3.0	− 2.8	4.5	+5	− 4.5	6	+7	− 6
	2.3	0	−1.1	3.8	0	− 2.0	4.2	0	− 2.4	7.6	0	− 4.6	14.5	0	− 9.5	20	0	−13
0.24 – 0.40	0.5	+1.4	−0.5	1.0	+2.2	− 1.0	1.6	+2.2	− 1.6	3.0	+3.5	− 3.0	5	+6	− 5	7	+9	− 7
	2.8	0	−1.4	4.6	0	− 2.4	5.2	0	− 3.0	8.7	0	− 5.2	17	0	−11	25	0	−16
0.40 – 0.71	0.6	+1.6	−0.6	1.2	+2.8	− 1.2	2.0	+2.8	− 2.0	3.5	+4.0	− 3.5	6	+7	− 6	8	+10	− 8
	3.2	0	−1.6	5.6	0	− 2.8	6.4	0	− 3.6	10.3	0	− 6.3	20	0	−13	28	0	−18
0.71 – 1.19	0.8	+2.0	−0.8	1.6	+3.5	− 1.6	2.5	+3.5	− 2.5	4.5	+5.0	− 4.5	7	+8	− 7	10	+12	−10
	4.0	0	−2.0	7.1	0	− 3.6	8.0	0	− 4.5	13.0	0	− 8.0	23	0	−15	34	0	−22
1.19 – 1.97	1.0	+2.5	−1.0	2.0	+4.0	− 2.0	3.6	+4.0	− 3.0	5.0	+6	− 5.0	8	+10	− 8	12	+16	−12
	5.1	0	−2.6	8.5	0	− 4.5	9.5	0	− 5.5	15.0	0	− 9.0	28	0	−18	44	0	−28
1.97 – 3.15	1.2	+3.0	−1.0	2.5	+4.5	− 2.5	4.0	+4.5	− 4.0	6.0	+7	− 6.0	10	+12	−10	14	+18	−14
	6.0	0	−3.0	10.0	0	− 5.5	11.5	0	− 7.0	17.5	0	−10.5	34	0	−22	50	0	−32
3.15 – 4.73	1.4	+3.5	−1.4	3.0	+5.0	− 3.0	5.0	+5.0	− 5.0	7	+9	− 7	11	+14	−11	16	+22	−16
	7.1	0	−3.6	11.5	0	− 6.5	13.5	0	− 8.5	21	0	−12	39	0	−25	60	0	−38
4.73 – 7.09	1.6	+4.0	−1.6	3.5	+6.0	− 3.5	6	+6	− 6	8	+10	− 8	12	+16	−12	18	+25	−18
	8.1	0	−4.1	13.5	0	− 7.5	16	0	−10	24	0	−14	44	0	−28	68	0	−43
7.09 – 9.85	2.0	+4.5	−2.0	4.0	+7.0	− 4.0	7	+7	− 7	10	+12	−10	16	+18	−16	22	+28	−22
	9.3	0	−4.8	15.5	0	− 8.5	18.5	0	−11.5	29	0	−17	52	0	−34	78	0	−50
9.85 – 12.41	2.2	+5.0	−2.2	4.5	+8.0	− 4.5	7	+8	− 7	12	+12	−12	20	+20	−20	28	+30	−28
	10.2	0	−5.2	17.5	0	− 9.5	20	0	−12	32	0	−20	60	0	−40	88	0	−58
12.41 – 15.75	2.5	+6.0	−2.5	5.0	+9.0	− 5.0	8	+9	− 8	14	+14	−14	22	+22	−22	30	+35	−30
	12.0	0	−6.0	20.0	0	−11.0	23	0	−14	37	0	−23	66	0	−44	100	0	−65
15.75 – 19.69	2.8	+6.0	−2.8	5.0	+10.0	− 5.0	9	+10	− 9	16	+16	−16	25	+25	−25	35	+40	−35
	12.8	0	−6.8	21.0	0	−11.0	25	0	−15	42	0	−26	75	0	−50	115	0	−75

Tolerance limits given in body of table are added or subtracted to basic size (as indicated by + or − sign) to obtain maximum and minimum sizes of mating parts.

All data above heavy lines are in accordance with American-British-Canadian (ABC) agreements. Symbols H6, H7, s6, etc. are hole and shaft designations in ABC system. Limits for sizes above 19.69 inches are not covered by ABC agreements but are given in the ANSI Standard.

Table 8. ANSI Standard Transition Locational Fits *ANSI B4.1-1967 (R1987)*

Nominal Size Range, Inches	Class LT 1			Class LT 2			Class LT 3			Class LT 4			Class LT 5			Class LT 6		
		Std. Tolerance Limits			Std. Tolerance Limits			Std. Tolerance Limits			Std. Tolerance Limits			Std. Tolerance Limits			Std. Tolerance Limits	
Over – To	Fit[a]	Hole H7	Shaft js6	Fit[a]	Hole H8	Shaft js7	Fit[a]	Hole H7	Shaft k6	Fit[a]	Hole H8	Shaft k7	Fit[a]	Hole H7	Shaft n6	Fit[a]	Hole H7	Shaft n7
	Values shown below are in thousandths of an inch																	
0 – 0.12	−0.12 +0.52	+0.4 0	+0.12 −0.12	−0.2 +0.8	+0.6 0	+0.2 −0.2							−0.5 +0.15	+0.4 0	+0.5 +0.25	−0.65 +0.15	+0.4 0	+0.65 +0.25
0.12 – 0.24	−0.15 +0.65	+0.5 0	+0.15 −0.15	−0.25 +0.95	+0.7 0	+0.25 −0.25							−0.6 +0.2	+0.5 0	+0.6 +0.3	−0.8 +0.2	+0.5 0	+0.8 +0.3
0.24 – 0.40	−0.2 +0.8	+0.6 0	+0.2 −0.2	−0.3 +1.2	+0.9 0	+0.3 −0.3	−0.5 +0.5	+0.6 0	+0.5 +0.1	−0.7 +0.8	+0.9 0	+0.7 +0.1	−0.8 +0.2	+0.6 0	+0.8 +0.4	−1.0 +0.2	+0.6 0	+1.0 +0.4
0.40 – 0.71	−0.2 +0.9	+0.7 0	+0.2 −0.2	−0.35 +1.35	+1.0 0	+0.35 −0.35	−0.5 +0.6	+0.7 0	+0.5 +0.1	−0.8 +0.9	+1.0 0	+0.8 +0.1	−0.9 +0.2	+0.7 0	+0.9 +0.5	−1.2 +0.2	+0.7 0	+1.2 +0.5
0.71 – 1.19	−0.25 +1.05	+0.8 0	+0.25 −0.25	−0.4 +1.6	+1.2 0	+0.4 −0.4	−0.6 +0.7	+0.8 0	+0.6 +0.1	−0.9 +1.1	+1.2 0	+0.9 +0.1	−1.1 +0.2	+0.8 0	+1.1 +0.6	−1.4 +0.2	+0.8 0	+1.4 +0.6
1.19 – 1.97	−0.3 +1.3	+1.0 0	+0.3 −0.3	−0.5 +2.1	+1.6 0	+0.5 −0.5	−0.7 +0.9	+1.0 0	+0.7 +0.1	−1.1 +1.5	+1.6 0	+1.1 +0.1	−1.3 +0.3	+1.0 0	+1.3 +0.7	−1.7 +0.3	+1.0 0	+1.7 +0.7
1.97 – 3.15	−0.3 +1.5	+1.2 0	+0.3 −0.3	−0.6 +2.4	+1.8 0	+0.6 −0.6	−0.8 +1.1	+1.2 0	+0.8 +0.1	−1.3 +1.7	+1.8 0	+1.3 +0.1	−1.5 +0.4	+1.2 0	+1.5 +0.8	−2.0 +0.4	+1.2 0	+2.0 +0.8
3.15 – 4.73	−0.4 +1.8	+1.4 0	+0.4 −0.4	−0.7 +2.9	+2.2 0	+0.7 −0.7	−1.0 +1.3	+1.4 0	+1.0 +0.1	−1.5 +2.1	+2.2 0	+1.5 +0.1	−1.9 +0.4	+1.4 0	+1.9 +1.0	−2.4 +0.4	+1.4 0	+2.4 +1.0
4.73 – 7.09	−0.5 +2.1	+1.6 0	+0.5 −0.5	−0.8 +3.3	+2.5 0	+0.8 −0.8	−1.1 +1.5	+1.6 0	+1.1 +0.1	−1.7 +2.4	+2.5 0	+1.7 +0.1	−2.2 +0.4	+1.6 0	+2.2 +1.2	−2.8 +0.4	+1.6 0	+2.8 +1.2
7.09 – 9.85	−0.6 +2.4	+1.8 0	+0.6 −0.6	−0.9 +3.7	+2.8 0	+0.9 −0.9	−1.4 +1.6	+1.8 0	+1.4 +0.2	−2.0 +2.6	+2.8 0	+2.0 +0.2	−2.6 +0.4	+1.8 0	+2.6 +1.4	−3.2 +0.4	+1.8 0	+3.2 +1.4
9.85 – 12.41	−0.6 +2.6	+2.0 0	+0.6 −6.6	−1.0 +4.0	+3.0 0	+1.0 −1.0	−1.4 +1.8	+2.0 0	+1.4 +0.2	−2.2 +2.8	+3.0 0	+2.2 +0.2	−2.6 +0.6	+2.0 0	+2.6 +1.4	−3.4 +0.6	+2.0 0	+3.4 +1.4
12.41 – 15.75	−0.7 +2.9	+2.2 0	+0.7 −0.7	−1.0 +4.5	+3.5 0	+1.0 −1.0	−1.6 +2.0	+2.2 0	+1.6 +0.2	−2.4 +3.3	+3.5 0	+2.4 +0.2	−3.0 +0.6	+2.2 0	+3.0 +1.6	−3.8 +0.6	+2.2 0	+3.8 +1.6
15.75 – 19.69	−0.8 +3.3	+2.5 0	+0.8 −0.8	−1.2 +5.2	+4.0 0	+1.2 −1.2	−1.8 +2.3	+2.5 0	+1.8 +0.2	−2.7 +3.8	+4.0 0	+2.7 +0.2	−3.4 +0.7	+2.5 0	+3.4 +1.8	−4.3 +0.7	+2.5 0	+4.3 +1.8

[a] Pairs of values shown represent maximum amount of interference (−) and maximum amount of clearance (+) resulting from application of standard tolerance limits.

All data above heavy lines are in accord with ABC agreements. Symbols H7, js6, etc., are hole and shaft designations in the ABC system.

Table 9. ANSI Standard Interference Location Fits ***ANSI B4.1-1967 (R1987)***

Nominal Size Range, Inches	Class LN 1			Class LN 2			Class LN 3		
	Limits of Interference	Standard Limits		Limits of Interference	Standard Limits		Limits of Interference	Standard Limits	
		Hole H6	Shaft n5		Hole H7	Shaft p6		Hole H7	Shaft r6
Over To	Values shown below are given in thousandths of an inch								
0 – 0.12	0	+0.25	+0.45	0	+0.4	+0.65	0.1	+0.4	+0.75
	0.45	0	+0.25	0.65	0	+0.4	0.75	0	+0.5
0.12 – 0.24	0	+0.3	+0.5	0	+0.5	+0.8	0.1	+0.5	+0.9
	0.5	0	+0.3	0.8	0	+0.5	0.9	0	+0.6
0.24 – 0.40	0	+0.4	+0.65	0	+0.6	+1.0	0.2	+0.6	+1.2
	0.65	0	+0.4	1.0	0	+0.6	1.2	0	+0.8
0.40 – 0.71	0	+0.4	+0.8	0	+0.7	+1.1	0.3	+0.7	+1.4
	0.8	0	+0.4	1.1	0	+0.7	1.4	0	+1.0
0.71 – 1.19	0	+0.5	+1.0	0	+0.8	+1.3	0.4	+0.8	+1.7
	1.0	0	+0.5	1.3	0	+0.8	1.7	0	+1.2
1.19 – 1.97	0	+0.6	+1.1	0	+1.0	+1.6	0.4	+1.0	+2.0
	1.1	0	+0.6	1.6	0	+1.0	2.0	0	+1.4
1.97 – 3.15	0.1	+0.7	+1.3	0.2	+1.2	+2.1	0.4	+1.2	+2.3
	1.3	0	+0.8	2.1	0	+1.4	2.3	0	+1.6
3.15 – 4.73	0.1	+0.9	+1.6	0.2	+1.4	+2.5	0.6	+1.4	+2.9
	1.6	0	+1.0	2.5	0	+1.6	2.9	0	+2.0
4.73 – 7.09	0.2	+1.0	+1.9	0.2	+1.6	+2.8	0.9	+1.6	+3.5
	1.9	0	+1.2	2.8	0	+1.8	3.5	0	+2.5
7.09 – 9.85	0.2	+1.2	+2.2	0.2	+1.8	+3.2	1.2	+1.8	+4.2
	2.2	0	+1.4	3.2	0	+2.0	4.2	0	+3.0
9.85 – 12.41	0.2	+1.2	+2.3	0.2	+2.0	+3.4	1.5	+2.0	+4.7
	2.3	0	+1.4	3.4	0	+2.2	4.7	0	+3.5
12.41 – 15.75	0.2	+1.4	+2.6	0.3	+2.2	+3.9	2.3	+2.2	+5.9
	2.6	0	+1.6	3.9	0	+2.5	5.9	0	+4.5
15.75 – 19.69	0.2	+1.6	+2.8	0.3	+2.5	+4.4	2.5	+2.5	+6.6
	2.8	0	+1.8	4.4	0	+2.8	6.6	0	+5.0

All data in this table are in accordance with American-British-Canadian (ABC) agreements.

Limits for sizes above 19.69 inches are not covered by ABC agreements but are given in the ANSI Standard.

Symbols H7, p6, etc., are hole and shaft designations in the ABC system.

Tolerance limits given in body of table are added or subtracted to basic size (as indicated by + or – sign) to obtain maximum and minimum sizes of mating parts.

Table 10. ANSI Standard Force and Shrink Fits *ANSI B4.1-1967 (R1987)*

Nominal Size Range, Inches	Class FN 1			Class FN 2			Class FN 3			Class FN 4			Class FN 5		
	Interference[a]	Standard Tolerance Limits		Interference[a]	Standard Tolerance Limits		Interference[a]	Standard Tolerance Limits		Interference[a]	Standard Tolerance Limits		Interference[a]	Standard Tolerance Limits	
Over To		Hole H6	Shaft		Hole H7	Shaft s6		Hole H7	Shaft t6		Hole H7	Shaft u6		Hole H8	Shaft x7
	Values shown below are in thousandths of an inch														
0 – 0.12	0.05	+0.25	+0.5	0.2	+0.4	+0.85				0.3	+0.4	+0.95	0.3	+0.6	+1.3
	0.5	0	+0.3	0.85	0	+0.6				0.95	0	+0.7	1.3	0	+0.9
0.12 – 0.24	0.1	+0.3	+0.6	0.2	+0.5	+1.0				0.4	+0.5	+1.2	0.5	+0.7	+1.7
	0.6	0	+0.4	1.0	0	+0.7				1.2	0	+0.9	1.7	0	+1.2
0.24 – 0.40	0.1	+0.4	+0.75	0.4	+0.6	+1.4				0.6	+0.6	+1.6	0.5	+0.9	+2.0
	0.75	0	+0.5	1.4	0	+1.0				1.6	0	+1.2	2.0	0	+1.4
0.40 – 0.56	0.1	+0.4	+0.8	0.5	+0.7	+1.6				0.7	+0.7	+1.8	0.6	+1.0	+2.3
	0.8	0	+0.5	1.6	0	+1.2				1.8	0	+1.4	2.3	0	+1.6
0.56 – 0.71	0.2	+0.4	+0.9	0.5	+0.7	+1.6				0.7	+0.7	+1.8	0.8	+1.0	+2.5
	0.9	0	+0.6	1.6	0	+1.2				1.8	0	+1.4	2.5	0	+1.8
0.71 – 0.95	0.2	+0.5	+1.1	0.6	+0.8	+1.9				0.8	+0.8	+2.1	1.0	+1.2	+3.0
	1.1	0	+0.7	1.9	0	+1.4				2.1	0	+1.6	3.0	0	+2.2
0.95 – 1.19	0.3	+0.5	+1.2	0.6	+0.8	+1.9	0.8	+0.8	+2.1	+1.0	+0.8	+2.3	1.3	+1.2	+3.3
	1.2	0	+0.8	1.9	0	+1.4	2.1	0	+1.6	2.3	0	+1.8	3.3	0	+2.5
1.19 – 1.58	0.3	+0.6	+1.3	0.8	+1.0	+2.4	1.0	+1.0	+2.6	1.5	+1.0	+3.1	1.4	+1.6	+4.0
	1.3	0	+0.9	2.4	0	+1.8	2.6	0	+2.0	3.1	0	+2.5	4.0	0	+3.0
1.58 – 1.97	0.4	+0.6	+1.4	0.8	+1.0	+2.4	1.2	+1.0	+2.8	1.8	+1.0	+3.4	2.4	+1.6	+5.0
	1.4	0	+1.0	2.4	0	+1.8	2.8	0	+2.2	3.4	0	+2.8	5.0	0	+4.0
1.97 – 2.56	0.6	+0.7	+1.8	0.8	+1.2	+2.7	1.3	+1.2	+3.2	2.3	+1.2	+4.2	3.2	+1.8	+6.2
	1.8	0	+1.3	2.7	0	+2.0	3.2	0	+2.5	4.2	0	+3.5	6.2	0	+5.0
2.56 – 3.15	0.7	+0.7	+1.9	1.0	+1.2	+2.9	1.8	+1.2	+3.7	2.8	+1.2	+4.7	4.2	+1.8	+7.2
	1.9	0	+1.4	2.9	0	+2.2	3.7	0	+3.0	4.7	0	+4.0	7.2	0	+6.0
3.15 – 3.94	0.9	+0.9	+2.4	1.4	+1.4	+3.7	2.1	+1.4	+4.4	3.6	+1.4	+5.9	4.8	+2.2	+8.4
	2.4	0	+1.8	3.7	0	+2.8	4.4	0	+3.5	5.9	0	+5.0	8.4	0	+7.0
3.94 – 4.73	1.1	+0.9	+2.6	1.6	+1.4	+3.9	2.6	+1.4	+4.9	4.6	+1.4	+6.9	5.8	+2.2	+9.4
	2.6	0	+2.0	3.9	0	+3.0	4.9	0	+4.0	6.9	0	+6.0	9.4	0	+8.0
4.73 – 5.52	1.2	+1.0	+2.9	1.9	+1.6	+4.5	3.4	+1.6	+6.0	5.4	+1.6	+8.0	7.5	+2.5	+11.6
	2.9	0	+2.2	4.5	0	+3.5	6.0	0	+5.0	8.0	0	+7.0	11.6	0	+10.0

Table 10. *(Continued)* **ANSI Standard Force and Shrink Fits** *ANSI B4.1-1967 (R1987)*

Nominal Size Range, Inches	Class FN 1			Class FN 2			Class FN 3			Class FN 4			Class FN 5		
	Interference[a]	Standard Tolerance Limits		Interference[a]	Standard Tolerance Limits		Interference[a]	Standard Tolerance Limits		Interference[a]	Standard Tolerance Limits		Interference[a]	Standard Tolerance Limits	
Over To		Hole H6	Shaft		Hole H7	Shaft s6		Hole H7	Shaft t6		Hole H7	Shaft u6		Hole H8	Shaft x7
	Values shown below are in thousandths of an inch														
5.52– 6.30	1.5	+1.0	+3.2	2.4	+1.6	+5.0	3.4	+1.6	+6.0	5.4	+1.6	+8.0	9.5	+2.5	+13.6
	3.2	0	+2.5	5.0	0	+4.0	6.0	0	+5.0	8.0	0	+7.0	13.6	0	+12.0
6.30– 7.09	1.8	+1.0	+3.5	2.9	+1.6	+5.5	4.4	+1.6	+7.0	6.4	+1.6	+9.0	9.5	+2.5	+13.6
	3.5	0	+2.8	5.5	0	+4.5	7.0	0	+6.0	9.0	0	+8.0	13.6	0	+12.0
7.09– 7.88	1.8	+1.2	+3.8	3.2	+1.8	+6.2	5.2	+1.8	+8.2	7.2	+1.8	+10.2	11.2	+2.8	+15.8
	3.8	0	+3.0	6.2	0	+5.0	8.2	0	+7.0	10.2	0	+9.0	15.8	0	+14.0
7.88– 8.86	2.3	+1.2	+4.3	3.2	+1.8	+6.2	5.2	+1.8	+8.2	8.2	+1.8	+11.2	13.2	+2.8	+17.8
	4.3	0	+3.5	6.2	0	+5.0	8.2	0	+7.0	11.2	0	+10.0	17.8	0	+16.0
8.86– 9.85	2.3	+1.2	+4.3	4.2	+1.8	+7.2	6.2	+1.8	+9.2	10.2	+1.8	+13.2	13.2	+2.8	+17.8
	4.3	0	+3.5	7.2	0	+6.0	9.2	0	+8.0	13.2	0	+12.0	17.8	0	+16.0
9.85– 11.03	2.8	+1.2	+4.9	4.0	+2.0	+7.2	7.0	+2.0	+10.2	10.0	+2.0	+13.2	15.0	+3.0	+20.0
	4.9	0	+4.0	7.2	0	+6.0	10.2	0	+9.0	13.2	0	+12.0	20.0	0	+18.0
11.03– 12.41	2.8	+1.2	+4.9	5.0	+2.0	+8.2	7.0	+2.0	+10.2	12.0	+2.0	+15.2	17.0	+3.0	+22.0
	4.9	0	+4.0	8.2	0	+7.0	10.2	0	+9.0	15.2	0	+14.0	22.0	0	+20.0
12.41– 13.98	3.1	+1.4	+5.5	5.8	+2.2	+9.4	7.8	+2.2	+11.4	13.8	+2.2	+17.4	18.5	+3.5	+24.2
	5.5	0	+4.5	9.4	0	+8.0	11.4	0	+10.0	17.4	0	+16.0	24.2	0	+22.0
13.98– 15.75	3.6	+1.4	+6.1	5.8	+2.2	+9.4	9.8	+2.2	+13.4	15.8	+2.2	+19.4	21.5	+3.5	+27.2
	6.1	0	+5.0	9.4	0	+8.0	13.4	0	+12.0	19.4	0	+18.0	27.2	0	+25.0
15.75– 17.72	4.4	+1.6	+7.0	6.5	+2.5	+10.6	+9.5	+2.5	+13.6	17.5	+2.5	+21.6	24.0	+4.0	+30.5
	7.0	0	+6.0	10.6	0	+9.0	13.6	0	+12.0	21.6	0	+20.0	30.5	0	+28.0
17.72– 19.69	4.4	+1.6	+7.0	7.5	+2.5	+11.6	11.5	+2.5	+15.6	19.5	+2.5	+23.6	26.0	+4.0	+32.5
	7.0	0	+6.0	11.6	0	+10.0	15.6	0	+14.0	23.6	0	+22.0	32.5	0	+30.0

[a] Pairs of values shown represent minimum and maximum amounts of interference resulting from application of standard tolerance limits.

All data above heavy lines are in accordance with American-British-Canadian (ABC) agreements. Symbols H6, H7, s6, etc., are hole and shaft designations in the ABC system. Limits for sizes above 19.69 inches are not covered by ABC agreements but are given in the ANSI standard.

Modified Standard Fits.—Fits having the same limits of clearance or interference as those shown in Tables 3 to 7 may sometimes have to be produced by using holes or shafts having limits of size other than those shown in these tables. These modifications may be accomplished by using either a *Bilateral Hole* (*System B*) or a *Basic Shaft System* (*Symbol S*). Both methods will result in nonstandard holes and shafts.

Bilateral Hole Fits: (*Symbol B*): The common situation is where holes are produced with fixed tools such as drills or reamers; to provide a longer wear life for such tools, a bilateral tolerance is desired.

The symbols used for these fits are identical with those used for standard fits except that they are followed by the letter B. Thus, LC 4B is a clearance locational fit, Class 4, except that it is produced with a bilateral hole.

The limits of clearance or interference are identical with those shown in Tables 3 to 7 for the corresponding fits.

The hole tolerance, however, is changed so that the plus limit is that for one grade finer than the value shown in the tables and the minus limit equals the amount by which the plus limit was lowered. The shaft limits are both lowered by the same amount as the lower limit of size of the hole. The finer grade of tolerance required to make these modifications may be obtained from Table 2. For example, an LC 4B fit for a 6-inch diameter hole would have tolerance limits of + 4.0, − 2.0 (+ 0.0040 inch, − 0.0020 inch); the shaft would have tolerance limits of − 2.0, − 6.0 (− 0.0020 inch, − 0.0060 inch).

Basic Shaft Fits: (*Symbol S*): For these fits, the maximum size of the shaft is basic. The limits of clearance or interference are identical with those shown in Tables 3 to 6 for the corresponding fits and the symbols used for these fits are identical with those used for standard fits except that they are followed by the letter S. Thus, LC 4S is a clearance locational fit, Class 4, except that it is produced on a basic shaft basis.

The limits for hole and shaft as given in Tables 3 to 6 are increased for clearance fits (*decreased* for transition or interference fits) by the value of the upper shaft limit; that is, by the amount required to change the maximum shaft to the basic size.

American National Standard Preferred Metric Limits and Fits.—This standard ANSI B4.2-1978 (R1994) describes the ISO system of metric limits and fits for mating parts as approved for general engineering usage in the United States.

It establishes: 1) the designation symbols used to define dimensional limits on drawings, material stock, related tools, gages, etc.; 2) the preferred basic sizes (first and second choices); 3) the preferred tolerance zones (first, second, and third choices); 4) the preferred limits and fits for sizes (first choice only) up to and including 500 millimeters; and 5) the definitions of related terms.

The general terms "hole" and "shaft" can also be taken to refer to the space containing or contained by two parallel faces of any part, such as the width of a slot, or the thickness of a key.

Definitions.—The most important terms relating to limits and fits are shown in Fig. 1 and are defined as follows:

Basic Size: The size to which limits of deviation are assigned. The basic size is the same for both members of a fit. For example, it is designated by the numbers 40 in 40H7.

Deviation: The algebraic difference between a size and the corresponding basic size.

Upper Deviation: The algebraic difference between the maximum limit of size and the corresponding basic size.

Lower Deviation: The algebraic difference between the minimum limit of size and the corresponding basic size.

Fundamental Deviation: That one of the two deviations closest to the basic size. For example, it is designated by the letter H in 40H7.

Tolerance: The difference between the maximum and minimum size limits on a part.

Tolerance Zone: A zone representing the tolerance and its position in relation to the basic size.

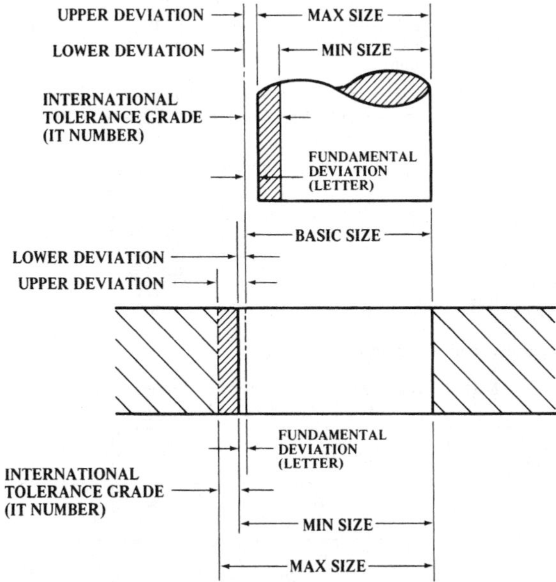

Fig. 1. Illustration of Definitions

International Tolerance Grade: (*IT*): A group of tolerances that vary depending on the basic size, but that provide the same relative level of accuracy within a given grade. For example, it is designated by the number 7 in 40H7 or as IT7.

Hole Basis: The system of fits where the minimum hole size is basic. The fundamental deviation for a hole basis system is H.

Shaft Basis: The system of fits where the maximum shaft size is basic. The fundamental deviation for a shaft basis system is h.

Clearance Fit: The relationship between assembled parts when clearance occurs under all tolerance conditions.

Interference Fit: The relationship between assembled parts when interference occurs under all tolerance conditions.

Transition Fit: The relationship between assembled parts when either a clearance or an interference fit can result, depending on the tolerance conditions of the mating parts.

Tolerances Designation.—An "International Tolerance grade" establishes the magnitude of the tolerance zone or the amount of part size variation allowed for external and internal dimensions alike (see Fig. 1). Tolerances are expressed in grade numbers that are consistent with International Tolerance grades identified by the prefix IT, such as IT6, IT11, etc. A smaller grade number provides a smaller tolerance zone.

A fundamental deviation establishes the position of the tolerance zone with respect to the basic size (see Fig. 1). Fundamental deviations are expressed by tolerance position letters.

Capital letters are used for internal dimensions and lowercase or small letters for external dimensions.

Symbols.—By combining the IT grade number and the tolerance position letter, the tolerance symbol is established that identifies the actual maximum and minimum limits of the part. The toleranced size is thus defined by the basic size of the part followed by a symbol composed of a letter and a number, such as 40H7, 40f7, etc.

A fit is indicated by the basic size common to both components, followed by a symbol corresponding to each component, the internal part symbol preceding the external part symbol, such as 40H8/f7.

Some methods of designating tolerances on drawings are:

A) 40H8

B) 40H8 $\begin{pmatrix}40.039\\40.000\end{pmatrix}$

C) $\begin{pmatrix}40.039\\40.000\end{pmatrix}$ 40H8

The values in parentheses indicate reference only.

Table 11. American National Standard Preferred Metric Sizes
ANSI B4.2-1978 (R1994)

Basic Size, mm		Basic Size, mm		Basic Size, mm		Basic Size, mm	
1st Choice	2nd Choice	1st Choice	2nd Choice	1st Choice	2nd Choice	1st Choice	2nd Choice
1	…	6	…	40	…	250	…
…	1.1	…	7	…	45	…	280
1.2	…	8	…	50	…	300	…
…	1.4	…	9	…	55	…	350
1.6	…	10	…	60	…	400	…
…	1.8	…	11	…	70	…	450
2	…	12	…	80	…	500	…
…	2.2	…	14	…	90	…	550
2.5	…	16	…	100	…	600	…
…	2.8	…	18	…	110	…	700
3	…	20	…	120	…	800	…
…	3.5	…	22	…	140	…	900
4	…	25	…	160	…	1000	…
…	4.5	…	28	…	180	…	…
5	…	30	…	200	…	…	…
…	5.5	…	35	…	220	…	…

Preferred Metric Sizes.—American National Standard ANSI B32.4M-1980 (R1994), presents series of preferred metric sizes for round, square, rectangular, and hexagonal metal products. Table 11 gives preferred metric diameters from 1 to 320 millimeters for round metal products. Wherever possible, sizes should be selected from the Preferred Series shown in the table. A Second Preference series is also shown. A Third Preference Series not shown in the table is: 1.3, 2.1, 2.4, 2.6, 3.2, 3.8, 4.2, 4.8, 7.5, 8.5, 9.5, 36, 85, and 95.

Most of the Preferred Series of sizes are derived from the American National Standard "10 series" of preferred numbers (see *American National Standard for Preferred Numbers* on page 19). Most of the Second Preference Series are derived from the "20 series" of preferred numbers. Third Preference sizes are generally from the "40 series" of preferred numbers.

For preferred metric diameters less than 1 millimeter, preferred across flat metric sizes of square and hexagon metal products, preferred across flat metric sizes of rectangular metal products, and preferred metric lengths of metal products, reference should be made to the Standard.

Preferred Fits.—First-choice tolerance zones are used to establish preferred fits in the Standard for Preferred Metric Limits and Fits, ANSI B4.2, as shown in Figs. 2 and 3. A complete listing of first-, second-, and third- choice tolerance zones is given in the Standard.

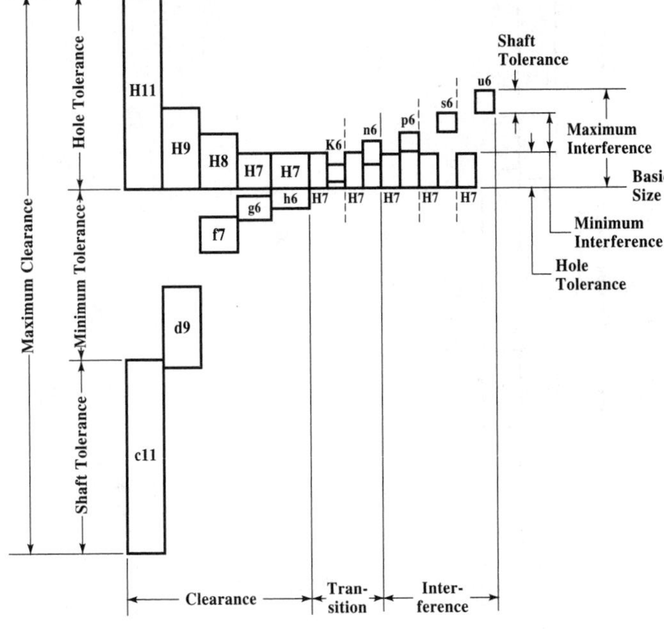

Fig. 2. Preferred Hole Basis Fits

Hole basis fits have a fundamental deviation of H on the hole, and shaft basis fits have a fundamental deviation of h on the shaft and are shown in Fig. 2 for hole basis and Fig. 3 for shaft basis fits. A description of both types of fits, that have the same relative fit condition, is given in Table 12. Normally, the hole basis system is preferred; however, when a common shaft mates with several holes, the shaft basis system should be used.

The hole basis and shaft basis fits shown in Table 12 are combined with the first-choice sizes shown in Table 11 to form Tables 13, 14, 15, and 16, where specific limits as well as the resultant fits are tabulated.

If the required size is not tabulated in Tables 13 through 16 then the preferred fit can be calculated from numerical values given in an appendix of ANSI B4.2-1978 (R1984). It is anticipated that other fit conditions may be necessary to meet special requirements, and a preferred fit can be loosened or tightened simply by selecting a standard tolerance zone as given in the Standard. Information on how to calculate limit dimensions, clearances, and interferences, for nonpreferred fits and sizes can also be found in an appendix of this Standard.

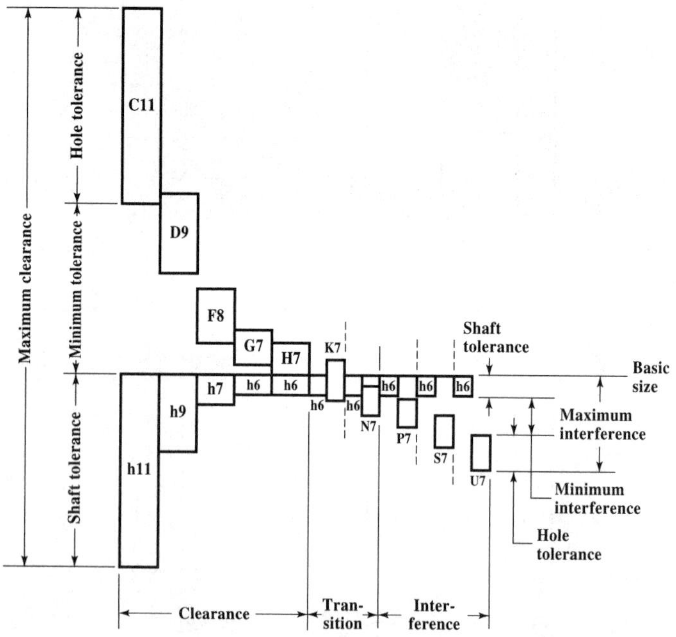

Fig. 3. Preferred Shaft Basis Fits

Table 12. Description of preferred fits

	ISO SYMBOL		DESCRIPTION	
	Hole Basis	**Shaft Basis**		
Clearance Fits	H11/c11	C11/h11	*Loose running* fit for wide commercial tolerances or allowances on external members.	↑ More Clearance
	H9/d9	D9/h9	*Free running* fit not for use where accuracy is essential, but good for large temperature variations, high running speeds, or heavy journal pressures.	
	H8/f7	F8/h7	*Close Running* fit for running on accurate machines and for accurate moderate speeds and journal pressures.	
	H7/g6	G7/h6	*Sliding fit* not intended to run freely, but to move and turn freely and locate accurately.	
	H7/h6	H7/h6	*Locational clearance* fit provides snug fit for locating stationary parts; but can be freely assembled and disassembled.	
Transition Fits	H7/k6	K7/h6	*Locational transition* fit for accurate location, a compromise between clearance and interference.	
	H7/n6	N7/h6	*Locational transition* fit for more accurate location where greater interference is permissible.	
Interference Fits	H7/p6[a]	P7/h6	*Locational interference* fit for parts requiring and alignment with prime accuracy of location but without special bore pressure requirements.	More Interference ↓
	H7/s6	S7/h6	*Medium drive* fit for ordinary steel parts or shrink fits on light sections, the tightest fit usable with cast iron.	
	H7/u6	U7/h6	*Force* fit suitable for parts which can be highly stressed or for shrink fits where the heavy pressing forces required are impractical.	

[a] Transition fit for basic sizes in range from 0 through 3 mm.

Table 13. American National Standard Preferred Hole Basis Metric Clearance Fits *ANSI B4.2-1978 (R1994)*

Basic Size[a]		Loose Running			Free Running			Close Running			Sliding			Locational Clearance		
		Hole H11	Shaft C11	Fit[b]	Hole H9	Shaft d9	Fit[b]	Hole H8	Shaft f7	Fit[b]	Hole H7	Shaft g6	Fit[b]	Hole H7	Shaft h6	Fit[b]
1	Max	1.060	0.940	0.180	1.025	0.980	0.070	1.014	0.994	0.030	1.010	0.998	0.018	1.010	1.000	0.016
	Min	1.000	0.880	0.060	1.000	0.995	0.020	1.000	0.984	0.006	1.000	0.992	0.002	1.000	0.994	0.000
1.2	Max	1.260	1.140	0.180	1.225	1.180	0.070	1.214	1.194	0.030	1.210	1.198	0.018	1.210	1.200	0.016
	Min	1.200	1.080	0.060	1.200	1.155	0.020	1.200	1.184	0.006	1.200	1.192	0.002	1.200	1.194	0.000
1.6	Max	1.660	1.540	0.180	1.625	1.580	0.070	1.614	1.594	0.030	1.610	1.598	0.018	1.610	1.600	0.016
	Min	1.600	1.480	0.060	1.600	1.555	0.020	1.600	1.584	0.006	1.600	1.592	0.002	1.600	1.594	0.000
2	Max	2.060	1.940	0.180	2.025	1.980	0.070	2.014	1.994	0.030	2.010	1.998	0.018	2.010	2.000	0.016
	Min	2.000	1.880	0.060	2.000	1.955	0.020	2.000	1.984	0.006	2.000	1.992	0.002	2.000	1.994	0.000
2.5	Max	2.560	2.440	0.180	2.525	2.480	0.070	2.514	2.494	0.030	2.510	2.498	0.018	2.510	2.500	0.016
	Min	2.500	2.380	0.060	2.500	2.455	0.020	2.500	2.484	0.006	2.500	2.492	0.002	2.500	2.494	0.000
3	Max	3.060	2.940	0.180	3.025	2.980	0.070	3.014	2.994	0.030	3.010	2.998	0.018	3.010	3.000	0.016
	Min	3.000	2.880	0.060	3.000	2.955	0.020	3.000	2.984	0.006	3.000	2.992	0.002	3.000	2.994	0.000
4	Max	4.075	3.930	0.220	4.030	3.970	0.090	4.018	3.990	0.040	4.012	3.996	0.024	4.012	4.000	0.020
	Min	4.000	3.855	0.070	4.000	3.940	0.030	4.000	3.978	0.010	4.000	3.988	0.004	4.000	3.992	0.000
5	Max	5.075	4.930	0.220	5.030	4.970	0.090	5.018	4.990	0.040	5.012	4.996	0.024	5.012	5.000	0.020
	Min	5.000	4.855	0.070	5.000	4.940	0.030	5.000	4.978	0.010	5.000	4.988	0.004	5.000	4.992	0.000
6	Max	6.075	5.930	0.220	6.030	5.970	0.090	6.018	5.990	0.040	6.012	5.996	0.024	6.012	6.000	0.020
	Min	6.000	5.855	0.070	6.000	5.940	0.030	6.000	5.978	0.010	6.000	5.988	0.004	6.000	5.992	0.000
8	Max	8.090	7.920	0.260	8.036	7.960	0.112	8.022	7.987	0.050	8.015	7.995	0.029	8.015	8.000	0.024
	Min	8.000	7.830	0.080	8.000	7.924	0.040	8.000	7.972	0.013	8.000	7.986	0.005	8.000	7.991	0.000
10	Max	10.090	9.920	0.260	10.036	9.960	0.112	10.022	9.987	0.050	10.015	9.995	0.029	10.015	10.000	0.024
	Min	10.000	9.830	0.080	10.000	9.924	0.040	10.000	9.972	0.013	10.000	9.986	0.005	10.000	9.991	0.000
12	Max	12.110	11.905	0.315	12.043	11.956	0.136	12.027	11.984	0.061	12.018	11.994	0.035	12.018	12.000	0.029
	Min	12.000	11.795	0.095	12.000	11.907	0.050	12.000	11.966	0.016	12.000	11.983	0.006	12.000	11.989	0.000
16	Max	16.110	15.905	0.315	16.043	15.950	0.136	16.027	15.984	0.061	16.018	15.994	0.035	16.018	16.000	0.029
	Min	16.000	15.795	0.095	16.000	15.907	0.050	16.000	15.966	0.016	16.000	15.983	0.006	16.000	15.989	0.000
20	Max	20.130	19.890	0.370	20.052	19.935	0.169	20.033	19.980	0.074	20.021	19.993	0.041	20.021	20.000	0.034
	Min	20.000	19.760	0.110	20.000	19.883	0.065	20.000	19.959	0.020	20.000	19.980	0.007	20.000	19.987	0.000
25	Max	25.130	24.890	0.370	25.052	24.935	0.169	25.033	24.980	0.074	25.021	24.993	0.041	25.021	25.000	0.034
	Min	25.000	24.760	0.110	25.000	24.883	0.065	25.000	24.959	0.020	25.000	24.980	0.007	25.000	24.987	0.000

Table 13. *(Continued)* **American National Standard Preferred Hole Basis Metric Clearance Fits** *ANSI B4.2-1978 (R1994)*

Basic Size[a]		Loose Running			Free Running			Close Running			Sliding			Locational Clearance		
		Hole H11	Shaft C11	Fit[b]	Hole H9	Shaft d9	Fit[b]	Hole H8	Shaft f7	Fit[b]	Hole H7	Shaft g6	Fit[b]	Hole H7	Shaft h6	Fit[b]
30	Max	30.130	29.890	0.370	30.052	29.935	0.169	30.033	29.980	0.074	30.021	29.993	0.041	30.021	30.000	0.034
	Min	30.000	29.760	0.110	30.000	29.883	0.065	30.000	29.959	0.020	30.000	29.980	0.007	30.000	29.987	0.000
40	Max	40.160	39.880	0.440	40.062	39.920	0.204	40.039	39.975	0.089	40.025	39.991	0.050	40.025	40.000	0.041
	Min	40.000	39.720	0.120	40.000	39.858	0.080	40.000	39.950	0.025	40.000	39.975	0.009	40.000	39.984	0.000
50	Max	50.160	49.870	0.450	50.062	49.920	0.204	50.039	49.975	0.089	50.025	49.991	0.050	50.025	50.000	0.041
	Min	50.000	49.710	0.130	50.000	49.858	0.080	50.000	49.950	0.025	50.000	49.975	0.009	50.000	49.984	0.000
60	Max	60.190	59.860	0.520	60.074	59.900	0.248	60.046	59.970	0.106	60.030	59.990	0.059	60.030	60.000	0.049
	Min	60.000	59.670	0.140	60.000	59.826	0.100	60.000	59.940	0.030	60.000	59.971	0.010	60.000	59.981	0.000
80	Max	80.190	79.850	0.530	80.074	79.900	0.248	80.046	79.970	0.106	80.030	79.990	0.059	80.030	80.000	0.049
	Min	80.000	79.660	0.150	80.000	79.826	0.100	80.000	79.940	0.030	80.000	79.971	0.010	80.000	79.981	0.000
100	Max	100.220	99.830	0.610	100.087	99.880	0.294	100.054	99.964	0.125	100.035	99.988	0.069	100.035	100.000	0.057
	Min	100.000	99.610	0.170	100.000	99.793	0.120	100.000	99.929	0.036	100.000	99.966	0.012	100.000	99.978	0.000
120	Max	120.220	119.820	0.620	120.087	119.880	0.294	120.054	119.964	0.125	120.035	119.988	0.069	120.035	120.000	0.057
	Min	120.000	119.600	0.180	120.000	119.793	0.120	120.000	119.929	0.036	120.000	119.966	0.012	120.000	119.978	0.000
160	Max	160.250	159.790	0.710	160.100	159.855	0.345	160.063	159.957	0.146	160.040	159.986	0.079	160.040	160.000	0.065
	Min	160.000	159.540	0.210	160.000	159.755	0.145	160.000	159.917	0.043	160.000	159.961	0.014	160.000	159.975	0.000
200	Max	200.290	199.760	0.820	200.115	199.830	0.400	200.072	199.950	0.168	200.046	199.985	0.090	200.046	200.000	0.075
	Min	200.000	199.470	0.240	200.000	199.715	0.170	200.000	199.904	0.050	200.000	199.956	0.015	200.000	199.971	0.000
250	Max	250.290	249.720	0.860	250.115	249.830	0.400	250.072	249.950	0.168	250.046	249.985	0.090	250.046	250.000	0.075
	Min	250.000	249.430	0.280	250.000	249.715	0.170	250.000	249.904	0.050	250.000	249.956	0.015	250.000	249.971	0.000
300	Max	300.320	299.670	0.970	300.130	299.810	0.450	300.081	299.944	0.189	300.052	299.983	0.101	300.052	300.000	0.084
	Min	300.000	299.350	0.330	300.000	299.680	0.190	300.000	299.892	0.056	300.000	299.951	0.017	300.000	299.968	0.000
400	Max	400.360	399.600	1.120	400.140	399.790	0.490	400.089	399.938	0.208	400.057	399.982	0.111	400.057	400.000	0.093
	Min	400.000	399.240	0.400	400.000	399.650	0.210	400.000	399.881	0.062	400.000	399.946	0.018	400.000	399.964	0.000
500	Max	500.400	499.520	1.280	500.155	499.770	0.540	500.097	499.932	0.228	500.063	499.980	0.123	500.063	500.000	0.103
	Min	500.000	499.120	0.480	500.000	499.615	0.230	500.000	499.869	0.068	500.000	499.940	0.020	500.000	499.960	0.000

[a] The sizes shown are first-choice basic sizes (see Table 11). Preferred fits for other sizes can be calculated from data given in ANSI B4.2-1978 (R1984).

[b] All fits shown in this table have clearance.

All dimensions are in millimeters.

Table 14. American National Standard Preferred Hole Basis Metric Transition and Interference Fits *ANSI B4.2-1978 (R1994)*

Basic Size[a]		Locational Transition			Locational Transition			Locational Interference			Medium Drive			Force		
		Hole H7	Shaft k6	Fit[b]	Hole H7	Shaft n6	Fit[b]	Hole H7	Shaft p6	Fit[b]	Hole H7	Shaft s6	Fit[b]	Hole H7	Shaft u6	Fit[b]
1	Max	1.010	1.006	+0.010	1.010	1.010	+0.006	1.010	1.012	+0.004	1.010	1.020	−0.004	1.010	1.024	−0.008
	Min	1.000	1.000	−0.006	1.000	1.004	−0.010	1.000	1.006	−0.012	1.000	1.014	−0.020	1.000	1.018	−0.024
1.2	Max	1.210	1.206	+0.010	1.210	1.210	+0.006	1.210	1.212	+0.004	1.210	1.220	−0.004	1.210	1.224	−0.008
	Min	1.200	1.200	−0.006	1.200	1.204	−0.010	1.200	1.206	−0.012	1.200	1.214	−0.020	1.200	1.218	−0.024
1.6	Max	1.610	1.606	+0.010	1.610	1.610	+0.006	1.610	1.612	+0.004	1.610	1.620	−0.004	1.610	1.624	−0.008
	Min	1.600	1.600	−0.006	1.600	1.604	−0.010	1.600	1.606	−0.012	1.600	1.614	−0.020	1.600	1.618	−0.024
2	Max	2.010	2.006	+0.010	2.010	2.010	+0.006	2.010	2.012	+0.004	2.010	2.020	−0.004	2.010	2.024	−0.008
	Min	2.000	2.000	−0.006	2.000	2.004	−0.010	2.000	2.006	−0.012	2.000	2.014	−0.020	2.000	2.018	−0.024
2.5	Max	2.510	2.506	+0.010	2.510	2.510	+0.006	2.510	2.512	+0.004	2.510	2.520	−0.004	2.510	2.524	−0.008
	Min	2.500	2.500	−0.006	2.500	2.504	−0.010	2.500	2.506	−0.012	2.500	2.514	−0.020	2.500	2.518	−0.024
3	Max	3.010	3.006	+0.010	3.010	3.010	+0.006	3.010	3.012	+0.004	3.010	3.020	−0.004	3.010	3.024	−0.008
	Min	3.000	3.000	−0.006	3.000	3.004	−0.010	3.000	3.006	−0.012	3.000	3.014	−0.020	3.000	3.018	−0.024
4	Max	4.012	4.009	+0.011	4.012	4.016	+0.004	4.012	4.020	0.000	4.012	4.027	−0.007	4.012	4.031	−0.011
	Min	4.000	4.001	−0.009	4.000	4.008	−0.016	4.000	4.012	−0.020	4.000	4.019	−0.027	4.000	4.023	−0.031
5	Max	5.012	5.009	+0.011	5.012	5.016	+0.004	5.012	5.020	0.000	5.012	5.027	−0.007	5.012	5.031	−0.011
	Min	5.000	5.001	−0.009	5.000	5.008	−0.016	5.000	5.012	−0.020	5.000	5.019	−0.027	5.000	5.023	−0.031
6	Max	6.012	6.009	+0.011	6.012	6.016	+0.004	6.012	6.020	0.000	6.012	6.027	−0.007	6.012	6.031	−0.011
	Min	6.000	6.001	−0.009	6.000	6.008	−0.016	6.000	6.012	−0.020	6.000	6.019	−0.027	6.000	6.023	−0.031
8	Max	8.015	8.010	+0.014	8.015	8.019	+0.005	8.015	8.024	0.000	8.015	8.032	−0.008	8.015	8.037	−0.013
	Min	8.000	8.001	−0.010	8.000	8.010	−0.019	8.000	8.015	−0.024	8.000	8.023	−0.032	8.000	8.028	−0.037
10	Max	10.015	10.010	+0.014	10.015	10.019	+0.005	10.015	10.024	0.000	10.015	10.032	−0.008	10.015	10.034	−0.013
	Min	10.000	10.001	−0.010	10.000	10.010	−0.019	10.000	10.015	−0.024	10.000	10.023	−0.032	10.000	10.028	−0.037
12	Max	12.018	12.012	+0.017	12.018	12.023	+0.006	12.018	12.029	0.000	12.018	12.039	−0.010	12.018	12.044	−0.015
	Min	12.000	12.001	−0.012	12.000	12.012	−0.023	12.000	12.018	−0.029	12.000	12.028	−0.039	12.000	12.033	−0.044
16	Max	16.018	16.012	+0.017	16.018	16.023	+0.006	16.018	16.029	0.000	16.018	16.039	−0.010	16.018	16.044	−0.015
	Min	16.000	16.001	−0.012	16.000	16.012	−0.023	16.000	16.018	−0.029	16.000	16.028	−0.039	16.000	16.033	−0.044
20	Max	20.021	20.015	+0.019	20.021	20.028	+0.006	20.021	20.035	−0.001	20.021	20.048	−0.014	20.021	20.054	−0.020
	Min	20.000	20.002	−0.015	20.000	20.015	−0.028	20.000	20.022	−0.035	20.000	20.035	−0.048	20.000	20.041	−0.054
25	Max	25.021	25.015	+0.019	25.021	25.028	+0.006	25.021	25.035	−0.001	25.021	25.048	−0.014	25.021	25.061	−0.027
	Min	25.000	25.002	−0.015	25.000	25.015	−0.028	25.000	25.022	−0.035	25.000	25.035	−0.048	25.000	25.048	−0.061

Table 14. *(Continued)* **American National Standard Preferred Hole Basis Metric Transition and Interference Fits** *ANSI B4.2-1978 (R1994)*

Basic Size[a]		Locational Transition			Locational Transition			Locational Interference			Medium Drive			Force		
		Hole H7	Shaft k6	Fit[b]	Hole H7	Shaft n6	Fit[b]	Hole H7	Shaft p6	Fit[b]	Hole H7	Shaft s6	Fit[b]	Hole H7	Shaft u6	Fit[b]
30	Max	30.021	30.015	+0.019	30.021	30.028	+0.006	30.021	30.035	−0.001	30.021	30.048	−0.014	30.021	30.061	−0.027
	Min	30.000	30.002	−0.015	30.000	30.015	−0.028	30.000	30.022	−0.035	30.000	30.035	−0.048	30.000	30.048	−0.061
40	Max	40.025	40.018	+0.023	40.025	40.033	+0.008	40.025	40.042	−0.001	40.025	40.059	−0.018	40.025	40.076	−0.035
	Min	40.000	40.002	−0.018	40.000	40.017	−0.033	40.000	40.026	−0.042	40.000	40.043	−0.059	40.000	40.060	−0.076
50	Max	50.025	50.018	+0.023	50.025	50.033	+0.008	50.025	50.042	−0.001	50.025	50.059	−0.018	50.025	50.086	−0.045
	Min	50.000	50.002	−0.018	50.000	50.017	−0.033	50.000	50.026	−0.042	50.000	50.043	−0.059	50.000	50.070	−0.086
60	Max	60.030	60.021	+0.028	60.030	60.039	+0.010	60.030	60.051	−0.002	60.030	60.072	−0.023	60.030	60.106	−0.057
	Min	60.000	60.002	−0.021	60.000	60.020	−0.039	60.000	60.032	−0.051	60.000	60.053	−0.072	60.000	60.087	−0.106
80	Max	80.030	80.021	+0.028	80.030	80.039	+0.010	80.030	80.051	−0.002	80.030	80.078	−0.029	80.030	80.121	−0.072
	Min	80.000	80.002	−0.021	80.000	80.020	−0.039	80.000	80.032	−0.051	80.000	80.059	−0.078	80.000	80.102	−0.121
100	Max	100.035	100.025	+0.032	100.035	100.045	+0.012	100.035	100.059	−0.002	100.035	100.093	−0.036	100.035	100.146	−0.089
	Min	100.000	100.003	−0.025	100.000	100.023	−0.045	100.000	100.037	−0.059	100.000	100.071	−0.093	100.000	100.124	−0.146
120	Max	120.035	120.025	+0.032	120.035	120.045	+0.012	120.035	120.059	−0.002	120.035	120.101	−0.044	120.035	120.166	−0.109
	Min	120.000	120.003	−0.025	120.000	120.023	−0.045	120.000	120.037	−0.059	120.000	120.079	−0.101	120.000	120.144	−0.166
160	Max	160.040	160.028	+0.037	160.040	160.052	+0.013	160.040	160.068	−0.003	160.040	160.125	−0.060	160.040	160.215	−0.150
	Min	160.000	160.003	−0.028	160.000	160.027	−0.052	160.000	160.043	−0.068	160.000	160.100	−0.125	160.000	160.190	−0.215
200	Max	200.046	200.033	+0.042	200.046	200.060	+0.015	200.046	200.079	−0.004	200.046	200.151	−0.076	200.046	200.265	−0.190
	Min	200.000	200.004	−0.033	200.000	200.031	−0.060	200.000	200.050	−0.079	200.000	200.122	−0.151	200.000	200.236	−0.265
250	Max	250.046	250.033	+0.042	250.046	250.060	+0.015	250.046	250.079	−0.004	250.046	250.169	−0.094	250.046	250.313	−0.238
	Min	250.000	250.004	−0.033	250.000	250.031	−0.060	250.000	250.050	−0.079	250.000	250.140	−0.169	250.000	250.284	−0.313
300	Max	300.052	300.036	+0.048	300.052	300.066	+0.018	300.052	300.088	−0.004	300.052	300.202	−0.118	300.052	300.382	−0.298
	Min	300.000	300.004	−0.036	300.000	300.034	−0.066	300.000	300.056	−0.088	300.000	300.170	−0.202	300.000	300.350	−0.382
400	Max	400.057	400.040	+0.053	400.057	400.073	+0.020	400.057	400.098	−0.005	400.057	400.244	−0.151	400.057	400.471	−0.378
	Min	400.000	400.004	−0.040	400.000	400.037	−0.073	400.000	400.062	−0.098	400.000	400.208	−0.244	400.000	400.435	−0.471
500	Max	500.063	500.045	+0.058	500.063	500.080	+0.023	500.063	500.108	−0.005	500.063	500.292	−0.189	500.063	500.580	−0.477
	Min	500.000	500.005	−0.045	500.000	500.040	−0.080	500.000	500.068	−0.108	500.000	500.252	−0.292	500.000	500.540	−0.580

[a] The sizes shown are first-choice basic sizes (see Table 11). Preferred fits for other sizes can be calculated from data given in ANSI B4.2-1978 (R1984).

[b] A plus sign indicates clearance; a minus sign indicates interference.

All dimensions are in millimeters.

Table 15. American National Standard Preferred Shaft Basis Metric Clearance Fits *ANSI B4.2-1978 (R1994)*

Basic Size[a]		Loose Running			Free Running			Close Running			Sliding			Locational Clearance		
		Hole C11	Shaft h11	Fit[b]	Hole D9	Shaft h9	Fit[b]	Hole F8	Shaft h7	Fit[b]	Hole G7	Shaft h6	Fit[b]	Hole H7	Shaft h6	Fit[b]
1	Max	1.120	1.000	0.180	1.045	1.000	0.070	1.020	1.000	0.030	1.012	1.000	0.018	1.010	1.000	0.016
	Min	1.060	0.940	0.060	1.020	0.975	0.020	1.006	0.990	0.006	1.002	0.994	0.002	1.000	0.994	0.000
1.2	Max	1.320	1.200	0.180	1.245	1.200	0.070	1.220	1.200	0.030	1.212	1.200	0.018	1.210	1.200	0.016
	Min	1.260	1.140	0.060	1.220	1.175	0.020	1.206	1.190	0.006	1.202	1.194	0.002	1.200	1.194	0.000
1.6	Max	1.720	1.600	0.180	1.645	1.600	0.070	1.620	1.600	0.030	1.612	1.600	0.018	1.610	1.600	0.016
	Min	1.660	1.540	0.060	1.620	1.575	0.020	1.606	1.590	0.006	1.602	1.594	0.002	1.600	1.594	0.000
2	Max	2.120	2.000	0.180	2.045	2.000	0.070	2.020	2.000	0.030	2.012	2.000	0.018	2.010	2.000	0.016
	Min	2.060	1.940	0.060	2.020	1.975	0.020	2.006	1.990	0.006	2.002	1.994	0.002	2.000	1.994	0.000
2.5	Max	2.620	2.500	0.180	2.545	2.500	0.070	2.520	2.500	0.030	2.512	2.500	0.018	2.510	2.500	0.016
	Min	2.560	2.440	0.060	2.520	2.475	0.020	2.506	2.490	0.006	2.502	2.494	0.002	2.500	2.494	0.000
3	Max	3.120	3.000	0.180	3.045	3.000	0.070	3.020	3.000	0.030	3.012	3.000	0.018	3.010	3.000	0.016
	Min	3.060	2.940	0.060	3.020	2.975	0.020	3.006	2.990	0.006	3.002	2.994	0.002	3.000	2.994	0.000
4	Max	4.145	4.000	0.220	4.060	4.000	0.090	4.028	4.000	0.040	4.016	4.000	0.024	4.012	4.000	0.020
	Min	4.070	3.925	0.070	4.030	3.970	0.030	4.010	3.988	0.010	4.004	3.992	0.004	4.000	3.992	0.000
5	Max	5.145	5.000	0.220	5.060	5.000	0.090	5.028	5.000	0.040	5.016	5.000	0.024	5.012	5.000	0.020
	Min	5.070	4.925	0.070	5.030	4.970	0.030	5.010	4.988	0.010	5.004	4.992	0.004	5.000	4.992	0.000
6	Max	6.145	6.000	0.220	6.060	6.000	0.090	6.028	6.000	0.040	6.016	6.000	0.024	6.012	6.000	0.020
	Min	6.070	5.925	0.070	6.030	5.970	0.030	6.010	5.988	0.010	6.004	5.992	0.004	6.000	5.992	0.000
8	Max	8.170	8.000	0.260	8.076	8.000	0.112	8.035	8.000	0.050	8.020	8.000	0.029	8.015	8.000	0.024
	Min	8.080	7.910	0.080	8.040	7.964	0.040	8.013	7.985	0.013	8.005	7.991	0.005	8.000	7.991	0.000
10	Max	10.170	10.000	0.260	10.076	10.000	0.112	10.035	10.000	0.050	10.020	10.000	0.029	10.015	10.000	0.024
	Min	10.080	9.910	0.080	10.040	9.964	0.040	10.013	9.985	0.013	10.005	9.991	0.005	10.000	9.991	0.000
12	Max	12.205	12.000	0.315	12.093	12.000	0.136	12.043	12.000	0.061	12.024	12.000	0.035	12.018	12.000	0.029
	Min	12.095	11.890	0.095	12.050	11.957	0.050	12.016	11.982	0.016	12.006	11.989	0.006	12.000	11.989	0.000
16	Max	16.205	16.000	0.315	16.093	16.000	0.136	16.043	16.000	0.061	16.024	16.000	0.035	16.018	16.000	0.029
	Min	16.095	15.890	0.095	16.050	15.957	0.050	16.016	15.982	0.016	16.006	15.989	0.006	16.000	15.989	0.000
20	Max	20.240	20.000	0.370	20.117	20.000	0.169	20.053	20.000	0.074	20.028	20.000	0.041	20.021	20.000	0.034
	Min	20.110	19.870	0.110	20.065	19.948	0.065	20.020	19.979	0.020	20.007	19.987	0.007	20.000	19.987	0.000
25	Max	25.240	25.000	0.370	25.117	25.000	0.169	25.053	25.000	0.074	25.028	25.000	0.041	25.021	25.000	0.034
	Min	25.110	24.870	0.110	25.065	24.948	0.065	25.020	24.979	0.020	25.007	24.987	0.007	25.000	24.987	0.000

Table 15. *(Continued)* **American National Standard Preferred Shaft Basis Metric Clearance Fits** *ANSI B4.2-1978 (R1994)*

Basic Size[a]		Loose Running			Free Running			Close Running			Sliding			Locational Clearance		
		Hole C11	Shaft h11	Fit[b]	Hole D9	Shaft h9	Fit[b]	Hole F8	Shaft h7	Fit[b]	Hole G7	Shaft h6	Fit[b]	Hole H7	Shaft h6	Fit[b]
30	Max	30.240	30.000	0.370	30.117	30.000	0.169	30.053	30.000	0.074	30.028	30.000	0.041	30.021	30.000	0.034
	Min	30.110	29.870	0.110	30.065	29.948	0.065	30.020	29.979	0.020	30.007	29.987	0.007	30.000	29.987	0.000
40	Max	40.280	40.000	0.440	40.142	40.000	0.204	40.064	40.000	0.089	40.034	40.000	0.050	40.025	40.000	0.041
	Min	40.120	39.840	0.120	40.080	39.938	0.080	40.025	39.975	0.025	40.009	39.984	0.009	40.000	39.984	0.000
50	Max	50.290	50.000	0.450	50.142	50.000	0.204	50.064	50.000	0.089	50.034	50.000	0.050	50.025	50.000	0.041
	Min	50.130	49.840	0.130	50.080	49.938	0.080	50.025	49.975	0.025	50.009	49.984	0.009	50.000	49.984	0.000
60	Max	60.330	60.000	0.520	60.174	60.000	0.248	60.076	60.000	0.106	60.040	60.000	0.059	60.030	60.000	0.049
	Min	60.140	59.810	0.140	60.100	59.926	0.100	60.030	59.970	0.030	60.010	59.981	0.010	60.000	59.981	0.000
80	Max	80.340	80.000	0.530	80.174	80.000	0.248	80.076	80.000	0.106	80.040	80.000	0.059	80.030	80.000	0.049
	Min	80.150	79.810	0.150	80.100	79.926	0.100	80.030	79.970	0.030	80.010	79.981	0.010	80.000	79.981	0.000
100	Max	100.390	100.000	0.610	100.207	100.000	0.294	100.090	100.000	0.125	100.047	100.000	0.069	100.035	100.000	0.057
	Min	100.170	99.780	0.170	100.120	99.913	0.120	100.036	99.965	0.036	100.012	99.978	0.012	100.000	99.978	0.000
120	Max	120.400	120.000	0.620	120.207	120.000	0.294	120.090	120.000	0.125	120.047	120.000	0.069	120.035	120.000	0.057
	Min	120.180	119.780	0.180	120.120	119.913	0.120	120.036	119.965	0.036	120.012	119.978	0.012	120.000	119.978	0.000
160	Max	160.460	160.000	0.710	160.245	160.000	0.345	160.106	160.000	0.146	160.054	160.000	0.079	160.040	160.000	0.065
	Min	160.210	159.750	0.210	160.145	159.900	0.145	160.043	159.960	0.043	160.014	159.975	0.014	160.000	159.975	0.000
200	Max	200.530	200.000	0.820	200.285	200.000	0.400	200.122	200.000	0.168	200.061	200.000	0.090	200.046	200.000	0.075
	Min	200.240	199.710	0.240	200.170	199.885	0.170	200.050	199.954	0.050	200.015	199.971	0.015	200.000	199.971	0.000
250	Max	250.570	250.000	0.860	250.285	250.000	0.400	250.122	250.000	0.168	250.061	250.000	0.090	250.046	250.000	0.075
	Min	250.280	249.710	0.280	250.170	249.885	0.170	250.050	249.954	0.050	250.015	249.971	0.015	250.000	249.971	0.000
300	Max	300.650	300.000	0.970	300.320	300.000	0.450	300.137	300.000	0.189	300.069	300.000	0.101	300.052	300.000	0.084
	Min	300.330	299.680	0.330	300.190	299.870	0.190	300.056	299.948	0.056	300.017	299.968	0.017	300.000	299.968	0.000
400	Max	400.760	400.000	1.120	400.350	400.000	0.490	400.151	400.000	0.208	400.075	400.000	0.111	400.057	400.000	0.093
	Min	400.400	399.640	0.400	400.210	399.860	0.210	400.062	399.943	0.062	400.018	399.964	0.018	400.000	399.964	0.000
500	Max	500.880	500.000	1.280	500.385	500.000	0.540	500.165	500.000	0.228	500.083	500.000	0.123	500.063	500.000	0.103
	Min	500.480	499.600	0.480	500.230	499.845	0.230	500.068	499.937	0.068	500.020	499.960	0.020	500.000	499.960	0.000

[a] The sizes shown are first-choice basic sizes (see Table 11). Preferred fits for other sizes can be calculated from data given in ANSI B4.2-1978 (R1984).

[b] All fits shown in this table have clearance.

All dimensions are in millimeters.

Table 16. American National Standard Preferred Shaft Basis Metric Transition and Interference Fits *ANSI B4.2-1978 (R1994)*

Basic Size[a]		Locational Transition			Locational Transition			Locational Interference			Medium Drive			Force		
		Hole K7	Shaft h6	Fit[b]	Hole N7	Shaft h6	Fit[b]	Hole P7	Shaft h6	Fit[b]	Hole S7	Shaft h6	Fit[b]	Hole U7	Shaft h6	Fit[b]
1	Max	1.000	1.000	+0.006	0.996	1.000	+0.002	0.994	1.000	0.000	0.986	1.000	−0.008	0.982	1.000	−0.012
	Min	0.990	0.994	−0.010	0.986	0.954	−0.014	0.984	0.994	−0.016	0.976	0.994	−0.024	0.972	0.994	−0.028
1.2	Max	1.200	1.200	+0.006	1.196	1.200	+0.002	1.194	1.200	0.000	1.186	1.200	−0.008	1.182	1.200	−0.012
	Min	1.190	1.194	−0.010	1.186	1.194	−0.014	1.184	1.194	−0.016	1.176	1.194	−0.024	1.172	1.194	−0.028
1.6	Max	1.600	1.600	+0.006	1.596	1.600	+0.002	1.594	1.600	0.000	1.586	1.600	−0.008	1.582	1.600	−0.012
	Min	1.590	1.594	−0.010	1.586	1.594	−0.014	1.584	1.594	−0.016	1.576	1.594	−0.024	1.572	1.594	−0.028
2	Max	2.000	2.000	+0.006	1.996	2.000	+0.002	1.994	2.000	0.000	1.986	2.000	−0.008	1.982	2.000	−0.012
	Min	1.990	1.994	−0.010	1.986	1.994	−0.014	1.984	1.994	−0.016	1.976	1.994	−0.024	1.972	1.994	−0.028
2.5	Max	2.500	2.500	+0.006	2.496	2.500	+0.002	2.494	2.500	0.000	2.486	2.500	−0.008	2.482	2.500	−0.012
	Min	2.490	2.494	−0.010	2.486	2.494	−0.014	2.484	2.494	−0.016	2.476	2.494	−0.024	2.472	2.494	−0.028
3	Max	3.000	3.000	+0.006	2.996	3.000	+0.002	2.994	3.000	0.000	2.986	3.000	−0.008	2.982	3.000	−0.012
	Min	2.990	2.994	−0.010	2.986	2.994	−0.014	2.984	2.994	−0.016	2.976	2.994	−0.024	2.972	2.994	−0.028
4	Max	4.003	4.000	+0.011	3.996	4.000	+0.004	3.992	4.000	0.000	3.985	4.000	−0.007	3.981	4.000	−0.011
	Min	3.991	3.992	−0.009	3.984	3.992	−0.016	3.980	3.992	−0.020	3.973	3.992	−0.027	3.969	3.992	−0.031
5	Max	5.003	5.000	+0.011	4.996	5.000	+0.004	4.992	5.000	0.000	4.985	5.000	−0.007	4.981	5.000	−0.011
	Min	4.991	4.992	−0.009	4.984	4.992	−0.016	4.980	4.992	−0.020	4.973	4.992	−0.027	4.969	4.992	−0.031
6	Max	6.003	6.000	+0.011	5.996	6.000	+0.004	5.992	6.000	0.000	5.985	6.000	−0.007	5.981	6.000	−0.011
	Min	5.991	5.992	−0.009	5.984	5.992	−0.016	5.980	5.992	−0.020	5.973	5.992	−0.027	5.969	5.992	−0.031
8	Max	8.005	8.000	+0.014	7.996	8.000	+0.005	7.991	8.000	0.000	7.983	8.000	−0.008	7.978	8.000	−0.013
	Min	7.990	7.991	−0.010	7.981	7.991	−0.019	7.976	7.991	−0.024	7.968	7.991	−0.032	7.963	7.991	−0.037
10	Max	10.005	10.000	+0.014	9.996	10.000	+0.005	9.991	10.000	0.000	9.983	10.000	−0.008	9.978	10.000	−0.013
	Min	9.990	9.991	−0.010	9.981	9.991	−0.019	9.976	9.991	−0.024	9.968	9.991	−0.032	9.963	9.991	−0.037
12	Max	12.006	12.000	+0.017	11.995	12.000	+0.006	11.989	12.000	0.000	11.979	12.000	−0.010	11.974	12.000	−0.015
	Min	11.988	11.989	−0.012	11.977	11.989	−0.023	11.971	11.989	−0.029	11.961	11.989	−0.039	11.956	11.989	−0.044
16	Max	16.006	16.000	+0.017	15.995	16.000	+0.006	15.989	16.000	0.000	15.979	16.000	−0.010	15.974	16.000	−0.015
	Min	15.988	15.989	−0.012	15.977	15.989	−0.023	15.971	15.989	−0.029	15.961	15.989	−0.039	15.956	15.989	−0.044
20	Max	20.006	20.000	+0.019	19.993	20.000	+0.006	19.986	20.000	−0.001	19.973	20.000	−0.014	19.967	20.000	−0.020
	Min	19.985	19.987	−0.015	19.972	19.987	−0.028	19.965	19.987	−0.035	19.952	19.987	−0.048	19.946	19.987	−0.054
25	Max	25.006	25.000	+0.019	24.993	25.000	+0.006	24.986	25.000	−0.001	24.973	25.000	−0.014	24.960	25.000	−0.027
	Min	24.985	24.987	−0.015	24.972	24.987	−0.028	24.965	24.987	−0.035	24.952	24.987	−0.048	24.939	24.987	−0.061

Table 16. *(Continued)* **American National Standard Preferred Shaft Basis Metric Transition and Interference Fits** *ANSI B4.2-1978 (R1994)*

Basic Size[a]		Locational Transition			Locational Transition			Locational Interference			Medium Drive			Force		
		Hole K7	Shaft h6	Fit[b]	Hole N7	Shaft h6	Fit[b]	Hole P7	Shaft h6	Fit[b]	Hole S7	Shaft h6	Fit[b]	Hole U7	Shaft h6	Fit[b]
30	Max	30.006	30.000	+0.019	29.993	30.000	+0.006	29.986	30.000	−0.001	29.973	30.000	−0.014	29.960	30.000	−0.027
	Min	29.985	29.987	−0.015	29.972	29.987	−0.028	29.965	29.987	−0.035	29.952	29.987	−0.048	29.939	29.987	−0.061
40	Max	40.007	40.000	+0.023	39.992	40.000	+0.008	39.983	40.000	−0.001	39.966	40.000	−0.018	39.949	40.000	−0.035
	Min	39.982	39.984	−0.018	39.967	39.984	−0.033	39.958	39.984	−0.042	39.941	39.984	−0.059	39.924	39.984	−0.076
50	Max	50.007	50.000	+0.023	49.992	50.000	+0.008	49.983	50.000	−0.001	49.966	50.000	−0.018	49.939	50.000	−0.045
	Min	49.982	49.984	−0.018	49.967	49.984	−0.033	49.958	49.984	−0.042	49.941	49.984	−0.059	49.914	49.984	−0.086
60	Max	60.009	60.000	+0.028	59.991	60.000	+0.010	59.979	60.000	−0.002	59.958	60.000	−0.023	59.924	60.000	−0.087
	Min	59.979	59.981	−0.021	59.961	59.981	−0.039	59.949	59.981	−0.051	59.928	59.981	−0.072	59.894	59.981	−0.106
80	Max	80.009	80.000	+0.028	79.991	80.000	+0.010	79.979	80.000	−0.002	79.952	80.000	−0.029	79.909	80.000	−0.072
	Min	79.979	79.981	−0.021	79.961	79.981	−0.039	79.949	79.981	−0.051	79.922	79.981	−0.078	79.879	79.981	−0.121
100	Max	100.010	100.000	+0.032	99.990	100.000	+0.012	99.976	100.000	−0.002	99.942	100.000	−0.036	99.889	100.000	−0.089
	Min	99.975	99.978	−0.025	99.955	99.978	−0.045	99.941	99.978	−0.059	99.907	99.978	−0.093	99.854	99.978	−0.146
120	Max	120.010	120.000	+0.032	119.990	120.000	+0.012	119.976	120.000	−0.002	119.934	120.000	−0.044	119.869	120.000	−0.109
	Min	119.975	119.978	−0.025	119.955	119.978	−0.045	119.941	119.978	−0.059	119.899	119.978	−0.101	119.834	119.978	−0.166
160	Max	160.012	160.000	+0.037	159.988	160.000	+0.013	159.972	160.000	−0.003	159.915	160.000	−0.060	159.825	160.000	−0.150
	Min	159.972	159.975	−0.028	159.948	159.975	−0.052	159.932	159.975	−0.068	159.875	159.975	−0.125	159.785	159.975	−0.215
200	Max	200.013	200.00	+0.042	199.986	200.000	+0.015	199.967	200.000	−0.004	199.895	200.000	−0.076	199.781	200.000	−0.190
	Min	199.967	199.971	−0.033	199.940	199.971	−0.060	199.921	199.971	−0.079	199.849	199.971	−0.151	199.735	199.971	−0.265
250	Max	250.013	250.000	+0.042	249.986	250.000	+0.015	249.967	250.000	−0.004	249.877	250.000	−0.094	249.733	250.000	−0.238
	Min	249.967	249.971	−0.033	249.940	249.971	−0.060	249.921	249.971	−0.079	249.831	249.971	−0.169	249.687	249.971	−0.313
300	Max	300.016	300.000	+0.048	299.986	300.000	+0.018	299.964	300.000	−0.004	299.850	300.000	−0.118	299.670	300.000	−0.298
	Min	299.964	299.968	−0.036	299.934	299.968	−0.066	299.912	299.968	−0.088	299.798	299.968	−0.202	299.618	299.968	−0.382
400	Max	400.017	400.000	+0.053	399.984	400.000	+0.020	399.959	400.000	−0.005	399.813	400.000	−0.151	399.586	400.000	−0.378
	Min	399.960	399.964	−0.040	399.927	399.964	−0.073	399.902	399.964	−0.098	399.756	399.964	−0.244	399.529	399.964	−0.471
500	Max	500.018	500.000	+0.058	499.983	500.000	+0.023	499.955	500.000	−0.005	499.771	500.000	−0.189	499.483	500.000	−0.477
	Min	499.955	499.960	−0.045	499.920	499.960	−0.080	499.892	499.960	−0.108	499.708	499.960	−0.292	499.420	499.960	−0.580

[a] The sizes shown are first-choice basic sizes (see Table 11). Preferred fits for other sizes can be calculated from data given in ANSI B4.2-1978 (R1984).

[b] A plus sign indicates clearance; a minus sign indicates interference.

All dimensions are in millimeters.

Table 17. American National Standard Gagemakers Tolerances
ANSI B4.4M-1981, R1987

Gagemakers Tolerance			Workpiece Tolerance	
	Class	ISO Symbol[a]	IT Grade	Recommended Gage Usage
Rejection of Good Parts Increase ↑	ZM	0.05 IT11	IT11	Low-precision gages recommended to be used to inspect workpieces held to internal (hole) tolerances C11 and H11 and to external (shaft) tolerances c11 and h11.
	YM	0.05 IT9	IT9	Gages recommended to be used to inspect workpieces held to internal (hole) tolerances D9 and H9 and to external (shaft) tolerances d9 and h9.
	XM	0.05 IT8	IT8	Precision gages recommended to be used to inspect workpieces held to internal (hole) tolerances F8 and H8.
Gage Cost Increase ↓	XXM	0.05 IT7	IT7	Recommended to be used for gages to inspect workpieces held to internal (hole) tolerances G7, H7, K7, N7, P7, S7, and U7, and to external (shaft) tolerances f7 and h7.
	XXX M	0.05 IT6	IT6	High-precision gages recommended to be used to inspect workpieces held to external (shaft) tolerances g6, h6, k6, n6, p6, s6, and u6.

[a] Gagemakers tolerance is equal to 5 per cent of workpiece tolerance or 5 per cent of applicable IT grade value. See table .

For workpiece tolerance class values, see previous Tables 13 through 16, incl.

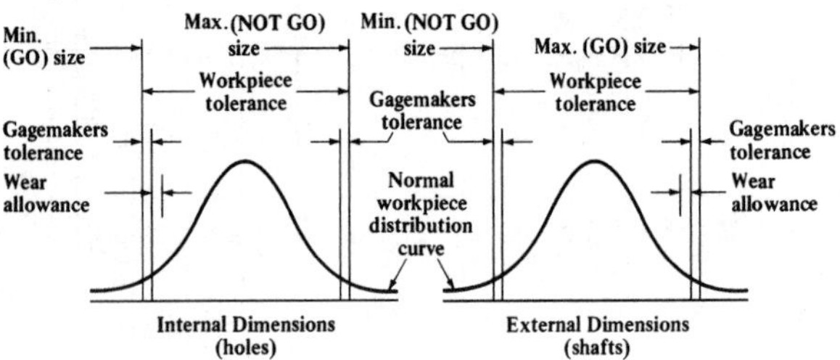

Fig. 4. Relationship between Gagemakers Tolerance, Wear Allowance and Workpiece Tolerance

Basic Size		Class ZM	Class YM	Class XM	Class XXM	Class XXXM
Over	To	(0.05 IT11)	(0.05 IT9)	(0.05 IT8)	(0.05 IT7)	(0.05 IT6)
0	3	0.0030	0.0012	0.0007	0.0005	0.0003
3	6	0.0037	0.0015	0.0009	0.0006	0.0004
6	10	0.0045	0.0018	0.0011	0.0007	0.0005
10	18	0.0055	0.0021	0.0013	0.0009	0.0006
18	30	0.0065	0.0026	0.0016	0.0010	0.0007
30	50	0.0080	0.0031	0.0019	0.0012	0.0008
50	80	0.0095	0.0037	0.0023	0.0015	0.0010
80	120	0.0110	0.0043	0.0027	0.0017	0.0011
120	180	0.0125	0.0050	0.0031	0.0020	0.0013
180	250	0.0145	0.0057	0.0036	0.0023	0.0015
250	315	0.0160	0.0065	0.0040	0.0026	0.0016
315	400	0.0180	0.0070	0.0044	0.0028	0.0018
400	500	0.0200	0.0077	0.0048	0.0031	0.0020

All dimensions are in millimeters. For closer gagemakers tolerance classes than Class XXXM, specify 5 per cent of IT5, IT4, or IT3 and use the designation 0.05 IT5, 0.05 IT4, etc.

Applications.—Many factors such as length of engagement, bearing load, speed, lubrication, operating temperatures, humidity, surface texture, and materials must be taken into account in fit selections for a particular application.

Choice of other than the preferred fits might be considered necessary to satisfy extreme conditions. Subsequent adjustments might also be desired as the result of experience in a particular application to suit critical functional requirements or to permit optimum manufacturing economy. Selection of a departure from these recommendations will depend upon consideration of the engineering and economic factors that might be involved; however, the benefits to be derived from the use of preferred fits should not be overlooked.

A general guide to machining processes that may normally be expected to produce work within the tolerances indicated by the IT grades given in ANSI B4.2-1978 (R1994) is shown in the chart in Table 18.

Table 18. Relation of Machining Processes to IT Tolerance Grades

	IT Grades							
	4	5	6	7	8	9	10	11
Lapping & Honing	■	■						
Cylindrical Grinding		■	■	■				
Surface Grinding		■	■	■	■			
Diamond Turning		■	■	■				
Diamond Boring		■	■	■				
Broaching		■	■	■	■			
Powder Metal sizes			■	■	■			
Reaming			■	■	■	■	■	
Turning				■	■	■	■	■
Powder Metal sintered				■	■	■	■	
Boring				■	■	■	■	■
Milling							■	■
Planing & Shaping							■	■
Drilling							■	■
Punching							■	■
Die Casting								■

PINS

Straight Pins

American National Standard Straight Pins.—The diameter of both chamfered and square end straight pins is that of the commercial wire or rod from which the pins are made. The tolerances shown in Table 1 are applicable to carbon steel and some deviations in the diameter limits may be necessary for pins made from other materials.

Table 1. American National Standard Chamfered and Square End Straight Pins *ANSI/ASME B18.8.2-1995*

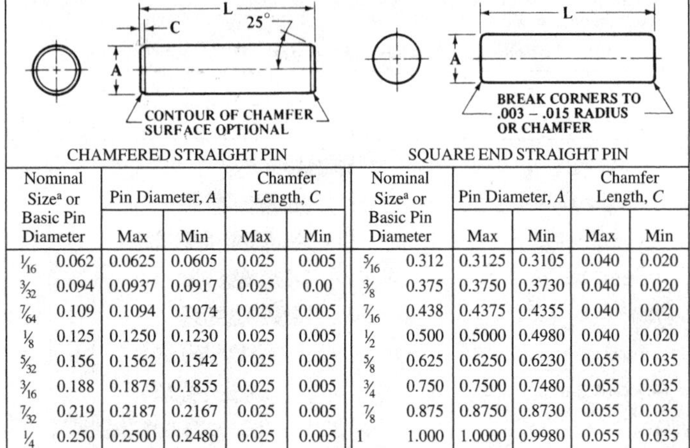

Nominal Size[a] or Basic Pin Diameter	Pin Diameter, A		Chamfer Length, C		Nominal Size[a] or Basic Pin Diameter	Pin Diameter, A		Chamfer Length, C	
	Max	Min	Max	Min		Max	Min	Max	Min
1/16 0.062	0.0625	0.0605	0.025	0.005	5/16 0.312	0.3125	0.3105	0.040	0.020
3/32 0.094	0.0937	0.0917	0.025	0.00	3/8 0.375	0.3750	0.3730	0.040	0.020
7/64 0.109	0.1094	0.1074	0.025	0.005	7/16 0.438	0.4375	0.4355	0.040	0.020
1/8 0.125	0.1250	0.1230	0.025	0.005	1/2 0.500	0.5000	0.4980	0.040	0.020
5/32 0.156	0.1562	0.1542	0.025	0.005	5/8 0.625	0.6250	0.6230	0.055	0.035
3/16 0.188	0.1875	0.1855	0.025	0.005	3/4 0.750	0.7500	0.7480	0.055	0.035
7/32 0.219	0.2187	0.2167	0.025	0.005	7/8 0.875	0.8750	0.8730	0.055	0.035
1/4 0.250	0.2500	0.2480	0.025	0.005	1 1.000	1.0000	0.9980	0.055	0.035

[a] Where specifying nominal size in decimals, zeros preceding decimal point are omitted.

All dimensions are in inches.

Length Increments: Lengths are as specified by the purchaser; however, it is recommended that nominal pin lengths be limited to increments of not less than 0.062 inch.

Material: Straight pins are normally made from cold drawn steel wire or rod having a maximum carbon content of 0.28 per cent. Where required, pins may also be made from corrosion resistant steel, brass, or other metals.

Designation: Straight pins are designated by the following data, in the sequence shown: Product name (noun first), nominal size (fraction or decimal equivalent), material, and protective finish, if required.

Examples: Pin, Chamfered Straight, 1/8 × 1.500, Steel

Pin, Square End Straight, 0.250 × 2.250, Steel, Zinc Plated

American National Standard Taper Pins.—Taper pins have a uniform taper over the pin length with both ends crowned. Most sizes are supplied in commercial and precision classes, the latter having generally tighter tolerances and being more closely controlled in manufacture.

Diameters: The major diameter of both commercial and precision classes of pins is the diameter of the large end and is the basis for pin size. The diameter at the small end is computed by multiplying the nominal length of the pin by the factor 0.02083 and subtracting the result from the basic pin diameter. See also Table 2.

Taper: The taper on commercial class pins is 0.250 ± 0.006 inch per foot and on the precision class pins is 0.250 ± 0.004 inch per foot of length.

Materials: Unless otherwise specified, taper pins are made from SAE 1211 steel or cold drawn SAE 1212 or 1213 steel or equivalents, and no mechanical property requirements apply.

Designation: Taper pins are designated by the following data in the sequence shown: Product name (noun first), class, size number (or decimal equivalent), length (fraction or three-place decimal equivalent), material, and protective finish, if required.

Examples: Pin, Taper (Commercial Class) No. 0 × ¾, Steel
Pin, Taper (Precision Class) 0.219 × 1.750, Steel, Zinc Plated

Table 2. Nominal Diameter at Small Ends of Standard Taper Pins

Pin Length in inches	Pin Number and Small End Diameter for Given Length										
	0	1	2	3	4	5	6	7	8	9	10
¾	0.140	0.156	0.177	0.203	0.235	0.273	0.325	0.393	0.476	0.575	0.690
1	0.135	0.151	0.172	0.198	0.230	0.268	0.320	0.388	0.471	0.570	0.685
1¼	0.130	0.146	0.167	0.192	0.224	0.263	0.315	0.382	0.466	0.565	0.680
1½	0.125	0.141	0.162	0.187	0.219	0.258	0.310	0.377	0.460	0.560	0.675
1¾	0.120	0.136	0.157	0.182	0.214	0.252	0.305	0.372	0.455	0.554	0.669
2	0.114	0.130	0.151	0.177	0.209	0.247	0.299	0.367	0.450	0.549	0.664
2¼	0.109	0.125	0.146	0.172	0.204	0.242	0.294	0.362	0.445	0.544	0.659
2½	0.104	0.120	0.141	0.166	0.198	0.237	0.289	0.356	0.440	0.539	0.654
2¾	0.099	0.115	0.136	0.161	0.193	0.232	0.284	0.351	0.434	0.534	0.649
3	0.094	0.110	0.131	0.156	0.188	0.227	0.279	0.346	0.429	0.528	0.643
3¼	…	…	…	0.151	0.182	0.221	0.273	0.340	0.424	0.523	0.638
3½	…	…	…	0.146	0.177	0.216	0.268	0.335	0.419	0.518	0.633
3¾	…	…	…	0.141	0.172	0.211	0.263	0.330	0.414	0.513	0.628
4	…	…	…	0.136	0.167	0.206	0.258	0.326	0.409	0.508	0.623
4¼	…	…	…	0.131	0.162	0.201	0.253	0.321	0.403	0.502	0.617
4½	…	…	…	0.125	0.156	0.195	0.247	0.315	0.398	0.497	0.612
5	…	…	…	…	0.146	0.185	0.237	0.305	0.389	0.487	0.602
5½	…	…	…	…	…	…	…	0.294	0.377	0.476	0.591
6	…	…	…	…	…	…	…	0.284	0.367	0.466	0.581

Drilling Specifications for Taper Pins.—When helically fluted taper pin reamers are used, the diameter of the through hole drilled prior to reaming is equal to the diameter at the small end of the taper pin. (See Table 2.) However, when straight fluted taper reamers are to be used, it may be necessary, for long pins, to step drill the hole before reaming, the number and sizes of the drills to be used depending on the depth of the hole (pin length).

To determine the number and sizes of step drills required: Find the length of pin to be used at the top of the chart on page 241 and follow this length down to the intersection with that heavy line which represents the size of taper pin (see taper pin numbers at the right-hand end of each heavy line). If the length of pin falls between the first and second dots, counting from the left, only one drill is required. Its size is indicated by following the nearest horizontal line from the point of intersection (of the pin length) on the heavy line over to the drill diameter values at the left. If the intersection of pin length comes between the second and third dots, then two drills are required. The size of the smaller drill then corresponds to the intersection of the pin length and the heavy line and the larger is the corresponding drill diameter for the intersection of one-half this length with the heavy line. Should the pin length fall between the third and fourth dots, three drills are required. The smallest drill will have a diameter corresponding to the intersection of the total pin length with the heavy line, the next in size will have a diameter corresponding to the intersection of two-thirds of this length with the heavy line and the largest will have a diameter corresponding to the intersection of one-third of this length with the heavy line. Where the intersection falls between two drill sizes, use the smaller.

Table 3. American National Standard Taper Pins *ANSI/ASME B18.8.2-1995*

Pin Size Number and Basic Pin Dia.[a]	Major Diameter (Large End), *A*				End Crown Radius, *R*		Range of Lengths,[b] *L*	
	Commercial Class		Precision Class					
	Max	Min	Max	Min	Max	Min	Stand. Reamer Avail.[c]	Other
7/0 0.0625	0.0638	0.0618	0.0635	0.0625	0.072	0.052	…	1/4–1
6/0 0.0780	0.0793	0.0773	0.0790	0.0780	0.088	0.068	…	1/4–1/2
5/0 0.0940	0.0953	0.0933	0.0950	0.0940	0.104	0.084	1/4–1	1 1/4, 1 1/2
4/0 0.1090	0.1103	0.1083	0.1100	0.1090	0.119	0.099	1/4–1	1 1/4–2
3/0 0.1250	0.1263	0.1243	0.1260	0.1250	0.135	0.115	1/4–1	1 1/4–2
2/0 0.1410	0.1423	0.1403	0.1420	0.1410	0.151	0.131	1/2–1 1/4	1 1/2–2 1/2
0 0.1560	0.1573	0.1553	0.1570	0.1560	0.166	0.146	1/2–1 1/4	1 1/2–3
1 0.1720	0.1733	0.1713	0.1730	0.1720	0.182	0.162	3/4–1 1/4	1 1/2–3
2 0.1930	0.1943	0.1923	0.1940	0.1930	0.203	0.183	3/4–1 1/2	1 3/4–3
3 0.2190	0.2203	0.2183	0.2200	0.2190	0.229	0.209	3/4–1 3/4	2–4
4 0.2500	0.2513	0.2493	0.2510	0.2500	0.260	0.240	3/4–2	2 1/4–4
5 0.2890	0.2903	0.2883	0.2900	0.2890	0.299	0.279	1–2 1/2	2 3/4–6
6 0.3410	0.3423	0.3403	0.3420	0.3410	0.351	0.331	1 1/4–3	3 1/4–6
7 0.4090	0.4103	0.4083	0.4100	0.4090	0.419	0.399	1 1/4–3 3/4	4–8
8 0.4920	0.4933	0.4913	0.4930	0.4920	0.502	0.482	1 1/4–4 1/2	4 3/4–8
9 0.5910	0.5923	0.5903	0.5920	0.5910	0.601	0.581	1 1/4–5 1/4	5 1/2–8
10 0.7060	0.7073	0.7053	0.7070	0.7060	0.716	0.696	1 1/2–6	6 1/4–8
11 0.8600	0.8613	0.8593	…	…	0.870	0.850	…	2–8
12 1.0320	1.0333	1.0313	…	…	1.042	1.022	…	2–9
13 1.2410	1.2423	1.2403	…	…	1.251	1.231	…	3–11
14 1.5210	1.5223	1.5203	…	…	1.531	1.511	…	3–13

[a] When specifying nominal pin size in decimals, zeros preceding the decimal and in the fourth decimal place are omitted.

[b] Lengths increase in 1/8-inch steps up to 1 inch and in 1/4-inch steps above 1 inch.

[c] Standard reamers are available for pin lengths in this column.

All dimensions are in inches.

For nominal diameters, B, see Table 2.

Chart to Facilitate Selection of Number and Sizes of Drills for Step-Drilling Prior to Taper Reaming

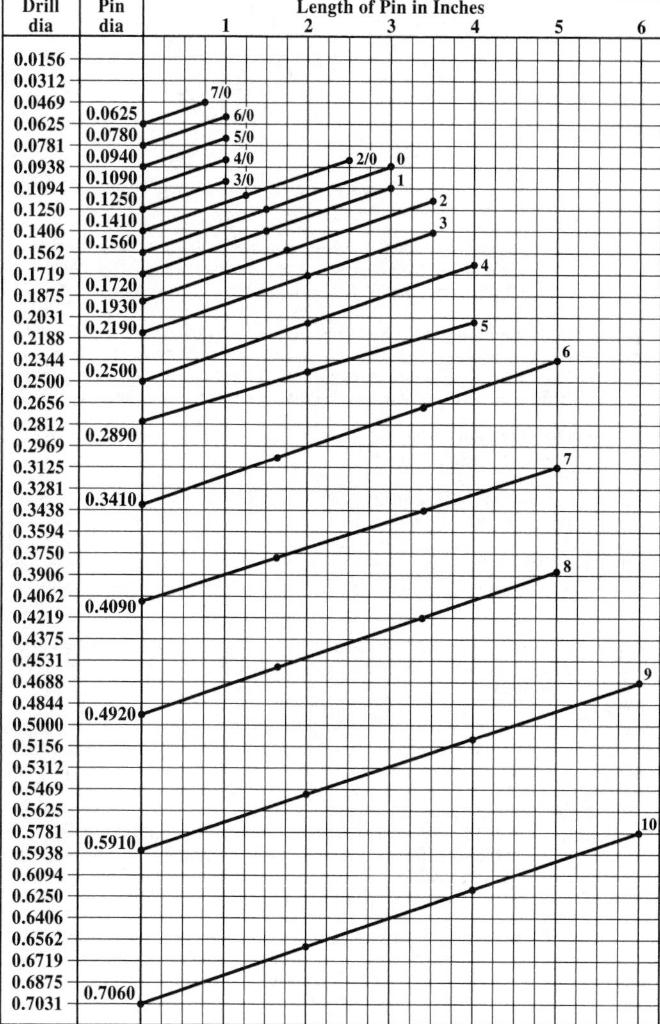

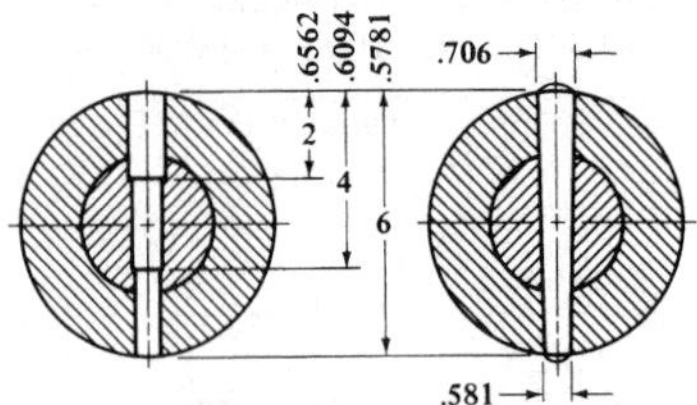

Examples: For a No. 10 taper pin 6-inches long, three drills would be used, of the sizes and for the depths shown in the accompanying diagram.

For a No. 10 taper pin 3-inches long, two drills would be used because the 3-inch length falls between the second and third dots. The first or through drill will be 0.6406 inch and the second drill, 0.6719 inch for a depth of 1½ inches.

American National Standard Grooved Pins.—These pins have three equally spaced longitudinal grooves and an expanded diameter over the crests of the ridges formed by the material displaced when the grooves are produced. The grooves are aligned with the axes of the pins. There are seven types of grooved pins as shown in the illustration on page 243.

Standard Sizes and Lengths: The standard sizes and lengths in which grooved pins are normally available are given in Table 4.

Materials: Grooved pins are normally made from cold drawn low carbon steel wire or rod. Where additional performance is required, carbon steel pins may be supplied surface hardened and heat treated to a hardness consistent with the performance requirements. Pins may also be made from alloy steel, corrosion resistant steel, brass, Monel and other non-ferrous metals having chemical properties as agreed upon between manufacturer and purchaser.

Performance Requirements: Grooved pins are required to withstand the minimum double shear loads given in Table 4 for the respective materials shown, when tested in accordance with the Double Shear Testing of Pins as set forth in ANSI/ASME B18.8.2-1995, Appendix B.

Hole Sizes: To obtain maximum product retention under average conditions, it is recommended that holes for the installation of grooved pins be held as close as possible to the limits shown in Table 4. The minimum limits correspond to the drill size, which is the same as the basic pin diameter. The maximum limits are generally suitable for length-diameter ratios of not less than 4 to 1 nor greater than 10 to 1. For smaller length-to-diameter ratios, the hole should be held closer to the minimum limits where retention is critical. Conversely for larger ratios where retention requirements are less important, it may be desirable to increase the hole diameters beyond the maximum limits shown.

Designation: Grooved pins are designated by the following data in the sequence shown: Product name (noun first) including type designation, nominal size (number, fraction or decimal equivalent), length (fraction or decimal equivalent), material, including specification or heat treatment where necessary, protective finish, if required.

Examples: Pin, Type A Grooved, $\frac{3}{32} \times \frac{3}{4}$, Steel, Zinc Plated

Pin, Type F Grooved, 0.250 × 1.500, Corrosion Resistant Steel

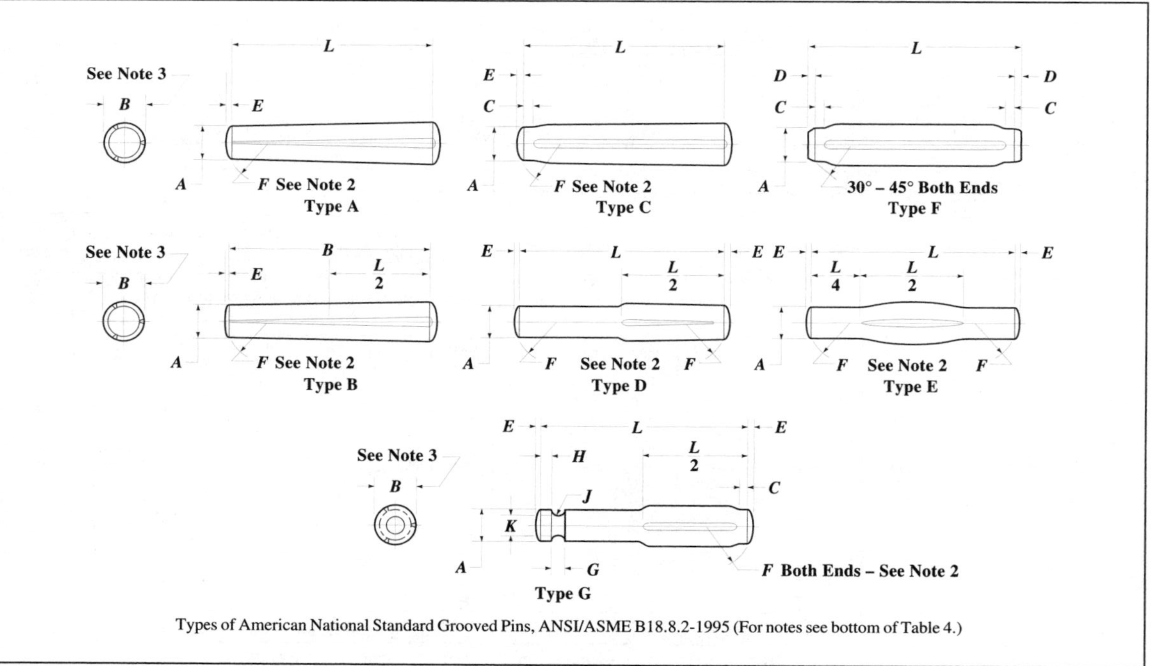

Types of American National Standard Grooved Pins, ANSI/ASME B18.8.2-1995 (For notes see bottom of Table 4.)

Table 4. American National Standard Grooved Pins *ANSI/ASME B18.8.2-1995*

Nominal Size or Basic Pin Diameter		Pin Diameter,[a] *A*		Pilot Length, *C*	Chamfer Length,[b] *D*	Crown Height,[b] *E*		Crown Radius,[b] *F*		Neck Width, *G*		Shoulder Length, *H*		Neck Radius, *J*	Neck Diameter, *K*		Range of Standard Lengths[c]
		Max	Min	Ref	Min	Max	Min	Max	Min	Max	Min	Max	Min	Ref	Max	Min	
1/32[d]	0.0312	0.0312	0.0302	0.015	…	…	…	…	…	…	…	…	…	…	…	…	1/8–1/2
3/64[d]	0.0469	0.0469	0.0459	0.031	…	…	…	…	…	…	…	…	…	…	…	…	1/8–5/8
1/16	0.0625	0.0625	0.0615	0.031	0.016	0.0115	0.0015	0.088	0.068	…	…	…	…	…	…	…	1/8–1
5/64[d]	0.0781	0.0781	0.0771	0.031	0.016	0.0137	0.0037	0.104	0.084	…	…	…	…	…	…	…	1/4–1
3/32	0.0938	0.0938	0.0928	0.031	0.016	0.0141	0.0041	0.135	0.115	0.038	0.028	0.041	0.031	0.016	0.067	0.057	1/4–1 1/4
7/64[d]	0.1094	0.1094	0.1074	0.031	0.016	0.0160	0.0060	0.150	0.130	0.038	0.028	0.041	0.031	0.016	0.082	0.072	1/4–1 1/4
1/8	0.1250	0.1250	0.1230	0.031	0.016	0.0180	0.0080	0.166	0.146	0.069	0.059	0.041	0.031	0.031	0.088	0.078	1/4–1 1/2
5/32	0.1563	0.1563	0.1543	0.062	0.031	0.0220	0.0120	0.198	0.178	0.069	0.059	0.057	0.047	0.031	0.109	0.099	3/8–2
3/16	0.1875	0.1875	0.1855	0.062	0.031	0.0230	0.0130	0.260	0.240	0.069	0.059	0.057	0.047	0.031	0.130	0.120	3/8–2 1/4
7/32	0.2188	0.2188	0.2168	0.062	0.031	0.0270	0.0170	0.291	0.271	0.101	0.091	0.072	0.062	0.047	0.151	0.141	1/2–3
1/4	0.2500	0.2500	0.2480	0.062	0.031	0.0310	0.0210	0.322	0.302	0.101	0.091	0.072	0.062	0.047	0.172	0.162	1/2–3 1/4
5/16	0.3125	0.3125	0.3105	0.094	0.047	0.0390	0.0290	0.385	0.365	0.132	0.122	0.104	0.094	0.062	0.214	0.204	5/8–3 1/2
3/8	0.3750	0.3750	0.3730	0.094	0.047	0.0440	0.0340	0.479	0.459	0.132	0.122	0.135	0.125	0.062	0.255	0.245	3/4–4 1/4
7/16	0.4375	0.4375	0.4355	0.094	0.047	0.0520	0.0420	0.541	0.521	0.195	0.185	0.135	0.125	0.094	0.298	0.288	7/8–4 1/2
1/2	0.5000	0.5000	0.4980	0.094	0.047	0.0570	0.0470	0.635	0.615	0.195	0.185	0.135	0.125	0.094	0.317	0.307	1–4 1/2

[a] For expanded diameters, *B*, see ANSI/ASME B18.8.2-1995.

[b] Pins in 1/32- and 3/64-inch sizes of any length and all sizes of 1/4-inch nominal length or shorter are not crowned or chamfered.

[c] Standard lengths increase in 1/8-inch steps from 1/8 to 1 inch, and in 1/4-inch steps above 1 inch. Standard lengths for the 1/32-, 3/64-, 1/16-, and 5/64-inch sizes and the 1/4-inch length for the 3/32-, 7/64-, and 1/8-inch sizes do not apply to Type G grooved pins.

[d] Non-stock items, not recommended for new designs.

Pin Material	Nominal Pin Size														
	1/32	3/64	1/16	5/64	3/32	7/64	1/8	5/32	3/16	7/32	1/4	5/16	3/8	7/16	1/2
Steels	Double Shear Load, Min, lb														
Low Carbon	100	220	410	620	890	1,220	1,600	2,300	3,310	4,510	5,880	7,660	11,000	15,000	19,600
Alloy (R_c 40 – 48 hardness)	180	400	720	1,120	1,600	2,180	2,820	4,520	6,440	8,770	11,500	17,900	26,000	35,200	46,000
Corrosion Resistant	140	300	540	860	1,240	1,680	2,200	3,310	4,760	6,480	8,460	12,700	18,200	24,800	32,400
Brass	60	140	250	390	560	760	990	1,540	2,220	3,020	3,950	6,170	9,050	12,100	15,800
	Recommended Hole Sizes for Unplated Pins (The minimum drill size is the same as the pin size. See also text on page 242.)														
Maximum Diameter	0.0324	0.0482	0.0640	0.0798	0.0956	0.1113	0.1271	0.1587	0.1903	0.2219	0.2534	0.3166	0.3797	0.4428	0.5060
Minimum Diameter	0.0312	0.0469	0.0625	0.0781	0.0938	0.1094	0.1250	0.1563	0.1875	0.2188	0.2500	0.3125	0.3750	0.4375	0.5000

All dimensions are in inches.

STANDARD TAPERS

Certain types of small tools and machine parts, such as twist drills, end mills, arbors, lathe centers, etc., are provided with taper shanks which fit into spindles or sockets of corresponding taper, thus providing not only accurate alignment between the tool or other part and its supporting member, but also more or less frictional resistance for driving the tool. There are several standards for "self-holding" tapers, but the American National, Morse, and the Brown & Sharpe are the standards most widely used by American manufacturers.

The name *self-holding* has been applied to the smaller tapers—like the Morse and the Brown & Sharpe—because, where the angle of the taper is only 2 or 3 degrees, the shank of a tool is so firmly seated in its socket that there is considerable frictional resistance to any force tending to turn or rotate the tool relative to the socket. The term "self-holding" is used to distinguish relatively small tapers from the larger or *self-releasing* type. A milling machine spindle having a taper of 3½ inches per foot is an example of a self-releasing taper. The included angle in this case is over 16 degrees and the tool or arbor requires a positive locking device to prevent slipping, but the shank may be released or removed more readily than one having a smaller taper of the self-holding type.

Morse Taper.—Dimensions relating to Morse standard taper shanks and sockets may be found in an accompanying table. The taper for different numbers of Morse tapers is slightly different, but it is approximately ⅝ inch per foot in most cases. The table gives the actual tapers, accurate to five decimal places. Morse taper shanks are used on a variety of tools, and exclusively on the shanks of twist drills. Dimensions for Morse Stub Taper Shanks are given in Table 1a.

Brown & Sharpe Taper.—This standard taper is used for taper shanks on tools such as end mills and reamers, the taper being approximately ½ inch per foot for all sizes except for taper No. 10, where the taper is 0.5161 inch per foot. Brown & Sharpe taper sockets are used for many arbors, collets, and machine tool spindles, especially milling machines and grinding machines. In many cases there are a number of different lengths of sockets corresponding to the same number of taper; all these tapers, however, are of the same diameter at the small end.

Jarno Taper.—The Jarno taper was originally proposed by Oscar J. Beale of the Brown & Sharpe Mfg. Co. This taper is based on such simple formulas that practically no calculations are required when the number of taper is known. The taper per foot of all Jarno taper sizes is 0.600 inch on the diameter. The diameter at the large end is as many eighths, the diameter at the small end is as many tenths, and the length as many half inches as are indicated by the number of the taper. For example, a No. 7 Jarno taper is ⅞ inch in diameter at the large end; $^7/_{10}$, or 0.700 inch at the small end; and $^7/_2$, or 3½ inches long; hence, diameter at large end = No. of taper ÷ 8; diameter at small end = No. of taper ÷ 10; length of taper = No. of taper ÷ 2. The Jarno taper is used on various machine tools, especially profiling machines and die-sinking machines. It has also been used for the headstock and tailstock spindles of some lathes.

American National Standard Machine Tapers: This standard includes a self-holding series (Table 8 and 6) and a steep taper series, Table 7. The self-holding taper series consists of 22 sizes which are listed in Table 8. The reference gage for the self-holding tapers is a plug gage. Table 2 gives the dimensions and tolerances for both plug and ring gages applying to this series. Tables 6 give the dimensions for self-holding taper shanks and sockets which are classified as to (1) means of transmitting torque from spindle to the tool shank, and (2) means of retaining the shank in the socket. The steep machine tapers consist of a preferred series (bold-face type, Table 7) and an intermediate series (light-face type). A self-holding taper is defined as "a taper with an angle small enough to hold a shank in place ordinarily by friction without holding means. (Sometimes referred to as slow taper.)"

A steep taper is defined as "a taper having an angle sufficiently large to insure the easy or self-releasing feature." The term "gage line" indicates the basic diameter at or near the large end of the taper.

Table 1a. Morse Stub Taper Shanks

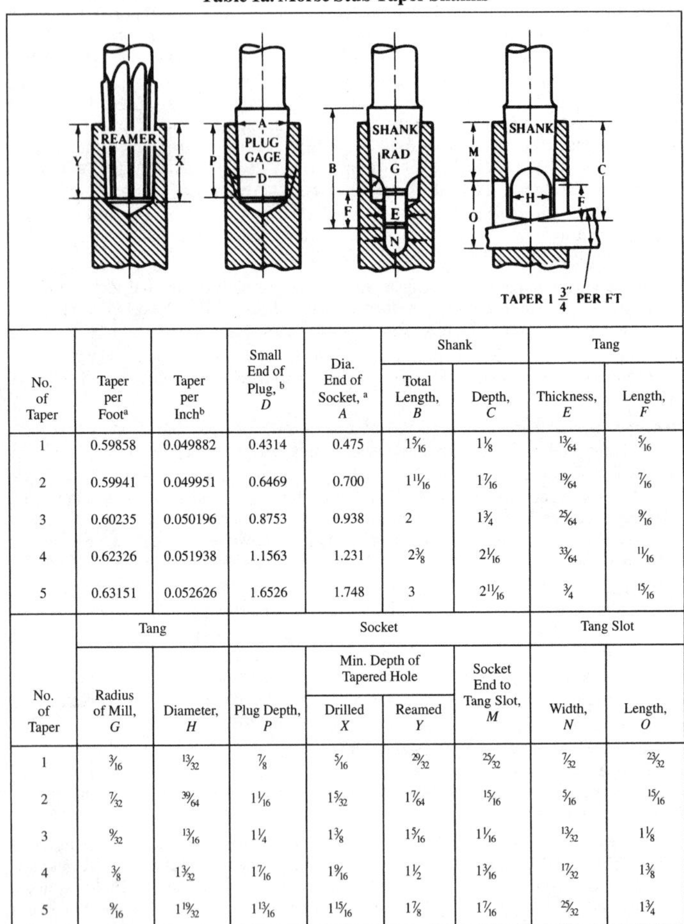

No. of Taper	Taper per Foot[a]	Taper per Inch[b]	Small End of Plug,[b] *D*	Dia. End of Socket,[a] *A*	Shank: Total Length, *B*	Shank: Depth, *C*	Tang: Thickness, *E*	Tang: Length, *F*
1	0.59858	0.049882	0.4314	0.475	1 5/16	1 1/8	13/64	5/16
2	0.59941	0.049951	0.6469	0.700	1 11/16	1 7/16	19/64	7/16
3	0.60235	0.050196	0.8753	0.938	2	1 3/4	25/64	9/16
4	0.62326	0.051938	1.1563	1.231	2 3/8	2 1/16	33/64	11/16
5	0.63151	0.052626	1.6526	1.748	3	2 11/16	3/4	15/16

No. of Taper	Tang: Radius of Mill, *G*	Tang: Diameter, *H*	Socket: Plug Depth, *P*	Socket: Min. Depth of Tapered Hole: Drilled *X*	Socket: Min. Depth of Tapered Hole: Reamed *Y*	Socket: Socket End to Tang Slot, *M*	Tang Slot: Width, *N*	Tang Slot: Length, *O*
1	3/16	13/32	7/8	5/16	29/32	25/32	7/32	23/32
2	7/32	39/64	1 1/16	1 5/32	1 7/64	15/16	5/16	15/16
3	9/32	13/16	1 1/4	1 3/8	1 5/16	1 1/16	13/32	1 1/8
4	3/8	1 3/32	1 7/16	1 9/16	1 1/2	1 3/16	17/32	1 3/8
5	9/16	1 19/32	1 13/16	1 15/16	1 7/8	1 7/16	25/32	1 3/4

[a] These are basic dimensions.

[b] These dimensions are calculated for reference only.

All dimensions in inches.

Radius *J* is 3/64, 1/16, 5/64, 3/32, and 1/8 inch respectively for Nos. 1, 2, 3, 4, and 5 tapers.

Table 1b. Morse Standard Taper Shanks

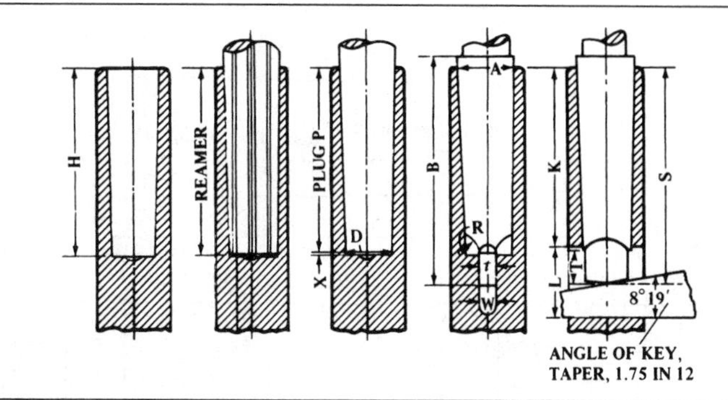

No. of Taper	Taper per Foot	Taper per Inch	Small End of Plug D	Diameter End of Socket A	Shank		Depth of Hole H
					Length B	Depth S	
0	0.62460	0.05205	0.252	0.3561	2 11/32	2 7/32	2 1/32
1	0.59858	0.04988	0.369	0.475	2 9/16	2 7/16	2 5/32
2	0.59941	0.04995	0.572	0.700	3 1/8	2 15/16	2 39/64
3	0.60235	0.05019	0.778	0.938	3 7/8	3 11/16	3 1/4
4	0.62326	0.05193	1.020	1.231	4 7/8	4 5/8	4 1/8
5	0.63151	0.05262	1.475	1.748	6 1/8	5 7/8	5 1/4
6	0.62565	0.05213	2.116	2.494	8 9/16	8 1/4	7 21/64
7	0.62400	0.05200	2.750	3.270	11 5/8	11 1/4	10 5/64

Plug Depth P	Tang or Tongue				Keyway		Keyway to End K
	Thickness t	Length T	Radius R	Dia.	Width W	Length L	
2	0.1562	1/4	5/32	0.235	11/64	9/16	1 15/16
2 1/8	0.2031	3/8	3/16	0.343	0.218	3/4	2 1/16
2 9/16	0.2500	7/16	1/4	17/32	0.266	7/8	2 1/2
3 3/16	0.3125	9/16	9/32	23/32	0.328	1 3/16	3 1/16
4 1/16	0.4687	5/8	5/16	31/32	0.484	1 1/4	3 7/8
5 3/16	0.6250	3/4	3/8	1 13/32	0.656	1 1/2	4 15/16
7 1/4	0.7500	1 1/8	1/2	2	0.781	1 3/4	7
10	1.1250	1 3/8	3/4	2 5/8	1.156	2 5/8	9 1/2

Table 2. Dimensions of Morse Taper Sleeves

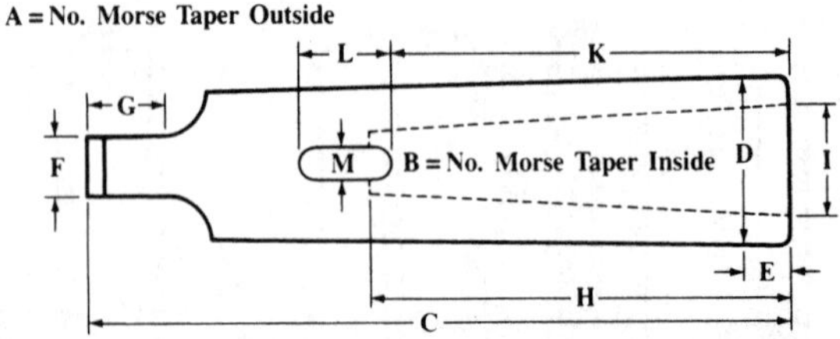

A = No. Morse Taper Outside

B = No. Morse Taper Inside

A	*B*	*C*	*D*	*E*	*F*	*G*	*H*	*I*	*K*	*L*	*M*
2	1	3 9/16	0.700	5/8	1/4	7/16	2 3/16	0.475	2 1/16	3/4	0.213
3	1	3 15/16	0.938	1/4	5/16	9/16	2 3/16	0.475	2 1/16	3/4	0.213
3	2	4 7/16	0.938	3/4	5/16	9/16	2 5/8	0.700	2 1/2	7/8	0.260
4	1	4 7/8	1.231	1/4	15/32	5/8	2 3/16	0.475	2 1/16	3/4	0.213
4	2	4 7/8	1.231	1/4	15/32	5/8	2 5/8	0.700	2 1/2	7/8	0.260
4	3	5 3/8	1.231	3/4	15/32	5/8	3 1/4	0.938	3 1/16	1 3/16	0.322
5	1	6 1/8	1.748	1/4	5/8	3/4	2 3/16	0.475	2 1/16	3/4	0.213
5	2	6 1/8	1.748	1/4	5/8	3/4	2 5/8	0.700	2 1/2	7/8	0.260
5	3	6 1/8	1.748	1/4	5/8	3/4	3 1/4	0.938	3 1/16	1 3/16	0.322
5	4	6 5/8	1.748	3/4	5/8	3/4	4 1/8	1.231	3 7/8	1 1/4	0.478
6	1	8 5/8	2.494	3/8	3/4	1 1/8	2 3/16	0.475	2 1/16	3/4	0.213
6	2	8 5/8	2.494	3/8	3/4	1 1/8	2 5/8	0.700	2 1/2	7/8	0.260
6	3	8 5/8	2.494	3/8	3/4	1 1/8	3 1/4	0.938	3 1/16	1 3/16	0.322
6	4	8 5/8	2.494	3/8	3/4	1 1/8	4 1/8	1.231	3 7/8	1 1/4	0.478
6	5	8 5/8	2.494	3/8	3/4	1 1/8	5 1/4	1.748	4 15/16	1 1/2	0.635
7	3	11 5/8	3.270	3/8	1 1/8	1 3/8	3 1/4	0.938	3 1/16	1 3/16	0.322
7	4	11 5/8	3.270	3/8	1 1/8	1 3/8	4 1/8	1.231	3 7/8	1 1/4	0.478
7	5	11 5/8	3.270	3/8	1 1/8	1 3/8	5 1/4	1.748	4 15/16	1 1/2	0.635
7	6	12 1/2	3.270	1 1/4	1 1/8	1 3/8	7 3/8	2.494	7	1 3/4	0.760

Table 3. Morse Taper Sockets — Hole and Shank Sizes

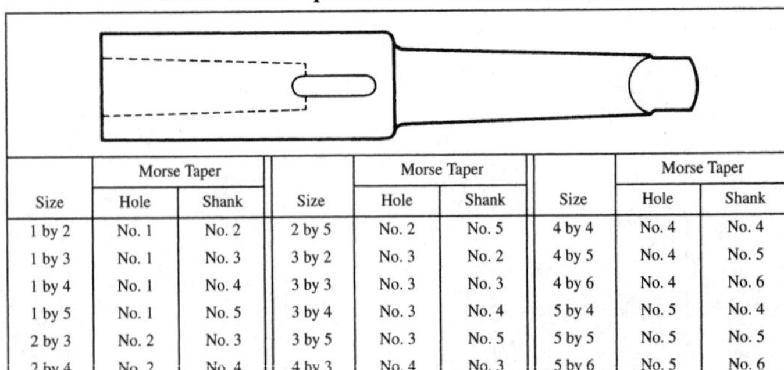

Size	Morse Taper		Size	Morse Taper		Size	Morse Taper	
	Hole	Shank		Hole	Shank		Hole	Shank
1 by 2	No. 1	No. 2	2 by 5	No. 2	No. 5	4 by 4	No. 4	No. 4
1 by 3	No. 1	No. 3	3 by 2	No. 3	No. 2	4 by 5	No. 4	No. 5
1 by 4	No. 1	No. 4	3 by 3	No. 3	No. 3	4 by 6	No. 4	No. 6
1 by 5	No. 1	No. 5	3 by 4	No. 3	No. 4	5 by 4	No. 5	No. 4
2 by 3	No. 2	No. 3	3 by 5	No. 3	No. 5	5 by 5	No. 5	No. 5
2 by 4	No. 2	No. 4	4 by 3	No. 4	No. 3	5 by 6	No. 5	No. 6

Table 4. Brown & Sharpe Taper Shanks

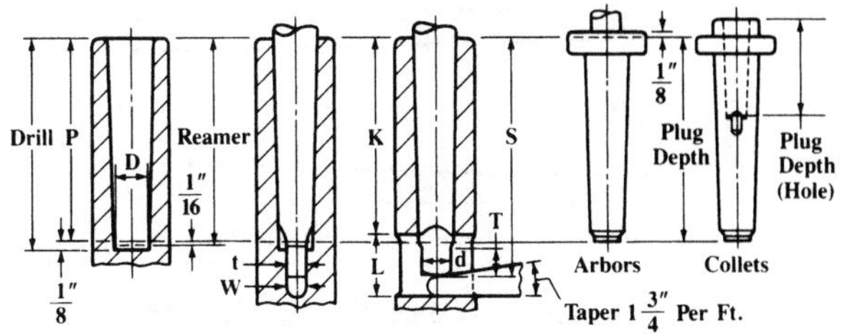

Number of Taper	Taper per Foot (inch)	Dia. of Plug at Small End	Plug Depth, P			Keyway from End of Spindle	Shank Depth	Length of Keyway[a]	Width of Keyway	Length of Arbor Tongue	Diameter of Arbor Tongue	Thickness of Arbor Tongue
		D	B & S[b] Standard	Mill. Mach. Standard	Miscell.	K	S	L	W	T	d	t
1[c]	.50200	.20000	15/16	…	…	15/16	1 3/16	3/8	.135	3/16	.170	1/8
2[c]	.50200	.25000	1 3/16	…	…	1 11/64	1 1/2	1/2	.166	1/4	.220	5/32
3[c]	.50200	.31250	1 1/2	…	…	1 15/32	1 7/8	5/8	.197	5/16	.282	3/16
			…	…	1 3/4	1 23/32	2 1/8	5/8	.197	5/16	.282	3/16
			…	…	2	1 31/32	2 3/8	5/8	.197	5/16	.282	3/16
4	.50240	.35000	…	1 1/4	…	1 13/64	1 21/32	11/16	.228	11/32	.320	7/32
			1 11/16	…	…	1 41/64	2 3/32	11/16	.228	11/32	.320	7/32
5	.50160	.45000	…	1 3/4	…	1 11/16	2 3/16	3/4	.260	3/8	.420	1/4
			…	…	2	1 15/16	2 7/16	3/4	.260	3/8	.420	1/4
			2 1/8	…	…	2 1/16	2 9/16	3/4	.260	3/8	.420	1/4
6	.50329	.50000	2 3/8	…	…	2 19/64	2 7/8	7/8	.291	7/16	.460	9/32
7	.50147	.60000	…	…	2 1/2	2 13/32	3 1/32	15/16	.322	15/32	.560	5/16
			2 7/8	…	…	2 25/32	3 13/32	15/16	.322	15/32	.560	5/16
			…	3	…	2 29/32	3 17/32	15/16	.322	15/32	.560	5/16
8	.50100	.75000	3 9/16	…	…	3 29/64	4 1/8	1	.353	1/2	.710	11/32
9	.50085	.90010	…	4	…	3 7/8	4 5/8	1 1/8	.385	9/16	.860	3/8
			4 1/4	…	…	4 1/8	4 7/8	1 1/8	.385	9/16	.860	3/8
10	.51612	1.04465	5	…	…	4 27/32	5 23/32	1 5/16	.447	21/32	1.010	7/16
			…	5 11/16	…	5 17/32	6 13/32	1 5/16	.447	21/32	1.010	7/16
			…	…	6 7/32	6 1/16	6 15/16	1 5/16	.447	21/32	1.010	7/16
11	.50100	1.24995	5 15/16	…	…	5 25/32	6 21/32	1 5/16	.447	21/32	1.210	7/16
			…	6 3/4	…	6 19/32	7 15/32	1 5/16	.447	21/32	1.210	7/16
12	.49973	1.50010	7 1/8	7 1/8	…	6 15/16	7 15/16	1 1/2	.510	3/4	1.460	1/2
			…	…	6 1/4	…	…	…	…	…	…	…
13	.50020	1.75005	7 3/4	…	…	7 9/16	8 9/16	1 1/2	.510	3/4	1.710	1/2
14	.50000	2.00000	8 1/4	8 1/4	…	8 1/32	9 5/32	1 11/16	.572	27/32	1.960	9/16
15	.5000	2.25000	8 3/4	…	…	8 17/32	9 21/32	1 11/16	.572	27/32	2.210	9/16
16	.50000	2.50000	9 1/4	…	…	9	10 1/4	1 7/8	.635	15/16	2.450	5/8
17	.50000	2.75000	9 3/4	…	…	…	…	…	…	…	…	…
18	.50000	3.00000	10 1/4	…	…	…	…	…	…	…	…	…

[a] Special lengths of keyway are used instead of standard lengths in some places. Standard lengths need not be used when keyway is for driving only and not for admitting key to force out tool.

[b] "B & S Standard" Plug Depths are not used in all cases.

[c] Adopted by American Standards Association.

Table 5. Jarno Taper Shanks

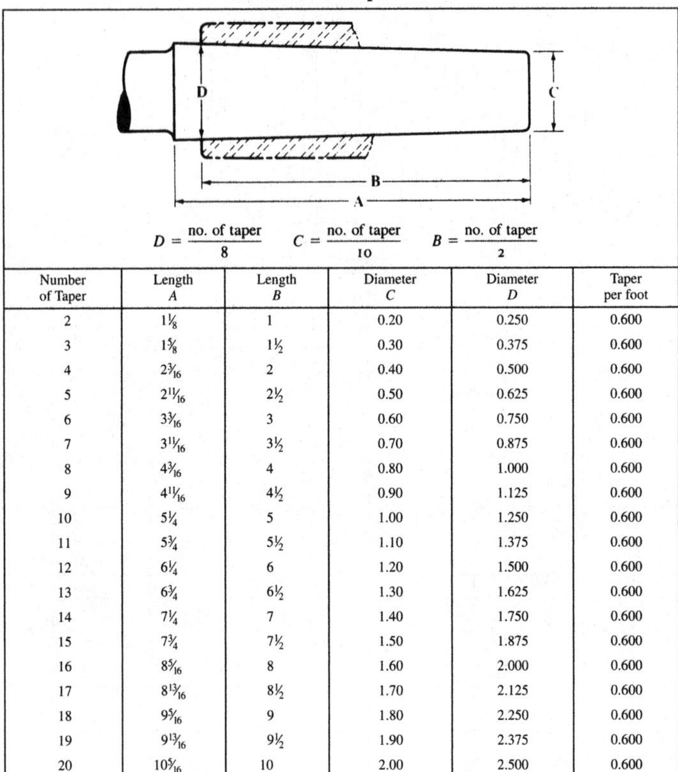

$D = \frac{\text{no. of taper}}{8}$ $C = \frac{\text{no. of taper}}{10}$ $B = \frac{\text{no. of taper}}{2}$

Number of Taper	Length *A*	Length *B*	Diameter *C*	Diameter *D*	Taper per foot
2	1 1/8	1	0.20	0.250	0.600
3	1 5/8	1 1/2	0.30	0.375	0.600
4	2 3/16	2	0.40	0.500	0.600
5	2 11/16	2 1/2	0.50	0.625	0.600
6	3 3/16	3	0.60	0.750	0.600
7	3 11/16	3 1/2	0.70	0.875	0.600
8	4 3/16	4	0.80	1.000	0.600
9	4 11/16	4 1/2	0.90	1.125	0.600
10	5 1/4	5	1.00	1.250	0.600
11	5 3/4	5 1/2	1.10	1.375	0.600
12	6 1/4	6	1.20	1.500	0.600
13	6 3/4	6 1/2	1.30	1.625	0.600
14	7 1/4	7	1.40	1.750	0.600
15	7 3/4	7 1/2	1.50	1.875	0.600
16	8 5/16	8	1.60	2.000	0.600
17	8 13/16	8 1/2	1.70	2.125	0.600
18	9 5/16	9	1.80	2.250	0.600
19	9 13/16	9 1/2	1.90	2.375	0.600
20	10 5/16	10	2.00	2.500	0.600

Tapers for Machine Tool Spindles.—Most lathe spindles have Morse tapers, most milling machine spindles have American Standard tapers, almost all smaller milling machine spindles have R8 tapers, and large vertical milling machine spindles have American Standard tapers. The spindles of drilling machines and the taper shanks of twist drills are made to fit the Morse taper. For lathes, the Morse taper is generally used, but lathes may have the Jarno, Brown & Sharpe, or a special taper. Of 33 lathe manufacturers, 20 use the Morse taper; 5, the Jarno; 3 use special tapers of their own; 2 use modified Morse (longer than the standard but the same taper); 2 use Reed (which is a short Jarno); 1 uses the Brown & Sharpe standard. For grinding machine centers, Jarno, Morse, and Brown & Sharpe tapers are used. Of ten grinding machine manufacturers, 3 use Brown & Sharpe; 3 use Morse; and 4 use Jarno. The Brown & Sharpe taper is used extensively for milling machine and dividing head spindles. The standard milling machine spindle adopted in 1927 by the milling machine manufacturers of the National Machine Tool Builders' Association (now The Association for Manufacturing Technology [AMT]), has a taper of 3 1/2 inches per foot. This comparatively steep taper was adopted to ensure easy release of arbors.

Table 6. American National Standard Taper Drive with Tang, Self-Holding Tapers *ANSI/ASME B5.10-1994*

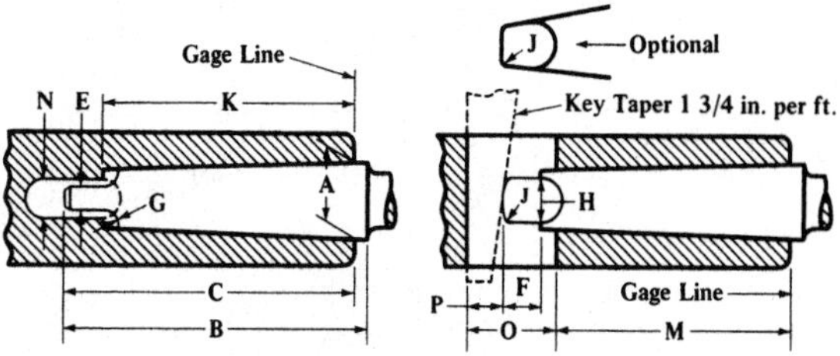

No. of Taper	Diameter at Gage Line (1) *A*	Shank: Total Length of Shank *B*	Shank: Gage Line to End of Shank *C*	Tang: Thickness *E*	Tang: Length *F*	Tang: Radius of Mill *G*	Tang: Diameter *H*
0.239	0.23922	1.28	1.19	0.125	0.19	0.19	0.18
0.299	0.29968	1.59	1.50	0.156	0.25	0.19	0.22
0.375	0.37525	1.97	1.88	0.188	0.31	0.19	0.28
1	0.47500	2.56	2.44	0.203	0.38	0.19	0.34
2	0.70000	3.13	2.94	0.250	0.44	0.25	0.53
3	0.93800	3.88	3.69	0.312	0.56	0.22	0.72
4	1.23100	4.88	4.63	0.469	0.63	0.31	0.97
4½	1.50000	5.38	5.13	0.562	0.69	0.38	1.20
5	1.74800	6.12	5.88	0.625	0.75	0.38	1.41
6	2.49400	8.25	8.25	0.750	1.13	0.50	2.00

No. of Taper	Radius *J*	Socket: Min. Depth of Hole *K*: Drilled	Socket: Min. Depth of Hole *K*: Reamed	Socket: Gage Line to Tang Slot *M*	Tang Slot: Width *N*	Tang Slot: Length *O*	Tang Slot: Shank End to Back of Tang Slot *P*
0.239	0.03	1.06	1.00	0.94	0.141	0.38	0.13
0.299	0.03	1.31	1.25	1.17	0.172	0.50	0.17
0.375	0.05	1.63	1.56	1.47	0.203	0.63	0.22
1	0.05	2.19	2.16	2.06	0.218	0.75	0.38
2	0.06	2.66	2.61	2.50	0.266	0.88	0.44
3	0.08	3.31	3.25	3.06	0.328	1.19	0.56
4	0.09	4.19	4.13	3.88	0.484	1.25	0.50
4½	0.13	4.62	4.56	4.31	0.578	1.38	0.56
5	0.13	5.31	5.25	4.94	0.656	1.50	0.56
6	0.16	7.41	7.33	7.00	0.781	1.75	0.50

All dimensions are in inches. (1) See Table 2 for plug and ring gage dimensions.

Tolerances: For shank diameter *A* at gage line, + 0.002 – 0.000; for hole diameter *A*, + 0.000 – 0.002. For tang thickness *E* up to No. 5 inclusive, + 0.000 – 0.006; No. 6, + 0.000 – 0.008. For width *N* of tang slot up to No. 5 inclusive, + 0.006; – 0.000; No. 6, + 0.008 – 0.000. For centrality of tang *E* with center line of taper, 0.0025 (0.005 total indicator variation). These centrality tolerances also apply to the tang slot *N*. On rate of taper, all sizes 0.002 per foot. This tolerance may be applied on *shanks* only in the direction which *increases* the rate of taper and on *sockets* only in the direction which *decreases* the rate of taper. Tolerances for two-decimal dimensions are plus or minus 0.010, unless otherwise specified.

Table 7. ANSI Standard Steep Machine Tapers *ANSI/ASME B5.10-1994*

No. of Taper	Taper per Foot[a]	Dia. at Gage Line[b]	Length Along Axis	No. of Taper	Taper per Foot[a]	Dia. at Gage Line[b]	Length Along Axis
5	3.500	0.500	0.6875	35	3.500	1.500	2.2500
10	**3.500**	**0.625**	**0.8750**	**40**	**3.500**	**1.750**	**2.5625**
15	3.500	0.750	1.0625	45	3.500	2.250	3.3125
20	**3.500**	**0.875**	**1.3125**	**50**	**3.500**	**2.750**	**4.0000**
25	3.500	1.000	1.5625	55	3.500	3.500	5.1875
30	**3.500**	**1.250**	**1.8750**	**60**	**3.500**	**4.250**	**6.3750**

[a] This taper corresponds to an included angle of 16°, 35′, 39.4″.
[b] The basic diameter at gage line is at large end of taper.

All dimensions given in inches.

The tapers numbered 10, 20, 30, 40, 50, and 60 that are printed in heavy-faced type are designated as the "Preferred Series." The tapers numbered 5, 15, 25, 35, 45, and 55 that are printed in light-faced type are designated as the "Intermediate Series."

Table 8. American National Standard Self-holding Tapers — Basic Dimensions ANSI/ASME B5.10-1994

No. of Taper	Taper per Foot	Dia.at Gage Line[a] *A*	Means of Driving and Holding[a]	Origin of Series
.239	0.50200	0.23922	Tang Drive With Shank Held in by Friction (See Table 6)	Brown & Sharpe Taper Series
.299	0.50200	0.29968		
.375	0.50200	0.37525		
1	0.59858	0.47500		Morse Taper Series
2	0.59941	0.70000		
3	0.60235	0.93800		
4	0.62326	1.23100	Tang Drive With Shank Held in by Key	
4½	0.62400	1.50000		
5	0.63151	1.74800		
6	0.62565	2.49400		
7	0.62400	3.27000		
200	0.750	2.000	Key Drive With Shank Held in by Key	¾ Inch per Foot Taper Series
250	0.750	2.500	Key Drive With Shank Heldin by Draw-bolt	
300	0.750	3.000		
350	0.750	3.500		
400	0.750	4.000		
450	0.750	4.500		
500	0.750	5.000		
600	0.750	6.000		
800	0.750	8.000		
1000	0.750	10.000		
1200	0.750	12.000		

[a] See illustrations above Tables 6.

All dimensions given in inches.

KEYS AND KEYSEATS

ANSI Standard Keys and Keyseats.—American National Standard, B17.1 Keys and Keyseats, based on current industry practice, was approved in 1967, and reaffirmed in 1989. This standard establishes a uniform relationship between shaft sizes and key sizes for parallel and taper keys as shown in Table 1. Other data in this standard are given in Tables 2 and 3 through 7. The sizes and tolerances shown are for single key applications only.

The following definitions are given in the standard:

Key: A demountable machinery part which, when assembled into keyseats, provides a positive means for transmitting torque between the shaft and hub.

Keyseat: An axially located rectangular groove in a shaft or hub.

This standard recognizes that there are two classes of stock for parallel keys used by industry. One is a close, plus toleranced key stock and the other is a broad, negative toleranced bar stock. Based on the use of two types of stock, two classes of fit are shown:

Class 1: A clearance or metal-to-metal side fit obtained by using bar stock keys and keyseat tolerances as given in Table 4. This is a relatively free fit and applies only to parallel keys.

Class 2: A side fit, with possible interference or clearance, obtained by using key stock and keyseat tolerances as given in Table 4. This is a relatively tight fit.

Class 3: This is an interference side fit and is not tabulated in Table 4 since the degree of interference has not been standardized. However, it is suggested that the top and bottom fit range given under Class 2 in Table 4, for parallel keys be used.

Table 1. Key Size Versus Shaft Diameter ***ANSI B17.1-1967 (R1998)***

Nominal Shaft Diameter		Nominal Key Size			Normal Keyseat Depth	
			Height, *H*		*H*/2	
Over	To (Incl.)	Width, *W*	Square	Rectangular	Square	Rectangular
5/16	7/16	3/32	3/32	…	3/64	…
7/16	9/16	1/8	1/8	3/32	1/16	3/64
9/16	7/8	3/16	3/16	1/8	3/32	1/16
7/8	1 1/4	1/4	1/4	3/16	1/8	3/32
1 1/4	1 3/8	5/16	5/16	1/4	5/32	1/8
1 3/8	1 3/4	3/8	3/8	1/4	3/16	1/8
1 3/4	2 1/4	1/2	1/2	3/8	1/4	3/16
2 1/4	2 3/4	5/8	5/8	7/16	5/16	7/32
2 3/4	3 1/4	3/4	3/4	1/2	3/8	1/4
3 1/4	3 3/4	7/8	7/8	5/8	7/16	5/16
3 3/4	4 1/2	1	1	3/4	1/2	3/8
4 1/2	5 1/2	1 1/4	1 1/4	7/8	5/8	7/16
5 1/2	6 1/2	1 1/2	1 1/2	1	3/4	1/2
Square Keys preferred for shaft diameters above this line; rectangular keys, below						
6 1/2	7 1/2	1 3/4	1 3/4	1 1/2[a]	7/8	3/4
7 1/2	9	2	2	1 1/2	1	3/4
9	11	2 1/2	2 1/2	1 3/4	1 1/4	7/8

[a] Some key standards show 1 1/4 inches; preferred height is 1 1/2 inches.

All dimensions are given in inches. For larger shaft sizes, see *ANSI Standard Woodruff Keys and Keyseats.*

Key Size vs. Shaft Diameter: Shaft diameters are listed in Table 1 for identification of various key sizes and are not intended to establish shaft dimensions, tolerances or selections. For a stepped shaft, the size of a key is determined by the diameter of the shaft at the point of location of the key. Up through 6½-inch diameter shafts square keys are preferred; rectangular keys are preferred for larger shafts.

If special considerations dictate the use of a keyseat in the hub shallower than the preferred nominal depth shown, it is recommended that the tabulated preferred nominal standard keyseat always be used in the shaft.

Keyseat Alignment Tolerances: A tolerance of 0.010 inch, max is provided for offset (due to parallel displacement of keyseat centerline from centerline of shaft or bore) of keyseats in shaft and bore. The following tolerances for maximum lead (due to angular displacement of keyseat centerline from centerline of shaft or bore and measured at right angles to the shaft or bore centerline) of keyseats in shaft and bore are specified: 0.002 inch for keyseat length up to and including 4 inches; 0.0005 inch per inch of length for keyseat lengths above 4 inches to and including 10 inches; and 0.005 inch for keyseat lengths above 10 inches. For the effect of keyways on shaft strength, see *Effect of Keyways on Shaft Strength* on page 283.

Table 2. Depth Control Values *S* and *T* for Shaft and Hub

Nominal Shaft Diameter	Parallel and Taper		Parallel		Taper	
	Square	Rectangular	Square	Rectangular	Square	Rectangular
	S	*S*	*T*	*T*	*T*	*T*
½	0.430	0.445	0.560	0.544	0.535	0.519
9/16	0.493	0.509	0.623	0.607	0.598	0.582
5/8	0.517	0.548	0.709	0.678	0.684	0.653
11/16	0.581	0.612	0.773	0.742	0.748	0.717
¾	0.644	0.676	0.837	0.806	0.812	0.781
13/16	0.708	0.739	0.900	0.869	0.875	0.844
7/8	0.771	0.802	0.964	0.932	0.939	0.907
15/16	0.796	0.827	1.051	1.019	1.026	0.994
1	0.859	0.890	1.114	1.083	1.089	1.058
1 1/16	0.923	0.954	1.178	1.146	1.153	1.121
1⅛	0.986	1.017	1.241	1.210	1.216	1.185
1 3/16	1.049	1.080	1.304	1.273	1.279	1.248
1¼	1.112	1.144	1.367	1.336	1.342	1.311
1 5/16	1.137	1.169	1.455	1.424	1.430	1.399
1⅜	1.201	1.232	1.518	1.487	1.493	1.462
1 7/16	1.225	1.288	1.605	1.543	1.580	1.518
1½	1.289	1.351	1.669	1.606	1.644	1.581
1 9/16	1.352	1.415	1.732	1.670	1.707	1.645
1⅝	1.416	1.478	1.796	1.733	1.771	1.708
1 11/16	1.479	1.541	1.859	1.796	1.834	1.771
1¾	1.542	1.605	1.922	1.860	1.897	1.835
1 13/16	1.527	1.590	2.032	1.970	2.007	1.945
1⅞	1.591	1.654	2.096	2.034	2.071	2.009
1 15/16	1.655	1.717	2.160	2.097	2.135	2.072
2	1.718	1.781	2.223	2.161	2.198	2.136
2 1/16	1.782	1.844	2.287	2.224	2.262	2.199
2⅛	1.845	1.908	2.350	2.288	2.325	2.263
2 3/16	1.909	1.971	2.414	2.351	2.389	2.326
2¼	1.972	2.034	2.477	2.414	2.452	2.389
2 5/16	1.957	2.051	2.587	2.493	2.562	2.468
2⅜	2.021	2.114	2.651	2.557	2.626	2.532

Table 2. *(Continued)* **Depth Control Values *S* and *T* for Shaft and Hub**

Nominal Shaft Diameter	Parallel and Taper		Parallel		Taper	
	Square	Rectangular	Square	Rectangular	Square	Rectangular
	S	*S*	*T*	*T*	*T*	*T*
2 7/16	2.084	2.178	2.714	2.621	2.689	2.596
2 1/2	2.148	2.242	2.778	2.684	2.753	2.659
2 9/16	2.211	2.305	2.841	2.748	2.816	2.723
2 5/8	2.275	2.369	2.905	2.811	2.880	2.786
2 11/16	2.338	2.432	2.968	2.874	2.943	2.849
2 3/4	2.402	2.495	3.032	2.938	3.007	2.913
2 13/16	2.387	2.512	3.142	3.017	3.117	2.992
2 7/8	2.450	2.575	3.205	3.080	3.180	3.055
2 15/16	2.514	2.639	3.269	3.144	3.244	3.119
3	2.577	2.702	3.332	3.207	3.307	3.182
3 1/16	2.641	2.766	3.396	3.271	3.371	3.246
3 1/8	2.704	2.829	3.459	3.334	3.434	3.309
3 3/16	2.768	2.893	3.523	3.398	3.498	3.373
3 1/4	2.831	2.956	3.586	3.461	3.561	3.436
3 5/16	2.816	2.941	3.696	3.571	3.671	3.546
3 3/8	2.880	3.005	3.760	3.635	3.735	3.610
3 7/16	2.943	3.068	3.823	3.698	3.798	3.673
3 1/2	3.007	3.132	3.887	3.762	3.862	3.737
3 9/16	3.070	3.195	3.950	3.825	3.925	3.800
3 5/8	3.134	3.259	4.014	3.889	3.989	3.864
3 11/16	3.197	3.322	4.077	3.952	4.052	3.927
3 3/4	3.261	3.386	4.141	4.016	4.116	3.991
3 13/16	3.246	3.371	4.251	4.126	4.226	4.101
3 7/8	3.309	3.434	4.314	4.189	4.289	4.164
3 15/16	3.373	3.498	4.378	4.253	4.353	4.228
4	3.436	3.561	4.441	4.316	4.416	4.291
4 3/16	3.627	3.752	4.632	4.507	4.607	4.482
4 1/4	3.690	3.815	4.695	4.570	4.670	4.545
4 3/8	3.817	3.942	4.822	4.697	4.797	4.672
4 7/16	3.880	4.005	4.885	4.760	4.860	4.735
4 1/2	3.944	4.069	4.949	4.824	4.924	4.799
4 3/4	4.041	4.229	5.296	5.109	5.271	5.084
4 7/8	4.169	4.356	5.424	5.236	5.399	5.211
4 15/16	4.232	4.422	5.487	5.300	5.462	5.275
5	4.296	4.483	5.551	5.363	5.526	5.338
5 3/16	4.486	4.674	5.741	5.554	5.716	5.529
5 1/4	4.550	4.737	5.805	5.617	5.780	5.592
5 7/16	4.740	4.927	5.995	5.807	5.970	5.782
5 1/2	4.803	4.991	6.058	5.871	6.033	5.846
5 3/4	4.900	5.150	6.405	6.155	6.380	6.130
5 15/16	5.091	5.341	6.596	6.346	6.571	6.321
6	5.155	5.405	6.660	6.410	6.635	6.385
6 1/4	5.409	5.659	6.914	6.664	6.889	6.639
6 1/2	5.662	5.912	7.167	6.917	7.142	6.892
6 3/4	5.760	[a]5.885	7.515	[a]7.390	7.490	[a]7.365
7	6.014	[a]6.139	7.769	[a]7.644	7.744	[a]7.619
7 1/4	6.268	[a]6.393	8.023	[a]7.898	7.998	[a]7.873
7 1/2	6.521	[a]6.646	8.276	[a]8.151	8.251	[a]8.126

Table 2. *(Continued)* **Depth Control Values *S* and *T* for Shaft and Hub**

Nominal Shaft Diameter	Parallel and Taper		Parallel		Taper	
	Square	Rectangular	Square	Rectangular	Square	Rectangular
	S	*S*	*T*	*T*	*T*	*T*
7¾	6.619	6.869	8.624	8.374	8.599	8.349
8	6.873	7.123	8.878	8.628	8.853	8.603
9	7.887	8.137	9.892	9.642	9.867	9.617
10	8.591	8.966	11.096	10.721	11.071	10.696
11	9.606	9.981	12.111	11.736	12.086	11.711
12	10.309	10.809	13.314	12.814	13.289	12.789
13	11.325	11.825	14.330	13.830	14.305	13.805
14	12.028	12.528	15.533	15.033	15.508	15.008
15	13.043	13.543	16.548	16.048	16.523	16.023

[a] 1¾ × 1½ inch key.

All dimensions are given in inches. See Table 4 for tolerances.

Definition.—The following definitions are given in this standard:

Woodruff Key: A Remountable machinery part which, when assembled into key-seats, provides a positive means for transmitting torque between the shaft and hub.

Woodruff Key Number: An identification number by which the size of key may be readily determined.

Woodruff Keyseat—Shaft: The circular pocket in which the key is retained.

Woodruff Keyseat—Hub: An axially located rectangular groove in a hub. (This has been referred to as a keyway.)

Woodruff Keyseat Milling Cutter: An arbor type or shank type milling cutter normally used for milling Woodruff keyseats in shafts.

Chamfered Keys and Filleted Keyseats.—In general practice, chamfered keys and filleted keyseats are not used. However, it is recognized that fillets in keyseats decrease stress concentration at corners. When used, fillet radii should be as large as possible without causing excessive bearing stresses due to reduced contact area between the key and its mating parts. Keys must be chamfered or rounded to clear fillet radii. Values in Table 5 assume general conditions and should be used only as a guide when critical stresses are encountered.

Depths for Milling Keyseats.—The above table has been compiled to facilitate the accurate milling of keyseats. This table gives the distance *M* (see illustration accompanying table) between the top of the shaft and a line passing through the upper corners or edges of the keyseat. Dimension *M* is calculated by the formula: $M = \frac{1}{2}(S - \sqrt{S^2 - E^2})$ where *S* is diameter of shaft, and *E* is width of keyseat. A simple approximate formula that gives *M* to within 0.001 inch is $M = E_2 \div 4S$.

Cotters.—A cotter is a form of key that is used to connect rods, etc., that are subjected either to tension or compression or both, the cotter being subjected to shearing stresses at two transverse cross-sections. When taper cotters are used for drawing and holding parts together, if the cotter is held in place by the friction between the bearing surfaces, the taper should not be too great. Ordinarily a taper varying from ¼ to ½ inch per foot is used for plain cotters. When a set-screw or other device is used to prevent the cotter from backing out of its slot, the taper may vary from 1½ to 2 inches per foot.

Table 3. ANSI Standard Plain and Gib Head Keys ***ANSI B17.1-1967 (R1998)***

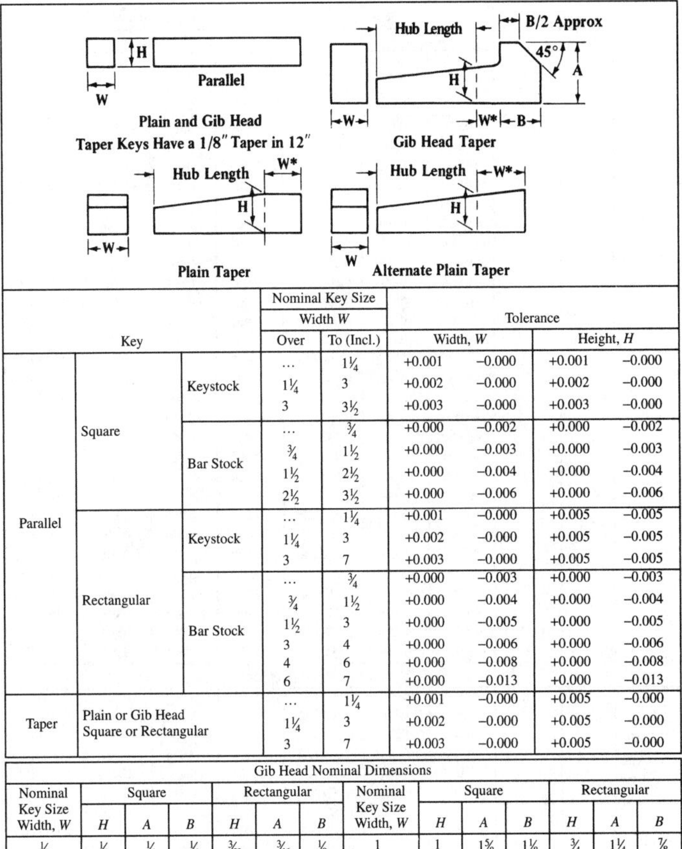

Key			Nominal Key Size Width *W* Over	Nominal Key Size Width *W* To (Incl.)	Tolerance Width, *W*		Tolerance Height, *H*	
Parallel	Square	Keystock	…	1¼	+0.001	−0.000	+0.001	−0.000
			1¼	3	+0.002	−0.000	+0.002	−0.000
			3	3½	+0.003	−0.000	+0.003	−0.000
		Bar Stock	…	¾	+0.000	−0.002	+0.000	−0.002
			¾	1½	+0.000	−0.003	+0.000	−0.003
			1½	2½	+0.000	−0.004	+0.000	−0.004
			2½	3½	+0.000	−0.006	+0.000	−0.006
	Rectangular	Keystock	…	1¼	+0.001	−0.000	+0.005	−0.005
			1¼	3	+0.002	−0.000	+0.005	−0.005
			3	7	+0.003	−0.000	+0.005	−0.005
		Bar Stock	…	¾	+0.000	−0.003	+0.000	−0.003
			¾	1½	+0.000	−0.004	+0.000	−0.004
			1½	3	+0.000	−0.005	+0.000	−0.005
			3	4	+0.000	−0.006	+0.000	−0.006
			4	6	+0.000	−0.008	+0.000	−0.008
			6	7	+0.000	−0.013	+0.000	−0.013
Taper	Plain or Gib Head Square or Rectangular		…	1¼	+0.001	−0.000	+0.005	−0.000
			1¼	3	+0.002	−0.000	+0.005	−0.000
			3	7	+0.003	−0.000	+0.005	−0.000

Gib Head Nominal Dimensions												
Nominal Key Size Width, *W*	Square			Rectangular			Nominal Key Size Width, *W*	Square			Rectangular	
	H	*A*	*B*	*H*	*A*	*B*		*H*	*A*	*B*	*H*	*A*	*B*
⅛	⅛	¼	¼	3/32	3/16	⅛	1	1	1⅝	1⅛	¾	1¼	⅞
3/16	3/16	5/16	5/16	⅛	¼	¼	1¼	1¼	2	1 7/16	⅞	1⅜	1
¼	¼	7/16	⅜	3/16	5/16	5/16	1½	1½	2⅜	1¾	1	1⅝	1⅛
5/16	5/16	½	7/16	¼	7/16	⅜	1¾	1¾	2¾	2	1½	2⅜	1¾
⅜	⅜	⅝	½	¼	7/16	⅜	2	2	3½	2¼	1½	2⅜	1¾
½	½	⅞	⅝	⅜	⅝	½	2½	2½	4	3	1¾	2¾	2
⅝	⅝	1	¾	7/16	¾	9/16	3	3	5	3½	2	3½	2¼
¾	¾	1¼	⅞	½	⅞	⅝	3½	3½	6	4	2½	4	3
⅞	⅞	1⅜	1	⅝	1	¾	…	…	…	…	…	…	…

All dimensions are given in inches.

*For locating position of dimension *H*. Tolerance does not apply.

For larger sizes the following relationships are suggested as guides for establishing *A* and *B*: A = 1.8*H* and *B* = 1.2*H*.

Table 4. ANSI Standard Fits for Parallel and Taper Keys *ANSI B17.1-1967 (R1998)*

Type of Key	Key Width		Side Fit			Top and Bottom Fit			
	Over	To (Incl.)	Width Tolerance		Fit Range[a]	Depth Tolerance			Fit Range[a]
			Key	Key-Seat		Key	Shaft Key-Seat	Hub Key-Seat	
Class 1 Fit for Parallel Keys									
Square	…	½	+0.000	+0.002	0.004 CL	+0.000	+0.000	+0.010	0.032 CL
			−0.002	−0.000	0.000	−0.002	−0.015	−0.000	0.005 CL
	½	¾	+0.000	+0.003	0.005 CL	+0.000	+0.000	+0.010	0.032 CL
			−0.002	−0.000	0.000	−0.002	−0.015	−0.000	0.005 CL
	¾	1	+0.000	+0.003	0.006 CL	+0.000	+0.000	+0.010	0.033 CL
			−0.003	−0.000	0.000	−0.003	−0.015	−0.000	0.005 CL
	1	1½	+0.000	+0.004	0.007 CL	+0.000	+0.000	+0.010	0.033 CL
			−0.003	−0.000	0.000	−0.003	−0.015	−0.000	0.005 CL
	1½	2½	+0.000	+0.004	0.008 CL	+0.000	+0.000	+0.010	0.034 CL
			−0.004	−0.000	0.000	−0.004	−0.015	−0.000	0.005 CL
	2½	3½	+0.000	+0.004	0.010 CL	+0.000	+0.000	+0.010	0.036 CL
			−0.006	−0.000	0.000	−0.006	−0.015	−0.000	0.005 CL
Rectangular	…	½	+0.000	+0.002	0.005 CL	+0.000	+0.000	+0.010	0.033 CL
			−0.003	−0.000	0.000	−0.003	−0.015	−0.000	0.005 CL
	½	¾	+0.000	+0.003	0.006 CL	+0.000	+0.000	+0.010	0.033 CL
			−0.003	−0.000	0.000	−0.003	−0.015	−0.000	0.005 CL
	¾	1	+0.000	+0.003	0.007 CL	+0.000	+0.000	+0.010	0.034 CL
			−0.004	−0.000	0.000	−0.004	−0.015	−0.000	0.005 CL
	1	1½	+0.000	+0.004	0.008 CL	+0.000	+0.000	+0.010	0.034 CL
			−0.004	−0.000	0.000	−0.004	−0.015	−0.000	0.005 CL
	1½	3	+0.000	+0.004	0.009 CL	+0.000	+0.000	+0.010	0.035 CL
			−0.005	−0.000	0.000	−0.005	−0.015	−0.000	0.005 CL
	3	4	+0.000	+0.004	0.010 CL	+0.000	+0.000	+0.010	0.036 CL
			−0.006	−0.000	0.000	−0.006	−0.015	−0.000	0.005 CL
	4	6	+0.000	+0.004	0.012 CL	+0.000	+0.000	+0.010	0.038 CL
			−0.008	−0.000	0.000	−0.008	−0.015	−0.000	0.005 CL
	6	7	+0.000	+0.004	0.017 CL	+0.000	+0.000	+0.010	0.043 CL
			−0.013	−0.000	0.000	−0.013	−0.015	−0.000	0.005 CL
Class 2 Fit for Parallel and Taper Keys									
Parallel Square	…	1¼	+0.001	+0.002	0.002 CL	+0.001	+0.000	+0.010	0.030 CL
			−0.000	−0.000	0.001 INT	−0.000	−0.015	−0.000	0.004 CL
	1¼	3	+0.002	+0.002	0.002 CL	+0.002	+0.000	+0.010	0.030 CL
			−0.000	−0.000	0.002 INT	−0.000	−0.015	−0.000	0.003 CL
	3	3½	+0.003	+0.002	0.002 CL	+0.003	+0.000	+0.010	0.030 CL
			−0.000	−0.000	0.003 INT	−0.000	−0.015	−0.000	0.002 CL
Parallel Rectangular	…	1¼	+0.001	+0.002	0.002 CL	+0.005	+0.000	+0.010	0.035 CL
			−0.000	−0.000	0.001 INT	−0.005	−0.015	−0.000	0.000 CL
	1¼	3	+0.002	+0.002	0.002 CL	+0.005	+0.000	+0.010	0.035 CL
			−0.000	−0.000	0.002 INT	−0.005	−0.015	−0.000	0.000 CL
	3	7	+0.003	+0.002	0.002 CL	+0.005	+0.000	+0.010	0.035 CL
			−0.000	−0.000	0.003 INT	−0.005	−0.015	−0.000	0.000 CL
Taper	…	1¼	+0.001	+0.002	0.002 CL	+0.005	+0.000	+0.010	0.005 CL
			−0.000	−0.000	0.001 INT	−0.000	−0.015	−0.000	0.025 INT
	1¼	3	+0.002	+0.002	0.002 CL	+0.005	+0.000	+0.010	0.005 CL
			−0.000	−0.000	0.002 INT	−0.000	−0.015	−0.000	0.025 INT
	3	[b]	+0.003	+0.002	0.002 CL	+0.005	+0.000	+0.010	0.005 CL
			−0.000	−0.000	0.003 INT	−0.000	−0.015	−0.000	0.025 INT

[a] Limits of variation. CL = Clearance; INT = Interference.

[b] To (Incl.) 3½-inch Square and 7-inch Rectangular key widths.

All dimensions are given in inches. See also text on page 253.

Table 5. Suggested Keyseat Fillet Radius and Key Chamfer *ANSI B17.1-1967 (R1998)*

Keyseat Depth, *H*/2		Fillet Radius	45 deg. Chamfer	Keyseat Depth, *H*2		Fillet Radius	45 deg. Chamfer
Over	To (Incl.)			Over	To (Incl.)		
1/8	1/4	1/32	3/64	7/8	1 1/4	3/16	7/32
1/4	1/2	1/16	5/64	1 1/4	1 3/4	1/4	9/32
1/2	7/8	1/8	5/32	1 3/4	2 1/2	3/8	13/32

All dimensions are given in inches.

Table 6. ANSI Standard Keyseat Tolerances for Electric Motor and Generator Shaft Extensions *ANSI B17.1-1967 (R1998)*

Keyseat Width		Width Tolerance	Depth Tolerance
Over	To (Incl.)		
…	1/4	+0.001	+0.000
		−0.001	−0.015
1/4	3/4	+0.000	+0.000
		−0.002	−0.015
3/4	1 1/4	+0.000	+0.000
		−0.003	−0.015

All dimensions are given in inches.

Table 7. Set Screws for Use Over Keys *ANSI B17.1-1967 (R1998)*

Nom. Shaft Diam.		Nom. Key Width	Set Screw Diam.	Nom. Shaft Diam.		Nom. Key Width	Set Screw Diam.
Over	To (Incl.)			Over	To (Incl.)		
5/16	7/16	3/32	No. 10	2 1/4	2 3/4	5/8	1/2
7/16	9/16	1/8	No. 10	2 3/4	3 1/4	3/4	5/8
9/16	7/8	3/16	1/4	3 1/4	3 3/4	7/8	3/4
7/8	1 1/4	1/4	5/16	3 3/4	4 1/2	1	3/4
1 1/4	1 3/8	5/16	3/8	4 1/2	5 1/2	1 1/4	7/8
1 3/8	1 3/4	3/8	3/8	5 1/2	6 1/2	1 1/2	1
1 3/4	2 1/4	1/2	1/2	…	…	…	…

All dimensions are given in inches.

These set screw diameter selections are offered as a guide but their use should be dependent upon design considerations.

ANSI Standard Woodruff Keys and Keyseats.—American National Standard B17.2 was approved in 1967, and reaffirmed in 1990. Data from this standard are shown in Tables 8, 9, and 10.

Table 8. ANSI Standard Woodruff Keys *ANSI B17.2-1967 (R1998)*

+0.000 −0.010 *B* · **+0.001 −0.000** *W* · *E* · *C* · *F* · **Break corners 0.020 max *R*** · **Full radius type**

+0.000 −0.010 *B* · **+0.001 −0.000** *W* · *E* · *D* · *F* · **Break corners 0.020 max *R*** · **Flat bottom type**

Key No.	Nominal Key Size *W* X *B*	Actual Length *F* +0.000 −0.010	Height of Key *C* Max.	*C* Min.	*D* Max.	*D* Min.	Distance Below Center *E*
202	1/16 × 1/4	0.248	0.109	0.104	0.109	0.104	1/64
202.5	1/16 × 5/16	0.311	0.140	0.135	0.140	0.135	1/64
302.5	3/32 × 5/16	0.311	0.140	0.135	0.140	0.135	1/64
203	1/16 × 3/8	0.374	0.172	0.167	0.172	0.167	1/64
303	3/32 × 3/8	0.374	0.172	0.167	0.172	0.167	1/64
403	1/8 × 3/8	0.374	0.172	0.167	0.172	0.167	1/64
204	1/16 × 1/2	0.491	0.203	0.198	0.194	0.188	3/64
304	3/32 × 1/2	0.491	0.203	0.198	0.194	0.188	3/64
404	1/8 × 1/2	0.491	0.203	0.198	0.194	0.188	3/64
305	3/32 × 5/8	0.612	0.250	0.245	0.240	0.234	1/16
405	1/8 × 5/8	0.612	0.250	0.245	0.240	0.234	1/16
505	5/32 × 5/8	0.612	0.250	0.245	0.240	0.234	1/16
605	3/16 × 5/8	0.612	0.250	0.245	0.240	0.234	1/16
406	1/8 × 3/4	0.740	0.313	0.308	0.303	0.297	1/16
506	5/32 × 3/4	0.740	0.313	0.308	0.303	0.297	1/16
606	3/16 × 3/4	0.740	0.313	0.308	0.303	0.297	1/16
806	1/4 × 3/4	0.740	0.313	0.308	0.303	0.297	1/16
507	5/32 × 7/8	0.866	0.375	0.370	0.365	0.359	1/16
607	3/16 × 7/8	0.866	0.375	0.370	0.365	0.359	1/16
707	7/32 × 7/8	0.866	0.375	0.370	0.365	0.359	1/16
807	1/4 × 7/8	0.866	0.375	0.370	0.365	0.359	1/16
608	3/16 × 1	0.992	0.438	0.433	0.428	0.422	1/16
708	7/32 × 1	0.992	0.438	0.433	0.428	0.422	1/16
808	1/4 × 1	0.992	0.438	0.433	0.428	0.422	1/16
1008	5/16 × 1	0.992	0.438	0.433	0.428	0.422	1/16
1208	3/8 × 1	0.992	0.438	0.433	0.428	0.422	1/16
609	3/16 × 1 1/8	1.114	0.484	0.479	0.475	0.469	5/64
709	7/32 × 1 1/8	1.114	0.484	0.479	0.475	0.469	5/64
809	1/4 × 1 1/8	1.114	0.484	0.479	0.475	0.469	5/64
1009	5/16 × 1 1/8	1.114	0.484	0.479	0.475	0.469	5/64
610	3/16 × 1 1/4	1.240	0.547	0.542	0.537	0.531	5/64
710	7/32 × 1 1/4	1.240	0.547	0.542	0.537	0.531	5/64
810	1/4 × 1 1/4	1.240	0.547	0.542	0.537	0.531	5/64
1010	5/16 × 1 1/4	1.240	0.547	0.542	0.537	0.531	5/64
1210	3/8 × 1 1/4	1.240	0.547	0.542	0.537	0.531	5/64
811	1/4 × 1 3/8	1.362	0.594	0.589	0.584	0.578	3/32
1011	5/16 × 1 3/8	1.362	0.594	0.589	0.584	0.578	3/32
1211	3/8 × 1 3/8	1.362	0.594	0.589	0.584	0.578	3/32
812	1/4 × 1 1/2	1.484	0.641	0.636	0.631	0.625	7/64
1012	5/16 × 1 1/2	1.484	0.641	0.636	0.631	0.625	7/64
1212	3/8 × 1 1/2	1.484	0.641	0.636	0.631	0.625	7/64

All dimensions are given in inches.

The Key numbers indicate normal key dimensions. The last two digits give the nominal diameter *B* in eighths of an inch and the digits preceding the last two give the nominal width *W* in thirty-seconds of an inch.

Table 9. ANSI Standard Woodruff Keys *ANSI B17.2-1967 (R1998)*

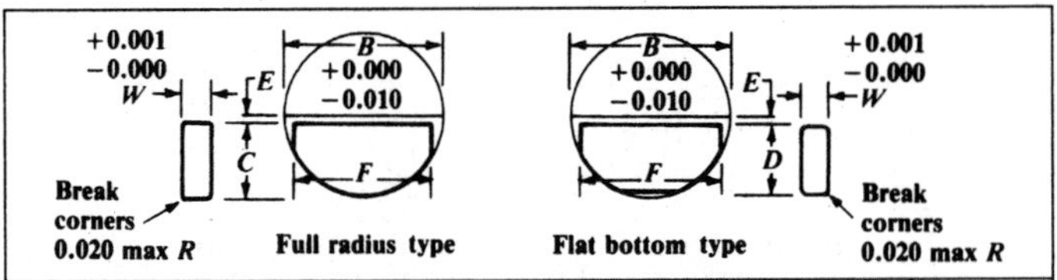

+0.001 −0.000 W
B +0.000 −0.010
E C F
Break corners 0.020 max R
Full radius type
Flat bottom type
D
Break corners 0.020 max R

Key No.	Nominal Key Size $W \times B$	Actual Length F +0.000 −0.010	Height of Key C Max.	Height of Key C Min.	Height of Key D Max.	Height of Key D Min.	Distance Below Center E
617-1	$\frac{3}{16} \times 2\frac{1}{8}$	1.380	0.406	0.401	0.396	0.390	$\frac{21}{32}$
817-1	$\frac{1}{4} \times 2\frac{1}{8}$	1.380	0.406	0.401	0.396	0.390	$\frac{21}{32}$
1017-1	$\frac{5}{16} \times 2\frac{1}{8}$	1.380	0.406	0.401	0.396	0.390	$\frac{21}{32}$
1217-1	$\frac{3}{8} \times 2\frac{1}{8}$	1.380	0.406	0.401	0.396	0.390	$\frac{21}{32}$
617	$\frac{3}{16} \times 2\frac{1}{8}$	1.723	0.531	0.526	0.521	0.515	$\frac{17}{32}$
817	$\frac{1}{4} \times 2\frac{1}{8}$	1.723	0.531	0.526	0.521	0.515	$\frac{17}{32}$
1017	$\frac{5}{16} \times 2\frac{1}{8}$	1.723	0.531	0.526	0.521	0.515	$\frac{17}{32}$
1217	$\frac{3}{8} \times 2\frac{1}{8}$	1.723	0.531	0.526	0.521	0.515	$\frac{17}{32}$
822-1	$\frac{1}{4} \times 2\frac{3}{4}$	2.000	0.594	0.589	0.584	0.578	$\frac{25}{32}$
1022-1	$\frac{5}{16} \times 2\frac{3}{4}$	2.000	0.594	0.589	0.584	0.578	$\frac{25}{32}$
1222-1	$\frac{3}{8} \times 2\frac{3}{4}$	2.000	0.594	0.589	0.584	0.578	$\frac{25}{32}$
1422-1	$\frac{7}{16} \times 2\frac{3}{4}$	2.000	0.594	0.589	0.584	0.578	$\frac{25}{32}$
1622-1	$\frac{1}{2} \times 2\frac{3}{4}$	2.000	0.594	0.589	0.584	0.578	$\frac{25}{32}$
822	$\frac{1}{4} \times 2\frac{3}{4}$	2.317	0.750	0.745	0.740	0.734	$\frac{5}{8}$
1022	$\frac{5}{16} \times 2\frac{3}{4}$	2.317	0.750	0.745	0.740	0.734	$\frac{5}{8}$
1222	$\frac{3}{8} \times 2\frac{3}{4}$	2.317	0.750	0.745	0.740	0.734	$\frac{5}{8}$
1422	$\frac{7}{16} \times 2\frac{3}{4}$	2.317	0.750	0.745	0.740	0.734	$\frac{5}{8}$
1622	$\frac{1}{2} \times 2\frac{3}{4}$	2.317	0.750	0.745	0.740	0.734	$\frac{5}{8}$
1228	$\frac{3}{8} \times 3\frac{1}{2}$	2.880	0.938	0.933	0.928	0.922	$\frac{13}{16}$
1428	$\frac{7}{16} \times 3\frac{1}{2}$	2.880	0.938	0.933	0.928	0.922	$\frac{13}{16}$
1628	$\frac{1}{2} \times 3\frac{1}{2}$	2.880	0.938	0.933	0.928	0.922	$\frac{13}{16}$
1828	$\frac{9}{16} \times 3\frac{1}{2}$	2.880	0.938	0.933	0.928	0.922	$\frac{13}{16}$
2028	$\frac{5}{8} \times 3\frac{1}{2}$	2.880	0.938	0.933	0.928	0.922	$\frac{13}{16}$
2228	$\frac{11}{16} \times 3\frac{1}{2}$	2.880	0.938	0.933	0.928	0.922	$\frac{13}{16}$
2428	$\frac{3}{4} \times 3\frac{1}{2}$	2.880	0.938	0.933	0.928	0.922	$\frac{13}{16}$

All dimensions are given in inches.

The key numbers indicate nominal key dimensions. The last two digits give the nominal diameter B in eighths of an inch and the digits preceding the last two give the nominal width W in thirty-seconds of an inch.

The key numbers with the −1 designation, while representing the nominal key size have a shorter length F and due to a greater distance below center E are less in height than the keys of the same number without the −1 designation.

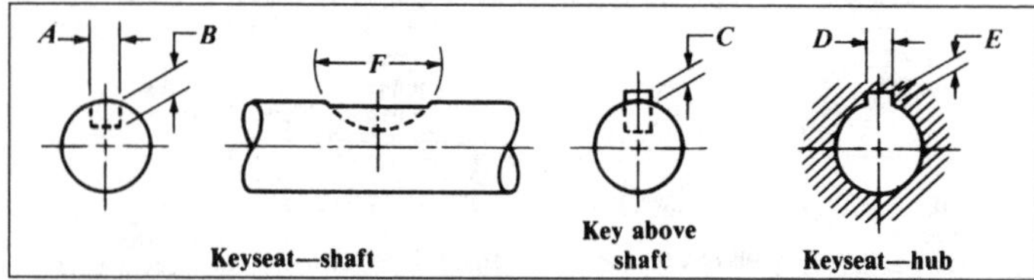

Table 10. ANSI Keyseat Dimensions for Woodruff Keys

Key No.	Nominal Size Key	Keyseat—Shaft					Key-Above Shaft	Keyseat—Hub	
		Width *A*[a]		Depth *B*	Diameter *F*		Height *C*	Width *D*	Depth *E*
		Min.	Max.	+0.005 −0.000	Min.	Max.	+0.005 −0.005	+0.002 −0.000	+0.005 −0.000
202	1/16 × 1/4	0.0615	0.0630	0.0728	0.250	0.268	0.0312	0.0635	0.0372
202.5	1/16 × 5/16	0.0615	0.0630	0.1038	0.312	0.330	0.0312	0.0635	0.0372
302.5	3/32 × 5/16	0.0928	0.0943	0.0882	0.312	0.330	0.0469	0.0948	0.0529
203	1/16 × 3/8	0.0615	0.0630	0.1358	0.375	0.393	0.0312	0.0635	0.0372
303	3/32 × 3/8	0.0928	0.0943	0.1202	0.375	0.393	0.0469	0.0948	0.0529
403	1/8 × 3/8	0.1240	0.1255	0.1045	0.375	0.393	0.0625	0.1260	0.0685
204	1/16 × 1/2	0.0615	0.0630	0.1668	0.500	0.518	0.0312	0.0635	0.0372
304	3/32 × 1/2	0.0928	0.0943	0.1511	0.500	0.518	0.0469	0.0948	0.0529
404	1/8 × 1/2	0.1240	0.1255	0.1355	0.500	0.518	0.0625	0.1260	0.0685
305	3/32 × 5/8	0.0928	0.0943	0.1981	0.625	0.643	0.0469	0.0948	0.0529
405	1/8 × 5/8	0.1240	0.1255	0.1825	0.625	0.643	0.0625	0.1260	0.0685
505	5/32 × 5/8	0.1553	0.1568	0.1669	0.625	0.643	0.0781	0.1573	0.0841
605	3/16 × 5/8	0.1863	0.1880	0.1513	0.625	0.643	0.0937	0.1885	0.0997
406	1/8 × 3/4	0.1240	0.1255	0.2455	0.750	0.768	0.0625	0.1260	0.0685
506	5/32 × 3/4	0.1553	0.1568	0.2299	0.750	0.768	0.0781	0.1573	0.0841
606	3/16 × 3/4	0.1863	0.1880	0.2143	0.750	0.768	0.0937	0.1885	0.0997
806	1/4 × 3/4	0.2487	0.2505	0.1830	0.750	0.768	0.1250	0.2510	0.1310
507	5/32 × 7/8	0.1553	0.1568	0.2919	0.875	0.895	0.0781	0.1573	0.0841
607	3/16 × 7/8	0.1863	0.1880	0.2763	0.875	0.895	0.0937	0.1885	0.0997
707	7/32 × 7/8	0.2175	0.2193	0.2607	0.875	0.895	0.1093	0.2198	0.1153
807	1/4 × 7/8	0.2487	0.2505	0.2450	0.875	0.895	0.1250	0.2510	0.1310
608	3/16 × 1	0.1863	0.1880	0.3393	1.000	1.020	0.0937	0.1885	0.0997
708	7/32 × 1	0.2175	0.2193	0.3237	1.000	1.020	0.1093	0.2198	0.1153
808	1/4 × 1	0.2487	0.2505	0.3080	1.000	1.020	0.1250	0.2510	0.1310
1008	5/16 × 1	0.3111	0.3130	0.2768	1.000	1.020	0.1562	0.3135	0.1622
1208	3/8 × 1	0.3735	0.3755	0.2455	1.000	1.020	0.1875	0.3760	0.1935
609	3/16 × 1 1/8	0.1863	0.1880	0.3853	1.125	1.145	0.0937	0.1885	0.0997
709	7/32 × 1 1/8	0.2175	0.2193	0.3697	1.125	1.145	0.1093	0.2198	0.1153
809	1/4 × 1 1/8	0.2487	0.2505	0.3540	1.125	1.145	0.1250	0.2510	0.1310
1009	5/16 × 1 1/8	0.3111	0.3130	0.3228	1.125	1.145	0.1562	0.3135	0.1622
610	3/16 × 1 1/4	0.1863	0.1880	0.4483	1.250	1.273	0.0937	0.1885	0.0997
710	7/32 × 1 1/4	0.2175	0.2193	0.4327	1.250	1.273	0.1093	0.2198	0.1153

Table 10. *(Continued)* **ANSI Keyseat Dimensions for Woodruff Keys**

Key No.	Nominal Size Key	Keyseat—Shaft					Key-Above Shaft	Keyseat—Hub	
		Width *A*[a]		Depth *B*	Diameter *F*		Height *C*	Width *D*	Depth *E*
		Min.	Max.	+0.005 −0.000	Min.	Max.	+0.005 −0.005	+0.002 −0.000	+0.005 −0.000
810	$\frac{1}{4} \times 1\frac{1}{4}$	0.2487	0.2505	0.4170	1.250	1.273	0.1250	0.2510	0.1310
1010	$\frac{5}{16} \times 1\frac{1}{4}$	0.3111	0.3130	0.3858	1.250	1.273	0.1562	0.3135	0.1622
1210	$\frac{3}{8} \times 1\frac{1}{4}$	0.3735	0.3755	0.3545	1.250	1.273	0.1875	0.3760	0.1935
811	$\frac{1}{4} \times 1\frac{3}{8}$	0.2487	0.2505	0.4640	1.375	1.398	0.1250	0.2510	0.1310
1011	$\frac{5}{16} \times 1\frac{3}{8}$	0.3111	0.3130	0.4328	1.375	1.398	0.1562	0.3135	0.1622
1211	$\frac{3}{8} \times 1\frac{3}{8}$	0.3735	0.3755	0.4015	1.375	1.398	0.1875	0.3760	0.1935
812	$\frac{1}{4} \times 1\frac{1}{2}$	0.2487	0.2505	0.5110	1.500	1.523	0.1250	0.2510	0.1310
1012	$\frac{5}{16} \times 1\frac{1}{2}$	0.3111	0.3130	0.4798	1.500	1.523	0.1562	0.3135	0.1622
1212	$\frac{3}{8} \times 1\frac{1}{2}$	0.3735	0.3755	0.4485	1.500	1.523	0.1875	0.3760	0.1935
617-1	$\frac{3}{16} \times 2\frac{1}{8}$	0.1863	0.1880	0.3073	2.125	2.160	0.0937	0.1885	0.0997
817-1	$\frac{1}{4} \times 2\frac{1}{8}$	0.2487	0.2505	0.2760	2.125	2.160	0.1250	0.2510	0.1310
1017-1	$\frac{5}{16} \times 2\frac{1}{8}$	0.3111	0.3130	0.2448	2.125	2.160	0.1562	0.3135	0.1622
1217-1	$\frac{3}{8} \times 2\frac{1}{8}$	0.3735	0.3755	0.2135	2.125	2.160	0.1875	0.3760	0.1935
617	$\frac{3}{16} \times 2\frac{1}{8}$	0.1863	0.1880	0.4323	2.125	2.160	0.0937	0.1885	0.0997
817	$\frac{1}{4} \times 2\frac{1}{8}$	0.2487	0.2505	0.4010	2.125	2.160	0.1250	0.2510	0.1310
1017	$\frac{5}{16} \times 2\frac{1}{8}$	0.3111	0.3130	0.3698	2.125	2.160	0.1562	0.3135	0.1622
1217	$\frac{3}{8} \times 2\frac{1}{8}$	0.3735	0.3755	0.3385	2.125	2.160	0.1875	0.3760	0.1935
822-1	$\frac{1}{4} \times 2\frac{3}{4}$	0.2487	0.2505	0.4640	2.750	2.785	0.1250	0.2510	0.1310
1022-1	$\frac{5}{16} \times 2\frac{3}{4}$	0.3111	0.3130	0.4328	2.750	2.785	0.1562	0.3135	0.1622
1222-1	$\frac{3}{8} \times 2\frac{3}{4}$	0.3735	0.3755	0.4015	2.750	2.785	0.1875	0.3760	0.1935
1422-1	$\frac{7}{16} \times 2\frac{3}{4}$	0.4360	0.4380	0.3703	2.750	2.785	0.2187	0.4385	0.2247
1622-1	$\frac{1}{2} \times 2\frac{3}{4}$	0.4985	0.5005	0.3390	2.750	2.785	0.2500	0.5010	0.2560
822	$\frac{1}{4} \times 2\frac{3}{4}$	0.2487	0.2505	0.6200	2.750	2.785	0.1250	0.2510	0.1310
1022	$\frac{5}{16} \times 2\frac{3}{4}$	0.3111	0.3130	0.5888	2.750	2.785	0.1562	0.3135	0.1622
1222	$\frac{3}{8} \times 2\frac{3}{4}$	0.3735	0.3755	0.5575	2.750	2.785	0.1875	0.3760	0.1935
1422	$\frac{7}{16} \times 2\frac{3}{4}$	0.4360	0.4380	0.5263	2.750	2.785	0.2187	0.4385	0.2247
1622	$\frac{1}{2} \times 2\frac{3}{4}$	0.4985	0.5005	0.4950	2.750	2.785	0.2500	0.5010	0.2560
1228	$\frac{3}{8} \times 3\frac{1}{2}$	0.3735	0.3755	0.7455	3.500	3.535	0.1875	0.3760	0.1935
1428	$\frac{7}{16} \times 3\frac{1}{2}$	0.4360	0.4380	0.7143	3.500	3.535	0.2187	0.4385	0.2247
1628	$\frac{1}{2} \times 3\frac{1}{2}$	0.4985	0.5005	0.6830	3.500	3.535	0.2500	0.5010	0.2560
1828	$\frac{9}{16} \times 3\frac{1}{2}$	0.5610	0.5630	0.6518	3.500	3.535	0.2812	0.5635	0.2872
2028	$\frac{5}{8} \times 3\frac{1}{2}$	0.6235	0.6255	0.6205	3.500	3.535	0.3125	0.6260	0.3185
2228	$\frac{11}{16} \times 3\frac{1}{2}$	0.6860	0.6880	0.5893	3.500	3.535	0.3437	0.6885	0.3497
2428	$\frac{3}{4} \times 3\frac{1}{2}$	0.7485	0.7505	0.5580	3.500	3.535	0.3750	0.7510	0.3810

[a] These Width *A* values were set with the maximum keyseat (shaft) width as that figure which will receive a key with the greatest amount of looseness consistent with assuring the key's sticking in the keyseat (shaft). Minimum keyseat width is that figure permitting the largest shaft distortion acceptable when assembling maximum key in minimum keyseat.Dimensions *A*, *B*, *C*, *D* are taken at side intersection.

All dimensions are in inches.

Table 11. Finding Depth of Keyseat and Distance from Top of Key to Bottom of Shaft

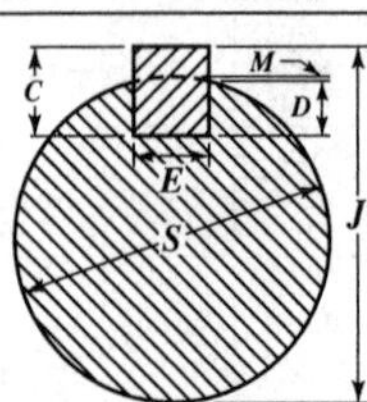

For milling keyseats, the total depth to feed cutter in from outside of shaft to bottom of keyseat is $M + D$, where D is depth of keyseat.

For checking an assembled key and shaft, caliper measurement J between top of key and bottom of shaft is used.

$$J = S - (M + D) + C$$

where C is depth of key. For Woodruff keys, dimensions C and D can be found in Tables 8 through 10. Assuming shaft diameter S is normal size, the tolerance on dimension J for Woodruff keys in keyslots are + 0.000, −0.010 inch.

Dia. of Shaft *S*, Inches	Width of Keyseat, *E*														
	1/16	3/32	1/8	5/32	3/16	7/32	1/4	5/16	3/8	7/16	1/2	9/16	5/8	11/16	3/4
	Dimension *M*, Inch														
0.3125	.0032	…	…	…	…	…	…	…	…	…	…	…	…	…	…
0.3437	.0029	.0065	…	…	…	…	…	…	…	…	…	…	…	…	…
0.3750	.0026	.0060	.0107	…	…	…	…	…	…	…	…	…	…	…	…
0.4060	.0024	.0055	.0099	…	…	…	…	…	…	…	…	…	…	…	…
0.4375	.0022	.0051	.0091	…	…	…	…	…	…	…	…	…	…	…	…
0.4687	.0021	.0047	.0085	.0134	…	…	…	…	…	…	…	…	…	…	…
0.5000	.0020	.0044	.0079	.0125	…	…	…	…	…	…	…	…	…	…	…
0.5625	…	.0039	.0070	.0111	.0161	…	…	…	…	…	…	…	…	…	…
0.6250	…	.0035	.0063	.0099	.0144	.0198	…	…	…	…	…	…	…	…	…
0.6875	…	.0032	.0057	.0090	.0130	.0179	.0235	…	…	…	…	…	…	…	…
0.7500	…	.0029	.0052	.0082	.0119	.0163	.0214	.0341	…	…	…	…	…	…	…
0.8125	…	.0027	.0048	.0076	.0110	.0150	.0197	.0312	…	…	…	…	…	…	…
0.8750	…	.0025	.0045	.0070	.0102	.0139	.0182	.0288	…	…	…	…	…	…	…
0.9375	…	…	.0042	.0066	.0095	.0129	.0170	.0263	.0391	…	…	…	…	…	…
1.0000	…	…	.0039	.0061	.0089	.0121	.0159	.0250	.0365	…	…	…	…	…	…
1.0625	…	…	.0037	.0058	.0083	.0114	.0149	.0235	.0342	…	…	…	…	…	…
1.1250	…	…	.0035	.0055	.0079	.0107	.0141	.0221	.0322	.0443	…	…	…	…	…
1.1875	…	…	.0033	.0052	.0074	.0102	.0133	.0209	.0304	.0418	…	…	…	…	…
1.2500	…	…	.0031	.0049	.0071	.0097	.0126	.0198	.0288	.0395	…	…	…	…	…
1.3750	…	…	…	.0045	.0064	.0088	.0115	.0180	.0261	.0357	.0471	…	…	…	…
1.5000	…	…	…	.0041	.0059	.0080	.0105	.0165	.0238	.0326	.0429	…	…	…	…
1.6250	…	…	…	.0038	.0054	.0074	.0097	.0152	.0219	.0300	.0394	.0502	…	…	…
1.7500	…	…	…	…	.0050	.0069	.0090	.0141	.0203	.0278	.0365	.0464	…	…	…
1.8750	…	…	…	…	.0047	.0064	.0084	.0131	.0189	.0259	.0340	.0432	.0536	…	…
2.0000	…	…	…	…	.0044	.0060	.0078	.0123	.0177	.0242	.0318	.0404	.0501	…	…
2.1250	…	…	…	…	…	.0056	.0074	.0116	.0167	.0228	.0298	.0379	.0470	.0572	.0684
2.2500	…	…	…	…	…	…	.0070	.0109	.0157	.0215	.0281	.0357	.0443	.0538	.0643
2.3750	…	…	…	…	…	…	…	.0103	.0149	.0203	.0266	.0338	.0419	.0509	.0608
2.5000	…	…	…	…	…	…	…	…	.0141	.0193	.0253	.0321	.0397	.0482	.0576
2.6250	…	…	…	…	…	…	…	…	.0135	.0184	.0240	.0305	.0377	.0457	.0547
2.7500	…	…	…	…	…	…	…	…	…	.0175	.0229	.0291	.0360	.0437	.0521
2.8750	…	…	…	…	…	…	…	…	…	.0168	.0219	.0278	.0344	.0417	.0498
3.0000	…	…	…	…	…	…	…	…	…	…	.0210	.0266	.0329	.0399	.0476

Table 12. British Preferred Lengths of Plain (Parallel or Taper) and Gib-head Keys, Rectangular and Square Section ***BS 46:Part 1:1958 (1985) Appendix***

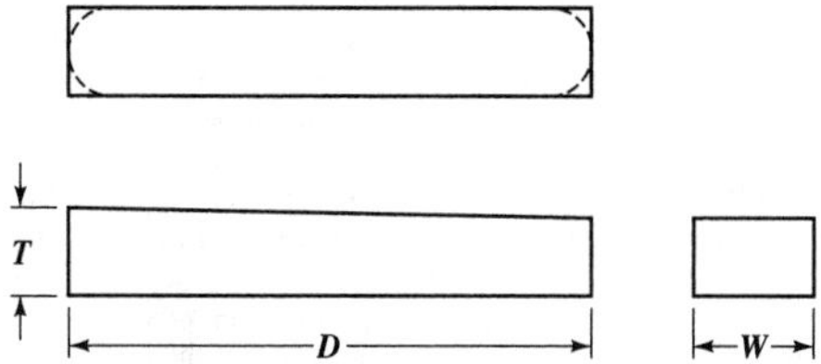

Plain Key Size $W \times T$	Overall Length, L														
	3/4	1	1 1/4	1 1/2	1 3/4	2	2 1/4	2 1/2	2 3/4	3	3 1/2	4	4 1/2	5	6
1/8 × 1/8	X	X													
3/16 × 3/16	X	X	X	X	X	X									
1/4 × 1/4	X	X	X	X	X	X	X	X	X	X					
5/16 × 1/4	X	X	X	X	X	X	X	X	X	X					
5/16 × 5/16	X	X	X	X	X	X	X	X	X	X					
3/8 × 1/4		X	X	X	X	X	X	X	X	X	X	X			
3/8 × 3/8		X	X	X	X	X	X	X	X	X	X	X			
7/16 × 5/16				X	X	X	X	X	X	X	X	X			
7/16 × 7/16					X	X	X	X	X	X	X	X			
1/2 × 5/16					X	X	X	X	X	X	X	X	X	X	
1/2 × 1/2						X	X	X	X	X	X	X	X	X	
5/8 × 7/16								X	X	X	X	X	X	X	
5/8 × 5/8								X		X	X	X	X	X	X
3/4 × 1/2										X	X	X	X	X	X
3/4 × 3/4											X	X	X	X	X
7/8 × 5/8											X	X	X	X	X

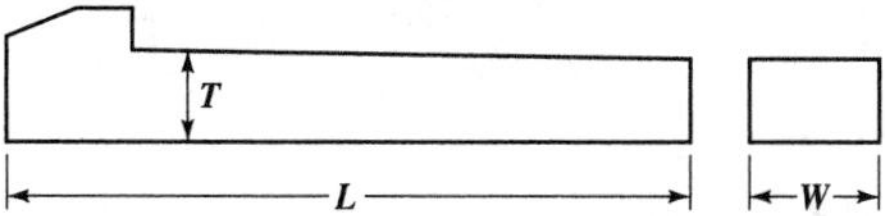

Gib-head Key Size, $W \times T$	Overall Length, L																
	1 1/2	1 3/4	2	2 1/4	2 1/2	2 3/4	3	3 1/2	4	4 1/2	5	5 1/2	6	6 1/2	7	7 1/2	8
3/16 × 3/16	X	X	X	X	X		X										
1/4 × 1/4	X	X	X	X	X	X	X	X	X								
5/16 × 1/4			X	X	X	X	X	X	X								
5/16 × 5/16			X	X	X	X	X	X	X	X							
3/8 × 1/4			X	X	X	X	X	X	X	X	X						
3/8 × 3/8			X	X	X	X	X	X	X	X	X	X	X				
7/16 × 5/16					X		X	X	X	X	X	X	X				
7/16 × 7/16					X		X	X	X	X	X	X	X				
1/2 × 5/16					X		X	X	X	X	X	X	X	X			
1/2 × 1/2					X		X	X	X	X	X	X	X	X			
5/8 × 7/16							X		X	X	X	X	X	X	X		
5/8 × 5/8									X	X	X	X	X	X	X		X
3/4 × 1/2									X	X	X	X	X	X	X	X	X
3/4 × 3/4									X		X		X		X		X
7/8 × 5/8											X		X	X	X	X	X
7/8 × 7/8													X		X		X
1 × 3/4													X		X	X	X
1 × 1													X		X		X

All dimension are in inches

SCREW THREAD SYSTEMS

Screw Thread Forms.—Of the various screw thread forms which have been developed, the most used are those having symmetrical sides inclined at equal angles with a vertical center line through the thread apex. Present-day examples of such threads would include the Unified, the Whitworth and the Acme forms. One of the early forms was the Sharp V which is now used only occasionally. Symmetrical threads are relatively easy to manufacture and inspect and hence are widely used on mass-produced general-purpose threaded fasteners of all types. In addition to general-purpose fastener applications, certain threads are used to repeatedly move or translate machine parts against heavy loads. For these so-called translation threads a stronger form is required. The most widely used translation thread forms are the square, the Acme, and the buttress. Of these, the square thread is the most efficient, but it is also the most difficult to cut owing to its parallel sides and it cannot be adjusted to compensate for wear. Although less efficient, the Acme form of thread has none of the disadvantages of the square form and has the advantage of being somewhat stronger. The buttress form is used for translation of loads in one direction only because of its non-symmetrical form and combines the high efficiency and strength of the square thread with the ease of cutting and adjustment of the Acme thread.

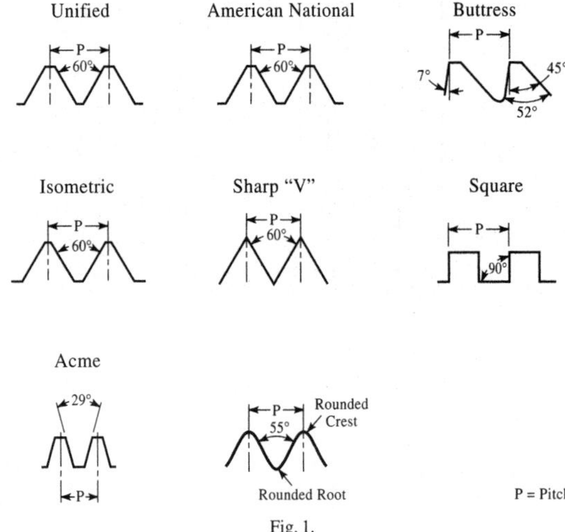

Fig. 1.

Parts of a Screw Thread

Major Diameter: The largest diameter of the thread. This is the distance between the crests of the thread measured perpendicular to the thread axis.

Pitch Diameter: The diameter of the thread used to establish the relationship, or fit, between an internal and external thread. The pitch diameter is the distance between the pitch points measured perpendicular to the thread axis. The pitch points are the points on the thread where the thread ridge and the space between the threads are the same width.

Minor Diameter: The smallest diameter of the thread. This is the distance between the roots of the thread measured perpendicular to the thread axis.

Thread Angle: The included angle of the thread form.

Helix Angle: The angle of the thread ridge measured from a plane perpendicular to the thread axis.

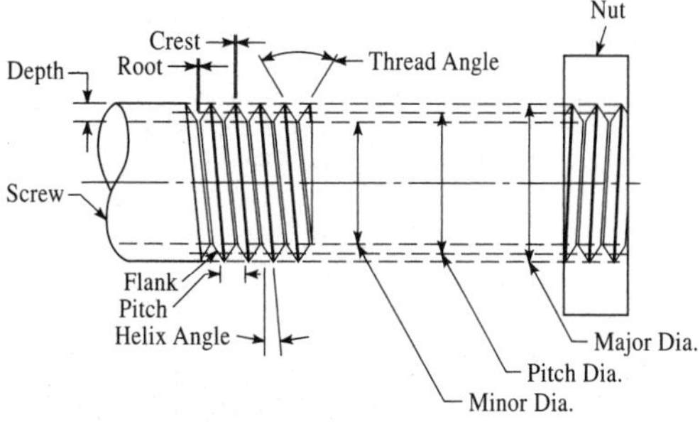

Fig. 2.

Pitch: The distance between the same points on adjacent threads. This is also the linear distance the thread will travel in one revolution.

Depth: The distance between the crest and root of a thread measured perpendicular to the axis of the thread on one side.

Flank: The sides of the thread that connect the root with the crest of each thread.

Root: The surface of the thread that joins the flanks of adjacent threads. The distance between the roots on opposite sides of the thread is called the root, or minor, diameter.

Crest: The surface of a thread that joins the flanks of the same thread. The distance between the crests on opposite sides of the thread is called the outside, or major, diameter.

Thread Designations.—Screw threads must conform to specific dimensions to work properly. To interpret thread dimensions correctly you must know how threads are specified. Since the Unified, Metric, Acme, and pipe are the most common thread forms, this discussion will cover just these four.

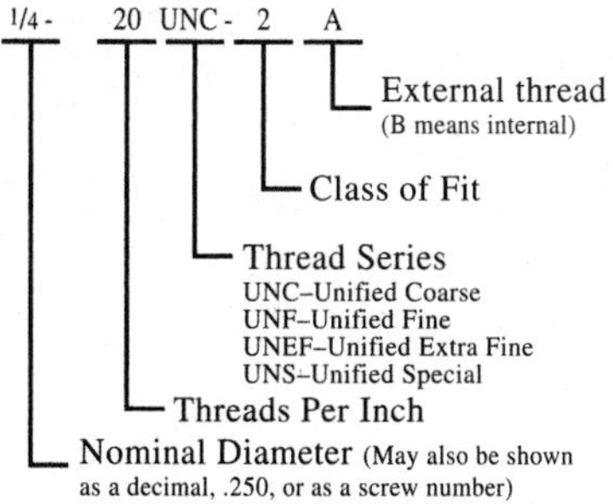

Fig. 3.

Unified: The size of a Unified thread is shown with a series of numbers and letters. This series follows the pattern shown in Fig. 3, and must be read from left to right in the sequence indicated. The first number in the series (¼) shows the nominal diameter of the thread. This number represents the fractional diameter, the decimal diameter. or the number of the screw size. The second number (20) indicates the number of threads per inch. The next group of letters (UNC) shows the thread series. In this example, Unified coarse is indicated. The last number and letter set (2A) indicates the thread class, or class of fit. The standard classes are 1, 2, and 3. Number is the loosest fit and number 3 is the tightest fit. The letters used here are A and B. The A means the thread is external and the B is used to designate internal threads.

Other information may be added to the standard designation, Fig. 4. Here the LH means left hand. All threads are considered to be right hand unless this LH designation is used. The last entry in this thread designation indicates the length of thread. This length may be shown in the thread designation or, as is more often the case, as a separate dimension on the print.

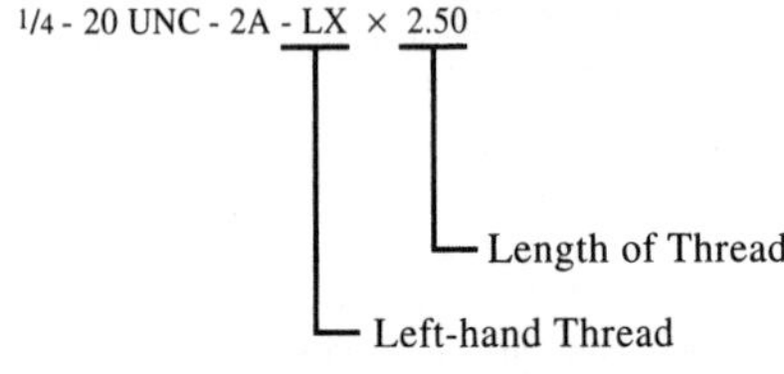

Fig. 4.

Metric: Metric threads are also designated by a series of numbers and letters. As shown in Fig. 5, the first letter and number (M10) indicate the nominal diameter of the thread. The letter M is used to identify all ISO (International Standards Organization) or American National Standard M Profile metric threads. The second number (1.5) specifies the pitch, or distance between adjacent threads, in millimeters. The last number and letter combination (4g) indicate the class of fit. In this designation lower case letters indicate external threads and upper case letters are used with internal threads. As with Unified threads, the abbreviation LH after the thread designation also means the thread is left handed.

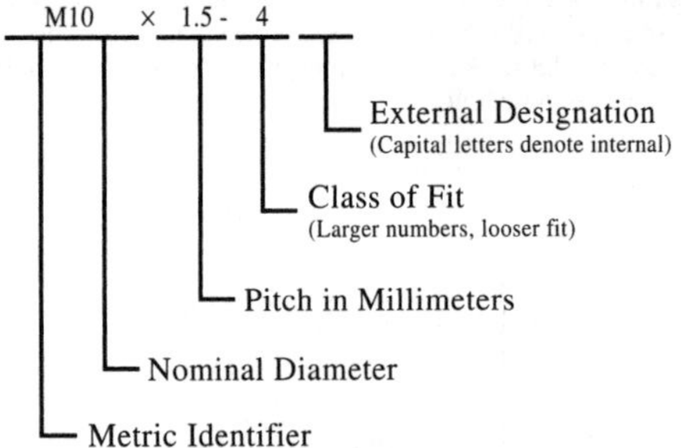

Fig. 5.

Acme: Acme threads are also designated by their nominal diameter, threads per inch, thread series, and fit, Fig. 6. The principal differences are the word ACME and the different thread classes. The thread classes commonly used for Acme threads are 2, 3, 4, and 5 with 2 the loosest and the tightest. The letter G indicates a general purpose thread. If the letter C is used, in place of the letter G, it indicates a centralizing

Acme thread: The LH, when used, means the thread is left handed. There is also a Stub Acme thread which is designated, for example, as ½-20 Stub Acme without a letter symbol.

Pipe: Pipe threads are specified on a print by the nominal size and threads per inch, Fig. 7. The letters NPT are used to specify taper pipe and NPS is used for straight pipe.

In addition to the thread designations discussed here there are also several other thread designations in common use today. To find specific information about these thread types, or more detailed information about those designations mentioned here, consult *Machinery's Handbook.*

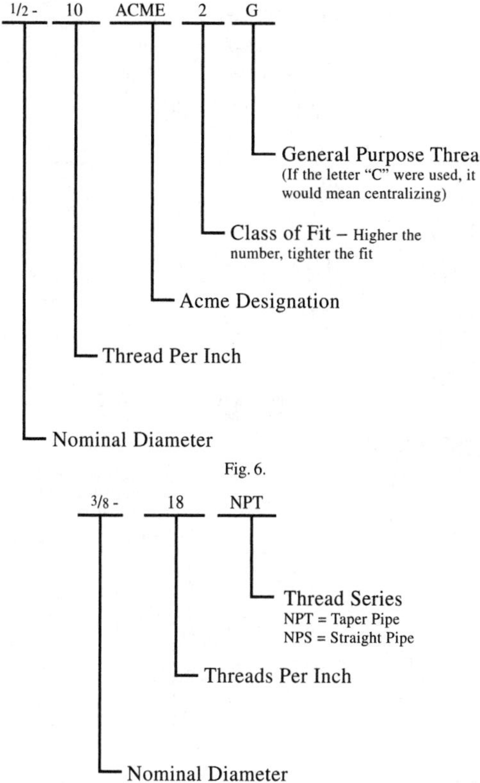

Fig. 6.

Fig. 7.

Cutting Screw Threads.—Right triangle calculations can be used to determine the depth of cut needed to cut a thread. The cutting tool is advanced along the angle set by the compound rest. This angle is approximately 30°. The depth of the thread is normally specified as the distance from the crest to the root of the thread measured perpendicular to the axis. The movement of the threading tool is along the hypotenuse formed by the side of the thread and the perpendicular depth. (Note, it is not along the perpendicular.) The length of the finished thread is the value that must be determined so the thread will be cut to the proper depth.

Hence, the angles that are known when cutting threads are the 30° threading angle and the right angle formed by the perpendicular. To find the depth of the thread, a thread table or the formula for sharp V threads may be used. This formula is

$$\text{thread depth} = 0.86603 \times \text{thread pitch}$$

$$\text{where thread pitch} = 1/\text{number of threads per inch}$$

For example, to cut 8 threads per inch on a 1.00-inch rod, the threading tool is set with its axis perpendicular to the axis of the rod to be threaded. (Fig. 8) The calculation is

$$\begin{aligned}\text{thread depth} &= 0.86603 \times \text{thread pitch}\\ &= 0.86603 \times \tfrac{1}{8}\text{ inch}\\ &= 0.86603 \times 0.125\text{ inch}\\ &= 0.10825\text{ inch}\end{aligned}$$

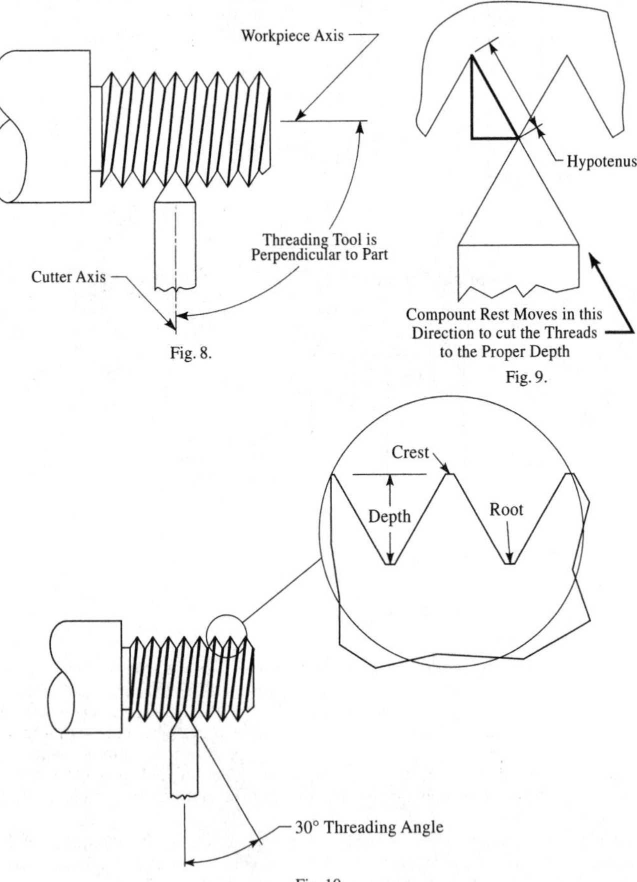

Fig. 8.

Fig. 9.

Fig. 10.

Table 1. Tap Drill Sizes and Percentage of Thread (Unified Threads)

Taps Size Threads per Inch	Drills				Percentage of Full Thread	Nut Minor Diameter, min[a] max
	No.	inch	mm	Decimal		
0-80	…	…	1.30	0.0512	54	0.0465
	…	…	1.25	0.0492	68	0.0514
	…	…	1.20	0.0472	79	
1-64	…	$\frac{1}{16}$	…	0.0625	51	0.0561
	…	…	1.55	0.0610	59	0.0623
	53	…	…	0.0595	66	
	…	…	1.50	0.0590	69	
	…	…	1.45	0.0571	77	
1-72	52	…	…	0.0635	53	0.0580
	…	$\frac{1}{16}$	…	0.0625	58	0.0635
	…	…	1.55	0.0610	66	
	53	…	…	0.0595	75	
	…	…	1.50	0.0590	77	
2-56	49	…	…	0.0730	56	0.0667
	…	…	1.80	0.0709	65	0.0737
	50	…	…	0.0700	69	
	…	…	1.75	0.0689	74	
2-64	…	…	1.90	0.0748	55	0.0691
	49	…	…	0.0730	64	0.0753
	…	…	1.85	0.0728	65	
	…	…	1.80	0.0709	74	
	50	…	…	0.0700	79	
3-48	…	…	2.15	0.0846	53	0.0764
	…	…	2.10	0.0827	60	0.0845
	45	…	…	0.0820	63	
	46	…	…	0.0810	66	
	47	…	…	0.0785	76	
	…	$\frac{5}{64}$	…	0.0781	77	
3-56	…	…	2.20	0.0866	53	0.0797
	44	…	…	0.0860	56	0.0865
	…	…	2.15	0.0846	62	
	…	…	2.10	0.0827	70	
	45	…	…	0.0820	73	
	46	…	…	0.0801	77	
	…	…	2.05	0.0807	79	
4-36	43	…	…	0.0890	64	0.0821[b]
	44	…	…	0.0860	72	0.0882[b]
4-40	…	…	2.40	0.0945	54	0.0849
	42	…	…	0.0935	57	0.0939
	…	…	2.35	0.0925	60	
	…	…	2.30	0.0905	66	

Table 1. Tap Drill Sizes and Percentage of Thread (Unified Threads) *(Continued)*

Taps Size Threads per Inch	Drills				Percentage of Full Thread	Nut Minor Diameter, min[a] max
	No.	inch	mm	Decimal		
4-40	43	…	…	0.0890	71	
	…	…	2.20	0.0866	78	
4-48	…	…	2.45	0.0964	57	0.0894
	41	…	…	0.0960	59	0.0968
	…	…	2.40	0.0945	64	
	…	3/32	…	0.0938	68	
	42	…	…	0.0935	68	
	…	…	2.35	0.0925	72	
	…	…	2.30	0.0905	79	
5-40	…	…	2.70	0.1063	57	0.0979
	37	…	…	0.1040	65	0.1062
	…	…	2.60	0.1024	70	
	38	…	…	0.1015	72	
	39	…	…	0.0995	78	
5-44	…	…	2.75	0.1083	57	0.1004
	36	…	…	0.1065	62	0.1079
	37	…	…	0.1040	71	
	…	…	2.6	0.1024	76	
	38	…	…	0.1015	79	
6-32	33	…	…	0.1130	62	0.1042
	34	…	…	0.1110	67	0.1130
	…	…	2.8	0.1102	68	
	35	…	…	0.1100	69	
	…	7/64	…	0.1094	70	
	…	…	2.75	0.1083	73	
	36	…	…	0.1065	77	
6-40	31	…	…	0.1200	55	0.1109
	…	…	3	0.1181	61	0.1186
	32	…	…	0.1160	68	
	…	…	2.9	0.1142	73	
	33	…	…	0.1130	77	
8-32	…	9/64	…	0.1406	57	0.1302
	28	…	…	0.1405	58	0.1389
	…	…	3.5	0.1378	65	
	29	…	…	0.1360	69	
	…	…	3.4	0.1339	74	
8-36	27	…	…	0.1440	55	0.1339
	…	…	3.6	0.1417	62	0.1416
	…	9/64	…	0.1406	68	
	28	…	…	0.1405	65	
		…	3.5	0.1378	73	
	29	…	…	0.1360	78	

Table 1. Tap Drill Sizes and Percentage of Thread (Unified Threads) *(Continued)*

Taps Size Threads per Inch	Drills				Percentage of Full Thread	Nut Minor Diameter, min[a] max
	No.	inch	mm	Decimal		
10-24	21	…	…	0.1590	57	0.1449
	22	…	…	0.1570	61	0.1555
	…	5⁄32	…	0.1563	62	
	23	…	…	0.1540	66	
	24	…	…	0.1520	67	
	25	…	…	0.1496	75	
	26	…	…	0.1470	79	
10-32	19	…	…	0.1660	59	0.1562
	…	…	4.1	0.1614	70	0.1641
	20	…	…	0.1610	71	
	21	…	…	0.1590	76	
12-24	14	…	…	0.1820	63	0.1709
	15	…	…	0.1800	67	0.1708
	16	…	…	0.1770	72	
	17	…	…	0.1730	79	
12-28	…	3⁄16	…	0.1875	61	0.1773
	13	…	…	0.1850	67	0.1857
	14	…	…	0.1820	73	
	15	…	…	0.1800	78	
1⁄4-20	4	…	…	0.2090	63	0.1959
	5	…	…	0.2055	69	0.2067
	6	…	…	0.2040	71	
	…	13⁄64	…	0.2031	72	
	7	…	…	0.2010	75	
	8	…	…	0.1990	79	
1⁄4-27	2	…	…	0.2210	61	0.2120
	…	…	…	0.2187	65	0.2180[b]
	3	…	…	0.2130	77	
1⁄4-28	2	…	…	0.2210	63	0.2113
	…	7⁄32	…	0.2187	67	0.2190
	…	…	5.5	0.2165	72	
	3	…	…	0.2130	80	
1⁄4-32	…	…	5.7	0.2244	63	0.2162
	2	…	…	0.2210	71	0.2208[b]
	…	7⁄28	…	0.2187	77	
5⁄16-18	…	N	…	0.2660	64	0.2524
	…	17⁄64	…	0.2656	65	0.2630
	…	G	…	0.2610	71	
	…	F	…	0.2570	77	

Table 1. Tap Drill Sizes and Percentage of Thread (Unified Threads) *(Continued)*

Taps Size Threads per Inch	Drills				Percentage of Full Thread	Nut Minor Diameter, min[a] max
	No.	inch	mm	Decimal		
5⁄16-24	…	J	…	0.2770	66	0.2674
	…	…	7	0.2756	68	0.2754
	…	I	…	0.2720	75	
5⁄16-27	…	9⁄32	…	0.2812	68	0.2718
	…	J	…	0.2770	74	0.2792[b]
5⁄16-32	…	…	7.3	0.2841	62	0.2787
	…	…	7.2	0.2835	75	0.2833[b]
	…	9⁄32	…	0.2812	77	
3⁄8-16	…	P	…	0.3230	64	0.3073
	…	…	8.1	0.3189	69	0.3182
	…	O	…	0.3160	72	
	…	5⁄16	…	0.3125	77	
3⁄8-24	…	R	…	0.3390	67	0.3299
	…	…	8.5	0.3345	74	0.3372
	…	Q	…	0.3320	79	
3⁄8-27	…	11⁄32	…	0.3437	65	0.3347
	…	R	…	0.3390	75	0.3416[b]
7⁄16-24	…	…	9.7	0.3818	60	0.3602
	…	V	…	0.3770	65	0.3717
	…	3⁄8	…	0.3750	67	
	…	U	…	0.3680	75	
7⁄16-20	…	X	…	0.3970	62	0.3834
	…	…	10	0.3937	67	0.3916
	…	25⁄64	…	0.3906	72	
	…	W	…	0.3860	79	
7⁄16-24	…	Y	…	0.4040	62	0.3925
	…	X	…	0.3970	74	0.3985[b]
7⁄16-27	…	13⁄32	…	0.4062	65	0.3982
	…	Y	…	0.4040	70	0.4043[b]
1⁄2-12	…	27⁄64	…	0.4219	72	0.4098
						0.4223
1⁄2-13	…	7⁄16	…	0.4375	62	0.4167
	…	…	11	0.4331	68	0.4284
	…	27⁄64	…	0.4219	78	
1⁄2-20	…	29⁄64	…	0.4531	72	0.4459
	…	[c]	…	0.4492	78	0.4537
1⁄2-27	…	15⁄32	…	0.4687	65	0.4618
	[c]	…	…	0.4640	75	0.4672[b]

Table 1. Tap Drill Sizes and Percentage of Thread (Unified Threads) *(Continued)*

Taps Size Threads per Inch	Drills				Percentage of Full Thread	Nut Minor Diameter, min[a] max
	No.	inch	mm	Decimal		
9⁄16-12	…	1⁄2	…	0.5000	58	0.4723
	…	…	12.5	0.4921	65	0.4843
	…	31⁄64	…	0.4844	72	
9⁄16-18	…	33⁄64	…	0.5156	65	0.5024
	…	…	13	0.5118	70	0.5106
9⁄16-27	…	17⁄32	…	0.5312	65	0.5234
	…	c	…	0.5265	75	0.5273[b]
5⁄8-11	…	…	14	0.5512	62	0.5266
	…	35⁄64	…	0.5469	66	0.5391
	…	17⁄32	…	0.5312	79	
5⁄8-18	…	37⁄64	…	0.5781	65	0.5649
	…	…	14.5	0.5709	75	0.5730
5⁄8-27	…	19⁄32	…	0.5937	65	0.5860
	…	c	…	0.5890	75	0.5912
3⁄4-10	…	…	17	0.6693	62	0.6417
	c	21⁄32	…	0.6563	72	0.6545[b]
	…	…	16.5	0.6496	77	
3⁄4-16	…	45⁄64	…	0.7031	58	0.6823
	…	…	17.5	0.6890	75	0.6908
	…	11⁄64	…	0.6875	77	
3⁄4-27	…	23⁄32	…	0.7187	65	0.7102
	c	…	…	0.7140	75	0.7164[c]
7⁄8-9	…	25⁄32	…	0.7813	65	0.7547
	…	49⁄64	…	0.7656	76	0.7679
7⁄8-14	…	13⁄16	…	0.8125	67	0.7977
	…	…	20.5	0.8071	73	0.8068
7⁄8-18	…	53⁄64	…	0.8281	65	0.8149
	c	…	…	0.8210	75	0.8223[b]
7⁄8-27	…	27⁄32	…	0.8437	65	0.8340
	c	…	…	0.8390	75	0.8406[c]
1-8	…	…	23.0	0.9055	58	0.8647
	…	57⁄64	…	0.8906	67	0.8797
	…	7⁄8	…	0.8750	77	
1-12	…	59⁄64	…	0.9219	72	0.9098
						0.9198

[a] Unified or American Standard Threads.
[b] Not based on Unified or American Thread Standards.
[c] Special drill required for this size as the next size smaller gives too great a percentage of thread.

Table 2. Tap Drill Sizes and Percentage of Thread (Metric Threads)

Metric Tap Size	Recommended Metric Drill				Closest Recommended Inch Drill				
	Tape Drill Size, mm	Theoretical Percentage of Thread	Probable Hole Size, mm	Percentage of Thread	Tap Drill Size	Tap Drill Equiv., in.	Theoretical Percentage of Thread	Probable Hole Size, in.	Percentage of Thread
M1.6 × 0.35	1.25	77	1.288	69	#55	0.052	61	0.0535	53
	1.30	66	1.339	59					
M1.8 × 0.35	1.45	77	1.488	69	#53	0.0575	64	0.061	55
	1.50	66	1.539	57					
M2 × 0.4	1.60	77	1.643	69	#52	0.0635	74	0.0652	66
M2.2 × 0.45	1.75	77	1.793	70	#50	0.0700	72	0.0717	65
M2.5 × 0.45	2.05	77	2.098	69	#45	0.0820	71	0.0839	63
M3 × 0.5	2.50	77	2.558	68	#39	0.0995	73	0.1018	64
M3.5 × 0.6	2.90	77	2.967	68	#32	0.1160	71	0.1186	63
M4 × 0.7	3.30	77	3.373	69	#29	0.1360	60	0.1389	52
	3.40	66	3.475	58					
M4.5 × 0.75	3.75	77	3.830	69	#25	0.1495	72	0.1527	64
M5 × 0.8	4.20	77	4.282	69	#18	0.1695	67	0.1730	58
M6 × 1	5.00	77	5.095	70	#8	0.1990	73	0.2028	65
M6.3 × 1	5.30	77	5.396	70	#4	0.2090	76	0.2128	69
M7 × 1	6.00	77	6.096	70	B	0.2380	74	0.2418	66
M8 × 1.25	6.75	77	6.853	71	I	0.2720	67	0.2761	61
	6.80	74	6.904	68					
M8 × 1	7.00	77	7.104	69	J	0.2770	74	0.2811	66
M10 × 1.5	8.50	77	8.611	71	K	0.3390	71	0.3434	66
M10 × 1.25	8.75	77	8.867	70	S	0.3480	71	0.3526	64
M12 × 1.75	10.20	79	10.319	74	13⁄32	0.4062	74	0.4109	69
	10.30	75	10.419	70	Z	0.4130	66	0.4177	61
M12 × 1.25	11.00	62	11.120	54	27⁄64	0.4129	79	0.4266	72
M14 × 2	12	77	12.121	72	31⁄64	0.4844	65	0.4892	61
M14 × 1.5	12.50	77	12.621	71	½	0.5	67	0.5048	60

Table 2. Tap Drill Sizes and Percentage of Thread (Metric Threads) *(Continued)*

Metric Tap Size	Recommended Metric Drill				Closest Recommended Inch Drill				
	Tape Drill Size, mm	Theoretical Percentage of Thread	Probable Hole Size, mm	Percentage of Thread	Tap Drill Size	Tap Drill Equiv., in.	Theoretical Percentage of Thread	Probable Hole Size, in.	Percentage of Thread
M16 × 2	14	77	14.125	72	9/16	0.5625	66	0.5674	61
M16 × 1.5	14.50	77	14.625	71	37/64	0.5781	68	0.5830	61
M18 × 2.5	15.50	77	15.626	73	5/8	0.6250	65	0.6300	62
M18 × 1.5	16.5	77	16.627	70	21/32	0.6562	68	0.6612	62
M20 × 2.5	17.50	77	17.633	73	45/64	0.7031	66	0.7083	62
M20 × 1.5	18.5	77	18.631	70	47/64	0.7344	69	0.7396	62
M22 × 2.5	19.5	77	19.632	73	25/32	0.7812	66	0.7864	62
M22 × 1.5	20.5	77	20.632	70	13/16	0.8125	70	0.8177	63
M24 × 3	21	77	21.151	73	27/32	0.8437	66	0.8496	62
M24 × 2	22	77	22.149	71	7/8	0.8750	68	0.8809	63
M27 × 3	24	77	24.158	73	61/64	0.9531	72	0.9593	68
M27 × 2	25	77	25.179	70	63/64	0.9844	77	0.9914	70
M30 × 3.5	26.5	77			1 1/16	1.0625	66		
M30 × 2	28	77			1 7/64	1.1094	70		
M33 × 3.5	29.5	77			1 11/64	1.1719	71		
M33 × 2	31	77			1 15/64	1.2344	63		
			Reaming Recommended					Reaming Recommended	
M36 × 4	32	77			1 17/64	1.2656	74		
M36 × 3	32	77			1 19/64	1.2969	78		
					1 5/16	1.3125	68		
M39 × 4	35	77			1 3/8	1.3750	78		
					1 25/64	1.3906	71		
M39 × 3	36	77			1 27/64	1.4219	74		

Courtesy of Cleveland Twist Drill Co.

Table 3. Tap Drill Sizes (Pipe Threads)

Taper Pipe				Straight Pipe			
Thread	Tap Drill	Thread	Tap Drill	Thread	Tap Drill	Thread	Tap Drill
1/8-27	R	1 1/2-11 1/2	1 47/64	1/8-27	S	1 1/2-11 1/2	1 3/4
1/4-18	7/16	2-11 1/2	2 7/32	1/4-18	29/64	2-11 1/2	2 7/32
3/8-18	37/64	2 1/2-8	2 5/8	3/8-18	19/32	2 1/2-8	2 21/32
1/2-14	23/32	3-8	3 1/4	1/2-14	47/64	3-8	3 9/32
3/4-14	59/64	3 1/2-8	3 3/4	3/4-14	15/16	3 1/2-8	3 25/32
1-11 1/2	1 5/32	4-8	4 1/4	1-11 1/2	1 3/16	4-8	4 9/32
1 1/4-11 1/2	1 1/2	1 1/4-11 1/2	1 1/2	1 1/4-11 1/2	1 33/64	...	...

Table 4. General Threading Formulas

Tap Drill Sizes	$S = D_m - 1.0825 \times P \times \%$ (Unified Threads) $S = D_m - 1.2990 \times P \times \%$ (American Standard Threads) $S = D_m - 1.0825 \times P \times \%$ (ISO Metric Threads)
Percentage of Full Thread	$\frac{D_m - S}{1.0825 \times P}$ = % of full thread (Unified Threads) $\frac{D_m - S}{1.2990 \times P}$ = % of full thread (American Standard Threads) $\frac{D_m - S}{1.0825 \times P}$ = % of full thread (ISO Metric Thread)
Determining Machine Screw Sizes	$N = \frac{D_m - 0.060}{0.013}$ $D_m = N \times 0.013 + 0.060$

All dimensions in mm

D_m = Major Diameter, P = Pitch, % = Percentage of Full Thread, S = Size of selected Tap Drill, N = Number of Machine Screw

Courtesy of the Society of Manufacturing Engineers

Table 5. Internal Metric Thread - M Profile Limiting Dimensions, ANSI/ASME B1.13M-1983 (R1995)

Basic Thread Designation	Toler. Class	Minor Diameter D_1		Pitch Diameter D_2			Major Diameter D	
		Min	Max	Min	Max	Tol	Min	Max[a]
M1.6 × 0.35	6H	1.221	1.321	1.373	1.458	0.085	1.600	1.736
M2 × 0.4	6H	1.567	1.679	1.740	1.830	0.090	2.000	2.148
M2.5 × 0.45	6H	2.013	2.138	2.208	2.303	0.095	2.500	2.660
M3 × 0.5	6H	2.459	2.599	2.675	2.775	0.100	3.000	3.172
M3.5 × 0.6	6H	2.850	3.010	3.110	3.222	0.112	3.500	3.699
M4 × 0.7	6H	3.242	3.422	3.545	3.663	0.118	4.000	4.219
M5 × 0.8	6H	4.134	4.334	4.480	4.605	0.125	5.000	5.240
M6 × 1	6H	4.917	5.153	5.350	5.500	0.150	6.000	6.294
M8 × 1.25	6H	6.647	6.912	7.188	7.348	0.160	8.000	8.340
M8 × 1	6H	6.917	7.153	7.350	7.500	0.150	8.000	8.294
M10 × 1.5	6H	8.376	8.676	9.026	9.206	0.180	10.000	10.396
M10 × 1.25	6H	8.647	8.912	9.188	9.348	0.160	10.000	10.340
M10 × 0.75	6H	9.188	9.378	9.513	9.645	0.132	10.000	10.240
M12 × 1.75	6H	10.106	10.441	10.863	11.063	0.200	12.000	12.453
M12 × 1.5	6H	10.376	10.676	11.026	11.216	0.190	12.000	12.406
M12 × 1.25	6H	10.647	10.912	11.188	11.368	0.180	12.000	12.360
M12 × 1	6H	10.917	11.153	11.350	11.510	0.160	12.000	12.304
M14 × 2	6H	11.835	12.210	12.701	12.913	0.212	14.000	14.501
M14 × 1.5	6H	12.376	12.676	13.026	13.216	0.190	14.000	14.406
M15 × 1	6H	13.917	14.153	14.350	14.510	0.160	15.000	15.304
M16 × 2	6H	13.835	14.210	14.701	14.913	0.212	16.000	16.501
M16 × 1.5	6H	14.376	14.676	15.026	15.216	0.190	16.000	16.406
M17 × 1	6H	15.917	16.153	16.350	16.510	0.160	17.000	17.304
M18 × 1.5	6H	16.376	16.676	17.026	17.216	0.190	18.000	18.406

Table 5. *(Continued)* **Internal Metric Thread - M Profile Limiting Dimensions, ANSI/ASME B1.13M-1983 (R1995)**

Basic Thread Designation	Toler. Class	Minor Diameter D_1		Pitch Diameter D_2			Major Diameter D	
		Min	Max	Min	Max	Tol	Min	Max[a]
M20 × 2.5	6H	17.294	17.744	18.376	18.600	0.224	20.000	20.585
M20 × 1.5	6H	18.376	18.676	19.026	19.216	0.190	20.000	20.406
M20 × 1	6H	18.917	19.153	19.350	19.510	0.160	20.000	20.304
M22 × 2.5	6H	19.294	19.744	20.376	20.600	0.224	22.000	22.585
M22 × 1.5	6H	20.376	20.676	21.026	21.216	0.190	22.000	22.406
M24 × 3	6H	20.752	21.252	22.051	22.316	0.265	24.000	24.698
M24 × 2	6H	21.835	22.210	22.701	22.925	0.224	24.000	24.513
M25 × 1.5	6H	23.376	23.676	24.026	24.226	0.200	25.000	25.416
M27 × 3	6H	23.752	24.252	25.051	25.316	0.265	27.000	27.698
M27 × 2	6H	24.835	25.210	25.701	25.925	0.224	27.000	27.513
M30 × 3.5	6H	26.211	26.771	27.727	28.007	0.280	30.000	30.785
M30 × 2	6H	27.835	28.210	28.701	28.925	0.224	30.000	30.513
M30 × 1.5	6H	28.376	28.676	29.026	29.226	0.200	30.000	30.416
M33 × 2	6H	30.835	31.210	31.701	31.925	0.224	33.000	33.513
M35 × 1.5	6H	33.376	33.676	34.026	34.226	0.200	35.000	35.416
M36 × 4	6H	31.670	32.270	33.402	33.702	0.300	36.000	36.877
M36 × 2	6H	33.835	34.210	34.701	34.925	0.224	36.000	36.513
M39 × 2	6H	36.835	37.210	37.701	37.925	0.224	39.000	39.513
M40 × 1.5	6H	38.376	38.676	39.026	39.226	0.200	40.000	40.416
M42 × 4.5	6H	37.129	37.799	39.077	39.392	0.315	42.000	42.965
M42 × 2	6H	39.835	40.210	40.701	40.925	0.224	42.000	42.513
M45 × 1.5	6H	43.376	43.676	44.026	44.226	0.200	45.000	45.416
M48 × 5	6H	42.587	43.297	44.752	45.087	0.335	48.000	49.057
M48 × 2	6H	45.835	46.210	46.701	46.937	0.236	48.000	48.525
M50 × 1.5	6H	48.376	48.676	49.026	49.238	0.212	50.000	50.428
M55 × 1.5	6H	53.376	53.676	54.026	54.238	0.212	55.000	55.428
M56 × 5.5	6H	50.046	50.796	52.428	52.783	0.355	56.000	57.149
M56 × 2	6H	53.835	54.210	54.701	54.937	0.236	56.000	56.525
M60 × 1.5	6H	58.376	58.676	59.026	59.238	0.212	60.000	60.428
M64 × 6	6H	57.505	58.305	60.103	60.478	0.375	64.000	65.241
M64 × 2	6H	61.835	62.210	62.701	62.937	0.236	64.000	64.525
M65 × 1.5	6H	63.376	63.676	64.026	64.238	0.212	65.000	65.428
M70 × 1.5	6H	68.376	68.676	69.026	69.238	0.212	70.000	70.428
M72 × 6	6H	65.505	66.305	68.103	68.478	0.375	72.000	73.241
M72 × 2	6H	69.835	70.210	70.701	70.937	0.236	72.000	72.525
M75 × 1.5	6H	73.376	73.676	74.026	74.238	0.212	75.000	75.428
M80 × 6	6H	73.505	74.305	76.103	76.478	0.375	80.000	81.241
M80 × 2	6H	77.835	78.210	78.701	78.937	0.236	80.000	80.525
M80 × 1.5	6H	78.376	78.676	79.026	79.238	0.212	80.000	80.428
M85 × 2	6H	82.835	83.210	83.701	83.937	0.236	85.000	85.525
M90 × 6	6H	83.505	84.305	86.103	86.478	0.375	90.000	91.241
M90 × 2	6H	87.835	88.210	88.701	88.937	0.236	90.000	90.525
M95 × 2	6H	92.835	93.210	93.701	93.951	0.250	95.000	95.539
M100 × 6	6H	93.505	94.305	96.103	96.503	0.400	100.000	101.266
M100 × 2	6H	97.835	98.210	98.701	98.951	0.250	100.000	100.539
M105 × 2	6H	102.835	103.210	103.701	103.951	0.250	105.000	105.539
M110 × 2	6H	107.835	108.210	108.701	108.951	0.250	110.000	110.539
M120 × 2	6H	117.835	118.210	118.701	118.951	0.250	120.000	120.539
M130 × 2	6H	127.835	128.210	128.701	128.951	0.250	130.000	130.539
M140 × 2	6H	137.835	138.210	138.701	138.951	0.250	140.000	140.539
M150 × 2	6H	147.835	148.210	148.701	148.951	0.250	150.000	150.539
M160 × 3	6H	156.752	157.252	158.051	158.351	0.300	160.000	160.733
M170 × 3	6H	166.752	167.252	168.051	168.351	0.300	170.000	170.733
M180 × 3	6H	176.752	177.252	178.051	178.351	0.300	180.000	180.733
M190 × 3	6H	186.752	187.252	188.051	188.386	0.335	190.000	190.768
M200 × 3	6H	196.752	197.252	198.051	198.386	0.335	200.000	200.768

[a] This reference dimension is used in design of tools, etc., and is not normally specified. Generally, major diameter acceptance is based upon maximum material condition gaging.

All dimensions are in millimeters.

Table 6. External Metric Thread—M Profile Limiting Dimensions ANSI/ASME B1.13M-1983 (R1995)

Basic Thread Desig.	Toler. Class	Allow. es^a	Major Diam.[b] d		Pitch Diam.[b] d_2			Minor-Diam., d_1[b]	Minor Diam., d_3[c]
			Max	Min	Max	Min	Tol.	Max	Min
M1.6 × 0.35	6g	0.019	1.581	1.496	1.354	1.291	0.063	1.202	1.075
M1.6 × 0.35	4g6g	0.019	1.581	1.496	1.354	1.314	0.040	1.202	1.098
M2 × 0.4	6g	0.019	1.981	1.886	1.721	1.654	0.067	1.548	1.408
M2 × 0.4	4g6g	0.019	1.981	1.886	1.721	1.679	0.042	1.548	1.433
M2.5 × 0.45	6g	0.020	2.480	2.380	2.188	2.117	0.071	1.993	1.840
M2.5 × 0.45	4g6g	0.020	2.480	2.380	2.188	2.143	0.045	1.993	1.866
M3 × 0.5	6g	0.020	2.980	2.874	2.655	2.580	0.075	2.439	2.272
M3 × 0.5	4g6g	0.020	2.980	2.874	2.655	2.607	0.048	2.439	2.299
M3.5 × 0.6	6g	0.021	3.479	3.354	3.089	3.004	0.085	2.829	2.635
M3.5 × 0.6	4g6g	0.021	3.479	3.354	3.089	3.036	0.053	2.829	2.667
M4 × 0.7	6g	0.022	3.978	3.838	3.523	3.433	0.090	3.220	3.002
M4 × 0.7	4g6g	0.022	3.978	3.838	3.523	3.467	0.056	3.220	3.036
M5 × 0.8	6g	0.024	4.976	4.826	4.456	4.361	0.095	4.110	3.869
M5 × 0.8	4g6g	0.024	4.976	4.826	4.456	4.396	0.060	4.110	3.904
M6 × 1	6g	0.026	5.974	5.794	5.324	5.212	0.112	4.891	4.596
M6 × 1	4g6g	0.026	5.974	5.794	5.324	5.253	0.071	4.891	4.637
M8 × 1.25	6g	0.028	7.972	7.760	7.160	7.042	0.118	6.619	6.272
M8 × 1.25	4g6g	0.028	7.972	7.760	7.160	7.085	0.075	6.619	6.315
M8 × 1	6g	0.026	7.974	7.794	7.324	7.212	0.112	6.891	6.596
M8 × 1	4g6g	0.026	7.974	7.794	7.324	7.253	0.071	6.891	6.637
M10 × 1.5	6g	0.032	9.968	9.732	8.994	8.862	0.132	8.344	7.938
M10 × 1.5	4g6g	0.032	9.968	9.732	8.994	8.909	0.085	8.344	7.985
M10 × 1.25	6g	0.028	9.972	9.760	9.160	9.042	0.118	8.619	8.272
M10 × 1.25	4g6g	0.028	9.972	9.760	9.160	9.085	0.075	8.619	8.315
M10 × 0.75	6g	0.022	9.978	9.838	9.491	9.391	0.100	9.166	8.929
M10 × 0.75	4g6g	0.022	9.978	9.838	9.491	9.428	0.063	9.166	8.966
M12 × 1.75	6g	0.034	11.966	11.701	10.829	10.679	0.150	10.072	9.601
M12 × 1.75	4g6g	0.034	11.966	11.701	10.829	10.734	0.095	10.072	9.656
M12 × 1.5	6g	0.032	11.968	11.732	10.994	10.854	0.140	10.344	9.930
M12 × 1.25	6g	0.028	11.972	11.760	11.160	11.028	0.132	10.619	10.258
M12 × 1.25	4g6g	0.028	11.972	11.760	11.160	11.075	0.085	10.619	10.305
M12 × 1	6g	0.026	11.974	11.794	11.324	11.206	0.118	10.891	10.590
M12 × 1	4g6g	0.026	11.974	11.794	11.324	11.249	0.075	10.891	10.633
M14 × 2	6g	0.038	13.962	13.682	12.663	12.503	0.160	11.797	11.271
M14 × 2	4g6g	0.038	13.962	13.682	12.663	12.563	0.100	11.797	11.331
M14 × 1.5	6g	0.032	13.968	13.732	12.994	12.854	0.140	12.344	11.930
M14 × 1.5	4g6g	0.032	13.968	13.732	12.994	12.904	0.090	12.344	11.980
M15 × 1	6g	0.026	14.974	14.794	14.324	14.206	0.118	13.891	13.590
M15 × 1	4g6g	0.026	14.974	14.794	14.324	14.249	0.075	13.891	13.633
M16 × 2	6g	0.038	15.962	15.682	14.663	14.503	0.160	13.797	13.271
M16 × 2	4g6g	0.038	15.962	15.682	14.663	14.563	0.100	13.797	13.331
M16 × 1.5	6g	0.032	15.968	15.732	14.994	14.854	0.140	14.344	13.930
M16 × 1.5	4g6g	0.032	15.968	15.732	14.994	14.904	0.090	14.344	13.980
M17 × 1	6g	0.026	16.974	16.794	16.324	16.206	0.118	15.891	15.590
M17 × 1	4g6g	0.026	16.974	16.794	16.324	16.249	0.075	15.891	15.633
M18 × 1.5	6g	0.032	17.968	17.732	16.994	16.854	0.140	16.344	15.930
M18 × 1.5	4g6g	0.032	17.968	17.732	16.994	16.904	0.090	16.344	15.980
M20 × 2.5	6g	0.042	19.958	19.623	18.334	18.164	0.170	17.252	16.624
M20 × 2.5	4g6g	0.042	19.958	19.623	18.334	18.228	0.106	17.252	16.688
M20 × 1.5	6g	0.032	19.968	19.732	18.994	18.854	0.140	18.344	17.930
M20 × 1.5	4g6g	0.032	19.968	19.732	18.994	18.904	0.090	18.344	17.980
M20 × 1	6g	0.026	19.974	19.794	19.324	19.206	0.118	18.891	18.590
M20 × 1	4g6g	0.026	19.974	19.794	19.324	19.249	0.075	18.891	18.633
M22 × 2.5	6g	0.042	21.958	21.623	20.334	20.164	0.170	19.252	18.624
M22 × 1.5	6g	0.032	21.968	21.732	20.994	20.854	0.140	20.344	19.930

Table 6. *(Continued)* **External Metric Thread—M Profile Limiting Dimensions ANSI/ASME B1.13M-1983 (R1995)**

Basic Thread Desig.	Toler. Class	Allow. es[a]	Major Diam.[b] d		Pitch Diam.[b] d_2			Minor-Diam., d_1[b]	Minor Diam., d_3[c]
			Max	Min	Max	Min	Tol.	Max	Min
M22 × 1.5	4g6g	0.032	21.968	21.732	20.994	20.904	0.090	20.344	19.980
M24 × 3	6g	0.048	23.952	23.577	22.003	21.803	0.200	20.704	19.955
M24 × 3	4g6g	0.048	23.952	23.557	22.003	21.878	0.125	20.704	20.030
M24 × 2	6g	0.038	23.962	23.682	22.663	22.493	0.170	21.797	21.261
M24 × 2	4g6g	0.038	23.962	23.682	22.663	22.557	0.106	21.797	21.325
M25 × 1.5	6g	0.032	24.968	24.732	23.994	23.844	0.150	23.344	22.920
M25 × 1.5	4g6g	0.032	24.968	24.732	23.994	23.899	0.095	23.344	22.975
M27 × 3	6g	0.048	26.952	26.577	25.003	24.803	0.200	23.704	22.955
M27 × 2	6g	0.038	26.962	26.682	25.663	25.493	0.170	24.797	24.261
M27 × 2	4g6g	0.038	29.962	26.682	25.663	25.557	0.106	24.797	24.325
M30 × 3.5	6g	0.053	29.947	29.522	27.674	27.462	0.212	26.158	25.306
M30 × 3.5	4g6g	0.053	29.947	29.522	27.674	27.542	0.132	26.158	25.386
M30 × 2	6g	0.038	29.962	29.682	28.663	28.493	0.170	27.797	27.261
M30 × 2	4g6g	0.038	29.962	29.682	28.663	28.557	0.106	27.797	27.325
M30 × 1.5	6g	0.032	29.968	29.732	28.994	28.844	0.150	28.344	27.920
M30 × 1.5	4g6g	0.032	29.968	29.732	28.994	28.899	0.095	28.344	27.975
M33 × 2	6g	0.038	32.962	32.682	31.663	31.493	0.170	30.797	30.261
M33 × 2	4g6g	0.038	32.962	32.682	31.663	31.557	0.106	30.797	30.325
M35 × 1.5	6g	0.032	34.968	34.732	33.994	33.844	0.150	33.344	33.920
M36 × 4	6g	0.060	35.940	35.465	33.342	33.118	0.224	31.610	30.654
M36 × 4	4g6g	0.060	35.940	35.465	33.342	33.202	0.140	31.610	30.738
M36 × 2	6g	0.038	35.962	35.682	34.663	34.493	0.170	33.797	33.261
M36 × 2	4g6g	0.038	35.962	35.682	34.663	34.557	0.106	33.797	33.325
M39 × 2	6g	0.038	38.962	38.682	37.663	37.493	0.170	36.797	36.261
M39 × 2	4g6g	0.038	38.962	38.682	37.663	37.557	0.106	36.797	36.325
M40 × 1.5	6g	0.032	39.968	39.732	38.994	38.844	0.150	38.344	37.920
M40 × 1.5	4g6g	0.032	39.968	39.732	38.994	38.899	0.095	38.344	37.975
M42 × 4.5	6g	0.063	41.937	41.437	39.014	38.778	0.236	37.066	36.006
M42 × 4.5	4g6g	0.063	41.937	41.437	39.014	38.864	0.150	37.066	36.092
M42 × 2	6g	0.038	41.962	41.682	40.663	40.493	0.170	39.797	39.261
M42 × 2	4g6g	0.038	41.962	41.682	40.663	40.557	0.106	39.797	39.325
M45 × 1.5	6g	0.032	44.968	44.732	43.994	43.844	0.150	43.344	42.920
M45 × 1.5	4g6g	0.032	44.968	44.732	43.994	43.899	0.095	43.344	42.975
M48 × 5	6g	0.071	47.929	47.399	44.681	44.431	0.250	42.516	41.351
M48 × 5	4g6g	0.071	47.929	47.399	44.681	44.521	0.160	42.516	41.441
M48 × 2	6g	0.038	47.962	47.682	46.663	46.483	0.180	45.797	45.251
M48 × 2	4g6g	0.038	47.962	47.682	46.663	46.551	0.112	45.797	45.319
M50 × 1.5	6g	0.032	49.968	49.732	48.994	48.834	0.160	48.344	47.910
M50 × 1.5	4g6g	0.032	49.968	49.732	48.994	48.894	0.100	48.344	47.970
M55 × 1.5	6g	0.032	54.968	54.732	53.994	53.834	0.160	53.344	52.910
M55 × 1.5	4g6g	0.032	54.968	54.732	53.994	53.894	0.100	53.344	52.970
M56 × 5.5	6g	0.075	55.925	55.365	52.353	52.088	0.265	49.971	48.700
M56 × 5.5	4g6g	0.075	55.925	55.365	52.353	52.183	0.170	49.971	48.795
M56 × 2	6g	0.038	55.962	55.682	54.663	54.483	0.180	53.797	53.251
M56 × 2	4g6g	0.038	55.962	55.682	54.663	54.551	0.112	53.797	53.319
M60 × 1.5	6	0.032	59.968	59.732	58.994	58.834	0.160	58.344	57.910
M60 × 1.5	4g6g	0.032	59.968	59.732	58.994	58.894	0.100	58.344	57.970
M64 × 6	6g	0.080	63.920	63.320	60.023	59.743	0.280	57.425	56.047
M64 × 6	4g6g	0.080	63.920	63.320	60.023	59.843	0.180	57.425	56.147
M64 × 2	6g	0.038	63.962	63.682	62.663	62.483	0.180	61.797	61.251
M64 × 2	4g6g	0.038	63.962	63.682	62.663	62.551	0.112	61.797	61.319
M65 × 1.5	6g	0.032	64.968	64.732	63.994	63.834	0.160	63.344	62.910
M65 × 1.5	4g6g	0.032	64.968	64.732	63.994	63.894	0.100	63.344	62.970
M70 × 1.5	6g	0.032	69.968	69.732	68.994	68.834	0.160	68.344	67.910
M70 × 1.5	4g6g	0.032	69.968	69.732	68.994	68.894	0.100	68.344	67.970
M72 × 6	6g	0.080	71.920	71.320	68.023	67.743	0.280	65.425	64.047

Table 6. *(Continued)* **External Metric Thread—M Profile Limiting Dimensions ANSI/ASME B1.13M-1983 (R1995)**

Basic Thread Desig.	Toler. Class	Allow. es[a]	Major Diam.[b] d		Pitch Diam.[b] d_2			Minor-Diam., d_1[b]	Minor Diam., d_3[c]
			Max	Min	Max	Min	Tol.	Max	Min
M72 × 6	4g6g	0.080	71.920	71.320	68.023	67.843	0.180	65.425	64.147
M72 × 2	6g	0.038	71.962	71.682	70.663	70.483	0.180	69.797	69.251
M72 × 2	4g6g	0.038	71.962	71.682	70.663	70.551	0.112	69.797	69.319
M75 × 1.5	6g	0.032	74.968	74.732	73.994	73.834	0.160	73.344	72.910
M75 × 1.5	4g6g	0.032	74.968	74.732	73.994	73.894	0.100	73.344	72.970
M80 × 6	6g	0.080	79.920	79.320	76.023	75.743	0.280	73.425	72.047
M80 × 6	4g6g	0.080	79.920	79.320	76.023	75.843	0.180	73.425	72.147
M80 × 2	6g	0.038	79.962	79.682	78.663	78.483	0.180	77.797	77.251
M80 × 2	4g6g	0.038	79.962	79.682	78.663	78.551	0.112	77.797	77.319
M80 × 1.5	6g	0.032	79.968	79.732	78.994	78.834	0.160	78.344	77.910
M80 × 1.5	4g6g	0.032	79.968	79.732	78.994	78.894	0.100	78.334	77.970
M85 × 2	6g	0.038	84.962	84.682	83.663	83.483	0.180	82.797	82.251
M85 × 2	4g6g	0.038	84.962	84.682	83.663	83.551	0.112	82.797	82.319
M90 × 6	6g	0.080	89.920	89.320	86.023	85.743	0.280	83.425	82.047
M90 × 6	4g6g	0.080	89.920	89.320	86.023	85.843	0.180	83.425	82.147
M90 × 2	6g	0.038	89.962	89.682	88.663	88.483	0.180	87.797	87.251
M90 × 2	4g6g	0.038	89.962	89.682	88.663	88.551	0.112	87.797	87.319
M95 × 2	6g	0.038	94.962	94.682	93.663	93.473	0.190	92.797	92.241
M95 × 2	4g6g	0.038	94.962	94.682	93.663	93.545	0.118	92.797	92.313
M100 × 6	6g	0.080	99.920	99.320	96.023	95.723	0.300	93.425	92.027
M100 × 6	4g6g	0.080	99.920	99.320	96.023	95.833	0.190	93.425	92.137
M100 × 2	6g	0.038	99.962	99.682	98.663	98.473	0.190	97.797	97.241
M100 × 2	4g6g	0.038	99.962	99.682	98.663	98.545	0.118	97.797	97.313
M105 × 2	6g	0.038	104.962	104.682	103.663	103.473	0.190	102.797	102.241
M105 × 2	4g6g	0.038	104.962	104.682	103.663	103.545	0.118	102.797	102.313
M110 × 2	6g	0.038	109.962	109.682	108.663	108.473	0.190	107.797	107.241
M110 × 2	4g6g	0.038	109.962	109.682	108.663	108.545	0.118	107.797	107.313
M120 × 2	6g	0.038	119.962	119.682	118.663	118.473	0.190	117.797	117.241
M120 × 2	4g6g	0.038	119.962	119.682	118.663	118.545	0.118	117.797	117.313
M130 × 2	6g	0.038	129.962	129.682	128.663	128.473	0.190	127.797	127.241
M130 × 2	4g6g	0.038	139.962	139.682	138.663	138.545	0.118	137.797	137.313
M140 × 2	6g	0.038	139.962	139.682	138.663	138.473	0.190	137.797	137.241
M140 × 2	4g6g	0.038	139.962	139.682	138.663	138.545	0.118	137.797	137.313
M150 × 2	6g	0.038	149.962	149.682	148.663	148.473	0.190	147.797	147.241
M150 × 2	4g6g	0.038	149.962	149.682	148.663	148.545	0.118	147.797	147.313
M160 × 3	6g	0.048	159.952	159.577	158.003	157.779	0.224	156.704	155.931
M160 × 3	4g6g	0.048	159.952	159.577	158.003	157.863	0.140	156.704	156.015
M170 × 3	6g	0.048	169.952	169.577	168.003	167.779	0.224	166.704	165.931
M170 × 3	4g6g	0.048	169.952	169.577	168.003	167.863	0.140	166.704	166.015
M180 × 3	6g	0.048	179.952	179.577	178.003	177.779	0.224	176.704	175.931
M180 × 3	4g6g	0.048	179.952	179.577	178.003	177.863	0.140	176.704	176.015
M190 × 3	6g	0.048	189.952	189.577	188.003	187.753	0.250	186.704	185.905
M190 × 3	4g6g	0.048	189.952	189.577	188.003	187.843	0.160	186.704	185.995
M200 × 3	6g	0.048	199.952	199.577	198.003	197.753	0.250	196.704	195.905
M200 × 3	4g6g	0.048	199.952	199.577	198.003	197.843	0.160	196.704	195.995

[a] *es* is an absolute value.

[b] (Flat form) For screw threads at maximum limits of tolerance position *h*, add the absolute value *es* to the maximum diameters required. For maximum major diameter this value is the basic thread size listed in Table 5 as Minimum Major Diameter (D_{min}; for maximum pitch diameter this value is the same as listed in Table 5 as Minimum Pitch Diameter ($D_{2\ min}$); and for maximum minor diameter this value is the same as listed in Table 5 as Minimum Minor Diameter ($D_{1\ min}$).

[c] (Rounded form) This reference dimension is used in the design of tools, etc. In dimensioning external threads it is not normally specified. Generally minor diameter acceptance is based upon maximum material condition gaging.

All dimensions are in millimeters.

Table 7. Standard Series and Selected Combinations — Unified Screw Threads

Nominal Size, Threads per Inch, and Series Designation[a]	External[b]								Internal[b]					
	Class	Allow-ance	Major Diameter			Pitch Diameter		UNR Minor Dia.,[c] Max (Ref.)	Class	Minor Diameter		Pitch Diameter		Major Diameter
			Max[d]	Min	Min[e]	Max[d]	Min			Min	Max	Min	Max	Min
0–80 UNF	2A	0.0005	0.0595	0.0563	—	0.0514	0.0496	0.0446	2B	0.0465	0.0514	0.0519	0.0542	0.0600
	3A	0.0000	0.0600	0.0568	—	0.0519	0.0506	0.0451	3B	0.0465	0.0514	0.0519	0.0536	0.0600
1–64 UNC	2A	0.0006	0.0724	0.0686	—	0.0623	0.0603	0.0538	2B	0.0561	0.0623	0.0629	0.0655	0.0730
	3A	0.0000	0.0730	0.0692	—	0.0629	0.0614	0.0544	3B	0.0561	0.0623	0.0629	0.0648	0.0730
1–72 UNF	2A	0.0006	0.0724	0.0689	—	0.0634	0.0615	0.0559	2B	0.0580	0.0635	0.0640	0.0665	0.0730
	3A	0.0000	0.0730	0.0695	—	0.0640	0.0626	0.0565	3B	0.0580	0.0635	0.0640	0.0659	0.0730
2–56 UNC	2A	0.0006	0.0854	0.0813	—	0.0738	0.0717	0.0642	2B	0.0667	0.0737	0.0744	0.0772	0.0860
	3A	0.0000	0.0860	0.0819	—	0.0744	0.0728	0.0648	3B	0.0667	0.0737	0.0744	0.0765	0.0860
2–64 UNF	2A	0.0006	0.0854	0.0816	—	0.0753	0.0733	0.0668	2B	0.0691	0.0753	0.0759	0.0786	0.0860
	3A	0.0000	0.0860	0.0822	—	0.0759	0.0744	0.0674	3B	0.0691	0.0753	0.0759	0.0779	0.0860
3–48 UNC	2A	0.0007	0.0983	0.0938	—	0.0848	0.0825	0.0734	2B	0.0764	0.0845	0.0855	0.0885	0.0990
	3A	0.0000	0.0990	0.0945	—	0.0855	0.0838	0.0741	3B	0.0764	0.0845	0.0855	0.0877	0.0990
3–56 UNF	2A	0.0007	0.0983	0.0942	—	0.0867	0.0845	0.0771	2B	0.0797	0.0865	0.0874	0.0902	0.0990
	3A	0.0000	0.0990	0.0949	—	0.0874	0.0858	0.0778	3B	0.0797	0.0865	0.0874	0.0895	0.0990
4–40 UNC	2A	0.0008	0.1112	0.1061	—	0.0950	0.0925	0.0814	2B	0.0849	0.0939	0.0958	0.0991	0.1120
	3A	0.0000	0.1120	0.1069	—	0.0958	0.0939	0.0822	3B	0.0849	0.0939	0.0958	0.0982	0.1120
4–48 UNF	2A	0.0007	0.1113	0.1068	—	0.0978	0.0954	0.0864	2B	0.0894	0.0968	0.0985	0.1016	0.1120
	3A	0.0000	0.1120	0.1075	—	0.0985	0.0967	0.0871	3B	0.0894	0.0968	0.0985	0.1008	0.1120
5–40 UNC	2A	0.0008	0.1242	0.1191	—	0.1080	0.1054	0.0944	2B	0.0979	0.1062	0.1088	0.1121	0.1250
	3A	0.0000	0.1250	0.1199	—	0.1088	0.1069	0.0952	3B	0.0979	0.1062	0.1088	0.1113	0.1250
5–44 UNF	2A	0.0007	0.1243	0.1195	—	0.1095	0.1070	0.0972	2B	0.1004	0.1079	0.1102	0.1134	0.1250
	3A	0.0000	0.1250	0.1202	—	0.1102	0.1083	0.0979	3B	0.1004	0.1079	0.1102	0.1126	0.1250
6–32 UNC	2A	0.0008	0.1372	0.1312	—	0.1169	0.1141	0.1000	2B	0.104	0.114	0.1177	0.1214	0.1380
	3A	0.0000	0.1380	0.1320	—	0.1177	0.1156	0.1008	3B	0.1040	0.1140	0.1177	0.1204	0.1380
6–40 UNF	2A	0.0008	0.1372	0.1321	—	0.1210	0.1184	0.1074	2B	0.111	0.119	0.1218	0.1252	0.1380
	3A	0.0000	0.1380	0.1329	—	0.1218	0.1198	0.1082	3B	0.1110	0.1186	0.1218	0.1243	0.1380
8–32 UNC	2A	0.0009	0.1631	0.1571	—	0.1428	0.1399	0.1259	2B	0.130	0.139	0.1437	0.1475	0.1640
	3A	0.0000	0.1640	0.1580	—	0.1437	0.1415	0.1268	3B	0.1300	0.1389	0.1437	0.1465	0.1640
8–36 UNF	2A	0.0008	0.1632	0.1577	—	0.1452	0.1424	0.1301	2B	0.134	0.142	0.1460	0.1496	0.1640
	3A	0.0000	0.1640	0.1585	—	0.1460	0.1439	0.1309	3B	0.1340	0.1416	0.1460	0.1487	0.1640
10–24 UNC	2A	0.0010	0.1890	0.1818	—	0.1619	0.1586	0.1394	2B	0.145	0.156	0.1629	0.1672	0.1900

Table 7. *(Continued)* **Standard Series and Selected Combinations — Unified Screw Threads**

Nominal Size, Threads per Inch, and Series Designation[a]	External[b]								Internal[b]					
	Class	Allow-ance	Major Diameter			Pitch Diameter		UNR Minor Dia.,[c] Max (Ref.)	Class	Minor Diameter		Pitch Diameter		Major Diameter
			Max[d]	Min	Min[e]	Max[d]	Min			Min	Max	Min	Max	Min
	3A	0.0000	0.1900	0.1828	—	0.1629	0.1604	0.1404	3B	0.1450	0.1555	0.1629	0.1661	0.1900
10–28 UNS	2A	0.0010	0.1890	0.1825	—	0.1658	0.1625	0.1464	2B	0.151	0.160	0.1668	0.1711	0.1900
10–32 UNF	2A	0.0009	0.1891	0.1831	—	0.1688	0.1658	0.1519	2B	0.156	0.164	0.1697	0.1736	0.1900
	3A	0.0000	0.1900	0.1840	—	0.1697	0.1674	0.1528	3B	0.1560	0.1641	0.1697	0.1726	0.1900
10–36 UNS	2A	0.0009	0.1891	0.1836	—	0.1711	0.1681	0.1560	2B	0.160	0.166	0.1720	0.1759	0.1900
10–40 UNS	2A	0.0009	0.1891	0.1840	—	0.1729	0.1700	0.1592	2B	0.163	0.169	0.1738	0.1775	0.1900
10–48 UNS	2A	0.0008	0.1892	0.1847	—	0.1757	0.1731	0.1644	2B	0.167	0.172	0.1765	0.1799	0.1900
10–56 UNS	2A	0.0007	0.1893	0.1852	—	0.1777	0.1752	0.1681	2B	0.171	0.175	0.1784	0.1816	0.1900
12–24 UNC	2A	0.0010	0.2150	0.2078	—	0.1879	0.1845	0.1654	2B	0.171	0.181	0.1889	0.1933	0.2160
	3A	0.0000	0.2160	0.2088	—	0.1889	0.1863	0.1664	3B	0.1710	0.1807	0.1889	0.1922	0.2160
12–28 UNF	2A	0.0010	0.2150	0.2085	—	0.1918	0.1886	0.1724	2B	0.177	0.186	0.1928	0.1970	0.2160
	3A	0.0000	0.2160	0.2095	—	0.1928	0.1904	0.1734	3B	0.1770	0.1857	0.1928	0.1959	0.2160
12–32 UNEF	2A	0.0009	0.2151	0.2091	—	0.1948	0.1917	0.1779	2B	0.182	0.190	0.1957	0.1998	0.2160
	3A	0.0000	0.2160	0.2100	—	0.1957	0.1933	0.1788	3B	0.1820	0.1895	0.1957	0.1988	0.2160
12–36 UNS	2A	0.0009	0.2151	0.2096	—	0.1971	0.1941	0.1821	2B	0.186	0.192	0.1980	0.2019	0.2160
12–40 UNS	2A	0.0009	0.2151	0.2100	—	0.1989	0.1960	0.1835	2B	0.189	0.195	0.1998	0.2035	0.2160
12–48 UNS	2A	0.0008	0.2152	0.2107	—	0.2017	0.1991	0.1904	2B	0.193	0.198	0.2025	0.2059	0.2160
12–56 UNS	2A	0.0007	0.2153	0.2112	—	0.2037	0.2012	0.1941	2B	0.197	0.201	0.2044	0.2076	0.2160
¼–20 UNC	1A	0.0011	0.2489	0.2367	—	0.2164	0.2108	0.1894	1B	0.196	0.207	0.2175	0.2248	0.2500
	2A	0.0011	0.2489	0.2408	0.2367	0.2164	0.2127	0.1894	2B	0.196	0.207	0.2175	0.2224	0.2500
	3A	0.0000	0.2500	0.2419	—	0.2175	0.2147	0.1905	3B	0.1960	0.2067	0.2175	0.2211	0.2500
¼–24 UNS	2A	0.0011	0.2489	0.2417	—	0.2218	0.2181	0.1993	2B	0.205	0.215	0.2229	0.2277	0.2500
¼–27 UNS	2A	0.0010	0.2490	0.2423	—	0.2249	0.2214	0.2049	2B	0.210	0.219	0.2259	0.2304	0.2500
¼–28 UNF	1A	0.0010	0.2490	0.2392	—	0.2258	0.2208	0.2064	1B	0.211	0.220	0.2268	0.2333	0.2500
	2A	0.0010	0.2490	0.2425	—	0.2258	0.2225	0.2064	2B	0.211	0.220	0.2268	0.2311	0.2500
	3A	0.0000	0.2500	0.2435	—	0.2268	0.2243	0.2074	3B	0.2110	0.2190	0.2268	0.2300	0.2500
¼–32 UNEF	2A	0.0010	0.2490	0.2430	—	0.2287	0.2255	0.2118	2B	0.216	0.224	0.2297	0.2339	0.2500
	3A	0.0000	0.2500	0.2440	—	0.2297	0.2273	0.2128	3B	0.2160	0.2229	0.2297	0.2328	0.2500
¼–36 UNS	2A	0.0009	0.2491	0.2436	—	0.2311	0.2280	0.2161	2B	0.220	0.226	0.2320	0.2360	0.2500
¼–40 UNS	2A	0.0009	0.2491	0.2440	—	0.2329	0.2300	0.2193	2B	0.223	0.229	0.2338	0.2376	0.2500

Table 7. *(Continued)* **Standard Series and Selected Combinations — Unified Screw Threads**

Nominal Size, Threads per Inch, and Series Designation[a]	External[b]								Internal[b]					
	Class	Allowance	Major Diameter			Pitch Diameter		UNR Minor Dia.,[c] Max (Ref.)	Class	Minor Diameter		Pitch Diameter		Major Diameter
			Max[d]	Min	Min[e]	Max[d]	Min			Min	Max	Min	Max	Min
1/4–48 UNS	2A	0.0008	0.2492	0.2447	—	0.2357	0.2330	0.2243	2B	0.227	0.232	0.2365	0.2401	0.2500
1/4–56 UNS	2A	0.0008	0.2492	0.2451	—	0.2376	0.2350	0.2280	2B	0.231	0.235	0.2384	0.2417	0.2500
5/16–18 UNC	1A	0.0012	0.3113	0.2982	—	0.2752	0.2691	0.2452	1B	0.252	0.265	0.2764	0.2843	0.3125
	2A	0.0012	0.3113	0.3026	0.2982	0.2752	0.2712	0.2452	2B	0.252	0.265	0.2764	0.2817	0.3125
	3A	0.0000	0.3125	0.3038	—	0.2764	0.2734	0.2464	3B	0.2520	0.2630	0.2764	0.2803	0.3125
5/16–20 UN	2A	0.0012	0.3113	0.3032	—	0.2788	0.2748	0.2518	2B	0.258	0.270	0.2800	0.2852	0.3125
	3A	0.0000	0.3125	0.3044	—	0.2800	0.2770	0.2530	3B	0.2580	0.2680	0.2800	0.2839	0.3125
5/16–24 UNF	1A	0.0011	0.3114	0.3006	—	0.2843	0.2788	0.2618	1B	0.267	0.277	0.2854	0.2925	0.3125
	2A	0.0011	0.3114	0.3042	—	0.2843	0.2806	0.2618	2B	0.267	0.277	0.2854	0.2902	0.3125
	3A	0.0000	0.3125	0.3053	—	0.2854	0.2827	0.2629	3B	0.2670	0.2754	0.2854	0.2890	0.3125
5/16–27 UNS	2A	0.0010	0.3115	0.3048	—	0.2874	0.2839	0.2674	2B	0.272	0.281	0.2884	0.2929	0.3125
5/16–28 UN	2A	0.0010	0.3115	0.3050	—	0.2883	0.2849	0.2689	2B	0.274	0.282	0.2893	0.2937	0.3125
	3A	0.0000	0.3125	0.3060	—	0.2893	0.2867	0.2699	3B	0.2740	0.2807	0.2893	0.2926	0.3125
5/16–32 UNEF	2A	0.0010	0.3115	0.3055	—	0.2912	0.2880	0.2743	2B	0.279	0.286	0.2922	0.2964	0.3125
	3A	0.0000	0.3125	0.3065	—	0.2922	0.2898	0.2753	3B	0.2790	0.2847	0.2922	0.2953	0.3125
5/16–36 UNS	2A	0.0009	0.3116	0.3061	—	0.2936	0.2905	0.2785	2B	0.282	0.289	0.2945	0.2985	0.3125
5/16–40 UNS	2A	0.0009	0.3116	0.3065	—	0.2954	0.2925	0.2818	2B	0.285	0.291	0.2963	0.3001	0.3125
5/16–48 UNS	2A	0.0008	0.3117	0.3072	—	0.2982	0.2955	0.2869	2B	0.290	0.295	0.2990	0.3026	0.3125
3/8–16 UNC	1A	0.0013	0.3737	0.3595	—	0.3331	0.3266	0.2992	1B	0.307	0.321	0.3344	0.3429	0.3750
	2A	0.0013	0.3737	0.3643	0.3595	0.3331	0.3287	0.2992	2B	0.307	0.321	0.3344	0.3401	0.3750
	3A	0.0000	0.3750	0.3656	—	0.3344	0.3311	0.3005	3B	0.3070	0.3182	0.3344	0.3387	0.3750
3/8–18 UNS	2A	0.0013	0.3737	0.3650	—	0.3376	0.3333	0.3076	2B	0.315	0.328	0.3389	0.3445	0.3750
3/8–20 UN	2A	0.0012	0.3738	0.3657	—	0.3413	0.3372	0.3143	2B	0.321	0.332	0.3425	0.3479	0.3750
	3A	0.0000	0.3750	0.3669	—	0.3425	0.3394	0.3155	3B	0.3210	0.3297	0.3425	0.3465	0.3750
3/8–24 UNF	1A	0.0011	0.3739	0.3631	—	0.3468	0.3411	0.3243	1B	0.330	0.340	0.3479	0.3553	0.3750
	2A	0.0011	0.3739	0.3667	—	0.3468	0.3430	0.3243	2B	0.330	0.340	0.3479	0.3528	0.3750
3/8–24 UNF	3A	0.0000	0.3750	0.3678	—	0.3479	0.3450	0.3254	3B	0.3300	0.3372	0.3479	0.3516	0.3750
3/8–27 UNS	2A	0.0011	0.3739	0.3672	—	0.3498	0.3462	0.3298	2B	0.335	0.344	0.3509	0.3556	0.3750

Table 7. *(Continued)* **Standard Series and Selected Combinations — Unified Screw Threads**

Nominal Size, Threads per Inch, and Series Designation[a]	External[b]								Internal[b]					
	Class	Allowance	Major Diameter			Pitch Diameter		UNR Minor Dia.,[c] Max (Ref.)	Class	Minor Diameter		Pitch Diameter		Major Diameter
			Max[d]	Min	Min[e]	Max[d]	Min			Min	Max	Min	Max	Min
3/8–28 UN	2A	0.0011	0.3739	0.3674	—	0.3507	0.3471	0.3313	2B	0.336	0.345	0.3518	0.3564	0.3750
	3A	0.0000	0.3750	0.3685	—	0.3518	0.3491	0.3324	3B	0.3360	0.3426	0.3518	0.3553	0.3750
3/8–32 UNEF	2A	0.0010	0.3740	0.3680	—	0.3537	0.3503	0.3368	2B	0.341	0.349	0.3547	0.3591	0.3750
	3A	0.0000	0.3750	0.3690	—	0.3547	0.3522	0.3378	3B	0.3410	0.3469	0.3547	0.3580	0.3750
3/8–36 UNS	2A	0.0010	0.3740	0.3685	—	0.3560	0.3528	0.3409	2B	0.345	0.352	0.3570	0.3612	0.3750
3/8–40 UNS	2A	0.0009	0.3741	0.3690	—	0.3579	0.3548	0.3443	2B	0.348	0.354	0.3588	0.3628	0.3750
0.390–27 UNS	2A	0.0011	0.3889	0.3822	—	0.3648	0.3612	0.3448	2B	0.350	0.359	0.3659	0.3706	0.3900
7/16–14 UNC	1A	0.0014	0.4361	0.4206	—	0.3897	0.3826	0.3511	1B	0.360	0.376	0.3911	0.4003	0.4375
	2A	0.0014	0.4361	0.4258	0.4206	0.3897	0.3850	0.3511	2B	0.360	0.376	0.3911	0.3972	0.4375
	3A	0.0000	0.4375	0.4272	—	0.3911	0.3876	0.3525	3B	0.3600	0.3717	0.3911	0.3957	0.4375
7/16–16 UN	2A	0.0014	0.4361	0.4267	—	0.3955	0.3909	0.3616	2B	0.370	0.384	0.3969	0.4028	0.4375
	3A	0.0000	0.4375	0.4281	—	0.3969	0.3935	0.3630	3B	0.3700	0.3800	0.3969	0.4014	0.4375
7/16–18 UNS	2A	0.0013	0.4362	0.4275	—	0.4001	0.3958	0.3701	2B	0.377	0.390	0.4014	0.4070	0.4375
7/16–20 UNF	1A	0.0013	0.4362	0.4240	—	0.4037	0.3975	0.3767	1B	0.383	0.395	0.4050	0.4131	0.4375
	2A	0.0013	0.4362	0.4281	—	0.4037	0.3995	0.3767	2B	0.383	0.395	0.4050	0.4104	0.4375
	3A	0.0000	0.4375	0.4294	—	0.4050	0.4019	0.3780	3B	0.3830	0.3916	0.4050	0.4091	0.4375
7/16–24 UNS	2A	0.0011	0.4364	0.4292	—	0.4093	0.4055	0.3868	2B	0.392	0.402	0.4104	0.4153	0.4375
7/16–27 UNS	2A	0.0011	0.4364	0.4297	—	0.4123	0.4087	0.3923	2B	0.397	0.406	0.4134	0.4181	0.4375
7/16–28 UNEF	2A	0.0011	0.4364	0.4299	—	0.4132	0.4096	0.3938	2B	0.399	0.407	0.4143	0.4189	0.4375
	3A	0.0000	0.4375	0.4310	—	0.4143	0.4116	0.3949	3B	0.3990	0.4051	0.4143	0.4178	0.4375
7/16–32 UN	2A	0.0010	0.4365	0.4305	—	0.4162	0.4128	0.3993	2B	0.404	0.411	0.4172	0.4216	0.4375
	3A	0.0000	0.4375	0.4315	—	0.4172	0.4147	0.4003	3B	0.4040	0.4094	0.4172	0.4205	0.4375
1/2–12 UNS	2A	0.0016	0.4984	0.4870	—	0.4443	0.4389	0.3992	2B	0.410	0.428	0.4459	0.4529	0.5000
	3A	0.0000	0.5000	0.4886	—	0.4459	0.4419	0.4008	3B	0.4100	0.4223	0.4459	0.4511	0.5000
1/2–13 UNC	1A	0.0015	0.4985	0.4822	—	0.4485	0.4411	0.4069	1B	0.417	0.434	0.4500	0.4597	0.5000
	2A	0.0015	0.4985	0.4876	0.4822	0.4485	0.4435	0.4069	2B	0.417	0.434	0.4500	0.4565	0.5000
	3A	0.0000	0.5000	0.4891	—	0.4500	0.4463	0.4084	3B	0.4170	0.4284	0.4500	0.4548	0.5000
1/2–14 UNS	2A	0.0015	0.4985	0.4882	—	0.4521	0.4471	0.4135	2B	0.423	0.438	0.4536	0.4601	0.5000
1/2–16 UN	2A	0.0014	0.4986	0.4892	—	0.4580	0.4533	0.4241	2B	0.432	0.446	0.4594	0.4655	0.5000

Table 7. *(Continued)* **Standard Series and Selected Combinations — Unified Screw Threads**

Nominal Size, Threads per Inch, and Series Designation[a]	External[b]								Internal[b]					
	Class	Allowance	Major Diameter			Pitch Diameter		UNR Minor Dia.,[c] Max (Ref.)	Class	Minor Diameter		Pitch Diameter		Major Diameter
			Max[d]	Min	Min[e]	Max[d]	Min			Min	Max	Min	Max	Min
	3A	0.0000	0.5000	0.4906	—	0.4594	0.4559	0.4255	3B	0.4320	0.4419	0.4594	0.4640	0.5000
½–18 UNS	2A	0.0013	0.4987	0.4900	—	0.4626	0.4582	0.4326	2B	0.440	0.453	0.4639	0.4697	0.5000
½–20 UNF	1A	0.0013	0.4987	0.4865	—	0.4662	0.4598	0.4392	1B	0.446	0.457	0.4675	0.4759	0.5000
	2A	0.0013	0.4987	0.4906	—	0.4662	0.4619	0.4392	2B	0.446	0.457	0.4675	0.4731	0.5000
	3A	0.0000	0.5000	0.4919	—	0.4675	0.4643	0.4405	3B	0.4460	0.4537	0.4675	0.4717	0.5000
½–24 UNS	2A	0.0012	0.4988	0.4916	—	0.4717	0.4678	0.4492	2B	0.455	0.465	0.4729	0.4780	0.5000
½–27 UNS	2A	0.0011	0.4989	0.4922	—	0.4748	0.4711	0.4548	2B	0.460	0.469	0.4759	0.4807	0.5000
½–28 UNEF	2A	0.0011	0.4989	0.4924	—	0.4757	0.4720	0.4563	2B	0.461	0.470	0.4768	0.4816	0.5000
	3A	0.0000	0.5000	0.4935	—	0.4768	0.4740	0.4574	3B	0.4610	0.4676	0.4768	0.4804	0.5000
½–32 UN	2A	0.0010	0.4990	0.4930	—	0.4787	0.4752	0.4618	2B	0.466	0.474	0.4797	0.4842	0.5000
	3A	0.0000	0.5000	0.4940	—	0.4797	0.4771	0.4628	3B	0.4660	0.4719	0.4797	0.4831	0.5000
9⁄16–12 UNC	1A	0.0016	0.5609	0.5437	—	0.5068	0.4990	0.4617	1B	0.472	0.490	0.5084	0.5186	0.5625
	2A	0.0016	0.5609	0.5495	0.5437	0.5068	0.5016	0.4617	2B	0.472	0.490	0.5084	0.5152	0.5625
	3A	0.0000	0.5625	0.5511	—	0.5084	0.5045	0.4633	3B	0.4720	0.4843	0.5084	0.5135	0.5625
9⁄16–14 UNS	2A	0.0015	0.5610	0.5507	—	0.5146	0.5096	0.4760	2B	0.485	0.501	0.5161	0.5226	0.5625
9⁄16–16 UN	2A	0.0014	0.5611	0.5517	—	0.5205	0.5158	0.4866	2B	0.495	0.509	0.5219	0.5280	0.5625
	3A	0.0000	0.5625	0.5531	—	0.5219	0.5184	0.4880	3B	0.4950	0.5040	0.5219	0.5265	0.5625
9⁄16–18 UNF	1A	0.0014	0.5611	0.5480	—	0.5250	0.5182	0.4950	1B	0.502	0.515	0.5264	0.5353	0.5625
	2A	0.0014	0.5611	0.5524	—	0.5250	0.5205	0.4950	2B	0.502	0.515	0.5264	0.5323	0.5625
	3A	0.0000	0.5625	0.5538	—	0.5264	0.5230	0.4964	3B	0.5020	0.5106	0.5264	0.5308	0.5625
9⁄16–20 UN	2A	0.0013	0.5612	0.5531	—	0.5287	0.5245	0.5017	2B	0.508	0.520	0.5300	0.5355	0.5625
	3A	0.0000	0.5625	0.5544	—	0.5300	0.5268	0.5030	3B	0.5080	0.5162	0.5300	0.5341	0.5625
9⁄16–24 UNEF	2A	0.0012	0.5613	0.5541	—	0.5342	0.5303	0.5117	2B	0.517	0.527	0.5354	0.5405	0.5625
	3A	0.0000	0.5625	0.5553	—	0.5354	0.5325	0.5129	3B	0.5170	0.5244	0.5354	0.5392	0.5625
9⁄16–27 UNS	2A	0.0011	0.5614	0.5547	—	0.5373	0.5336	0.5173	2B	0.522	0.531	0.5384	0.5432	0.5625
9⁄16–28 UN	2A	0.0011	0.5614	0.5549	—	0.5382	0.5345	0.5188	2B	0.524	0.532	0.5393	0.5441	0.5625
	3A	0.0000	0.5625	0.5560	—	0.5393	0.5365	0.5199	3B	0.5240	0.5301	0.5393	0.5429	0.5625
9⁄16–32 UN	2A	0.0010	0.5615	0.5555	—	0.5412	0.5377	0.5243	2B	0.529	0.536	0.5422	0.5467	0.5625
	3A	0.0000	0.5625	0.5565	—	0.5422	0.5396	0.5253	3B	0.5290	0.5344	0.5422	0.5456	0.5625

Table 7. *(Continued)* **Standard Series and Selected Combinations — Unified Screw Threads**

Nominal Size, Threads per Inch, and Series Designation[a]	External[b]								Internal[b]					
	Class	Allowance	Major Diameter			Pitch Diameter		UNR Minor Dia.,[c] Max (Ref.)	Class	Minor Diameter		Pitch Diameter		Major Diameter
			Max[d]	Min	Min[e]	Max[d]	Min			Min	Max	Min	Max	Min
5/8–11 UNC	1A	0.0016	0.6234	0.6052	—	0.5644	0.5561	0.5152	1B	0.527	0.546	0.5660	0.5767	0.6250
	2A	0.0016	0.6234	0.6113	0.6052	0.5644	0.5589	0.5152	2B	0.527	0.546	0.5660	0.5732	0.6250
	3A	0.0000	0.6250	0.6129	—	0.5660	0.5619	0.5168	3B	0.5270	0.5391	0.5660	0.5714	0.6250
5/8–12 UN	2A	0.0016	0.6234	0.6120	—	0.5693	0.5639	0.5242	2B	0.535	0.553	0.5709	0.5780	0.6250
	3A	0.0000	0.6250	0.6136	—	0.5709	0.5668	0.5258	3B	0.5350	0.5463	0.5709	0.5762	0.6250
5/8–14 UNS	2A	0.0015	0.6235	0.6132	—	0.5771	0.5720	0.5385	2B	0.548	0.564	0.5786	0.5852	0.6250
5/8–16 UN	2A	0.0014	0.6236	0.6142	—	0.5830	0.5782	0.5491	2B	0.557	0.571	0.5844	0.5906	0.6250
	3A	0.0000	0.6250	0.6156	—	0.5844	0.5808	0.5505	3B	0.5570	0.5662	0.5844	0.5890	0.6250
5/8–18 UNF	1A	0.0014	0.6236	0.6105	—	0.5875	0.5805	0.5575	1B	0.565	0.578	0.5889	0.5980	0.6250
	2A	0.0014	0.6236	0.6149	—	0.5875	0.5828	0.5575	2B	0.565	0.578	0.5889	0.5949	0.6250
	3A	0.0000	0.6250	0.6163	—	0.5889	0.5854	0.5589	3B	0.5650	0.5730	0.5889	0.5934	0.6250
5/8–20 UN	2A	0.0013	0.6237	0.6156	—	0.5912	0.5869	0.5642	2B	0.571	0.582	0.5925	0.5981	0.6250
	3A	0.0000	0.6250	0.6169	—	0.5925	0.5893	0.5655	3B	0.5710	0.5787	0.5925	0.5967	0.6250
5/8–24 UNEF	2A	0.0012	0.6238	0.6166	—	0.5967	0.5927	0.5742	2B	0.580	0.590	0.5979	0.6031	0.6250
	3A	0.0000	0.6250	0.6178	—	0.5979	0.5949	0.5754	3B	0.5800	0.5869	0.5979	0.6018	0.6250
5/8–27 UNS	2A	0.0011	0.6239	0.6172	—	0.5998	0.5960	0.5798	2B	0.585	0.594	0.6009	0.6059	0.6250
5/8–28 UN	2A	0.0011	0.6239	0.6174	—	0.6007	0.5969	0.5813	2B	0.586	0.595	0.6018	0.6067	0.6250
	3A	0.0000	0.6250	0.6185	—	0.6018	0.5990	0.5824	3B	0.5860	0.5926	0.6018	0.6055	0.6250
5/8–32 UN	2A	0.0011	0.6239	0.6179	—	0.6036	0.6000	0.5867	2B	0.591	0.599	0.6047	0.6093	0.6250
	3A	0.0000	0.6250	0.6190	—	0.6047	0.6020	0.5878	3B	0.5910	0.5969	0.6047	0.6082	0.6250
11/16–12 UN	2A	0.0016	0.6859	0.6745	—	0.6318	0.6264	0.5867	2B	0.597	0.615	0.6334	0.6405	0.6875
	3A	0.0000	0.6875	0.6761	—	0.6334	0.6293	0.5883	3B	0.5970	0.6085	0.6334	0.6387	0.6875
11/16–16 UN	2A	0.0014	0.6861	0.6767	—	0.6455	0.6407	0.6116	2B	0.620	0.634	0.6469	0.6531	0.6875
	3A	0.0000	0.6875	0.6781	—	0.6469	0.6433	0.6130	3B	0.6200	0.6284	0.6469	0.6515	0.6875
11/16–20 UN	2A	0.0013	0.6862	0.6781	—	0.6537	0.6494	0.6267	2B	0.633	0.645	0.6550	0.6606	0.6875
	3A	0.0000	0.6875	0.6794	—	0.6550	0.6518	0.6280	3B	0.6330	0.6412	0.6550	0.6592	0.6875
11/16–24 UNEF	2A	0.0012	0.6863	0.6791	—	0.6592	0.6552	0.6367	2B	0.642	0.652	0.6604	0.6656	0.6875
	3A	0.0000	0.6875	0.6803	—	0.6604	0.6574	0.6379	3B	0.6420	0.6494	0.6604	0.6643	0.6875
11/16–28 UN	2A	0.0011	0.6864	0.6799	—	0.6632	0.6594	0.6438	2B	0.649	0.657	0.6643	0.6692	0.6875

Table 7. *(Continued)* **Standard Series and Selected Combinations — Unified Screw Threads**

Nominal Size, Threads per Inch, and Series Designation[a]	External[b]								Internal[b]					
	Class	Allowance	Major Diameter			Pitch Diameter		UNR Minor Dia.,[c] Max (Ref.)	Class	Minor Diameter		Pitch Diameter		Major Diameter
			Max[d]	Min	Min[e]	Max[d]	Min			Min	Max	Min	Max	Min
	3A	0.0000	0.6875	0.6810	—	0.6643	0.6615	0.6449	3B	0.6490	0.6551	0.6643	0.6680	0.6875
11/16–32 UN	2A	0.0011	0.6864	0.6804	—	0.6661	0.6625	0.6492	2B	0.654	0.661	0.6672	0.6718	0.6875
	3A	0.0000	0.6875	0.6815	—	0.6672	0.6645	0.6503	3B	0.6540	0.6594	0.6672	0.6707	0.6875
3/4–10 UNC	1A	0.0018	0.7482	0.7288	—	0.6832	0.6744	0.6291	1B	0.642	0.663	0.6850	0.6965	0.7500
	2A	0.0018	0.7482	0.7353	0.7288	0.6832	0.6773	0.6291	2B	0.642	0.663	0.6850	0.6927	0.7500
	3A	0.0000	0.7500	0.7371	—	0.6850	0.6806	0.6309	3B	0.6420	0.6545	0.6850	0.6907	0.7500
3/4–12 UN	2A	0.0017	0.7483	0.7369	—	0.6942	0.6887	0.6491	2B	0.660	0.678	0.6959	0.7031	0.7500
	3A	0.0000	0.7500	0.7386	—	0.6959	0.6918	0.6508	3B	0.6600	0.6707	0.6959	0.7013	0.7500
3/4–14 UNS	2A	0.0015	0.7485	0.7382	—	0.7021	0.6970	0.6635	2B	0.673	0.688	0.7036	0.7103	0.7500
3/4–16 UNF	1A	0.0015	0.7485	0.7343	—	0.7079	0.7004	0.6740	1B	0.682	0.696	0.7094	0.7192	0.7500
	2A	0.0015	0.7485	0.7391	—	0.7079	0.7029	0.6740	2B	0.682	0.696	0.7094	0.7159	0.7500
	3A	0.0000	0.7500	0.7406	—	0.7094	0.7056	0.6755	3B	0.6820	0.6908	0.7094	0.7143	0.7500
3/4–18 UNS	2A	0.0014	0.7486	0.7399	—	0.7125	0.7079	0.6825	2B	0.690	0.703	0.7139	0.7199	0.7500
3/4–20 UNEF	2A	0.0013	0.7487	0.7406	—	0.7162	0.7118	0.6892	2B	0.696	0.707	0.7175	0.7232	0.7500
	3A	0.0000	0.7500	0.7419	—	0.7175	0.7142	0.6905	3B	0.6960	0.7037	0.7175	0.7218	0.7500
3/4–24 UNS	2A	0.0012	0.7488	0.7416	—	0.7217	0.7176	0.6992	2B	0.705	0.715	0.7229	0.7282	0.7500
3/4–24 UNS	2A	0.0012	0.7488	0.7421	—	0.7247	0.7208	0.7047	2B	0.710	0.719	0.7259	0.7310	0.7500
3/4–28 UN	2A	0.0012	0.7488	0.7423	—	0.7256	0.7218	0.7062	2B	0.711	0.720	0.7268	0.7318	0.7500
	3A	0.0000	0.7500	0.7435	—	0.7268	0.7239	0.7074	3B	0.7110	0.7176	0.7268	0.7305	0.7500
3/4–32 UN	2A	0.0011	0.7489	0.7429	—	0.7286	0.7250	0.7117	2B	0.716	0.724	0.7297	0.7344	0.7500
	3A	0.0000	0.7500	0.7440	—	0.7297	0.7270	0.7128	3B	0.7160	0.7219	0.7297	0.7333	0.7500
13/16–12 UN	2A	0.0017	0.8108	0.7994	—	0.7567	0.7512	0.7116	2B	0.722	0.740	0.7584	0.7656	0.8125
	3A	0.0000	0.8125	0.8011	—	0.7584	0.7543	0.7133	3B	0.7220	0.7329	0.7584	0.7638	0.8125
13/16–16 UN	2A	0.0015	0.8110	0.8016	—	0.7704	0.7655	0.7365	2B	0.745	0.759	0.7719	0.7782	0.8125
	3A	0.0000	0.8125	0.8031	—	0.7719	0.7683	0.7380	3B	0.7450	0.7533	0.7719	0.7766	0.8125
13/16–20 UNEF	2A	0.0013	0.8112	0.8031	—	0.7787	0.7743	0.7517	2B	0.758	0.770	0.7800	0.7857	0.8125
	3A	0.0000	0.8125	0.8044	—	0.7800	0.7767	0.7530	3B	0.7580	0.7662	0.7800	0.7843	0.8125
13/16–28 UN	2A	0.0012	0.8113	0.8048	—	0.7881	0.7843	0.7687	2B	0.774	0.782	0.7893	0.7943	0.8125
	3A	0.0000	0.8125	0.8060	—	0.7893	0.7864	0.7699	3B	0.7740	0.7801	0.7893	0.7930	0.8125

Table 7. *(Continued)* **Standard Series and Selected Combinations — Unified Screw Threads**

Nominal Size, Threads per Inch, and Series Designation[a]	External[b]								Internal[b]					
	Class	Allow-ance	Major Diameter			Pitch Diameter		UNR Minor Dia.,[c] Max (Ref.)	Class	Minor Diameter		Pitch Diameter		Major Diameter
			Max[d]	Min	Min[e]	Max[d]	Min			Min	Max	Min	Max	Min
13/16–32 UN	2A	0.0011	0.8114	0.8054	—	0.7911	0.7875	0.7742	2B	0.779	0.786	0.7922	0.7969	0.8125
	3A	0.0000	0.8125	0.8065	—	0.7922	0.7895	0.7753	3B	0.7790	0.7844	0.7922	0.7958	0.8125
7/8–9 UNC	1A	0.0019	0.8731	0.8523	—	0.8009	0.7914	0.7408	1B	0.755	0.778	0.8028	0.8151	0.8750
	2A	0.0019	0.8731	0.8592	0.8523	0.8009	0.7946	0.7408	2B	0.755	0.778	0.8028	0.8110	0.8750
	3A	0.0000	0.8750	0.8611	—	0.8028	0.7981	0.7427	3B	0.7550	0.7681	0.8028	0.8089	0.8750
7/8–10 UNS	2A	0.0018	0.8732	0.8603	—	0.8082	0.8022	0.7542	2B	0.767	0.788	0.8100	0.8178	0.8750
7/8–12 UN	2A	0.0017	0.8733	0.8619	—	0.8192	0.8137	0.7741	2B	0.785	0.803	0.8209	0.8281	0.8750
	3A	0.0000	0.8750	0.8636	—	0.8209	0.8168	0.7758	3B	0.7850	0.7948	0.8209	0.8263	0.8750
7/8–14 UNF	1A	0.0016	0.8734	0.8579	—	0.8270	0.8189	0.7884	1B	0.798	0.814	0.8286	0.8392	0.8750
	2A	0.0016	0.8734	0.8631	—	0.8270	0.8216	0.7884	2B	0.798	0.814	0.8286	0.8356	0.8750
	3A	0.0000	0.8750	0.8647	—	0.8286	0.8245	0.7900	3B	0.7980	0.8068	0.8286	0.8339	0.8750
7/8–16 UN	2A	0.0015	0.8735	0.8641	—	0.8329	0.8280	0.7900	2B	0.807	0.821	0.8344	0.8407	0.8750
	3A	0.0000	0.8750	0.8656	—	0.8344	0.8308	0.8005	3B	0.8070	0.8158	0.8344	0.8391	0.8750
7/8–18 UNS	2A	0.0014	0.8736	0.8649	—	0.8375	0.8329	0.8075	2B	0.815	0.828	0.8389	0.8449	0.8750
7/8–20 UNEF	2A	0.0013	0.8737	0.8656	—	0.8412	0.8368	0.8142	2B	0.821	0.832	0.8425	0.8482	0.8750
	3A	0.0000	0.8750	0.8669	—	0.8425	0.8392	0.8155	3B	0.8210	0.8287	0.8425	0.8468	0.8750
7/8–24 UNS	2A	0.0012	0.8738	0.8666	—	0.8467	0.8426	0.8242	2B	0.830	0.840	0.8479	0.8532	0.8750
7/8–27 UNS	2A	0.0012	0.8738	0.8671	—	0.8497	0.8458	0.8297	2B	0.835	0.844	0.8509	0.8560	0.8750
7/8–28 UN	2A	0.0012	0.8738	0.8673	—	0.8506	0.8468	0.8312	2B	0.836	0.845	0.8518	0.8568	0.8750
	3A	0.0000	0.8750	0.8685	—	0.8518	0.8489	0.8324	3B	0.8360	0.8426	0.8518	0.8555	0.8750
7/8–32 UN	2A	0.0011	0.8739	0.8679	—	0.8536	0.8500	0.8367	2B	0.841	0.849	0.8547	0.8594	0.8750
	3A	0.0000	0.8750	0.8690	—	0.8547	0.8520	0.8378	3B	0.8410	0.8469	0.8547	0.8583	0.8750
15/16–12 UN	2A	0.0017	0.9358	0.9244	—	0.8817	0.8760	0.8366	2B	0.847	0.865	0.8834	0.8908	0.9375
	3A	0.0000	0.9375	0.9261	—	0.8834	0.8793	0.8383	3B	0.8470	0.8575	0.8834	0.8889	0.9375
15/16–16 UN	2A	0.0015	0.9360	0.9266	—	0.8954	0.8904	0.8615	2B	0.870	0.884	0.8969	0.9034	0.9375
	3A	0.0000	0.9375	0.9281	—	0.8969	0.8932	0.8630	3B	0.8700	0.8783	0.8969	0.9018	0.9375
15/16–20 UNEF	2A	0.0014	0.9361	0.9280	—	0.9036	0.8991	0.8766	2B	0.883	0.895	0.9050	0.9109	0.9375
	3A	0.0000	0.9375	0.9294	—	0.9050	0.9016	0.8780	3B	0.8830	0.8912	0.9050	0.9094	0.9375
15/16–28 UN	2A	0.0012	0.9363	0.9298	—	0.9131	0.9091	0.8937	2B	0.899	0.907	0.9143	0.9195	0.9375

Table 7. *(Continued)* **Standard Series and Selected Combinations — Unified Screw Threads**

Nominal Size, Threads per Inch, and Series Designation[a]	External[b]								Internal[b]					
	Class	Allowance	Major Diameter			Pitch Diameter		UNR Minor Dia.,[c] Max (Ref.)	Class	Minor Diameter		Pitch Diameter		Major Diameter
			Max[d]	Min	Min[e]	Max[d]	Min			Min	Max	Min	Max	Min
	3A	0.0000	0.9375	0.9310	—	0.9143	0.9113	0.8949	3B	0.8990	0.9051	0.9143	0.9182	0.9375
$^{15}/_{16}$–32 UN	2A	0.0011	0.9364	0.9304	—	0.9161	0.9123	0.8992	2B	0.904	0.911	0.9172	0.9221	0.9375
	3A	0.0000	0.9375	0.9315	—	0.9172	0.9144	0.9003	3B	0.9040	0.9094	0.9172	0.9209	0.9375
1–8 UNC	1A	0.0020	0.9980	0.9755	—	0.9168	0.9067	0.8492	1B	0.865	0.890	0.9188	0.9320	1.0000
	2A	0.0020	0.9980	0.9830	0.9755	0.9168	0.9100	0.8492	2B	0.865	0.890	0.9188	0.9276	1.0000
	3A	0.0000	1.0000	0.9850	—	0.9188	0.9137	0.8512	3B	0.8650	0.8797	0.9188	0.9254	1.0000
1–10 UNS	2A	0.0018	0.9982	0.9853	—	0.9332	0.9270	0.8792	2B	0.892	0.913	0.9350	0.9430	1.0000
1–12 UNF	1A	0.0018	0.9982	0.9810	—	0.9441	0.9353	0.8990	1B	0.910	0.928	0.9459	0.9573	1.0000
	2A	0.0018	0.9982	0.9868	—	0.9441	0.9382	0.8990	2B	0.910	0.928	0.9459	0.9535	1.0000
	3A	0.0000	1.0000	0.9886	—	0.9459	0.9415	0.9008	3B	0.9100	0.9198	0.9459	0.9516	1.0000
1–14 UNS$_f$	1A	0.0017	0.9983	0.9828	—	0.9519	0.9435	0.9132	1B	0.923	0.938	0.9536	0.9645	1.0000
	2A	0.0017	0.9983	0.9880	—	0.9519	0.9463	0.9132	2B	0.923	0.938	0.9536	0.9609	1.0000
	3A	0.0000	1.0000	0.9897	—	0.9536	0.9494	0.9149	3B	0.9230	0.9315	0.9536	0.9590	1.0000
1–16 UN	2A	0.0015	0.9985	0.9891	—	0.9579	0.9529	0.9240	2B	0.932	0.946	0.9594	0.9659	1.0000
	3A	0.0000	1.0000	0.9906	—	0.9594	0.9557	0.9255	3B	0.9320	0.9408	0.9594	0.9643	1.0000
1–18 UNS	2A	0.0014	0.9986	0.9899	—	0.9625	0.9578	0.9325	2B	0.940	0.953	0.9639	0.9701	1.0000
1–20 UNEF	2A	0.0014	0.9986	0.9905	—	0.9661	0.9616	0.9391	2B	0.946	0.957	0.9675	0.9734	1.0000
	3A	0.0000	1.0000	0.9919	—	0.9675	0.9641	0.9405	3B	0.9460	0.9537	0.9675	0.9719	1.0000
1–24 UNS	2A	0.0013	0.9987	0.9915	—	0.9716	0.9674	0.9491	2B	0.955	0.965	0.9729	0.9784	1.0000
1–27 UNS	2A	0.0012	0.9988	0.9921	—	0.9747	0.9707	0.9547	2B	0.960	0.969	0.9759	0.9811	1.0000
1–28 UN	2A	0.0012	0.9988	0.9923	—	0.9756	0.9716	0.9562	2B	0.961	0.970	0.9768	0.9820	1.0000
	3A	0.0000	1.0000	0.9935	—	0.9768	0.9738	0.9574	3B	0.9610	0.9676	0.9768	0.9807	1.0000
1–32 UN	2A	0.0011	0.9989	0.9929	—	0.9786	0.9748	0.9617	2B	0.966	0.974	0.9797	0.9846	1.0000
	3A	0.0000	1.0000	0.9940	—	0.9797	0.9769	0.9628	3B	0.9660	0.9719	0.9797	0.9834	1.0000
1$^{1}/_{16}$–8 UN	2A	0.0020	1.0605	1.0455	—	0.9793	0.9725	0.9117	2B	0.927	0.952	0.9813	0.9902	1.0625
	3A	0.0000	1.0625	1.0475	—	0.9813	0.9762	0.9137	3B	0.9270	0.9422	0.9813	0.9880	1.0625
1$^{1}/_{16}$–12 UN	2A	0.0017	1.0608	1.0494	—	1.0067	1.0010	0.9616	2B	0.972	0.990	1.0084	1.0158	1.0625
	3A	0.0000	1.0625	1.0511	—	1.0084	1.0042	0.9633	3B	0.9720	0.9823	1.0084	1.0139	1.0625
1$^{1}/_{16}$–16 UN	2A	0.0015	1.0610	1.0516	—	1.0204	1.0154	0.9865	2B	0.995	1.009	1.0219	1.0284	1.0625
	3A	0.0000	1.0625	1.0531	—	1.0219	1.0182	0.9880	3B	0.9950	1.0033	1.0219	1.0268	1.0625
1$^{1}/_{16}$–18 UNEF	2A	0.0014	1.0611	1.0524	—	1.0250	1.0203	0.9950	2B	1.002	1.015	1.0264	1.0326	1.0625

Table 7. *(Continued)* **Standard Series and Selected Combinations — Unified Screw Threads**

Nominal Size, Threads per Inch, and Series Designation[a]	External[b]								Internal[b]					
	Class	Allowance	Major Diameter			Pitch Diameter		UNR Minor Dia.,[c] Max (Ref.)	Class	Minor Diameter		Pitch Diameter		Major Diameter
			Max[d]	Min	Min[e]	Max[d]	Min			Min	Max	Min	Max	Min
	3A	0.0000	1.0625	1.0538	—	1.0264	1.0228	0.9964	3B	1.0020	1.0105	1.0264	1.0310	1.0625
1 1/16–20 UN	2A	0.0014	1.0611	1.0530	—	1.0286	1.0241	1.0016	2B	1.008	1.020	1.0300	1.0359	1.0625
	3A	0.0000	1.0625	1.0544	—	1.0300	1.0266	1.0030	3B	1.0080	1.0162	1.0300	1.0344	1.0625
1 1/16–28 UN	2A	0.0012	1.0613	1.0548	—	1.0381	1.0341	1.0187	2B	1.024	1.032	1.0393	1.0445	1.0625
	3A	0.0000	1.0625	1.0560	—	1.0393	1.0363	1.0199	3B	1.0240	1.0301	1.0393	1.0432	1.0625
1 1/8–7 UNC	1A	0.0022	1.1228	1.0982	—	1.0300	1.0191	0.9527	1B	0.970	0.998	1.0322	1.0463	1.1250
	2A	0.0022	1.1228	1.1064	1.0982	1.0300	1.0228	0.9527	2B	0.970	0.998	1.0322	1.0416	1.1250
	3A	0.0000	1.1250	1.1086	—	1.0322	1.0268	0.9549	3B	0.9700	0.9875	1.0322	1.0393	1.1250
1 1/8–8 UN	2A	0.0021	1.1229	1.1079	1.1004	1.0417	1.0348	0.9741	2B	0.990	1.015	1.0438	1.0528	1.1250
	3A	0.0000	1.1250	1.1100	—	1.0438	1.0386	0.9762	3B	0.9900	1.0047	1.0438	1.0505	1.1250
1 1/8–10 UNS	2A	0.0018	1.1232	1.1103	—	1.0582	1.0520	1.0042	2B	1.017	1.038	1.0600	1.0680	1.1250
1 1/8–12 UNF	1A	0.0018	1.1232	1.1060	—	1.0691	1.0601	1.0240	1B	1.035	1.053	1.0709	1.0826	1.1250
	2A	0.0018	1.1232	1.1118	—	1.0691	1.0631	1.0240	2B	1.035	1.053	1.0709	1.0787	1.1250
	3A	0.0000	1.1250	1.1136	—	1.0709	1.0664	1.0258	3B	1.0350	1.0448	1.0709	1.0768	1.1250
1 1/8–14 UNS	2A	0.0016	1.1234	1.1131	—	1.0770	1.0717	1.0384	2B	1.048	1.064	1.0786	1.0855	1.1250
1 1/8–16 UN	2A	0.0015	1.1235	1.1141	—	1.0829	1.0779	1.0490	2B	1.057	1.071	1.0844	1.0909	1.1250
	3A	0.0000	1.1250	1.1156	—	1.0844	1.0807	1.0505	3B	1.0570	1.0658	1.0844	1.0893	1.1250
1 1/8–18 UNEF	2A	0.0014	1.1236	1.1149	—	1.0875	1.0828	1.0575	2B	1.065	1.078	1.0889	1.0951	1.1250
	3A	0.0000	1.1250	1.1163	—	1.0889	1.0853	1.0589	3B	1.0650	1.0730	1.0889	1.0935	1.1250
1 1/8–20 UN	2A	0.0014	1.1236	1.1155	—	1.0911	1.0866	1.0641	2B	1.071	1.082	1.0925	1.0984	1.1250
	3A	0.0000	1.1250	1.1169	—	1.0925	1.0891	1.0655	3B	1.0710	1.0787	1.0925	1.0969	1.1250
1 1/8–24 UNS	2A	0.0013	1.1237	1.1165	—	1.0966	1.0924	1.0742	2B	1.080	1.090	1.0979	1.1034	1.1250
1 1/8–28 UN	2A	0.0012	1.1238	1.1173	—	1.1006	1.0966	1.0812	2B	1.086	1.095	1.1018	1.1070	1.1250
	3A	0.0000	1.1250	1.1185	—	1.1018	1.0988	1.0824	3B	1.0860	1.0926	1.1018	1.1057	1.1250
1 3/16–8 UN	2A	0.0021	1.1854	1.1704	—	1.1042	1.0972	1.0366	2B	1.052	1.077	1.1063	1.1154	1.1875
	3A	0.0000	1.1875	1.1725	—	1.1063	1.1011	1.0387	3B	1.0520	1.0672	1.1063	1.1131	1.1875
1 3/16–12 UN	2A	0.0017	1.1858	1.1744	—	1.1317	1.1259	1.0866	2B	1.097	1.115	1.1334	1.1409	1.1875
	3A	0.0000	1.1875	1.1761	—	1.1334	1.1291	1.0883	3B	1.0970	1.1073	1.1334	1.1390	1.1875
1 3/16–16 UN	2A	0.0015	1.1860	1.1766	—	1.1454	1.1403	1.1115	2B	1.120	1.134	1.1469	1.1535	1.1875

Table 7. *(Continued)* **Standard Series and Selected Combinations — Unified Screw Threads**

Nominal Size, Threads per Inch, and Series Designation[a]	External[b]								Internal[b]					
	Class	Allowance	Major Diameter			Pitch Diameter		UNR Minor Dia.,[c] Max (Ref.)	Class	Minor Diameter		Pitch Diameter		Major Diameter
			Max[d]	Min	Min[e]	Max[d]	Min			Min	Max	Min	Max	Min
	3A	0.0000	1.1875	1.1781	—	1.1469	1.1431	1.1130	3B	1.1200	1.1283	1.1469	1.1519	1.1875
1 3/16–18 UNEF	2A	0.0015	1.1860	1.1773	—	1.1499	1.1450	1.1199	2B	1.127	1.140	1.1514	1.1577	1.1875
	3A	0.0000	1.1875	1.1788	—	1.1514	1.1478	1.1214	3B	1.1270	1.1355	1.1514	1.1561	1.1875
1 3/16–20 UN	2A	0.0014	1.1861	1.1780	—	1.1536	1.1489	1.1266	2B	1.133	1.145	1.1550	1.1611	1.1875
	3A	0.0000	1.1875	1.1794	—	1.1550	1.1515	1.1280	3B	1.1330	1.1412	1.1550	1.1595	1.1875
1 3/16–28 UN	2A	0.0012	1.1863	1.1798	—	1.1631	1.1590	1.1437	2B	1.149	1.157	1.1643	1.1696	1.1875
	3A	0.0000	1.1875	1.1810	—	1.1643	1.1612	1.1449	3B	1.1490	1.1551	1.1643	1.1683	1.1875
1 1/4–7 UNC	1A	0.0022	1.2478	1.2232	—	1.1550	1.1439	1.0777	1B	1.095	1.123	1.1572	1.1716	1.2500
	2A	0.0022	1.2478	1.2314	1.2232	1.1550	1.1476	1.0777	2B	1.095	1.123	1.1572	1.1668	1.2500
	3A	0.0000	1.2500	1.2336	—	1.1572	1.1517	1.0799	3B	1.0950	1.1125	1.1572	1.1644	1.2500
1 1/4–8 UN	2A	0.0021	1.2479	1.2329	1.2254	1.1667	1.1597	1.0991	2B	1.115	1.140	1.1688	1.1780	1.2500
	3A	0.0000	1.2500	1.2350	—	1.1688	1.1635	1.1012	3B	1.1150	1.1297	1.1688	1.1757	1.2500
1 1/4–10 UNS	2A	0.0019	1.2481	1.2352	—	1.1831	1.1768	1.1291	2B	1.142	1.163	1.1850	1.1932	1.2500
1 1/4–12 UNF	1A	0.0018	1.2482	1.2310	—	1.1941	1.1849	1.1490	1B	1.160	1.178	1.1959	1.2079	1.2500
	2A	0.0018	1.2482	1.2368	—	1.1941	1.1879	1.1490	2B	1.160	1.178	1.1959	1.2039	1.2500
	3A	0.0000	1.2500	1.2386	—	1.1959	1.1913	1.1508	3B	1.1600	1.1698	1.1959	1.2019	1.2500
1 1/4–14 UNS	2A	0.0016	1.2484	1.2381	—	1.2020	1.1966	1.1634	2B	1.173	1.188	1.2036	1.2106	1.2500
1 1/4–16 UN	2A	0.0015	1.2485	1.2391	—	1.2079	1.2028	1.1740	2B	1.182	1.196	1.2094	1.2160	1.2500
	3A	0.0000	1.2500	1.2406	—	1.2094	1.2056	1.1755	3B	1.1820	1.1908	1.2094	1.2144	1.2500
1 1/4–18 UNEF	2A	0.0015	1.2485	1.2398	—	1.2124	1.2075	1.1824	2B	1.190	1.203	1.2139	1.2202	1.2500
	3A	0.0000	1.2500	1.2413	—	1.2139	1.2103	1.1839	3B	1.1900	1.1980	1.2139	1.2186	1.2500
1 1/4–20 UN	2A	0.0014	1.2486	1.2405	—	1.2161	1.2114	1.1891	2B	1.196	1.207	1.2175	1.2236	1.2500
	3A	0.0000	1.2500	1.2419	—	1.2175	1.2140	1.1905	3B	1.1960	1.2037	1.2175	1.2220	1.2500
1 1/4–24 UNS	2A	0.0013	1.2487	1.2415	—	1.2216	1.2173	1.1991	2B	1.205	1.215	1.2229	1.2285	1.2500
1 1/4–28 UN	2A	0.0012	1.2488	1.2423	—	1.2256	1.2215	1.2062	2B	1.211	1.220	1.2268	1.2321	1.2500
	3A	0.0000	1.2500	1.2435	—	1.2268	1.2237	1.2074	3B	1.2110	1.2176	1.2268	1.2308	1.2500
1 5/16–8 UN	2A	0.0021	1.3104	1.2954	—	1.2292	1.2221	1.1616	2B	1.177	1.202	1.2313	1.2405	1.3125
	3A	0.0000	1.3125	1.2975	—	1.2313	1.2260	1.1637	3B	1.1770	1.1922	1.2313	1.2382	1.3125
1 5/16–12 UN	2A	0.0017	1.3108	1.2994	—	1.2567	1.2509	1.2116	2B	1.222	1.240	1.2584	1.2659	1.3125

Table 7. *(Continued)* **Standard Series and Selected Combinations — Unified Screw Threads**

Nominal Size, Threads per Inch, and Series Designation[a]	External[b]								Internal[b]					
	Class	Allow-ance	Major Diameter			Pitch Diameter		UNR Minor Dia.,[c] Max (Ref.)	Class	Minor Diameter		Pitch Diameter		Major Diameter
			Max[d]	Min	Min[e]	Max[d]	Min			Min	Max	Min	Max	Min
	3A	0.0000	1.3125	1.3011	—	1.2584	1.2541	1.2133	3B	1.2220	1.2323	1.2584	1.2640	1.3125
1 5/16–16 UN	2A	0.0015	1.3110	1.3016	—	1.2704	1.2653	1.2365	2B	1.245	1.259	1.2719	1.2785	1.3125
	3A	0.0000	1.3125	1.3031	—	1.2719	1.2681	1.2380	3B	1.2450	1.2533	1.2719	1.2769	1.3125
1 5/16–18 UNEF	2A	0.0015	1.3110	1.3023	—	1.2749	1.2700	1.2449	2B	1.252	1.265	1.2764	1.2827	1.3125
	3A	0.0000	1.3125	1.3038	—	1.2764	1.2728	1.2464	3B	1.2520	1.2605	1.2764	1.2811	1.3125
1 5/16–20 UN	2A	0.0014	1.3111	1.3030	—	1.2786	1.2739	1.2516	2B	1.258	1.270	1.2800	1.2861	1.3125
	3A	0.0000	1.3125	1.3044	—	1.2800	1.2765	1.2530	3B	1.2580	1.2662	1.2800	1.2845	1.3125
1 5/16–28 UN	2A	0.0012	1.3113	1.3048	—	1.2881	1.2840	1.2687	2B	1.274	1.282	1.2893	1.2946	1.3125
	3A	0.0000	1.3125	1.3060	—	1.2893	1.2862	1.2699	3B	1.2740	1.2801	1.2893	1.2933	1.3125
1 3/8–6 UNC	1A	0.0024	1.3726	1.3453	—	1.2643	1.2523	1.1742	1B	1.195	1.225	1.2667	1.2822	1.3750
	2A	0.0024	1.3726	1.3544	1.3453	1.2643	1.2563	1.1742	2B	1.195	1.225	1.2667	1.2771	1.3750
	3A	0.0000	1.3750	1.3568	—	1.2667	1.2607	1.1766	3B	1.1950	1.2146	1.2667	1.2745	1.3750
1 3/8–8 UN	2A	0.0022	1.3728	1.3578	1.3503	1.2916	1.2844	1.2240	2B	1.240	1.265	1.2938	1.3031	1.3750
	3A	0.0000	1.3750	1.3600	—	1.2938	1.2884	1.2262	3B	1.2400	1.2547	1.2938	1.3008	1.3750
1 3/8–10 UNS	2A	0.0019	1.3731	1.3602	—	1.3081	1.3018	1.2541	2B	1.267	1.288	1.3100	1.3182	1.3750
1 3/8–12 UNF	1A	0.0019	1.3731	1.3559	—	1.3190	1.3096	1.2739	1B	1.285	1.303	1.3209	1.3332	1.3750
	2A	0.0019	1.3731	1.3617	—	1.3190	1.3127	1.2739	2B	1.285	1.303	1.3209	1.3291	1.3750
	3A	0.0000	1.3750	1.3636	—	1.3209	1.3162	1.2758	3B	1.2850	1.2948	1.3209	1.3270	1.3750
1 3/8–14 UNS	2A	0.0016	1.3734	1.3631	—	1.3270	1.3216	1.2884	2B	1.298	1.314	1.3286	1.3356	1.3750
1 3/8–16 UN	2A	0.0015	1.3735	1.3641	—	1.3329	1.3278	1.2990	2B	1.307	1.321	1.3344	1.3410	1.3750
	3A	0.0000	1.3750	1.3656	—	1.3344	1.3306	1.3005	3B	1.3070	1.3158	1.3344	1.3394	1.3750
1 3/8–18 UNEF	2A	0.0015	1.3735	1.3648	—	1.3374	1.3325	1.3074	2B	1.315	1.328	1.3389	1.3452	1.3750
	3A	0.0000	1.3750	1.3663	—	1.3389	1.3353	1.3089	3B	1.3150	1.3230	1.3389	1.3436	1.3750
1 3/8–20 UN	2A	0.0014	1.3736	1.3655	—	1.3411	1.3364	1.3141	2B	1.321	1.332	1.3425	1.3486	1.3750
	3A	0.0000	1.3750	1.3669	—	1.3425	1.3390	1.3155	3B	1.3210	1.3287	1.3425	1.3470	1.3750
1 3/8–24 UNS	2A	0.0013	1.3737	1.3665	—	1.3466	1.3423	1.3241	2B	1.330	1.340	1.3479	1.3535	1.3750
1 3/8–28 UN	2A	0.0012	1.3738	1.3673	—	1.3506	1.3465	1.3312	2B	1.336	1.345	1.3518	1.3571	1.3750
	3A	0.0000	1.3750	1.3685	—	1.3518	1.3487	1.3324	3B	1.3360	1.3426	1.3518	1.3558	1.3750
1 7/16–6 UN	2A	0.0024	1.4351	1.4169	—	1.3268	1.3188	1.2367	2B	1.257	1.288	1.3292	1.3396	1.4375

Table 7. *(Continued)* **Standard Series and Selected Combinations — Unified Screw Threads**

Nominal Size, Threads per Inch, and Series Designation[a]	External[b]								Internal[b]					
	Class	Allowance	Major Diameter			Pitch Diameter		UNR Minor Dia.,[c] Max (Ref.)	Class	Minor Diameter		Pitch Diameter		Major Diameter
			Max[d]	Min	Min[e]	Max[d]	Min			Min	Max	Min	Max	Min
	3A	0.0000	1.4375	1.4193	—	1.3292	1.3232	1.2391	3B	1.2570	1.2771	1.3292	1.3370	1.4375
1⁷⁄₁₆–8 UN	2A	0.0022	1.4353	1.4203	—	1.3541	1.3469	1.2865	2B	1.302	1.327	1.3563	1.3657	1.4375
	3A	0.0000	1.4375	1.4225	—	1.3563	1.3509	1.2887	3B	1.3020	1.3172	1.3563	1.3634	1.4375
1⁷⁄₁₆–12 UN	2A	0.0018	1.4357	1.4243	—	1.3816	1.3757	1.3365	2B	1.347	1.365	1.3834	1.3910	1.4375
	3A	0.0000	1.4375	1.4261	—	1.3834	1.3790	1.3383	3B	1.3470	1.3573	1.3834	1.3891	1.4375
1⁷⁄₁₆–16 UN	2A	0.0016	1.4359	1.4265	—	1.3953	1.3901	1.3614	2B	1.370	1.384	1.3969	1.4037	1.4375
	3A	0.0000	1.4375	1.4281	—	1.3969	1.3930	1.3630	3B	1.3700	1.3783	1.3969	1.4020	1.4375
1⁷⁄₁₆–18 UNEF	2A	0.0015	1.4360	1.4273	—	1.3999	1.3949	1.3699	2B	1.377	1.390	1.4014	1.4079	1.4375
	3A	0.0000	1.4375	1.4288	—	1.4014	1.3977	1.3714	3B	1.3770	1.3855	1.4014	1.4062	1.4375
1⁷⁄₁₆–20 UN	2A	0.0014	1.4361	1.4280	—	1.4036	1.3988	1.3766	2B	1.383	1.395	1.4050	1.4112	1.4375
	3A	0.0000	1.4375	1.4294	—	1.4050	1.4014	1.3780	3B	1.3830	1.3912	1.4050	1.4096	1.4375
1⁷⁄₁₆–28 UN	2A	0.0013	1.4362	1.4297	—	1.4130	1.4088	1.3936	2B	1.399	1.407	1.4143	1.4198	1.4375
	3A	0.0000	1.4375	1.4310	—	1.4143	1.4112	1.3949	3B	1.3990	1.4051	1.4143	1.4184	1.4375
1½–6	1A	0.0024	1.4976	1.4703	—	1.3893	1.3772	1.2992	1B	1.320	1.350	1.3917	1.4075	1.5000
	2A	0.0024	1.4976	1.4794	1.4703	1.3893	1.3812	1.2992	2B	1.320	1.350	1.3917	1.4022	1.5000
	3A	0.0000	1.5000	1.4818	—	1.3917	1.3856	1.3016	3B	1.3200	1.3396	1.3917	1.3996	1.5000
1½–8 UN	2A	0.0022	1.4978	1.4828	1.4753	1.4166	1.4093	1.3490	2B	1.365	1.390	1.4188	1.4283	1.5000
	3A	0.0000	1.5000	1.4850	—	1.4188	1.4133	1.3512	3B	1.3650	1.3797	1.4188	1.4259	1.5000
1½–10 UNS	2A	0.0019	1.4981	1.4852	—	1.4331	1.4267	1.3791	2B	1.392	1.413	1.4350	1.4433	1.5000
1½–12 UNF	1A	0.0019	1.4981	1.4809	—	1.4440	1.4344	1.3989	1B	1.410	1.428	1.4459	1.4584	1.5000
	2A	0.0019	1.4981	1.4867	—	1.4440	1.4376	1.3989	2B	1.410	1.428	1.4459	1.4542	1.5000
	3A	0.0000	1.5000	1.4886	—	1.4459	1.4411	1.4008	3B	1.4100	1.4198	1.4459	1.4522	1.5000
1½–14 UNS	2A	0.0017	1.4983	1.4880	—	1.4519	1.4464	1.4133	2B	1.423	1.438	1.4536	1.4608	1.5000
1½–16 UN	2A	0.0016	1.4984	1.4890	—	1.4578	1.4526	1.4239	2B	1.432	1.446	1.4594	1.4662	1.5000
	3A	0.0000	1.5000	1.4906	—	1.4594	1.4555	1.4255	3B	1.4320	1.4408	1.4594	1.4645	1.5000
1½–18 UNEF	2A	0.0015	1.4985	1.4898	—	1.4624	1.4574	1.4324	2B	1.440	1.452	1.4639	1.4704	1.5000
	3A	0.0000	1.5000	1.4913	—	1.4639	1.4602	1.4339	3B	1.4400	1.4480	1.4639	1.4687	1.5000
1½–20 UN	2A	0.0014	1.4986	1.4905	—	1.4661	1.4613	1.4391	2B	1.446	1.457	1.4675	1.4737	1.5000
	3A	0.0000	1.5000	1.4919	—	1.4675	1.4639	1.4405	3B	1.4460	1.4537	1.4675	1.4721	1.5000

Table 7. *(Continued)* **Standard Series and Selected Combinations — Unified Screw Threads**

Nominal Size, Threads per Inch, and Series Designation[a]	External[b]								Internal[b]					
	Class	Allowance	Major Diameter			Pitch Diameter		UNR Minor Dia.,[c] Max (Ref.)	Class	Minor Diameter		Pitch Diameter		Major Diameter
			Max[d]	Min	Min[e]	Max[d]	Min			Min	Max	Min	Max	Min
1½–24 UNS	2A	0.0013	1.4987	1.4915	—	1.4716	1.4672	1.4491	2B	1.455	1.465	1.4729	1.4787	1.5000
1½–28 UN	2A	0.0013	1.4987	1.4922	—	1.4755	1.4713	1.4561	2B	1.461	1.470	1.4768	1.4823	1.5000
	3A	0.0000	1.5000	1.4935	—	1.4768	1.4737	1.4574	3B	1.4610	1.4676	1.4768	1.4809	1.5000
1 9/16–6 UN	2A	0.0024	1.5601	1.5419	—	1.4518	1.4436	1.3617	2B	1.382	1.413	1.4542	1.4648	1.5625
	3A	0.0000	1.5625	1.5443	—	1.4542	1.4481	1.3641	3B	1.3820	1.4021	1.4542	1.4622	1.5625
1 9/16–8 UN	2A	0.0022	1.5603	1.5453	—	1.4791	1.4717	1.4115	2B	1.427	1.452	1.4813	1.4909	1.5625
	3A	0.0000	1.5625	1.5475	—	1.4813	1.4758	1.4137	3B	1.4270	1.4422	1.4813	1.4885	1.5625
1 9/16–12 UN	2A	0.0018	1.5607	1.5493	—	1.5066	1.5007	1.4615	2B	1.472	1.490	1.5084	1.5160	1.5625
	3A	0.0000	1.5625	1.5511	—	1.5084	1.5040	1.4633	3B	1.4720	1.4823	1.5084	1.5141	1.5625
1 9/16–16 UN	2A	0.0016	1.5609	1.5515	—	1.5203	1.5151	1.4864	2B	1.495	1.509	1.5219	1.5287	1.5625
	3A	0.0000	1.5625	1.5531	—	1.5219	1.5180	1.4880	3B	1.4950	1.5033	1.5219	1.5270	1.5625
1 9/16–18 UNEF	2A	0.0015	1.5610	1.5523	—	1.5249	1.5199	1.4949	2B	1.502	1.515	1.5264	1.5329	1.5625
	3A	0.0000	1.5625	1.5538	—	1.5264	1.5227	1.4964	3B	1.5020	1.5105	1.5264	1.5312	1.5625
1 9/16–20 UN	2A	0.0014	1.5611	1.5530	—	1.5286	1.5238	1.5016	2B	1.508	1.520	1.5300	1.5362	1.5625
	3A	0.0000	1.5625	1.5544	—	1.5300	1.5264	1.5030	3B	1.5080	1.5162	1.5300	1.5346	1.5625
1⅝–6 UN	2A	0.0025	1.6225	1.6043	—	1.5142	1.5060	1.4246	2B	1.445	1.475	1.5167	1.5274	1.6250
	3A	0.0000	1.6250	1.6068	—	1.5167	1.5105	1.4271	3B	1.4450	1.4646	1.5167	1.5247	1.6250
1⅝–8 UN	2A	0.0022	1.6228	1.6078	1.6003	1.5416	1.5342	1.4784	2B	1.490	1.515	1.5438	1.5535	1.6250
	3A	0.0000	1.6250	1.6100	—	1.5438	1.5382	1.4806	3B	1.4900	1.5047	1.5438	1.5510	1.6250
1⅝–10 UNS	2A	0.0019	1.6231	1.6102	—	1.5581	1.5517	1.5041	2B	1.517	1.538	1.5600	1.5683	1.6250
1⅝–12 UN	2A	0.0018	1.6232	1.6118	—	1.5691	1.5632	1.5240	2B	1.535	1.553	1.5709	1.5785	1.6250
	3A	0.0000	1.6250	1.6136	—	1.5709	1.5665	1.5258	3B	1.5350	1.5448	1.5709	1.5766	1.6250
1⅝–14 UNS	2A	0.0017	1.6233	1.6130	—	1.5769	1.5714	1.5383	2B	1.548	1.564	1.5786	1.5858	1.6250
1⅝–16 UN	2A	0.0016	1.6234	1.6140	—	1.5828	1.5776	1.5489	2B	1.557	1.571	1.5844	1.5912	1.6250
	3A	0.0000	1.6250	1.6156	—	1.5844	1.5805	1.5505	3B	1.5570	1.5658	1.5844	1.5895	1.6250
1⅝–18 UNEF	2A	0.0015	1.6235	1.6148	—	1.5874	1.5824	1.5574	2B	1.565	1.578	1.5889	1.5954	1.6250
	3A	0.0000	1.6250	1.6163	—	1.5889	1.5852	1.5589	3B	1.5650	1.5730	1.5889	1.5937	1.6250
1⅝–20 UN	2A	0.0014	1.6236	1.6155	—	1.5911	1.5863	1.5641	2B	1.571	1.582	1.5925	1.5987	1.6250
	3A	0.0000	1.6250	1.6169	—	1.5925	1.5889	1.5655	3B	1.5710	1.5787	1.5925	1.5971	1.6250

Table 7. *(Continued)* **Standard Series and Selected Combinations — Unified Screw Threads**

Nominal Size, Threads per Inch, and Series Designation[a]	External[b]								Internal[b]					
	Class	Allowance	Major Diameter			Pitch Diameter		UNR Minor Dia.,[c] Max (Ref.)	Class	Minor Diameter		Pitch Diameter		Major Diameter
			Max[d]	Min	Min[e]	Max[d]	Min			Min	Max	Min	Max	Min
1 5/8–24 UNS	2A	0.0013	1.6237	1.6165	—	1.5966	1.5922	1.5741	2B	1.580	1.590	1.5979	1.6037	1.6250
1 11/16–6 UN	2A	0.0025	1.6850	1.6668	—	1.5767	1.5684	1.4866	2B	1.507	1.538	1.5792	1.5900	1.6875
	3A	0.0000	1.6875	1.6693	—	1.5792	1.5730	1.4891	3B	1.5070	1.5271	1.5792	1.5873	1.6875
1 11/16–8 UN	2A	0.0022	1.6853	1.6703	—	1.6041	1.5966	1.5365	2B	1.552	1.577	1.6063	1.6160	1.6875
	3A	0.0000	1.6875	1.6725	—	1.6063	1.6007	1.5387	3B	1.5520	1.5672	1.6063	1.6136	1.6875
1 11/16–12 UN	2A	0.0018	1.6857	1.6743	—	1.6316	1.6256	1.5865	2B	1.597	1.615	1.6334	1.6412	1.6875
	3A	0.0000	1.6875	1.6761	—	1.6334	1.6289	1.5883	3B	1.5970	1.6073	1.6334	1.6392	1.6875
1 11/16–16 UN	2A	0.0016	1.6859	1.6765	—	1.6453	1.6400	1.6114	2B	1.620	1.634	1.6469	1.6538	1.6875
	3A	0.0000	1.6875	1.6781	—	1.6469	1.6429	1.6130	3B	1.6200	1.6283	1.6469	1.6521	1.6875
1 11/16–18 UNEF	2A	0.0015	1.6860	1.6773	—	1.6499	1.6448	1.6199	2B	1.627	1.640	1.6514	1.6580	1.6875
	3A	0.0000	1.6875	1.6788	—	1.6514	1.6476	1.6214	3B	1.6270	1.6355	1.6514	1.6563	1.6875
1 11/16–20 UN	2A	0.0015	1.6860	1.6779	—	1.6535	1.6487	1.6265	2B	1.633	1.645	1.6550	1.6613	1.6875
	3A	0.0000	1.6875	1.6794	—	1.6550	1.6514	1.6280	3B	1.6330	1.6412	1.6550	1.6597	1.6875
1 3/4–5 UNC	1A	0.0027	1.7473	1.7165	—	1.6174	1.6040	1.5092	1B	1.534	1.568	1.6201	1.6375	1.7500
	2A	0.0027	1.7473	1.7268	1.7165	1.6174	1.6085	1.5092	2B	1.534	1.568	1.6201	1.6317	1.7500
	3A	0.0000	1.7500	1.7295	—	1.6201	1.6134	1.5119	3B	1.5340	1.5575	1.6201	1.6288	1.7500
1 3/4–6 UN	2A	0.0025	1.7475	1.7293	—	1.6392	1.6309	1.5491	2B	1.570	1.600	1.6417	1.6525	1.7500
	3A	0.0000	1.7500	1.7318	—	1.6417	1.6354	1.5516	3B	1.5700	1.5896	1.6417	1.6498	1.7500
1 3/4–8 UN	2A	0.0023	1.7477	1.7327	1.7252	1.6665	1.6590	1.5989	2B	1.615	1.640	1.6688	1.6786	1.7500
	3A	0.0000	1.7500	1.7350	—	1.6688	1.6632	1.6012	3B	1.6150	1.6297	1.6688	1.6762	1.7500
1 3/4–10 UNS	2A	0.0019	1.7481	1.7352	—	1.6831	1.6766	1.6291	2B	1.642	1.663	1.6850	1.6934	1.7500
1 3/4–12 UN	2A	0.0018	1.7482	1.7368	—	1.6941	1.6881	1.6490	2B	1.660	1.678	1.6959	1.7037	1.7500
	3A	0.0000	1.7500	1.7386	—	1.6959	1.6914	1.6508	3B	1.6600	1.6698	1.6959	1.7017	1.7500
1 3/4–14 UNS	2A	0.0017	1.7483	1.7380	—	1.7019	1.6963	1.6632	2B	1.673	1.688	1.7036	1.7109	1.7500
1 3/4–16 UN	2A	0.0016	1.7484	1.7390	—	1.7078	1.7025	1.6739	2B	1.682	1.696	1.7094	1.7163	1.7500
	3A	0.0000	1.7500	1.7406	—	1.7094	1.7054	1.6755	3B	1.6820	1.6908	1.7094	1.7146	1.7500
1 3/4–18 UNS	2A	0.0015	1.7485	1.7398	—	1.7124	1.7073	1.6824	2B	1.690	1.703	1.7139	1.7205	1.7500
1 3/4–20 UN	2A	0.0015	1.7485	1.7404	—	1.7160	1.7112	1.6890	2B	1.696	1.707	1.7175	1.7238	1.7500
	3A	0.0000	1.7500	1.7419	—	1.7175	1.7139	1.6905	3B	1.6960	1.7037	1.7175	1.7222	1.7500

Table 7. *(Continued)* **Standard Series and Selected Combinations — Unified Screw Threads**

Nominal Size, Threads per Inch, and Series Designation[a]	External[b]								Internal[b]					
	Class	Allow-ance	Major Diameter			Pitch Diameter		UNR Minor Dia.,[c] Max (Ref.)	Class	Minor Diameter		Pitch Diameter		Major Diameter
			Max[d]	Min	Min[e]	Max[d]	Min			Min	Max	Min	Max	Min
1 13/16–6 UN	2A	0.0025	1.8100	1.7918	—	1.7017	1.6933	1.6116	2B	1.632	1.663	1.7042	1.7151	1.8125
	3A	0.0000	1.8125	1.7943	—	1.7042	1.6979	1.6141	3B	1.6320	1.6521	1.7042	1.7124	1.8125
1 13/16–8 UN	2A	0.0023	1.8102	1.7952	—	1.7290	1.7214	1.6614	2B	1.677	1.702	1.7313	1.7412	1.8125
	3A	0.0000	1.8125	1.7975	—	1.7313	1.7256	1.6637	3B	1.6770	1.6922	1.7313	1.7387	1.8125
1 13/16–12 UN	2A	0.0018	1.8107	1.7993	—	1.7566	1.7506	1.7115	2B	1.722	1.740	1.7584	1.7662	1.8125
	3A	0.0000	1.8125	1.8011	—	1.7584	1.7539	1.7133	3B	1.7220	1.7323	1.7584	1.7642	1.8125
1 13/16–16 UN	2A	0.0016	1.8109	1.8015	—	1.7703	1.7650	1.7364	2B	1.745	1.759	1.7719	1.7788	1.8125
	3A	0.0000	1.8125	1.8031	—	1.7719	1.7679	1.7380	3B	1.7450	1.7533	1.7719	1.7771	1.8125
1 13/16–20 UN	2A	0.0015	1.8110	1.8029	—	1.7785	1.7737	1.7515	2B	1.758	1.770	1.7800	1.7863	1.8125
	3A	0.0000	1.8125	1.8044	—	1.7800	1.7764	1.7530	3B	1.7580	1.7662	1.7800	1.7847	1.8125
1 7/8–6 UN	2A	0.0025	1.8725	1.8543	—	1.7642	1.7558	1.6741	2B	1.695	1.725	1.7667	1.7777	1.8750
	3A	0.0000	1.8750	1.8568	—	1.7667	1.7604	1.6766	3B	1.6950	1.7146	1.7667	1.7749	1.8750
1 7/8–8 UN	2A	0.0023	1.8727	1.8577	1.8502	1.7915	1.7838	1.7239	2B	1.740	1.765	1.7938	1.8038	1.8750
	3A	0.0000	1.8750	1.8600	—	1.7938	1.7881	1.7262	3B	1.7400	1.7547	1.7938	1.8013	1.8750
1 7/8–10 UNS	2A	0.0019	1.8731	1.8602	—	1.8081	1.8016	1.7541	2B	1.767	1.788	1.8100	1.8184	1.8750
1 7/8–12 UN	2A	0.0018	1.8732	1.8618	—	1.8191	1.8131	1.7740	2B	1.785	1.803	1.8209	1.8287	1.8750
	3A	0.0000	1.8750	1.8636	—	1.8209	1.8164	1.7758	3B	1.7850	1.7948	1.8209	1.8267	1.8750
1 7/8–14 UNS	2A	0.0017	1.8733	1.8630	—	1.8269	1.8213	1.7883	2B	1.798	1.814	1.8286	1.8359	1.8750
1 7/8–16 UN	2A	0.0016	1.8734	1.8640	—	1.8328	1.8275	1.7989	2B	1.807	1.821	1.8344	1.8413	1.8750
	3A	0.0000	1.8750	1.8656	—	1.8344	1.8304	1.8005	3B	1.8070	1.8158	1.8344	1.8396	1.8750
1 7/8–18 UNS	2A	0.0015	1.8735	1.8648	—	1.8374	1.8323	1.8074	2B	1.815	1.828	1.8389	1.8455	1.8750
1 7/8–20 UN	2A	0.0015	1.8735	1.8654	—	1.8410	1.8362	1.8140	2B	1.821	1.832	1.8425	1.8488	1.8750
	3A	0.0000	1.8750	1.8669	—	1.8425	1.8389	1.8155	3B	1.8210	1.8287	1.8425	1.8472	1.8750
1 15/16–6 UN	2A	0.0026	1.9349	1.9167	—	1.8266	1.8181	1.7365	2B	1.757	1.788	1.8292	1.8403	1.9375
	3A	0.0000	1.9375	1.9193	—	1.8292	1.8228	1.7391	3B	1.7570	1.7771	1.8292	1.8375	1.9375
1 15/16–8 UN	2A	0.0023	1.9352	1.9202	—	1.8540	1.8463	1.7864	2B	1.802	1.827	1.8563	1.8663	1.9375
	3A	0.0000	1.9375	1.9225	—	1.8563	1.8505	1.7887	3B	1.8020	1.8172	1.8563	1.8638	1.9375
1 15/16–12 UN	2A	0.0018	1.9357	1.9243	—	1.8816	1.8755	1.8365	2B	1.847	1.865	1.8834	1.8913	1.9375
	3A	0.0000	1.9375	1.9261	—	1.8834	1.8789	1.8383	3B	1.8470	1.8573	1.8834	1.8893	1.9375

Table 7. *(Continued)* **Standard Series and Selected Combinations — Unified Screw Threads**

Nominal Size, Threads per Inch, and Series Designation[a]	External[b]								Internal[b]					
	Class	Allowance	Major Diameter			Pitch Diameter		UNR Minor Dia.,[c] Max (Ref.)	Class	Minor Diameter		Pitch Diameter		Major Diameter
			Max[d]	Min	Min[e]	Max[d]	Min			Min	Max	Min	Max	Min
1 15/16–16 UN	2A	0.0016	1.9359	1.9265	—	1.8953	1.8899	1.8614	2B	1.870	1.884	1.8969	1.9039	1.9375
	3A	0.0000	1.9375	1.9281	—	1.8969	1.8929	1.8630	3B	1.8700	1.8783	1.8969	1.9021	1.9375
1 15/16–20 UN	2A	0.0015	1.9360	1.9279	—	1.9035	1.8986	1.8765	2B	1.883	1.895	1.9050	1.9114	1.9375
	3A	0.0000	1.9375	1.9294	—	1.9050	1.9013	1.8780	3B	1.8830	1.8912	1.9050	1.9098	1.9375
2–4½ UNC	1A	0.0029	1.9971	1.9641	—	1.8528	1.8385	1.7324	1B	1.759	1.795	1.8557	1.8743	2.0000
	2A	0.0029	1.9971	1.9751	1.9641	1.8528	1.8433	1.7324	2B	1.759	1.795	1.8557	1.8681	2.0000
	3A	0.0000	2.0000	1.9780	—	1.8557	1.8486	1.7353	3B	1.7590	1.7861	1.8557	1.8650	2.0000
2–6 UN	2A	0.0026	1.9974	1.9792	—	1.8891	1.8805	1.7990	2B	1.820	1.850	1.8917	1.9028	2.0000
	3A	0.0000	2.0000	1.9818	—	1.8917	1.8853	1.8016	3B	1.8200	1.8396	1.8917	1.9000	2.0000
2–8 UN	2A	0.0023	1.9977	1.9827	1.9752	1.9165	1.9087	1.8489	2B	1.865	1.890	1.9188	1.9289	2.0000
	3A	0.0000	2.0000	1.9850	—	1.9188	1.9130	1.8512	3B	1.8650	1.8797	1.9188	1.9264	2.0000
2–10 UNS	2A	0.0020	1.9980	1.9851	—	1.9330	1.9265	1.8790	2B	1.892	1.913	1.9350	1.9435	2.0000
2–12 UN	2A	0.0018	1.9982	1.9868	—	1.9441	1.9380	1.8990	2B	1.910	1.928	1.9459	1.9538	2.0000
	3A	0.0000	2.0000	1.9886	—	1.9459	1.9414	1.9008	3B	1.9100	1.9198	1.9459	1.9518	2.0000
2–14 UNS	2A	0.0017	1.9983	1.9880	—	1.9519	1.9462	1.9133	2B	1.923	1.938	1.9536	1.9610	2.0000
2–16 UN	2A	0.0016	1.9984	1.9890	—	1.9578	1.9524	1.9239	2B	1.932	1.946	1.9594	1.9664	2.0000
	3A	0.0000	2.0000	1.9906	—	1.9594	1.9554	1.9255	3B	1.9320	1.9408	1.9594	1.9646	2.0000
2–18 UNS	2A	0.0015	1.9985	1.9898	—	1.9624	1.9573	1.9324	2B	1.940	1.953	1.9639	1.9706	2.0000
2–20 UN	2A	0.0015	1.9985	1.9904	—	1.9660	1.9611	1.9390	2B	1.946	1.957	1.9675	1.9739	2.0000
	3A	0.0000	2.0000	1.9919	—	1.9675	1.9638	1.9405	3B	1.9460	1.9537	1.9675	1.9723	2.0000
2 1/16–16 UNS	2A	0.0016	2.0609	2.0515	—	2.0203	2.0149	1.9864	2B	1.995	2.009	2.0219	2.0289	2.0625
	3A	0.0000	2.0625	2.0531	—	2.0219	2.0179	1.9880	3B	1.9950	2.0033	2.0219	2.0271	2.0625
2⅛–6 UN	2A	0.0026	2.1224	2.1042	—	2.0141	2.0054	1.9240	2B	1.945	1.975	2.0167	2.0280	2.1250
	3A	0.0000	2.1250	2.1068	—	2.0167	2.0102	1.9266	3B	1.9450	1.9646	2.0167	2.0251	2.1250
2⅛–8 UN	2A	0.0024	2.1226	2.1076	2.1001	2.0414	2.0335	1.9738	2B	1.990	2.015	2.0438	2.0540	2.1250
	3A	0.0000	2.1250	2.1100	—	2.0438	2.0379	1.9762	3B	1.9900	2.0047	2.0438	2.0515	2.1250
2⅛–12 UN	2A	0.0018	2.1232	2.1118	—	2.0691	2.0630	2.0240	2B	2.035	2.053	2.0709	2.0788	2.1250
	3A	0.0000	2.1250	2.1136	—	2.0709	2.0664	2.0258	3B	2.0350	2.0448	2.0709	2.0768	2.1250
2⅛–16 UN	2A	0.0016	2.1234	2.1140	—	2.0828	2.0774	2.0489	2B	2.057	2.071	2.0844	2.0914	2.1250
	3A	0.0000	2.1250	2.1156	—	2.0844	2.0803	2.0505	3B	2.0570	2.0658	2.0844	2.0896	2.1250

Table 7. *(Continued)* **Standard Series and Selected Combinations — Unified Screw Threads**

Nominal Size, Threads per Inch, and Series Designation[a]	External[b]								Internal[b]					
	Class	Allowance	Major Diameter			Pitch Diameter		UNR Minor Dia.,[c] Max (Ref.)	Class	Minor Diameter		Pitch Diameter		Major Diameter
			Max[d]	Min	Min[e]	Max[d]	Min			Min	Max	Min	Max	Min
2 1/8–20 UN	2A	0.0015	2.1235	2.1154	—	2.0910	2.0861	2.0640	2B	2.071	2.082	2.0925	2.0989	2.1250
	3A	0.0000	2.1250	2.1169	—	2.0925	2.0888	2.0655	3B	2.0710	2.0787	2.0925	2.0973	2.1250
2 3/16–16 UNS	2A	0.0016	2.1859	2.1765	—	2.1453	2.1399	2.1114	2B	2.120	2.134	2.1469	2.1539	2.1875
	3A	0.0000	2.1875	2.1781	—	2.1469	2.1428	2.1130	3B	2.1200	2.1283	2.1469	2.1521	2.1875
2 1/4–4 1/2 UNC	1A	0.0029	2.2471	2.2141	—	2.1028	2.0882	1.9824	1B	2.009	2.045	2.1057	2.1247	2.2500
	2A	0.0029	2.2471	2.2251	2.2141	2.1028	2.0931	1.9824	2B	2.009	2.045	2.1057	2.1183	2.2500
	3A	0.0000	2.2500	2.2280	—	2.1057	2.0984	1.9853	3B	2.0090	2.0361	2.1057	2.1152	2.2500
2 1/4–6 UN	2A	0.0026	2.2474	2.2292	—	2.1391	2.1303	2.0490	2B	2.070	2.100	2.1417	2.1531	2.2500
	3A	0.0000	2.2500	2.2318	—	2.1417	2.1351	2.0516	3B	2.0700	2.0896	2.1417	2.1502	2.2500
2 1/4–8 UN	2A	0.0024	2.2476	2.2326	2.2251	2.1664	2.1584	2.0988	2B	2.115	2.140	2.1688	2.1792	2.2500
	3A	0.0000	2.2500	2.2350	—	2.1688	2.1628	2.1012	3B	2.1150	2.1297	2.1688	2.1766	2.2500
2 1/4–10 UNS	2A	0.0020	2.2480	2.2351	—	2.1830	2.1765	2.1290	2B	2.142	2.163	2.1850	2.1935	2.2500
2 1/4–12 UN	2A	0.0018	2.2482	2.2368	—	2.1941	2.1880	2.1490	2B	2.160	2.178	2.1959	2.2038	2.2500
	3A	0.0000	2.2500	2.2386	—	2.1959	2.1914	2.1508	3B	2.1600	2.1698	2.1959	2.2018	2.2500
2 1/4–14 UNS	2A	0.0017	2.2483	2.2380	—	2.2019	2.1962	2.1633	2B	2.173	2.188	2.2036	2.2110	2.2500
2 1/4–16 UN	2A	0.0016	2.2484	2.2390	—	2.2078	2.2024	2.1739	2B	2.182	2.196	2.2094	2.2164	2.2500
	3A	0.0000	2.2500	2.2406	—	2.2094	2.2053	2.1755	3B	2.1820	2.1908	2.2094	2.2146	2.2500
2 1/4–18 UNS	2A	0.0015	2.2485	2.2398	—	2.2124	2.2073	2.1824	2B	2.190	2.203	2.2139	2.2206	2.2500
2 1/4–20 UN	2A	0.0015	2.2485	2.2404	—	2.2160	2.2111	2.1890	2B	2.196	2.207	2.2175	2.2239	2.2500
	3A	0.0000	2.2500	2.2419	—	2.2175	2.2137	2.1905	3B	2.1960	2.2037	2.2175	2.2223	2.2500
2 5/16–16 UNS	2A	0.0017	2.3108	2.3014	—	2.2702	2.2647	2.2363	2B	2.245	2.259	2.2719	2.2791	2.3125
	3A	0.0000	2.3125	2.3031	—	2.2719	2.2678	2.2380	3B	2.2450	2.2533	2.2719	2.2773	2.3125
2 3/8–6 UN	2A	0.0027	2.3723	2.3541	—	2.2640	2.2551	2.1739	2B	2.195	2.226	2.2667	2.2782	2.3750
	3A	0.0000	2.3750	2.3568	—	2.2667	2.2601	2.1766	3B	2.1950	2.2146	2.2667	2.2753	2.3750
2 3/8–8 UN	2A	0.0024	2.3726	2.3576	—	2.2914	2.2833	2.2238	2B	2.240	2.265	2.2938	2.3043	2.3750
	3A	0.0000	2.3750	2.3600	—	2.2938	2.2878	2.2262	3B	2.2400	2.2547	2.2938	2.3017	2.3750
2 3/8–12 UN	2A	0.0019	2.3731	2.3617	—	2.3190	2.3128	2.2739	2B	2.285	2.303	2.3209	2.3290	2.3750
	3A	0.0000	2.3750	2.3636	—	2.3209	2.3163	2.2758	3B	2.2850	2.2948	2.3209	2.3269	2.3750
2 3/8–16 UN	2A	0.0017	2.3733	2.3639	—	2.3327	2.3272	2.2988	2B	2.307	2.321	2.3344	2.3416	2.3750

Table 7. *(Continued)* **Standard Series and Selected Combinations — Unified Screw Threads**

Nominal Size, Threads per Inch, and Series Designation[a]	External[b]								Internal[b]					
	Class	Allowance	Major Diameter			Pitch Diameter		UNR Minor Dia.,[c] Max (Ref.)	Class	Minor Diameter		Pitch Diameter		Major Diameter
			Max[d]	Min	Min[e]	Max[d]	Min			Min	Max	Min	Max	Min
	3A	0.0000	2.3750	2.3656	—	2.3344	2.3303	2.3005	3B	2.3070	2.3158	2.3344	2.3398	2.3750
2⅜–20 UN	2A	0.0015	2.3735	2.3654	—	2.3410	2.3359	2.3140	2B	2.321	2.332	2.3425	2.3491	2.3750
	3A	0.0000	2.3750	2.3669	—	2.3425	2.3387	2.3155	3B	2.3210	2.3287	2.3425	2.3475	2.3750
2 7/16–16 UNS	2A	0.0017	2.4358	2.4264	—	2.3952	2.3897	2.3613	2B	2.370	2.384	2.3969	2.4041	2.4375
	3A	0.0000	2.4375	2.4281	—	2.3969	2.3928	2.3630	3B	2.3700	2.3783	2.3969	2.4023	2.4375
2½–4 UNC	1A	0.0031	2.4969	2.4612	—	2.3345	2.3190	2.1992	1B	2.229	2.267	2.3376	2.3578	2.5000
	2A	0.0031	2.4969	2.4731	2.4612	2.3345	2.3241	2.1992	2B	2.229	2.267	2.3376	2.3511	2.5000
	3A	0.0000	2.5000	2.4762	—	2.3376	2.3298	2.2023	3B	2.2290	2.2594	2.3376	2.3477	2.5000
2½–6 UN	2A	0.0027	2.4973	2.4791	—	2.3890	2.3800	2.2989	2B	2.320	2.350	2.3917	2.4033	2.5000
	3A	0.0000	2.5000	2.4818	—	2.3917	2.3850	2.3016	3B	2.3200	2.3396	2.3917	2.4004	2.5000
2½–8 UN	2A	0.0024	2.4976	2.4826	2.4751	2.4164	2.4082	2.3488	2B	2.365	2.390	2.4188	2.4294	2.5000
	3A	0.0000	2.5000	2.4850	—	2.4188	2.4127	2.3512	3B	2.3650	2.3797	2.4188	2.4268	2.5000
2½–10 UNS	2A	0.0020	2.4980	2.4851	—	2.4330	2.4263	2.3790	2B	2.392	2.413	2.4350	2.4437	2.5000
2½–12 UN	2A	0.0019	2.4981	2.4867	—	2.4440	2.4378	2.3989	2B	2.410	2.428	2.4459	2.4540	2.5000
	3A	0.0000	2.5000	2.4886	—	2.4459	2.4413	2.4008	3B	2.4100	2.4198	2.4459	2.4519	2.5000
2½–14 UNS	2A	0.0017	2.4983	2.4880	—	2.4519	2.4461	2.4133	2B	2.423	2.438	2.4536	2.4612	2.5000
2½–16 UN	2A	0.0017	2.4983	2.4889	—	2.4577	2.4522	2.4238	2B	2.432	2.446	2.4594	2.4666	2.5000
	3A	0.0000	2.5000	2.4906	—	2.4594	2.4553	2.4255	3B	2.4320	2.4408	2.4594	2.4648	2.5000
2½–18 UNS	2A	0.0016	2.4984	2.4897	—	2.4623	2.4570	2.4323	2B	2.440	2.453	2.4639	2.4708	2.5000
2½–20 UN	2A	0.0015	2.4985	2.4904	—	2.4660	2.4609	2.4390	2B	2.446	2.457	2.4675	2.4741	2.5000
	3A	0.0000	2.5000	2.4919	—	2.4675	2.4637	2.4405	3B	2.4460	2.4537	2.4675	2.4725	2.5000
2⅝–6 UN	2A	0.0027	2.6223	2.6041	—	2.5140	2.5050	2.4239	2B	2.445	2.475	2.5167	2.5285	2.6250
	3A	0.0000	2.6250	2.6068	—	2.5167	2.5099	2.4266	3B	2.4450	2.4646	2.5167	2.5255	2.6250
2⅝–8 UN	2A	0.0025	2.6225	2.6075	—	2.5413	2.5331	2.4737	2B	2.490	2.515	2.5438	2.5545	2.6250
	3A	0.0000	2.6250	2.6100	—	2.5438	2.5376	2.4762	3B	2.4900	2.5047	2.5438	2.5518	2.6250
2⅝–12 UN	2A	0.0019	2.6231	2.6117	—	2.5690	2.5628	2.5239	2B	2.535	2.553	2.5709	2.5790	2.6250
	3A	0.0000	2.6250	2.6136	—	2.5709	2.5663	2.5258	3B	2.5350	2.5448	2.5709	2.5769	2.6250
2⅝–16 UN	2A	0.0017	2.6233	2.6139	—	2.5827	2.5772	2.5488	2B	2.557	2.571	2.5844	2.5916	2.6250
	3A	0.0000	2.6250	2.6156	—	2.5844	2.5803	2.5505	3B	2.5570	2.5658	2.5844	2.5898	2.6250

Table 7. *(Continued)* **Standard Series and Selected Combinations — Unified Screw Threads**

Nominal Size, Threads per Inch, and Series Designation[a]	External[b]								Internal[b]					
	Class	Allowance	Major Diameter			Pitch Diameter		UNR Minor Dia.,[c] Max (Ref.)	Class	Minor Diameter		Pitch Diameter		Major Diameter
			Max[d]	Min	Min[e]	Max[d]	Min			Min	Max	Min	Max	Min
2⅝–20 UN	2A	0.0015	2.6235	2.6154	—	2.5910	2.5859	2.5640	2B	2.571	2.582	2.5925	2.5991	2.6250
	3A	0.0000	2.6250	2.6169	—	2.5925	2.5887	2.5655	3B	2.5710	2.5787	2.5925	2.5975	2.6250
2¾–4 UNC	1A	0.0032	2.7468	2.7111	—	2.5844	2.5686	2.4491	1B	2.479	2.517	2.5876	2.6082	2.7500
	2A	0.0032	2.7468	2.7230	2.7111	2.5844	2.5739	2.4491	2B	2.479	2.517	2.5876	2.6013	2.7500
	3A	0.0000	2.7500	2.7262	—	2.5876	2.5797	2.4523	3B	2.4790	2.5094	2.5876	2.5979	2.7500
2¾–6 UN	2A	0.0027	2.7473	2.7291	—	2.6390	2.6299	2.5489	2B	2.570	2.600	2.6417	2.6536	2.7500
	3A	0.0000	2.7500	2.7318	—	2.6417	2.6349	2.5516	3B	2.5700	2.5896	2.6417	2.6506	2.7500
2¾–8 UN	2A	0.0025	2.7475	2.7325	2.7250	2.6663	2.6580	2.5987	2B	2.615	2.640	2.6688	2.6796	2.7500
	3A	0.0000	2.7500	2.7350	—	2.6688	2.6625	2.6012	3B	2.6150	2.6297	2.6688	2.6769	2.7500
2¾–10 UNS	2A	0.0020	2.7480	2.7351	—	2.6830	2.6763	2.6290	2B	2.642	2.663	2.6850	2.6937	2.7500
2¾–12 UN	2A	0.0019	2.7481	2.7367	—	2.6940	2.6878	2.6489	2B	2.660	2.678	2.6959	2.7040	2.7500
	3A	0.0000	2.7500	2.7386	—	2.6959	2.6913	2.6508	3B	2.6600	2.6698	2.6959	2.7019	2.7500
2¾–14 UNS	2A	0.0017	2.7483	2.7380	—	2.7019	2.6961	2.6633	2B	2.673	2.688	2.7036	2.7112	2.7500
2¾–16 UN	2A	0.0017	2.7483	2.7389	—	2.7077	2.7022	2.6738	2B	2.682	2.696	2.7094	2.7166	2.7500
	3A	0.0000	2.7500	2.7406	—	2.7094	2.7053	2.6755	3B	2.6820	2.6908	2.7094	2.7148	2.7500
2¾–18 UNS	2A	0.0016	2.7484	2.7397	—	2.7123	2.7070	2.6823	2B	2.690	2.703	2.7139	2.7208	2.7500
2¾–20 UN	2A	0.0015	2.7485	2.7404	—	2.7160	2.7109	2.6890	2B	2.696	2.707	2.7175	2.7241	2.7500
	3A	0.0000	2.7500	2.7419	—	2.7175	2.7137	2.6905	3B	2.6960	2.7037	2.7175	2.7225	2.7500
2⅞–6 UN	2A	0.0028	2.8722	2.8540	—	2.7639	2.7547	2.6738	2B	2.695	2.725	2.7667	2.7787	2.8750
	3A	0.0000	2.8750	2.8568	—	2.7667	2.7598	2.6766	3B	2.6950	2.7146	2.7667	2.7757	2.8750
2⅞–8 UN	2A	0.0025	2.8725	2.8575	—	2.7913	2.7829	2.7237	2B	2.740	2.765	2.7938	2.8048	2.8750
	3A	0.0000	2.8750	2.8600	—	2.7938	2.7875	2.7262	3B	2.7400	2.7547	2.7938	2.8020	2.8750
2⅞–12 UN	2A	0.0019	2.8731	2.8617	—	2.8190	2.8127	2.7739	2B	2.785	2.803	2.8209	2.8291	2.8750
	3A	0.0000	2.8750	2.8636	—	2.8209	2.8162	2.7758	3B	2.7850	2.7948	2.8209	2.8271	2.8750
2⅞–16 UN	2A	0.0017	2.8733	2.8639	—	2.8327	2.8271	2.7988	2B	2.807	2.821	2.8344	2.8417	2.8750
	3A	0.0000	2.8750	2.8656	—	2.8344	2.8302	2.8005	3B	2.8070	2.8158	2.8344	2.8399	2.8750
2⅞–20 UN	2A	0.0016	2.8734	2.8653	—	2.8409	2.8357	2.8139	2B	2.821	2.832	2.8425	2.8493	2.8750
	3A	0.0000	2.8750	2.8669	—	2.8425	2.8386	2.8155	3B	2.8210	2.8287	2.8425	2.8476	2.8750
3–4 UNC	1A	0.0032	2.9968	2.9611	—	2.8344	2.8183	2.6991	1B	2.729	2.767	2.8376	2.8585	3.0000

Table 7. *(Continued)* **Standard Series and Selected Combinations — Unified Screw Threads**

Nominal Size, Threads per Inch, and Series Designation[a]	External[b]								Internal[b]					
	Class	Allowance	Major Diameter			Pitch Diameter		UNR Minor Dia.,[c] Max (Ref.)	Class	Minor Diameter		Pitch Diameter		Major Diameter
			Max[d]	Min	Min[e]	Max[d]	Min			Min	Max	Min	Max	Min
	2A	0.0032	2.9968	2.9730	2.9611	2.8344	2.8237	2.6991	2B	2.729	2.767	2.8376	2.8515	3.0000
	3A	0.0000	3.0000	2.9762	—	2.8376	2.8296	2.7023	3B	2.7290	2.7594	2.8376	2.8480	3.0000
3–6 UN	2A	0.0028	2.9972	2.9790	—	2.8889	2.8796	2.7988	2B	2.820	2.850	2.8917	2.9038	3.0000
	3A	0.0000	3.0000	2.9818	—	2.8917	2.8847	2.8016	3B	2.8200	2.8396	2.8917	2.9008	3.0000
3–8 UN	2A	0.0026	2.9974	2.9824	2.9749	2.9162	2.9077	2.8486	2B	2.865	2.890	2.9188	2.9299	3.0000
	3A	0.0000	3.0000	2.9850	—	2.9188	2.9124	2.8512	3B	2.8650	2.8797	2.9188	2.9271	3.0000
3–10 UNS	2A	0.0020	2.9980	2.9851	—	2.9330	2.9262	2.8790	2B	2.892	2.913	2.9350	2.9439	3.0000
3–12 UN	2A	0.0019	2.9981	2.9867	—	2.9440	2.9377	2.8989	2B	2.910	2.928	2.9459	2.9541	3.0000
	3A	0.0000	3.0000	2.9886	—	2.9459	2.9412	2.9008	3B	2.9100	2.9198	2.9459	2.9521	3.0000
3–14 UNS	2A	0.0018	2.9982	2.9879	—	2.9518	2.9459	2.9132	2B	2.923	2.938	2.9536	2.9613	3.0000
3–16 UN	2A	0.0017	2.9983	2.9889	—	2.9577	2.9521	2.9238	2B	2.932	2.946	2.9594	2.9667	3.0000
	3A	0.0000	3.0000	2.9906	—	2.9594	2.9552	2.9255	3B	2.9320	2.9408	2.9594	2.9649	3.0000
3–18 UNS	2A	0.0016	2.9984	2.9897	—	2.9623	2.9569	2.9323	2B	2.940	2.953	2.9639	2.9709	3.0000
3–20 UN	2A	0.0016	2.9984	2.9903	—	2.9659	2.9607	2.9389	2B	2.946	2.957	2.9675	2.9743	3.0000
	3A	0.0000	3.0000	2.9919	—	2.9675	2.9636	2.9405	3B	2.9460	2.9537	2.9675	2.9726	3.0000
3⅛–6 UN	2A	0.0028	3.1222	3.1040	—	3.0139	3.0045	2.9238	2B	2.945	2.975	3.0167	3.0289	3.1250
	3A	0.0000	3.1250	3.1068	—	3.0167	3.0097	2.9266	3B	2.9450	2.9646	3.0167	3.0259	3.1250
3⅛–8 UN	2A	0.0026	3.1224	3.1074	—	3.0412	3.0326	2.9736	2B	2.990	3.015	3.0438	3.0550	3.1250
	3A	0.0000	3.1250	3.1100	—	3.0438	3.0374	2.9762	3B	2.9900	3.0047	3.0438	3.0522	3.1250
3⅛–12 UN	2A	0.0019	3.1231	3.1117	—	3.0690	3.0627	3.0239	2B	3.035	3.053	3.0709	3.0791	3.1250
	3A	0.0000	3.1250	3.1136	—	3.0709	3.0662	3.0258	3B	3.0350	3.0448	3.0709	3.0771	3.1250
3⅛–16 UN	2A	0.0017	3.1233	3.1139	—	3.0827	3.0771	3.0488	2B	3.057	3.071	3.0844	3.0917	3.1250
	3A	0.0000	3.1250	3.1156	—	3.0844	3.0802	3.0505	3B	3.0570	3.0658	3.0844	3.0899	3.1250
3¼–4 UNC	1A	0.0033	3.2467	3.2110	—	3.0843	3.0680	2.9490	1B	2.979	3.017	3.0876	3.1088	3.2500
	2A	0.0033	3.2467	3.2229	3.2110	3.0843	3.0734	2.9490	2B	2.979	3.017	3.0876	3.1017	3.2500
	3A	0.0000	3.2500	3.2262	—	3.0876	3.0794	2.9523	3B	2.9790	3.0094	3.0876	3.0982	3.2500
3¼–6 UN	2A	0.0028	3.2472	3.2290	—	3.1389	3.1294	3.0488	2B	3.070	3.100	3.1417	3.1540	3.2500
	3A	0.0000	3.2500	3.2318	—	3.1417	3.1346	3.0516	3B	3.0700	3.0896	3.1417	3.1509	3.2500
3¼–8 UN	2A	0.0026	3.2474	3.2324	3.2249	3.1662	3.1575	3.0986	2B	3.115	3.140	3.1688	3.1801	3.2500
	3A	0.0000	3.2500	3.2350	—	3.1688	3.1623	3.1012	3B	3.1150	3.1297	3.1688	3.1773	3.2500

Table 7. *(Continued)* **Standard Series and Selected Combinations — Unified Screw Threads**

Nominal Size, Threads per Inch, and Series Designation[a]	External[b]								Internal[b]					
	Class	Allow-ance	Major Diameter			Pitch Diameter		UNR Minor Dia.,[c] Max (Ref.)	Class	Minor Diameter		Pitch Diameter		Major Diameter
			Max[d]	Min	Min[e]	Max[d]	Min			Min	Max	Min	Max	Min
3¼–10 UNS	2A	0.0020	3.2480	3.2351	—	3.1830	3.1762	3.1290	2B	3.142	3.163	3.1850	3.1939	3.2500
3¼–12 UN	2A	0.0019	3.2481	3.2367	—	3.1940	3.1877	3.1489	2B	3.160	3.178	3.1959	3.2041	3.2500
	3A	0.0000	3.2500	3.2386	—	3.1959	3.1912	3.1508	3B	3.1600	3.1698	3.1959	3.2041	3.2500
3¼–14 UNS	2A	0.0018	3.2482	3.2379	—	3.2018	3.1959	3.1632	2B	3.173	3.188	3.2036	3.2113	3.2500
3¼–16 UN	2A	0.0017	3.2483	3.2389	—	3.2077	3.2021	3.1738	2B	3.182	3.196	3.2094	3.2167	3.2500
	3A	0.0000	3.2500	3.2406	—	3.2094	3.2052	3.1755	3B	3.1820	3.1908	3.2094	3.2149	3.2500
3¼–18 UNS	2A	0.0016	3.2484	3.2397	—	3.2123	3.2069	3.1823	2B	3.190	3.203	3.2139	3.2209	3.2500
3⅜–6 UN	2A	0.0029	3.3721	3.3539	—	3.2638	3.2543	3.1737	2B	3.195	3.225	3.2667	3.2791	3.3750
	3A	0.0000	3.3750	3.3568	—	3.2667	3.2595	3.1766	3B	3.1950	3.2146	3.2667	3.2760	3.3750
3⅜–8 UN	2A	0.0026	3.3724	3.3574	—	3.2912	3.2824	3.2236	2B	3.240	3.265	3.2938	3.3052	3.3750
	3A	0.0000	3.3750	3.3600	—	3.2938	3.2872	3.2262	3B	3.2400	3.2547	3.2938	3.3023	3.3750
3⅜–12 UN	2A	0.0019	3.3731	3.3617	—	3.3190	3.3126	3.2739	2B	3.285	3.303	3.3209	3.3293	3.3750
	3A	0.0000	3.3750	3.3636	—	3.3209	3.3161	3.2758	3B	3.2850	3.2948	3.3209	3.3272	3.3750
3⅜–16 UN	2A	0.0017	3.3733	3.3639	—	3.3327	3.3269	3.2988	2B	3.307	3.321	3.3344	3.3419	3.3750
	3A	0.0000	3.3750	3.3656	—	3.3344	3.3301	3.3005	3B	3.3070	3.3158	3.3344	3.3400	3.3750
3½–4 UNC	1A	0.0033	3.4967	3.4610	—	3.3343	3.3177	3.1990	1B	3.229	3.267	3.3376	3.3591	3.5000
	2A	0.0033	3.4967	3.4729	3.4610	3.3343	3.3233	3.1990	2B	3.229	3.267	3.3376	3.3519	3.5000
	3A	0.0000	3.5000	3.4762	—	3.3376	3.3293	3.2023	3B	3.2290	3.2594	3.3376	3.3484	3.5000
3½–6 UN	2A	0.0029	3.4971	3.4789	—	3.3888	3.3792	3.2987	2B	3.320	3.350	3.3917	3.4042	3.5000
	3A	0.0000	3.5000	3.4818	—	3.3917	3.3845	3.3016	3B	3.3200	3.3396	3.3917	3.4011	3.5000
3½–8 UN	2A	0.0026	3.4974	3.4824	3.4749	3.4162	3.4074	3.3486	2B	3.365	3.390	3.4188	3.4303	3.5000
	3A	0.0000	3.5000	3.4850	—	3.4188	3.4122	3.3512	3B	3.3650	3.3797	3.4188	3.4274	3.5000
3½–10 UNS	2A	0.0021	3.4979	3.4850	—	3.4329	3.4260	3.3789	2B	3.392	3.413	3.4350	3.4440	3.5000
3½–12 UN	2A	0.0019	3.4981	3.4867	—	3.4440	3.4376	3.3989	2B	3.410	3.428	3.4459	3.4543	3.5000
	3A	0.0000	3.5000	3.4886	—	3.4459	3.4411	3.4008	3B	3.4100	3.4198	3.4459	3.4522	3.5000
3½–14 UNS	2A	0.0018	3.4982	3.4879	—	3.4518	3.4457	3.4132	2B	3.423	3.438	3.4536	3.4615	3.5000
3½–16 UN	2A	0.0017	3.4983	3.4889	—	3.4577	3.4519	3.4238	2B	3.432	3.446	3.4594	3.4669	3.5000
	3A	0.0000	3.5000	3.4906	—	3.4594	3.4551	3.4255	3B	3.4320	3.4408	3.4594	3.4650	3.5000
3½–18 UNS	2A	0.0017	3.4983	3.4896	—	3.4622	3.4567	3.4322	2B	3.440	3.453	3.4639	3.4711	3.5000

Table 7. *(Continued)* **Standard Series and Selected Combinations — Unified Screw Threads**

Nominal Size, Threads per Inch, and Series Designation[a]	External[b]								Internal[b]					
	Class	Allowance	Major Diameter			Pitch Diameter		UNR Minor Dia.,[c] Max (Ref.)	Class	Minor Diameter		Pitch Diameter		Major Diameter
			Max[d]	Min	Min[e]	Max[d]	Min			Min	Max	Min	Max	Min
3⅝–6 UN	2A	0.0029	3.6221	3.6039	—	3.5138	3.5041	3.4237	2B	3.445	3.475	3.5167	3.5293	3.6250
	3A	0.0000	3.6250	3.6068	—	3.5167	3.5094	3.4266	3B	3.4450	3.4646	3.5167	3.5262	3.6250
3⅝–8 UN	2A	0.0027	3.6223	3.6073	—	3.5411	3.5322	3.4735	2B	3.490	3.515	3.5438	3.5554	3.6250
	3A	0.0000	3.6250	3.6100	—	3.5438	3.5371	3.4762	3B	3.4900	3.5047	3.5438	3.5525	3.6250
3⅝–12 UN	2A	0.0019	3.6231	3.6117	—	3.5690	3.5626	3.5239	2B	3.535	3.553	3.5709	3.5793	3.6250
	3A	0.0000	3.6250	3.6136	—	3.5709	3.5661	3.5258	3B	3.5350	3.5448	3.5709	3.5772	3.6250
3⅝–16 UN	2A	0.0017	3.6233	3.6139	—	3.5827	3.5769	3.5488	2B	3.557	3.571	3.5844	3.5919	3.6250
	3A	0.0000	3.6250	3.6156	—	3.5844	3.5801	3.5505	3B	3.5570	3.5658	3.5844	3.5900	3.6250
3¾–4 UNC	1A	0.0034	3.7466	3.7109	—	3.5842	3.5674	3.4489	1B	3.479	3.517	3.5876	3.6094	3.7500
	2A	0.0034	3.7466	3.7228	3.7109	3.5842	3.5730	3.4489	2B	3.479	3.517	3.5876	3.6021	3.7500
	3A	0.0000	3.7500	3.7262	—	3.5876	3.5792	3.4523	3B	3.4790	3.5094	3.5876	3.5985	3.7500
3¾–6 UN	2A	0.0029	3.7471	3.7289	—	3.6388	3.6290	3.5487	2B	3.570	3.600	3.6417	3.6544	3.7500
	3A	0.0000	3.7500	3.7318	—	3.6417	3.6344	3.5516	3B	3.5700	3.5896	3.6417	3.6512	3.7500
3¾–8 UN	2A	0.0027	3.7473	3.7323	3.7248	3.6661	3.6571	3.5985	2B	3.615	3.640	3.6688	3.6805	3.7500
	3A	0.0000	3.7500	3.7350	—	3.6688	3.6621	3.6012	3B	3.6150	3.6297	3.6688	3.6776	3.7500
3¾–10 UNS	2A	0.0021	3.7479	3.7350	—	3.6829	3.6760	3.6289	2B	3.642	3.663	3.6850	3.6940	3.7500
3¾–12 UN	2A	0.0019	3.7481	3.7367	—	3.6940	3.6876	3.6489	2B	3.660	3.678	3.6959	3.7043	3.7500
	3A	0.0000	3.7500	3.7386	—	3.6959	3.6911	3.6508	3B	3.6600	3.6698	3.6959	3.7022	3.7500
3¾–14 UNS	2A	0.0018	3.7482	3.7379	—	3.7018	3.6957	3.6632	2B	3.673	3.688	3.7036	3.7115	3.7500
3¾–16 UN	2A	0.0017	3.7483	3.7389	—	3.7077	3.7019	3.6738	2B	3.682	3.696	3.7094	3.7169	3.7500
	3A	0.0000	3.7500	3.7406	—	3.7094	3.7051	3.6755	3B	3.6820	3.6908	3.7094	3.7150	3.7500
3¾–18 UNS	2A	0.0017	3.7483	3.7396	—	3.7122	3.7067	3.6822	2B	3.690	3.703	3.7139	3.7211	3.7500
3⅞–6 UN	2A	0.0030	3.8720	3.8538	—	3.7637	3.7538	3.6736	2B	3.695	3.725	3.7667	3.7795	3.8750
	3A	0.0000	3.8750	3.8568	—	3.7667	3.7593	3.6766	3B	3.6950	3.7146	3.7667	3.7763	3.8750
3⅞–8 UN	2A	0.0027	3.8723	3.8573	—	3.7911	3.7820	3.7235	2B	3.740	3.765	3.7938	3.8056	3.8750
	3A	0.0000	3.8750	3.8600	—	3.7938	3.7870	3.7262	3B	3.7400	3.7547	3.7938	3.8026	3.8750
3⅞–12 UN	2A	0.0020	3.8730	3.8616	—	3.8189	3.8124	3.7738	2B	3.785	3.803	3.8209	3.8294	3.8750
	3A	0.0000	3.8750	3.8636	—	3.8209	3.8160	3.7758	3B	3.7850	3.7948	3.8209	3.8273	3.8750
3⅞–16 UN	2A	0.0018	3.8732	3.8638	—	3.8326	3.8267	3.7987	2B	3.807	3.821	3.8344	3.8420	3.8750

Table 7. *(Continued)* **Standard Series and Selected Combinations — Unified Screw Threads**

Nominal Size, Threads per Inch, and Series Designation[a]	External[b]								Internal[b]					
	Class	Allowance	Major Diameter			Pitch Diameter		UNR Minor Dia.,[c] Max (Ref.)	Class	Minor Diameter		Pitch Diameter		Major Diameter
			Max[d]	Min	Min[e]	Max[d]	Min			Min	Max	Min	Max	Min
	3A	0.0000	3.8750	3.8656	—	3.8344	3.8300	3.8005	3B	3.8070	3.8158	3.8344	3.8401	3.8750
4–4 UNC	1A	0.0034	3.9966	3.9609	—	3.8342	3.8172	3.6989	1B	3.729	3.767	3.8376	3.8597	4.0000
	2A	0.0034	3.9966	3.9728	3.9609	3.8342	3.8229	3.6989	2B	3.729	3.767	3.8376	3.8523	4.0000
	3A	0.0000	4.0000	3.9762	—	3.8376	3.8291	3.7023	3B	3.7290	3.7594	3.8376	3.8487	4.0000
4–6 UN	2A	0.0030	3.9970	3.9788	—	3.8887	3.8788	3.7986	2B	3.820	3.850	3.8917	3.9046	4.0000
	3A	0.0000	4.0000	3.9818	—	3.8917	3.8843	3.8016	3B	3.8200	3.8396	3.8917	3.9014	4.0000
4–8 UN	2A	0.0027	3.9973	3.9823	3.9748	3.9161	3.9070	3.8485	2B	3.865	3.890	3.9188	3.9307	4.0000
	3A	0.0000	4.0000	3.9850	—	3.9188	3.9120	3.8512	3B	3.8650	3.8797	3.9188	3.9277	4.0000
4–10 UNS	2A	0.0021	3.9979	3.9850	—	3.9329	3.9259	3.8768	2B	3.892	3.913	3.9350	3.9441	4.0000
4–12 UN	2A	0.0020	3.9980	3.9866	—	3.9439	3.9374	3.8988	2B	3.910	3.928	3.9459	3.9544	4.0000
	3A	0.0000	4.0000	3.9886	—	3.9459	3.9410	3.9008	3B	3.9100	3.9198	3.9459	3.9523	4.0000
4–14 UNS	2A	0.0018	3.9982	3.9879	—	3.9518	3.9456	3.9132	2B	3.923	3.938	3.9536	3.9616	4.0000
4–16 UN	2A	0.0018	3.9982	3.9888	—	3.9576	3.9517	3.9237	2B	3.932	3.946	3.9594	3.9670	4.0000
	3A	0.0000	4.0000	3.9906	—	3.9594	3.9550	3.9255	3B	3.9320	3.9408	3.9594	3.9651	4.0000

[a] Use UNR designation instead of UN wherever UNR thread form is desired for external use.
[b] Regarding combinations of thread classes.
[c] UN series external thread maximum minor diameter is basic for Class3A and basic minus allowance for Classes 1A and 2A.
[d] For Class 2A threads having an additive finish the maximum is increased, by the allowance, to the basic size, the value being the same as for Class 3A.
[e] For unfinished hot-rolled material not including standard fasteners with rolled threads.

All dimensions in inches.
Use UNS threads only if Standard Series do not meet requirements . For sizes above 4 inches see ASME/ANSI B1.1-1989.

Measuring Screw Threads

Pitch and Lead of Screw Threads.—The *pitch* of a screw thread is the distance from the center of one thread to the center of the next thread. This applies no matter whether the screw has a single, double, triple or quadruple thread. The *lead* of a screw thread is the distance the nut will move forward on the screw if it is turned around one full revolution. In a single-threaded screw, the pitch and lead are equal, because the nut would move forward the distance from one thread to the next, if turned around once. In a double-threaded screw, the nut will move forward two threads, or twice the pitch, so that in this case the lead equals twice the pitch. In a triple-threaded screw, the lead equals three times the pitch, and so on.

The word "pitch" is often, although improperly, used to denote the *number of threads per inch.* Screws are spoken of as having a 12-pitch thread, when twelve threads per inch is what is really meant. The number of threads per inch equals 1 divided by the pitch, or expressed as a formula:

$$\text{Number of threads per inch} = \frac{1}{\text{pitch}}$$

The pitch of a screw equals 1 divided by the number of threads per inch, or:

$$\text{Pitch} = \frac{1}{\text{number of threads per inch}}$$

If the number of threads per inch equals 16, the pitch = $\frac{1}{16}$. If the pitch equals 0.05, the number of threads equals $1 \div 0.05 = 20$. If the pitch is $\frac{2}{5}$ inch, the number of threads per inch equals $1 \div \frac{2}{5} = 2\frac{1}{2}$.

Confusion is often caused by the indefinite designation of multiple-thread screws (double, triple, quadruple, etc.). The expression, "four threads per inch, triple," for example, is not to be recommended. It means that the screw is cut with four triple threads or with twelve threads per inch, if the threads are counted by placing a scale alongside the screw. To cut this screw, the lathe would be geared to cut four threads per inch, but they would be cut only to the depth required for twelve threads per inch. The best expression, when a multiple-thread is to be cut, is to say, in this case, "$\frac{1}{4}$ inch lead, $\frac{1}{12}$ inch pitch, triple thread." For single-threaded screws, only the number of threads per inch and the form of the thread are specified. The word "single" is not required.

Measuring Screw Thread Pitch Diameters by Thread Micrometers.—As the pitch or angle diameter of a tap or screw is the most important dimension, it is necessary that the pitch diameter of screw threads be measured, in addition to the outside diameter.

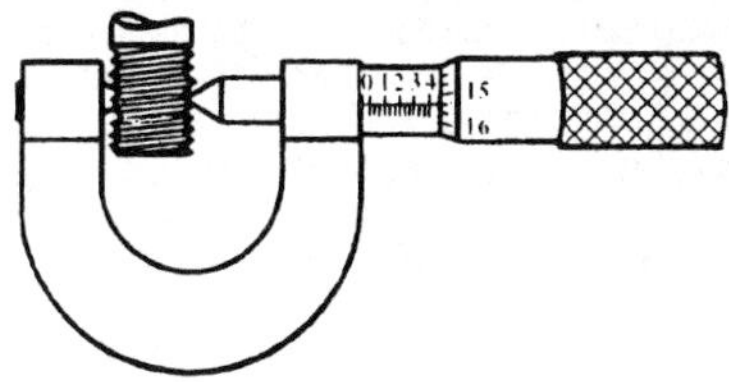

Fig. 1.

One method of measuring in the angle of a thread is by means of a special screw thread micrometer, as shown in the accompanying engraving, Fig. 1. The fixed anvil is W-shaped to engage two thread flanks, and the movable point is cone-shaped so as to enable it to enter the space between two threads, and at the same time be at liberty to revolve. The contact points are on the sides of the thread, as they necessarily must be in order that the pitch diameter may be determined. The cone-shaped point of the measuring screw is slightly rounded so that it will not bear in the bottom of the thread. There is also sufficient clearance at the bottom of the V-shaped anvil to prevent it from bearing on the top of the thread. The movable point is adapted to measuring all pitches, but the fixed anvil is limited in its capacity. To cover the whole range of pitches, from the finest to the coarsest, a number of fixed anvils are therefore required.

To find the theoretical pitch diameter, which is measured by the micrometer, subtract twice the addendum of the thread from the standard outside diameter. The addendum of the thread for the American and other standard threads is given in the section on screw thread systems.

Ball-point Micrometers.—If standard plug gages are available, it is not necessary to actually measure the pitch diameter, but merely to compare it with the standard gage. In this case, a ball-point micrometer, as shown in Fig. 2, may be employed. Two types of ball-point micrometers are ordinarily used. One is simply a regular plain micrometer with ball points made to slip over both measuring points. (See B, Fig. 2.) This makes a kind of combination plain and ball-point micrometer, the ball points being easily removed. These ball points, however, do not fit solidly on their seats, even if they are split, as shown, and are apt to cause errors in measurements. The best, and, in the long run, the cheapest, method is to use a regular micrometer arranged as shown at *A*. Drill and ream out both the end of the measuring screw or spindle and the anvil, and fit ball points into them as shown. Care should be taken to have the ball point in the spindle run true. The holes in the micrometer spindle and anvil and the shanks on the points are tapered to insure a good fit. The hole *H* in spindle *G* is provided so that the ball point can be easily driven out when a change for a larger or smaller size of ball point is required.

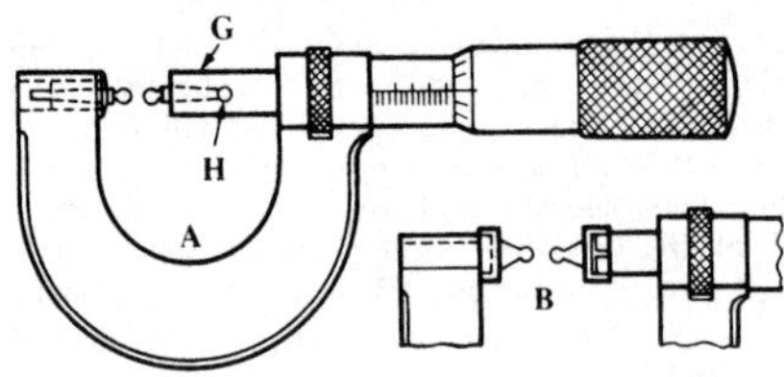

Fig. 2.

A ball-point micrometer may be used for comparing the *angle* of a screw thread, with that of a gage. This can be done by using different sizes of ball points, comparing the size first near the root of the thread, then (using a larger ball point) at about the point of the pitch diameter, and finally near the top of the thread (using in the latter case, of course, a much larger ball point). If the gage and thread measurements are the same at each of the three points referred to, this indicates that the thread angle is correct.

Measuring Screw Threads by Three-wire Method.—The *effective* or *pitch diameter* of a screw thread may be measured very accurately by means of some form of micrometer and three wires of equal diameter. This method is extensively used in checking the accuracy of threaded plug gages and other precision screw threads. Two of the wires are placed in contact with the thread on one side and the third wire in a position diametrically opposite as illustrated by the diagram, (see table "Formulas for Checking Pitch Diameters of Screw Threads") and the dimension over the wires is determined by means of a micrometer. An ordinary micrometer is commonly used but some form of "floating micrometer" is preferable, especially for measuring thread gages and other precision work. The floating micrometer is mounted upon a compound slide so that it can move freely in directions parallel or at right angles to the axis of the screw, which is held in a horizontal position between adjustable centers. With this arrangement the micrometer is held constantly at right angles to the axis of the screw so that only one wire on each side may be used instead of having two on one side and one on the other, as is necessary when using an ordinary micrometer. The pitch diameter may be determined accurately if the correct micrometer reading for wires of a given size is known.

Classes of Formulas for Three-Wire Measurement.—Various formulas have been established for checking the pitch diameters of screw threads by measurement over wires of known size. These formulas differ with regard to their simplicity or complexity and resulting accuracy. They also differ in that some show what measurement M over the wires should be to obtain a given pitch diameter E, whereas others show the value of the pitch diameter E for a given measurement M.

Formulas for Finding Measurement M: In using a formula for finding the value of measurement M, the required pitch diameter E is inserted in the formula. Then, in cutting or grinding a screw thread, the actual measurement M is made to conform to the calculated value of M. Formulas for finding measurement M may be modified so that the basic major or outside diameter is inserted in the formula instead of the pitch diameter; however, the pitch-diameter type of formula is preferable because the pitch diameter is a more important dimension than the major diameter.

Formulas for Finding Pitch Diameters E: Some formulas are arranged to show the value of the pitch diameter E when measurement M is known. Thus, the value of M is first determined by measurement and then is inserted in the formula for finding the corresponding pitch diameter E. This type of formula is useful for determining the pitch diameter of an existing thread gage or other screw thread in connection with inspection work. The formula for finding measurement M is more convenient to use in the shop or tool room in cutting or grinding new threads, because the pitch diameter is specified on the drawing and the problem is to find the value of measurement M for obtaining that pitch diameter.

General Classes of Screw Thread Profiles.—Thread profiles may be divided into three general classes or types as follows:

Screw Helicoid: Represented by a screw thread having a straight-line profile in the axial plane. Such a screw thread may be cut in a lathe by using a straight-sided single-point tool, provided the top surface lies in the axial plane.

Involute Helicoid: Represented either by a screw thread or a helical gear tooth having an involute profile in a plane perpendicular to the axis. A rolled screw thread, theoretically at least, is an exact involute helicoid.

Intermediate Profiles: An intermediate profile that lies somewhere between the screw helicoid and the involute helicoid will be formed on a screw thread either by milling or grinding with a straight-sided wheel set in alignment with the thread groove. The resulting form will approach closely the involute helicoid form. In milling or grinding a thread, the included cutter or wheel angle may either equal the standard thread angle (which is always measured in the axial plane) or the cutter or wheel angle may be reduced to approximate, at least, the thread angle in the normal plane. In practice, all these variations affect the three-wire measurement.

Accuracy of Formulas for Checking Pitch Diameters by Three-Wire Method.—The exact measurement M for a given pitch diameter depends upon the lead angle, the thread angle, and the profile or cross-sectional shape of the thread. As pointed out in the preceding paragraph, the profile depends upon the method of cutting or forming the thread. In a milled or ground thread, the profile is affected not only by the cutter or wheel angle, but also by the diameter of the cutter or wheel; hence, because of these variations, an absolutely exact and reasonably simple general formula for measurement M cannot be established; however, if the lead angle is low, as with a standard single-thread screw, and especially if the thread angle is high like a 60-degree thread, simple formulas that are not arranged to compensate for the lead angle are used ordinarily and meet most practical requirements, particularly in measuring 60-degree threads. If lead angles are large enough to greatly affect the result, as with most multiple threads (especially Acme or 29-degree worm threads), a formula should be used that compensates for the lead angle sufficiently to obtain the necessary accuracy.

The formulas that follow include 1) a very simple type in which the effect of the lead angle on measurement M is entirely ignored. This simple formula usually is applicable to the measurement of 60-degree single-thread screws, except possibly when gage-making accuracy is required; 2) formulas that do include the effect of the lead angle but, nevertheless, are approximations and not always suitable for the higher lead angles when extreme accuracy is required; and 3) formulas for the higher lead angles and the most precise classes of work.

Where approximate formulas are applied consistently in the measurement of both thread plug gages and the thread "setting plugs" for ring gages, interchangeability might be secured, assuming that such approximate formulas were universally employed.

Wire Sizes for Checking Pitch Diameters of Screw Threads.—In checking screw threads by the 3-wire method, the general practice is to use measuring wires of the so-called "best size." The "best-size" wire is one that contacts at the pitch line or midslope of the thread because then the measurement of the pitch diameter is least affected by an error in the thread angle. In the following formula for determining approximately the "best-size" wire or the diameter for pitch-line contact, A = one-half included angle of thread in the axial plane.

$$\text{Best-size wire} = \frac{0.5 \times \text{pitch}}{\cos A} = 0.5\ \text{pitch} \times \sec A$$

For 60-degree threads, this formula reduces to

$$\text{Best-size wire} = 0.57735 \times \text{pitch}$$

Table 8. Diameters of Wires for Measuring American Standard and British Standard Whitworth Screw Threads

Threads per Inch	Pitch, Inch	Wire Diameters for American Standard Threads			Wire Diameters for Whitworth Standard Threads		
		Max.	Min.	Pitch-Line Contact	Max.	Min.	Pitch-Line Contact
4	0.2500	0.2250	0.1400	0.1443	0.1900	0.1350	0.1409
4½	0.2222	0.2000	0.1244	0.1283	0.1689	0.1200	0.1253
5	0.2000	0.1800	0.1120	0.1155	0.1520	0.1080	0.1127
5½	0.1818	0.1636	0.1018	0.1050	0.1382	0.0982	0.1025
6	0.1667	0.1500	0.0933	0.0962	0.1267	0.0900	0.0939
7	0.1428	0.1283	0.0800	0.0825	0.1086	0.0771	0.0805
8	0.1250	0.1125	0.0700	0.0722	0.0950	0.0675	0.0705
9	0.1111	0.1000	0.0622	0.0641	0.0844	0.0600	0.0626
10	0.1000	0.0900	0.0560	0.0577	0.0760	0.0540	0.0564
11	0.0909	0.0818	0.0509	0.0525	0.0691	0.0491	0.0512
12	0.0833	0.0750	0.0467	0.0481	0.0633	0.0450	0.0470
13	0.0769	0.0692	0.0431	0.0444	0.0585	0.0415	0.0434
14	0.0714	0.0643	0.0400	0.0412	0.0543	0.0386	0.0403
16	0.0625	0.0562	0.0350	0.0361	0.0475	0.0337	0.0352
18	0.0555	0.0500	0.0311	0.0321	0.0422	0.0300	0.0313

Table 8. *(Continued)* **Diameters of Wires for Measuring American Standard and British Standard Whitworth Screw Threads**

Threads per Inch	Pitch, Inch	Wire Diameters for American Standard Threads			Wire Diameters for Whitworth Standard Threads		
		Max.	Min.	Pitch-Line Contact	Max.	Min.	Pitch-Line Contact
20	0.0500	0.0450	0.0280	0.0289	0.0380	0.0270	0.0282
22	0.0454	0.0409	0.0254	0.0262	0.0345	0.0245	0.0256
24	0.0417	0.0375	0.0233	0.0240	0.0317	0.0225	0.0235
28	0.0357	0.0321	0.0200	0.0206	0.0271	0.0193	0.0201
32	0.0312	0.0281	0.0175	0.0180	0.0237	0.0169	0.0176
36	0.0278	0.0250	0.0156	0.0160	0.0211	0.0150	0.0156
40	0.0250	0.0225	0.0140	0.0144	0.0190	0.0135	0.0141

These formulas are based upon a thread groove of zero lead angle because ordinary variations in the lead angle have little effect on the wire diameter and it is desirable to use one wire size for a given pitch regardless of the lead angle. A theoretically correct solution for finding the *exact* size for pitch-line contact involves the use of cumbersome indeterminate equations with solution by successive trials. The accompanying table gives the wire sizes for both American Standard (formerly, U.S. Standard) and the Whitworth Standard Threads. The following formulas for determining wire diameters do not give the extreme theoretical limits, but the smallest and largest practicable sizes. The diameters in the table are based upon these approximate formulas.

Measuring Wire Accuracy.—A set of three measuring wires should have the same diameter within 0.0002 inch. To measure the pitch diameter of a screw-thread gage to an accuracy of 0.0001 inch by means of wires, it is necessary to know the wire diameters to 0.00002 inch. If the diameters of the wires are known only to an accuracy of 0.0001 inch, an accuracy better than 0.0003 inch in the measurement of pitch diameter cannot be expected. The wires should be accurately finished hardened steel cylinders of the maximum possible hardness without being brittle. The hardness should not be less than that corresponding to a Knoop indentation number of 630. A wire of this hardness can be cut with a file only with difficulty. The surface should not be rougher than the equivalent of a deviation of 3 microinches from a true cylindrical surface.

Measuring or Contact Pressure.—In measuring screw threads or screw-thread gages by the 3-wire method, variations in contact pressure will result in different readings. The effect of a variation in contact pressure in measuring threads of fine pitches is indicated by the difference in readings obtained with pressures of 2 and 5 pounds in checking a thread plug gage having 24 threads per inch. The reading over the wires with 5 pounds pressure was 0.00013 inch less than with 2 pounds pressure. For pitches finer than 20 threads per inch, a pressure of 16 ounces is recommended by the National Bureau of Standards, now National Institute of Standards and Technology (NIST). For pitches of 20 threads per inch and coarser, a pressure of 2 ½ pounds is recommended.

For Acme threads, the wire presses against the sides of the thread with a pressure of approximately twice that of the measuring instrument. To limit the tendency of the wires to wedge in between the sides of an Acme thread, it is recommended that pitch-diameter measurements be made at 1 pound on 8 threads per inch and finer, and at 2 ½ pounds for pitches coarser than 8 threads per inch.

Table 9. Formulas for Checking Pitch Diameters of Screw Threads

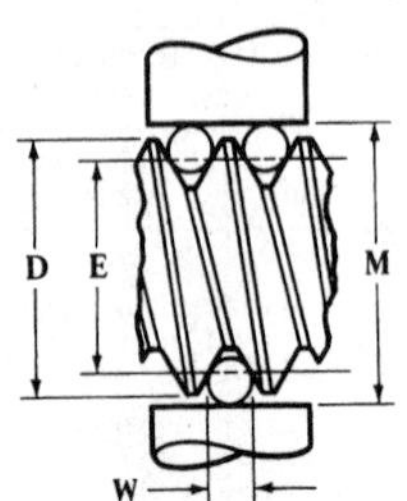

The formulas below do not compensate for the effect of the lead angle upon measurement *M*, but they are sufficiently accurate for checking standard single-thread screws unless exceptional accuracy is required. See accompanying information on effect of lead angle; also matter relating to measuring wire sizes, accuracy required for such wires, and contact or measuring pressure. The approximate best wire size for pitch-line contact may be obtained by the formula

$$W = 0.5 \times \text{pitch} \times \sec \tfrac{1}{2} \text{ included thread angle}$$

For 60-degree threads, $W = 0.57735 \times \text{pitch}$.

Form of Thread	Formulas for determining measurement *M* corresponding to correct pitch diameter and the pitch diameter *E* corresponding to a given measurement over wires.[a]
American National Standard Unified	When measurement *M* is known. $E = M + 0.86603P - 3W$ When pitch diameter *E* is used in formula. $M = E - 0.86603P + 3W$ The American Standard formerly was known as U.S. Standard.
British Standard Whitworth	When measurement *M* is known. $E = M + 0.9605P - 3.1657W$ When pitch diameter *E* is used in formula. $M = E - 0.9605P + 3.1657W$
British Association Standard	When measurement *M* is known. $E = M + 1.1363P - 3.4829W$ When pitch diameter *E* is used in formula. $M = E - 1.1363P + 3.4829W$
Lowenherz Thread	When measurement *M* is known. $E = M + P - 3.2359W$ When pitch diameter *E* is used in formula. $M = E - P + 3.2359W$
Sharp V-Thread	When measurement *M* is known. $E = M + 0.86603P - 3W$ When pitch diameter *E* is used in formula. $M = E - 0.86603P + 3W$
International Standard	Use the formula given above for the American National Standard Unified Thread.

[a] The wires must be lapped to a uniform diameter and it is very important to insert in the rule or formula the wire diameter as determined by precise means of measurement. Any error will be multiplied. See paragraph on *Wire Sizes for Checking Pitch Diameters of Screw Threads*.

COMMON HARDWARE

Grade Marks and Material Properties for Bolts and Screws.—Bolts, screws, and other fasteners are marked on the head with a symbol that identifies the grade of the fastener. The grade specification establishes the minimum mechanical properties that the fastener must meet. Additionally, industrial fasteners must be stamped with a registered head mark that identifies the manufacturer. The grade identification table identifies the grade markings and gives mechanical properties for some commonly used ASTM and SAE steel fasteners. Metric fasteners are identified by property grade marks, which are specified in ISO and SAE standards. These marks are discussed with metric fasteners.

Table 1. Grade Identification Marks and Mechanical Properties of Bolts and Screws

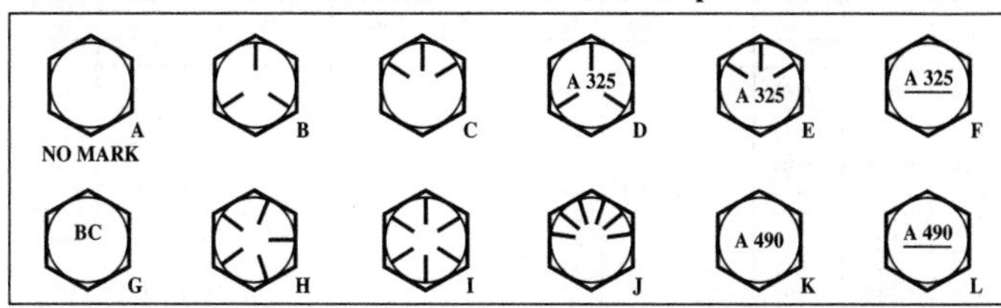

Identifier	Grade	Size (in.)	Min. Strength (10^3 psi) Proof	Tensile	Yield	Material & Treatment
A	SAE Grade 1	$\frac{1}{4}$ to $1\frac{1}{2}$	33	60	36	1
	ASTM A307	$\frac{1}{4}$ to $1\frac{1}{2}$	33	60	36	3
	SAE Grade 2	$\frac{1}{4}$ to $\frac{3}{4}$	55	74	57	1
		$\frac{7}{8}$ to $1\frac{1}{2}$	33	60	36	
	SAE Grade 4	$\frac{1}{4}$ to $1\frac{1}{2}$	65	115	100	2, a
B	SAE Grade 5	$\frac{1}{4}$ to 1	85	120	92	2,b
	ASTM A449	$1\frac{1}{8}$ to $1\frac{1}{2}$	74	105	81	
	ASTM A449	$1\frac{3}{4}$ to 3	55	90	58	
C	SAE Grade 5.2	$\frac{1}{4}$ to 1	85	120	92	4, b
D	ASTM A325, Type 1	$\frac{1}{2}$ to 1	85	120	92	2,b
		$1\frac{1}{8}$ to $1\frac{1}{2}$	74	105	81	
E	ASTM A325, Type 2	$\frac{1}{2}$ to 1	85	120	92	4, b
		$1\frac{1}{8}$ to $1\frac{1}{2}$	74	105	81	
F	ASTM A325, Type 3	$\frac{1}{2}$ to 1	85	120	92	5, b
		$1\frac{1}{8}$ to $1\frac{1}{2}$	74	105	81	
G	ASTM A354, Grade BC	$\frac{1}{4}$ to $2\frac{1}{2}$	105	125	109	5,b
		$2\frac{3}{4}$ to 4	95	115	99	
H	SAE Grade 7	$\frac{1}{4}$ to $1\frac{1}{2}$	105	133	115	7, b
I	SAE Grade 8	$\frac{1}{4}$ to $1\frac{1}{2}$	120	150	130	7, b
	ASTM A354, Grade BD	$\frac{1}{4}$ to $1\frac{1}{2}$	120	150	130	6, b
J	SAE Grade 8.2	$\frac{1}{4}$ to 1	120	150	130	4, b
K	ASTM A490, Type 1	$\frac{1}{2}$ to $1\frac{1}{2}$	120	150	130	6, b
L	ASTM A490, Type 3					5, b

Material Steel: 1—low or medium carbon; 2—medium carbon; 3—low carbon; 4—low-carbon martensite; 5—weathering steel; 6—alloy steel; 7— medium-carbon alloy. Treatment: a—cold drawn; b—quench and temper.

Table 2. American National Standard Hexagon and Spline Socket Head Cap Screws *ANSI/ASME B18.3-1998*

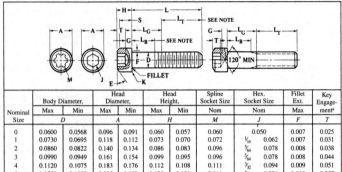

Nominal Size	Body Diameter, D Max	Body Diameter, D Min	Head Diameter, A Max	Head Diameter, A Min	Head Height, H Max	Head Height, H Min	Spline Socket Size M Nom	Hex. Socket Size J Nom	Fillet Ext. F Max	Key Engagement[a] T
0	0.0600	0.0568	0.096	0.091	0.060	0.057	0.060	0.050	0.007	0.025
1	0.0730	0.0695	0.118	0.112	0.073	0.070	0.072	1/16 0.062	0.007	0.031
2	0.0860	0.0822	0.140	0.134	0.086	0.083	0.096	5/64 0.078	0.008	0.038
3	0.0990	0.0949	0.161	0.154	0.099	0.095	0.096	5/64 0.078	0.008	0.044
4	0.1120	0.1075	0.183	0.176	0.112	0.108	0.111	3/32 0.094	0.009	0.051
5	0.1250	0.1202	0.205	0.198	0.125	0.121	0.111	3/32 0.094	0.010	0.057
6	0.1380	0.1329	0.226	0.218	0.138	0.134	0.133	7/64 0.109	0.010	0.064
8	0.1640	0.1585	0.270	0.262	0.164	0.159	0.168	9/64 0.141	0.012	0.077
10	0.1900	0.1840	0.312	0.303	0.190	0.185	0.183	5/32 0.156	0.014	0.090
1/4	0.2500	0.2435	0.375	0.365	0.250	0.244	0.216	3/16 0.188	0.014	0.120
5/16	0.3125	0.3053	0.469	0.457	0.312	0.306	0.291	1/4 0.250	0.017	0.151
3/8	0.3750	0.3678	0.562	0.550	0.375	0.368	0.372	5/16 0.312	0.020	0.182
7/16	0.4375	0.4294	0.656	0.642	0.438	0.430	0.454	3/8 0.375	0.023	0.213
1/2	0.5000	0.4919	0.750	0.735	0.500	0.492	0.454	3/8 0.375	0.026	0.245
5/8	0.6250	0.6163	0.938	0.921	0.625	0.616	0.595	1/2 0.500	0.032	0.307
3/4	0.7500	0.7406	1.125	1.107	0.750	0.740	0.620	5/8 0.625	0.039	0.370
7/8	0.8750	0.8647	1.312	1.293	0.875	0.864	0.698	3/4 0.750	0.044	0.432
1	1.0000	0.9886	1.500	1.479	1.000	0.988	0.790	3/4 0.750	0.050	0.495
1 1/8	1.1250	1.1086	1.688	1.665	1.125	1.111	…	7/8 0.875	0.055	0.557
1 1/4	1.2500	1.2336	1.875	1.852	1.250	1.236	…	7/8 0.875	0.060	0.620
1 3/8	1.3750	1.3568	2.062	2.038	1.375	1.360	…	1 1.000	0.065	0.682
1 1/2	1.5000	1.4818	2.250	2.224	1.500	1.485	…	1 1.000	0.070	0.745
1 3/4	1.7500	1.7295	2.625	2.597	1.750	1.734	…	1 1/4 1.250	0.080	0.870
2	2.0000	1.9780	3.000	2.970	2.000	1.983	…	1 1/2 1.500	0.090	0.995
2 1/4	2.2500	2.2280	3.375	3.344	2.250	2.232	…	1 3/4 1.750	0.100	1.120
2 1/2	2.5000	2.4762	3.750	3.717	2.500	2.481	…	1 3/4 1.750	0.110	1.245
2 3/4	2.7500	2.7262	4.125	4.090	2.750	2.730	…	2 2.000	0.120	1.370
3	3.0000	2.9762	4.500	4.464	3.000	2.979	…	2 1/4 2.250	0.130	1.495
3 1/4	3.2500	3.2262	4.875	4.837	3.250	3.228	…	2 1/4 2.250	0.140	1.620
3 1/2	3.5000	3.4762	5.250	5.211	3.500	3.478	…	2 3/4 2.750	0.150	1.745
3 3/4	3.7500	3.7262	5.625	5.584	3.750	3.727	…	2 3/4 2.750	0.160	1.870
4	4.0000	3.9762	6.000	5.958	4.000	3.976	…	3 3.000	0.170	1.995

[a] (Key engagement depths are minimum.)

All dimensions in inches. The body length L_B of the screw is the length of the unthreaded cylindrical portion of the shank. The length of thread, L_T, is the distance from the extreme point to the last complete (full form) thread. Standard length increments for screw diameters up to 1 inch are 1/16 inch for lengths 1/8 through 1/4 inch, 1/8 inch for lengths 1/4 through 1 inch, 1/4 inch for lengths 1 through 3 1/2 inches, 1/2 inch for lengths 3 1/2 through 7 inches, 1 inch for lengths 7 through 10 inches and for diameters over 1 inch are 1/2 inch for lengths 1 through 7 inches, 1 inch for lengths 7 through 10 inches, and 2 inches for lengths over 10 inches. Heads may be plain or knurled, and chamfered to an angle *E* of 30 to 45 degrees with the surface of the flat. The thread conforms to the Unified Standard with radius root, Class 3A UNRC and UNRF for screw sizes No. 0 through 1 inch inclusive, Class 2A UNRC and UNRF for over 1 inch through 1 1/2 inches inclusive, and Class 2A UNRC for larger sizes. For details not shown, including materials, see ANSI/ASME B18.3-1998.

Table 3. Drill and Counterbore Sizes For Socket Head Cap Screws (1960 Series)

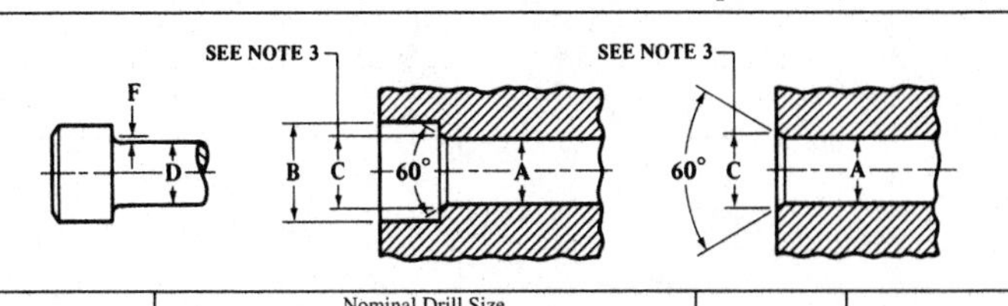

Nominal Size or Basic Screw Diameter		Nominal Drill Size				Counterbore Diameter	Countersink Diameter[a]
		Close Fit[b]		Normal Fit[c]			
		Number or Fractional Size	Decimal Size	Number or Fractional Size	Decimal Size		
		A				B	C
0	0.0600	51	0.067	49	0.073	1/8	0.074
1	0.0730	46	0.081	43	0.089	5/32	0.087
2	0.0860	3/32	0.094	36	0.106	3/16	0.102
3	0.0990	36	0.106	31	0.120	7/32	0.115
4	0.1120	1/8	0.125	29	0.136	7/32	0.130
5	0.1250	9/64	0.141	23	0.154	1/4	0.145
6	0.1380	23	0.154	18	0.170	9/32	0.158
8	0.1640	15	0.180	10	0.194	5/16	0.188
10	0.1900	5	0.206	2	0.221	3/8	0.218
1/4	0.2500	17/64	0.266	9/32	0.281	7/16	0.278
5/16	0.3125	21/64	0.328	11/32	0.344	17/32	0.346
3/8	0.3750	25/64	0.391	13/32	0.406	5/8	0.415
7/16	0.4375	29/64	0.453	15/32	0.469	23/32	0.483
1/2	0.5000	33/64	0.516	17/32	0.531	13/16	0.552
5/8	0.6250	41/64	0.641	21/32	0.656	1	0.689
3/4	0.7500	49/64	0.766	25/32	0.781	1 3/16	0.828
7/8	0.8750	57/64	0.891	29/32	0.906	1 3/8	0.963
1	1.0000	1 1/64	1.016	1 1/32	1.031	1 5/8	1.100
1 1/4	1.2500	1 9/32	1.281	1 5/16	1.312	2	1.370
1 1/2	1.5000	1 17/32	1.531	1 9/16	1.562	2 3/8	1.640
1 3/4	1.7500	1 25/32	1.781	1 13/16	1.812	2 3/4	1.910
2	2.0000	2 1/32	2.031	2 1/16	2.062	3 1/8	2.180

[a] *Countersink:* It is considered good practice to countersink or break the edges of holes which are smaller than (D Max + $2F$ Max) in parts having a hardness which approaches, equals or exceeds the screw hardness. If such holes are not countersunk, the heads of screws may not seat properly or the sharp edges on holes may deform the fillets on screws thereby making them susceptible to fatigue in applications involving dynamic loading. The countersink or corner relief, however, should not be larger than is necessary to insure that the fillet on the screw is cleared.

[b] *Close Fit:* The close fit is normally limited to holes for those lengths of screws which are threaded to the head in assemblies where only one screw is to be used or where two or more screws are to be used and the mating holes are to be produced either at assembly or by matched and coordinated tooling.

[c] *Normal Fit:* The normal fit is intended for screws of relatively long length or for assemblies involving two or more screws where the mating holes are to be produced by conventional tolerancing methods. It provides for the maximum allowable eccentricity of the longest standard screws and for certain variations in the parts to be fastened, such as: deviations in hole straightness, angularity between the axis of the tapped hole and that of the hole for the shank, differences in center distances of the mating holes, etc.

All dimensions in inches.

Source: Appendix to American National Standard ANSI/ASME B18.3-1998.

Table 4. Applicability of Hexagon and Spline Keys and Bits

Nominal Key or Bit Size		Cap Screws 1960 Series	Flat Countersunk Head Cap Screws	Button Head Cap Screws	Shoulder Screws	Set Screws
		Nominal Screw Sizes				
HEXAGON KEYS AND BITS						
	0.028	…	…	…	…	0
	0.035	…	0	0	…	1 & 2
	0.050	0	1 & 2	1 & 2	…	3 & 4
1/16	0.062	1	3 & 4	3 & 4	…	5 & 6
5/64	0.078	2 & 3	5 & 6	5 & 6	…	8
3/32	0.094	4 & 5	8	8	…	10
7/64	0.109	6	…	…	…	…
1/8	0.125	…	10	10	1/4	1/4
9/64	0.141	8	…	…	…	…
5/32	0.156	10	1/4	1/4	5/16	5/16
3/16	0.188	1/4	5/16	5/16	3/8	3/8
7/32	0.219	…	3/8	3/8	…	7/16
1/4	0.250	5/16	7/16	…	1/2	1/2
5/16	0.312	3/8	1/2	1/2	5/8	5/8
3/8	0.375	7/16 & 1/2	5/8	5/8	3/4	3/4
7/16	0.438	…	…	…	…	…
1/2	0.500	5/8	3/4	…	1	7/8
9/16	0.562	…	7/8	…	…	1 & 1 1/8
5/8	0.625	3/4	1	…	1 1/4	1 1/4 & 1 3/8
3/4	0.750	7/8 & 1	1 1/8	…	…	1 1/2
7/8	0.875	1 1/8 & 1 1/4	1 1/4 & 1 3/8	…	1 1/2	…
1	1.000	1 3/8 & 1 1/2	1 1/2	…	1 3/4	1 3/4 & 2
1 1/4	1.250	1 3/4	…	…	2	…
1 1/2	1.500	2	…	…	…	…
1 3/4	1.750	2 1/4 & 2 1/2	…	…	…	…
2	2.000	2 3/4	…	…	…	…
2 1/4	2.250	3 & 3 1/4	…	…	…	…
2 3/4	2.750	3 1/2 & 3 3/4	…	…	…	…
3	3.000	4	…	…	…	…
SPLINE KEYS AND BITS						
	0.033	…	…	…	…	0 & 1
	0.048	…	0	0	…	2 & 3
	0.060	0	1 & 2	1 & 2	…	4
	0.072	1	3 & 4	3 & 4	…	5 & 6
	0.096	2 & 3	5 & 6	5 & 6	…	8
	0.111	4 & 5	8	8	…	10
	0.133	6	…	…	…	…
	0.145	…	10	10	…	1/4
	0.168	8	…	…	…	…
	0.183	10	1/4	1/4	…	5/16
	0.216	1/4	5/16	5/16	…	3/8
	0.251	…	3/8	3/8	…	7/16
	0.291	5/16	7/16	…	…	1/2
	0.372	3/8	1/2	1/2	…	5/8
	0.454	7/16 & 1/2	5/8 & 3/4	5/8	…	3/4
	0.595	5/8	…	…	…	7/8
	0.620	3/4	…	…	…	…
	0.698	7/8	…	…	…	…
	0.790	1	…	…	…	…

Source: Appendix to American National StandardANSI/ASME B18.3-1998.

Table 5. American National Standard Socket Head Cap Screws—Metric Series
ANSI/ASME B18.3.1M-1986

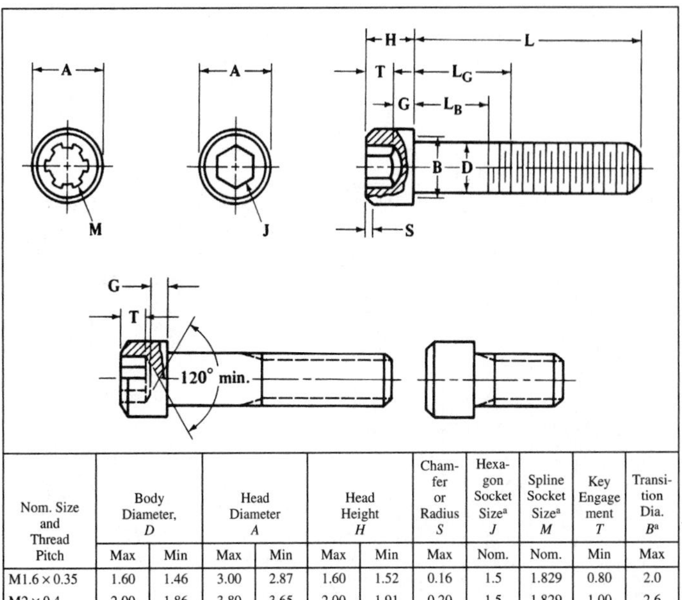

Nom. Size and Thread Pitch	Body Diameter, *D*		Head Diameter *A*		Head Height *H*		Chamfer or Radius *S*	Hexagon Socket Size[a] *J*	Spline Socket Size[a] *M*	Key Engagement *T*	Transition Dia. *B*[a]
	Max	Min	Max	Min	Max	Min	Max	Nom.	Nom.	Min	Max
M1.6 × 0.35	1.60	1.46	3.00	2.87	1.60	1.52	0.16	1.5	1.829	0.80	2.0
M2 × 0.4	2.00	1.86	3.80	3.65	2.00	1.91	0.20	1.5	1.829	1.00	2.6
M2.5 × 0.45	2.50	2.36	4.50	4.33	2.50	2.40	0.25	2.0	2.438	1.25	3.1
M3 × 0.5	3.00	2.86	5.50	5.32	3.00	2.89	0.30	2.5	2.819	1.50	3.6
M4 × 0.7	4.00	3.82	7.00	6.80	4.00	3.88	0.40	3.0	3.378	2.00	4.7
M5 × 0.8	5.00	4.82	8.50	8.27	5.00	4.86	0.50	4.0	4.648	2.50	5.7
M6 × 1	6.00	5.82	10.00	9.74	6.00	5.85	0.60	5.0	5.486	3.00	6.8
M8 × 1.25	8.00	7.78	13.00	12.70	8.00	7.83	0.80	6.0	7.391	4.00	9.2
M10 × 1.5	10.00	9.78	16.00	15.67	10.00	9.81	1.00	8.0	…	5.00	11.2
M12 × 1.75	12.00	11.73	18.00	17.63	12.00	11.79	1.20	10.0	…	6.00	14.2
M14 × 2[b]	14.00	13.73	21.00	20.60	14.00	13.77	1.40	12.0	…	7.00	16.2
M16 × 2	16.00	15.73	24.00	23.58	16.00	15.76	1.60	14.0	…	8.00	18.2
M20 × 2.5	20.00	19.67	30.00	29.53	20.00	19.73	2.00	17.0	…	10.00	22.4
M24 × 3	24.00	23.67	36.00	35.48	24.00	23.70	2.40	19.0	…	12.00	26.4
M30 × 3.5	30.00	29.67	45.00	44.42	30.00	29.67	3.00	22.0	…	15.00	33.4
M36 × 4	36.00	35.61	54.00	53.37	36.00	35.64	3.60	27.0	…	18.00	39.4
M42 × 4.5	42.00	41.61	63.00	62.31	42.00	41.61	4.20	32.0	…	21.00	45.6
M48 × 5	48.00	47.61	72.00	71.27	48.00	47.58	4.80	36.0	…	24.00	52.6

[a] See also Table 7.
[b] The M14 × 2 size is not recommended for use in new designs.

All dimensions are in millimeters

L_G is grip length and L_B is body length (see Table 6).

For length of complete thread, see Table 6.

For additional manufacturing and acceptance specifications, see ANSI/ASME B18.3.1M-1986.

Table 6. Socket Head Cap Screws (Metric Series)—Length of Complete Thread *ANSI/ASME B18.3.1M-1986*

Nominal Size	Length of Complete Thread, L_T	Nominal Size	Length of Complete Thread, L_T	Nominal Size	Length of Complete Thread, L_T
M1.6	15.2	M6	24.0	M20	52.0
M2	16.0	M8	28.0	M24	60.0
M2.5	17.0	M10	32.0	M30	72.0
M3	18.0	M12	36.0	M36	84.0
M4	20.0	M14	40.0	M42	96.0
M5	22.0	M16	44.0	M48	108.0

Grip length, L_G equals screw length, L, minus L_T. Total length of thread L_{TT} equals L_T plus 5 times the pitch of the coarse thread for the respective screw size. Body length L_B equals L minus L_{TT}.

Nominal Length, L	Nominal Size													
	M1.6	M2	M2.5	M3	M4	M5	M6	M8	M10	M12	M14	M16	M20	M24
20	X	X												
25	X	X	X	X										
30	X	X	X	X	X									
35	…	X	X	X	X	X	X							
40	…	X	X	X	X	X	X							
45	…	…	X	X	X	X	X	X						
50	…	…	X	X	X	X	X	X	X					
55	…	…	…	X	X	X	X	X	X					
60	…	…	…	X	X	X	X	X	X	X				
65	…	…	…	X	X	X	X	X	X	X	X			
70	…	…	…	…	X	X	X	X	X	X	X	X		
80	…	…	…	…	X	X	X	X	X	X	X	X		
90	…	…	…	…	…	X	X	X	X	X	X	X	X	
100	…	…	…	…	…	X	X	X	X	X	X	X	X	X
110	…	…	…	…	…	…	X	X	X	X	X	X	X	X
120	…	…	…	…	…	…	X	X	X	X	X	X	X	X
130	…	…	…	…	…	…	…	X	X	X	X	X	X	X
140	…	…	…	…	…	…	…	X	X	X	X	X	X	X
150	…	…	…	…	…	…	…	X	X	X	X	X	X	X
160	…	…	…	…	…	…	…	X	X	X	X	X	X	X
180	…	…	…	…	…	…	…	…	X	X	X	X	X	X
200	…	…	…	…	…	…	…	…	X	X	X	X	X	X
220	…	…	…	…	…	…	…	…	…	X	X	X	X	X
240	…	…	…	…	…	…	…	…	…	X	X	X	X	X
260	…	…	…	…	…	…	…	…	…	…	X	X	X	X
300	…	…	…	…	…	…	…	…	…	…	…	X	X	X

All dimensions are in millimeters. Screws with lengths above heavy cross lines are threaded full length. Diameter-length combinations are indicated by the symbol X. Standard sizes for government use. In addition to the lengths shown, the following lengths are standard: 3, 4, 5, 6, 8, 10, 12, and 16 mm. No diameter-length combinations are given in the Standard for these lengths. Screws larger than M24 with lengths equal to or shorter than L_{TT} (see Table 6 footnote) are threaded full length.

Table 7. American National Standard Hexagon and Spline Sockets for Socket Head Cap Screws—Metric Series *ANSI/ASME B18.3.1M-1986*

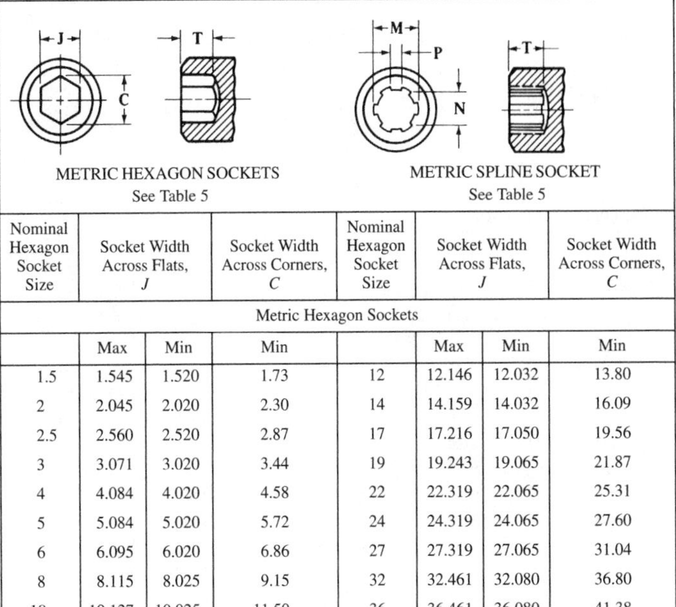

Nominal Hexagon Socket Size	Socket Width Across Flats, *J*		Socket Width Across Corners, *C*	Nominal Hexagon Socket Size	Socket Width Across Flats, *J*		Socket Width Across Corners, *C*
Metric Hexagon Sockets							
	Max	Min	Min		Max	Min	Min
1.5	1.545	1.520	1.73	12	12.146	12.032	13.80
2	2.045	2.020	2.30	14	14.159	14.032	16.09
2.5	2.560	2.520	2.87	17	17.216	17.050	19.56
3	3.071	3.020	3.44	19	19.243	19.065	21.87
4	4.084	4.020	4.58	22	22.319	22.065	25.31
5	5.084	5.020	5.72	24	24.319	24.065	27.60
6	6.095	6.020	6.86	27	27.319	27.065	31.04
8	8.115	8.025	9.15	32	32.461	32.080	36.80
10	10.127	10.025	11.50	36	36.461	36.080	41.38

Metric Spline Sockets[a]						
Nominal Spline Socket Size	Socket Major Diameter, *M*		Socket Minor Diameter, *N*		Width of Tooth, *P*	
	Max	Min	Max	Min	Max	Min
1.829	1.8796	1.8542	1.6256	1.6002	0.4064	0.3810
2.438	2.4892	2.4638	2.0828	2.0320	0.5588	0.5334
2.819	2.9210	2.8702	2.4892	2.4384	0.6350	0.5842
3.378	3.4798	3.4290	2.9972	2.9464	0.7620	0.7112
4.648	4.7752	4.7244	4.1402	4.0894	0.9906	0.9398
5.486	5.6134	5.5626	4.8260	4.7752	1.2700	1.2192
7.391	7.5692	7.5184	6.4516	6.4008	1.7272	2.6764

[a] The tabulated dimensions represent direct metric conversions of the equivalent inch size spline sockets shown in American National Standard Socket Cap, Shoulder and Set Screws—Inch Series ANSI B18.3. Therefore, the spline keys and bits shown therein are applicable for wrenching the corresponding size metric spline sockets.

Table 8. Drill and Counterbore Sizes for Metric Socket Head Cap Screws

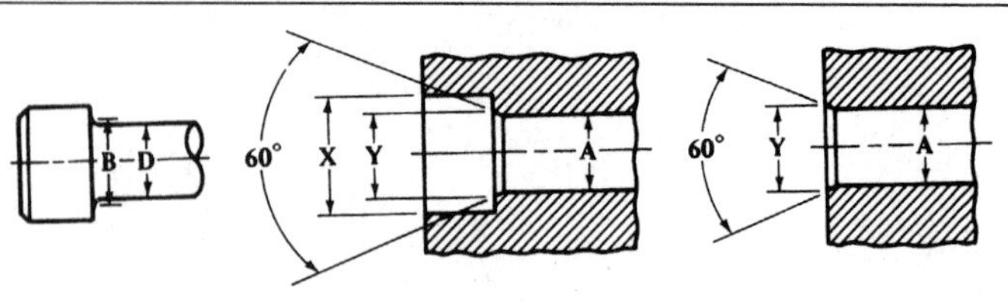

Nominal Size or Basic Screw Diameter	Nominal Drill Size, *A*		Counterbore Diameter, *X*	Countersink Diameter,[a] *Y*
	Close Fit[b]	Normal Fit[c]		
M1.6	1.80	1.95	3.50	2.0
M2	2.20	2.40	4.40	2.6
M2.5	2.70	3.00	5.40	3.1
M3	3.40	3.70	6.50	3.6
M4	4.40	4.80	8.25	4.7
M5	5.40	5.80	9.75	5.7
M6	6.40	6.80	11.25	6.8
M8	8.40	8.80	14.25	9.2
M10	10.50	10.80	17.25	11.2
M12	12.50	12.80	19.25	14.2
M14	14.50	14.75	22.25	16.2
M16	16.50	16.75	25.50	18.2
M20	20.50	20.75	31.50	22.4
M24	24.50	24.75	37.50	26.4
M30	30.75	31.75	47.50	33.4
M36	37.00	37.50	56.50	39.4
M42	43.00	44.00	66.00	45.6
M48	49.00	50.00	75.00	52.6

[a] *Countersink:* It is considered good practice to countersink or break the edges of holes which are smaller than *B* Max. (see Table 5) in parts having a hardness which approaches, equals, or exceeds the screw hardness. If such holes are not countersunk, the heads of screws may not seat properly or the sharp edges on holes may deform the fillets on screws, thereby making them susceptible to fatigue in applications involving dynamic loading. The countersink or corner relief, however, should not be larger than is necessary to ensure that the fillet on the screw is cleared. Normally, the diameter of countersink does not have to exceed *B* Max. Countersinks or corner reliefs in excess of this diameter reduce the effective bearing area and introduce the possibility of embedment where the parts to be fastened are softer than the screws or of brinnelling or flaring the heads of the screws where the parts to be fastened are harder than the screws.

[b] *Close Fit:* The close fit is normally limited to holes for those lengths of screws which are threaded to the head in assemblies where only one screw is to be used or where two or more screws are to be used and the mating holes are to be produced either at assembly or by matched and coordinated tooling.

[c] *Normal Fit:* The normal fit is intended for screws of relatively long length or for assemblies involving two or more screws where the mating holes are to be produced by conventional tolerancing methods. It provides for the maximum allowable eccentricity of the longest standard screws and for certain variations in the parts to be fastened, such as: deviations in hole straightness, angularity between the axis of the tapped hole and that of the hole for shank, differences in center distances of the mating holes, etc.

All dimensions are in millimeters.

Table 9. American National Standard Hardened Ground Machine Dowel Pins *ANSI/ASME B18.8.2-1995*

Nominal Size[a] or Nominal Pin Diameter		Pin Diameter, A						Point Diameter, B		Crown Height, C	Crown Radius, R	Range of Preferred Lengths,[b] L	Single Shear Load, for Carbon or Alloy Steel, Calculated lb	Suggested Hole Diameter[c]	
		Standard Series Pins			Oversize Series Pins										
		Basic	Max	Min	Basic	Max	Min	Max	Min	Max	Min			Max	Min
1/16	0.0625	0.0627	0.0628	0.0626	0.0635	0.0636	0.0634	0.058	0.048	0.020	0.008	3/16 – 3/4	400	0.0625	0.0620
5/64[d]	0.0781	0.0783	0.0784	0.0782	0.0791	0.0792	0.0790	0.074	0.064	0.026	0.010	…	620	0.0781	0.0776
3/32	0.0938	0.0940	0.0941	0.0939	0.0948	0.0949	0.0947	0.089	0.079	0.031	0.012	5/16 – 1	900	0.0937	0.0932
1/8	0.1250	0.1252	0.1253	0.1251	0.1260	0.1261	0.1259	0.120	0.110	0.041	0.016	3/8 – 2	1,600	0.1250	0.1245
5/32[d]	0.1562	0.1564	0.1565	0.1563	0.1572	0.1573	0.1571	0.150	0.140	0.052	0.020	…	2,500	0.1562	0.1557
3/16	0.1875	0.1877	0.1878	0.1876	0.1885	0.1886	0.1884	0.180	0.170	0.062	0.023	1/2 – 2	3,600	0.1875	0.1870
1/4	0.2500	0.2502	0.2503	0.2501	0.2510	0.2511	0.2509	0.240	0.230	0.083	0.031	1/2 – 2 1/2	6,400	0.2500	0.2495
5/16	0.3125	0.3127	0.3128	0.3126	0.3135	0.3136	0.3134	0.302	0.290	0.104	0.039	1/2 – 2 1/2	10,000	0.3125	0.3120
3/8	0.3750	0.3752	0.3753	0.3751	0.3760	0.3761	0.3759	0.365	0.350	0.125	0.047	1/2 – 3	14,350	0.3750	0.3745
7/16	0.4375	0.4377	0.4378	0.4376	0.4385	0.4386	0.4384	0.424	0.409	0.146	0.055	7/8 – 3	19,550	0.4375	0.4370
1/2	0.5000	0.5002	0.5003	0.5001	0.5010	0.5011	0.5009	0.486	0.471	0.167	0.063	3/4, 1 – 4	25,500	0.5000	0.4995
5/8	0.6250	0.6252	0.6253	0.6251	0.6260	0.6261	0.6259	0.611	0.595	0.208	0.078	1 1/4 – 5	39,900	0.6250	0.6245
3/4	0.7500	0.7502	0.7503	0.7501	0.7510	0.7511	0.7509	0.735	0.715	0.250	0.094	1 1/2 – 6	57,000	0.7500	0.7495
7/8	0.8750	0.8752	0.8753	0.8751	0.8760	0.8761	0.8759	0.860	0.840	0.293	0.109	2, 2 1/2 – 6	78,000	0.8750	0.8745
1	1.0000	1.0002	1.0003	1.0001	1.0010	1.0011	1.0009	0.980	0.960	0.333	0.125	2, 2 1/2 – 5, 6	102,000	1.0000	0.9995

[a] Where specifying nominal size as basic diameter, zeros preceding decimal and in the fourth decimal place are omitted.

[b] Lengths increase in 1/16-inch steps up to 3/8 inch, in 1/8-inch steps from 3/8 inch to 1 inch, in 1/4-inch steps from 1 inch to 2 1/2 inches, and in 1/2-inch steps above 2 1/2 inches. Tolerance on length is ±0.010 inch.

[c] These hole sizes have been commonly used for press fitting Standard Series machine dowel pins into materials such as mild steels and cast iron. In soft materials such as aluminum or zinc die castings, hole size limits are usually decreased by 0.0005 inch to increase the press fit.

[d] Nonpreferred sizes, not recommended for use in new designs.

All dimensions are in inches.

GEARS AND GEARING

Gears.—Gears are a very important part of many mechanical assemblies. They transmit power and torque, increase or decrease speed, and provide a constant slip-free rate of speed. The most common types of gears in use today are spur gears, helical gears, bevel gears, and worm gears, Fig. 1a, Fig. 1b, Fig. 1c, and Fig. 1d. In addition to these basic gear forms there are also several variations, or modified forms, of these basic gear types. These include herringbone, hypoid, spiral bevel, and elliptical gears.

Type of Gears

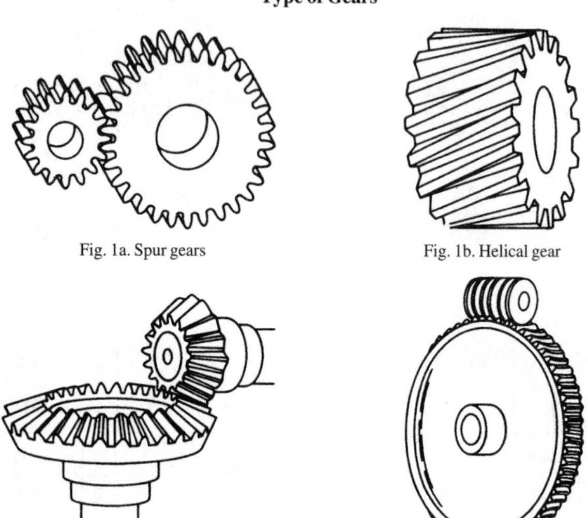

Fig. 1a. Spur gears

Fig. 1b. Helical gear

Fig. 1c. Bevel gears

Fig. 1d. Worm gear and worm wheel

Spur Gears: Spur gears are the simplest and most common type of gears used in industry today. These gears have straight teeth cut parallel with the axis of rotation of the gear body. Spur gear dimensions, like other gear dimensions on a shop print, are generally divided into two groups: gear blank dimensions and gear cutting dimensions. Gear blank dimensions deal directly with the gear blank and are usually shown on the drawn detail. The gear cutting dimensions refer to the actual cutting of the gear and are generally shown in a separate data block on the print. Many times the gear cutting dimensions will include the note (REF) which indicates a calculated dimension that cannot be measured directly.

Spur gears are usually shown in prints by a front view and a full section side view. Rarely are complete gear profiles shown on prints. Generally the addendum circle and the dedendum circle are shown with phantom lines and the pitch circle is shown with a center line. Occasionally a few gear teeth may be drawn for clarity and to avoid confusion. In addition to the external spur gears, there are also two other spur gear forms: the internal gear and the rack. An internal spur gear is basically the same as an external gear with the exception of the location and the shape of the teeth. Internal spur gear teeth have the same shape as the spaces between the teeth of the mating external spur gear. A rack gear is simply a flat gear. A rack and spur gear are often used to transfer rotary motion into linear motion or vice versa.

Helical Gears: Helical gears are a variation of the spur gear that have teeth cut at an angle to the axis of rotation. Helical gears are cut by advancing a cutter across the gear blank while the blank slowly revolves. This method produces a tooth form which is smoother running and quieter than normal spur gears. Spur gears engage across the entire face of the tooth while helical gears engage on a small area which moves across the face of the tooth.

Helical gears may be used for parallel gear shafts, or shafts that are at an angle to each other. Helical gears are made as either right hand or left hand. The hand of a gear may be determined by looking at the tooth. If the tooth runs down on the right side it is a right-hand helix, likewise, if it runs down on the left side it is a left-hand helical gear. Helical gears are usually specified in much the same way as spur gears. The only major difference is in the additions to the gear cutting data block. The drawn view of the helical gear is usually the same as that used for a spur gear.

Bevel Gears: Bevel gears are characterized by a variable pitch diameter, tooth depth, and tooth thickness. Like helical gears, bevel gears are also used for applications where the gears must be positioned at an angle to each other. The most common type of bevel gears run at a 90° angle. These gears are called miter gears. Bevel gears can best be compared to cones that have teeth. Unlike spur gears and helical gears which engage on their outer surfaces, bevel gears engage, or mesh, along the angular side of their conical form. Bevel gears are normally shown as full sections on most prints. When two mating bevel gears are required, they may both be shown in their operational position. The gear cutting data block is then divided into two sections, one for the driver and one for the driven gear. Depending on their construction they may also be called a pinion (driver) and ring (driven) gear.

Worm Gears: Worm gearing is primarily used to obtain great reductions in speed and an increase in power. The two basic parts that are included in a worm drive are the worm and the worm gear. Owing to their basic design, the worm is always the driver and the worm gear is always the driven gear in this system. Worm gearing can be compared to an endless screw. As the worm rotates the worm gear revolves, much the same as the movement of a nut along a bolt. The helix angle of the thread is what causes the worm to drive the worm gear. So the greater the helix angle the faster the movement. Worms and worm gears are sometimes shown as individual details in a print. In some cases only the worm gear will be shown and the cutting data of the worm will be shown in the data block along with the data necessary to make the worm gear. As with other forms of gears, the drawn views are normally intended to show the dimensions needed to make the blank and all relevant gear cutting data is contained in a special data block on the print.

Metric Gears: Metric gears use basically the same values as inch type gears. The only major difference is the use of the term module in place of diametral pitch and circular pitch. Where the term diametral pitch is defined as the ratio of the number of teeth per inch of pitch diameter, the module is equal to the length of pitch diameter per tooth expressed in millimeters. So, rather than a ratio, the module is a dimension.

Definitions of Gear Terms.—The following terms are commonly applied to the various classes of gears:

Active face width is the dimension of the tooth face width that makes contact with a mating gear.

Addendum is the radial or perpendicular distance between the pitch circle and the top of the tooth.

Arc of action is the arc of the pitch circle through which a tooth travels from the first point of contact with the mating tooth to the point where contact ceases.

Arc of approach is the arc of the pitch circle through which a tooth travels from the first point of contact with the mating tooth to the pitch point.

Arc of recession is the arc of the pitch circle through which a tooth travels from its contact with a mating tooth at the pitch point until contact ceases.

Axial pitch is the distance parallel to the axis between corresponding sides of adjacent teeth.

Axial plane is the plane that contains the two axes in a pair of gears. In a single gear the axial plane is any plane containing the axis and any given point.

Axial thickness is the distance parallel to the axis between two pitch line elements of the same tooth.

Backlash is the shortest distance between the non-driving surfaces of adjacent teeth when the working flanks are in contact.

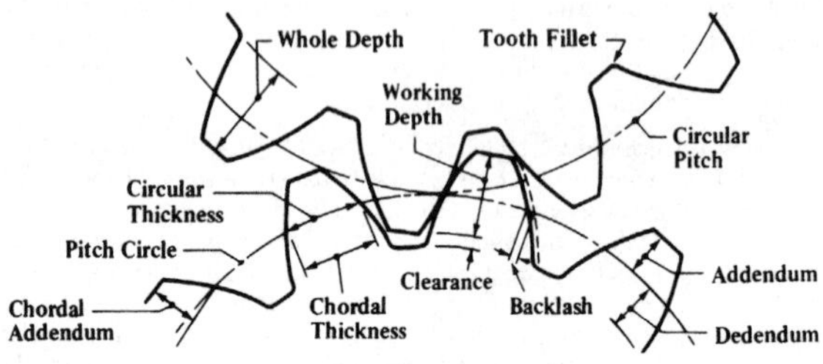

Gear Tooth Parts

Base circle is the circle from which the involute tooth curve is generated or developed.

Base helix angle is the angle at the base cylinder of an involute gear that the tooth makes with the gear axis.

Base pitch is the circular pitch taken on the circumference of the base circles, or the distance along the line of action between two successive and corresponding involute tooth profiles. The *normal base pitch* is the base pitch in the normal plane and the *axial base pitch* is the base pitch in the axial plane.

Base tooth thickness is the distance on the base circle in the plane of rotation between involutes of the same pitch.

Bottom land is the surface of the gear between the flanks of adjacent teeth.

Center distance is the shortest distance between the non-intersecting axes of mating gears, or between the parallel axes of spur gears and parallel helical gears, or the crossed axes of crossed helical gears or worm gears.

Central plane is the plane perpendicular to the gear axis in a worm gear, which contains the common perpendicular of the gear and the worm axes. In the usual arrangement with the axes at right angles, it contains the worm axis.

Chordal addendum is the radial distance from the circular thickness chord to the top of the tooth, or the height from the top of the tooth to the chord subtending the circular thickness arc.

Chordal thickness is the length of the chord subtended by the circular thickness arc. The dimension obtained when a gear tooth caliper is used to measure the tooth thickness at the pitch circle.

Circular pitch is the distance on the circumference of the pitch circle, in the plane of rotation, between corresponding points of adjacent teeth. The length of the arc of the pitch circle between the centers or other corresponding points of adjacent teeth.

Circular thickness is the thickness of the tooth on the pitch circle in the plane of rotation, or the length of arc between the two sides of a gear tooth measured on the pitch circle.

Clearance is the radial distance between the top of a tooth and the bottom of a mating tooth space, or the amount by which the dedendum in a given gear exceeds the addendum of its mating gear.

Contact diameter is the smallest diameter on a gear tooth with which the mating gear makes contact.

Contact ratio is the ratio of the arc of action in the plane of rotation to the circular pitch, and is sometimes thought of as the average number of teeth in contact. This ratio is obtained most directly as the ratio of the length of action to the base pitch.

Contact ratio – face is the ratio of the face advance to the circular pitch in helical gears.

Contact ratio – total is the ratio of the sum of the arc of action and the face advance to the circular pitch.

Contact stress is the maximum compressive stress within the contact area between mating gear tooth profiles. Also called the Hertz stress.

Cycloid is the curve formed by the path of a point on a circle as it rolls along a straight line. When such a circle rolls along the outside of another circle the curve is called an *epicycloid*, and when it rolls along the inside of another circle it is called a *hypocycloid*. These curves are used in defining the former American Standard composite Tooth Form.

Dedendum is the radial or perpendicular distance between the pitch circle and the bottom of the tooth space.

Diametral pitch is the ratio of the number of teeth to the number of inches in the pitch diameter in the plane of rotation, or the number of gear teeth to each inch of pitch diameter. Normal diametral pitch is the diametral pitch as calculated in the normal plane, or the diametral pitch divided by the cosine of the helix angle.

Efficiency is the torque ratio of a gear set divided by its gear ratio.

Equivalent pitch radius is the radius of curvature of the pitch surface at the pitch point in a plane normal to the pitch line element.

Face advance is the distance on the pitch circle that a gear tooth travels from the time pitch point contact is made at one end of the tooth until pitch point contact is made at the other end.

Fillet radius is the radius of the concave portion of the tooth profile where it joins the bottom of the tooth space.

Fillet stress is the maximum tensile stress in the gear tooth fillet.

Flank of tooth is the surface between the pitch circle and the bottom land, including the gear tooth fillet.

Gear ratio is the ratio between the numbers of teeth in mating gears.

Helical overlap is the effective face width of a helical gear divided by the gear axial pitch.

Helix angle is the angle that a helical gear tooth makes with the gear axis at the pitch circle, unless specified otherwise.

Hertz stress, see *Contact stress.*

Highest point of single tooth contact (HPSTC) is the largest diameter on a spur gear at which a single tooth is in contact with the mating gear.

Interference is the contact between mating teeth at some point other than along the line of action.

Internal diameter is the diameter of a circle that coincides with the tops of the teeth of an internal gear.

Internal gear is a gear with teeth on the inner cylindrical surface.

Involute is the curve generally used as the profile of gear teeth. The curve is the path of a point on a straight line as it rolls along a convex base curve, usually a circle.

Land: The top land is the top surface of a gear tooth and the *bottom land* is the surface of the gear between the fillets of adjacent teeth.

Lead is the axial advance of the helix in one complete turn, or the distance along its own axis on one revolution if the gear were free to move axially.

Length of action is the distance on an involute line of action through which the point of contact moves during the action of the tooth profile.

Line of action is the portion of the common tangent to the base cylinders along which contact between mating involute teeth occurs.

Lowest point of single tooth contact (LPSTC) is the smallest diameter on a spur gear at which a single tooth is in contact with its mating gear. Gear set contact stress is determined with a load placed on the pinion at this point.

Module is the ratio of the pitch diameter to the number of teeth, normally the ratio of pitch diameter in mm to the number of teeth. Module in the inch system is the ratio of the pitch diameter in inches to the number of teeth.

Normal plane is a plane normal to the tooth surfaces at a point of contact and perpendicular to the pitch plane.

Number of teeth is the number of teeth contained in a gear.

Outside diameter is the diameter of the circle that contains the tops of the teeth of external gears.

Pitch is the distance between similar, equally-spaced tooth surfaces in a given direction along a given curve or line.

Pitch circle is the circle through the pitch point having its center at the gear axis.

Pitch diameter is the diameter of the pitch circle. The operating pitch diameter is the pitch diameter at which the gear operates.

Pitch plane is the plane parallel to the axial plane and tangent to the pitch surfaces in any pair of gears. In a single gear, the pitch plane may be any plane tangent to the pitch surfaces.

Pitch point is the intersection between the axes of the line of centers and the line of action.

Plane of rotation is any plane perpendicular to a gear axis.

Pressure angle is the angle between a tooth profile and a radial line at its pitch point. In involute teeth, the pressure angle is often described as the angle between the line of action and the line tangent to the pitch circle. *Standard pressure angles* are established in connection with standard tooth proportions. A given pair of involute profiles will transmit smooth motion at the same velocity ratio when the center distance is changed. Changes in center distance in gear design and gear manufacturing operations may cause changes in pitch diameter, pitch and pressure angle in the same gears under different conditions. Unless otherwise specified, the pressure angle is the *standard pressure angle at the standard pitch diameter*. The *operating pressure angle* is determined by the center distance at which a pair of gears operate. In oblique teeth such as helical and spiral designs, the pressure angle is specified in the transverse, normal or axial planes.

Principle reference planes are pitch plane, axial plane and transverse plane, all intersecting at a point and mutually perpendicular.

Rack: A rack is a gear with teeth spaced along a straight line, suitable for straight line motion. A basic rack is a rack that is adopted as the basis of a system of interchangeable gears. Standard gear tooth dimensions are often illustrated on an outline of a basic rack.

Roll angle is the angle subtended at the center of a base circle from the origin of an involute to the point of tangency of a point on a straight line from any point on the same involute. The radian measure of this angle is the tangent of the pressure angle of the point on the involute.

Root diameter is the diameter of the circle that contains the roots or bottoms of the tooth spaces.

Tangent plane is a plane tangent to the tooth surfaces at a point or line of contact.

Tip relief is an arbitrary modification of a tooth profile where a small amount of material is removed from the involute face of the tooth surface near the tip of the gear tooth.

Tooth face is the surface between the pitch line element and the tooth tip.

Tooth surface is the total tooth area including the flank of the tooth and the tooth face.

Comparative Sizes and Shapes of Gear Teeth

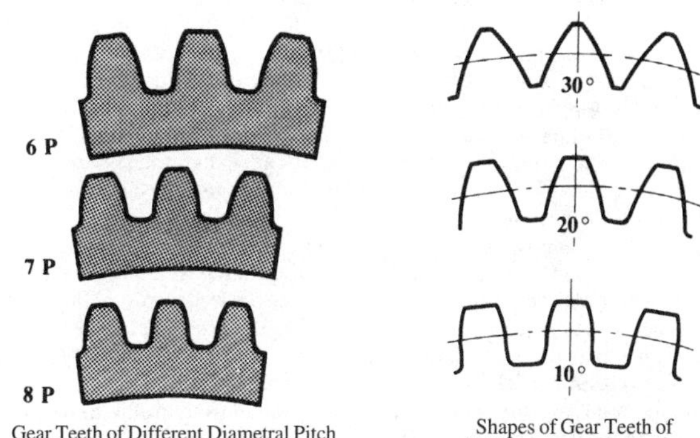

Gear Teeth of Different Diametral Pitch

Shapes of Gear Teeth of Different Pressure Angles

Nomenclature of Gear TeethTerms Used in Gear Geometry from Table on page 337

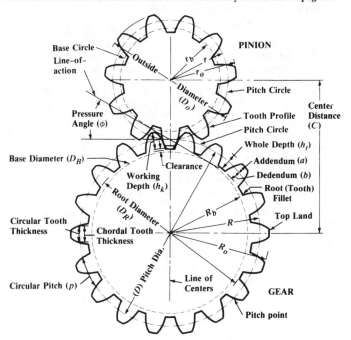

PINION
Base Circle
Line-of-action
Outside Diameter (D_o)
r_b
r
r_o
Pitch Circle
Pressure Angle (ϕ)
Tooth Profile
Pitch Circle
Center Distance (C)
Whole Depth (h_t)
Base Diameter (D_B)
Clearance
Addendum (a)
Dedendum (b)
Working Depth (h_k)
Root (Tooth) Fillet
Root Diameter (D_R)
R_b
Top Land
R
Circular Tooth Thickness
Chordal Tooth Thickness
R_o
(D) Pitch Dia.
Circular Pitch (p)
Line of Centers
GEAR
Pitch point

Total face width is the dimensional width of a gear blank and may exceed the effective face width as with a double-helical gear where the total face width includes any distance separating the right-hand and left-hand helical gear teeth.

Transverse plane is a plane that is perpendicular to the axial plane and to the pitch plane. In gears with parallel axes, the transverse plane and the plane of rotation coincide.

Trochoid is the curve formed by the path of a point on the extension of a radius of a circle as it rolls along a curve or line. A trochoid is also the curve formed by the path of a point on a perpendicular to a straight line as the straight line rolls along the convex side of a base curve. By the first definition, a trochoid is derived from the *cycloid*, by the second definition it is derived from the *involute*.

True involute form diameter is the smallest diameter on the tooth at which the point of tangency of the involute tooth profile exists. Usually this position is the point of tangency of the involute tooth profile and the fillet curve, and is often referred to as the TIF diameter.

Undercut is a condition in generated gear teeth when any part of the fillet curve lies inside a line drawn at a tangent to the working profile at its lowest point. Undercut may be introduced deliberately to facilitate shaving operations, as in pre-shaving.

Whole depth is the total depth of a tooth space, equal to the addendum plus the dedendum and equal to the working depth plus clearance.

Working depth is the depth of engagement of two gears, or the sum of their addendums. The standard working distance is the depth to which a tooth extends into the tooth space of a mating gear when the center distance is standard.

Definitions of gear terms are given in AGMA Standards 112.05, 115.01, and 116.01 entitled "Terms, Definitions, Symbols and Abbreviations," "Reference Information—Basic Gear Geometry," and "Glossary—Terms Used in Gearing," respectively; obtainable from American Gear Manufacturers Assn., 1500 King. St., Alexandria, VA 22314.

Properties of the Involute Curve.—The involute curve is used almost exclusively for gear-tooth profiles, because of the following important properties.

1) The form or shape of an involute curve depends upon the diameter of the base circle from which it is derived. (If a taut line were unwound from the circumference of a circle—the *base circle* of the involute—the end of that line or any point on the unwound portion, would describe an involute curve.)

2) If a gear tooth of involute curvature acts against the involute tooth of a mating gear while rotating at a uniform rate, the angular motion of the driven gear will also be uniform, even though the center-to-center distance is varied.

3) The relative rate of motion between driving and driven gears having involute tooth curves is established by the diameters of their base circles.

4) Contact between intermeshing involute teeth on a driving and driven gear is along a straight line that is tangent to the two base circles of these gears. This is the *line of action*.

5) The point where the line of action intersects the common center-line of the mating involute gears, establishes the radii of the pitch circles of these gears; hence true pitch circle diameters are affected by a change in the center distance. (Pitch diameters obtained by dividing the number of teeth by the diametral pitch apply when the center distance equals the total number of teeth on both gears divided by twice the diametral pitch.)

6) The pitch diameters of mating involute gears are directly proportional to the diameters of their respective base circles; thus, if the base circle of one mating gear is three times as large as the other, the pitch circle diameters will be in the same ratio.

7) The angle between the line of action and a line perpendicular to the common center-line of mating gears, is the *pressure angle*; hence the pressure angle is affected by any change in the center distance.

8) When an involute curve acts against a straight line (as in the case of an involute pinion acting against straight-sided rack teeth), the straight line is tangent to the involute and perpendicular to its line of action.

9) The pressure angle, in the case of an involute pinion acting against straight-sided rack teeth, is the angle between the line of action and the line of the rack's motion. If the involute pinion rotates at a uniform rate, movement of the rack will also be uniform.

Diametral and Circular Pitch Systems.—Gear tooth system standards are established by specifying the tooth proportions of the basic rack. The diametral pitch system is applied to most of the gearing produced in the United States. If gear teeth are larger than about one diametral pitch, it is common practice to use the circular pitch system. The circular pitch system is also applied to cast gearing and it is commonly used in connection with the design and manufacture of worm gearing.

Pitch Diameters Obtained with Diametral Pitch System.—The diametral pitch system is arranged to provide a series of standard tooth sizes, the principle being similar to the standardization of screw thread pitches. Inasmuch as there must be a whole number of teeth on each gear, the increase in pitch diameter per tooth varies according to the pitch. For example, the pitch diameter of a gear having, say, 20 teeth of 4 diametral pitch, will be 5 inches; 21 teeth, 5¼ inches; and so on, the increase in diameter for each additional tooth being equal to ¼ inch for 4 diametral pitch. Similarly, for 2 diametral pitch the variations for successive numbers of teeth would equal ½ inch, and for 10 diametral pitch the variations would equal $\frac{1}{10}$ inch, etc. Where a given center distance must be maintained and no standard diametral pitch can be used.

Tables for Chordal Thicknesses and Chordal Addenda of Milled, Full-depth Teeth.—Two convenient tables for checking gears with milled, full-depth teeth are given on pages 331. The first shows chordal thicknesses and chordal addenda for the lowest number of teeth cut by gear cutters Nos. 1 through 8, and for the commonly used diametral pitches. The second gives similar data for commonly used circular pitches. In each case the data shown are accurate for the number of gear teeth indicated, but are approximate for other numbers of teeth within the range of the cutter under which they appear in the table. For the higher diametral pitches and lower circular pitches, the error introduced by using the data for any tooth number within the range of the cutter under which it appears is comparatively small. The chordal thicknesses and chordal addenda for gear cutters Nos. 1 through 8 of the more commonly used diametral and circular pitches can be obtained from the table and formulas on pages 331.

Caliper Measurement of Gear Tooth.—In cutting gear teeth, the general practice is to adjust the cutter or hob until it grazes the outside diameter of the blank; the cutter is then sunk to the total depth of the tooth space plus whatever slight additional amount may be required to provide the necessary play or backlash between the teeth. If the outside diameter of the gear blank is correct, the tooth thickness should also be correct after the cutter has been sunk to the depth required for a given pitch and backlash. However, it is advisable to check the tooth thickness by measuring it, and the vernier gear-tooth caliper (see following illustration) is commonly used in measuring the thickness.

The vertical scale of this caliper is set so that when it rests upon the top of the tooth as shown, the lower ends of the caliper jaws will be at the height of the pitch circle; the horizontal scale then shows the chordal thickness of the tooth at this point. If the gear is being cut on a milling machine or with the type of gear-cutting machine employing a formed milling cutter, the tooth thickness is checked by first taking a trial cut for a short distance at one side of the blank; then the gear blank is indexed for the next space and another cut is taken far enough to mill the full outline of the tooth. The tooth thickness is then measured.

Before the gear-tooth caliper can be used, it is necessary to determine the correct chordal thickness and also the chordal addendum (or "corrected addendum" as it is sometimes called). The vertical scale is set to the chordal addendum, thus locating the ends of the jaws at the height of the pitch circle. The rules or formulas to use in determining the chordal thickness and chordal addendum will depend upon the outside diameter of the gear; for example, if the outside diameter of a small pinion is enlarged to avoid undercut and improve the tooth action, this must be taken into account in figuring the chordal thickness and chordal addendum as shown by the accompanying rules. The detail of a gear tooth included with the gear-tooth caliper illustration, represents the chordal thickness *T*, the addendum *S*, and the chordal addendum *H*.

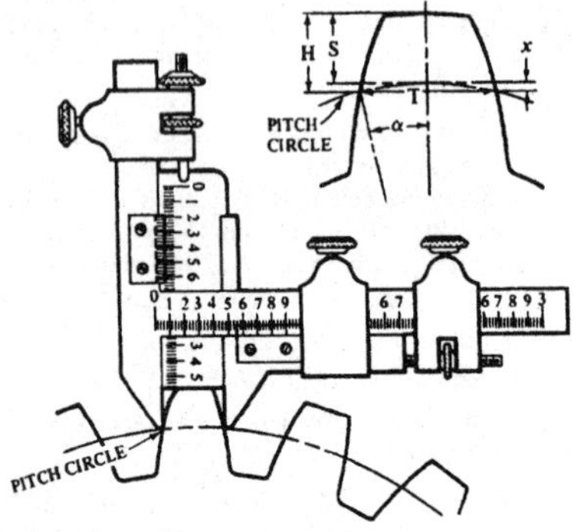

Method of setting a gear tooth caliper

Selection of Involute Gear Milling Cutter for a Given Diametral Pitch and Number of Teeth.—When gear teeth are cut by using formed milling cutters, the cutter must be selected to suit both the pitch and the number of teeth, because the shapes of the tooth spaces vary according to the number of teeth. For instance, the tooth spaces of a small pinion are not of the same shape as the spaces of a large gear of equal pitch. Theoretically, there should be a different formed cutter for every tooth number, but such refinement is unnecessary in practice. The involute formed cutters commonly used are made in series of eight cutters for each diametral pitch (see). The shape of each cutter in this series is correct for a certain number of teeth only, but it can be used for other numbers within the limits given. For instance, a No. 6 cutter may be used for gears having from 17 to 20 teeth, but the tooth outline is correct only for 17 teeth or the lowest number in the range, which is also true of the other cutters listed. When this cutter is used for a gear having, say, 19 teeth, too much material is removed from the upper surfaces of the teeth, although the gear meets ordinary requirements. When greater accuracy of tooth shape is desired to ensure smoother or quieter operation, an intermediate series of cutters having half-numbers may be used provided the number of gear teeth is between the number listed for the regular cutters (see).

Involute gear milling cutters are designed to cut a composite tooth form, the center portion being a true involute while the top and bottom portions are cycloidal. This composite form is necessary to prevent tooth interference when milled mating gears are meshed with each other. Because of their composite form, milled gears will not mate satisfactorily enough for high grade work with those of generated, full-involute form. Composite form hobs are available, however, which will produce generated gears that mesh with those cut by gear milling cutters.

Metric Module Gear Cutters: The accompanying table for selecting the cutter number to be used to cut a given number of teeth may be used also to select metric module gear cutters except that the numbers are designated in reverse order. For example, cutter No. 1, in the metric module system, is used for 12–13 teeth, cutter No. 2 for 14–16 teeth, etc.

Chordal Thicknesses and Chordal Addenda of Milled, Full-depth Gear Teeth and of Gear Milling Cutters

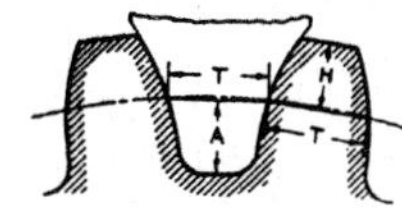

T = chordal thickness of gear tooth and cutter tooth at pitch line;
H = chordal addendum for full-depth gear tooth;
A = chordal addendum of cutter = (2.157 ÷ diametral pitch) − H
= (0.6866 × circular pitch) − H.

Diametral Pitch	Dimension	Number of Gear Cutter, and Corresponding Number of Teeth							
		No. 1 135 Teeth	No. 2 55 Teeth	No. 3 35 Teeth	No. 4 26 Teeth	No. 5 21 Teeth	No. 6 17 Teeth	No. 7 14 Teeth	No. 8 12 Teeth
1	T	1.5707	1.5706	1.5702	1.5698	1.5694	1.5686	1.5675	1.5663
	H	1.0047	1.0112	1.0176	1.0237	1.0294	1.0362	1.0440	1.0514
1½	T	1.0471	1.0470	1.0468	1.0465	1.0462	1.0457	1.0450	1.0442
	H	0.6698	0.6741	0.6784	0.6824	0.6862	0.6908	0.6960	0.7009
2	T	0.7853	0.7853	0.7851	0.7849	0.7847	0.7843	0.7837	0.7831
	H	0.5023	0.5056	0.5088	0.5118	0.5147	0.5181	0.5220	0.5257
2½	T	0.6283	0.6282	0.6281	0.6279	0.6277	0.6274	0.6270	0.6265
	H	0.4018	0.4044	0.4070	0.4094	0.4117	0.4144	0.4176	0.4205
3	T	0.5235	0.5235	0.5234	0.5232	0.5231	0.5228	0.5225	0.5221
	H	0.3349	0.3370	0.3392	0.3412	0.3431	0.3454	0.3480	0.3504
3½	T	0.4487	0.4487	0.4486	0.4485	0.4484	0.4481	0.4478	0.4475
	H	0.2870	0.2889	0.2907	0.2919	0.2935	0.2954	0.2977	0.3004
4	T	0.3926	0.3926	0.3926	0.3924	0.3923	0.3921	0.3919	0.3915
	H	0.2511	0.2528	0.2544	0.2559	0.2573	0.2590	0.2610	0.2628
5	T	0.3141	0.3141	0.3140	0.3139	0.3138	0.3137	0.3135	0.3132
	H	0.2009	0.2022	0.2035	0.2047	0.2058	0.2072	0.2088	0.2102
6	T	0.2618	0.2617	0.2617	0.2616	0.2615	0.2614	0.2612	0.2610
	H	0.1674	0.1685	0.1696	0.1706	0.1715	0.1727	0.1740	0.1752
7	T	0.2244	0.2243	0.2243	0.2242	0.2242	0.2240	0.2239	0.2237
	H	0.1435	0.1444	0.1453	0.1462	0.1470	0.1480	0.1491	0.1502
8	T	0.1963	0.1963	0.1962	0.1962	0.1961	0.1960	0.1959	0.1958
	H	0.1255	0.1264	0.1272	0.1279	0.1286	0.1295	0.1305	0.1314
9	T	0.1745	0.1745	0.1744	0.1744	0.1743	0.1743	0.1741	0.1740
	H	0.1116	0.1123	0.1130	0.1137	0.1143	0.1151	0.1160	0.1168
10	T	0.1570	0.1570	0.1570	0.1569	0.1569	0.1568	0.1567	0.1566
	H	0.1004	0.1011	0.1017	0.1023	0.1029	0.1036	0.1044	0.1051
11	T	0.1428	0.1428	0.1427	0.1427	0.1426	0.1426	0.1425	0.1424
	H	0.0913	0.0919	0.0925	0.0930	0.0935	0.0942	0.0949	0.0955
12	T	0.1309	0.1309	0.1308	0.1308	0.1308	0.1307	0.1306	0.1305
	H	0.0837	0.0842	0.0848	0.0853	0.0857	0.0863	0.0870	0.0876
14	T	0.1122	0.1122	0.1121	0.1121	0.1121	0.1120	0.1119	0.1118
	H	0.0717	0.0722	0.0726	0.0731	0.0735	0.0740	0.0745	0.0751
16	T	0.0981	0.0981	0.0981	0.0981	0.0980	0.0980	0.0979	0.0979
	H	0.0628	0.0632	0.0636	0.0639	0.0643	0.0647	0.0652	0.0657
18	T	0.0872	0.0872	0.0872	0.0872	0.0872	0.0871	0.0870	0.0870
	H	0.0558	0.0561	0.0565	0.0568	0.0571	0.0575	0.0580	0.0584
20	T	0.0785	0.0785	0.0785	0.0785	0.0784	0.0784	0.0783	0.0783
	H	0.0502	0.0505	0.0508	0.0511	0.0514	0.0518	0.0522	0.0525

Chordal Thicknesses and Chordal Addenda of Milled, Full-depth Gear Teeth and of Gear Milling Cutters

Circular Pitch	Dimension	Number of Gear Cutter, and Corresponding Number of Teeth							
		No. 1 135 Teeth	No. 2 55 Teeth	No. 3 35 Teeth	No. 4 26 Teeth	No. 5 21 Teeth	No. 6 17 Teeth	No. 7 14 Teeth	No. 8 12 Teeth
1/4	T	0.1250	0.1250	0.1249	0.1249	0.1249	0.1248	0.1247	0.1246
	H	0.0799	0.0804	0.0809	0.0814	0.0819	0.0824	0.0830	0.0836
5/16	T	0.1562	0.1562	0.1562	0.1561	0.1561	0.1560	0.1559	0.1558
	H	0.0999	0.1006	0.1012	0.1018	0.1023	0.1030	0.1038	0.1045
3/8	T	0.1875	0.1875	0.1874	0.1873	0.1873	0.1872	0.1871	0.1870
	H	0.1199	0.1207	0.1214	0.1221	0.1228	0.1236	0.1245	0.1254
7/16	T	0.2187	0.2187	0.2186	0.2186	0.2185	0.2184	0.2183	0.2181
	H	0.1399	0.1408	0.1416	0.1425	0.1433	0.1443	0.1453	0.1464
1/2	T	0.2500	0.2500	0.2499	0.2498	0.2498	0.2496	0.2495	0.2493
	H	0.1599	0.1609	0.1619	0.1629	0.1638	0.1649	0.1661	0.1673
9/16	T	0.2812	0.2812	0.2811	0.2810	0.2810	0.2808	0.2806	0.2804
	H	0.1799	0.1810	0.1821	0.1832	0.1842	0.1855	0.1868	0.1882
5/8	T	0.3125	0.3125	0.3123	0.3123	0.3122	0.3120	0.3118	0.3116
	H	0.1998	0.2012	0.2023	0.2036	0.2047	0.2061	0.2076	0.2091
11/16	T	0.3437	0.3437	0.3436	0.3435	0.3434	0.3432	0.3430	0.3427
	H	0.2198	0.2213	0.2226	0.2239	0.2252	0.2267	0.2283	0.2300
3/4	T	0.3750	0.3750	0.3748	0.3747	0.3747	0.3744	0.3742	0.3740
	H	0.2398	0.2414	0.2428	0.2443	0.2457	0.2473	0.2491	0.2509
13/16	T	0.4062	0.4062	0.4060	0.4059	0.4059	0.4056	0.4054	0.4050
	H	0.2598	0.2615	0.2631	0.2647	0.2661	0.2679	0.2699	0.2718
7/8	T	0.4375	0.4375	0.4373	0.4372	0.4371	0.4368	0.4366	0.4362
	H	0.2798	0.2816	0.2833	0.2850	0.2866	0.2885	0.2906	0.2927
15/16	T	0.4687	0.4687	0.4685	0.4684	0.4683	0.4680	0.4678	0.4674
	H	0.2998	0.3018	0.3035	0.3054	0.3071	0.3092	0.3114	0.3137
1	T	0.5000	0.5000	0.4998	0.4997	0.4996	0.4993	0.4990	0.4986
	H	0.3198	0.3219	0.3238	0.3258	0.3276	0.3298	0.3322	0.3346
1 1/8	T	0.5625	0.5625	0.5623	0.5621	0.5620	0.5617	0.5613	0.5610
	H	0.3597	0.3621	0.3642	0.3665	0.3685	0.3710	0.3737	0.3764
1 1/4	T	0.6250	0.6250	0.6247	0.6246	0.6245	0.6241	0.6237	0.6232
	H	0.3997	0.4023	0.4047	0.4072	0.4095	0.4122	0.4152	0.4182
1 3/8	T	0.6875	0.6875	0.6872	0.6870	0.6869	0.6865	0.6861	0.6856
	H	0.4397	0.4426	0.4452	0.4479	0.4504	0.4534	0.4567	0.4600
1 1/2	T	0.7500	0.7500	0.7497	0.7495	0.7494	0.7489	0.7485	0.7480
	H	0.4797	0.4828	0.4857	0.4887	0.4914	0.4947	0.4983	0.5019
1 3/4	T	0.8750	0.8750	0.8746	0.8744	0.8743	0.8737	0.8732	0.8726
	H	0.5596	0.5633	0.5666	0.5701	0.5733	0.5771	0.5813	0.5855
2	T	1.0000	1.0000	0.9996	0.9994	0.9992	0.9986	0.9980	0.9972
	H	0.6396	0.6438	0.6476	0.6516	0.6552	0.6596	0.6644	0.6692
2 1/4	T	1.1250	1.1250	1.1246	1.1242	1.1240	1.1234	1.1226	1.1220
	H	0.7195	0.7242	0.7285	0.7330	0.7371	0.7420	0.7474	0.7528
2 1/2	T	1.2500	1.2500	1.2494	1.2492	1.2490	1.2482	1.2474	1.2464
	H	0.7995	0.8047	0.8095	0.8145	0.8190	0.8245	0.8305	0.8365
3	T	1.5000	1.5000	1.4994	1.4990	1.4990	1.4978	1.4970	1.4960
	H	0.9594	0.9657	0.9714	0.9774	0.9828	0.9894	0.9966	1.0038

Series of Involute, Finishing Gear Milling Cutters for Each Pitch

Number of Cutter	Will cut Gears from	Number of Cutter	Will cut Gears from
1	135 teeth to a rack	5	21 to 25 teeth
2	55 to 134 teeth	6	17 to 20 teeth
3	35 to 54 teeth	7	14 to 16 teeth
4	26 to 34 teeth	8	12 to 13 teeth
The regular cutters listed above are used ordinarily. The cutters listed below (an intermediate series having half numbers) may be used when greater accuracy of tooth shape is essential in cases where the number of teeth is between the numbers for which the regular cutters are intended.			

Series of Involute, Finishing Gear Milling Cutters for Each Pitch *(Continued)*

Number of Cutter	Will cut Gears from	Number of Cutter	Will cut Gears from
1½	80 to 134 teeth	5½	19 to 20 teeth
2½	42 to 54 teeth	6½	15 to 16 teeth
3½	30 to 34 teeth	7½	13 teeth
4½	23 to 25 teeth	…	…

Roughing cutters are made with No. 1 form only.

Gear Design Based upon Module System.—The *module* of a gear is equal to the pitch diameter divided by the number of teeth, whereas *diametral pitch* is equal to the number of teeth divided by the pitch diameter. The module system (see accompanying table and diagram) is in general use in countries that have adopted the metric system; hence, the term module is usually understood to mean the pitch diameter *in millimeters* divided by the number of teeth. The module system, however, may also be based on inch measurements and then it is known as the English module to avoid confusion with the metric module. Module is an actual dimension, whereas diametral pitch is only a ratio. Thus, if the pitch diameter of a gear is 50 millimeters and the number of teeth 25, the module is 2, which means that there are 2 millimeters of pitch diameter for each tooth. The table *Tooth Dimensions Based Upon Module System* shows the relation among module, diametral pitch, and circular pitch.

German Standard Tooth Form for Spur and Bevel Gears ***DIN 867***

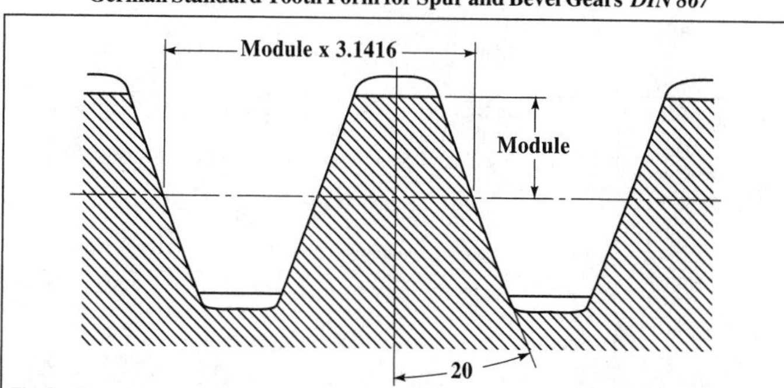

The flanks or sides are straight (involute system) and the pressure angle is 20 degrees. The shape of the root clearance space and the amount of clearance depend upon the method of cutting and special requirements. The amount of clearance may vary from 0.1 × module to 0.3 × module.

To Find	Module Known	Circular Pitch Known
Addendum	Equals module	0.31823 × Circular pitch
Dedendum	1.157 × module* 1.167 × module**	0.3683 × Circular pitch* 0.3714 × Circular pitch**
Working Depth	2 × module	0.6366 × Circular pitch
Total Depth	2.157 × module* 2.167 × module**	0.6866 × Circular pitch* 0.6898 × Circulate pitch**
Tooth Thickness on Pitch Line	1.5708 × module	0.5 × Circular pitch

Formulas for dedendum and total depth, marked (*) are used when clearance equals 0.157 × module. Formulas marked (**) are used when clearance equals one-sixth module. It is common practice among American cutter manufacturers to make the clearance of metric or module cutters equal to 0.157 × module.

Rules for Module System of Gearing

To Find	Rule
Metric Module	*Rule 1:* To find the metric module, divide the pitch diameter in millimeters by the number of teeth. *Example 1:* The pitch diameter of a gear is 200 millimeters and the number of teeth, 40; then $\text{Module} = \frac{200}{40} = 5$ *Rule 2:* Multiply circular pitch in millimeters by 0.3183. *Example 2:* (Same as Example 1. Circular pitch of this gear equals 15.708 millimeters.) $\text{Module} = 15.708 \times 0.3183 = 5$ *Rule 3:* Divide outside diameter in millimeters by the number of teeth plus 2.
English Module	*Note:* The module system is usually applied when gear dimensions are expressed in millimeters, but module may also be based on inch measurements. *Rule:* To find the English module, divide pitch diameter in inches by the number of teeth. *Example:* A gear has 48 teeth and a pitch diameter of 12 inches. $\text{Module} = \frac{12}{48} = \frac{1}{4}$ module or 4 diametral pitch
Metric Module Equivalent to Diametral Pitch	*Rule:* To find the metric module equivalent to a given diametral pitch, divide 25.4 by the diametral pitch. *Example:* Determine metric module equivalent to 10 diameteral pitch. $\text{Equivalent module} = \frac{25.4}{10} = 2.54$ *Note:* The nearest standard module is 2.5.
Diametral Pitch Equivalent to Metric Module	*Rule:* To find the diametral pitch equivalent to a given module, divide 25.4 by the module. (25.4 = number of millimeters per inch.) *Example:* The module is 12; determine equivalent diametral pitch. $\text{Equivalent diametral pitch} = \frac{25.4}{12} = 2.117$ *Note:* A diametral pitch of 2 is the nearest *standard* equivalent.
Pitch Diameter	*Rule:* Multiply number of teeth by module. *Example:* The metric module is 8 and the gear has 40 teeth; then $D = 40 \times 8 = 320 \text{ millimeters} = 12.598 \text{ inches}$
Outside Diameter	*Rule:* Add 2 to the number of teeth and multiply sum by the module. *Example:* A gear has 40 teeth and module is 6. Find outside or blank diameter. $\text{Outside diameter} = (40 + 2) \times 6 = 252 \text{ millimeters}$

For tooth dimensions, see table *Tooth Dimensions Based Upon Module System*; also formulas in *German Standard Tooth Form for Spur and Bevel Gears DIN 867*.

Tooth Dimensions Based Upon Module System

Module, *DIN* Standard Series	Equivalent Diametral Pitch	Circular Pitch		Addendum, Millimeters	Dedendum, Millimeters[a]	Whole Depth,[a] Millimeters	Whole Depth,[b] Millimeters
		Millimeters	Inches				
0.3	84.667	0.943	0.0371	0.30	0.35	0.650	0.647
0.4	63.500	1.257	0.0495	0.40	0.467	0.867	0.863
0.5	50.800	1.571	0.0618	0.50	0.583	1.083	1.079
0.6	42.333	1.885	0.0742	0.60	0.700	1.300	1.294
0.7	36.286	2.199	0.0865	0.70	0.817	1.517	1.510
0.8	31.750	2.513	0.0989	0.80	0.933	1.733	1.726
0.9	28.222	2.827	0.1113	0.90	1.050	1.950	1.941
1	25.400	3.142	0.1237	1.00	1.167	2.167	2.157
1.25	20.320	3.927	0.1546	1.25	1.458	2.708	2.697
1.5	16.933	4.712	0.1855	1.50	1.750	3.250	3.236
1.75	14.514	5.498	0.2164	1.75	2.042	3.792	3.774
2	12.700	6.283	0.2474	2.00	2.333	4.333	4.314
2.25	11.289	7.069	0.2783	2.25	2.625	4.875	4.853
2.5	10.160	7.854	0.3092	2.50	2.917	5.417	5.392
2.75	9.236	8.639	0.3401	2.75	3.208	5.958	5.932
3	8.466	9.425	0.3711	3.00	3.500	6.500	6.471
3.25	7.815	10.210	0.4020	3.25	3.791	7.041	7.010
3.5	7.257	10.996	0.4329	3.50	4.083	7.583	7.550
3.75	6.773	11.781	0.4638	3.75	4.375	8.125	8.089
4	6.350	12.566	0.4947	4.00	4.666	8.666	8.628
4.5	5.644	14.137	0.5566	4.50	5.25	9.750	9.707
5	5.080	15.708	0.6184	5.00	5.833	10.833	10.785
5.5	4.618	17.279	0.6803	5.50	6.416	11.916	11.864
6	4.233	18.850	0.7421	6.00	7.000	13.000	12.942
6.5	3.908	20.420	0.8035	6.50	7.583	14.083	14.021
7	3.628	21.991	0.8658	7.	8.166	15.166	15.099
8	3.175	25.132	0.9895	8.	9.333	17.333	17.256
9	2.822	28.274	1.1132	9.	10.499	19.499	19.413
10	2.540	31.416	1.2368	10.	11.666	21.666	21.571
11	2.309	34.558	1.3606	11.	12.833	23.833	23.728
12	2.117	37.699	1.4843	12.	14.000	26.000	25.884
13	1.954	40.841	1.6079	13.	15.166	28.166	28.041
14	1.814	43.982	1.7317	14.	16.332	30.332	30.198
15	1.693	47.124	1.8541	15.	17.499	32.499	32.355
16	1.587	50.266	1.9790	16.	18.666	34.666	34.512
18	1.411	56.549	2.2263	18.	21.000	39.000	38.826
20	1.270	62.832	2.4737	20.	23.332	43.332	43.142
22	1.155	69.115	2.7210	22.	25.665	47.665	47.454
24	1.058	75.398	2.9685	24.	28.000	52.000	51.768
27	0.941	84.823	3.339	27.	31.498	58.498	58.239
30	0.847	94.248	3.711	30.	35.000	65.000	64.713
33	0.770	103.673	4.082	33.	38.498	71.498	71.181
36	0.706	113.097	4.453	36.	41.998	77.998	77.652
39	0.651	122.522	4.824	39.	45.497	84.497	84.123
42	0.605	131.947	5.195	42.	48.997	90.997	90.594
45	0.564	141.372	5.566	45.	52.497	97.497	97.065
50	0.508	157.080	6.184	50.	58.330	108.330	107.855
55	0.462	172.788	6.803	55.	64.163	119.163	118.635
60	0.423	188.496	7.421	60.	69.996	129.996	129.426
65	0.391	204.204	8.040	65.	75.829	140.829	140.205
70	0.363	219.911	8.658	70.	81.662	151.662	150.997
75	0.339	235.619	9.276	75.	87.495	162.495	161.775

[a] Dedendum and total depth when clearance = 0.1666 × module, or one-sixth module.

[b] Total depth equivalent to American standard full-depth teeth. (Clearance = 0.157 × module.)

Equivalent Diametral Pitches, Circular Pitches, and Metric Modules
Commonly Used Pitches and Modules in Bold Type

Diametral Pitch	Circular Pitch, Inches	Module Millimeters	Diametral Pitch	Circular Pitch, Inches	Module Millimeters	Diametral Pitch	Circular Pitch, Inches	Module Millimeters
1/2	6.2832	50.8000	2.2848	**1 3/8**	11.1170	10.0531	**5/16**	2.5266
0.5080	6.1842	**50**	2.3091	1.3605	**11**	10.1600	0.3092	**2 1/2**
0.5236	**6**	48.5104	**2 1/2**	1.2566	10.1600	**11**	0.2856	2.3091
0.5644	5.5658	**45**	2.5133	**1 1/4**	10.1063	**12**	0.2618	2.1167
0.5712	**5 1/2**	44.4679	2.5400	1.2368	**10**	12.5664	**1/4**	2.0213
0.6283	**5**	40.4253	**2 3/4**	1.1424	9.2364	12.7000	0.2474	**2**
0.6350	4.9474	**40**	2.7925	**1 1/8**	9.0957	**13**	0.2417	1.9538
0.6981	**4 1/2**	36.3828	2.8222	1.1132	**9**	**14**	0.2244	1.8143
0.7257	4.3290	**35**	**3**	1.0472	8.4667	**15**	0.2094	1.6933
3/4	4.1888	33.8667	3.1416	**1**	8.0851	**16**	0.1963	1.5875
0.7854	**4**	32.3403	3.1750	0.9895	**8**	16.7552	**3/16**	1.5160
0.8378	**3 3/4**	30.3190	3.3510	**15/16**	7.5797	16.9333	0.1855	**1 1/2**
0.8467	3.7105	**30**	**3 1/2**	0.8976	7.2571	**17**	0.1848	1.4941
0.8976	**3 1/2**	28.2977	3.5904	7/8	7.0744	**18**	0.1745	1.4111
0.9666	**3 1/4**	26.2765	3.6286	0.8658	**7**	**19**	0.1653	1.3368
1	3.1416	25.4000	3.8666	**13/16**	6.5691	**20**	0.1571	1.2700
1.0160	3.0921	**25**	3.9078	0.8040	**6 1/2**	**22**	0.1428	1.1545
1.0472	**3**	24.2552	**4**	0.7854	6.3500	**24**	0.1309	1.0583
1.1424	**2 3/4**	22.2339	4.1888	**3/4**	6.0638	**25**	0.1257	1.0160
1 1/4	2.5133	20.3200	4.2333	0.7421	**6**	25.1328	**1/8**	1.0106
1.2566	**2 1/2**	20.2127	4.5696	**11/16**	5.5585	25.4000	0.1237	**1**
1.2700	2.4737	**20**	4.6182	0.6803	5 1/2	**26**	0.1208	0.9769
1.3963	**2 1/4**	18.1914	**5**	0.6283	5.0800	**28**	0.1122	0.9071
1.4111	2.2263	**18**	5.0265	**5/8**	5.0532	**30**	0.1047	0.8467
1 1/2	2.0944	16.9333	5.0800	0.6184	**5**	**32**	0.0982	0.7937
1.5708	**2**	16.1701	5.5851	**9/16**	4.5478	**34**	0.0924	0.7470
1.5875	1.9790	**16**	5.6443	0.5566	**4 1/2**	**36**	0.0873	0.7056
1.6755	**1 7/8**	15.1595	**6**	0.5236	4.2333	**38**	0.0827	0.6684
1.6933	1.8553	**15**	6.2832	**1/2**	4.0425	**40**	0.0785	0.6350
1 3/4	1.7952	14.5143	6.3500	0.4947	**4**	**42**	0.0748	0.6048
1.7952	**1 3/4**	14.1489	**7**	0.4488	3.6286	**44**	0.0714	0.5773
1.8143	1.7316	**14**	7.1808	**7/16**	3.5372	**46**	0.0683	0.5522
1.9333	**1 5/8**	13.1382	7.2571	0.4329	**3 1/2**	**48**	0.0654	0.5292
1.9538	1.6079	**13**	**8**	0.3927	3.1750	**50**	0.0628	0.5080
2	1.5708	12.7000	8.3776	**3/8**	3.0319	50.2656	**1/16**	0.5053
2.0944	**1 1/2**	12.1276	8.4667	0.3711	**3**	50.8000	0.0618	**1/2**
2.1167	1.4842	**12**	**9**	0.3491	2.8222	**56**	0.0561	0.4536
2 1/4	1.3963	11.2889	**10**	0.3142	2.5400	**60**	0.0524	0.4233

The module of a gear is the pitch diameter divided by the number of teeth. The module may be expressed in any units; but when no units are stated, it is understood to be in millimeters. The metric module, therefore, equals the pitch diameter in millimeters divided by the number of teeth. To find the metric module equivalent to a given diametral pitch, divide 25.4 by the diametral pitch. To find the diametral pitch equivalent to a given module, divide 25.4 by the module. (25.4 = number of millimeters per inch.)

Rules and Formulas for Spur Gear Calculations (Circular Pitch)

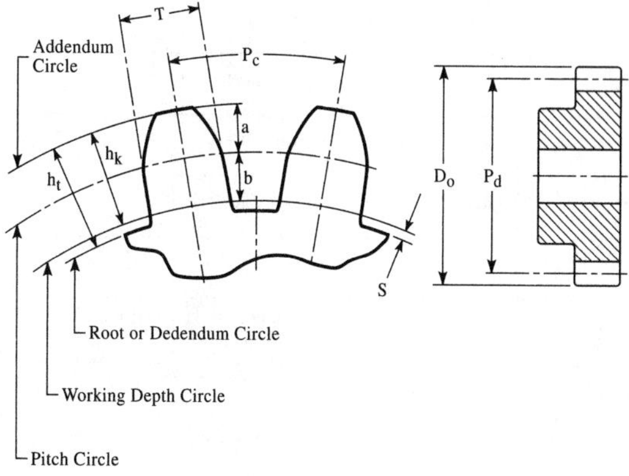

Fig. 2. Spur Gear

The following symbols are used in conjunction with the formulas for determining the proportions of spur gear teeth

P = Diametral pitch
P_C = Circular pitch
P_d = Pitch diameter
D_o = Outside diameter
N = Number of teeth in the gear
a = Addendum
b = Dedendum
h_k = Working depth
h_t = Whole depth
S = Clearance
C = Center distance
L = Length of rack

To Find	Rule	Formula
Diametral pitch P	Divide 3.1416 by the circular pitch.	$P = \frac{3.1416}{P_c}$
Circular pitch P_C	Divide 3.1416 by the diametral pitch.	$P_c = \frac{3.1416}{P}$
Pitch diameter P_d	Divide the number of teeth by the diametral pitch.	$P_d = \frac{N}{P}$
Outside diameter D_o	Add 2 to the number of teeth and divide the sum by the diametral pitch.	$D_o = \frac{N+2}{P}$
Number of teeth N	Multiply the pitch diameter by the diametral pitch.	$N = P_d \cdot P$
Tooth circular thickness T	Divide 1.5708 by the diametral pitch.	$T = \frac{1.5708}{P}$

To Find	Rule	Formula
Addendum a	Divide 1.0 by the diametral pitch.	$a = \frac{1.0}{P}$
Dedendum b	Divide 1.157 by the diametral pitch.	$b = \frac{1.157}{P}$
Working depth h_k	Divide 2 by the diametral pitch.	$h_k = \frac{2}{P}$
Whole depth h_t	Divide 2.157 by the diametral pitch.	$h_t = \frac{2.157}{P}$
Clearance S	Divide 0.157 by the diametral pitch.	$S = \frac{0.157}{P}$
Center distance C	Add the number of teeth in both gears and divide the sum by two times the diametral pitch.	$C = \frac{N_1 + N_2}{2P}$
Length of rack L	Multiply the number of teeth in the rack by the circular pitch.	$L = NP_c$

Rules and Formulas for Helical Gear Calculations

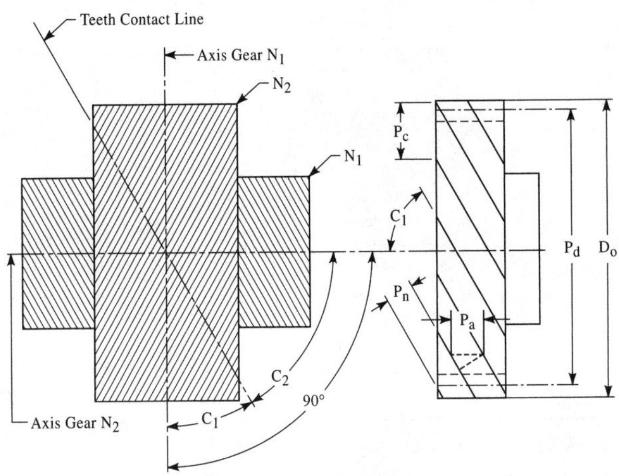

Fig. 3.

The following symbols are used in conjunction with the formulas for determining the proportions of helical gear teeth

P_{nd} = Normal diametral pitch (pitch of the cutter)

P_c = Circular pitch

P_a = Axial pitch

P_n = Normal pitch

P_d = Pitch diameter

S = Center distance

C, C_1, C_2 = Helix angle of the gears

L = Lead of tooth helix

T_n = Normal tooth thickness at pitch line

a = Addendum

h_t = Whole depth of tooth

N, N_1, N_2 = Number of thhth in the gears

D_o = Outside diameter

N_c = Hypothetical number of teeth for which gear cutter should be selected

To Find	Rule	Formula
Normal diametral pitch P_{nd}	Divide the number of teeth by the product of the pitch diameter and the cosine of the helix angle.	$P_{nd} = \frac{N}{P_d \cos C_1}$
Circular pitch P_c	Multiply the pitch diameter of the gear by 3.1416, and divide the product by the number of teeth in the gear.	$P_c = \frac{3.1416 P_d}{N}$
Axial pitch P_a	Multiply the circular pitch by the cotangent of the helix angle.	$P_a = P_c \cot C_1$
Normal pitch P_n	Divide 3.1416 by the normal diametral pitch.	$P_n = \frac{3.1416}{P_{nd}}$
Pitch diameter P_d	Divide the number of teeth by the product of the normal pitch and the cosine of the helix angle.	$P_d = \frac{N}{P_{nd} \cos C_1}$
Center distance S	Divide the sum of the pitch diameters of the mating gears by 2.	$S = \frac{P_{d1} + P_{d2}}{2}$
Checking Formulas (shafts at right angles)	Multiply the number of teeth in the first gear by the tangent of the tooth angle of that gear, and add the number of teeth in the second gear to the product. The sum should equal twice the product of the center distance multiplied by the normal diametral pitch, multiplied by the sine of the helix angle.	$N_1 + N_2 \tan C_2$ $= 2SP_{nd} \sin C_1$
Lead of tooth helix L	Multiply the pitch diameter by 3.1416 times the cotangent of the helix angle.	L $= 3.14 \cdot P_d \cot C_1$
Normal circular tooth thickness at pitch line T_n	Divide 1.571 by the normal diametral pitch.	$T_n = \frac{1.571}{P_{nd}}$
Addendum a	Divide the normal pitch by 3.1416.	$a = \frac{P_n}{3.1416}$
Whole depth of tooth h_t	Divide 2.157 by the normal diametral pitch.	$h_t = \frac{2.157}{P_{nd}}$
Outside diameter D_o	Add twice the addendum to the pitch diameter.	$D_o = P_d + 2a$
Hypothetical number of teeth for which gear cutter should be selected N_c	Divide the number of teeth in the gear by the cube of the cosine of the helix angle.	$N_C = \frac{N_1}{(\cos C_1)^3}$

Rules and Formulas for Bevel Gear Calculations

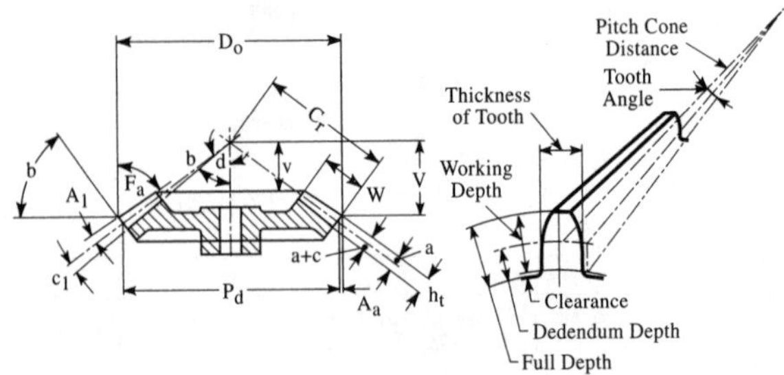

The following symbols are used in conjunction with the formulas for determining the proportions of bevel gear teeth

P = Diametral pitch
P_C = Circular pitch
P_d = Pitch diameter
b = Pitch angle
Cr = Pitch cone diatance
a = Addendum
A_l = Addendum angle
A_a = Angular addendum
D_o = Outside diameter
C_l = Dedendum angle
$a + c$= Addendum plus clearance
a_s = Addendum of small end of tooth
T_L = Thickness of tooth at pitch line
T_s = Thickness of tooth at pitch line at small end of gear
F_a = Face width
h_t = Whole depth of tooth space
V = Apex distance at large end of tooth
v = Apex distance at small end of tooth
m_g = Gear ratio
N = Number of teeth
N_g = Number of teeth in gear
N_p = Number of teeth in pinion
d = Root angle
W = Width of gear tooth face
N_c = Number of teeth ofimaginary spur gear forwhich cutter is selected

To Find	Rule	Formula
Diametral pitch P	Divide the number of teeth by the pitch diameter.	$P = \frac{N}{P_d}$
Circular pitch P_C	Divide 3.1416 by the diametral pitch.	$P_c = \frac{3.1416}{P}$
Pitch diameter P_d	Divide the number of teeth by the diametral pitch.	$P_d = \frac{N}{P}$

To Find	Rule	Formula
Pitch angle of pinion tan b_p	Divide the number of teeth in the pinion by the number of teeth in the gear to obtain the tangent.	$\tan b_p = \frac{N_p}{N_g}$
Pitch angle of gear tan b_g	Divide the number of teeth in the gear by the number of teeth in the pinion to obtain the tangent.	$\tan b_g = \frac{N_g}{N_p}$
Pitch cone diatance C_r	Divide the pitch diameter by twice the sine of the pitch angle.	$C_r = \frac{P_d}{2\sin b}$
Addendum a	Divide 1.0 by the diametral pitch.	$a = \frac{1.0}{p}$
Addendum angle tan A_l	Divide the addendum by the pitch cone distance to obtain the tangent.	$\tan A_1 = \frac{a}{C_r}$
Angular addendum A_a	Multiply the addendum by the cosine of the pitch angle.	$A_a = a\cos b$
Outside diameter D_o	Add twice the angular addendum to the pitch diameter.	$D_o = P_d + 2A_a$
Dedendum angle tan C_l	Divide the dedendum by the pitch cone distance to obtain the tangent.	$\tan c_1 = \frac{a+c}{C_r}$
Addendum of small end of tooth a_s	Subtract the width of face from the pitch cone distance, divide the remainder by the pitch cone distance and multiply by the addendum.	$a_s = a\left(\frac{C_r - W}{C_r}\right)$
Thickness of tooth at pitch line T_L	Divide the circular pitch by 2.	$T_l = \frac{P_c}{2}$
Thickness of tooth at pitch line at small end of gear T_s	Subtract the width of face from the pitch cone distance, divide the remainder by the pitch cone distance and multiply by the thickness of the tooth at the pitch line.	$T_s = T_L\left(\frac{C_r - W}{C_r}\right)$
Face width F_a	Face cone of blank turned parallel to root cone of mating gear.	$F_a = b + c_l$
Whole depth of tooth space h_t	Divide 2.157 by the diametral pitch.	$h_t = \frac{2.157}{P}$
Apex distance at large end of tooth V	Multiply one-half the outside diameter by the tangent of the face angle.	$V = \left(\frac{D_o}{2}\right)\tan F_a$
Apex distance at small end of tooth v	Subtract the width of face from the pitch cone distance, divide the remainder by the pitch cone distance and multiply by the apex distance.	$v = V\left(\frac{C_r - W}{C_r}\right)$
Gear ratio m_g	Divide the number of teeth in the gear by the number of teeth in the pinion.	$m_g = \frac{N_g}{N_p}$
Number of teeth in gear and/or pinion N_g, N_p	Multiply the pitch diameter by the diametral pitch.	$N_g = P_d P$ $N_p = P_d P$
Cutting angle d	Subtract the addendum plus clearance angle from the pitch angle.	$d = b - c_1$
Number of teeth of imaginary spur gear for which cutter is selected N_o	Divide the number of teeth in actual gear by the cosine of the pitch angle.	$N_c = \frac{N}{\cos b}$

Milling Bevel Gears.—Once the teeth of a bevel gear have been milled to their proper depth, only the back, or large end of the teeth, have the proper size and form. To achieve the proper form on the small end of the teeth, the gear blank must be adjusted and the excess material, shown in Fig. 4, must be removed. To determine the proper position of the blank for this cutting operation, the following formulas can be used to calculate the set over and roll of the blank.

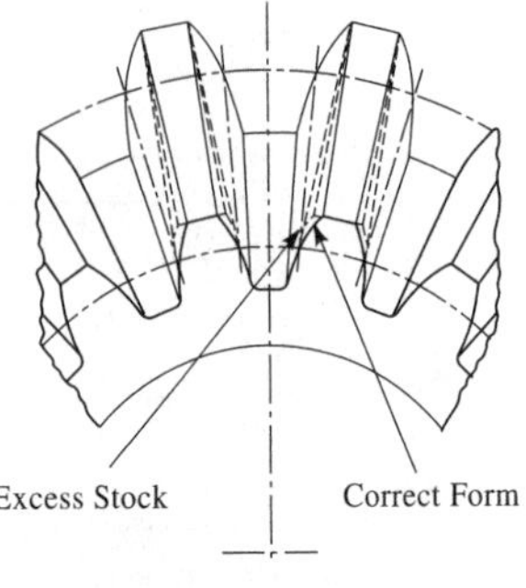

Fig. 4.

Calculating The Angle of Roll: The roll of the blank needed to produce the proper tooth form is calculated using the roll formula shown below. Once calculated, the dividing head is adjusted the required angular amount to produce the required tooth form.

Calculating The Set Over: In addition to roll, the gear blank must also be moved to produce the required form. The amount of this lateral movement is called the set over and is found using the set over formula shown on the next page.

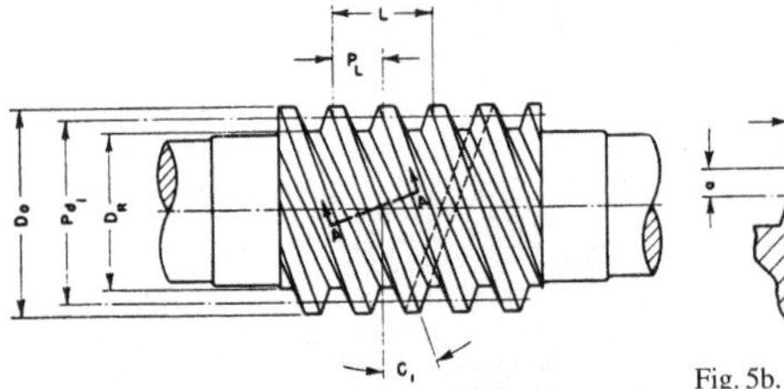

Fig. 5a. Section A-A Double size normal to helix angle

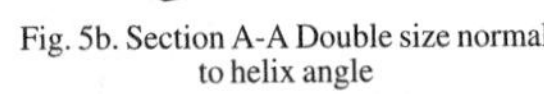

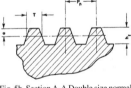

Fig. 5b. Section A-A Double size normal to helix angle

The following symbols are used in conjunction with the formulas for determining the proportions of bevel gear teeth

P_l = Linear pitch	C_l = Helix angle
P_{dl} = Pitch diameter	P_n = Normal pitch
D_o = Outside diameter	a = Addendum
N_W = Number of threads	L = Lead
D_R = Root diameter	T = Normal tooth thickness
h_t = Whole depth of tooth	t = Width of thread tool at end

Roll $$C = \frac{57.3}{P_d}\left(\frac{P_c}{2} - \frac{C_r}{W}(T_l - T_s)\right)$$

Set over $$n = \frac{T_l}{2} - \frac{(T_l - T_s)}{2}\frac{C_r}{W}$$

Where:

C = angle of roll, degrees.

P_d = pitch diameter at large end of gear, inches.

P_c = circular pitch at large end of gear, inches.

C_r = pitch cone distance at large end of gear, inches.

T_s, T_L = chordal thickness ofgear cutter tooth corresponding to pitch line at small and large ends of gear, respectively, inches.

57.3 = degrees per radian.

W = width of gear tooth face, inches

Rules and Formulas for Worm Gear Calculations (Solid Type) Single and Double Thread—14½° Pressure Angle

To Find	Rule	Formula
Linear pitch P_l	Divide the lead by the number of threads in the whole worm: i.e., one if single-threaded or four if quadrupled threaded.	$P_l = \frac{L}{N_w}$
Pitch diameter Pd_l	Subtract twice the addendum from the outside diameter.	$Pd_1 = D_o - 2a$
Outside diameter D_o	Add twice the addendum of the worm to the pitch diameter of the worm wheel.	$D_o = Pd_1 + 2a$
Root diameter D_R	Subtract twice the whole depth of the tooth from the outside diameter.	$D_R = D_o - 2h_t$
Whole depth of tooth h_t	Multiply the linear pitch by 0.6866.	$h_t = 0.6866P_L$
Helix angle C_l	Multiply the pitch diameter of the worm by 3.1416, and divide the product by the lead. The quotient is the cotangent of the helix angle.	$\cot C_l = \frac{3.1416Pd_2}{L}$
Normal pitch P_n	Multiply the linear pitch by the cosine of the helix angle of the worm.	$P_n = P_L \cos C_l$
Addendum a	Multiply the linear pitch by 0.3183.	$a = 0.3183P_L$
Lead L	Multiply the linear pitch by the number of threads.	$L = P_L N_w$
Normal tooth thickness T	Multiply one-half the linear pitch by the cosine of the helix angle.	$T = \frac{P_L}{2}\cos C_l$
Width of thread tool at end t	Multiply the linear pitch by 0.31.	$t = 0.31P_L$

Rules and Formulas for Worm Wheel Calculations (Single and Double Thread—14½° Pressure Angle)

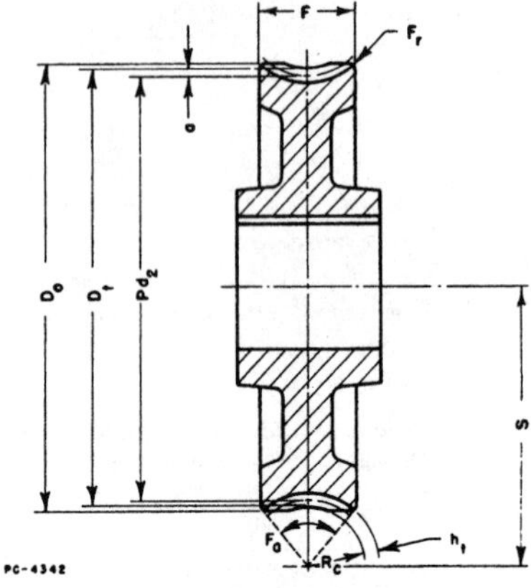

Fig. 6.

The following symbols are used in conjunction with the formulas for determining the proportions of worm wheel teeth.

P_c = Circular pitch

Pd_2 = Pitch diameter

N = Number of Teeth

D_o = Outside diameter

D_t = Throat diameter

R_c = Radius of curvature of worm wheel throat

D = Diameter to sharp corners

F_a = Face angle

F = Face width of rim

F_r = Radius at edge of face

a = Addendum

h_t = Whole depth of tooth

S = Center distance between worm and worm wheel

G = Gashing angle

Rules and Formulas for Worm Wheel Calculations (Single and Double Thread—14½° Pressure Angle)

To Find	Rule	Formula
Circular pitch P_c	Divide the pitch diameter by the product of 0.3883 and the number of teeth.	$P_c = \frac{Pd_2}{0.3183N}$
Pitch diameter Pd_2	Multiply the number of teeth in the worm wheel by the linear pitch of the worm, and divide the product by 3.1416.	$Pd_2 = \frac{NP_l}{3.1416}$
Outside diameter D_o	Multiply the circular pitch by 0.4775 and add the product to the throat diameter.	$D_o = D_t + 0.4775P_c$
Throat diameter D_t	Add twice the addendum of the worm tooth to the pitch diameter of the worm wheel.	$D_t = Pd_2 + 2a$
Radius of curvature of worm wheel throat R_c	Subtract twice the addendum of the worm tooth from half the outside diameter of the worm.	$R_c = \frac{D_o}{2} - 2a$
Diameter to sharp corners D	Multiply the radius of curvature of the worm-wheel throat by the cosine of half the face angle, subtract this quantity from the radius of curvature. Multiply the remainder by 2, and add the product to the throat diameter of the worm wheel.	$D = 2R_C$ $- 2R_C \times \cos\frac{F_a}{2} + D_t$
Face width of rim F	Multiply the circular pitch by 2.38 and add 0.25 to the product.	$F = 2.38P_c + 0.25$
Radius at edge of face F_r	Divide the circular pitch by 4.	$F_r = \frac{P_c}{4}$
Addendum a	Multiply the circular pitch by 0.3183.	$a = 0.3183P_c$
Whole depth of tooth h_t	Multiply the circular pitch by 0.6866.	$h_t = 0.6866P_c$
Center distance between worm and worm wheel S	Add the pitch diameter of the worm to the pitch diameter of the worm wheel and divide the sum by 2.	$S = \frac{Pd_1 + Pd_2}{2}$
Gashing angle G	Divide the lead of the worm by the circumference of the pitch circle. The result will be the co-tangent of the gashing angle.	$\cot G = \frac{L}{3.1416d}$

INDEXING

Milling Machine Indexing.—Positioning a workpiece at a precise angle or interval of rotation for a machining operation is called indexing. A dividing head is a milling machine attachment that provides this fine control of rotational positioning through a combination of a crank-operated worm and worm gear, and one or more indexing plates with several circles of evenly spaced holes to measure partial turns of the worm crank. The indexing crank carries a movable indexing pin that can be inserted into and withdrawn from any of the holes in a given circle with an adjustment provided for changing the circle that the indexing pin tracks.

Hole Circles.—The Brown & Sharpe dividing head has three standard indexing plates, each with six circles of holes as listed in the table below.

Numbers of Holes in Brown & Sharpe Standard Indexing Plates

Plate Number	Numbers of Holes					
1	15	16	17	18	19	20
2	21	23	27	29	31	33
3	37	39	41	43	47	49

Dividing heads of Cincinnati Milling Machine design have two-sided, standard, and high-number plates with the numbers of holes shown in the following table.

Numbers of Holes in Cincinnati Milling Machine Standard Indexing Plates

Side	Standard Plate										
1	24	25	28	30	34	37	38	39	41	42	43
2	46	47	49	51	53	54	57	58	59	62	66
	High-Number Plates										
A	30	48	69	91	99	117	129	147	171	177	189
B	36	67	81	97	111	127	141	157	169	183	199
C	34	46	79	93	109	123	139	153	167	181	197
D	32	44	77	89	107	121	137	151	163	179	193
E	26	42	73	87	103	119	133	149	161	175	191
F	28	38	71	83	101	113	131	143	159	173	187

Some dividing heads provide for *Direct Indexing* through the attachment of a special indexing plate directly to the main spindle where a separate indexing pin engages indexing holes in the plate. The worm is disengaged from the worm gear during this quick method of indexing, which is mostly used for common, small-numbered divisions such as six, used in machining hexagonal forms for bolt heads and nuts, for instance.

Simple Indexing.—Also called *Plain Indexing* or *Indirect Indexing*, simple indexing is based on the ratio between the worm and the worm gear, which is usually, but not always, 40:1. The number of turns of the indexing crank needed for each indexing movement to produce a specified number of evenly spaced divisions is equal to the number of turns of the crank that produce exactly one full turn of the main spindle, divided by the specified number of divisions required for the workpiece. The accompanying tables provide data for the indexing movements to meet most division requirements, and include the simple indexing movements along with the more complex movements for divisions that are not available through simple indexing. The fractional entries in the tables are deliberately not reduced to lowest terms. Thus, the numerator represents the number of holes to be moved on the circle of holes specified by the denominator.

Indexing Movements for Standard Index Plate—Cincinnati Milling Machine

The standard index plate indexes all numbers up to and including 60; all even numbers and those divisible by 5 up to 120; and all divisions listed below up to 400. The plate is drilled on both sides, and has holes as follows: First side: 24, 25, 28, 30, 34, 37, 38, 39, 41, 42, 43.
Second side: 46, 47, 49, 51, 53, 54, 57, 58, 59, 62, 66.

No of divisions	Circle	Turns	Holes	No of divisions	Circle	Holes	No of divisions	Circle	Holes	No of divisions	Circle	Holes
2	Any	20	…	44	66	60	104	39	15	205	41	8
3	24	13	8	45	54	48	105	42	16	210	42	8
4	Any	10	…	46	46	40	106	53	20	212	53	10
5	Any	8	…	47	47	40	108	54	20	215	43	8
6	24	6	16	48	24	20	110	66	24	216	54	10
7	28	5	20	49	49	40	112	112	40	220	66	12
8	Any	5	…	50	25	20	114	57	20	224	28	5
9	54	4	24	51	51	40	115	46	16	228	57	10
10	Any	4	…	52	39	30	116	58	20	232	58	10
11	66	3	42	53	53	40	118	59	20	235	47	8
12	24	3	8	54	54	40	120	66	22	236	59	10
13	39	3	3	55	66	48	124	62	20	245	49	8
14	49	2	42	56	28	20	125	25	8	248	62	10
15	24	2	16	57	57	40	130	39	12	250	25	4
16	24	2	12	58	58	40	132	66	20	255	51	8
17	34	2	12	59	59	40	135	54	16	260	39	6
18	54	2	12	60	42	28	136	34	10	264	66	10
19	38	2	4	62	62	40	140	28	8	270	54	8
20	Any	2	…	64	24	15	144	54	15	272	34	5
21	42	1	38	65	39	24	145	58	16	280	28	4
22	66	1	54	66	66	40	148	37	10	290	58	8
23	46	1	34	68	34	20	150	30	8	296	37	5
24	24	1	16	70	28	16	152	38	10	300	30	4
25	25	1	15	72	54	30	155	62	16	304	38	5
26	39	1	21	74	37	20	156	39	10	310	62	8
27	54	1	26	75	30	16	160	28	7	312	39	5
28	42	1	18	76	38	20	164	41	10	320	24	3
29	58	1	22	78	39	20	165	66	16	328	41	5
30	24	1	8	80	34	17	168	42	10	330	66	8
31	62	1	18	82	41	20	170	34	8	336	42	5
32	28	1	7	84	42	20	172	43	10	340	34	4
33	66	1	14	85	34	16	176	66	15	344	43	5
34	34	1	6	86	43	20	180	54	12	360	54	6
35	28	1	4	88	66	30	184	46	10	368	46	5
36	54	1	6	90	54	24	185	37	8	370	37	4
37	37	1	3	92	46	20	188	47	10	376	47	5
38	38	1	2	94	47	20	190	38	8	380	38	4
39	39	1	1	95	38	16	192	24	5	390	39	4
40	Any	1	…	96	24	10	195	39	8	392	49	5
41	41	…	40	98	49	20	196	49	10	400	30	3
42	42	…	40	100	25	10	200	30	6	…	…	…
43	43	…	40	102	51	20	204	51	10	…	…	…

Setting up for an indexing job includes setting the sector arms to the fractional part of a turn required for each indexing movement to avoid the need to count holes each time. The current location of the indexing pin in the circle of holes to be used is always hole zero when counting the number of holes to be moved. The wormshaft hub carrying the dividing plate may also carry one or two sets of sector arms, each of which can be used to define two arcs of holes.

Example: If 40 turns of the index crank are required for one revolution of the spindle, and 12 divisions are required, the number of turns of the index crank for each indexing would equal $40 \div 12 = 3\frac{1}{3}$ turns.

Angular Indexing.—The plain dividing head with a 40:1 gear ratio will rotate the main spindle and the workpiece 9 degrees for each full turn of the indexing crank, and therefore 1 degree for movements of 2/18 or 3/27 on Brown & Sharpe dividing heads and 6/54 on heads of Cincinnati design. To find the indexing movement for an angle, divide that angle, in degrees, by 9 to get the number of full turns and the remainder, if any. If the remainder, expressed in minutes, is evenly divisible by 36, 33.75, 30, 27, or 20, then the quotient is the number of holes to be moved on the 15-, 16-, 18-, 20-, or 27-hole circles, respectively, to obtain the fractional turn required (or evenly divisible by 22.5, 21.6, 18, 16.875, 15, 11.25, or 10 for the number of holes to be moved on the 24-, 25-, 30-, 32-, 36-, 48-, or 54-hole circles, respectively, for the standard and high number plates of a Cincinnati dividing head). If none of these divisions is even, it is not possible to index the angle (exactly) by this method.

Example: An angle of 61° 48′ is required. Expressed in degrees, this angle is 61.8°, which when divided by 9 equals 6 with a remainder of 7.8°, or 468′. Division of 468 by 20, 27, 30, 33.75, and 36 reveals an even division by 36, yielding 13. The indexing movement for 61° 48′ is six full turns plus 13 holes on the 15-hole circle.

The number of degrees left to turn after having completed the full turns are indexed by taking two holes in the 18-hole circle for each degree. In this case, $35/9 = 3\frac{8}{9}$, which indicates that the index crank must be turned three full revolutions, and then 8 degrees more are indexed by moving 16 holes in the 18-hole circle.

To index for $11\frac{1}{2}$ degrees, for example, first turn the index crank one revolution, this being a 9-degree movement. Then to index $2\frac{1}{2}$ degrees, move the index cranks holes in the 18-hole circle (4 holes for the two whole degrees and one hole for the $\frac{1}{2}$ degree equals the total movement of 5 holes).

Below is shown how this calculation may be carried out to plainly indicate the movement required for this angle:

$$11\tfrac{1}{2}\text{ deg.} = 9\text{ deg.} + 2\text{ deg.} + \tfrac{1}{2}\text{ deg.}$$

1 turn + 4 holes + 1 hole in the 18-hole circle.

Should it be required to index only $\frac{1}{3}$ degree, this may be done by using the 27-hole circle. In this circle a three-hole movement equals one degree, and a one-hole movement in that circle thus equals $\frac{1}{3}$ degree, or 20 minutes. Assume that it is required to index the work through an angle of 48 degrees 40 minutes. Below is plainly shown how this calculation may be carried out:

$$48\text{ deg. }40\text{ mm} = 45\text{ deg.} + 3\text{ deg.} + 40\text{ mm.}$$

5 turns + 9 holes + 2 holes in the 27-hole circle.

Angular Indexing

Angle in Degs.	Turns of Index Crank	Angle in Degs.	Turns of Index Crank	Angle in Degs.	Turns of Index Crank	Angle in Degs.	Turns of Index Crank
1	2/18	10 2/3	1 5/27	23	2 10/18	35	3 16/18
1 1/3	4/27	11	1 4/18	23 1/3	2 16/27	35 1/3	3 25/27
1 1/2	3/18	11 1/3	1 7/27	23 2/3	2 17/27	35 2/3	3 26/27
1 2/3	5/27	11 2/3	1 8/27	24	2 12/18	36	4
2	6/27	12	1 6/18	24 1/3	2 19/27	36 1/3	4 1/27
2 1/3	7/27	12 1/3	1 10/27	24 2/3	2 20/27	36 2/3	4 2/27
2 1/2	5/18	12 2/3	1 11/27	25	2 14/18	37	4 2/18
2 2/3	8/27	13	1 4/9	25 1/3	2 22/27	37 1/3	4 4/27
3	6/18	13 1/3	1 13/27	25 2/3	2 23/27	37 2/3	4 5/27
3 1/3	10/27	13 2/3	1 14/27	26	2 16/18	38	4 2/9
3 1/2	7/18	14	1 10/18	26 1/3	2 25/27	38 1/3	4 7/27
3 2/3	11/27	14 1/3	1 16/27	26 2/3	2 26/27	38 2/3	4 8/27
4	8/18	14 2/3	1 17/27	27	3	39	4 6/18
4 1/3	13/27	15	1 12/18	27 1/3	3 1/27	39 1/3	4 10/27
4 1/2	9/18	15 1/3	1 19/27	27 2/3	3 2/27	39 2/3	4 11/27
4 2/3	14/27	15 2/3	1 20/27	28	3 2/18	40	4 8/18
5	10/18	16	1 14/18	28 1/3	3 4/27	40 1/3	4 13/27
5 1/3	16/27	16 1/3	1 22/27	28 2/3	3 5/27	40 2/3	4 14/27
5 1/2	11/18	16 2/3	1 23/27	29	3 4/18	41	4 10/18
5 2/3	17/27	17	1 16/18	29 1/3	3 7/27	41 1/3	4 16/27
6	12/18	17 1/3	1 25/27	29 2/3	3 8/27	41 2/3	4 17/27
6 1/3	19/27	17 2/3	1 26/27	30	3 6/18	42	4 12/18
6 1/2	13/18	18	2	30 1/3	3 10/27	42 1/3	4 19/27
6 2/3	20/27	18 1/3	2 1/27	30 2/3	3 11/27	42 2/3	4 20/27
7	14/18	18 2/3	2 2/27	31	3 8/18	43	4 14/18
7 1/3	22/27	19	2 2/18	31 1/3	3 13/27	43 1/3	4 22/27
7 1/2	15/18	19 1/3	2 4/27	31 2/3	3 14/27	43 2/3	4 23/27
7 2/3	23/27	19 2/3	2 5/27	32	3 10/18	44	4 16/18
8	16/18	20	2 4/18	32 1/3	3 16/27	44 1/3	4 25/27
8 1/3	25/27	20 1/3	2 7/27	32 2/3	3 17/27	44 2/3	4 26/27
8 2/3	26/27	20 2/3	2 8/27	33	3 18/27	45	5
9	1	21	2 6/18	33 1/3	3 19/27	45 1/3	5 1/27
9 1/3	1 1/27	21 1/3	2 10/27	33 2/3	3 20/27	45 2/3	5 2/27
9 2/3	1 2/27	21 2/3	2 11/27	34	3 14/18	46	5 2/18
10	1 2/18	22	2 8/18	34 1/3	3 22/27	46 1/3	5 4/27
10 1/3	1 4/27	22 1/3	2 13/27	34 2/3	3 23/27	46 2/3	5 5/27
10 2/3	1 5/27	22 2/3	2 14/27	35	3 16/18	47	5 4/18

Tables for Angular Indexing: The table, "Angular Indexing," gives the number of turns of the index crank for indexing various angles. In the column headed, "Turns of Index Crank," the whole number (where given) indicates the number of full revolutions; the numerator of the fraction, the number of holes additional; and the denominator, the number of holes in the index circle to be used. The angular movement obtained for a movement of one hole, in various index plates is given in the table, "Angular Values of One-Hole Moves."

Approximate Indexing for Angles: The following general rule for *approximate indexing* of small angles is applicable to any index head requiring 40 revolutions of the index crank for one revolution of the work.

Rule: Divide 540 by the total number of minutes to be indexed. If the quotient is approximately equal to the number of holes in any index circle available, the angular movement is obtained by moving the crank one hole in this index circle; but if the quotient is not approximately equal, multiply it by any trial number which will give a product equal to the number of holes in an available index circle and move the index crank as many holes as are indicated by the trial number. (If the quotient of 540 divided by the total number of minutes is greater than the number of holes in any of the index circles, it is not possible to obtain the required movement for the angle by simple indexing.)

Example: Assume that it is required to index to an angle of 2 degrees 46 minutes. Changing this to minutes gives a total of 166 minutes. Dividing 540 by 166 we have $540 \div 166 = 3.253$. This quotient is next multiplied by some trial number to obtain a product which equals the number of holes in an available index circle. Multiplying by 12, we have $3.253 \times 12 = 39.036$. Therefore, for indexing 2 degrees 46 minutes, the 39-hole circle can be used and the index crank would be moved 12 holes.

Angular Values of One-Hole Moves – B.& S. Index Plates

15-hole circle =	36	minutes	29-hole circle =	18.621	minutes
16-hole circle =	33.750	minutes	31-hole circle =	17.419	minutes
17-hole circle =	31.765	minutes	33-hole circle =	16.364	minutes
18-hole circle =	30	minutes	37-hole circle =	14.595	minutes
19-hole circle =	28.421	minutes	39-hole circle =	13.846	minutes
20-hole circle =	27	minutes	41-hole circle =	13.171	minutes
21-hole circle =	25.714	minutes	43-hole circle =	12.558	minutes
23-hole circle =	23.478	minutes	47-hole circle =	11.489	minutes
27-hole circle =	20	minutes	49-hole circle =	11.020	minutes

CUTTING SPEEDS AND FEEDS

Work Materials.—The large number of work materials that are commonly machined vary greatly in their basic structure and the ease with which they can be machined. Yet it is possible to group together certain materials having similar machining characteristics, for the purpose of recommending the cutting speed at which they can be cut. Most materials that are machined are metals and it has been found that the most important single factor influencing the ease with which a metal can be cut is its microstructure, followed by any cold work that may have been done to the metal, which increases its hardness. Metals that have a similar, but not necessarily the same microstructure, will tend to have similar machining characteristics. Thus, the grouping of the metals in the accompanying tables has been done on the basis of their microstructure.

With the exception of a few soft and gummy metals, experience has shown that harder metals are more difficult to cut than softer metals. Furthermore, any given metal is more difficult to cut when it is in a harder form than when it is softer. It is more difficult to penetrate the harder metal and more power is required to cut it. These factors in turn will generate a higher cutting temperature at any given cutting speed, thereby making it necessary to use a slower speed, for the cutting temperature must always be kept within the limits that can be sustained by the cutting tool without failure. Hardness, then, is an important property that must be considered when machining a given metal. Hardness alone, however, cannot be used as a measure of cutting speed. For example, if pieces of AISI 11L17 and AISI 1117 steel both have a hardness of 150 Bhn, their recommended cutting speeds for high-speed steel tools will be 140 fpm and 130 fpm, respectively. In some metals, two entirely different microstructures can produce the same hardness. As an example, a fine pearlite microstructure and a tempered martensite microstructure can result in the same hardness in a steel. These microstructures will not machine alike. For practical purposes, however, information on hardness is usually easier to obtain than information on microstructure; thus, hardness alone is usually used to differentiate between different cutting speeds for machining a metal. In some situations, the hardness of a metal to be machined is not known. When the hardness is not known, the material condition can be used as a guide.

The surface of ferrous metal castings has a scale that is more difficult to machine than the metal below. Some scale is more difficult to machine than others, depending on the foundry sand used, the casting process, the method of cleaning the casting, and the type of metal cast. Special electrochemical treatments sometimes can be used that almost entirely eliminate the effect of the scale on machining, although castings so treated are not frequently encountered. Usually, when casting scale is encountered, the cutting speed is reduced approximately 5 or 10 per cent. Difficult-to-machine surface scale can also be encountered when machining hot-rolled or forged steel bars.

Metallurgical differences that affect machining characteristics are often found within a single piece of metal. The occurrence of hard spots in castings is an example. Different microstructures and hardness levels may occur within a casting as a result of variations in the cooling rate in different parts of the casting. Such variations are less severe in castings that have been heat treated. Steel bar stock is usually harder toward the outside than toward the center of the bar. Sometimes there are slight metallurgical differences along the length of a bar that can affect its cutting characteristics.

Cutting Speed, Feed, Depth of Cut: The cutting conditions that determine the rate of metal removal are the cutting speed, the feed rate, and the depth of cut. These cutting conditions and the nature of the material to be cut determine the power required to take the cut. The cutting conditions must be adjusted to stay within the power available on the machine tool to be used.

The cutting conditions must also be considered in relation to the tool life. Tool life can be defined as the length of time that a cutting tool will cut before becoming dull or before it must be replaced. The end of tool life is defined as a given amount of wear on the flank of the tool by the ANSI Standard Specification For Tool Life Testing With Single-Point Tools-ANSI B94.34-1946, R1971 and B94.36-1956,R1971. These standards are followed when making scientific machinability tests with single-point cutting tools in order to achieve uniformity in testing procedure so that results from different machinability laboratories can readily be compared. It is not practicable or necessary to follow this standard in the shop; however, it should be understood that the cutting conditions and tool life are related. For example, the cutting speed and the feed may be increased if a shorter tool life is accepted. Furthermore, the decrease in the tool life will be proportionately greater than the increase in the cutting speed or the feed. Conversely, if the cutting speed or the feed is decreased, the increase in the tool life will be proportionately greater than the decrease in the cutting speed or the feed.

Tool life is influenced most by cutting speed, then by the feed rate, and least by the depth of cut. After the depth of cut is about 10 times greater than the feed rate, a further increase in the depth of cut will have no significant effect on the tool life. This characteristic of the performance of cutting tools is very important in determining the operating or cutting conditions for machining metals.

The first step in selecting the cutting conditions is to select the depth of cut. The depth of cut will be limited by the amount of metal that is to be machined from the workpiece, by the power available on the machine tool, by the rigidity of the workpiece and the cutting tool, and by the rigidity of the setup. Since the depth of cut has the least effect upon the tool life, always use the heaviest depth of cut that is possible.

The second step is to select the feed rate. In selecting the feed rate, consideration must be given to the power available on the machine tool, to the rigidity of the workpiece and the cutting tool, to the rigidity of the setup, and to the surface finish required on the finished workpiece. The available power must be considered in relation to the depth of cut previously selected in considering the feed. Select the maximum feed possible; however, it must not be greater than that which will produce an acceptable surface finish.

The third step is to select the cutting speed. The accompanying tables provide recommended cutting speeds. If previous experience has been had in machining a certain material, this may form the basis for selecting the cutting speed. In either case, however, the depth of cut should be selected first, followed by the feed, and the last to be selected should be the cutting speed.

Use of Cutting Speed Tables.—On the following pages tables of recommended cutting speeds are provided. The values in these tables are for average conditions and serve as a basis from which to start. In many cases they will be found to be the most satisfactory cutting speeds, while in others, modifications may offer advantages as experience is gained on a particular job. It is not possible to specify a single optimum cutting speed for a material that will fit every situation. Many factors unique to each job may make a modification desirable. These factors include: the size, type, model, and make of the machine tool; its power, rigidity, and the foundation on which it is standing; the workpiece configuration; the rigidity of the workpiece setup, or fix-turing; safety aspects of the setup; the particular grade of cemented carbide used; the influence of a cutting fluid; and, the tool life desired. Except for certain difficult-to-machine materials, most materials can be cut successfully over a rather wide range of cutting speeds with a particular type of cutting tool material; however, not just any speed within this range should be used. A cutting speed that is too slow will result in a loss of production and in an increase in the cost of the part. Likewise, a cutting speed that is too fast can have the same result because the tool life will be too short and the production must frequently be interrupted to change tools. Moreover, the cost of sharpening or replacing the cutting tool must be considered. There is usually a narrow-range of cutting speeds within which the most economical results will be obtained. When making a modification to the cutting speed the fundamental behavior of cutting tools must be kept in mind; i.e., increasing the cutting speed will result in a proportionately larger reduction in the tool life; reducing the cutting speed will result in a proportionately larger increase in the tool life.

Feed and Depth of Cut Factors: Factors used to modify the cutting speeds in compensation for different feed rates and depths of cut are given in Table 1. These factors should be used only with the cutting speeds listed in Tables 2 through Table 5 for turning. Moreover, they do not apply when the hardness of a material exceeds approximately 350 to 400 HB. They should, however, be used in the other hardness ranges, at which most materials are machined. Experience in production and scientific investigations conducted in machinability laboratories have shown that a change in the feed or in the depth of cut will require a compensating change in the cutting speed if the tool life is to remain unchanged. For this reason a modification in the cutting speed should be made if the feed or depth of cut is changed significantly when machining the materials listed in the other tables.

Cutting Tool Material: The cutting speeds in the tables listed under cemented carbide are based on the assumption that a correct grade of uncoated straight tungsten carbide or an uncoated crater resisting carbide is used. Most carbide producers will be able to recommend a grade of carbide that can be used at the cutting speed given. An incorrect grade of carbide may, however, require a modification to the cutting speed, and in some instances it may result in an extremely short tool life. High speed steels are less sensitive in this respect; i.e., any type of high speed steel can cut most of the materials listed in the tables at the recommended cutting speed for high speed steel. It is true, however, that certain types of high speed steels do offer advantages in some applications, such as a longer tool life. Very hard and tough materials, such as the superalloys, should be cut with T15 high speed steel, or with one of the M40 types, as for example M42. These premium high speed steels should not be used to cut most other materials, since they can be cut as well by other types of high speed steel.

Coated Carbides: A faster cutting speed can be used with coated carbide cutting tools than with uncoated tools when machining materials that can be cut by the coated grades. When using titanium carbide or titanium nitride coated cutting tools, the recommended cutting speed in the tables should be increased by 20 to 30 per cent. in some cases an increase up to 50 per cent is possible. The cutting speed should be increased even more when using aluminum oxide coated carbides.

Titanium Carbides: The cutting speed should be increased by 30 to 60 per cent over the values given in the tables when using titanium carbide cutting tools.

Milling: The recommended cutting speed for milling given in the tables can be used for all face milling operations, slab milling operations, slab milling type cuts taken with end milling cutters, and for shallow slotting cuts taken with either end milling cutters or with side milling cutters. When milling deep slots with end milling cutters or with side milling cutters, the cutting speed should be reduced approximately 10 per cent. Likewise, the cutting speed should be reduced about 10 per cent when taking wide side milling cuts with side milling cutters; no reduction in the cutting speed is required if the width of the side milling cut is small.

Planing and Shaping: The cutting speeds in Tables 2 through 8, and Table 12 can be used for planing and shaping. The feed and depth of cut factors in Table 1 should also be used as explained previously. Very often other factors relating to the machine or the setup will require a reduction in the actual cutting speed used on the job.

Surface Scale: Certain heavy and abrasive surface scales encountered on castings and on some wrought metals require a reduction in the cutting speed, especially when drilling.

Cutting Fluids: Many cutting fluids permit a somewhat higher cutting speed to be used. It is not possible, however, to provide specific recommendations because each proprietary cutting fluid exhibits its own characteristics.

Cutting Speed, Feed, Depth of Cut, Tool Wear, and Tool Life.—The cutting conditions that determine the rate of metal removal are the cutting speed, the feed rate, and the depth of cut. These cutting conditions and the nature of the material to be cut determine the power required to take the cut. The cutting conditions must be adjusted to stay within the power available on the machine tool to be used. Power requirements are discussed in Estimating Machining Power later in this section.

The cutting conditions must also be considered in relation to the tool life. Tool life is defined as the cutting time to reach a predetermined amount of wear, usually flank wear. Tool life is determined by assessing the time—the tool life—at which a given predetermined flank wear is reached (0.01, 0.015, 0.025, 0.03 inch, for example). This amount of wear is called the tool wear criterion, and its size depends on the tool grade used. Usually, a tougher grade can be used with a bigger flank wear, but for finishing operations, where close tolerances are required, the wear criterion is relatively small. Other wear criteria are a predetermined value of the machined surface roughness and the depth of the crater that develops on the rake face of the tool.

The ANSI standard, Specification For Tool Life Testing With Single-Point Tools (ANSI B94.55M-1985), defines the end of tool life as a given amount of wear on the flank of a tool. This standard is followed when making scientific machinability tests with single-point cutting tools in order to achieve uniformity in testing procedures so that results from different machinability laboratories can be readily compared. It is not practicable or necessary to follow this standard in the shop; however, it should be understood that the cutting conditions and tool life are related.

Tool life is influenced most by cutting speed, then by the feed rate, and least by the depth of cut. When the depth of cut is increased to about 10 times greater than the feed, a further increase in the depth of cut will have no significant effect on the tool life. This characteristic of the cutting tool performance is very important in determining the operating or cutting conditions for machining metals. Conversely, if the cutting speed or the feed is decreased, the increase in the tool life will be proportionately greater than the decrease in the cutting speed or the feed.

Tool life is reduced when either feed or cutting speed is increased. For example, the cutting speed and the feed may be increased if a shorter tool life is accepted; furthermore, the reduction in the tool life will be proportionately greater than the increase in the cutting speed or the feed. However, it is less well understood that a higher feed rate (feed/rev × speed) may result in a longer tool life if a higher feed/rev is used in combination with a lower cutting speed. This principle is well illustrated in the speed tables of this section, where two sets of feed and speed data are given (labeled *optimum* and *average*) that result in the same tool life. The *optimum* set results in a greater feed rate (i.e., increased productivity) although the feed/rev is higher and cutting speed lower than the *average* set. Complete instructions for using the speed tables and for estimating tool life are given in *Use of Cutting Speed Tables* starting on page 352.

Selecting Cutting Conditions.—The first step in establishing the cutting conditions is to select the depth of cut. The depth of cut will be limited by the amount of metal that is to be machined from the workpiece, by the power available on the machine tool, by the rigidity of the workpiece and the cutting tool, and by the rigidity of the setup. The depth of cut has the least effect upon the tool life, so the heaviest possible depth of cut should always be used.

The second step is to select the feed (feed/rev for turning, drilling, and reaming, or feed/tooth for milling). The available power must be sufficient to make the required depth of cut at the selected feed. The maximum feed possible that will produce an acceptable surface finish should be selected.

The third step is to select the cutting speed. Although the accompanying tables provide recommended cutting speeds and feeds for many materials, experience in machining a certain material may form the best basis for adjusting the given cutting speeds to a particular job. However, in general, the depth of cut should be selected first, followed by the feed, and last the cutting speed.

Principal Feeds and Speeds Tables for Turning, Milling, Drilling, Reaming, and Threading

Table 1	*Cutting-Speed Adjustment Factors for Turning with HSS Tools*
Table 2	*Recommended Cutting Speeds in Feet per Minute for Turning, Milling, Drilling and Reaming Plain Carbon and Alloy Steels*
Table 3	*Recommended Cutting Speeds in Feet per Minute for Turning, Milling, Drilling and Reaming Ferrous Cast Metals*
Table 4	*Recommended Cutting Speeds in Feet per Minute for Turning, Milling, Drilling and Reaming Stainless Steels*
Table 5	*Recommended Cutting Speeds in Feet per Minute for Turning, Milling, Drilling and Reaming Tool Steels*
Table 6	*Recommended Cutting Speeds in Feet per Minute for Turning, Milling, Drilling and Reaming Light Metals*
Table 7	*Recommended Cutting Speeds in Feet per Minute for Turning and Drilling Titanium and Titanium Alloys*
Table 8	*Recommended Cutting Speeds in Feet per Minute for Turning, Milling and Drilling Superalloys*
Table 9	*Recommended Feed in Inches per Tooth (ft) for Milling with High Speed Steel Cutters*
Table 10	*Recommended Feed in Inch per Tooth (ft) for Milling with Cemented Carbide Cutters*
Table 11	*Representative Cutting Conditions for Rough Turning Hard-to-Machine Materials with Single Point Cubic Boron Nitride (CBN) Cutting Tools*
Table 12	*Recommended Cutting Speeds in Feet per Minute for Turning and Milling Copper Alloys*

Metric Units: In metric practice, the units used for the cutting speed is meters per minute, abbreviated m/min. or rpm. The cutting speeds in the tables are given in inch units, or feet per minute, which is abbreviated fpm. These units can be converted as follows: to obtain meters per minute, multiply feet per minute by 0.3; to obtain feet per minute, multiply meters per minute by 3.28.

Cutting Speed Formulas.—Most machining operations are conducted on machine tools having a rotating spindle. Cutting speeds are usually given in feet or meters per minute and these speeds must be converted to spindle speeds, in revolutions per minute, to operate the machine. Conversion is accomplished by use of the following formulas:

<u>For U.S. units:</u>

$$N = \frac{12V}{\pi D} = 3.82\frac{V}{D} \text{ rpm}$$

<u>For metric units:</u>

$$N = \frac{1000V}{\pi D} = 318.3\frac{V}{D} \text{ rpm}$$

where N is the spindle speed in revolutions per minute (rpm); V is the cutting speed in feet per minute (fpm) for U.S. units and meters per minute (m/min.) for metric units. In turning, D is the diameter of the workpiece; in milling, drilling, reaming, and other operations that use a rotating tool, D is the cutter diameter in inches for U.S. units and in millimeters for metric units. $\pi = 3.1417$.

The feed and depth-of-cut factors in Table 1 are only to be used together with Table 2 through 5. While the values in Table 1 are for inch units, they can be used with metric units by converting the metric feed and depth of cut into inch units. First select the cutting speed from the appropriate table and then apply the factors from Tables, using the formula given below:

$$V = V_0 F_f F_d$$

V = Cutting speed to be used; fpm, or m/min.
V_0 = Cutting speed from tables; fpm, or m/min.
F_f = Feed Factor
F_d = Depth-of-cut factor

Example: Using both the inch and the metric formulas, calculate the spindle speed for turning a 1¼ inch (31.75 mm) bar of 200-220 HB AISI 1040 steel using depth of cut of 0.100 in. (2.54 mm) and a feed rate of 0.015 in. (0.38 mm/rev.). (A small insignificant difference in the answers may occur, which is caused by the conversion of the units and the rounding of numbers.)

The cutting speed for turning a 4-inch (102-mm) diameter bar has been found to be 575 fpm (175.3 m/min.). Using both the inch and metric formulas, calculate the lathe spindle speed.

From Table 2: $V_0 = 85$ fpm.

From Table 1: $F_f = 0.91$; $F_d = 1.03$

$$V = V_0 F_f F_d = 85 \times 0.91 \times 1.03 = 80 \text{ fpm}$$

$$N = \frac{12V}{\pi D} = \frac{12 \times 80}{\pi \times 1.250} = 244 \text{ rpm}$$

$$V_o = 85 \times 0.3 = 25.5 \text{ m/min}$$

$$V = V_0 F_f F_d = 25.5 \times 0.91 \times 1.03 = 24 \text{ fpm}$$

$$N = \frac{1000V}{\pi D} = \frac{1000 \times 24}{\pi \times 31.75} = 241 \text{ rpm}$$

It is often necessary to calculate the cutting speed in feet per minute or in meters per minute, when the diameter of the workpiece or of the cutting tool and the spindle speed is known. In this event, the following formulas are used.

For U.S. units:

$$V = \frac{\pi DN}{12} \text{ fpm}$$

For metric units:

$$V = \frac{\pi DN}{1000} \text{ m/min}$$

As in the previous formulas, N is the rpm and D is the diameter in inches for the U.S. unit formula and in millimeters for the metric formula.

Example: Calculate the cutting speed in feet per minute and in meters per minute if the spindle speed of a ¾-inch (19.05-mm) drill is 400 rpm.

$$V = \frac{\pi DN}{12} = \frac{\pi \times 0.75 \times 400}{12} = 78.5 \text{ fpm}$$

$$V = \frac{\pi DN}{1000} = \frac{\pi \times 19.05 \times 400}{1000} = 24.9 \text{ m/min}$$

Cutting Time for Turning, Boring, and Facing.—The time required to turn a length of metal can be determined by the following formula in which T = time in minutes, L = length of cut in inches, f = feed in inches per revolution, and N = lathe spindle speed in revolutions per minute.

$$T = \frac{L}{fN}$$

When making job estimates, the time required to load and to unload the workpiece on the machine, and the machine handling time must be added to the cutting time for each length cut to obtain the floor-to-floor time.

Table 1. Cutting-Speed Adjustment Factors for Turning with HSS Tools

Feed		Feed Factor	Depth of Cut		Depth-of-Cut Factor
in.	mm	F_f	in.	mm	F_d
0.002	0.05	1.50	0.005	0.13	1.50
0.003	0.08	1.50	0.010	0.25	1.42
0.004	0.10	1.50	0.016	0.41	1.33
0.005	0.13	1.44	0.031	0.79	1.21
0.006	0.15	1.34	0.047	1.19	1.15
0.007	0.18	1.25	0.062	1.57	1.10
0.008	0.20	1.18	0.078	1.98	1.07
0.009	0.23	1.12	0.094	2.39	1.04
0.010	0.25	1.08	0.100	2.54	1.03
0.011	0.28	1.04	0.125	3.18	1.00
0.012	0.30	1.00	0.150	3.81	0.97
0.013	0.33	0.97	0.188	4.78	0.94
0.014	0.36	0.94	0.200	5.08	0.93
0.015	0.38	0.91	0.250	6.35	0.91
0.016	0.41	0.88	0.312	7.92	0.88
0.018	0.46	0.84	0.375	9.53	0.86
0.020	0.51	0.80	0.438	11.13	0.84
0.022	0.56	0.77	0.500	12.70	0.82
0.025	0.64	0.73	0.625	15.88	0.80
0.028	0.71	0.70	0.688	17.48	0.78
0.030	0.76	0.68	0.750	19.05	0.77
0.032	0.81	0.66	0.812	20.62	0.76
0.035	0.89	0.64	0.938	23.83	0.75
0.040	1.02	0.60	1.000	25.40	0.74
0.045	1.14	0.57	1.250	31.75	0.73
0.050	1.27	0.55	1.250	31.75	0.72
0.060	1.52	0.50	1.375	34.93	0.71

For use with HSS tool data only from Tables 2 through 8. Adjusted cutting speed $V = V_{HSS} \times F_f \times F_d$, where V_{HSS} is the tabular speed for turning with high-speed tools.

Cutting Speed for Tapping.—A table of cutting speeds for tapping is not given. Several factors, singly or in combination, can cause very great differences in the permissible tapping speed. The principal factors affecting the tapping speed are the pitch of the thread, the chamfer length on the tap, the percentage of full thread to be cut, the length of the hole to be tapped, the cutting fluid used, whether the threads are straight or tapered, the machine tool used to perform the operation, and the material to be tapped.

The cutting speed for coarse-pitch taps must be lower than for fine-pitch taps with the same diameter. Usually, the difference in pitch becomes more pronounced as the diameter of the tap becomes larger and slight differences in the pitch of smaller-diameter taps have little significant effect on the cutting speed. Unlike all other cutting tools, the feed per revolution of a tap cannot be independently adjusted—it is always equal to the lead of the thread and is always greater for coarse pitches than for fine pitches. Furthermore, the thread form of a coarse-pitch thread is larger than that of a fine-pitch thread; therefore, it is necessary to remove more metal when cutting a coarse-pitch thread.

Taps with a long chamfer, such as starting or tapper taps, can cut faster in a short hole than short chamfer taps, such as plug taps. In deep holes, however, short chamfer or plug taps can run faster than long chamfer taps. Bottoming taps must be run more slowly than either starting or plug taps. The chamfer helps to start the tap in the hole. It also functions to involve more threads, or thread form cutting edges, on the tap in cutting the thread in the hole, thus reducing the cutting load on any one set of thread form cutting edges. In so doing, more chips and thinner chips are produced that are difficult to remove from deeper holes. Shortening the chamfer length causes fewer thread form cutting edges to cut, thereby producing fewer and thicker chips that can easily be disposed of. Only one or two sets of thread form cutting edges are cut on bottoming taps, causing these cutting edges to assume a heavy cutting load and produce very thick chips.

Spiral-pointed taps can operate at a faster cutting speed than taps with normal flutes. These taps are made with supplementary angular flutes on the end that push the chips ahead of the tap and prevent the tapped hole from becoming clogged with chips. They are used primarily to tap open or through holes although some are made with shorter supplementary flutes for tapping blind holes.

The tapping speed must be reduced as the percentage of full thread to be cut is increased. Experiments have shown that the torque required to cut a 100 per cent thread form is more than twice that required to cut a 50 per cent thread form. An increase in the percentage of full thread will also produce a greater volume of chips.

The tapping speed must be lowered as the length of the hole to be tapped is increased. More friction must be overcome in turning the tap and more chips accumulate in the hole. It will be more difficult to apply the cutting fluid at the cutting edges and to lubricate the tap to reduce friction. This problem becomes greater when the hole is being tapped in a horizontal position.

Cutting fluids have a very great effect on the cutting speed for tapping. Although other operating conditions when tapping frequently cannot be changed, a free selection of the cutting fluid usually can be made. When planning the tapping operation, the selection of a cutting fluid warrants a very careful consideration and perhaps an investigation.

Taper threaded taps, such as pipe taps, must be operated at a slower speed than straight thread taps with a comparable diameter. All the thread form cutting edges of a taper threaded tap that are engaged in the work cut and produce a chip, but only those cutting edges along the chamfer length cut on straight thread taps. Pipe taps often are required to cut the tapered thread from a straight hole, adding to the cutting burden.

The machine tool used for the tapping operation must be considered in selecting the tapping speed. Tapping machines and other machines that are able to feed the tap at a rate of advance equal to the lead of the tap, and that have provisions for quickly reversing the spindle, can be operated at high cutting speeds. On machines where the feed of the tap is controlled manually—such as on drill presses and turret lathes—the tapping speed must be reduced to allow the operator to maintain safe control of the operation.

There are other special considerations in selecting the tapping speed. Very accurate threads are usually tapped more slowly than threads with a commercial grade of accuracy. Thread forms that require deep threads for which a large amount of metal must be removed, producing a large volume of chips, require special techniques and slower cutting speeds. Acme, buttress, and square threads, therefore, are generally cut at lower speeds.

Not the least important consideration is the material to be tapped. Like other metal cutting operations, tapping is affected by the basic structure of the metal being tapped and its hardness. Because of all of the factors mentioned above, it is not practical to tabulate cutting speeds for tapping as has been done for other operations. On a relative basis, the cutting speeds for tapping can be compared to the cutting speed for drilling with high speed steel drills. The speeds may be used as a starting point but normally must be reduced, and must not be applied as such without weighing every factor that affects the tapping operation.

Cutting Speed for Broaching.—Broaching offers many advantages in manufacturing metal parts, including high production rates, excellent surface finishes, and close dimensional tolerances. These advantages are not derived from the use of high cutting speeds;

they are derived from the large number of cutting teeth that can be applied consecutively in a given period of time, from their configuration and precise dimensions, and from the width or diameter of the surface that can be machined in a single stroke. Most broaching cutters are expensive in their initial cost and are expensive to sharpen. For these reasons, a long tool life is desirable, and to obtain a long tool life, relatively slow cutting speeds are used. In many instances, slower cutting speeds are used because of the limitations of the machine in accelerating and stopping heavy broaching cutters. At other times, the available power on the machine places a limit on the cutting speed that can be used; i.e., the cubic inches of metal removed per minute must be within the power capacity of the machine.

The cutting speeds for high-speed steel broaches range from 3 to 50 feet per minute, although faster speeds have been used. In general, the harder and more difficult to machine materials are cut at a slower cutting speed and those that are easier to machine are cut at a faster speed. Some typical recommendations for high-speed steel broaches are: AISI 1040, 10 to 30 fpm; AISI 1060, 10 to 25 fpm; AISI 4140, 10 to 25 fpm; AISI 41L40, 20 to 30 fpm; 201 austenitic stainless steel, 10 to 20 fpm; Class 20 gray cast iron, 20 to 30 fpm; Class 40 gray cast iron, 15 to 25 fpm; aluminum and magnesium alloys, 30 to 50 fpm; copper alloys, 20 to 30 fpm; commercially pure titanium, 20 to 25 fpm; alpha and beta titanium alloys, 5 fpm; and the superalloys, 3 to 10 fpm. Surface broaching operations on gray iron castings have been conducted at a cutting speed of 150 fpm, using indexable insert cemented carbide broaching cutters. In selecting the speed for broaching, the cardinal principle of the performance of all metal cutting tools should be kept in mind; i.e., increasing the cutting speed may result in a proportionately larger reduction in tool life, and reducing the cutting speed may result in a proportionately larger increase in the tool life. When broaching most materials, a suitable cutting fluid should be used to obtain a good surface finish and a better tool life. Gray cast iron can be broached without using a cutting fluid although some shops prefer to use a soluble oil.

Thread Cutting with Single Point Cutting Tools.—Whenever possible the cutting speed recommended for turning should be used to cut internal and external threads with single point thread cutting tools. This cutting speed can frequently be used on numerically controlled lathes, using either cemented carbide or high speed steel tools. There are occasions, however, when a slower cutting speed must be used as a result of the workpiece configuration, the setup, or when cutting certain difficult-to-machine threads, such as coarse pitch Acme threads. A slightly reduced cutting speed is sometimes used on numerically controlled lathes to obtain a longer tool life.

Thread cutting on an engine lathe is not necessarily a slow speed operation, although on these machines the operation is controlled manually. However, there must never be a compromise with safety; the operator must always be sure that he has control over the machine to the extent that he or others will not be injured, and that the machine or the workpiece will not be damaged. On some jobs a skilled operator can safely manipulate a lathe with such skill that a fast spindle speed can be used to cut the thread, in which case the cutting speed may be equal to that recommended for turning with high speed steel, or with cemented carbide in the case of the more difficult-to-machine materials. Other jobs require using a slower cutting speed, even when the thread cutting operation is performed by a highly skilled operator. Some of the reasons for cutting threads at a slower speed have been given in the previous paragraph. Other reasons involve the ability to safely manipulate the machine such as when cutting a thread close to a large shoulder, or when cutting an internal thread with the cutting tool feeding into the bore.

Cutting Speed for Thread Chasing.—Cutting threads with a self-opening die head is called thread chasing. The die head contains a set of thread chasers that cut the thread and feed the die head in a nut-and-screw-like action. The feed rate is determined entirely by the lead or pitch of the thread. Since the feed of the die head must not be too rapid, the thread lead and pitch place a limit on the spindle speed, and thereby on the cutting speed. Other factors affecting the cutting speed are the work material, the type and size of the thread, the thread tolerance, and the finish required. A cutting fluid should be used in most cases, which may also have an effect on the cutting speed. Much slower cutting speeds are recommended for thread chasing, as compared to turning. A cutting speed that is too fast will reduce the life of the thread chasers and may cause the threads cut to be rough or torn. Some typical cutting speeds recommended by one manufacturer of self-opening die heads are given below. These cutting speeds may have to be modified somewhat to suit existing conditions on each job.

Material	Threads per Inch			
	3-7½	8-15	16-24	25-Up
Cutting Speed (fpm) for Threads per Inch				
AISI 1010-1035 Steel	20	30	40	50
AISI 1112-1340 Steel	20	30	40	50
AISI 1040-1095 Steel	15	20	25	30
AISI 4130-4820 Steel	8	10	15	20
AISI 5120-52100 Steel	8	10	15	20
Stainless Steel	8	10	15	20
Gray Cast Iron	25	40	50	80
Aluminum Alloys	50	100	150	200
Brass Bar Stock	50	100	150	200
Phosphor Bronze	40	80	100	150
Zinc Die Castings	50	100	150	200

Feed Rate for Milling.—Whenever the power feed is to be used to perform a milling operation, the table feed rate, in inches per minute, should always be calculated in order to achieve the best results. The table feed rate governs the production rate may result in overloading the milling cutter which can have serious consequences. Aside from the possible breakage of equipment, overloading the cutter will cause the tool life of the cutter to decrease. The tool life can also be decreased if the feed rate is too slow, which, in addition, certainly leads to a loss of production.

Using the Feed and Speed Tables for Milling: The basic feed for milling cutters is the feed per tooth (*f*), which is expressed in inches per tooth. There are many factors to consider in selecting the feed per tooth and no formula is available to resolve these factors. Among the factors to consider are the cutting tool material; the work material and its hardness; the width and the depth of the cut to be taken; the type of milling cutter to be used and its size; the surface finish to be produced; the power available on the milling machine; and the rigidity of the milling machine, the workpiece, the workpiece setup, the milling cutter, and the cutter mounting.

The cardinal principle is to always use the maximum feed that conditions will permit. Avoid, if possible, using a feed that is less than 0.001 inch per tooth because such low feeds reduce the tool life of the cutter. When milling hard materials with small-diameter end mills, such small feeds may be necessary, but otherwise use as much feed as possible. Harder materials in general will require lower feeds than softer materials. The width and the depth of cut also affect the feeds. Wider and deeper cuts must be fed somewhat more slowly than narrow and shallow cuts. A slower feed rate will result in a better surface finish; however, always use the heaviest feed that will produce the surface finish desired. Fine chips produced by fine feeds are dangerous when milling magnesium because spontaneous combustion can occur. Thus, when milling magnesium, a fast feed that will produce a relatively thick chip should be used. Cutting stainless steel produces a work-hardened layer on the surface that has been cut. Thus, when milling this material, the feed should be large enough to allow each cutting edge on the cutter to penetrate below the work-hardened layer produced by the previous cutting edge. The heavy feeds recommended for face milling cutters are to be used primarily with larger cutters on milling machines having an adequate amount of power. For smaller face milling cutters, start with smaller feeds and increase as indicated by the performance of the cutter and the machine.

When planning a milling operation that requires a high cutting speed and a fast feed, always check to determine if the power required to take the cut is within the capacity of the milling machine. Excessive power requirements are often encountered when milling with cemented carbide cutters. The large metal removal rates that can be attained require a high horsepower output. An example of this type of calculation is given in the section on Machining Power that follows this section. If the size of the cut must be reduced in order to stay within the power capacity of the machine, start by reducing the cutting speed rather than the feed in inches per tooth.

The formula for calculating the table feed rate, when the feed in inches per tooth is known, is as follows:

$$f_m = f_t n_t N$$

where f_m = milling machine table feed rate in inches per minute (ipm)

f_t = feed in inch per tooth (ipt)

n_t = number of teeth in the milling cutter

N = spindle speed of the milling machine in revolutions per minute (rpm)

Example: Calculate the feed rate for milling a piece of AISI 1040 steel having a hardness of 180 Bhn. The cutter is a 3-inch diameter high-speed steel plain or slab milling cutter with 8 teeth. The width of the cut is 2 inches, the depth of cut is 0.062 inch, and the cutting speed from Table 3 is 85 fpm. From Table 10, the feed rate selected is 0.008 inch per tooth.

$$N = \frac{12V}{\pi D} = \frac{12 \times 85}{3.14 \times 3} = 108 \text{ rpm}$$

$$f_m = f_t n_t N = 0.008 \times 8 \times 108$$

$$= 7 \text{ ipm (approximately)}$$

Feed Rates for Drilling.—The feed rate for drilling is governed primarily by the size of the drill and by the material to be drilled. Other factors that also affect the feed rate that can be used are the workpiece configuration, the rigidity of the machine tool and the workpiece setup, and the length of the chisel edge. A chisel edge that is too long will result in a very significant increase in the thrust force, which may cause large deflections to occur on the machine tool and drill breakage. For ordinary twist drills the feed rate used is .001 to .003 in./rev. for drills smaller than ⅛ in.; .002 to .006 in./rev. for ⅛ to ¼ in. drills; .004 to .010 in./rev. for ¼ to ½ in. drills; .007 to .015 in./rev. for ½ to 1 in. drills; and, .010 to .025 in./rev. for drills larger than 1 inch. The lower values in the feed ranges should be used for hard materials such as tool steels, superalloys, and work hardening stainless steels; the higher values in the feed ranges should be used to drill soft materials such as aluminum and brass.

Drilling Difficulties: A drill split at the web is evidence of too much feed or insufficient lip clearance at the center due to improper grinding. Rapid wearing away of the extreme outer corners of the cutting edges indicates that the speed is too high. A drill chipping or breaking out at the cutting edges indicates that either the feed is too heavy or the drill has been ground with too much lip clearance. Nothing will "check" a high-speed steel drill quicker than to turn a stream of cold water on it after it has been heated while in use. It is equally bad to plunge it in cold water after the point has been heated in grinding. The small checks or cracks resulting from this practice will eventually chip out and cause rapid wear or breakage. Insufficient speed in drilling small holes with hand feed greatly increases the risk of breakage, especially at the moment the drill is breaking through the farther side of the work, due to the operator's inability to gage the feed when the drill is running too slowly.

Small drills have heavier webs and smaller flutes in proportion to their size than do larger drills, so breakage due to clogging of chips in the flutes is more likely to occur. When drilling holes deeper than three times the diameter of the drill, it is advisable to withdraw the drill (peck feed) at intervals to remove the chips and permit coolant to reach the tip of the drill.

Drilling Holes in Glass: The simplest method of drilling holes in glass is to use a standard, tungsten-carbide-tipped masonry drill of the appropriate diameter, in a gun-drill. The edges of the carbide in contact with the glass should be sharp. Kerosene or other liquid may be used as a lubricant, and a light force is maintained on the drill until just before the point breaks through. The hole should then be started from the other side if possible, or a very light force applied for the remainder of the operation, to prevent excessive breaking of material from the sides of the hole. As the hard particles of glass are abraded, they accumulate and act to abrade the hole, so it may be advisable to use a slightly smaller drill than the required diameter of the finished hole.

Alternatively, for holes of medium and large size, use brass or copper tubing, having an outside diameter equal to the size of hole required. Revolve the tube at a peripheral speed of about 100 feet per minute, and use carborundum (80 to 100 grit) and light machine oil between the end of the pipe and the glass. Insert the abrasive under the drill with a thin piece of soft wood, to avoid scratching the glass. The glass should be supported by a felt or rubber cushion, not much larger than the hole to be drilled. If practicable, it is advisable to drill about halfway through, then turn the glass over, and drill down to meet the first cut. Any fin that may be left in the hole can be removed with a round second-cut file wetted with turpentine.

Smaller-diameter holes may also be drilled with triangular-shaped cemented carbide drills that can be purchased in standard sizes. The end of the drill is shaped into a long tapering triangular point. The other end of the cemented carbide bit is brazed onto a steel shank. A glass drill can be made to the same shape from hardened drill rod or an old three-cornered file. The location at which the hole is to be drilled is marked on the workpiece. A dam of putty or glazing compound is built up on the work surface to contain the cutting fluid, which can be either kerosene or turpentine mixed with camphor. Chipping on the back edge of the hole can be prevented by placing a scrap plate of glass behind the area to be drilled and drilling into the backup glass. This procedure also provides additional support to the workpiece and is essential for drilling very thin plates. The hole is usually drilled with an electric hand drill. When the hole is being produced, the drill should be given a small circular motion using the point as a fulcrum, thereby providing a clearance for the drill in the hole.

Very small round or intricately shaped holes and narrow slots can be cut in glass by the ultrasonic machining process or by the abrasive jet cutting process.

Table 2. Recommended Cutting Speeds in Feet per Minute for Turning, Milling, Drilling and Reaming Plain Carbon and Alloy Steels

Material AISI and SAE Steels	Hardness HB[a]	Material Condition	Cutting Speed, fpm HSS			
			Turning	Milling	Drilling	Reaming
Free Machining Plain Carbon Steels (Resulfurized)						
1212, 1213, 1215	100-150	HR, A	150	140	120	80
	150-200	CD	160	130	125	80
1108, 1109, 1115, 1117, 1118, 1120, 1126, 1211	100-150	HR, A	130	130	110	75
	150-200	CD	120	115	120	80
1132, 1137, 1139, 1140, 1144, 1146, 1151	175-225	HR, A, N, CD	120	115	100	65
	275-325	Q and T	75	70	70	45
	325-375	Q and T	50	45	45	30
	375-425	Q and T	40	35	35	20
Free Machining Plain Carbon Steels (Leaded)						
11L17, 11L18, 12L13, 12L14	100-150	HR, A, N, CD	140	140	130	85
	150-200	HR, A, N, CD	145	130	120	80
	200-250	N, CD	110	110	90	60
Plain Carbon Steels						
1006, 1008, 1009, 1010, 1012, 1015, 1016, 1017, 1018, 1019, 1020, 1021, 1022, 1023, 1024, 1025, 1026, 1513, 1514	100-125	HR, A, N, CD	120	110	100	65
	125-175	HR, A, N, CD	110	110	90	60
	175-225	HR, N, CD	90	90	70	45
	225-275	CD	70	65	60	40
1027, 1030, 1033, 1035, 1036, 1037, 1038, 1039, 1040, 1041, 1042, 1043, 1045, 1046, 1048, 1049, 1050, 1052, 1152, 1524, 1526, 1527, 1541	125-175	HR, A, N, CD	100	100	90	60
	175-225	HR, A, N, CD	85	85	75	50
	225-275	N, CD, Q and T	70	70	60	40
	275-325	Q and T	60	55	50	30
	325-375	Q and T	40	35	35	20
	375-425	Q and T	30	25	25	15
1055, 1060, 1064, 1065, 1070, 1074, 1078, 1080, 1084, 1086, 1090, 1095, 1548, 1551, 1552, 1561, 1566	125-175	HR, A, N, CD	100	90	85	55
	175-225	HR, A, N, CD	80	75	70	45
	225-275	N, CD, Q and T	65	60	50	30
	275-325	Q and T	50	45	40	25
	325-375	Q and T	35	30	30	20
	375-425	Q and T	30	15	15	10
Free Machining Alloy Steels (Resulfurized)						
4140, 4150	175-200	HR, A, N, CD	110	100	90	60
	200-250	HR, N, CD	90	90	80	50
	250-300	Q and T	65	60	55	30
	300-375	Q and T	50	45	40	25
	375-425	Q and T	40	35	30	15
Free Machining Alloy Steels (Leaded)						
41L30, 41L40, 41L47, 41L50, 43L47, 51L32, 52L100, 86L20, 86L40	150-200	HR, A, N, CD	120	115	100	65
	200-250	HR, N, CD	100	95	90	60
	250-300	Q and T	75	70	65	40
	300-375	Q and T	55	50	45	30
	375-425	Q and T	50	40	30	15

Table 2. *(Continued)* **Recommended Cutting Speeds in Feet per Minute for Turning, Milling, Drilling and Reaming Plain Carbon and Alloy Steels**

Material AISI and SAE Steels	Hardness HB[a]	Material Condition	Cutting Speed, fpm HSS			
			Turning	Milling	Drilling	Reaming
Alloy Steels						
4012, 4023, 4024, 4028, 4118, 4320, 4419, 4422, 4427, 4615, 4620, 4621, 4626, 4718, 4720, 4815, 4817, 4820, 5015, 5117, 5120, 6118, 8115, 8615, 8617, 8620, 8622, 8625, 8627, 8720, 8822, 94B17	125-175	HR, A, N, CD	100	100	85	55
	175-225	HR, A, N, CD	90	90	70	45
	225-275	CD, N, Q and T	70	60	55	35
	275-325	Q and T	60	50	50	30
	325-375	Q and T	50	40	35	25
	375-425	Q and T	35	25	25	15
1330, 1335, 1340, 1345, 4032, 4037, 4042, 4047, 4130, 4135, 4137, 4140, 4142, 4145, 4147, 4150, 4161, 4337, 4340, 50B44, 50B46, 50B50, 50B60, 5130, 5132, 5140, 5145, 5147, 5150, 5160, 51B60, 6150, 81B45, 8630, 8635, 8637, 8640, 8642, 8645, 8650, 8655, 8660, 8740, 9254, 9255, 9260, 9262, 94B30	175-225	HR, A, N, CD	85	75	75	50
	225-275	N, CD, Q and T	70	60	60	40
	275-325	N, Q and T	60	50	45	30
	325-375	N, Q and T	40	35	30	15
	375-425	Q and T	30	20	20	15
E51100, E52100	175-225	HR, A, CD	70	65	60	40
	225-275	N, CD, Q and T	65	60	50	30
	275-325	N, Q and T	50	40	35	25
	325-375	N, Q and T	30	30	30	20
	375-425	Q and T	20	20	20	10
Ultra High Strength Steels (Not AISI)						
AMS 6421 (98B37 Mod.), AMS 6422 (98BV40), AMS 6424, AMS 6427, AMS 6428, AMS 6430, AMS 6432, AMS 6433, AMS 6434, AMS 6436, AMS 6442, 300M, D6ac	220-300	A	65	60	50	30
	300-350	N	50	45	35	20
	350-400	N	35	20	20	10
	43-48 HRC	Q and T	25	…	…	…
	48-52 HRC	Q and T	10	…	…	…
Maraging Steels (Not AISI)						
18% Ni Grade 200, 18% Ni Grade 250, 18% Ni Grade 300, 18% Ni Grade 350	250-325	A	60	50	50	30
	50-52 HRC	Maraged	10	…	…	…
Nitriding Steels (Not AISI)						
Nitralloy 125, Nitralloy 135, Nitralloy 135 Mod., Nitralloy 225, Nitralloy 230, Nitralloy N, Nitralloy EZ, Nitrex I	200-250	A	70	60	60	40
	300-350	N, Q and T	30	25	35	20

[a] Abbreviations designate: HR, hot rolled; CD, cold drawn; A, annealed; N, normalized; Q and T, quenched and tempered; and HB, Brinell hardness number.

Speeds for turning based on a feed rate of 0.012 inch per revolution and a depth of cut of 0.125 inch.

Table 3. Recommended Cutting Speeds in Feet per Minute for Turning, Milling, Drilling and Reaming Ferrous Cast Metals

Material	Hardness HB[a]	Material Condition	Cutting Speed, fpm HSS			
			Turning	Milling	Drilling	Reaming
Gray Cast Iron						
ASTM Class 20	120-150	A	120	100	100	65
ASTM Class 25	160-200	AC	90	80	90	60
ASTM Class 30, 35, and 40	190-220	AC	80	70	80	55
ASTM Class 45 and 50	220-260	AC	60	50	60	40
ASTM Class 55 and 60	250-320	AC, HT	35	30	30	20
ASTM Type 1, 1b, 5 (Ni Resist)	100-215	AC	70	50	50	30
ASTM Type 2, 3, 6 (Ni Resist)	120-175	AC	65	40	40	25
ASTM Type 2b, 4 (Ni Resist)	150-250	AC	50	30	30	20
Malleable Iron						
(Ferritic), 32510, 35018	110-160	MHT	130	110	110	75
(Pearlitic), 40010, 43010, 45006, 45008, 48005, 50005	160-200	MHT	95	80	80	55
	200-240	MHT	75	65	70	45
(Martensitic), 53004, 60003, 60004	200-255	MHT	70	55	55	35
(Martensitic), 70002, 70003	220-260	MHT	60	50	50	30
(Martensitic), 80002	240-280	MHT	50	45	45	30
(Martensitic), 90001	250-320	MHT	30	25	25	15
Nodular (Ductile) Iron						
(Ferritic), 60-40-18, 65-45-12	140-190	A	100	75	100	65
(Ferritic-Pearlitic), 80-55-06	190-225	AC	80	60	70	45
	225-260	AC	65	50	50	30
(Pearlitic-Martensitic), 100-70-03	240-300	HT	45	40	40	25
(Martensitic), 120-90-02	270-330	HT	30	25	25	15
	330-400	HT	15	–	10	5
Cast Steels						
(Low Carbon), 1010, 1020	100-150	AC, A, N	110	100	100	65
(Medium Carbon), 1030, 1040, 1050	125-175	AC, A, N	100	95	90	60
	175-225	AC, A, N	90	80	70	45
	225-300	AC, HT	70	60	55	35
(Low Carbon Alloy), 1320, 2315, 2320, 4110, 4120, 4320, 8020, 8620	150-200	AC, A, N	90	85	75	50
	200-250	AC, A, N	80	75	65	40
	250-300	AC, HT	60	50	50	30
(Medium Carbon Alloy), 1330, 1340, 2325, 2330, 4125, 4130, 4140, 4330, 4340, 8030, 80B30, 8040, 8430, 8440, 8630, 8640, 9525, 9530, 9535	175-225	AC, A, N	80	70	70	45
	225-250	AC, A, N	70	65	60	35
	250-300	AC, HT	55	50	45	30
	300-350	AC, HT	45	30	30	20
	350-400	HT	30	…	20	10

[a] Abbreviations designate: A, annealed; AC, as cast; N, normalized; HT, heat treated; MHT, malleabilizing heat treatment; and HB, Brinell hardness number.

Speeds for turning based on a feed rate of 0.012 inch per revolution and a depth of cut of 0.125 inch.

Table 4. Recommended Cutting Speeds in Feet per Minute for Turning, Milling, Drilling and Reaming Stainless Steels

Material	Hardness HB[a]	Material Condition	Cutting Speed, fpm HSS			
			Turning	Milling	Drilling	Reaming
Free Machining Stainless Steels (Ferritic)						
430F, 430F Se	135-185	A	110	95	90	60
(Austenitic), 203EZ, 303, 303Se, 303MA, 303Pb, 303Cu, 303 Plus X	135-185	A	100	90	85	55
	225-275	CD	80	75	70	45
(Martensitic), 416, 416Se, 416Plus X, 420F, 420FSe, 440F, 440FSe	135-185	A	110	95	90	60
	185-240	A,CD	100	80	70	45
	275-325	Q and T	60	50	40	25
	375-425	Q and T	30	20	20	10
Stainless Steels						
(Ferritic), 405, 409, 429, 430, 434, 436, 442, 446, 502	135-185	A	90	75	65	45
(Austenitic), 201, 202, 301, 302, 304, 304L, 305, 308, 321, 347, 348	135-185	A	75	60	55	35
	225-275	CD	65	50	50	30
(Austenitic), 302B, 309, 309S, 310, 310S, 314, 316, 316L, 317, 330	135-185	A	70	50	50	30
(Martensitic), 403, 410, 420, 501	135-175	A	95	75	75	50
	175-225	A	85	65	65	45
	275-325	Q and T	55	40	40	25
	375-425	Q and T	35	25	25	15
(Martensitic), 414, 431, Greek Ascoloy	225-275	A	60	55	50	30
	275-325	Q and T	50	45	40	25
	375-425	Q and T	30	25	25	15
(Martensitic), 440A, 440B, 440C	225-275	A	55	50	45	30
	275-325	Q and T	45	40	40	25
	375-425	Q and T	30	20	20	10
(Precipitation Hardening), 15-5PH, 17-4PH, 17-7PH, AF-71, 17-14CuMo, AFC-77, AM-350, AM-355, AM-362, Custom 455, HNM, PH13-8, PH14-8Mo, PH15-7Mo, Stainless W	150-200	A	60	60	50	30
	275-325	H	50	50	45	25
	325-375	H	40	40	35	20
	375-450	H	25	25	20	10

[a] Abbreviations designate: A, annealed; CD, cold drawn: N, normalized; H, precipitation hardened; Q and T, quenched and tempered; and HB, Brinell hardness number.

Speeds for turning based on a feed rate of 0.012 inch per revolution and a depth of cut of 0.125 inch.

Table 5. Recommended Cutting Speeds in Feet per Minute for Turning, Milling, Drilling and Reaming Tool Steels

Material Tool Steels (AISI Types)	Hardness HB[a]	Material Condition	Cutting Speed, fpm HSS			
			Turning	Milling	Drilling	Reaming
Water Hardening W1, W2, W5	150-200	A	100	85	85	55
Shock Resisting S1, S2, S5, S6, S7	175-225	A	70	55	50	35
Cold Work, Oil Hardening O1, O2, O6, O7	175-225	A	70	50	45	30
Cold Work, High Carbon High Chromium D2, D3, D4, D5, D7	200-250	A	45	40	30	20
Cold Work, Air Hardening A2, A3, A8, A9, A10	200-250	A	70	50	50	35
A4, A6	200-250	A	55	45	45	30
A7	225-275	A	45	40	30	20
Hot Work, Chromium Type H10, H11, H12, H13, H14, H19	150-200	A	80	60	60	40
	200-250	A	65	50	50	30
	325-375	Q and T	50	30	30	20
	48-50 HRC	Q and T	20	…	…	…
	50-52 HRC	Q and T	10	…	…	…
	52-54 HRC	Q and T	…	…	…	…
	54-56 HRC	Q and T	…	…	…	…
Hot Work, Tungsten Type H21, H22, H23, H24, H25, H26	150-200	A	60	55	55	35
	200-250	A	50	45	40	25
Hot Work, Molybdenum Type H41, H42, H43	150-200	A	55	55	45	30
	200-250	A	45	45	35	20
Special Purpose, Low Alloy L2, L3, L6	150-200	A	75	65	60	40
Mold P2, P3, P4, P5, P6	100-150	A	90	75	75	50
P20, P21	150-200	A	80	60	60	40
High Speed Steel M1, M2, M6, M10, T1, T2, T6	200-250	A	65	50	45	30
M3-1, M4, M7, M30, M33, M34, M36, M41, M42, M43, M44, M46, M47, T5, T8	225-275	A	55	40	35	20
T15, M3-2	225-275	A	45	30	25	15

[a] Abbreviations designate: A, annealed; Q and T, quenched and tempered; HB, Brinell hardness number; and HRC, Rockwell C scale hardness number.

Speeds for turning based on a feed rate of 0.012 inch per revolution and a depth of cut of 0.125 inch.

Table 6. Recommended Cutting Speeds in Feet per Minute for Turning, Milling, Drilling and Reaming Light Metals

Material Light Metals	Material Condition[a]	Cutting Speed, fpm HSS			
		Turning	Milling	Drilling	Reaming
All Wrought Aluminum Alloys	CD	600	600	400	400
	ST and A	500	500	350	350
All Aluminum Sand and Permanent Mold Casting Alloys	AC	750	750	500	500
	ST and A	600	600	350	350
All Aluminum Die Casting Alloys	AC	125	125	300	300
	ST and A	100	100	70	70
except Alloys 390.0 and 392.0	AC	80	80	125	100
	ST and A	60	60	45	40
All Wrought Magnesium Alloys	A, CD, ST, and A	800	800	500	500
All Cast Magnesium Alloys	A, AC, ST, and A	800	800	450	450

[a] Abbreviations designate: A, annealed; AC, as cast; CD, cold drawn; ST and A, solution treated as aged.

Table 7. Recommended Cutting Speeds in Feet per Minute for Turning and Drilling Titanium and Titanium Alloys

Material Titanium and Titanium Alloys	Material Condition[a]	Hardness, HB[a]	Cutting Speed, fpm HSS
Commercially Pure			
99.5 Ti	A	110-150	110
99.1 Ti, 99.2 Ti	A	180-240	90
99.0 Ti	A	250-275	70
Low Alloyed			
99.5 Ti - 0.15 Pd	A	110-150	100
99.2 Ti - 0.15 Pd, 98.9 Ti - 0.8 Ni - 0.3 Mo	A	180-250	85
Alpha Alloys and Alpha-Beta Alloys			
5Al-2.5 Sn, 8Mn, 2Al-11Sn-5Zr-1Mo, 4Al-3Mo-1V, 5Al-6Sn-2Zr-1Mo, 6Al-2Sn-4Zr-2Mo, 6Al-2Sn-4Zr-6Mo, 6Al-2Sn-4Zr-2Mo-0.25Si	A	300-350	50
6Al-4V	A	310-350	40
6Al-6V-2Sn, 7Al-4Mo, 8Al-1Mo-1V	A	320-370	30
8V-5Fe-1Al	A	320-380	20
6Al-4V, 6Al-2Sn-4Zr-2Mo, 6Al-2Sn-4Zr-6Mo, 6Al-2Sn-4Zr-2Mo-0.25Si	ST and A	320-380	40
4Al-3Mo-1V, 6Al-6V-2Sn, 7Al-4Mo	ST and A	375-420	20
1Al-8V-5Fe	ST and A	375-440	20
Beta Alloys			
13V-11Cr-3Al, 8Mo-8V-2Fe-3Al, 3Al-8V-6Cr-4Mo-4Zr, 11.5Mo-6Zr-4.5Sn	A, ST	275-350	25
	ST and A	350-440	20

[a] Abbreviations designate: A, annealed; ST, solution treated; ST and A, solution treated as aged; and HB, Brinell hardness number.

Table 8. Recommended Cutting Speeds in Feet per Minute for Turning, Milling and Drilling* Superalloys

Material Superalloys	Cutting Speed, fpm HSS		Material Superalloys	Cutting Speed, fpm HSS	
	Roughing	Finishing		Roughing	Finishing
A-286	30-35	35-40	Mar-M200, M246, M421, and M432	8-10	10-12
AF_2-1DA	8-10	10-15	Mar-M905, and M918	15-20	20-25
Air Resist 213	15-20	20-25	Mar-M302, M322, and M509	10-12	10-15
Air Resist 13, and 215	10-12	10-15	N-12M	8-12	10-15
Astroloy	5-10	5-15	N-155	15-20	15-25
B-1900	8-10	8-10	Nasa C-W-Re	10-12	10-15
CW-12M	8-12	10-15	Nimonic 75, and 80	15-20	20-25
Discalloy	15-35	35-40	Nimonic 90, and 95	10-12	12-15
FSH-H14	10-12	10-15	Refractaloy 26	15-20	20-25
GMR-235, and 235D	8-10	8-10	Rene 41	10-15	12-20
Hastelloy B, C, G, and X (wrought)	15-20	20-25	Rene 80, and 95	8-10	10-15
Hastelloy B, and C (cast)	8-12	10-15	S-590	10-20	15-30
Haynes 25, and 188	15-20	20-25	S-816	10-15	15-20
Haynes 36, and 151	10-12	10-15	T-D Nickel	70-80	80-100
HS 6, 21, 2, 31(X40), 36, and 151	10-12	10-15	Udimet 500, 700, and 710	10-15	12-20
IN 100, and 738	8-10	8-10	Udimet 630	10-20	20-25
Incoloy 800, 801, and 802	30-35	35-40	Unitemp 1753	8-10	10-15
Incoloy 804, and 825	15-20	20-25	V-36	10-15	15-20
Incoloy 901	10-20	20-35	V-57	30-35	35-40
Inconel 625, 702, 706, 718 (wrought), 721, 722, X750, 751, 901, 600, and 604	15-20	20-25	W-545	25-35	30-40
Inconel 700, and 702	10-12	12-15	WI-52	10-12	10-15
Inconel 713C, and 718 (cast)	8-10	8-10	Waspaloy	10-30	25-35
J1300	15-25	20-30	X-45	10-12	10-15
J1570	15-20	20-25	16-25-6	30-35	35-40
M252 (wrought)	15-20	20-25	19-9DL	25-35	30-40
M252 (cast)	8-10	8-10			

* For milling and drilling, use the cutting speeds recommended under roughing.

Table 9. Recommended Feed in Inches per Tooth (f_t) for Milling with High Speed Steel Cutters

Material	Hardness, HB	End Mills							Plain or Slab Mills	Form Relieved Cutters	Face Mills and Shell End Mills	Slotting and Side Mills
		Depth of Cut, .250 in			Depth of Cut, .050 in							
		Cutter Diam., in			Cutter Diam., in							
		½	¾	1 and up	¼	½	¾	1 and up				
		Feed per Tooth, inch										
Free-machining plain carbon steels	100–185	.001	.003	.004	.001	.002	.003	.004	.003–.008	.005	.004–.012	.002–.008
Plain carbon steels, AISI 1006 to 1030; 1513 to 1522	100–150	.001	.003	.003	.001	.002	.003	.004	.003–.008	.004	.004–.012	.002–.008
	150–200	.001	.002	.003	.001	.002	.002	.003	.003–.008	.004	.003–.012	.002–.008
AISI 1033 to 1095; 1524 to 1566	120–180	.001	.003	.003	.001	.002	.003	.004	.003–.008	.004	.004–.012	.002–.008
	180–220	.001	.002	.003	.001	.002	.002	.003	.003–.008	.004	.003–.012	.002–.008
	220–300	.001	.002	.002	.001	.001	.002	.003	.002–.006	.003	.002–.008	.002–.006
Alloy steels having less than 3% carbon. Typical examples: AISI 4012, 4023, 4027, 4118, 4320 4422, 4427, 4615, 4620, 4626, 4720, 4820, 5015, 5120, 6118, 8115, 8620 8627, 8720, 8820, 8822, 9310, 93B17	125–175	.001	.003	.003	.001	.002	.003	.004	.003–.008	.004	.004–.012	.002–.008
	175–225	.001	.002	.003	.001	.002	.003	.003	.003–.008	.004	.003–.012	.002–.008
	225–275	.001	.002	.003	.001	.001	.002	.003	.002–.006	.003	.003–.008	.002–.006
	275–325	.001	.002	.002	.001	.001	.002	.002	.002–.005	.003	.002–.008	.002–.005
Alloy steels having 3% carbon or more. Typical examples: AISI 1330, 1340, 4032, 4037, 4130, 4140, 4150, 4340, 50B40, 50B60, 5130, 51B60, 6150, 81B45, 8630, 8640, 86B45, 8660, 8740, 94B30	175–225	.001	.002	.003	.001	.002	.003	.004	.003–.008	.004	.003–.012	.002–.008
	225–275	.001	.002	.003	.001	.001	.002	.003	.002–.006	.003	.003–.010	.002–.006
	275–325	.001	.002	.002	.001	.001	.002	.003	.002–.005	.003	.002–.008	.002–.005
	325–375	.001	.002	.002	.001	.001	.002	.002	.002–.004	.002	.002–.008	.002–.005
Tool steel	150–200	.001	.002	.002	.001	.002	.003	.003	.003–.008	.004	.003–.010	.002–.006
	200–250	.001	.002	.002	.001	.002	.002	.003	.002–.006	.003	.003–.008	.002–.005
Gray cast iron	120–180	.001	.003	.004	.002	.003	.004	.004	.004–.012	.005	.005–.016	.002–.010
	180–225	.001	.002	.003	.001	.002	.003	.003	.003–.010	.004	.004–.012	.002–.008
	225–300	.001	.002	.002	.001	.001	.002	.002	.002–.006	.003	.002–.008	.002–.005
Free malleable iron	110–160	.001	.003	.004	.002	.003	.004	.004	.003–.010	.005	.005–.016	.002–.010

Table 9. Recommended Feed in Inches per Tooth (f_t) for Milling with High Speed Steel Cutters

Material(*Continued*)	Hard-ness, HB	End Mills							Plain or Slab Mills	Form Relieved Cutters	Face Mills and Shell End Mills	Slotting and Side Mills
		Depth of Cut,.250 in			Depth of Cut,.050 in							
		Cutter Diam., in			Cutter Diam., in							
		½	¾	1 and up	¼	½	¾	1 and up				
		Feed per Tooth, inch										
Pearlitic-Martensitic malleable iron	160–200	.001	.003	.004	.001	.002	.003	.004	.003–.010	.004	.004–.012	.002–.018
	200–240	.001	.002	.003	.001	.002	.003	.003	.003–.007	.004	.003–.010	.002–.006
	240–300	.001	.002	.002	.001	.001	.002	.002	.002–.006	.003	.002–.008	.002–.005
Cast steel	100–180	.001	.003	.003	.001	.002	.003	.004	.003–.008	.004	.003–.012	.002–.008
	180–240	.001	.002	.003	.001	.002	.003	.003	.003–.008	.004	.003–.010	.002–.006
	240–300	.001	.002	.002	.005	.002	.002	.002	.002–.006	.003	.003–.008	.002–.005
Zinc alloys (die castings)	…	.002	.003	.004	.001	.003	.004	.006	.003–.010	.005	.004–.015	.002–.012
Copper alloys (brasses & bronzes)	100–150	.002	.004	.005	.002	.003	.005	.006	.003–.015	.004	.004–.020	.002–.010
	150–250	.002	.003	.004	.001	.003	.004	.005	.003–.015	.004	.003–.012	.002–.008
Free cutting brasses & bronzes	80–100	.002	.004	.005	.002	.003	.005	.006	.003–.015	.004	.004–.015	.002–.010
Cast aluminum alloys—as cast	…	.003	.004	.005	.002	.004	.005	.006	.005–.016	.006	.005–.020	.004–.012
Cast aluminum alloys—hardened	…	.003	.004	.005	.002	.003	.004	.005	.004–.012	.005	.005–.020	.004–.012
Wrought aluminum alloys— cold drawn	…	.003	.004	.005	.002	.003	.004	.005	.004–.014	.005	.005–.020	.004–.012
Wrought aluminum alloys—hardened	…	.002	.003	.004	.001	.002	.003	.004	.003–.012	.004	.005–.020	.004–.012
Magnesium alloys	…	.003	.004	.005	.003	.004	.005	.007	.005–.016	.006	.008–.020	.005–.012
Ferritic stainless steel	135–185	.001	.002	.003	.001	.002	.003	.003	.002–.006	.004	.004–.008	.002–.007
Austenitic stainless steel	135–185	.001	.002	.003	.001	.002	.003	.003	.003–.007	.004	.005–.008	.002–.007
	185–275	.001	.002	.003	.001	.002	.002	.002	.003–.006	.003	.004–.006	.002–.007
Martensitic stainless steel	135–185	.001	.002	.002	.001	.002	.003	.003	.003–.006	.004	.004–.010	.002–.007
	185–225	.001	.002	.002	.001	.002	.002	.003	.003–.006	.004	.003–.008	.002–.007
	225–300	.0005	.002	.002	.0005	.001	.002	.002	.002–.005	.003	.002–.006	.002–.005
Monel	100–160	.001	.003	.004	.001	.002	.003	.004	.002–.006	.004	.002–.008	.002–.006

Table 10. Recommended Feed in Inch per Tooth (f_t) for Milling with Cemented Carbide Cutters

Material	Hardness HB	Face Mills	Slotting and Side Mills
		Feed per Tooth, inch	
Free Machining Plain Carbon Steels	100-185	.008-.020	.003-.010
Plain Carbon Steels, AISI 1006 to 1030, 1513 to 1522	100-150 150-200	.008-.020 .008-.020	.003-.010 .003-.010
Plain Carbon Steels, AISI 1033 to 1095, 1524 to 1566	120-180 180-220 220-300	.005-.020 .005-.020 005-.020	.003-.010 .003-.010 .003-.010
Alloy Steels having less than .3% Carbon content. Typical examples: AISI 4012, 4023, 4027, 4118, 4320, 4422, 4427, 4615, 4620, 4626, 4720, 4820, 5015, 5120, 6118, 8115, 8620, 8627, 8720, 8822, 9310, 93B17	125-175 175-225 225-2 75 275-325 325-375	.006-020 .006-.020 .006-.016 .004-.012 .003-.008	.003-.010 .003-.010 .003-.010 .003-.008 .003-.007
Alloy Steels having .3% Carbon content, or more. Typical examples: AISI 1330, 1340, 4032, 4037, 4130, 4140, 4150, 4340, 50B40, 50B60, 5130, 51B60, 6150, 81B45, 8630, 8640, 86B45, 8660, 8740, 94B30	175-225 225-275 275-325 325-375	.005-.020 .004-.012 .003-010 .003-.008	.003-010 .003-.008 .003-.008 .003-.007
Tool Steels	200-275 275-325 36-45HRC 45-55HRC	.004-.012 .003-010 .003-.006 .003-005	.003-.007 .003-.006 .002-.005 .002-003
Ferritic Stainless Steels	110-160	.005-.015	.003-010
Austenitic Stainless Steels	135-185 185-275	.005-.012 .005-.010	.003-.010 .003-.008
Martensitic Stainless Steel	135-185 185-2 25 2 25-300	.005-.015 .005-.010 .004-.008	.003-.010 .003-008 .003-007
Precipitation Hardening Stainless Steels	Annealed 275-350 350-450	.004-.012 .003-.008 .002-.005	.003-010 .002-005 .002-.004
Cast Steel	100-180 180-240 240-300	.008-.020 .005-.016 .004-.012	.003-.010 .003-010 .003-.008
Gray Cast Iron	140-185 185-225 225-300	.008-020 .008-.016 .005-.012	.005-012 .005-010 .004-.008
Ferritic Malleable Iron	110-160	.005-.020	.004-.012
Pearlitic-Martensitic Malleable Iron	160-200 200-240 240-300	.005-.020 .005-.016 .004-.010	.003-.010 .003-010 .003-.008
Nodular (Ductile) Iron	140-200 200-275 275-325 325-400	.008-.020 .006-014 .005-.012 .003-.008	.003-010 .003-.008 .003-.007 .002-.004
Copper Alloys (Brasses and Bronzes)	100-150 150-250	.005-.020 .004-.014	.003-012 .003-.010
Wrought and Cast Magnesium Alloys	…	.005-.020	.005-.020
Superalloys	…	.003-.010	.003-.006
Titanium Alloys	…	.003-.010	.003-.006
Nickel Alloys	…	.003-.010	.003-.006
Monel	…	.003-.010	.003-.006
Plastics, Hard Rubber, Etc.	…	.003-.015	.003-.012

Table 11. Representative Cutting Conditions for Rough Turning Hard-to-Machine Materials with Single Point Cubic Boron Nitride (CBN) Cutting Tools

Material	Hardness. HRC	Cutting Speed. fpm	Feed. in./rev.	Depth of Cut, in.
Hardened Ferrous Alloys				
AISI 8620	63	250	.005	.060
AISI 52100	70	270	.020	.050
A2, A6. Cold Work Tool Steel	58	250	.008	.060
D2, Cold Work Tool Steel	54	250	.008	.060
H10, Hot Work Tool Steel	56	200	.005	.060
S5, Shock Resisting Tool Steel	60	400	.008	.060
O1, Oil Hardening Tool Steel	58	250	.008	.060
M2, High Speed Steel	62	250	.008	.060
Chilled Gray Iron	60	400	.010	.200
Mechanite Cast Iron	56	600	.008	.250
Superalloys				
Colmonoy	…	600	.006	.125
Incoloy 901	…	800	.006	.125
Inconel 600	…	600	.006	.125
Inconel 718	…	600	.006	.125
K-Monel	…	600	.006	.125
Rene 41	…	600	.006	.125
Rene 77	…	500	.006	.015
Rene 95 (Hot Isostatic Pressed)	…	900	.005	.125
Rene 95 (Forged)	…	450	.005	.125
Stellite	…	600	.006	.125
Waspaloy	…	600	.003	.060

Table 12. Recommended Cutting Speeds in Feet per Minute for Turning and Milling Copper Alloys

Group A
314 Leaded Commercial Bronze, 332 High Leaded Brass, 340 Medium Leaded Brass, 342 High Leaded Brass, 353 High Leaded Brass, 356 Extra High Leaded Brass, 360 Free Cutting Brass, 370 Free Cutting Muntz Metal, 377 Forging Brass, 385 Architectural Bronze, 485 Leaded Naval Brass, 544 Free Cutting Phosphor Bronze
Group B
226 Jewelry Bronze, 230 Red Brass, 240 Low Brass, 260 Cartridge Brass 70%, 268 Yellow Brass, 280 Muntz Metal, 335 Low Leaded Brass , 365 Leaded Muntz Metal, 368 Leaded Muntz Metal, 443 Admiralty Brass (inhibited), 445 Admiralty Brass (inhibited), 651 Low Silicon Bronze, 655 High Silicon Bronze, 675 Manganese Bronze, 687 Aluminum Brass, 770 Nickel Silver, 796 Leaded Nickel Silver
Group C
102 Oxygen Free Copper, 110 Electrolytic Tough Pitch Copper, 122 Phosphorus Deoxidized Copper, 170 Beryllium Copper, 172 Beryllium Copper, 175 Beryllium Copper, 210 Gilding, 95%, 220 Commercial Bronze, 502 Phosphor Bronze 1.25% , 510 Phosphor Bronze 5%, 521 Phosphor Bronze 8%, 524 Phosphor Bronze 10%, 614 Aluminum Bronze,706 Copper Nickel 10%, 715 Copper Nickel 30%, 745 Nickel Silver, 752 Nickel Silver, 754 Nickel Silver,757 Nickel Silver

Material Copper Alloys (Copper Alloy Nos. as per the Copper Development Assn. Inc.)	Material Condition[a]	Cutting Speed, fpm	
		HSS	Carbide
Group A	A	300	650
	CD	350	600
Group B	A	200	500
	CD	250	550
Group C	A	100	200
	CD	110	225

[a] Abbreviations used in this column are as follows: A, annealed; CD, cold drawn.

Cutting Speeds and Equivalent RPM for Drills of Number and Letter Sizes

Size No.	Cutting Speed, Feet per Minute										
	30′	40′	50′	60′	70′	80′	90′	100′	110′	130′	150′
	Revolutions per Minute for Number Sizes										
1	503	670	838	1005	1173	1340	1508	1675	1843	2179	2513
2	518	691	864	1037	1210	1382	1555	1728	1901	2247	2593
4	548	731	914	1097	1280	1462	1645	1828	2010	2376	2741
6	562	749	936	1123	1310	1498	1685	1872	2060	2434	2809
8	576	768	960	1151	1343	1535	1727	1919	2111	2495	2879
10	592	790	987	1184	1382	1579	1777	1974	2171	2566	2961
12	606	808	1010	1213	1415	1617	1819	2021	2223	2627	3032
14	630	840	1050	1259	1469	1679	1889	2099	2309	2728	3148
16	647	863	1079	1295	1511	1726	1942	2158	2374	2806	3237
18	678	904	1130	1356	1582	1808	2034	2260	2479	2930	3380
20	712	949	1186	1423	1660	1898	2135	2372	2610	3084	3559
22	730	973	1217	1460	1703	1946	2190	2433	2676	3164	3649
24	754	1005	1257	1508	1759	2010	2262	2513	2764	3267	3769
26	779	1039	1299	1559	1819	2078	2338	2598	2858	3378	3898
28	816	1088	1360	1631	1903	2175	2447	2719	2990	3534	4078
30	892	1189	1487	1784	2081	2378	2676	2973	3270	3864	4459
32	988	1317	1647	1976	2305	2634	2964	3293	3622	4281	4939
34	1032	1376	1721	2065	2409	2753	3097	3442	3785	4474	5162
36	1076	1435	1794	2152	2511	2870	3228	3587	3945	4663	5380
38	1129	1505	1882	2258	2634	3010	3387	3763	4140	4892	5645
40	1169	1559	1949	2339	2729	3118	3508	3898	4287	5067	5846
42	1226	1634	2043	2451	2860	3268	3677	4085	4494	5311	6128
44	1333	1777	2221	2665	3109	3554	3999	4442	4886	5774	6662
46	1415	1886	2358	2830	3301	3773	4244	4716	5187	6130	7074
48	1508	2010	2513	3016	3518	4021	4523	5026	5528	6534	7539
50	1637	2183	2729	3274	3820	4366	4911	5457	6002	7094	8185
52	1805	2406	3008	3609	4211	4812	5414	6015	6619	7820	9023
54	2084	2778	3473	4167	4862	5556	6251	6945	7639	9028	10417
Size	Revolutions per Minute for Letter Sizes										
A	491	654	818	982	1145	1309	1472	1636	1796	2122	2448
B	482	642	803	963	1124	1284	1445	1605	1765	2086	2407
C	473	631	789	947	1105	1262	1420	1578	1736	2052	2368
D	467	622	778	934	1089	1245	1400	1556	1708	2018	2329
E	458	611	764	917	1070	1222	1375	1528	1681	1968	2292
F	446	594	743	892	1040	1189	1337	1486	1635	1932	2229
G	440	585	732	878	1024	1170	1317	1463	1610	1903	2195
H	430	574	718	862	1005	1149	1292	1436	1580	1867	2154
I	421	562	702	842	983	1123	1264	1404	1545	1826	2106
J	414	552	690	827	965	1103	1241	1379	1517	1793	2068
K	408	544	680	815	951	1087	1223	1359	1495	1767	2039
L	395	527	659	790	922	1054	1185	1317	1449	1712	1976
M	389	518	648	777	907	1036	1166	1295	1424	1683	1942
N	380	506	633	759	886	1012	1139	1265	1391	1644	1897
O	363	484	605	725	846	967	1088	1209	1330	1571	1813
P	355	473	592	710	828	946	1065	1183	1301	1537	1774
Q	345	460	575	690	805	920	1035	1150	1266	1496	1726
R	338	451	564	676	789	902	1014	1127	1239	1465	1690
S	329	439	549	659	769	878	988	1098	1207	1427	1646
T	320	426	533	640	746	853	959	1066	1173	1387	1600
U	311	415	519	623	727	830	934	1038	1142	1349	1557
V	304	405	507	608	709	810	912	1013	1114	1317	1520
W	297	396	495	594	693	792	891	989	1088	1286	1484
X	289	385	481	576	672	769	865	962	1058	1251	1443
Y	284	378	473	567	662	756	851	945	1040	1229	1418
Z	277	370	462	555	647	740	832	925	1017	1202	1387

For fractional drill sizes, use the following table.

Revolutions per Minute for Various Cutting Speeds and Diameters

Dia., Inches	Cutting Speed, Feet per Minute											
	40	50	60	70	80	90	100	120	140	160	180	200
	Revolutions per Minute											
1/4	611	764	917	1070	1222	1376	1528	1834	2139	2445	2750	3056
5/16	489	611	733	856	978	1100	1222	1466	1711	1955	2200	2444
3/8	408	509	611	713	815	916	1018	1222	1425	1629	1832	2036
7/16	349	437	524	611	699	786	874	1049	1224	1398	1573	1748
1/2	306	382	459	535	611	688	764	917	1070	1222	1375	1528
9/16	272	340	407	475	543	611	679	813	951	1086	1222	1358
5/8	245	306	367	428	489	552	612	736	857	979	1102	1224
11/16	222	273	333	389	444	500	555	666	770	888	999	1101
3/4	203	254	306	357	408	458	508	610	711	813	914	1016
13/16	190	237	284	332	379	427	474	569	664	758	853	948
7/8	175	219	262	306	349	392	438	526	613	701	788	876
15/16	163	204	244	285	326	366	407	488	570	651	733	814
1	153	191	229	267	306	344	382	458	535	611	688	764
1 1/16	144	180	215	251	287	323	359	431	503	575	646	718
1 1/8	136	170	204	238	272	306	340	408	476	544	612	680
1 3/16	129	161	193	225	258	290	322	386	451	515	580	644
1 1/4	123	153	183	214	245	274	306	367	428	490	551	612
1 5/16	116	146	175	204	233	262	291	349	407	466	524	582
1 3/8	111	139	167	195	222	250	278	334	389	445	500	556
1 7/16	106	133	159	186	212	239	265	318	371	424	477	530
1 1/2	102	127	153	178	204	230	254	305	356	406	457	508
1 9/16	97.6	122	146	171	195	220	244	293	342	390	439	488
1 5/8	93.9	117	141	165	188	212	234	281	328	374	421	468
1 11/16	90.4	113	136	158	181	203	226	271	316	362	407	452
1 3/4	87.3	109	131	153	175	196	218	262	305	349	392	436
1 7/8	81.5	102	122	143	163	184	204	244	286	326	367	408
2	76.4	95.5	115	134	153	172	191	229	267	306	344	382
2 1/8	72.0	90.0	108	126	144	162	180	216	252	288	324	360
2 1/4	68.0	85.5	102	119	136	153	170	204	238	272	306	340
2 3/8	64.4	80.5	96.6	113	129	145	161	193	225	258	290	322
2 1/2	61.2	76.3	91.7	107	122	138	153	184	213	245	275	306
2 5/8	58.0	72.5	87.0	102	116	131	145	174	203	232	261	290
2 3/4	55.6	69.5	83.4	97.2	111	125	139	167	195	222	250	278
2 7/8	52.8	66.0	79.2	92.4	106	119	132	158	185	211	238	264
3	51.0	63.7	76.4	89.1	102	114	127	152	178	203	228	254
3 1/8	48.8	61.0	73.2	85.4	97.6	110	122	146	171	195	219	244
3 1/4	46.8	58.5	70.2	81.9	93.6	105	117	140	164	188	211	234
3 3/8	45.2	56.5	67.8	79.1	90.4	102	113	136	158	181	203	226
3 1/2	43.6	54.5	65.5	76.4	87.4	98.1	109	131	153	174	196	218
3 5/8	42.0	52.5	63.0	73.5	84.0	94.5	105	126	147	168	189	210
3 3/4	40.8	51.0	61.2	71.4	81.6	91.8	102	122	143	163	184	205
3 7/8	39.4	49.3	59.1	69.0	78.8	88.6	98.5	118	138	158	177	197
4	38.2	47.8	57.3	66.9	76.4	86.0	95.6	115	134	153	172	191
4 1/4	35.9	44.9	53.9	62.9	71.8	80.8	89.8	108	126	144	162	180
4 1/2	34.0	42.4	51.0	59.4	67.9	76.3	84.8	102	119	136	153	170
4 3/4	32.2	40.2	48.2	56.3	64.3	72.4	80.4	96.9	113	129	145	161
5	30.6	38.2	45.9	53.5	61.1	68.8	76.4	91.7	107	122	138	153
5 1/4	29.1	36.4	43.6	50.9	58.2	65.4	72.7	87.2	102	116	131	145
5 1/2	27.8	34.7	41.7	48.6	55.6	62.5	69.4	83.3	97.2	111	125	139
5 3/4	26.6	33.2	39.8	46.5	53.1	59.8	66.4	80.0	93.0	106	120	133
6	25.5	31.8	38.2	44.6	51.0	57.2	63.6	76.3	89.0	102	114	127
6 1/4	24.4	30.6	36.7	42.8	48.9	55.0	61.1	73.3	85.5	97.7	110	122
6 1/2	23.5	29.4	35.2	41.1	47.0	52.8	58.7	70.4	82.2	93.9	106	117
6 3/4	22.6	28.3	34.0	39.6	45.3	50.9	56.6	67.9	79.2	90.6	102	113
7	21.8	27.3	32.7	38.2	43.7	49.1	54.6	65.5	76.4	87.4	98.3	109
7 1/4	21.1	26.4	31.6	36.9	42.2	47.4	52.7	63.2	73.8	84.3	94.9	105
7 1/2	20.4	25.4	30.5	35.6	40.7	45.8	50.9	61.1	71.0	81.4	91.6	102
7 3/4	19.7	24.6	29.5	34.4	39.4	44.3	49.2	59.0	68.9	78.7	88.6	98.4
8	19.1	23.9	28.7	33.4	38.2	43.0	47.8	57.4	66.9	76.5	86.0	95.6

Revolutions per Minute for Various Cutting Speeds and Diameters

Dia., Inches	Cutting Speed, Feet per Minute											
	225	250	275	300	325	350	375	400	425	450	500	550
	Revolutions per Minute											
1/4	3438	3820	4202	4584	4966	5348	5730	6112	6493	6875	7639	8403
5/16	2750	3056	3362	3667	3973	4278	4584	4889	5195	5501	6112	6723
3/8	2292	2546	2801	3056	3310	3565	3820	4074	4329	4584	5093	5602
7/16	1964	2182	2401	2619	2837	3056	3274	3492	3710	3929	4365	4802
1/2	1719	1910	2101	2292	2483	2675	2866	3057	3248	3439	3821	4203
9/16	1528	1698	1868	2037	2207	2377	2547	2717	2887	3056	3396	3736
5/8	1375	1528	1681	1834	1987	2139	2292	2445	2598	2751	3057	3362
11/16	1250	1389	1528	1667	1806	1941	2084	2223	2362	2501	2779	3056
3/4	1146	1273	1401	1528	1655	1783	1910	2038	2165	2292	2547	2802
13/16	1058	1175	1293	1410	1528	1646	1763	1881	1998	2116	2351	2586
7/8	982	1091	1200	1310	1419	1528	1637	1746	1855	1965	2183	2401
15/16	917	1019	1120	1222	1324	1426	1528	1630	1732	1834	2038	2241
1	859	955	1050	1146	1241	1337	1432	1528	1623	1719	1910	2101
1 1/16	809	899	988	1078	1168	1258	1348	1438	1528	1618	1798	1977
1 1/8	764	849	933	1018	1103	1188	1273	1358	1443	1528	1698	1867
1 3/16	724	804	884	965	1045	1126	1206	1287	1367	1448	1609	1769
1 1/4	687	764	840	917	993	1069	1146	1222	1299	1375	1528	1681
1 5/16	654	727	800	873	946	1018	1091	1164	1237	1309	1455	1601
1 3/8	625	694	764	833	903	972	1042	1111	1181	1250	1389	1528
1 7/16	598	664	730	797	863	930	996	1063	1129	1196	1329	1461
1 1/2	573	636	700	764	827	891	955	1018	1082	1146	1273	1400
1 9/16	550	611	672	733	794	855	916	978	1039	1100	1222	1344
1 5/8	528	587	646	705	764	822	881	940	999	1057	1175	1293
1 11/16	509	566	622	679	735	792	849	905	962	1018	1132	1245
1 3/4	491	545	600	654	709	764	818	873	927	982	1091	1200
1 13/16	474	527	579	632	685	737	790	843	895	948	1054	1159
1 7/8	458	509	560	611	662	713	764	815	866	917	1019	1120
1 15/16	443	493	542	591	640	690	739	788	838	887	986	1084
2	429	477	525	573	620	668	716	764	811	859	955	1050
2 1/8	404	449	494	539	584	629	674	719	764	809	899	988
2 1/4	382	424	468	509	551	594	636	679	721	764	849	933
2 3/8	362	402	442	482	522	563	603	643	683	724	804	884
2 1/2	343	382	420	458	496	534	573	611	649	687	764	840
2 5/8	327	363	400	436	472	509	545	582	618	654	727	800
2 3/4	312	347	381	416	451	486	520	555	590	625	694	763
2 7/8	299	332	365	398	431	465	498	531	564	598	664	730
3	286	318	350	381	413	445	477	509	541	572	636	700
3 1/8	274	305	336	366	397	427	458	488	519	549	611	672
3 1/4	264	293	323	352	381	411	440	470	499	528	587	646
3 3/8	254	283	311	339	367	396	424	452	481	509	566	622
3 1/2	245	272	300	327	354	381	409	436	463	490	545	600
3 5/8	237	263	289	316	342	368	395	421	447	474	527	579
3 3/4	229	254	280	305	331	356	382	407	433	458	509	560
3 7/8	221	246	271	295	320	345	369	394	419	443	493	542
4	214	238	262	286	310	334	358	382	405	429	477	525
4 1/4	202	224	247	269	292	314	337	359	383	404	449	494
4 1/2	191	212	233	254	275	297	318	339	360	382	424	466
4 3/4	180	201	221	241	261	281	301	321	341	361	402	442
5	171	191	210	229	248	267	286	305	324	343	382	420
5 1/4	163	181	199	218	236	254	272	290	308	327	363	399
5 1/2	156	173	190	208	225	242	260	277	294	312	347	381
5 3/4	149	166	182	199	215	232	249	265	282	298	332	365
6	143	159	174	190	206	222	238	254	270	286	318	349
6 1/4	137	152	168	183	198	213	229	244	259	274	305	336
6 1/2	132	146	161	176	190	205	220	234	249	264	293	322
6 3/4	127	141	155	169	183	198	212	226	240	254	283	311
7	122	136	149	163	177	190	204	218	231	245	272	299
7 1/4	118	131	144	158	171	184	197	210	223	237	263	289
7 1/2	114	127	139	152	165	178	190	203	216	229	254	279
7 3/4	111	123	135	148	160	172	185	197	209	222	246	271
8	107	119	131	143	155	167	179	191	203	215	238	262

Revolutions per Minute for Various Cutting Speeds and Diameters (Metric Units)

Dia., mm	Cutting Speed, Meters per Minute											
	5	6	8	10	12	16	20	25	30	35	40	45
	Revolutions per Minute											
5	318	382	509	637	764	1019	1273	1592	1910	2228	2546	2865
6	265	318	424	530	637	849	1061	1326	1592	1857	2122	2387
8	199	239	318	398	477	637	796	995	1194	1393	1592	1790
10	159	191	255	318	382	509	637	796	955	1114	1273	1432
12	133	159	212	265	318	424	531	663	796	928	1061	1194
16	99.5	119	159	199	239	318	398	497	597	696	796	895
20	79.6	95.5	127	159	191	255	318	398	477	557	637	716
25	63.7	76.4	102	127	153	204	255	318	382	446	509	573
30	53.1	63.7	84.9	106	127	170	212	265	318	371	424	477
35	45.5	54.6	72.8	90.9	109	145	182	227	273	318	364	409
40	39.8	47.7	63.7	79.6	95.5	127	159	199	239	279	318	358
45	35.4	42.4	56.6	70.7	84.9	113	141	177	212	248	283	318
50	31.8	38.2	51	63.7	76.4	102	127	159	191	223	255	286
55	28.9	34.7	46.3	57.9	69.4	92.6	116	145	174	203	231	260
60	26.6	31.8	42.4	53.1	63.7	84.9	106	133	159	186	212	239
65	24.5	29.4	39.2	49	58.8	78.4	98	122	147	171	196	220
70	22.7	27.3	36.4	45.5	54.6	72.8	90.9	114	136	159	182	205
75	21.2	25.5	34	42.4	51	68	84.9	106	127	149	170	191
80	19.9	23.9	31.8	39.8	47.7	63.7	79.6	99.5	119	139	159	179
90	17.7	21.2	28.3	35.4	42.4	56.6	70.7	88.4	106	124	141	159
100	15.9	19.1	25.5	31.8	38.2	51	63.7	79.6	95.5	111	127	143
110	14.5	17.4	23.1	28.9	34.7	46.2	57.9	72.3	86.8	101	116	130
120	13.3	15.9	21.2	26.5	31.8	42.4	53.1	66.3	79.6	92.8	106	119
130	12.2	14.7	19.6	24.5	29.4	39.2	49	61.2	73.4	85.7	97.9	110
140	11.4	13.6	18.2	22.7	27.3	36.4	45.5	56.8	68.2	79.6	90.9	102
150	10.6	12.7	17	21.2	25.5	34	42.4	53.1	63.7	74.3	84.9	95.5
160	9.9	11.9	15.9	19.9	23.9	31.8	39.8	49.7	59.7	69.6	79.6	89.5
170	9.4	11.2	15	18.7	22.5	30	37.4	46.8	56.2	65.5	74.9	84.2
180	8.8	10.6	14.1	17.7	21.2	28.3	35.4	44.2	53.1	61.9	70.7	79.6
190	8.3	10	13.4	16.8	20.1	26.8	33.5	41.9	50.3	58.6	67	75.4
200	8	39.5	12.7	15.9	19.1	25.5	31.8	39.8	47.7	55.7	63.7	71.6
220	7.2	8.7	11.6	14.5	17.4	23.1	28.9	36.2	43.4	50.6	57.9	65.1
240	6.6	8	10.6	13.3	15.9	21.2	26.5	33.2	39.8	46.4	53.1	59.7
260	6.1	7.3	9.8	12.2	14.7	19.6	24.5	30.6	36.7	42.8	49	55.1
280	5.7	6.8	9.1	11.4	13.6	18.2	22.7	28.4	34.1	39.8	45.5	51.1
300	5.3	6.4	8.5	10.6	12.7	17	21.2	26.5	31.8	37.1	42.4	47.7
350	4.5	5.4	7.3	9.1	10.9	14.6	18.2	22.7	27.3	31.8	36.4	40.9
400	4	4.8	6.4	8	9.5	12.7	15.9	19.9	23.9	27.9	31.8	35.8
450	3.5	4.2	5.7	7.1	8.5	11.3	14.1	17.7	21.2	24.8	28.3	31.8
500	3.2	3.8	5.1	6.4	7.6	10.2	12.7	15.9	19.1	22.3	25.5	28.6

Revolutions per Minute for Various Cutting Speeds and Diameters (Metric Units)

Dia., mm	Cutting Speed, Meters per Minute											
	50	55	60	65	70	75	80	85	90	95	100	200
	Revolutions per Minute											
5	3183	3501	3820	4138	4456	4775	5093	5411	5730	6048	6366	12,732
6	2653	2918	3183	3448	3714	3979	4244	4509	4775	5039	5305	10,610
8	1989	2188	2387	2586	2785	2984	3183	3382	3581	3780	3979	7958
10	1592	1751	1910	2069	2228	2387	2546	2706	2865	3024	3183	6366
12	1326	1459	1592	1724	1857	1989	2122	2255	2387	2520	2653	5305
16	995	1094	1194	1293	1393	1492	1591	1691	1790	1890	1989	3979
20	796	875	955	1034	1114	1194	1273	1353	1432	1512	1592	3183
25	637	700	764	828	891	955	1019	1082	1146	1210	1273	2546
30	530	584	637	690	743	796	849	902	955	1008	1061	2122
35	455	500	546	591	637	682	728	773	819	864	909	1818
40	398	438	477	517	557	597	637	676	716	756	796	1592
45	354	389	424	460	495	531	566	601	637	672	707	1415
50	318	350	382	414	446	477	509	541	573	605	637	1273
55	289	318	347	376	405	434	463	492	521	550	579	1157
60	265	292	318	345	371	398	424	451	477	504	530	1061
65	245	269	294	318	343	367	392	416	441	465	490	979
70	227	250	273	296	318	341	364	387	409	432	455	909
75	212	233	255	276	297	318	340	361	382	403	424	849
80	199	219	239	259	279	298	318	338	358	378	398	796
90	177	195	212	230	248	265	283	301	318	336	354	707
100	159	175	191	207	223	239	255	271	286	302	318	637
110	145	159	174	188	203	217	231	246	260	275	289	579
120	133	146	159	172	186	199	212	225	239	252	265	530
130	122	135	147	159	171	184	196	208	220	233	245	490
140	114	125	136	148	159	171	182	193	205	216	227	455
150	106	117	127	138	149	159	170	180	191	202	212	424
160	99.5	109	119	129	139	149	159	169	179	189	199	398
170	93.6	103	112	122	131	140	150	159	169	178	187	374
180	88.4	97.3	106	115	124	133	141	150	159	168	177	354
190	83.8	92.1	101	109	117	126	134	142	151	159	167	335
200	79.6	87.5	95.5	103	111	119	127	135	143	151	159	318
220	72.3	79.6	86.8	94	101	109	116	123	130	137	145	289
240	66.3	72.9	79.6	86.2	92.8	99.5	106	113	119	126	132	265
260	61.2	67.3	73.4	79.6	85.7	91.8	97.9	104	110	116	122	245
280	56.8	62.5	68.2	73.9	79.6	85.3	90.9	96.6	102	108	114	227
300	53.1	58.3	63.7	69	74.3	79.6	84.9	90.2	95.5	101	106	212
350	45.5	50	54.6	59.1	63.7	68.2	72.8	77.3	81.8	99.1	91	182
400	39.8	43.8	47.7	51.7	55.7	59.7	63.7	67.6	71.6	75.6	79.6	159
450	35.4	38.9	42.4	46	49.5	53.1	56.6	60.1	63.6	67.2	70.7	141
500	31.8	35	38.2	41.4	44.6	47.7	50.9	54.1	57.3	60.5	63.6	127

Using the Feed and Speed Tables for Tapping and Threading.—The feed used in tapping and threading is always equal to the pitch of the screw thread being formed. The threading data contained in the tables for drilling, reaming, and threading (Tables 2 through 8) are primarily for tapping and thread chasing, and do not apply to thread cutting with single-point tools.

The threading data in Tables 2 through 8 give two sets of feed (pitch) and speed values, for 12 and 50 threads/inch, but these values can be used to obtain the cutting speed for any other thread pitches. If the desired pitch falls between the values given in the tables, i.e., between 0.020 inch (50 tpi) and 0.083 inch (12 tpi), the required cutting speed is obtained by interpolation between the given speeds. If the pitch is less than 0.020 inch (more than 50 tpi), use the *average* speed, i.e., the largest of the two given speeds. For pitches greater than 0.083 inch (fewer than 12 tpi), the *optimum* speed should be used. Tool life using the given feed/speed data is intended to be approximately 45 minutes, and should be about the same for threads between 12 and 50 threads per inch.

Example: Determine the cutting speed required for tapping 303 stainless steel with a ½–20 coated HSS tap.

The two feed/speed pairs for 303 stainless steel, in Table 8, are 83/35 (0.083 in./rev at 35 fpm) and 20/45 (0.020 in./rev at 45 fpm). The pitch of a ½–20 thread is 1/20 = 0.05 inch, so the required feed is 0.05 in./rev. Because 0.05 is between the two given feeds (Table 4), the cutting speed can be obtained by interpolation between the two given speeds as follows:

$$V = 35 + \frac{0.05 - 0.02}{0.083 - 0.02}(45 - 35) = 40 \text{ fpm}$$

The cutting speed for coarse-pitch taps must be lower than for fine-pitch taps with the same diameter. Usually, the difference in pitch becomes more pronounced as the diameter of the tap becomes larger and slight differences in the pitch of smaller-diameter taps have little significant effect on the cutting speed. Unlike all other cutting tools, the feed per revolution of a tap cannot be independently adjusted—it is always equal to the lead of the thread and is always greater for coarse pitches than for fine pitches. Furthermore, the thread form of a coarse-pitch thread is larger than that of a fine-pitch thread; therefore, it is necessary to remove more metal when cutting a coarse-pitch thread.

Taps with a long chamfer, such as starting or tapper taps, can cut faster in a short hole than short chamfer taps, such as plug taps. In deep holes, however, short chamfer or plug taps can run faster than long chamfer taps. Bottoming taps must be run more slowly than either starting or plug taps. The chamfer helps to start the tap in the hole. It also functions to involve more threads, or thread form cutting edges, on the tap in cutting the thread in the hole, thus reducing the cutting load on any one set of thread form cutting edges. In so doing, more chips and thinner chips are produced that are difficult to remove from deeper holes. Shortening the chamfer length causes fewer thread form cutting edges to cut, thereby producing fewer and thicker chips that can easily be disposed of. Only one or two sets of thread form cutting edges are cut on bottoming taps, causing these cutting edges to assume a heavy cutting load and produce very thick chips.

Spiral-pointed taps can operate at a faster cutting speed than taps with normal flutes. These taps are made with supplementary angular flutes on the end that push the chips ahead of the tap and prevent the tapped hole from becoming clogged with chips. They are used primarily to tap open or through holes although some are made with shorter supplementary flutes for tapping blind holes.

The tapping speed must be reduced as the percentage of full thread to be cut is increased. Experiments have shown that the torque required to cut a 100 per cent thread form is more than twice that required to cut a 50 per cent thread form. An increase in the percentage of full thread will also produce a greater volume of chips.

The tapping speed must be lowered as the length of the hole to be tapped is increased. More friction must be overcome in turning the tap and more chips accumulate in the hole. It will be more difficult to apply the cutting fluid at the cutting edges and to lubricate the tap to reduce friction. This problem becomes greater when the hole is being tapped in a horizontal position.

Cutting fluids have a very great effect on the cutting speed for tapping. Although other operating conditions when tapping frequently cannot be changed, a free selection of the cutting fluid usually can be made. When planning the tapping operation, the selection of a cutting fluid warrants a very careful consideration and perhaps an investigation.

Taper threaded taps, such as pipe taps, must be operated at a slower speed than straight thread taps with a comparable diameter. All the thread form cutting edges of a taper threaded tap that are engaged in the work cut and produce a chip, but only those cutting edges along the chamfer length cut on straight thread taps. Pipe taps often are required to cut the tapered thread from a straight hole, adding to the cutting burden.

The machine tool used for the tapping operation must be considered in selecting the tapping speed. Tapping machines and other machines that are able to feed the tap at a rate of advance equal to the lead of the tap, and that have provisions for quickly reversing the spindle, can be operated at high cutting speeds. On machines where the feed of the tap is controlled manually—such as on drill presses and turret lathes—the tapping speed must be reduced to allow the operator to maintain safe control of the operation.

There are other special considerations in selecting the tapping speed. Very accurate threads are usually tapped more slowly than threads with a commercial grade of accuracy. Thread forms that require deep threads for which a large amount of metal must be removed, producing a large volume of chips, require special techniques and slower cutting speeds. Acme, buttress, and square threads, therefore, are generally cut at lower speeds.

Speeds for Metal-cutting Saws.—The following speeds and feeds for metal-cutting saws are recommended by Henry Disston & Sons, Inc. Speeds and feeds vary over a wide range depending upon the kind of machine as well as the size, shape, and kind of metal to be cut; hence, the following speeds are intended as a general guide only.

Solid-tooth Circular Saws: Low-carbon steel: Speed, 60—90 feet per minute; feed, 3—6 inches per minute. Tool and alloy steel: Speed, 40—60 feet per minute; feed, $\frac{1}{32}$—2½ inches per minute. Cast iron: Speed, 60—70 feet per minute; feed, inches per minute. Brass: Speed, 500—800 feet per minute; feed, 80 inches per minute. Copper: Speed, 600—700 feet per minute; feed, 60 inches per minute. Aluminum: Speed, 800-1000 feet per minute; feed, 90 inches per minute.

"Hot saws" for sawing hot metal (structural shapes, rails, billets, etc., in rolling mills) should operate at speeds of 22,000 to 25,000 feet per minute.

Inserted-tooth Metal-cutting Saws: Generally speaking, an inserted-tooth saw is not recommended smaller than 18 inches in diameter, although smaller sizes are made. Small saws usually are applied to work requiring fine or relatively fine teeth which are impossible when the teeth are of the inserted type. The following speeds are not suitable for all compositions but represent general practice. Steel: Speed, 40 feet per minute; feed, 2½—10 inches per minute. Cast iron: Speed, 40 feet per minute; feed, inches per minute. Brass: Speed, 500 feet per minute; feed, 80 inches per minute. Copper: Speed, 750 feet per minute; feed, 60 inches per minute. Aluminum: Speed, 1200 feet per minute; feed, inches per minute.

Metal-cutting Band Saws: The speeds which follow apply to Disston "hard-edge" saws which have milled teeth. Aluminum sheets: Speed of blade, 1000—3000feet per minute. Bakelite: Speed, 800—1000 feet per minute. Brass sheets and tubing: Speed, 700—1500 feet per minute. Carbon tool steel: Speed, 100—150 feet per minute. Cast iron: Speed, 100-150 feet per minute. Cold-rolled steel: Speed, 150—200 feet per minute. High-speed steel:

Speed, 90—125 feet per minute. Malleable iron: Speed, 150—200 feet per minute. Hard rubber: Speed, 150—200 feet per minute. Slate: Speed, 100—150 feet per minute. The number of saw teeth per inch recommended for these different materials ranges from 8 to 14, and the saw user should be guided by the recommendations of the manufacturer.

Hacksaw Speeds: The following hacksaw speeds are given on the authority of a leading manufacturer: The average rate of travel of the cutting blade in feet per minute, including forward and return strokes, should be as follows: For mild steel, 130; for annealed tool steel, 90; and for unannealed tool steel, 60 feet per minute. Thus in the case of a 6-inch stroke, for example, the revolutions per minute of the driving crank should be 130 for mild steel, 90 for annealed tool steel, and 60 for unannealed tool steel. All of these steels are cut with the use of a cutting compound. Bronze can ordinarily be cut at the same speed as mild steel when a suitable compound is used. Brass heats the blade very rapidly if cut dry, and must be cut with a cooling compound adapted to brass. It also fills up the teeth of the saw if not used with the right kind of compound, but with a suitable compound, it may be cut at the same speed as machine steel.

Speeds for Turning Unusual Materials.—Slate, on account of its peculiarly stratified formation, is rather difficult to turn, but if handled carefully, can be machined in an ordinary lathe. The cutting speed should be about the same as for cast iron. A sheet of fiber or pressed paper should be interposed between the chuck or steadyrest jaws and the slate, to protect the latter. Slate rolls must not be centered and run on the tailstock. A satisfactory method of supporting a slate roll having journals at the ends is to bore a piece of lignum vitae to receive the turned end of the roll, and center it for the tailstock spindle.

Rubber can be turned at a peripheral speed of 200 feet per minute, although it is much easier to grind it with an abrasive wheel that is porous and soft. For cutting a rubber roll in two, the ordinary parting tool should not be used, but a tool shaped like a knife; such a tool severs the rubber without removing any material.

Gutta percha can be turned as easily as wood, but the tools must be sharp and a good soap-and-water lubricant used.

Copper can be turned easily at 200 feet per minute.

Lime-stone such as is used in the construction of pillars for balconies, etc., can be turned at 150 feet per minute, and the formation of ornamental contours is quite easy.

Marble is a treacherous material to turn. It should be cut with a tool such as would be used for brass, but at a speed suitable for cast iron. It must be handled very carefully to prevent flaws in the surface.

The foregoing speeds are for high-speed steel tools. Tools tipped with tungsten carbide are adapted for cutting various non-metallic products which cannot be machined readily with steel tools, such as slate, marble, synthetic plastic materials, etc. In drilling slate and marble, use flat drills; and for plastic materials, tungsten-carbide-tipped twist drills. Cutting speeds ranging from 75 to 150 feet per minute have been used for drilling slate (without coolant) and a feed of 0.025 inch per revolution for drills $\frac{3}{4}$ and inch in diameter.

CUTTING TOOLS

Tool Contour.—Tools for turning, planing, etc., are made in straight, bent, offset, and other forms to place the cutting edges in convenient positions for operating on differently located surfaces. The contour or shape of the cutting edge may also be varied to suit different classes of work. Tool shapes, however, are not only related to the kind of operation, but, in roughing tools particularly, the contour may have a decided effect upon the cutting efficiency of the tool. To illustrate, an increase in the side cutting-edge angle of a roughing tool, or in the nose radius, tends to permit higher cutting speeds because the chip will be thinner for a given feed rate. Such changes, however, may result in chattering or vibrations unless the work and the machine are rigid; hence, the most desirable contour may be a compromise between the ideal form and one that is needed to meet practical requirements.

Terms and Definitions.—The terms and definitions relating to single-point tools vary somewhat in different plants, but the following are in general use.

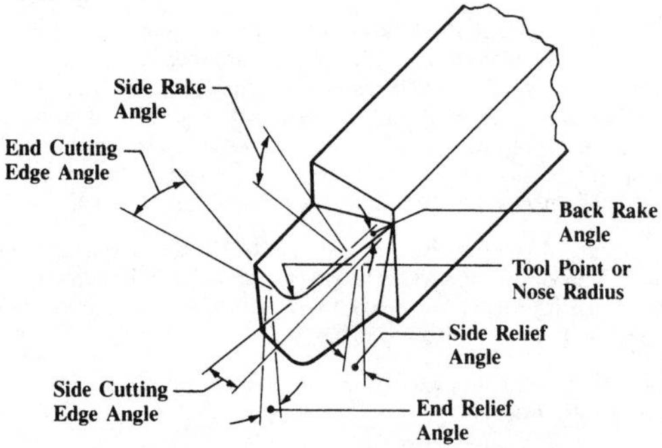

Fig. 1. Terms Applied to Single-point Turning Tools

Single-point Tool: This term is applied to tools for turning, planing, boring, etc., which have a cutting edge at one end. This cutting edge may be formed on one end of a solid piece of steel, or the cutting part of the tool may consist of an insert or tip which is held to the body of the tool by brazing, welding, or mechanical means.

Shank: The shank is the main body of the tool. If the tool is an inserted cutter type, the shank supports the cutter or bit. (See diagram, Fig. 1.)

Nose: A general term sometimes used to designate the cutting end but usually relating more particularly to the rounded tip of the cutting end.

Face: The surface against which the chips bear, as they are severed in turning or planing operations, is called the face.

Flank: The flank is that end surface adjacent to the cutting edge and below it when the tool is in a horizontal position as for turning.

Base: The base is the surface of the tool shank that bears against the supporting toolholder or block.

Side Cutting Edge: The side cutting edge is the cutting edge on the side of the tool. Tools such as shown in Fig. 1 do the bulk of the cutting with this cutting edge and are, therefore, sometimes called side cutting edge tools.

End Cutting Edge: The end cutting edge is the cutting edge at the end of the tool.

On side cutting edge tools, the end cutting edge can be used for light plunging and facing cuts. Cutoff tools and similar tools have only one cutting edge located on the end. These

Lathe Tools— Standard Shapes

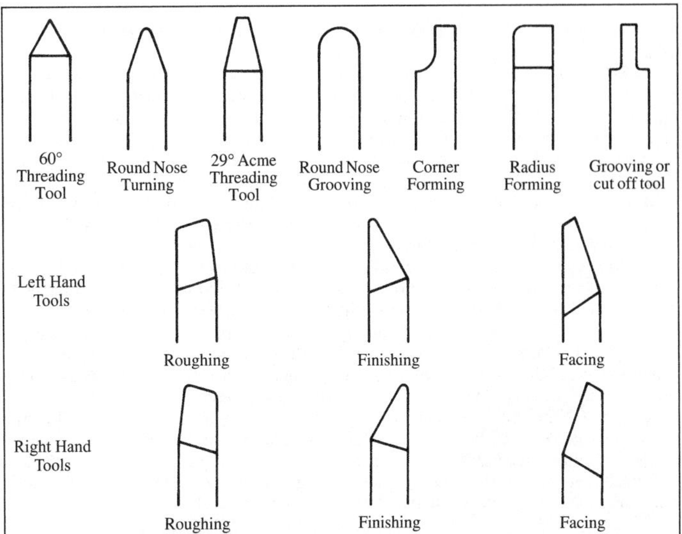

tools and other tools that are intended to cut primarily with the end cutting edge are sometimes called end cutting edge tools.

Rake: A metal-cutting tool is said to have rake when the tool face or surface against which the chips bear as they are being severed, is inclined for the purpose of either increasing or diminishing the keenness or bluntness of the edge. The magnitude of the rake is most conveniently measured by two angles called the back rake angle and the side rake angle. The tool shown in Fig. 1 has rake. If the face of the tool did not incline but was parallel to the base, there would be no rake; the rake angles would be zero.

Positive Rake: If the inclination of the tool face is such as to make the cutting edge keener or more acute than when the rake angle is zero, the rake angle is defined as positive.

Negative Rake: If the inclination of the tool face makes the cutting edge less keen or more blunt than when the rake angle is zero, the rake is defined as negative.

Back Rake: The back rake is the inclination of the face toward or away from the end or the end cutting edge of the tool. When the inclination is away from the end cutting edge, as shown in Fig. 1, the back rake is positive. If the inclination is downward toward the end cutting edge the back rake is negative.

Side Rake: The side rake is the inclination of the face toward or away from the side cutting edge. When the inclination is away from the side cutting edge, as shown in Fig. 1, the side rake is positive. If the inclination is toward the side cutting edge the side rake is negative.

Relief: The flanks below the side cutting edge and the end cutting edge must be relieved to allow these cutting edges to penetrate into the workpiece when taking a cut. If the flanks are not provided with relief, the cutting edges will rub against the workpiece and be unable to penetrate in order to form the chip. Relief is also provided below the nose of the tool to allow it to penetrate into the workpiece. The relief at the nose is usually a blend of the side relief and the end relief.

End Relief Angle: The end relief angle is a measure of the relief below the end cutting edge.

Side Relief Angle: The side relief angle is a measure of the relief below the side cutting edge.

Back Rake Angle: The back rake angle is a measure of the back rake. It is measured in a plane that passes through the side cutting edge and is perpendicular to the base. Thus, the back rake angle can be defined by measuring the inclination of the side cutting edge with respect to a line or plane that is parallel to the base. The back rake angle may be positive, negative, or zero depending upon the magnitude and direction of the back rake.

Side Rake Angle: The side rake angle is a measure of the side rake. This angle is always measured in a plane that is perpendicular to the side cutting edge and perpendicular to the base. Thus, the side rake angle is the angle of inclination of the face perpendicular to the side cutting edge with reference to a line or a plane that is parallel to the base.

End Cutting Edge Angle: The end cutting edge angle is the angle made by the end cutting edge with respect to a plane perpendicular to the axis of the tool shank. It is provided to allow the end cutting edge to clear the finish machined surface on the workpiece.

Side Cutting Edge Angle: The side cutting edge angle is the angle made by the side cutting edge and a plane that is parallel to the side of the shank.

Nose Radius: The nose radius is the radius of the nose of the tool. The performance of the tool, in part, is influenced by nose radius so that it must be carefully controlled.

Lead Angle: The lead angle, shown in Fig. 2, is not ground on the tool. It is a tool setting angle which has a great influence on the performance of the tool. The lead angle is bounded by the side cutting edge and a plane perpendicular to the workpiece surface when the tool is in position to cut; or, more exactly, the lead angle is the angle between the side cutting edge and a plane perpendicular to the direction of the feed travel.

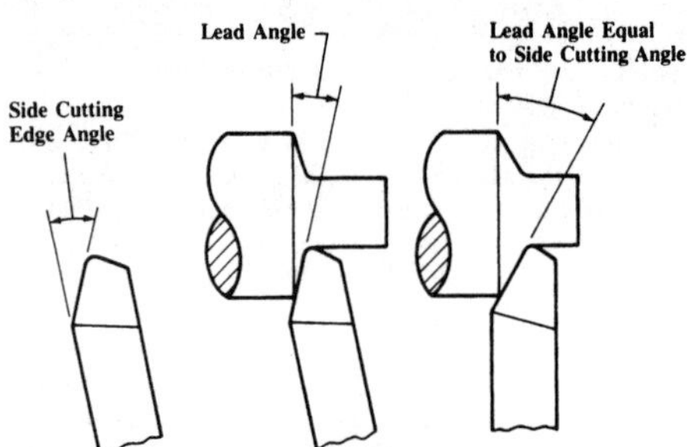

Fig. 2. Lead Angle on Single-point Turning Tool

Solid Tool: A solid tool is a cutting tool made from one piece of tool material.

Brazed Tool: A brazed tool is a cutting tool having a blank of cutting-tool material permanently brazed to a steel shank.

Blank: A blank is an unground piece of cutting-tool material from which a brazed tool is made.

Tool Bit: A tool bit is a relatively small cutting tool that is clamped in a holder in such a way that it can readily be removed and replaced. It is intended primarily to be reground when dull and not indexed.

Tool-bit Blank: The tool-bit blank is an unground piece of cutting-tool material from which a tool bit can be made by grinding. It is available in standard sizes and shapes.

Tool-bit Holder: Usually made from forged steel, the tool-bit holder is used to hold the tool bit, to act as an extended shank for the tool bit, and to provide a means for clamping in the tool post.

Straight-shank Tool-bit Holder: A straight-shank tool-bit holder has a straight shank when viewed from the top. The axis of the tool bit is held parallel to the axis of the shank.

Offset-shank Tool-bit Holder: An offset-shank tool-bit holder has the shank bent to the right or left, as seen in Fig. 3. The axis of the tool bit is held at an angle with respect to the axis of the shank.

Side cutting Tool: A side cutting tool has its major cutting edge on the side of the cutting part of the tool. The major cutting edge may be parallel or at an angle with respect to the axis of the tool.

Indexable Inserts: An indexable insert is a relatively small piece of cutting-tool material that is geometrically shaped to have two or several cutting edges that are used until dull. The insert is then indexed on the holder to apply a sharp cutting edge. When all the cutting edges have been dulled, the insert is discarded. The insert is held in a pocket or against other locating surfaces on an indexable insert holder by means of a mechanical clamping device that can be tightened or loosened easily.

Indexable Insert Holder: Made of steel, an indexable insert holder is used to hold indexable inserts. It is equipped with a mechanical clamping device that holds the inserts firmly in a pocket or against other seating surfaces.

Straight-shank Indexable Insert Holder: A straight-shank indexable insert tool-holder is essentially straight when viewed from the top, although the cutting edge of the insert may be oriented parallel, or at an angle to, the axis of the holder.

Offset-shank Indexable Insert Holder: An offset-shank indexable insert holder has the head end, or the end containing the insert pocket, offset to the right or left, as shown in Fig. 3.

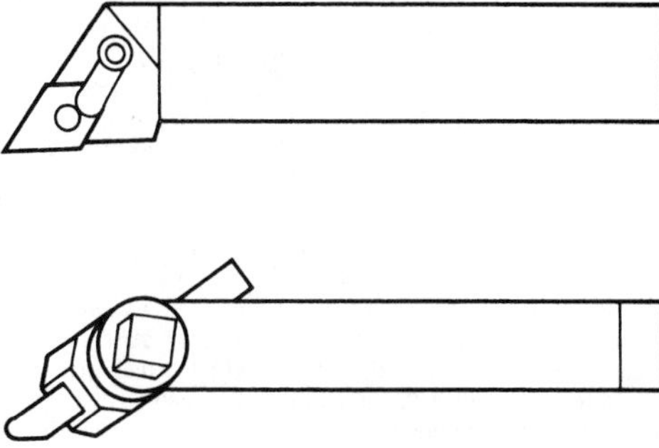

Fig. 3. Top: Right-hand Offset-shank, Indexable Insert Holder
Bottom: Right-hand Offset-shank Tool-bit Holder

End cutting Tool: An end cutting tool has its major cutting edge on the end of the cutting part of the tool. The major cutting edge may be perpendicular or at an angle, with respect to the axis of the tool.

Curved Cutting-edge Tool: A curved cutting-edge tool has a continuously variable side cutting edge angle. The cutting edge is usually in the form of a smooth, continuous curve along its entire length, or along a large portion of its length.

Right-hand Tool: A right-hand tool has the major, or working, cutting edge on the right-hand side when viewed from the cutting end with the face up. As used in a lathe, such a tool is usually fed into the work from right to left, when viewed from the shank end.

Left-hand Tool: A left-hand tool has the major or working cutting edge on the left-hand side when viewed from the cutting end with the face up. As used in a lathe, the tool is usually fed into the work from left to right, when viewed from the shank end.

Neutral-hand Tool: A neutral-hand tool is a tool to cut either left to right or right to left; or the cut may be parallel to the axis of the shank as when plunge cutting.

Chipbreaker: A groove formed in or on a shoulder on the face of a turning tool back of the cutting edge to break up the chips and prevent the formation of long, continuous chips which would be dangerous to the operator and also bulky and cumbersome to handle. A chipbreaker of the shoulder type may be formed directly on the tool face or it may consist of a separate piece that is held either by brazing or by clamping.

Relief Angles.—The end relief angle and the side relief angle on single-point cutting tools are usually, though not invariably, made equal to each other. The relief angle under the nose of the tool is a blend of the side and end relief angles.

The size of the relief angles has a pronounced effect on the performance of the cutting tool. If the relief angles are too large, the cutting edge will be weakened and in danger of breaking when a heavy cutting load is placed on it by a hard and tough material. On finish cuts, rapid wear of the cutting edge may cause problems with size control on the part. Relief angles that are too small will cause the rate of wear on the flank of the tool below the cutting edge to increase, thereby significantly reducing the tool life. In general, when cutting hard and tough materials, the relief angles should be 6 to 8 degrees for high-speed steel tools and 5 to 7 degrees for carbide tools. For medium steels, mild steels, cast iron, and other average work the recommended values of the relief angles are 8 to 12 degrees for high-speed steel tools and 5 to 10 degrees for carbides. Ductile materials having a relatively low modulus of elasticity should be cut using larger relief angles. For example, the relief angles recommended for turning copper, brass, bronze, aluminum, ferritic malleable iron, and similar metals are 12 to 16 degrees for high-speed steel tools and 8 to 14 degrees for carbides.

Larger relief angles generally tend to produce a better finish on the finish machined surface because less surface of the worn flank of the tool rubs against the workpiece. For this reason, single-point thread-cutting tools should be provided with relief angles that are as large as circumstances will permit. Problems encountered when machining stainless steel may be overcome by increasing the size of the relief angle. The relief angles used should never be smaller than necessary.

Rake Angles.—Machinability tests have confirmed that when the rake angle along which the chip slides, called the true rake angle, is made larger in the positive direction, the cutting force and the cutting temperature will decrease. Also, the tool life for a given cutting speed will increase with increases in the true rake angle up to an optimum value, after which it will decrease again. For turning tools which cut primarily with the side cutting edge, the true rake angle corresponds rather closely with the side rake angle except when taking shallow cuts. Increasing the side rake angle in the positive direction lowers the cutting force and the cutting temperature, while at the same time it results in a longer tool life or a higher permissible cutting speed up to an optimum value of the side rake angle. After the optimum value is exceeded, the cutting force and the cutting temperature will continue to drop; however, the tool life and the permissible cutting speed will decrease.

As an approximation, the magnitude of the cutting force will decrease about one per cent per degree increase in the side rake angle. While not exact, this rule of thumb does correspond approximately to test results and can be used to make rough estimates. Of course, the cutting force also increases about one per cent per degree decrease in the side rake angle. The limiting value of the side rake angle for optimum tool life or cutting speed depends upon the work material and the cutting tool material. In general, lower values can be used for hard and tough work materials. Cemented carbides are harder and more brittle than high-speed steel; therefore, the rake angles usually used for cemented carbides are less positive than for high-speed steel.

Negative rake angles cause the face of the tool to slope in the opposite direction from positive rake angles and, as might be expected, they have an opposite effect. For side cutting edge tools, increasing the side rake angle in a negative direction will result in an increase in the cutting force and an increase in the cutting temperature of approximately one per cent per degree change in rake angle. For example, if the side rake angle is changed from 5 degrees positive to 5 degrees negative, the cutting force will be about 10 per cent larger. Usually the tool life will also decrease when negative side rake angles are used, although the tool life will sometimes increase when the negative rake angle is not too large and when a fast cutting speed is used.

Negative side rake angles are usually used in combination with negative back rake angles on single-point cutting tools. The negative rake angles strengthen the cutting edges enabling them to sustain heavier cutting loads and shock loads. They are recommended for turning very hard materials and for heavy interrupted cuts. There is also an economic advantage in favor of using negative rake indexable inserts and tool holders inasmuch as the cutting edges provided on both the top and bottom of the insert can be used.

On turning tools that cut primarily with the side cutting edge, the effect of the back rake angle alone is much less than the effect of the side rake angle although the direction of the change in cutting force, cutting temperature, and tool life is the same. The effect that the back rake angle has can be ignored unless, of course, extremely large changes in this angle are made. A positive back rake angle does improve the performance of the nose of the tool somewhat and is helpful in taking light finishing cuts. A negative back rake angle strengthens the nose of the tool and is helpful when interrupted cuts are taken. The back rake angle has a very significant effect on the performance of end cutting edge tools, such as cut-off tools. For these tools, the effect of the back rake angle is very similar to the effect of the side rake angle on side cutting edge tools.

Side Cutting Edge and Lead Angles.—These angles are considered together because the side cutting edge angle is usually designed to provide the desired lead angle when the tool is being used. The side cutting edge angle and the lead angle will be equal when the shank of the cutting tool is positioned perpendicular to the workpiece, or, more correctly, perpendicular to the direction of the feed. When the shank is not perpendicular, the lead angle is determined by the side cutting edge and an imaginary line perpendicular to the feed direction.

The flow of the chips over the face of the tool is approximately perpendicular to the side cutting edge except when shallow cuts are taken. The thickness of the undeformed chip is measured perpendicular to the side cutting edge. As the lead angle is increased, the length of chip in contact with the side cutting edge is increased, and the chip will become longer and thinner. This effect is the same as increasing the depth of cut and decreasing the feed, although the actual depth of cut and feed remain the same and the same amount of metal is removed. The effect of lengthening and thinning the chip by increasing the lead angle is very beneficial as it increases the tool life for a given cutting speed or that speed can be increased. Increasing the cutting speed while the feed and the tool life remain the same leads to faster production.

However, an adverse effect must be considered. Chatter can be caused by a cutting edge that is oriented at a high lead angle when turning and sometimes, when turning long and slender shafts, even a small lead angle can cause chatter. In fact, an unsuitable lead angle of the side cutting edge is one of the principal causes of chatter. When chatter occurs, often simply reducing the lead angle will cure it. Sometimes, very long and slender shafts can be turned successfully with a tool having a zero degree lead angle (and having a small nose radius). Boring bars, being usually somewhat long and slender, are also susceptible to chatter if a large lead angle is used. The lead angle for boring bars should be kept small, and for very long and slender boring bars a zero degree lead angle is recommended. It is impossible to provide a rule that will determine when chatter caused by a lead angle will occur and when it will not. In making a judgment, the first consideration is the length to diameter ratio of the part to be turned, or of the boring bar. Then the method of holding the workpiece must be considered — a part that is firmly held is less apt to chatter. Finally, the overall condition and rigidity of the machine must be considered because they may be the real cause of chatter.

Although chatter can be a problem, the advantages gained from high lead angles are such that the lead angle should be as large as possible at all times.

End Cutting Edge Angle.—The size of the end cutting edge angle is important when tool wear by cratering occurs. Frequently, the crater will enlarge until it breaks through the end cutting edge just behind the nose, and tool failure follows shortly. Reducing the size of the end cutting edge angle tends to delay the time of crater breakthrough. When cratering takes place, the recommended end cutting edge angle is 8 to 15 degrees. If there is no cratering, the angle can be made larger. Larger end cutting edge angles may be required to enable profile turning tools to plunge into the work without interference from the end cutting edge.

Nose Radius.—The tool nose is a very critical part of the cutting edge since it cuts the finished surface on the workpiece. If the nose is made to a sharp point, the finish machined surface will usually be unacceptable and the life of the tool will be short. Thus, a nose radius is required to obtain an acceptable surface finish and tool life. The surface finish obtained is determined by the feed rate and by the nose radius if other factors such as the work material, the cutting speed, and cutting fluids are not considered. A large nose radius will give a better surface finish and will permit a faster feed rate to be used.

Machinability tests have demonstrated that increasing the nose radius will also improve the tool life or allow a faster cutting speed to be used. For example, high-speed steel tools were used to turn an alloy steel in one series of tests where complete or catastrophic tool failure was used as a criterion for the end of tool life. The cutting speed for a 60-minute tool life was found to be 125 fpm when the nose radius was $\frac{1}{16}$ inch and 160 fpm when the nose radius was $\frac{1}{4}$ inch.

A very large nose radius can often be used but a limit is sometimes imposed because the tendency for chatter to occur is increased as the nose radius is made larger. A nose radius that is too large can cause chatter and when it does, a smaller nose radius must be used on the tool. It is always good practice to make the nose radius as large as is compatible with the operation being performed.

Chipbreakers.—Many steel turning tools are equipped with chipbreaking devices to prevent the formation of long continuous chips in connection with the turning of steel at the high speeds made possible by high-speed steel and especially cemented carbide tools. Long steel chips are dangerous to the operator, and cumbersome to handle, and they may twist around the tool and cause damage. Broken chips not only occupy less space, but permit a better flow of coolant to the cutting edge. Several different forms of chipbreakers are illustrated in Fig. 4.

Angular Shoulder Type: The angular shoulder type shown at *A* is one of the commonly used forms. As the enlarged sectional view shows, the chipbreaking shoulder is located back of the cutting edge. The angle *a* between the shoulder and cutting edge may vary from 6 to 15 degrees or more, 8 degrees being a fair average. The ideal angle, width *W* and depth *G*, depend upon the speed and feed, the depth of cut, and the material. As a general rule, width *W*, at the end of the tool, varies from $\frac{3}{32}$ to $\frac{7}{32}$ inch, and the depth *G* may range from $\frac{1}{64}$ to $\frac{1}{16}$ inch. The shoulder radius equals depth *G*. If the tool has a large nose radius, the corner of the shoulder at the nose end may be beveled off, as illustrated at *B*, to prevent it from coming into contact with the work. The width *K* for type *B* should equal approximately 1.5 times the nose radius.

Parallel Shoulder Type: Diagram C shows a design with a chipbreaking shoulder that is parallel with the cutting edge. With this form, the chips are likely to come off in short curled sections. The parallel form may also be applied to straight tools which do not have a

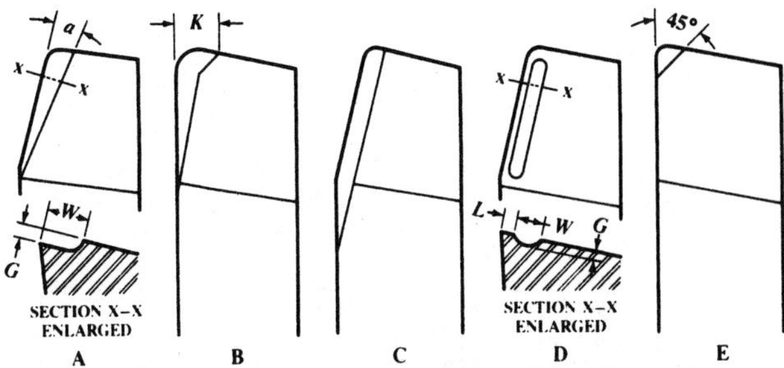

Fig. 4. Different Forms of Chipbreakers for Turning Tools

side cutting-edge angle. The tendency with this parallel shoulder form is to force the chips against the work and damage it.

Groove Type: This type (diagram D) has a groove in the face of the tool produced by grinding. Between the groove and the cutting edge, there is a land *L*. Under ideal conditions, this width *L*, the groove width *W*, and the groove depth *G*, would be varied to suit the feed, depth of cut and material. For average use, *L* is about $\frac{1}{32}$ inch; *G*, $\frac{1}{32}$ inch; and *W*, $\frac{1}{16}$ inch. There are differences of opinion concerning the relative merits of the groove type and the shoulder type. Both types have proved satisfactory when properly proportioned for a given class of work.

Chipbreaker for Light Cuts: Diagram E illustrates a form of chipbreaker that is sometimes used on tools for finishing cuts having a maximum depth of about $\frac{1}{32}$ inch. This chipbreaker is a shoulder type having an angle of 45 degrees and a maximum width of about $\frac{1}{16}$ inch. It is important in grinding all chipbreakers to give the chip-bearing surfaces a fine finish, such as would be obtained by honing. This finish greatly increases the life of the tool.

Planing Tools.—Many of the principles which govern the shape of turning tools also apply in the grinding of tools for planing. The amount of rake depends upon the hardness of the material, and the direction of the rake should be away from the *working part* of the cutting edge. The angle of clearance should be about 4 or 5 degrees for planer tools, which is less than for lathe tools. This small clearance is allowable because a planer tool is held about square with the platen, whereas a lathe tool, the height and inclination of which can be varied, may not always be clamped in the same position.

Carbide Tools: Carbide tools for planing usually have negative rake. Round-nose and square-nose end-cutting tools should have a "negative back rake" (or front rake) of 2 or 3 degrees. Side cutting tools may have a negative back rake of 10 degrees, a negative side rake of 5 degrees, and a side cutting-edge angle of 8 degrees.

Indexable Inserts.—A large proportion of the cemented carbide, single-point cutting tools are indexable inserts and indexable insert tool holders. Dimensional specifications for solid sintered carbide indexable inserts are given in American National Standard ANSI B212.12-1991. Samples of the many insert shapes are shown in Table 3. Most modern, cemented carbide, face milling cutters are of the indexable insert type. Larger size end milling cutters, side milling or slotting cutters, boring tools, and a wide variety of special tools are made to use indexable inserts. These inserts are primarily made from cemented carbide, although most of the cemented oxide cutting tools are also indexable inserts.

The objective of this type of tooling is to provide an insert with several cutting edges. When an edge is worn, the insert is indexed in the tool holder until all the cutting edges are used up, after which it is discarded. The insert is not intended to be reground. The advantages are that the cutting edges on the tool can be rapidly changed without removing the tool holder from the machine, tool-grinding costs are eliminated, and the cost of the insert is less than the cost of a similar, brazed carbide tool. Of course, the cost of the tool holder must be added to the cost of the insert; however, one tool holder will usually last for a long time before it, too, must be replaced.

Indexable inserts and tool holders are made with a negative rake or with a positive rake. Negative rake inserts have the advantage of having twice as many cutting edges available as comparable positive rake inserts, because the cutting edges on both the top and bottom of negative rake inserts can be used, while only the top cutting edges can be used on positive rake inserts. Positive rake inserts have a distinct advantage when machining long and slender parts, thin-walled parts, or other parts that are subject to bending or chatter when the cutting load is applied to them, because the cutting force is significantly lower as compared to that for negative rake inserts. Indexable inserts can be obtained in the following forms: utility ground, or ground on top and bottom only; precision ground, or ground on all surfaces; prehoned to produce a slight rounding of the cutting edge; and precision molded, which are unground. Positive-negative rake inserts also are available. These inserts are held on a negative-rake tool holder and have a chipbreaker groove that is formed to produce an effective positive-rake angle while cutting. Cutting edges may be available on the top surface only, or on both top and bottom surfaces. The positive-rake chipbreaker surface may be ground or precision molded on the insert.

Many materials, such as gray cast iron, form a discontinuous chip. For these materials an insert that has plain faces without chipbreaker grooves should always be used. Steels and other ductile materials form a continuous chip that must be broken into small segments when machined on lathes and planers having single-point, cemented-carbide and cemented-oxide cutting tools; otherwise, the chips can cause injury to the operator. In this case a chipbreaker must be used. Some inserts are made with chipbreaker grooves molded or ground directly on the insert. When inserts with plain faces are used, a cemented-carbide plate-type chipbreaker is clamped on top of the insert.

Identification System for Indexable Inserts.—The size of indexable inserts is determined by the diameter of an inscribed circle (I.C.), except for rectangular and parallelogram inserts where the length and width dimensions are used. To describe an insert in its entirety, a standard ANSI B212.4-1986 identification system is used where each position number designates a feature of the insert. The ANSI Standard includes items now commonly used and facilitates identification of items not in common use. Identification consists of up to ten positions; each position defines a characteristic of the insert as shown below:

1	2	3	4	5	6	7	8[a]	9[a]	10[a]
T	**N**	**M**	**G**	**5**	**4**	**3**			**A**

[a] Eighth, Ninth, and Tenth Positions are used only when required.

1) *Shape:* The shape of an insert is designated by a letter: **R** for round; **S,** square; **T,** triangle; **A,** 85° parallelogram; **B,** 82° parallelogram; **C,** 80° diamond; **D,** 55° diamond; **E,** 75° diamond; **H,** hexagon; **K,** 55° parallelogram; **L,** rectangle; **M,** 86° diamond; **O,** octagon; **P,** pentagon; **V,** 35° diamond; and **W,** 80° trigon.

2) *Relief Angle (Clearances):* The second position is a letter denoting the relief angles; **N** for 0°; **A,** 3°; **B,** 5°; **C,** 7°; **P,** 11°; **D,** 15°; **E,** 20°; **F,** 25°; **G,** 30°; **H,** 0° & 11°*; **J,** 0° & 14°*; **K,** 0° & 17°*; **L,** 0° & 20°*; **M,** 11° & 14°*; **R,** 11° & 17°*; **S,** 11° & 20°*. When mounted on a holder, the actual relief angle may be different from that on the insert.

3) *Tolerances:* The third position is a letter and indicates the tolerances which control the indexability of the insert. Tolerances specified do not imply the method of manufacture.

* Second angle is secondary facet angle, which may vary by ± 1°.

Symbol	Tolerance (± from nominal)		Symbol	Tolerance (± from nominal)	
	Inscribed Circle, Inch	Thickness, Inch		Inscribed Circle, Inch	Thickness, Inch
A	0.001	0.001	**H**	0.0005	0.001
B	0.001	0.005	**J**	0.002–0.005	0.001
C	0.001	0.001	**K**	0.002–0.005	0.001
D	0.001	0.005	**L**	0.002–0.005	0.001
E	0.001	0.001	**M**	0.002–0.004[a]	0.005
F	0.0005	0.001	**U**	0.005–0.010[a]	0.005
G	0.001	0.005	**N**	0.002–0.004[a]	0.001

[a] Exact tolerance is determined by size of insert. See ANSI B94.25.

4) *Type:* The type of insert is designated by a letter. **A,** with hole; **B,** with hole and countersink; **C,** with hole and two countersinks; **F,** chip grooves both surfaces, no hole; **G,** same as **F** but with hole; **H,** with hole, one countersink, and chip groove on one rake surface; **J,** with hole, two countersinks and chip grooves on two rake surfaces; **M,** with hole and chip groove on one rake surface; **N,** without hole; **Q,** with hole and two countersinks; **R,** without hole but with chip groove on one rake surface; **T,** with hole, one countersink, and chip groove on one rake face; **U,** with hole, two countersinks, and chip grooves on two rake faces; and **W,** with hole and one countersink. *Note:* a dash may be used after position 4 to separate the shape-describing portion from the following dimensional description of the insert and is not to be considered a position in the standard description.

5) *Size:* The size of the insert is designated by a one- or a two-digit number. For regular polygons and diamonds, it is the number of eighths of an inch in the nominal size of the inscribed circle, and will be a one- or two-digit number when the number of eighths is a whole number. It will be a two-digit number, including one decimal place, when it is not a whole number. Rectangular and parallelogram inserts require two digits: the first digit indicates the number of eighths of an inch width and the second digit, the number of quarters of an inch length.

6) *Thickness:* The thickness is designated by a one- or two-digit number, which indicates the number of sixteenths of an inch in the thickness of the insert. It is a one-digit number when the number of sixteenths is a whole number; it is a two-digit number carried to one decimal place when the number of sixteenths of an inch is not a whole number.

7) *Cutting Point Configuration:* The cutting point, or nose radius, is designated by a number representing $\frac{1}{64}$ths of an inch; a flat at the cutting point or nose, is designated by a letter: **0** for sharp corner; **1,** $\frac{1}{64}$ inch radius; **2,** $\frac{1}{32}$ inch radius; **3,** $\frac{3}{64}$inch radius; **4,** $\frac{1}{16}$ inch radius; **5,** $\frac{5}{64}$ inch radius; **6,** $\frac{3}{32}$ inch radius; **7,** $\frac{7}{64}$ inch radius; **8,** $\frac{1}{8}$ inch radius; **A,** square insert with 45° chamfer; **D,** square insert with 30° chamfer; **E,** square insert with 15° chamfer; **F,** square insert with 3° chamfer; **K,** square insert with 30° double chamfer; **L,** square insert with 15° double chamfer; **M,** square insert with 3° double chamfer; **N,** truncated triangle insert; and **P,** flatted corner triangle insert.

8) *Special Cutting Point Definition:* The eighth position, if it follows a letter in the 7th position, is a number indicating the number of $\frac{1}{64}$ths of an inch measured parallel to the edge of the facet.

9) *Hand:* **R,** right; **L,** left; to be used when required in ninth position.

10) *Other Conditions:* The tenth position defines special conditions (such as edge treatment, surface finish) as follows: **A,** honed, 0.0005 inch to less than 0.003 inch; **B,** honed, 0.003 inch to less than 0.005 inch; **C,** honed, 0.005 inch to less than 0.007 inch; **J,** polished,

4 microinch arithmetic average (AA) on rake surfaces only; **T,** chamfered, manufacturer's standard negative land, rake face only.

Indexable Insert Tool Holders.—Indexable insert tool holders are made from a good grade of steel which is heat treated to a hardness of 44 to 48 Rc for most normal applications. Accurate pockets that serve to locate the insert in position and to provide surfaces against which the insert can be clamped are machined in the ends of tool holders. A cemented carbide seat usually is provided, and is held in the bottom of the pocket by a screw or by the clamping pin, if one is used. The seat is necessary to provide a flat bearing surface upon which the insert can rest and, in so doing, it adds materially to the ability of the insert to withstand the cutting load. The seating surface of the holder may provide a positive-, negative-, or a neutral-rake orientation to the insert when it is in position on the holder. Holders, therefore, are classified as positive, negative, or neutral rake.

Four basic methods are used to clamp the insert on the holder: 1) Clamping, usually top clamping; 2) Pin-lock clamping; 3) Multiple clamping using a clamp, usually a top clamp, and a pin lock; and 4) Clamping the insert with a machine screw.

Table 1. Standard Shank Sizes for Indexable Insert Holders

Basic Shank Size	Shank Dimensions for Indexable Insert Holders					
	A		B		C[a]	
	In.	mm	In.	mm	In.	mm
$\frac{1}{2} \times \frac{1}{2} \times 4\frac{1}{2}$	0.500	12.70	0.500	12.70	4.500	114.30
$\frac{5}{8} \times \frac{5}{8} \times 4\frac{1}{2}$	0.625	15.87	0.625	15.87	4.500	114.30
$\frac{5}{8} \times 1\frac{1}{4} \times 6$	0.625	15.87	1.250	31.75	6.000	152.40
$\frac{3}{4} \times \frac{3}{4} \times 4\frac{1}{2}$	0.750	19.05	0.750	19.05	4.500	114.30
$\frac{3}{4} \times 1 \times 6$	0.750	19.05	1.000	25.40	6.000	152.40
$\frac{3}{4} \times 1\frac{1}{4} \times 6$	0.750	19.05	1.250	31.75	6.000	152.40
$1 \times 1 \times 6$	1.000	25.40	1.000	25.40	6.000	152.40
$1 \times 1\frac{1}{4} \times 6$	1.000	25.40	1.250	31.75	6.000	152.40
$1 \times 1\frac{1}{2} \times 6$	1.000	25.40	1.500	38.10	6.000	152.40
$1\frac{1}{4} \times 1\frac{1}{4} \times 7$	1.250	31.75	1.250	31.75	7.000	177.80
$1\frac{1}{4} \times 1\frac{1}{2} \times 8$	1.250	31.75	1.500	38.10	8.000	203.20
$1\frac{3}{8} \times 2\frac{1}{16} \times 6\frac{3}{8}$	1.375	34.92	2.062	52.37	6.380	162.05
$1\frac{1}{2} \times 1\frac{1}{2} \times 7$	1.500	38.10	1.500	38.10	7.000	177.80
$1\frac{3}{4} \times 1\frac{3}{4} \times 9\frac{1}{2}$	1.750	44.45	1.750	44.45	9.500	241.30
$2 \times 2 \times 8$	2.000	50.80	2.000	50.80	8.000	203.20

[a] Holder length; may vary by manufacturer. Actual shank length depends on holder style.

All top clamps are actuated by a screw that forces the clamp directly against the insert. When required, a cemented-carbide, plate-type chipbreaker is placed between the clamp and the insert. Pin-lock clamps require an insert having a hole: the pin acts against the walls of the hole to clamp the insert firmly against the seating surfaces of the holder. Multiple or combination clamping, simultaneously using both a pin-lock and a top clamp, is recommended when taking heavier or interrupted cuts. Holders are available on which all the above-mentioned methods of clamping may be used. Other holders are made with only a top clamp or a pin lock. Screw-on type holders use a machine screw to hold the insert in the pocket. Most standard indexable insert holders are either straight-shank or offset-shank, although special holders are made having a wide variety of configurations.

The common shank sizes of indexable insert tool holders are shown in Table 1. Not all styles are available in every shank size. Positive- and negative-rake tools are also not available in every style or shank size. Some manufacturers provide additional shank sizes for certain tool holder styles. For more complete details the manufacturers' catalogs must be consulted.

Identification System for Indexable Insert Holders.—The following identification system conforms to the American National Standard, ANSI B212.5-1986, Metric Holders for Indexable Inserts.

Each position in the system designates a feature of the holder in the following sequence:

1	2	3	4	5	—	6	—	7	—	8[a]	—	9	—	10[a]
C	**T**	**N**	**A**	**R**	**—**	**85**	**—**	**25**	**—**	**D**	**—**	**16**	**—**	**Q**

1) *Method of Holding Horizontally Mounted Insert:* The method of holding or clamping is designated by a letter: **C,** top clamping, insert without hole; **M,** top and hole clamping, insert with hole; **P,** hole clamping, insert with hole; **S,** screw clamping through hole, insert with hole; **W,** wedge clamping.

2) *Insert Shape:* The insert shape is identified by a letter: **H,** hexagonal; **O,** octagonal; **P,** pentagonal; **S,** square; **T,** triangular; **C,** rhombic, 80° included angle; **D,** rhombic, 55° included angle; **E,** rhombic, 75° included angle; **M,** rhombic, 86° included angle; **V,** rhombic, 35° included angle; **W,** hexagonal, 80° included angle; **L,** rectangular; **A,** parallelogram, 85° included angle; **B,** parallelogram, 82° included angle; **K,** parallelogram, 55° included angle; **R,** round. The included angle is always the smaller angle.

3) *Holder Style:* The holder style designates the shank style and the side cutting edge angle, or end cutting edge angle, or the purpose for which the holder is used. It is designated by a letter: **A,** for straight shank with 0° side cutting edge angle; **B,** straight shank with 15° side cutting edge angle; **C,** straight-shank end cutting tool with 0° end cutting edge angle; **D,** straight shank with 45° side cutting edge angle; **E,** straight shank with 30° side cutting edge angle; **F,** offset shank with 0° end cutting edge angle; **G,** offset shank with 0° side cutting edge angle; **J,** offset shank with negative 3° side cutting edge angle; **K,** offset shank with 15° end cutting edge angle; **L,** offset shank with negative 5° side cutting edge angle and 5° end cutting edge angle; **M,** straight shank with 40° side cutting edge angle; **N,** straight shank with 27° side cutting edge angle; **R,** offset shank with 15° side cutting edge angle; **S,** offset shank with 45° side cutting edge angle; **T,** offset shank with 30° side cutting edge angle; **U,** offset shank with negative 3° end cutting edge angle; **V,** straight shank with 17½° side cutting edge angle; **W,** offset shank with 30° end cutting edge angle; **Y,** offset shank with 5° end cutting edge angle.

4) *Normal Clearances:* The normal clearances of inserts are identified by letters: **A,** 3°; **B,** 5°; **C,** 7°; **D,** 15°; **E,** 20°; **F,** 25°; **G,** 30°; **N,** 0°; **P,** 11°.

5) *Hand of tool:* The hand of the tool is designated by a letter: **R** for right-hand; **L,** left-hand; and **N,** neutral, or either hand.

6) *Tool Height for Rectangular Shank Cross Sections:* The tool height for tool holders with a rectangular shank cross section and the height of cutting edge equal to shank height is given as a two-digit number representing this value in millimeters. For example, a height of 32 mm would be encoded as 32; 8 mm would be encoded as 08, where the one-digit value is preceded by a zero.

7) *Tool Width for Rectangular Shank Cross Sections:* The tool width for tool holders with a rectangular shank cross section is given as a two-digit number representing this value in millimeters. For example, a width of 25 mm would be encoded as 25; 8 mm would be encoded as 08, where the one-digit value is preceded by a zero.

8) *Tool Length:* The tool length is designated by a letter: **A,** 32 mm; **B,** 40 mm; **C,** 50 mm; **D,** 60 mm; **E,** 70 mm; **F,** 80 mm; **G,** 90 mm; **H,** 100 mm; **J,** 110 mm; **K,** 125 mm; **L,** 140 mm; **M,** 150 mm; **N,** 160 mm; **P,** 170 mm; **Q,** 180 mm; **R,** 200 mm; **S,** 250 mm; **T,** 300 mm; **U,** 350 mm; **V,** 400 mm; **W,** 450 mm; **X,** special length to be specified; **Y,** 500 mm.

9) *Indexable Insert Size:* The size of indexable inserts is encoded as follows: For insert shapes **C, D, E, H. M, O, P, R, S, T, V**, the side length (the diameter for **R** inserts) in millimeters is used as a two-digit number, with decimals being disregarded. For example, the symbol for a side length of 16.5 mm is 16. For insert shapes **A, B, K, L**, the length of the main cutting edge or of the longer cutting edge in millimeters is encoded as a two-digit number, disregarding decimals. If the symbol obtained has only one digit, then it should be preceded by a zero. For example, the symbol for a main cutting edge of 19.5 mm is 19; for an edge of 9.5 mm, the symbol is 09.

10) *Special Tolerances:* Special tolerances are indicated by a letter: **Q,** back and end qualified tool; **F,** front and end qualified tool; **B,** back, front, and end qualified tool. A qualified tool is one that has tolerances of ± 0.08 mm for dimensions *F*, *G*, and *C*. (See Table 3.)

Table 2. Qualified Indexable Insert Tool Holders: (left) Letter Designations for Shank Qualification and (right) Insert Nose Radius for Corresponding Inscribed Circle of Gage Insert

Length and Width Control		Length and Side Control		Insert Nose Radius, Inch	Insert Inscribed Circle (I.C.) Diam., Inch
Letter	Length, Inch	Letter	Length, Inch		
A	4.000	M	4.000		¼
B	4.500	N	4.500	.015	or
C	5.000	P	5.000		5⁄16
D	6.000	R	6.000		⅜
E	7.000	S	7.000	.031	or
F	8.000	T	8.000		½
G	5.500	U	5.500		⅝
H	9.000	V	9.000	.047	or
J	Unassigned	W	Unassigned		¾
K	12.000	X	12.000		
L	14.000	Y	14.000	.094	1
		Z	Length Only		

Table 3. Letter Symbols for Qualification of Tool Holders — Position 10
ANSI B212.5-1986

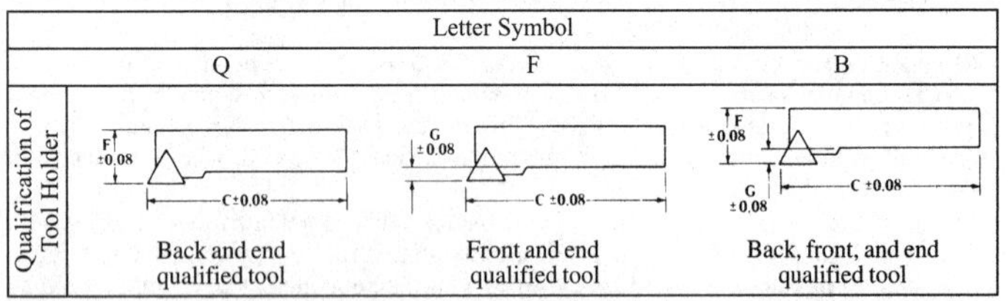

	Letter Symbol		
	Q	F	B
Qualification of Tool Holder	F ±0.08; C±0.08 — Back and end qualified tool	G ±0.08; C ±0.08 — Front and end qualified tool	F ±0.08; G ±0.08; C ±0.08 — Back, front, and end qualified tool

Selecting Indexable Insert Holders.—A guide for selecting indexable insert holders is provided by Table 4b. Some operations such as deep grooving, cut-off, and threading are not given in this table. However, tool holders designed specifically for these operations are available. The boring operations listed in Table 4b refer primarily to larger holes, into which the holders will fit. Smaller holes are bored using boring bars. An examination of this table shows that several tool-holder styles can be used and frequently are used for each operation. Selection of the best holder for a given job depends largely on the job and there are certain basic facts that should be considered in making the selection.

Rake Angle: A negative-rake insert has twice as many cutting edges available as a comparable positive-rake insert. Sometimes the tool life obtained when using the second face may be less than that obtained on the first face because the tool wear on the cutting edges of the first face may reduce the insert strength. Nevertheless, the advantage of negative-rake inserts and holders is such that they should be considered first in making any choice. Positive-rake holders should be used where lower cutting forces are required, as when machining slender or small-diameter parts, when chatter may occur, and for machining some materials, such as aluminum, copper, and certain grades of stainless steel, when positive-negative rake inserts can sometimes be used to advantage. These inserts are held on negative-rake holders that have their rake surfaces ground or molded to form a positive-rake angle.

Insert Shape: The configuration of the workpiece, the operation to be performed, and the lead angle required often determine the insert shape. When these factors need not be considered, the insert shape should be selected on the basis of insert strength and the maximum number of cutting edges available. Thus, a round insert is the strongest and has a maximum number of available cutting edges. It can be used with heavier feeds while producing a good surface finish. Round inserts are limited by their tendency to cause chatter, which may preclude their use. The square insert is the next most effective shape, providing good corner strength and more cutting edges than all other inserts except the round insert. The only limitation of this insert shape is that it must be used with a lead angle. Therefore, the square insert cannot be used for turning square shoulders or for back-facing. Triangle inserts are the most versatile and can be used to perform more operations than any other insert shape. The 80-degree diamond insert is designed primarily for heavy turning and facing operations, using the 100-degree corners, and for turning and back-facing square shoulders using the 80-degree corners. The 55- and 35-degree diamond inserts are intended primarily for tracing.

Lead Angle: Tool holders should be selected to provide the largest possible lead angle, although limitations are sometimes imposed by the nature of the job. For example, when tuning and back-facing a shoulder, a negative lead angle must be used. Slender or small-diameter parts may deflect, causing difficulties in holding size, or chatter when the lead angle is too large.

Table 4a. Maximum Plunge Angle for Tracing or Contour Turning

Tool Holder Style	Insert Shape	Maximum Plunge Angle	Tool Holder Style	Insert Shape	Maximum Plunge Angle
E	T	58°	J	D	30°
D and S	S	43°	J	V	50°
H	D	71°	N	T	55°
J	T	25°	N	D	58°–60°

End Cutting Edge Angle: When tracing or contour turning, the plunge angle is determined by the end cutting edge angle. A 2-deg minimum clearance angle should be provided between the workpiece surface and the end cutting edge of the insert. Table 4a provides the maximum plunge angle for holders commonly used to plunge when tracing where insert shape identifiers are S = square; T = triangle; D = 55-deg diamond, V = 35-deg diamond. When severe cratering cannot be avoided, an insert having a small, end cutting edge angle is desirable to delay the crater breakthrough behind the nose. For very heavy cuts a small, end cutting edge angle will strengthen the corner of the tool.

Table 4b. Indexable Insert Holder Application Guide

Tool	Tool Holder Style	Insert Shape	Rake N-Negative P-Positive	Application: Turn	Face	Turn and Face	Turn and Backface	Trace	Groove	Chamfer	Bore	Plane
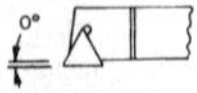 0°	A	T	N	●	●						●	
			P	●	●						●	
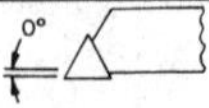 0°	A	T	N	●	●			●				
			P	●	●			●				
	A	R	N	●	●	●						●
	A	R	N	●	●	●		●				●
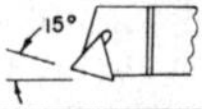 15°	B	T	N	●	●						●	
			P	●	●						●	
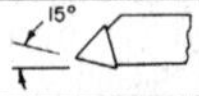 15°	B	T	N	●	●			●			●	
			P	●	●			●			●	
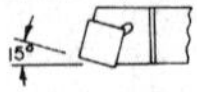 15°	B	S	N	●	●						●	
			P	●	●						●	
 5° 15°	B	C	N	●	●	●					●	●
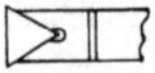	C	T	N	●	●				●	●		
			P	●	●				●	●		
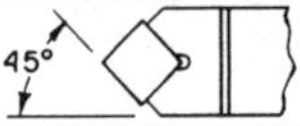 45°	D	S	N	●	●	●		●		●	●	●
			P	●	●	●		●		●	●	●
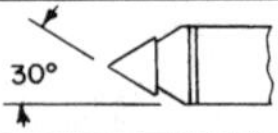 30°	E	T	N	●	●			●	●	●		
			P	●	●			●	●	●		
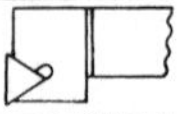	F	T	N	●	●						●	
			P	●	●						●	
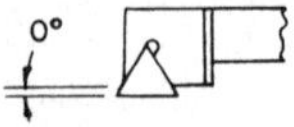 0°	G	T	N	●	●						●	
			P	●	●						●	
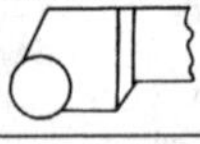	G	R	N	●	●	●						
 10° 0°	G	C	N	●	●	●						
			P	●	●	●						

Table 4b. *(Continued)* **Indexable Insert Holder Application Guide**

Tool	Tool Holder Style	Insert Shape	Rake N-Negative P-Positive	Application								
				Turn	Face	Turn and Face	Turn and Backface	Trace	Groove	Chamfer	Bore	Plane
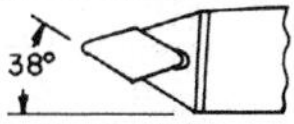 38°	H	D	N	●	●			●				
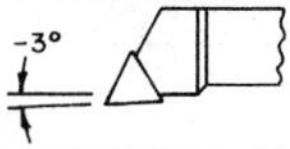 −3°	J	T	N				●	●				
			P				●	●				
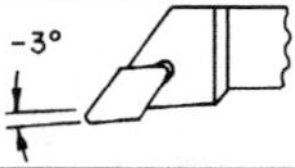 −3°	J	D	N				●	●				
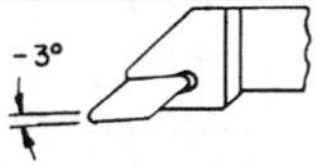 −3°	J	V	N				●	●				
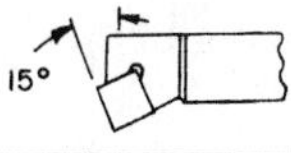 15°	K	S	N	●	●						●	
			P	●	●						●	
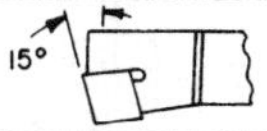 15°	K	C	N	●	●						●	
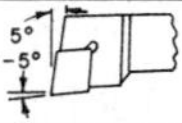 5° −5°	L	C	N			●	●					
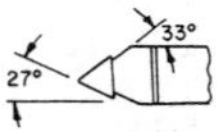 27° 33°	N	T	N	●	●			●				
			P	●	●			●				
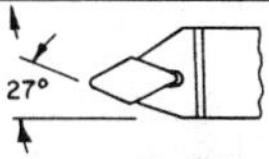 27°	N	D	N	●	●			●				
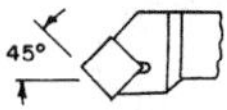 45°	S	S	N	●	●	●		●		●	●	●
			P	●	●	●		●		●	●	●
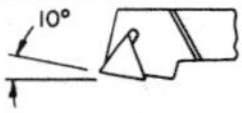 10°	W	S	N	●	●							

Table 5. Single-Point Turning and Boring Tools— Rake and Relief Angles-1

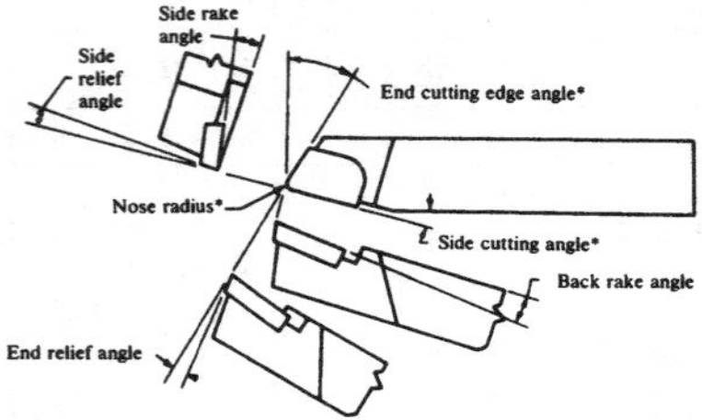

Side rake angle
Side relief angle
End cutting edge angle*
Nose radius*
Side cutting angle*
Back rake angle
End relief angle

Note: All rake and relief angles are measured in normal direction.

* Use the largest nose radius and the largest side cutting edge angle or end cutting edge angle that are consistent with part requirements.

Material	Hardness Bhn	High Speed Steel				Carbide					
						Brazed			Indexable		
		Back Rake Angle degrees	Side Rake Angle degrees	End Relief Angle degrees	Side Relief Angle degrees	Back Rake Angle degrees	Side Rake Angle degrees	Relief Angles degrees	Back Rake Angle degrees	Side Rake Angle degrees	Relief Angles degrees
Free Machining Carbon Steels-Wrought	85-225	10	12	5	5	0	6	7	0	5	5
Carbon Steels-Wrought and Cast	225-325	8	10	5	5	0	6	7	0	5	5
Free Machining Alloy Steels-Wrought Alloy Steels-Wrought and Cast; High Strength Steels-Wrought; Maraging Steels-Wrought; Tool Steels-Wrought; Nitriding Steels-Wrought; Armor Plate-Wrought; Structural Steels-Wrought	325-52Rc	0	10	5	5	0	6	7	−5	−5	5
	52Rc-58Rc	…	…	…	…	…	…	…	−5	−5	5
Free Machining Stainless Steels-Wrought	135-275	5	8	5	5	0	6	7	−5	−5	5
	275-425	0	10	5	5	0	6	7	−5	−5	5
Stainless Steels, Ferritic-Wrought and Cast	135-185	5	8	5	5	0	6	7	0	5	5
Stainless Steels, Austenitic-Wrought and Cast	135-275	0	10	5	5	0	6	7	0	5	5
Stainless Steels, Martensitic-Wrought and Cast	135-325	0	10	5	5	0	6	7	0	5	5
	325-425 48Rc-52Rc	0	10	5	5	0	6	7	−5	−5	5

Table 6. Single-Point Turning and Boring Tools—Rake and Relief Angles-2

Material	Hardness Bhn	High Speed Steel				Carbide					
						Brazed			Indexable		
		Back Rake Angle degrees	Side Rake Angle degrees	End Relief Angle degrees	Side Relief Angle degrees	Back Rake Angle degrees	Side Rake Angle degrees	Relief Angles degrees	Back Rake Angle degrees	Side Rake Angle degrees	Relief Angles degrees
Precipitation Hardening Stainless Steels-Wrought and Cast	150-450	0	10	5	5	0	6	7	–5	– 5	5
Gray Cast Irons Ductile Cast Irons Malleable Cast Irons Compacted Graphite Cast Irons White Cast Irons	100-200 200-300 300-400	5 5 5	10 8 5	5 5 5	5 5 5	0 0 –5	6 6 –5	7 7 7	–5 –5 –5	–5 –5 –5	5 5 5
Aluminum Alloys-Wrought and Cast	30-150 500 kg	20	15	12	10	3	15	7	0	5	5
Magnesium Alloys-Wrought and Cast	40-90 500 kg	20	15	12	10	3	15	7	0	5	5
Titanium Alloys-Wrought and Cast	110-440	0	5	5	5	0	6	7	–5	–5	5
Copper Alloys-Wrought and Cast	40-200 500 kg	5	10	8	8	0	8	7	0	5	5
Nickel Alloys-Wrought and Cast Chrome-Nickel Alloys Beryllium-Nickel Alloys	80-360	8	10	12	12	0	6	7	–5	–5	5
Nitinol Alloys-Wrought	210-340 48Rc-52Rc	…	…	…	…	0	5	7	0	5	5
High Temperature Alloys-Wrought and Cast	140-475	0	10	5	5	0	6	7	5	0	5

Table 7. Single-Point Turning and Boring Tools—Rake and Relief Angles-3

Material		Hardness Bhn	High Speed Steel				Carbide					
							Brazed			Indexable		
			Back Rake Angle degrees	Side Rake Angle degrees	End Relief Angle degrees	Side Relief Angle degrees	Back Rake Angle degrees	Side Rake Angle degrees	Relief Angles degrees	Back Rake Angle degrees	Side Rake Angle degrees	Relief Angles degrees
Columbium Alloys-Wrought, Cast, P/M Molybdenum Alloys-Wrought, Cast, P/M Tantalum Alloys-Wrought, Cast, P/M		170-290	0	20	5	5	0	20	7	…	…	5
Tungsten Alloys-Wrought Cast P/M		180-320	…	…	…	…	– 15	0	7	…	…	…
Zinc Alloys-Cast		80-100	10	10	12	4	5	5	7	0	5	5
Uranium Alloys-Wrought		190-2 10	…	…	…	…	0	0	7	–5	0	7
Zirconium Alloys-Wrought		140-280	15	10	10	10	5	5	7	5	5	6
Thermoplastics		All	0	0	20 to 30	15 to 20	0	0	20 to 30	0	0	20 to 30
Thermosetting Plastics		All	0	0	20 to 30	15 to 20	0	15	7	0	15	5
Magnetic Alloys, Nickel- and Cobalt-Base Controlled Expansion Alloys		125-250	10	8	8	8	…	…	…	…	…	…
Powder Metal Alloys	Copper	All	10	8	8	8	6	12	7	…	…	…
	Iron	All	0	0	8	8	6	16	7	0	0	5
Magnetic Core Iron		185-240	15	30	8	8	20	0	7	5	5	5
Carbon and Graphite		All	0	0	20	20	0	0	20	0	0	15
Machinable Carbide (Ferro-Tic)		40Rc-51Rc	–5	–5	5	5	–5	–5	7	–5	–5	5
Machinable Glass Ceramic		250 Knoop 100 g	0	15	5	5	1	0	7	0	0	5

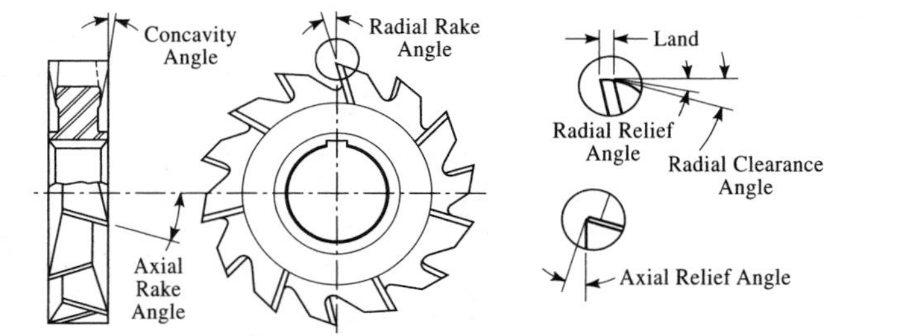

Table 8. Arbor-Mounted Side and Slot Mills— Rake and Relief Angles

Material	Hardness Bhn	High Speed Steel				Carbide			
		Axial Rake Angle degrees	Radial Rake Angle degrees	Axial Relief Angle degrees	Radial Relief Angle degrees	Axial Rake Angle degrees	Radial Rake Angle degrees	Axial Relief Angle degrees	Radial Relief Angle degrees
Free Machining Carbon Steels-Wrought Carbon Steels-Wrought and Cast Free Machining Alloy Steels-Wrought	85 - 325	10 to 15	10 to 15	3 to 5	4 to 8	0 to −5	−5 to 5	2 to 4	5 to 8
	325 - 425	10 to 12	5 to 12	3 to 5	4 to 8	0 to −5	−5 to 5	2 to 4	5 to 8
	45Rc-52Rc	10 to 12	5 to 12	2 to 4	3 to 7	−5 to − 10	0 to − 10	2 to 4	5 to 8
Alloy Steels-Wrought and Cast	125 - 425	10 to 12	5 to 12	3 to 5	4 to 8	−5 to −10	0 to −10	2 to 5	5 to 8
	45Rc-52Rc	10 to 12	5 to 12	2 to 4	3 to 7	−5 to −10	0 to −10	2 to 4	3 to 6

Table 8. *(Continued)* **Arbor-Mounted Side and Slot Mills— Rake and Relief Angles**

Material	Hardness Bhn	High Speed Steel				Carbide			
		Axial Rake Angle degrees	Radial Rake Angle degrees	Axial Relief Angle degrees	Radial Relief Angle degrees	Axial Rake Angle degrees	Radial Rake Angle degrees	Axial Relief Angle degrees	Radial Relief Angle degrees
High Strength Steels, Maraging Steels and Tool Steels-Wrought	100 - 52Rc	10 to 12	5 to 12	2 to 4	3 to 7	−5 to −10	0 to −10	2 to 4	5 to 8
Nitriding Steels-Wrought	200 - 350	10 to 12	5 to 12	2 to 4	3 to 7	−5 to −10	0 to −10	2 to 4	3 to 6
Armor Plate-Wrought	250 - 320	0 to 5	0 to 5	2 to 4	3 to 7	−5 to −10	−5 to −10	2 to 4	3 to 6
Structural Steels-Wrought	100 - 50Rc	10 to 12	5 to 12	3 to 5	4 to 8	0 to −5	0 to −5	2 to 4	5 to 8
Free Machining Stainless Steels-Wrought	135 - 425	10 to 12	5 to 12	3 to 5	4 to 8	0 to 5	−5 to 5	2 to 4	5 to 8
Stainless Steels, Ferritic-Wrought and Cast Stainless Steels, Austenitic-Wrought and Cast	135 - 52Rc	10 to 12	5 to 12	3 to 5	4 to 8	0 to 5	−5 to 5	2 to 4	5 to 8
Stainless Steels, Martensitic-Wrought and Cast	135 - 52Rc	10 to 12	5 to 12	2 to 4	3 to 7	−5 to −10	0 to – 10	2 to 4	5 to 8
Precipitation Hardening Stainless Steels-Wrought and Cast	150 - 450	10 to 12	5 to 12	2 to 4	4 to 8	0 to −5	0 to −10	2 to 4	5 to 8
Gray Cast Irons Ductile Cast Irons Malleable Cast Irons	100 - 400	10 to 12	10 to 12	2 to 4	3 to 7	0 to −10	5 to −10	3 to 5	5 to 8
Aluminium Alloys-Wrought and Cast	30 - 150 500 kg	12 to 25	10 to 20	5 to 7	5 to 11	10 to 20	5 to 15	5 to 7	7 to 10

Table 8. *(Continued)* **Arbor-Mounted Side and Slot Mills— Rake and Relief Angles**

Material	Hardness Bhn	High Speed Steel				Carbide			
		Axial Rake Angle degrees	Radial Rake Angle degrees	Axial Relief Angle degrees	Radial Relief Angle degrees	Axial Rake Angle degrees	Radial Rake Angle degrees	Axial Relief Angle degrees	Radial Relief Angle degrees
Magnesium Alloys-Wrought and Cast	40 - 90 500kg	12 to 25	10 to 20	5 to 7	5 to 11	10 to 20	5 to 15	5 to 7	7 to 10
Titanium Alloys-Wrought	110 - 440	10 to 15	5 to 10	5 to 7	5 to 11	0 to −10	0 to −10	5 to 7	5 to 8
Copper Alloys-Wrought and Cast	40 - 200 500kg	12 to 25	10 to 20	5 to 7	5 to 11	10 to 20	5 to 10	4 to 7	5 to 8
Nickel Alloys-Wrought and Cast	80 - 360	10 to 20	10 to 15	3 to 5	4 to 8	−5 to −10	0 to − 10	3 to 5	5 to 8
High Temperature Alloys-Wrought and Cast	140 - 300	10 to 15	10 to 15	1 to 5	5 to 10	−5 to − 10	0 to − 10	3 to 5	5 to 8
	300 - 475	10 to 12	5 to 12	1 to 5	4 to 8	−5 to − 10	0 to − 10	3 to 5	5 to 8
Columbium, Molybdenum Alloys-Wrought, Cast, P/M	170 - 290	0	15 to 20	3 to 5	5 to 10	0	5 to 15	7 to 10	7 to 10
Tantalum Alloys-Wrought, Cast, P/M	200 - 250	0	15 to 20	3 to 5	5 to 10	0	5 to 15	7 to 10	7 to 10
Tungsten Alloys-Wrought, Cast, P/M	180 - 320	…	…	…	…	−10 to −15	5 to 15	10 to 15	10 to 15
Zinc Alloys-Cast	80 - 100	10 to 20	10 to 20	5 to 7	8 to 11	10 to 15	10 to 15	7 to 10	7 to 10

Reprinted from the MACHINING DATA HANDBOOK. 3rd Edition, by permission of the Machinability Data Center. © 1980 by Metcut Research Associates Inc.

Table 9. High-Speed-Steel Peripheral and Slotting End Mills—Tool Geometry

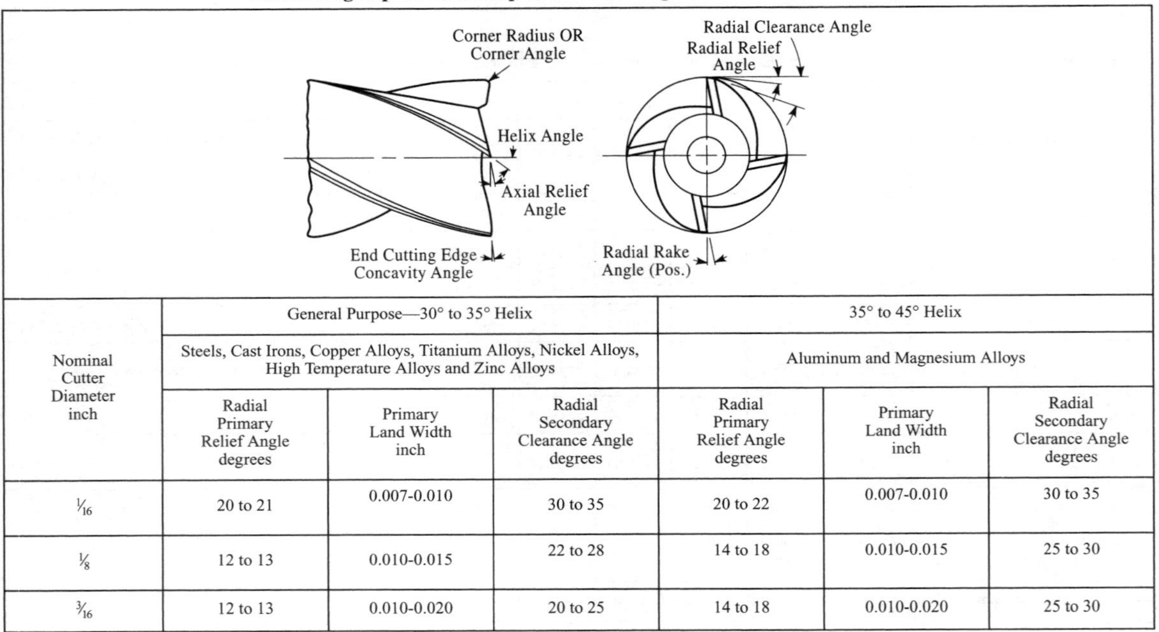

Nominal Cutter Diameter inch	General Purpose—30° to 35° Helix			35° to 45° Helix		
	Steels, Cast Irons, Copper Alloys, Titanium Alloys, Nickel Alloys, High Temperature Alloys and Zinc Alloys			Aluminum and Magnesium Alloys		
	Radial Primary Relief Angle degrees	Primary Land Width inch	Radial Secondary Clearance Angle degrees	Radial Primary Relief Angle degrees	Primary Land Width inch	Radial Secondary Clearance Angle degrees
1⁄16	20 to 21	0.007-0.010	30 to 35	20 to 22	0.007-0.010	30 to 35
1⁄8	12 to 13	0.010-0.015	22 to 28	14 to 18	0.010-0.015	25 to 30
3⁄16	12 to 13	0.010-0.020	20 to 25	14 to 18	0.010-0.020	25 to 30

Table 10. High-Speed-Steel Peripheral and Slotting End Mills—Tool Geometry

mm	degrees	mm	degrees	degrees	mm	degrees
¼	10 to 11	0.010-0.020	20 to 25	12 to 15	0.010-0.020	22 to 28
5⁄16	10 to 11	0.015-0.025	20 to 25	12 to 14	0.015-0.025	21 to 28
⅜	10 to 11	0.015-0.025	17 to 20	12 to 14	0.015-0.025	19 to 26
7⁄16	9 to 10	0.020-0.030	17 to 20	11 to 13	0.020-0.030	18 to 25
½	9 to 10	0.020-0.030	17 to 20	11 to 13	0.020-0.030	18 to 25
⅝	9 to 10	0.025-0.035	17 to 20	11 to 13	0.025-0.035	18 to 25
¾	8 to 9	0.030-0.040	15 to 18	10 to 12	0.030-0.040	17 to 24
⅞	8 to 9	0.030-0.040	15 to 18	10 to 12	0.030-0.040	17 to 24
1	8 to 9	0.035-0.050	15 to 18	10 to 12	0.035-0.050	16 to 23
1¼	7 to 8	0.040-0.060	13 to 18	9 to 11	0.040-0.060	14 to 22
1½	7 to 8	0.040-0.060	11 to 17	9 to 11	0.040-0.060	13 to 21
1¾	7 to 8	0.040-0.060	10 to 16	8 to 10	0.040-0.060	12 to 20
2	6 to 7	0.040-0.060	9 to 15	8 to 10	0.040-0.060	12 to 20
M1.6	20 to 21	0.200-0.250	30 to 35	20 to 22	0.200-0.250	30 to 35
M3	14 to 15	0.250-0.350	24 to 30	15 to 20	0.250-0.350	25 to 30
M4	12 to 13	0.250-0.400	22 to 28	14 to t8	0.250-0.400	25 to 30
M6	10 to 11	0.250-0.500	20 to 25	12 to 15	0.250-0.500	22 to 28
M7	10 to 11	0.250-0.500	20 to 25	12 to 15	0.250-0.500	22 to 28
M8	10 to 11	0.400-0.650	20 to 25	12 to 14	0.400-0.650	19 to 26
M10	10 to 11	0.400-0.650	17 to 20	12 to 14	0.400-0.650	19 to 26
M12	9 to 10	0.500-0.750	17 to 20	11 to 13	0.500-0.750	18 to 25
M14	9 to 10	0.500-0.750	17 to 20	11 to 13	0.500-0.750	18 to 25
M16	9 to 10	0.650-0.900	17 to 20	11 to 13	0.650-0.900	18 to 25
M20	8 to 9	0.750-1.000	15 to 18	10 to 12	0.750-1.000	17 to 24
M22	8 to 9	0.750-1.000	15 to 18	10 to 12	0.750-1.000	17 to 24
M25	8 to 9	0.900-1.250	15 to 18	10 to 12	0.900-1.250	16 to 23
M32	7 to 8	1.000-1.500	13 to 18	9 to 11	1.000-1.500	14 to 22
M40	7 to 8	1.000-1.500	11 to 17	9 to 11	1.000-1.500	13 to 21
M45	7 to 8	1.000-1.500	10 to 16	8 to 10	1.000-1.500	12 to 20

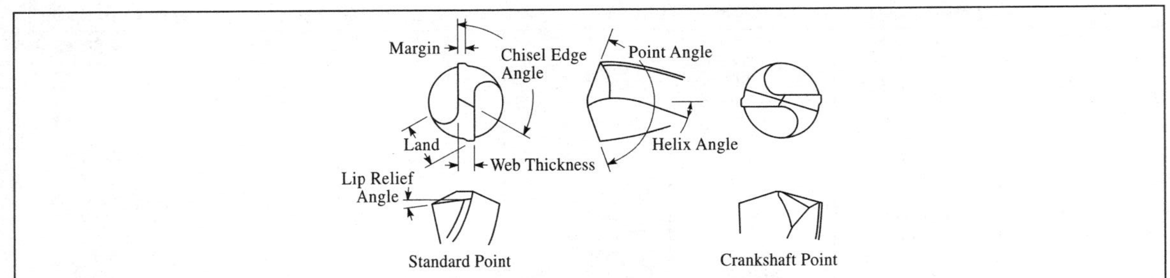

Table 11. High-Speed-Steel Twist Drill—Tool Geometry

<table>
<tr><th>Material</th><th>Hardness Bhn</th><th>Drill Type</th><th>Point Angle[a] degrees</th><th>Lip Relief Angle[b] degrees</th><th>Helix Angle</th><th>Point Grind</th></tr>
<tr><td rowspan="4">Free Machining Carbon Steels-Wrought
Carbon Steels-Wrought, Cast and P/M
Free Machining Alloy Steels-Wrought
Alloy Steels-Wrought, Cast and P/M
Maraging Steels-Wrought
Tool Steels-Wrought, Cast and P/M
Nitriding Steels-Wrought
Armor Plate-Wrought
Structural Steels-Wrought</td><td>85 - 225</td><td>General Purpose</td><td>118</td><td>A</td><td>Standard</td><td>Standard</td></tr>
<tr><td>225 - 325</td><td>General Purpose</td><td>118</td><td>A</td><td>Standard</td><td>Standard</td></tr>
<tr><td>325 - 425</td><td>General Purpose</td><td>118 to 135</td><td>B</td><td>Standard</td><td>Crankshaft</td></tr>
<tr><td>45Rc-52Rc</td><td>Heavy Web</td><td>118 to 135</td><td>B</td><td>Standard</td><td>Crankshaft</td></tr>
<tr><td>High Strength Steels-Wrought</td><td>175 - 325
325 - 52Rc</td><td>Heavy Web</td><td>118
118 to 135</td><td>A
B</td><td>Standard</td><td>Crankshaft</td></tr>
<tr><td>Austenitic Manganese Steels-Cast</td><td>150-220</td><td>Rail Drill</td><td>135</td><td>B</td><td>Low</td><td>Split</td></tr>
<tr><td>Free Machining Stainless Steels-Wrought</td><td>135-425</td><td>General Purpose</td><td>118</td><td>A</td><td>Standard</td><td>Crankshaft</td></tr>
</table>

Table 11. High-Speed-Steel Twist Drill—Tool Geometry

Material	Hardness Bhn	Drill Type	Point Angle[a] degrees	Lip Relief Angle[b] degrees	Helix Angle	Point Grind
Stainless Steels, Ferritic-Wrought and Cast Stainless Steels, Austenitic-Wrought, Cast and P/M Stainless Steels, Martensitic-Wrought, Cast and P/M Precipitation Hardening Stainless Steels-Wrought and Cast	135-200	General Purpose	118 to 135	A	Standard	Standard
	200-325	General Purpose	118 to 135	A	Standard	Crankshaft
	325-425 48Rc-52Rc	Heavy Web	118 to 135	B	Standard	Crankshaft
Gray Cast Irons Ductile Cast Irons Malleable Cast Irons Compacted Graphite Cast Iron White Cast Iron	110-220	General Purpose	118	A	Standard	Standard
	220-400	Heavy Web	118	A	Standard	Standard
Aluminum Alloys-Wrought and Cast	30-150 500 kg	Polished Flutes	90 to 118	C	High	Standard
Magnesium Alloys-Wrought and Cast	40-90 500 kg	Polished Flutes	70 to 118	C	High	Standard
Titanium Alloys-Wrought and Cast	110-275	General Purpose	118 to 135	B	Standard	Crankshaft
	275-440	Heavy Web	118 to 135	B	Standard	Crankshaft
Copper Alloys-Wrought, Cast and P/M	40-200 500 kg	Polished Flutes	118	C	Low	Standard
Nickel Alloys-Wrought, Cast and P/M Chromium-Nickel Alloys-Cast	80-360	General Purpose	118	B	Standard	Crankshaft
Nitinol Alloys-Wrought	210-360 48Rc-52Rc	General Purpose	118	B	Standard	Crankshaft
High Temperature Alloys-Wrought and Cast	140-475	Heavy Web	118 to 135	B	Standard	Crankshaft
Columbium, Molybdenum and Tantalum Alloys-Wrought, Cast. P/M	170-290	General Purpose	118	B	Standard	Standard

Table 11. High-Speed-Steel Twist Drill—Tool Geometry

Material	Hardness Bhn	Drill Type	Point Angle[a] degrees	Lip Relief Angle[b] degrees	Helix Angle	Point Grind
Tungsten Alloys (Anviloy)[c]	290-320	General Purpose	118	B	Standard	Standard
Zinc Alloys-Cast	80-100	General Purpose	118	C	Standard	Standard
Uranium Alloys-Wrought[d]	190-210	Special Carbide	118	5-8	20°	−5° Land on Drill Lip
Zirconium Alloys-Wrought	140-280	General Purpose	118	A	Standard	Crankshaft
Thermoplastics and Thermosetting Plastics[c]	All	Special Polished	60 to 90	C	Low	Standard
Magnetic Core Iron	185-240	General Purpose	100 to 118	A	High	Standard
Controlled Expansion Alloys	125-250	General Purpose	118	A	Standard	Crankshaft
Magnetic Alloys (Hi Perm 49, HyMu 80)	185-240	General Purpose	118	A	Standard	Crankshaft
Carbon and Graphite	8-100 Shore	General Purpose	90 to 118	C	Standard	Crankshaft
Machinable Carbide (Ferro-Tic)	40Rc-51Rc	General Purpose	118	B	Standard	Crankshaft

[a] Chisel edge angle: 115° to 135°
[b] See following chart.
[c] For both hss and carbide drills.
[d] For carbide drills.
Note: Use stub-length drills whenever possible on high strength materials.

Relief Angles at Periphery	Drill Size							
	#80 to #61	#60 to #41	#40 to #31	⅛ to ¼	¼ to ⅜	⅜ to ½	½ to ¾	1 inch up
A	24°	21°	18°	16°	14°	12°	10°	8°
B	20°	18°	16°	14°	12°	10°	8°	7°
C	26°	24°	22°	20°	18°	16°	14°	12°

Table 12. Length of Point on Twist Drills and Centering Tools

Size of Drill	Decimal Equivalent	Length of Point when Included Angle =90°	Length of Point when Included Angle =118°	Size of Drill	Decimal Equivalent	Length of Point when Included Angle =90°	Length of Point when Included Angle =118°	Size or Dia. of Drill	Decimal Equivalent	Length of Point when Included Angle =90°	Length of Point when Included Angle =118°	Dia. of Drill	Decimal Equivalent	Length of Point when Included Angle =90°	Length of Point when Included Angle =118°
60	0.0400	0.020	0.012	37	0.1040	0.052	0.031	14	0.1820	0.091	0.055	3/8	0.3750	0.188	0.113
59	0.0410	0.021	0.012	36	0.1065	0.054	0.032	13	0.1850	0.093	0.056	25/64	0.3906	0.195	0.117
58	0.0420	0.021	0.013	35	0.1100	0.055	0.033	12	0.1890	0.095	0.057	13/32	0.4063	0.203	0.122
57	0.0430	0.022	0.013	34	0.1110	0.056	0.033	11	0.1910	0.096	0.057	27/64	0.4219	0.211	0.127
56	0.0465	0.023	0.014	33	0.1130	0.057	0.034	10	0.1935	0.097	0.058	7/16	0.4375	0.219	0.131
55	0.0520	0.026	0.016	32	0.1160	0.058	0.035	9	0.1960	0.098	0.059	29/64	0.4531	0.227	0.136
54	0.0550	0.028	0.017	31	0.1200	0.060	0.036	8	0.1990	0.100	0.060	15/32	0.4688	0.234	0.141
53	0.0595	0.030	0.018	30	0.1285	0.065	0.039	7	0.2010	0.101	0.060	31/64	0.4844	0.242	0.145
52	0.0635	0.032	0.019	29	0.1360	0.068	0.041	6	0.2040	0.102	0.061	1/2	0.5000	0.250	0.150
51	0.0670	0.034	0.020	28	0.1405	0.070	0.042	5	0.2055	0.103	0.062	33/64	0.5156	0.258	0.155
50	0.0700	0.035	0.021	27	0.1440	0.072	0.043	4	0.2090	0.105	0.063	17/32	0.5313	0.266	0.159
49	0.0730	0.037	0.022	26	0.1470	0.074	0.044	3	0.2130	0.107	0.064	35/64	0.5469	0.273	0.164
48	0.0760	0.038	0.023	25	0.1495	0.075	0.045	2	0.2210	0.111	0.067	9/16	0.5625	0.281	0.169
47	0.0785	0.040	0.024	24	0.1520	0.076	0.046	1	0.2280	0.114	0.068	37/64	0.5781	0.289	0.173
46	0.0810	0.041	0.024	23	0.1540	0.077	0.046	15/64	0.2344	0.117	0.070	19/32	0.5938	0.297	0.178
45	0.0820	0.041	0.025	22	0.1570	0.079	0.047	1/4	0.2500	0.125	0.075	39/64	0.6094	0.305	0.183
44	0.0860	0.043	0.026	21	0.1590	0.080	0.048	17/64	0.2656	0.133	0.080	5/8	0.6250	0.313	0.188
43	0.0890	0.045	0.027	20	0.1610	0.081	0.048	9/32	0.2813	0.141	0.084	41/64	0.6406	0.320	0.192
42	0.0935	0.047	0.028	19	0.1660	0.083	0.050	19/64	0.2969	0.148	0.089	21/32	0.6563	0.328	0.197
41	0.0960	0.048	0.029	18	0.1695	0.085	0.051	5/16	0.3125	0.156	0.094	43/64	0.6719	0.336	0.202
40	0.0980	0.049	0.029	17	0.1730	0.087	0.052	21/64	0.3281	0.164	0.098	11/16	0.6875	0.344	0.206
39	0.0995	0.050	0.030	16	0.1770	0.089	0.053	11/32	0.3438	0.171	0.103	23/32	0.7188	0.359	0.216
38	0.1015	0.051	0.030	15	0.1800	0.090	0.054	23/64	0.3594	0.180	0.108	3/4	0.7500	0.375	0.225

Table 13. Tool Troubleshooting Check List

Problem	Tool Material	Remedy
Excessive flank wear—Tool life too short	Carbide	1. Change to harder, more wear-resistant grade
		2. Reduce the cutting speed
		3. Reduce the cutting speed and increase the feed to maintain production
		4. Reduce the feed
		5. For work-hardenable materials—increase the feed
		6. Increase the lead angle
		7. Increase the relief angles
	HSS	1. Use a coolant
		2. Reduce the cutting speed
		3. Reduce the cutting speed and increase the feed to maintain production
		4. Reduce the feed
		5. For work-hardenable materials—increase the feed
		6. Increase the lead angle
		7. Increase the relief angle
Excessive cratering	Carbide	1. Use a crater-resistant grade
		2. Use a harder, more wear-resistant grade
		3. Reduce the cutting speed
		4. Reduce the feed
		5. Widen the chip breaker groove
	HSS	1. Use a coolant
		2. Reduce the cutting speed
		3. Reduce the feed
		4. Widen the chip breaker groove
Cutting edge chipping	Carbide	1. Increase the cutting speed
		2. Lightly hone the cutting edge
		3. Change to a tougher grade
		4. Use negative-rake tools
		5. Increase the lead angle
		6. Reduce the feed
		7. Reduce the depth of cut
		8. Reduce the relief angles
		9. If low cutting speed must be used, use a high-additive EP cutting fluid

Table 13. *(Continued)* **Tool Troubleshooting Check List**

Problem	Tool Material	Remedy
Cutting edge chipping	HSS	1. Use a high additive EP cutting fluid
		2. Lightly hone the cutting edge before using
		3. Increase the lead angle
		4. Reduce the feed
		5. Reduce the depth of cut
		6. Use a negative rake angle
		7. Reduce the relief angles
Cutting edge chipping	Carbide and HSS	1. Check the setup for cause if chatter occurs
		2. Check the grinding procedure for tool overheating
		3. Reduce the tool overhang
Cutting edge deformation	Carbide	1. Change to a grade containing more tantalum
		2. Reduce the cutting speed
		3. Reduce the feed
Poor surface finish	Carbide	1. Increase the cutting speed
		2. If low cutting speed must be used, use a high additive EP cutting fluid
		4. For light cuts, use straight titanium carbide grade
		5. Increase the nose radius
		6. Reduce the feed
		7. Increase the relief angles
		8. Use positive rake tools
	HSS	1. Use a high additive EP cutting fluid
		2. Increase the nose radius
		3. Reduce the feed
		4. Increase the relief angles
		5. Increase the rake angles
	Diamond	1. Use diamond tool for soft materials
Notching at the depth of cut line	Carbide and HSS	1. Increase the lead angle
		2. Reduce the feed

Machining Methods*.— 1) Good machining practice requires a rigid setup in addition to the selection of proper cutting speed, feed, tool material, tool geometry and cutting fluid.

2) The machine tool must be capable of providing the rigidity required for the machining conditions used. If the size of the machine tool is not adequate or if looseness exists in the moving parts, such as spindle bearings or gibs, chatter will occur and poor tool life will result. When a rigid setup cannot be made, the feed and/or depth of cut must be reduced accordingly.

3) Excessive tool overhang is a common source of trouble in a machining operation. When this condition exists, poor tool life and surface finish result, and dimensional accuracy is difficult to maintain. Stub-length drills should be used instead of jobbers-length drills where the depth of hole permits. Milling cutters should be mounted as close to the spindle as the job will allow. The length of end mills should be kept at a minimum. Climb milling usually gives better tool life and surface finish than does conventional milling if the machine tool and setup have sufficient rigidity and the feed mechanism is free from backlash.

4) Misalignment and tool runout cause other machining problems, such as oversize and bellmouthed holes when reaming and rough and torn threads when tapping and die threading.

5) Maintenance of cutting tools must be given careful consideration in the development of good machining practice.

6) When dimensional accuracy and surface integrity are not critical, high speed steel tools should be removed when the wearland on the flank of the tool reaches approximately 0.060 in. (1.5 mm) width. With of carbide tools, the maximum width should not be allowed to exceed 0.030 in. (0.75 mm), otherwise complete tool failure may occur. On components where dimensional accuracy and surface integrity are critical, the tool wear must be carefully limited.

7) Wearland measurement is not always possible on the tool; however, it is practical to instruct the operator to change tools after a predetermined number of pieces have been machined. The number of parts to be machined per tool should be set conservatively so that the cutter will not fail. Occasionally by this procedure, a cutter may be removed before it is dull; therefore, the resharpening time will be short, but catastrophic failure will have been avoided.

8) The cutting fluid system should provide a copious flow of cutting fluid to the area where the chip is being formed. In the case of machining operations where cutting fluids are used with carbide tools, a continuous flow of the cutting fluid is imperative. Interrupted or intermittent flow can cause thermal shock and breakage of the carbide tool.

9) The concept of good machining practice involves consideration of all factors associated with the machining operation. Each detail—workpiece, fixturing, speed, feed, tool material, tool geometry, cutting fluid, and the machine tool itself—must be given careful attention to ensure success for the machining operation under consideration.

10) The *machinability* of a work material must be determined in order to select the proper machining conditions. The machinability of a material can be defined in terms of three major factors: surface integrity, tool life, and power or force requirements.

* Reprinted from the MACHINING DATA HANDBOOK, 3rd Edition, by permission of the Machinability Data Center. © 1980 by Metcut Research Associates Inc.

TOOL WEAR

Sharpening Twist Drills.—Twist drills are cutting tools designed to perform concurrently several functions, such as penetrating directly into solid material, ejecting the removed chips outside the cutting area, maintaining the essentially straight direction of the advance movement and controlling the size of the drilled hole. The geometry needed for these multiple functions is incorporated into the design of the twist drill in such a manner that it can be retained even after repeated sharpening operations. Twist drills are resharpened many times during their service life, with the practically complete restitution of their original operational characteristics. However, in order to assure all the benefits which the design of the twist drill is capable of providing, the surfaces generated in the sharpening process must agree with the original form of the tool's operating surfaces, unless a change of shape is required for use on a different work material.

The principal elements of the tool geometry which are essential for the adequate cutting performance of twist drills are shown in Fig. 1. The generally used values for these dimensions are the following:

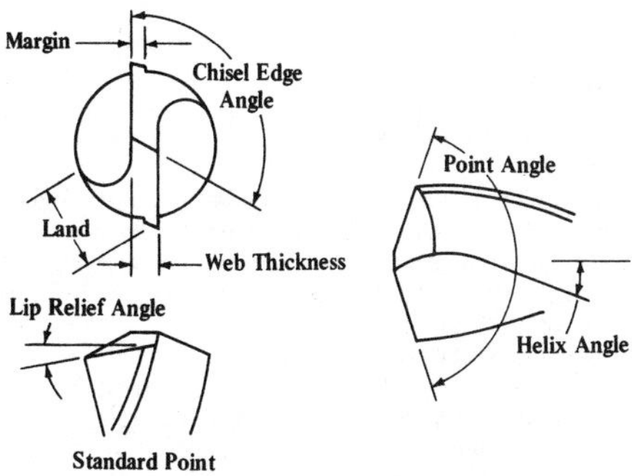

Fig. 1. The principal elements of tool geometry on twist drills.

Point angle: Commonly 118°, except for high strength steels, 118° to 135°; aluminum alloys, 90° to 140°; and magnesium alloys, 70° to 118°.

Helix angle: Commonly 24° to 32°, except for magnesium and copper alloys, 10° to 30°.

Lip relief angle: Commonly 10° to 15°, except for high strength or tough steels, 7° to 12°. The lower values of these angle ranges are used for drills of larger diameter, the higher values for the smaller diameters. For drills of diameters less than ¼ inch, the lip relief angles are increased beyond the listed maximum values up to 24°. For soft and free machining materials, 12° to 18° except for diameters less than ¼ inch, 20° to 26°.

Relief Grinding of the Tool Flanks.—In sharpening twist drills the tool flanks containing the two cutting edges are ground. Each flank consists of a curved surface which provides the relief needed for the easy penetration and free cutting of the tool edges. In grinding the flanks, Fig. 2, the drill is swung around the axis *A* of an imaginary cone while resting in a support which holds the drill at one-half the point angle *B* with respect to the face of the grinding wheel. Feed *f* for stock removal is in the direction of the drill axis. The relief angle is usually measured at the periphery of the twist drill and is also specified by that value. It is not a constant but should increase toward the center of the drill.

The relief grinding of the flank surfaces will generate the chisel angle on the web of the twist drill. The value of that angle, typically 55°, which can be measured, for example, with the protractor of an optical projector, is indicative of the correctness of the relief grinding.

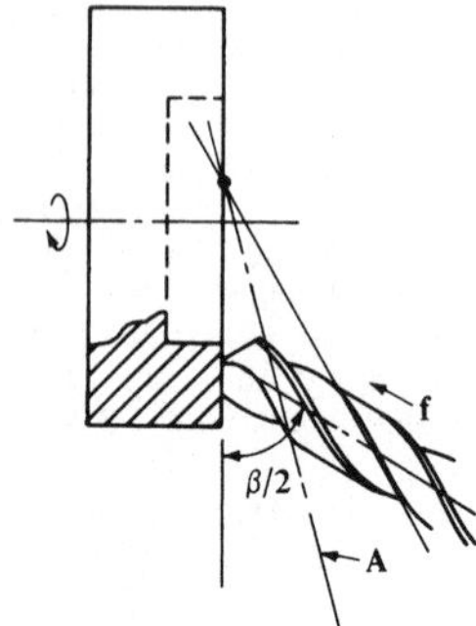

Fig. 2. In grinding the face of the twist drill the tool is swung around the axis *A* of an imaginary cone, while resting in a support tilted by half of the point angle β with respect to the face of the grinding wheel. Feed *f* for stock removal is in the direction of the drill axis.

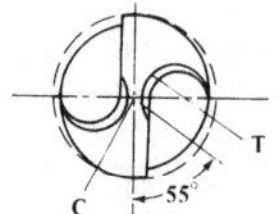

Fig. 3. The chisel edge *C* after thinning the web by grinding off area *T*.

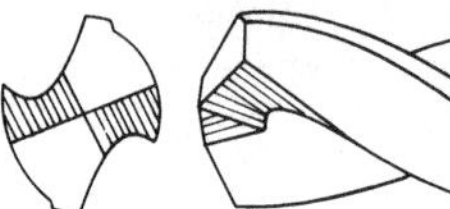

Fig. 4. Split point or "crankshaft" type web thinning.

Drill Point Thinning.—The chisel edge is the least efficient operating surface element of the twist drill because it does not cut, but actually squeezes or extrudes the work material. To improve the inefficient cutting conditions caused by the chisel edge, the point width is often reduced in a drill-point thinning operation, resulting in a condition such as that shown in Fig. 3. Point thinning is particularly desirable on larger size drills and also on those which become shorter in usage, because the thickness of the web increases toward the shaft of the twist drill, thereby adding to the length of the chisel edge. The extent of point thinning is limited by the minimum strength of the web needed to avoid splitting of the drill point under the influence of cutting forces.

Both sharpening operations—the relieved face grinding and the point thinning—should be carried out in special drill grinding machines or with twist drill grinding fixtures mounted on general-purpose tool grinding machines, designed to assure the essential accuracy of the required tool geometry. Off-hand grinding may be used for the important web thinning when a special machine is not available; however, such operation requires skill and experience.

Improperly sharpened twist drills, e.g. those with unequal edge length or asymmetrical point angle, will tend to produce holes with poor diameter and directional control.

For deep holes and also drilling into stainless steel, titanium alloys, high temperature alloys, nickel alloys, very high strength materials and in some cases tool steels, split point grinding, resulting in a "crankshaft" type drill point, is recommended. In this type of pointing, see Fig. 4, the chisel edge is entirely eliminated, extending the positive rake cutting edges to the center of the drill, thereby greatly reducing the required thrust in drilling. Points on modified-point drills must be restored after sharpening to maintain their increased drilling efficiency.

Sharpening Carbide Tools.—Cemented carbide indexable inserts are usually not resharpened but sometimes they require a special grind in order to form a contour on the cutting edge to suit a special purpose. Brazed type carbide cutting tools are resharpened after the cutting edge has become worn. On brazed carbide tools the cutting-edge wear should not be allowed to become excessive before the tool is re-sharpened. One method of determining when brazed carbide tools need resharpening is by periodic inspection of the flank wear and the condition of the face. Another method is to determine the amount of production which is normally obtained before excessive wear has taken place, or to determine the equivalent period of time. One disadvantage of this method is that slight variations in the work material will often cause the wear rate not to be uniform and the number of parts machined before regrinding will not be the same each time. Usually, sharpening should not require the removal of more than 0.005 to 0.010 inch of carbide.

General Procedure in Carbide Tool Grinding: The general procedure depends upon the kind of grinding operation required. If the operation is to resharpen a dull tool, a diamond wheel of 100 to 120 grain size is recommended although a finer wheel—up to 150 grain size—is sometimes used to obtain a better finish. If the tool is new or is a "standard" design and changes in shape are necessary, a 100-grit diamond wheel is recommended for roughing and a finer grit diamond wheel can be used for finishing. Some shops prefer to rough grind the carbide with a vitrified silicon carbide wheel, the finish grinding being done with a diamond wheel. A final operation commonly designated as lapping may or may not be employed for obtaining an extra-fine finish.

Wheel Speeds: The speed of silicon carbide wheels usually is about 5000 feet per minute. The speeds of diamond wheels generally range from 5000 to 6000 feet per minute; yet lower speeds (550 to 3000 fpm) can be effective.

Offhand Grinding: In grinding single-point tools (excepting chip breakers) the common practice is to hold the tool by hand, press it against the wheel face and traverse it continuously across the wheel face while the tool is supported on the machine rest or table which is adjusted to the required angle. This is known as "offhand grinding" to distinguish it from the machine grinding of cutters as in regular cutter grinding practice. The selection of wheels adapted to carbide tool grinding is very important.

Silicon Carbide Wheels.—The green colored silicon carbide wheels generally are preferred to the dark gray or gray-black variety, although the latter are sometimes used.

Grain or Grit Sizes: For roughing, a grain size of 60 is very generally used. For finish grinding with silicon carbide wheels, a finer grain size of 100 or 120 is common. A silicon carbide wheel such as C60-I-7V may be used for grinding both the steel shank and carbide tip. However, for under-cutting steel shanks up to the carbide tip, it may be advantageous to use an aluminum oxide wheel suitable for grinding softer, carbon steel.

Grade: According to the standard system of marking, different grades from soft to hard are indicated by letters from A to Z. For carbide tool grinding fairly soft grades such as G, H, I, and J are used. The usual grades for roughing are I or J and for finishing H, I, and J. The grade should be such that a sharp free-cutting wheel will be maintained without excessive grinding pressure. Harder grades than those indicated tend to overheat and crack the carbide.

Structure: The common structure numbers for carbide tool grinding are 7 and 8. The larger cup-wheels (10 to 14 inches) may be of the porous type and be designated as 12P. The standard structure numbers range from 1 to 15 with progressively higher numbers indicating less density and more open wheel structure.

Diamond Wheels.—Wheels with diamond-impregnated grinding faces are fast and cool cutting and have a very low rate of wear. They are used extensively both for resharpening and for finish grinding of carbide tools when preliminary roughing is required. Diamond wheels are also adapted for sharpening multi-tooth cutters such as milling cutters, reamers, etc., which are ground in a cutter grinding machine.

Resinoid bonded: wheels are commonly used for grinding chip breakers, milling cutters, reamers or other multi-tooth cutters. They are also applicable to precision grinding of carbide dies, gages, and various external, internal and surface grinding operations. Fast, cool cutting action is characteristic of these wheels.

Metal bonded: wheels are often used for offhand grinding of single-point tools especially when durability or long life and resistance to grooving of the cutting face, are considered more important than the rate of cutting. *Vitrified bonded* wheels are used both for roughing of chipped or very dull tools and for ordinary resharpening and finishing. They provide rigidity for precision grinding, a porous structure for fast cool cutting, sharp cutting action and durability.

Diamond Wheel Grit Sizes.—For roughing with diamond wheels a grit size of 100 is the most common both for offhand and machine grinding.

Grit sizes of 120 and 150 are frequently used in offhand grinding of single point tools 1) for resharpening; 2) for a combination roughing and finishing wheel; and 3) for chip-breaker grinding.

Grit sizes of 220 or 240 are used for ordinary finish grinding all types of tools (offhand and machine) and also for cylindrical, internal and surface finish grinding. Grits of 320 and 400 are used for "lapping" to obtain very fine finishes, and for hand hones. A grit of 500 is for lapping to a mirror finish on such work as carbide gages and boring or other tools for exceptionally fine finishes.

Diamond Wheel Grades.—Diamond wheels are made in several different grades to better adapt them to different classes of work. The grades vary for different types and shapes of wheels. Standard Norton grades are H, J, and L, for resinoid bonded wheels, grade N for metal bonded wheels and grades J, L, N, and P, for vitrified wheels. Harder and softer grades than standard may at times be used to advantage.

Diamond Concentration.—The relative amount (by carat weight) of diamond in the diamond section of the wheel is known as the "diamond concentration." Concentrations of 100 (high), 50 (medium) and 25 (low) ordinarily are supplied. A concentration of 50 represents one-half the diamond content of 100 (if the depth of the diamond is the same in each case) and 25 equals one-fourth the content of 100 or one-half the content of 50 concentration.

100 Concentration: Generally interpreted to mean 72 carats of diamond/in.^{3} of abrasive section. (A 75 concentration indicates 54 carats/in.^{3}.) Recommended (especially in grit sizes up to about 220) for general machine grinding of carbides, and for grinding cutters and chip breakers. Vitrified and metal bonded wheels usually have 100 concentration.

50 Concentration: In the finer grit sizes of 220, 240, 320, 400, and 500, a 50 concentration is recommended for offhand grinding with resinoid bonded cup-wheels.

25 Concentration: A low concentration of 25 is recommended for offhand grinding with resinoid bonded cup-wheels with grit sizes of 100, 120 and 150.

Depth of Diamond Section: The radial depth of the diamond section usually varies from $\frac{1}{16}$ to $\frac{1}{4}$ inch. The depth varies somewhat according to the wheel size and type of bond.

Dry Versus Wet Grinding of Carbide Tools.—In using silicon carbide wheels, grinding should be done either absolutely dry or with enough coolant to flood the wheel and tool. Satisfactory results may be obtained either by the wet or dry method. However, dry grinding is the most prevalent usually because, in wet grinding, operators tend to use an inadequate supply of coolant to obtain better visibility of the grinding operation and avoid getting wet; hence checking or cracking in many cases is more likely to occur in wet grinding than in dry grinding.

Wet Grinding with Silicon Carbide Wheels: One advantage commonly cited in connection with wet grinding is that an ample supply of coolant permits using wheels about one grade harder than in dry grinding thus increasing the wheel life. Plenty of coolant also prevents thermal stresses and the resulting cracks, and there is less tendency for the wheel to load. A dust exhaust system also is unnecessary.

Wet Grinding with Diamond Wheels: In grinding with diamond wheels the general practice is to use a coolant to keep the wheel face clean and promote free cutting. The amount of coolant may vary from a small stream to a coating applied to the wheel face by a felt pad.

Peripheral Versus Flat Side Grinding.—In grinding single point carbide tools with silicon carbide wheels, the roughing preparatory to finishing with diamond wheels may be done either by using the flat face of a cup-shaped wheel (side grinding) or the periphery of a "straight" or disk-shaped wheel. Even where side grinding is preferred, the periphery of a straight wheel may be used for heavy roughing as in grinding back chipped or broken tools (see left-hand diagram). Reasons for preferring peripheral grinding include faster cutting with less danger of localized heating and checking especially in grinding broad surfaces. The advantages usually claimed for side grinding are that proper rake or relief angles are easier to obtain and the relief or land is ground flat. The diamond wheels used for tool sharpening are designed for side grinding. (See right-hand diagram.)

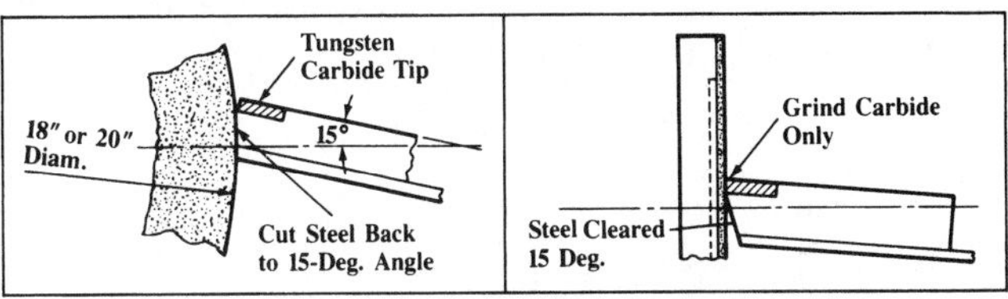

Lapping Carbide Tools.—Carbide tools may be finished by lapping, especially if an exceptionally fine finish is required on the work as, for example, tools used for precision boring or turning non-ferrous metals. If the finishing is done by using a diamond wheel of very fine grit (such as 240, 320, or 400), the operation is often called "lapping." A second lapping method is by means of a power-driven lapping disk charged with diamond dust, Norbide powder, or silicon carbide finishing compound. A third method is by using a hand lap or hone usually of 320 or 400 grit. In many plants the finishes obtained with carbide tools meet requirements without a special lapping operation. In all cases any feather edge which may be left on tools should be removed and it is good practice to bevel the edges of roughing tools at 45 degrees to leave a chamfer 0.005 to 0.010 inch wide. This is done by hand honing and the object is to prevent crumbling or flaking off at the edges when hard scale or heavy chip pressure is encountered.

Hand Honing: The cutting edge of carbide tools, and tools made from other tool materials, is sometimes hand honed before it is used in order to strengthen the cutting edge. When interrupted cuts or heavy roughing cuts are to be taken, or when the grade of carbide is slightly too hard, hand honing is beneficial because it will prevent chipping, or even possibly, breakage of the cutting edge. Whenever chipping is encountered, hand honing the cutting edge before use will be helpful. It is important, however, to hone the edge lightly and only when necessary. Heavy honing will always cause a reduction in tool life. Normally, removing 0.002 to 0.004 inch from the cutting edge is sufficient. When indexable inserts are used, the use of pre-honed inserts is preferred to hand honing although sometimes an additional amount of honing is required. Hand honing of carbide tools in between cuts is sometimes done to defer grinding or to increase the life of a cutting edge on an indexable insert. If correctly done, so as not to change the relief angle, this procedure is sometimes helpful. If improperly done, it can result in a reduction in tool life.

Chip Breaker Grinding.—For this operation a straight diamond wheel is used on a universal tool and cutter grinder, a small surface grinder, or a special chipbreaker grinder. A resinoid bonded wheel of the grade J or N commonly is used and the tool is held rigidly in an adjustable holder or vise. The width of the diamond wheel usually varies from ⅛ to ¼ inch. A vitrified bond may be used for wheels as thick as $\frac{1}{4}$ inch, and a resinoid bond for relatively narrow wheels.

Summary of Miscellaneous Points.—In grinding a single-point carbide tool, traverse it across the wheel face continuously to avoid localized heating. This traverse movement should be quite rapid in using silicon carbide wheels and comparatively slow with diamond wheels. A hand traversing and feeding movement, whenever practicable, is generally recommended because of greater sensitivity. In grinding, maintain a constant, moderate pressure. Manipulating the tool so as to keep the contact area with the wheel as small as possible will reduce heating and increase the rate of stock removal. Never cool a hot tool by dipping it in a liquid, as this may crack the tip. Wheel rotation should preferably be *against* the cutting edge or from the front face toward the back. If the grinder is driven by a reversing motor, opposite sides of a cup wheel can be used for grinding right-and left-hand tools and with rotation against the cutting edge. If it is necessary to grind the top face of a single-point tool, this should precede the grinding of the side and front relief, and top-face grinding should be minimized to maintain the tip thickness. In machine grinding with a diamond wheel, limit the feed per traverse to 0.001 inch for 100 to 120 grit; 0.0005 inch for 150 to 240 grit; and 0.0002 inch for 320 grit and finer.

CUTTING FLUIDS FOR MACHINING

The goal in all conventional metal-removal operations is to raise productivity and reduce costs by machining at the highest practical speed consistent with long tool life, fewest rejects, and minimum downtime, and with the production of surfaces of satisfactory accuracy and finish. Many machining operations can be performed "dry," but the proper application of a cutting fluid generally makes possible: higher cutting speeds, higher feed rates, greater depths of cut, lengthened tool life, decreased surface roughness, increased dimensional accuracy, and reduced power consumption. Selecting the proper cutting fluid for a specific machining situation requires knowledge of fluid functions, properties, and limitations. Cutting fluid selection deserves as much attention as the choice of machine tool, tooling, speeds, and feeds.

To understand the action of a cutting fluid it is important to realize that almost all the energy expended in cutting metal is transformed into heat, primarily by the deformation of the metal into the chip and, to a lesser degree, by the friction of the chip sliding against the tool face. With these factors in mind it becomes clear that the primary functions of any cutting fluid are: cooling of the tool, workpiece, and chip; reducing friction at the sliding contacts; and reducing or preventing welding or adhesion at the contact surfaces, which forms the "built-up edge" on the tool. Two other functions of cutting fluids are flushing away chips from the cutting zone and protecting the workpiece and tool from corrosion.

The relative importance of the functions is dependent on the material being machined, the cutting tool and conditions, and the finish and accuracy required on the part. For example, cutting fluids with greater lubricity are generally used in low-speed machining and on most difficult-to-cut materials. Cutting fluids with greater cooling ability are generally used in high-speed machining on easier-to-cut materials.

Types of Cutting and Grinding Fluids.—In recent years a wide range of cutting fluids has been developed to satisfy the requirements of new materials of construction and new tool materials and coatings.

There are four basic types of cutting fluids; each has distinctive features, as well as advantages and limitations. Selection of the right fluid is made more complex because the dividing line between types is not always clear. Most machine shops try to use as few different fluids as possible and prefer fluids that have long life, do not require constant changing or modifying, have reasonably pleasant odors, do not smoke or fog in use, and, most important, are neither toxic nor cause irritation to the skin. Other issues in selection are the cost and ease of disposal.

The major divisions and subdivisions used in classifying cutting fluids are:

Cutting Oils, including straight and compounded mineral oils plus additives.

Water-Miscible Fluids, including emulsifiable oils; chemical or synthetic fluids; and semi chemical fluids.

Gases.

Paste and Solid Lubricants.

Since the cutting oils and water-miscible types are the most commonly used cutting fluids in machine shops, discussion will be limited primarily to these types. It should be noted, however, that compressed air and inert gases, such as carbon dioxide, nitrogen, and Freon, are sometimes used in machining. Paste, waxes, soaps, graphite, and molybdenum disulfide may also be used, either applied directly to the workpiece or as an impregnant in the tool, such as in a grinding wheel.

Cutting Oils.—Cutting oils are generally compounds of mineral oil with the addition of animal, vegetable, or marine oils to improve the wetting and lubricating properties. Sulfur, chlorine, and phosphorous compounds, sometimes called extreme pressure (EP) additives, provide for even greater lubricity. In general, these cutting oils do not cool as well as water-miscible fluids.

Water-Miscible Fluids.—*Emulsions or soluble oils* are a suspension of oil droplets in water. These suspensions are made by blending the oil with emulsifying agents (soap and soap like materials) and other materials. These fluids combine the lubricating and rust-prevention properties of oil with water's excellent cooling properties. Their properties are affected by the emulsion concentration, with "lean" concentrations providing better cooling but poorer lubrication, and with "rich" concentrations having the opposite effect. Additions of sulfur, chlorine, and phosphorus, as with cutting oils, yield "extreme pressure" (EP) grades.

Chemical fluids are true solutions composed of organic and inorganic materials dissolved in water. Inactive types are usually clear fluids combining high rust inhibition, high cooling, and low lubricity characteristics with high surface tension. Surface-active types include wetting agents and possess moderate rust inhibition, high cooling, and moderate lubricating properties with low surface tension. They may also contain chlorine and/or sulfur compounds for extreme pressure properties.

Semichemical fluids are combinations of chemical fluids and emulsions. These fluids have a lower oil content but a higher emulsifier and surface-active-agent content than emulsions, producing oil droplets of much smaller diameter. They possess low surface tension, moderate lubricity and cooling properties, and very good rust inhibition. Sulfur, chlorine, and phosphorus also are sometimes added.

Selection of Cutting Fluids for Different Materials and Operations.—The choice of a cutting fluid depends on many complex interactions including the machinability of the metal; the severity of the operation; the cutting tool material; metallurgical, chemical, and human compatibility; fluid properties, reliability, and stability; and finally cost. Other factors affect results. Some shops standardize on a few cutting fluids which have to serve all purposes. In other shops, one cutting fluid must be used for all the operations performed on a machine. Sometimes, a very severe operating condition may be alleviated by applying the "right" cutting fluid manually while the machine supplies the cutting fluid for other operations through its coolant system. Several voluminous textbooks are available with specific recommendations for the use of particular cutting fluids for almost every combination of machining operation and workpiece and tool material. In general, when experience is lacking, it is wise to consult the material supplier and/or any of the many suppliers of different cutting fluids for advice and recommendations. Another excellent source is the Machinability Data Center, one of the many information centers supported by the U.S. Department of Defense. While the following recommendations represent good practice, they are to serve as a guide only, and it is not intended to say that other cutting fluids will not, in certain specific cases, also be effective.

Steels: Caution should be used when using a cutting fluid on steel that is being turned at a high cutting speed with cemented carbide cutting tools. See *Application of Cutting Fluids to Carbides* later. Frequently this operation is performed dry. If a cutting fluid is used, it should be a soluble oil mixed to a consistency of about 1 part oil to 20 to 30 parts water. A sulfurized mineral oil is recommended for reaming with carbide tipped reamers although a heavy-duty soluble oil has also been used successfully.

The cutting fluid recommended for machining steel with high speed cutting tools depends largely on the severity of the operation. For ordinary turning, boring, drilling, and milling on medium and low strength steels, use a soluble oil having a consistency of 1 part oil to 10 to 20 parts water. For tool steels and tough alloy steels, a heavy-duty soluble oil having a consistency of 1 part oil to 10 parts water is recommended for turning and milling. For drilling and reaming these materials, a light sulfurized mineral-fatty oil is used. For tough operations such as tapping, threading, and broaching, a sulfochlorinated mineral-fatty oil is recommended for tool steels and high-strength steels, and a heavy sulfurized mineral-fatty oil or a sulfochlorinated mineral oil can be used for medium- and low-strength steels. Straight sulfurized mineral oils are often recommended for machining tough, stringy low carbon steels to reduce tearing and produce smooth surface finishes.

Stainless Steel: For ordinary turning and milling a heavy-duty soluble oil mixed to a consistency of 1 part oil to 5 parts water is recommended. Broaching, threading, drilling, and reaming produce best results using a sulfochlorinated mineral-fatty oil.

Copper Alloys: Most brasses, bronzes, and copper are stained when exposed to cutting oils containing active sulfur and chlorine; thus, sulfurized and sulfochlorinated oils should not be used. For most operations a straight soluble oil, mixed to 1 part oil and 20 to 25 parts water is satisfactory. For very severe operations and for automatic screw machine work a mineral-fatty oil is used. A typical mineral-fatty oil might contain 5 to 10 percent lard oil with the remainder mineral oil.

Monel Metal: When turning this material, an emulsion gives a slightly longer tool life than a sulfurized mineral oil, but the latter aids in chip breakage, which is frequently desirable.

Aluminum Alloys: Aluminum and aluminum alloys are frequently machined dry. When a cutting fluid is used it should be selected for its ability to act as a coolant. Soluble oils mixed to a consistency of 1 part oil to 20 to 30 parts water can be used. Mineral oil-base cutting fluids, when used to machine aluminum alloys, are frequently cut back to increase their viscosity so as to obtain good cooling characteristics and to make them flow easily to cover the tool and the work. For example, a mineral-fatty oil or a mineral plus a sulfurized fatty oil can be cut back by the addition of as much as 50 Percent kerosene.

Cast Iron: Ordinarily, cast iron is machined dry. Some increase in tool life can be obtained or a faster cutting speed can be used with a chemical cutting fluid or a soluble oil mixed to consistency of 1 part oil and 20 to 40 parts water. A soluble oil is sometimes used to reduce the amount of dust around the machine.

Magnesium: Magnesium may be machined dry, or with an air blast for cooling. A light mineral oil of low acid content may be used on difficult cuts. Coolants containing water should not be used on magnesium because of the danger of releasing hydrogen caused by reaction of the chips with water. Proprietary water-soluble oil emulsions containing inhibitors that reduce the rate of hydrogen generation are available.

Grinding: Soluble oil emulsions or emulsions made from paste compounds are used extensively in precision grinding operations. For cylindrical grinding, 1 part oil to 40 to 50 parts water is used. Solution type fluids and translucent grinding emulsions are particularly suited for many fine-finish grinding applications. Mineral oil-base grinding fluids are recommended for many applications where a fine surface finish is required on the ground surface. Mineral oils are used with vitrified wheels but are not recommended for wheels with rubber or shellac bonds. Under certain conditions the oil vapor mist caused by the action of the grinding wheel can be ignited by the grinding sparks and explode. To quench the grinding spark a secondary coolant line to direct a flow of grinding oil below the grinding wheel is recommended.

Broaching: For steel, a heavy mineral oil such as sulfurized oil of 300 to 500 Saybolt viscosity at 100 degrees F can be used to provide both adequate lubricating effect and a dampening of the shock loads. Soluble oil emulsions may be used for the lighter broaching operations.

Cutting Fluids for Turning, Milling, Drilling and Tapping.—The following table, *Cutting Fluids Recommended for Machining Operations*, gives specific cutting oil recommendations for common machining operations.

Soluble Oils: Types of oils paste compounds that form emulsions when mixed with water: Soluble oils are used extensively in machining both ferrous and non-ferrous metals when the cooling quality is paramount and the chip-bearing pressure is not excessive. Care should be taken in selecting the proper soluble oil for precision grinding operations. Grinding coolants should be free from fatty materials that tend to load the wheel, thus affecting the finish on the machined part. Soluble coolants should contain rust preventive constituents to prevent corrosion.

Base Oils: Various types of highly sulfurized and chlorinated oils containing inorganic, animal, or fatty materials. This "base stock" usually is "cut back" or blended with a lighter oil, unless the chip-bearing pressures are high, as when cutting alloy steel. Base oils usually have a viscosity range of from 300 to 900 seconds at 100 degrees F.

Mineral Oils: This group includes all types of oils extracted from petroleum such as paraffin oil, mineral seal oil, and kerosene. Mineral oils are often blended with base stocks, but they are generally used in the original form for light machining operations on both free-machining steels and non-ferrous metals. The coolants in this class should be of a type that has a relatively high flash point. Care should be taken to see that they are nontoxic, so that they will not be injurious to the operator. The heavier mineral oils (paraffin oils) usually have a viscosity of about 100 seconds at 100 degrees F. Mineral seal oil and kerosene have a viscosity of 35 to 60 seconds at 100 degrees F.

Cutting Fluids Recommended for Machining Operations

Material to be Cut	Turning	Milling
Aluminum[a]	Mineral Oil with 10 Percent Fat (or) Soluble Oil	Soluble Oil (96 Percent Water) (or) Mineral Seal Oil (or) Mineral Oil
Alloy Steels[b]	25 Percent Sulfur base Oil[b] with 75 Percent Mineral Oil	10 Percent Lard Oil with 90 Percent Mineral Oil
Brass	Mineral Oil with 10 Percent Fat	Soluble Oil (96 Percent Water)
Tool Steels and Low-carbon Steels	25 Percent Lard Oil with 75 Percent Mineral Oil	Soluble Oil
Copper	Soluble Oil	Soluble Oil
Monel Metal	Soluble Oil	Soluble Oil
Cast Iron[c]	Dry	Dry
Malleable Iron	Soluble Oil	Soluble Oil
Bronze	Soluble Oil	Soluble Oil
Magnesium[d]	10 Percent Lard Oil with 90 Percent Mineral Oil	Mineral Seal Oil
Material to be Cut	**Drilling**	**Tapping**
Aluminum[e]	Soluble Oil (75 to 90 Percent Water) (or) 10 Percent Lard Oil with 90 Percent Mineral Oil	Lard Oil (or) Sperm Oil (or) Wool Grease (or) 25 Percent Sulfur-base Oil[b] Mixed with Mineral Oil
Alloy Steels[b]	Soluble Oil	30 Percent Lard Oil with 70 Percent Mineral Oil
Brass	Soluble Oil (75 to 90 Percent Water) (or) 30 Percent Lard Oil with 70 Percent Mineral Oil	10 to 20 Percent Lard Oil with Mineral Oil
Tool Steels and Low-carbon Steels	Soluble Oil	25 to 40 Percent Lard Oil with Mineral Oil (or) 25 Percent Sulfur-base Oil[b] with 75 Percent Mineral Oil
Copper	Soluble Oil	Soluble Oil
Monel Metal	Soluble Oil	25 to 40 Percent Lard Oil Mixed with Mineral Oil (or) Sulfur-base Oil[b] Mixed with Mineral Oil
Cast Iron[c]	Dry	Dry (or) 25 Percent Lard Oil with 75 Percent Mineral Oil
Malleable Iron	Soluble Oil	Soluble Oil
Bronze	Soluble Oil	20 Percent Lard Oil with 80 Percent Mineral Oil
Magnesium[d]	60-second Mineral Oil	20 Percent Lard Oil with 80 Percent Mineral Oil

[a] In machining aluminum, several varieties of coolants may be used. For rough machining, where the stock removal is sufficient to produce heat, water soluble mixtures can be used with good results to dissipate the heat. Other oils that may be recommended are straight mineral seal oil; a 50–50 mixture of mineral seal oil and kerosene; a mixture of 10 Percent lard oil with 90 Percent kerosene; and a 100-second mineral oil cut back with mineral seal oil or kerosene.

[b] The sulfur-base oil referred to contains 4½ Percent sulfur compound. Base oils are usually dark in color. As a rule, they contain sulfur compounds resulting from a thermal or catalytic refinery process. When so processed, they are more suitable for industrial coolants than when they have had such compounds as flowers of sulfur added by hand. The adding of sulfur compounds by hand to the coolant reservoir is of temporary value only, and the non-uniformity of the solution may affect the machining operation.

[c] A soluble oil or low-viscosity mineral oil may be used in machining cast iron to prevent excessive metal dust.

[d] When a cutting fluid is needed for machining magnesium, low or nonacid mineral seal or lard oils are recommended. Coolants containing water should not be used because of the fire danger when magnesium chips react with water, forming hydrogen gas.

[e] Sulfurized oils ordinarily are not recommended for tapping aluminum; however, for some tapping operations they have proved very satisfactory, although the work should be rinsed in a solvent right after machining to prevent discoloration.

Application of Cutting Fluids to Carbides.—Turning, boring, and similar operations on lathes using carbides are performed dry or with the help of soluble oil or chemical cutting fluids. The effectiveness of cutting fluids in improving tool life or by permitting higher cutting speeds to be used, is less with carbides than with high-speed steel tools. Furthermore, the effectiveness of the cutting fluid is reduced as the cutting speed is increased. Cemented carbides are very sensitive to sudden changes in temperature and to temperature gradients within the carbide. Thermal shocks to the carbide will cause thermal cracks to form near the cutting edge, which are a prelude to tool failure. An unsteady or interrupted flow of the coolant reaching the cutting edge will generally cause these thermal cracks. The flow of the chip over the face of the tool can cause an interruption to the flow of the coolant reaching the cutting edge even though a steady stream of coolant is directed at the tool. When a cutting fluid is used and frequent tool breakage is encountered, it is often best to cut dry. When a cutting fluid must be used to keep the workpiece cool for size control or to allow it to be handled by the operator, special precautions must be used. Sometimes applying the coolant from the front and the side of the tool simultaneously is helpful. On lathes equipped with overhead shields, it is very effective to apply the coolant from below the tool into the space between the shoulder of the work and the tool flank, in addition to applying the coolant from the top. Another method is not to direct the coolant stream at the cutting tool at all but to direct it at the workpiece above or behind the cutting tool.

The danger of thermal cracking is great when milling with carbide cutters. The nature of the milling operation itself tends to promote thermal cracking because the cutting edge is constantly heated to a high temperature and rapidly cooled as it enters and leaves the workpiece. For this reason, carbide milling operations should be performed dry.

Lower cutting-edge temperatures diminish the danger of thermal cracking. The cutting-edge temperatures usually encountered when reaming with solid carbide or carbide-tipped reamers are generally such that thermal cracking is not apt to occur except when reaming certain difficult-to-machine metals. Therefore, cutting fluids are very effective when used on carbide reamers. Practically every kind of cutting fluid has been used, depending on the job material encountered. For difficult surface-finish problems in holes, heavy duty soluble oils, sulfurized mineral-fatty oils, and sulfochlorinated mineral-fatty oils have been used successfully. On some work, the grade and the hardness of the carbide also have an effect on the surface finish of the hole.

Cutting fluids should be applied where the cutting action is taking place and at the highest possible velocity without causing splashing. As a general rule, it is preferable to supply from 3 to 5 gallons per minute for each single-point tool on a machine such as a turret or automatic lathe. The temperature of the cutting fluid should be kept below 110 degrees F. If the volume of fluid used is not sufficient to maintain the proper temperature, means of cooling the fluid should be provided.

Cutting Fluids for Machining Magnesium.—In machining magnesium, it is the general but not invariable practice in the United States to use a cutting fluid. In other places, magnesium usually is machined dry except where heat generated by high cutting speeds would not be dissipated rapidly enough without a cutting fluid. This condition may exist when, for example, small tools without much heat-conducting capacity are employed on automatics.

The cutting fluid for magnesium should be an anhydrous oil having, at most, a very low acid content. Various mineral-oil cutting fluids are used for magnesium.

To secure adequate cooling, the supply of fluid should be large (4 to 5 gallons per minute) and the viscosity low; however, to avoid too low a flash point, a compromise between cooling capacity and flash point is necessary. Soluble oils or emulsions should never be used for machining magnesium. Compressed air may be preferable to a fluid because it leaves the chips or swarf clean and dry.

A cutting fluid serves primarily to cool the work and also eliminate a possible fire hazard, especially when dull tools are operated at high speeds with fine feeds. Even when using sharp tools, the cut should not be less than 0.001 inch because fine chips are more likely to become ignited at high speeds. While a variety of mineral oils may be used, the following properties are recommended: Specific gravity 0.79 to 0.86; viscosity (Saybolt) at 100 degrees F., up to 55 seconds; flash point, minimum value (closed cup), 160 degrees F.; saponification No., 16 (max.); free acid (max.) 0.2 Percent. Oil-water emulsions, while good coolants, are objectionable because water will greatly intensify any accidental chip fire.

Coolants for Carbide Tool Grinding.—In grinding either with silicon carbide or diamond wheels a coolant that is used extensively consists of water plus a small amount either of soluble oil, sal soda, or soda ash to prevent corrosion. One prominent manufacturer recommends for silicon carbide wheels about 1 ounce of soda ash per gallon of water and for diamond wheels kerosene. The use of kerosene is quite general for diamond wheels and usually it is applied to the wheel face by a felt pad. Another coolant recommended for diamond wheels consists of 80 per cent water and 20 per cent soluble oil.

MACHINING NONFERROUS METALS

Machining Magnesium.—Magnesium alloys are readily machined and with relatively low power consumption per cubic inch of metal removed. The usual practice is to employ high cutting speeds with relatively coarse feeds and deep cuts. Exceptionally fine finishes can be obtained so that grinding to improve the finish usually is unnecessary. The horsepower normally required in machining magnesium varies from 0.15 to 0.30 per cubic inch per minute. While this value is low, especially in comparison with power required for cast iron and steel, the total amount of power for machining magnesium usually is high because of the exceptionally rapid rate at which metal is removed.

Carbide tools are recommended for maximum efficiency, although high-speed steel frequently is employed. Tools should be designed so as to dispose of chips readily or without excessive friction, by employing polished chip-bearing surfaces, ample chip spaces, large clearances, and small contact areas. *Keen-edged tools should always be used.*

Feeds and Speeds for Magnesium: Speeds ordinarily range up to 5000 feet per minute for rough- and finish-turning, up to 3000 feet per minute for rough-milling, and up to 9000 feet per minute for finish-milling. For rough-turning, the following combinations of speed in feet per minute, feed per revolution, and depth of cut are recommended: Speed 300 to 600 feet per minute — feed 0.030 to 0.100 inch, depth of cut 0.5 inch; speed 600 to 1000 — feed 0.020 to 0.080, depth of cut 0.4; speed 1000 to 1500 — feed 0.010 to 0.060, depth of cut 0.3; speed 1500 to 2000 — feed 0.010 to 0.040, depth of cut 0.2; speed 2000 to 5000 — feed 0.010 to 0.030, depth of cut 0.15.

Lathe Tool Angles for Magnesium: The true or actual rake angle resulting from back and side rakes usually varies from 10 to 15 degrees. Back rake varies from 10 to 20, and side rake from 0 to 10 degrees. Reduced back rake may be employed to obtain better chip breakage. The back rake may also be reduced to from 2 to 8 degrees on form tools or other broad tools to prevent chatter.

Parting Tools: For parting tools, the back rake varies from 15 to 20 degrees, the front end relief 8 to 10 degrees, the side relief measured perpendicular to the top face 8 degrees, the side relief measured in the plane of the top face from 3 to 5 degrees.

Milling Magnesium: In general, the coarse-tooth type of cutter is recommended. The number of teeth or cutting blades may be one-third to one-half the number normally used; however, the two-blade fly-cutter has proved to be very satisfactory. As a rule, the land relief or primary peripheral clearance is 10 degrees followed by secondary clearance of 20 degrees. The lands should be narrow, the width being about $\frac{3}{64}$ to $\frac{1}{16}$ inch. The rake, which is positive, is about 15 degrees.

For rough-milling and speeds in feet per minute up to 900 — feed, inch per tooth, 0.005 to 0.025, depth of cut up to 0.5; for speeds 900 to 1500 — feed 0.005 to 0.020, depth of cut up to 0.375; for speeds 1500 to 3000 — feed 0.005 to 0.010, depth of cut up to 0.2.

Drilling Magnesium: If the depth of a hole is less than five times the drill diameter, an ordinary twist drill with highly polished flutes may be used. The included angle of the point may vary from 70 degrees to the usual angle of 118 degrees. The relief angle is about 12 degrees. The drill should be kept sharp and the outer corners rounded to produce a smooth finish and prevent burr formation. For deep hole drilling, use a drill having a helix angle of 40 to 45 degrees with large polished flutes of uniform cross-section throughout the drill length to facilitate the flow of chips. A pyramid-shaped "spur" or "pilot point" at the tip of the drill will reduce the "spiraling or run-off."

Drilling speeds vary from 300 to 2000 feet per minute with feeds per revolution ranging from 0.015 to 0.050 inch.

Reaming Magnesium: Reamers up to 1 inch in diameter should have four flutes; larger sizes, six flutes. These flutes may be either parallel with the axis or have a negative helix angle of 10 degrees. The positive rake angle varies from 5 to 8 degrees, the relief angle from 4 to 7 degrees, and the clearance angle from 15 to 20 degrees.

Tapping Magnesium: Standard taps may be used unless Class 3B tolerances are required, in which case the tap should be designed for use in magnesium. A high-speed steel concentric type with a ground thread is recommended. The concentric form, which eliminates the radial thread relief, prevents jamming of chips while the tap is being backed out of the hole. The positive rake angle at the front may vary from 10 to 25 degrees and the "heel rake angle" at the back of the tooth from 3 to 5 degrees. The chamfer extends over two to three threads. For holes up to $\frac{1}{4}$ inch in diameter, two-fluted taps are recommended; for sizes from $\frac{1}{2}$ to $\frac{3}{4}$ inch, three flutes; and for larger holes, four flutes. Tapping speeds ordinarily range from 75 to 200 feet per minute, and mineral oil cutting fluid should be used.

Threading Dies for Magnesium: Threading dies for use on magnesium should have about the same cutting angles as taps. Narrow lands should be used to provide ample chip space. Either solid or self-opening dies may be used. The latter type is recommended when maximum smoothness is required. Threads may be cut at speeds up to 1000 feet per minute.

Grinding Magnesium: As a general rule, magnesium is ground dry. The highly inflammable dust should be formed into a sludge by means of a spray of water or low-viscosity mineral oil. Accumulations of dust or sludge should be avoided. For surface grinding, when a fine finish is desirable, a low-viscosity mineral oil may be used.

Machining Aluminum.—Some of the alloys of aluminum have been machined successfully without any lubricant or cutting compound, but in order to obtain the best results, some form of lubricant is desirable. For many purposes, a soluble cutting oil is good.

Tools for aluminum and aluminum alloys should have larger relief and rake angles than tools for cutting steel. For high-speed steel turning tools the following angles are recommended: relief angles, 14 to 16 degrees; back rake angle, 5 to 20 degrees; side rake angle, 15 to 35 degrees. For very soft alloys even larger side rake angles are sometimes used. High silicon aluminum alloys and some others have a very abrasive effect on the cutting tool. While these alloys can be cut successfully with high-speed-steel tools, cemented carbides are recommended because of their superior abrasion resistance. The tool angles recommended for cemented carbide turning tools are: relief angles, 12 to 14 degrees; back rake angle, 0 to 15 degrees; side rake angle, 8 to 30 degrees.

Cut-off tools and necking tools for machining aluminum and its alloys should have from 12 to 20 degrees back rake angle and the end relief angle should be from 8 to 12 degrees. Excellent threads can be cut with single-point tools in even the softest aluminum. Experience seems to vary somewhat regarding the rake angle for single-point thread cutting tools. Some prefer to use a rather large back and side rake angle although this requires a modification in the included angle of the tool to produce the correct thread contour. When both rake angles are zero, the included angle of the tool is ground equal to the included angle of the thread. Excellent threads have been cut in aluminum with zero rake angle thread-cutting tools using large relief angles, which are 16 to 18 degrees opposite the front side of the thread and 12 to 14 degrees opposite the back side of the thread. In either case, the cutting edges should be ground and honed to a keen edge. It is sometimes advisable to give the face of the tool a few strokes with a hone between cuts when chasing the thread to remove any built-up edge on the cutting edge.

Fine surface finishes are often difficult to obtain on aluminum and aluminum alloys, particularly the softer metals. When a fine finish is required, the cutting tool should be honed to a keen edge and the surfaces of the face and the flank will also benefit by being honed smooth. Tool wear is inevitable, but it should not be allowed to progress too far before the tool is changed or sharpened. A sulphurized mineral oil or a heavy-duty soluble oil will sometimes be helpful in obtaining a satisfactory surface finish. For best results, however, a diamond cutting tool is recommended. Excellent surface finishes can be obtained on even the softest aluminum and aluminum alloys with these tools.

Although ordinary milling cutters can be used successfully in shops where aluminum parts are only machined occasionally, the best results are obtained with coarse-tooth, large helix-angle cutters having large rake and clearance angles. Clearance angles up to 10 to 12 degrees are recommended. When slab milling and end milling a profile, using the peripheral teeth on the end mill, climb milling (also called down milling) will generally produce a better finish on the machined surface than conventional (or up) milling. Face milling cutters should have a large axial rake angle. Standard twist drills can be used without difficulty in drilling aluminum and aluminum alloys although high helix-angle drills are preferred. The wide flutes and high helix-angle in these drills helps to clear the chips. Sometimes split-point drills are preferred. Carbide tipped twist drills can be used for drilling aluminum and its alloys and may afford advantages in some production applications. Ordinary hand and machine taps can be used to tap aluminum and its alloys although spiral-fluted ground thread taps give superior results. Experience has shown that such taps should have a right-hand ground flute when intended to cut right-hand threads and the helix angle should be similar to that used in an ordinary twist drill.

Machining Plastics.—Molded plastic parts do not require machining, as general rule, unless the tolerances are exceptionally small or there are undercuts, angular holes, or other openings difficult or impracticable to reproduce in a mold. It is common practice, however, to machine laminated phenolic plastics and also cast phenolic plastics, as well as sheet, bar, and tube stock such as is commonly used for making parts such as pinions or gears. The machining characteristics of different plastics vary somewhat so that the general recommendations given may require some modification to obtain the best results for a given class of work. Although plastics are poor conductors of heat, they usually are machined dry or without a cutting fluid. In some cases, either an air jet, water, or a soap solution is used. The maximum speed at which a cutting tool can operate without excessive heating should be determined by actual test.

Turning and Boring Plastics: Tools of high-speed steel, Stellite and carbide are commonly used in machining plastics. The general practice in turning and boring is similar to that for brass so far as the feed and speed are concerned. Speeds usually vary from 250 to 500 feet per minute for high-speed steel tools and from 500 to 1500 for Stellite and carbide tools. According to the Haynes Stellite Co., hot-set molded parts require tools having less clearance and more rake than those used for steel or other metals. The cold-set acetate and polystyrene molded parts should be turned with tools having no rake and plenty of clearance. Parts molded of "Vinylite" plastic can usually be machined satisfactorily with ordinary metal-cutting tools, provided the front and side clearances are about double those required for machine steel.

Cutting Off Plastics: Tools for cutting off should have greater front and side clearances than for steel. The cutting speed should be about half that employed for a turning operation.

Drilling Plastics: Standard twist drills may be used for small holes up to ⅛ or $\frac{3}{16}$ inch, but for larger sizes particularly, it is preferable to use the commercial high-speed steel drills designed especially for plastics. These drills are made both in wire gage and fractional sizes. They have relatively large flutes to provide greater space for chips, and polished flutes are preferable. For large-quantity production, carbide-tipped drills are recommended. Extra clearance back of the edges of the flutes tends to reduce friction and heating. Frequent removal of the drill may be necessary, especially in drilling comparatively deep holes. A feed of 0.007 to 0.015 inch per revolution is a common range. Drilling speeds usually vary from 150 to 300 feet per minute. With carbide-tipped drills, the speeds may be as high as 12,000 to 15,000 rpm. The drill point should have an included angle of 55 to 60 degrees for thin sections and 90 degrees for the thicker sections. The clearance angle usually is 15 degrees. To avoid excessive heating and aid in chip removal, an air jet may be directed into the hole. In some cases, a soap solution is effective as a cutting fluid. If the drill-spindle movement is cam-operated, design the cam to advance the drill tip slowly at the beginning or for about 0.010 inch, then continue at the full rate of speed and allow a dwell of about 2 to 5 revolutions at the bottom of the hole; then withdraw the drill rapidly.

Tapping and Threading Plastics: For tapping small holes in hot- or cold-set molded parts, a high-speed steel nitrided chromium-plated tap is recommended. The rake may vary from 0 to 5 degrees negative. The size of the hole should allow for about three-fourths of the standard thread depth. Small holes generally are tapped dry. If a coolant is used, water is preferable to oil. Tapping speeds usually vary from 40 to 55 per minute. Threaded brass inserts or bushings are often used, in which case they may be tapped either before or after insertion. In cutting a 60-degree thread, such as the Unified Standard, use a tool that is ground to cut on one side only and feed it in at an angle of 30 degrees by setting the compound rest at this angle.

Drill Jigs for Plastics: In the design of jigs for drilling, close-fitting drill bushings should be avoided. They may increase not only the friction on the drill but also the tendency of the chips to plug up the drill flutes. If the operation is such that a drill bushing is absolutely essential, a floating leaf or templet should be employed. When using a templet, the hole should be spotted with the templet in place, using the drill size corresponding to the final hole size; then the templet should be removed and the hole completed. Pilot holes should be avoided, except in special instances when the hole is to be reamed or counterbored.

Sawing Hot-set Molded Parts: The sawing of molded hot-set parts is done chiefly on circular and band saws. Band saws are to be recommended at times for straight cutting because they run cooler than circular saws. Band saw manufacturers advocate saw teeth set to clear, some advocating one-half the thickness of the blade on each side so that saws give a width of cut double their thickness. Narrower saws and more set are needed for cutting curves than for straight cuts. Band saws just soft enough to permit filing are recommended, but saws must be kept sharp. Dull saws cause chipping and might result in saw breakage. Sawing is usually done dry but some recommend water for cooling. Saw teeth should have little set and eight to nine teeth per inch. The speed is 1800 to 2500 feet per minute.

Sawing Cold-set Molded Parts: Sawing of cold-set polystyrene or cellulose acetate can be done with circular saws having nine to twelve teeth per inch for thin sheets and six teeth per inch for thickness over ¼ inch. Saws six to nine inches in diameter are run at speeds of 3000 to 3600 rpm and should be hollow ground. They usually are 1/32 to 1/16 inch thick. The use of a water spray gives a cleaner cut. One large saw manufacturer recommends that pieces be cut with a stream of water running in the kerf while the saw is cutting. This applies to both circular and band saws; otherwise, the cold-set type of material will fuse. The circular saws recommended are 14 inches, 12 and 9 gage, 130 teeth, 10 degrees rake to be operated at 3000 rpm and made of a special alloy steel stock.

For band sawing, manufacturers suggest a band saw which is 19 to 20 gage, having twenty points to the inch and hardened and tempered. The saw should be operated at 4000 to 4500 feet per minute.

Milling Hot-set and Cold-set Molded Parts: Milling of molded parts is not as a rule feasible, but where it is required, milling speeds and feeds of the range used for brass are recommended. A speed of 400 fpm with carbon steel cutters and 1200–1600 fpm is recommended with carbide cutters. Single- and double-bladed fly cutters are sometimes used at high speed with fine cuts. Where little material has to be removed, a high-speed woodworking shaper with a carbide-tipped tool can be used to advantage. It is desirable and may be necessary to use an air blast to assure proper chip removal from the milling cutter. Wherever possible, it is recommended that spiral milling cutters be used, and that the number of teeth in the cutter head be such that at least two of them are in contact with the work at all times.

The same general rules that apply to turning, facing, and boring operations also hold for milling Vinyl molded parts. Standard cutters can be used, but higher speeds are feasible if extra clearances are ground on the cutter blades.

Machining Non-metallic Gear Blanks.—Laminated phenolic plastics are extensively used for non-metallic gears and pinions. A non-metallic pinion should preferably be used in conjunction with either a hardened steel or a cast-iron gear, thus providing a durable and comparatively noiseless drive. Small- or medium-sized gears may be formed by molding, provided the quantity is large enough to warrant the cost of a mold. Most non-metallic gears, however, are machined from blanks cut from laminated phenolic plastics which have physical properties superior to the molded gears. These blanks may be cut either by punching in a die, by shearing, or by sawing. An efficient method of producing small blanks is to punch them from thin sheet stock. If necessary, the face width of the gear or pinion may be increased by riveting together two or more of these blanks. Gear blanks may also be cut either from a laminated plastic bar or thick tube. Before cutting the gear teeth, the blanks are machined, as by grinding, to obtain a concentric blank of the required diameter. The larger gears are sawed from sheet stock and then turned to size. Gear blanks which are ready for the gear-cutting operation are supplied by manufacturers of laminated plastic materials. The gear teeth are cut with standard gear-cutting equipment. Metal reinforcing end plates and bushings are commonly employed in connection with laminated gears.

Punching Operations: Most grades of laminated phenolics can be punched either hot or cold. The die must be kept sharp, however, in order to produce good results. The minimum clearance between individual punchings, and also between punchings and the edge of the material strips, should be about three times the thickness of the material. For hot punching or shearing, the material is heated in a steam or electric oven, designed to give a uniform temperature throughout the heating chamber. The material is left in the oven just long enough to be uniformly heated to oven temperature. Further heating causes brittleness. Temperatures of 100 to 120 degrees C (212 to 248 degrees F) are recommended. The heating time ranges from five minutes for $\frac{1}{16}$-inch material to thirty minutes for $\frac{1}{4}$-inch material.

Dies for punching laminated plastics are designed the same as for punching metal, except that smaller clearances are allowed between punch and die. In cold punching, this clearance is small, approaching a "sliding fit." The strippers are close fitting and backed with strong springs.

Because these plastic materials expand after being compressed in a die, blanks will be larger than the die diameter and holes will be smaller. On hot punchings, allowance should be made for shrinkage of the material after punching. This shrinkage varies with the grade of material, thickness of piece, and the temperature of the material during the punching operation. For very small holes and blanks, allowances for shrinkage are often neglected, while for large pieces and accurate work, they must be carefully considered. As an example, suppose a 1-inch diameter hole is to be punched hot in $\frac{3}{32}$-inch thick stock; this would require a die of 1.009 inches and a punch of 1.007 inches in diameter. If this piece is punched cold, however, the die should be 1.005 inches in diameter and the punch 1.003 inches in diameter.

Shearing Laminated Plastics: Shears suitable for thin metal sheet are used for cutting laminated plastics. The knife must be kept sharp. In trimming paper-base grades cold, clean-cut edges are obtained with thicknesses of $\frac{1}{32}$ inch and under, and when trimmed hot, up to $\frac{1}{8}$ inch. Fabric-base grades are trimmed cold up to $\frac{1}{16}$ inch, and hot up to $\frac{1}{8}$ inch. Greater thicknesses can be sheared if the condition of the edge is not important.

Sawing Plastics: Material up to 1 inch thick may be cut with a 12- to 16-inch circular saw at about 3000 rpm, and material 1 inch thick and over, with a 16-inch saw at about 2400 rpm. A saw used for roughing cuts has bevel teeth, seven to the inch, while a smooth saw, with no set — similar to that for metal — is used for finishing cuts. For use on all thick material, the saws should be hollow-ground to prevent binding. The smaller the projection of the saw above the material on the sawing table, the better will be the sawed edges. A thin sheet of plastic or other material placed under the piece to be sawed, is of advantage when extreme smoothness of cut is desired.

A band saw is used for sawing round blanks from plate stock. The usual band saw is of the bevel-tooth type, with some set, and has three to seven teeth per inch. It is run at 3000 feet per minute.

Machining Zinc Alloy Die-Castings.—Machining of zinc alloy die-castings is mostly done without a lubricant. For particular work, especially deep drilling and tapping, a lubricant such as lard oil and kerosene (about half and half) or a 50-50 mixture of kerosene and machine oil may be used to advantage. A mixture of turpentine and kerosene has been found effective on certain difficult jobs.

In drilling, standard carbon steel drills are used for shallow holes and high-speed drills of the high-spiral type are recommended for deep holes. The standard 118degree angle of point between cutting edges is recommended. A lip clearance of 12 degrees is satisfactory for most drilling, but in some cases may be increased up to 15 degrees. Flutes that are larger than normal offer the advantage of providing plenty of chip clearance. Straight flute drills have been found useful in enlarging existing holes. Peripheral speeds of from 200 to 300 feet per minute are generally found satisfactory for high-speed steel drills and about half this speed for carbon steel drills.

Threading: Button or acorn dies are satisfactory for threading small diameters of work. Either radial or tangent type chasers may be employed for the larger diameters. For radial type chasers, one manufacturer recommends a 10-degree radial hook for straight threads, a 7-degree radial hook for tapered threads; and a surface speed of 50 feet per minute for 3½ to 7½ threads per inch, 100 feet per minute for 8 to 11 threads per inch, and 200 feet per minute for 12 to 32 threads per inch. In using tangent type chasers with zinc alloys, the cutting edge must be on or very near the center to avoid rapid wearing of the chasers just behind the cutting edge. A 5-degree positive rake is recommended.

Reaming: In reaming, tools with six straight flutes are commonly used, although tools with eight flutes irregularly spaced have been found to yield better results by one manufacturer. Many standard reamers have a land that is too wide for best results. A land about 0.015 inch wide is recommended but this may often be ground down to around 0.007 or even 0.005 inch to obtain freer cutting, less tendency to loading, and reduced heating.

Turning: Tools of high-speed steel are commonly employed although the application of Stellite and carbide tools, even on short runs, is feasible. For steel or Stellite, a positive top rake of from 0 to 20 degrees and an end clearance of about 15 degrees are commonly recommended. Where side cutting is involved, a side clearance of about 4 degrees minimum is recommended. With carbide tools, the end clearance should not exceed 6 to 8 degrees and the top rake should be from 5 to 10 degrees positive. For boring, facing, and other lathe operations, rake and clearance angles are about the same as for tools used in turning.

Machining Monel and Nickel Alloys.—These alloys are machined with high-speed steel and with cemented carbide cutting tools. High-speed steel lathe tools usually have a back rake of 6 to 8 degrees, a side rake of 10 to 15 degrees, and relief angles of 8 to 12 degrees. Broad-nose finishing tools have a back rake of 20 to 25 degrees and an end relief angle of 12 to 15 degrees. In most instances, standard commercial cemented-carbide tool holders and tool shanks can be used which provide an acceptable tool geometry. Honing the cutting edge lightly will help if chipping is encountered.

The most satisfactory tool materials for machining Monel and the softer nickel alloys, such as Nickel 200 and Nickel 230, are M2 and T5 for high-speed steel and crater resistant grades of cemented carbides. For the harder nickel alloys such as K Monel, Permanickel, Duranickel, and Nitinol alloys, the recommended tool materials are T15, M41, M42, M43, and for high-speed steel, M42. For carbides, a grade of crater resistant carbide is recommended when the hardness is less than 300 Bhn, and when the hardness is more than 300 Bhn, a grade of straight tungsten carbide will often work best, although some crater resistant grades will also work well.

A sulfurized oil or a water-soluble oil is recommended for rough and finish turning. A sulfurized oil is also recommended for milling, threading, tapping, reaming, and broaching. Recommended cutting speeds for Monel and the softer nickel alloys are 70 to 100 fpm for high-speed steel tools and 200 to 300 fpm for cemented carbide tools. For the harder nickel alloys, the recommended speed for high-speed steel is 40 to 70 fpm for a hardness up to 300 Bhn and for a higher hardness, 10 to 20 fpm; for cemented carbides, 175 to 225 fpm when the hardness is less than 300 Bhn and for a higher hardness, 30 to 70 fpm.

Nickel alloys have a high tendency to work harden. To minimize work hardening caused by machining, the cutting tools should be provided with adequate relief angles and positive rake angles. Furthermore, the cutting edges should be kept sharp and replaced when dull to prevent burnishing of the work surface. The depth of cut and feed should be sufficiently large to ensure that the tool penetrates the work without rubbing.

In all phases of steel production, various practices are employed which determine the quality and types of the finished material.

Quality Classifications: The term "quality" as it technically relates to steel products may be indicative of many conditions such as the degree of internal soundness, relative uniformity of composition, relative freedom from injurious surface imperfections, and finish. Steel quality also relates to general suitability for particular applications. Sheet steel surface requirements may be broadly identified as to the end use by the suffix E for exposed parts requiring a good painted surface and suffix U for unexposed parts for which surface finish is unimportant.

Carbon Steel may be obtained in a number of fundamental qualities which reflect various degrees of the quality conditions mentioned above. Some of those qualities may be modified by such requirements as limited Austenitic Grain Size, Specified Discard, Macrotech Test, Special Heat Treating, Maximum Incidental Alloy Elements, Restricted Chemical Composition and Nonmetallic Inclusions. In addition, several of the products have special qualities which are intended for specific end uses or fabricating practices.

Alloy Steels also may be obtained in special qualities with such requirements as Extensometer Test, Fracture Test, Impact Test, Macrotech Test, Nonmetallic Indusion Test, Special Hardenability Test, and Grain Size Test.

For complete descriptions of the qualities and supplementary requirements for carbon and alloy steels, reference should be made to the latest applicable American Iron and Steel Institute (AISI) Steel Products Manual Section.

High Strength, Low Alloy Steel, SAE 950.—High strength, low alloy steel represents a specific type of steel in which enhanced mechanical properties and, in most cases, good resistance to atmospheric corrosion are obtained by the addition of moderate amounts of one or more alloying elements other than carbon.

Steels of this type are normally furnished in the hot rolled or annealed condition to minimum mechanical properties. They are not intended for quenching and tempering. The user should not subject them to such treatment without assuming responsibility for the ensuing mechanical properties. There these steels are used for fabrication by welding, no preheat or post heat is required. In certain complex structures, stress relieving may be desirable. These steels may be obtained in the standard shapes or forms normally available in carbon steel.

Application: These steels, because of their enhanced strength, corrosion and erosion resistance, and their high strength-to-weight ratio and service life, are adapted particularly for use in mobile equipment and other structures where substantial weight savings are generally desirable. Typical applications are automotive bumper face bars, truck bodies, frames and structural members, scrapers, dump wagons, cranes, shovels, booms, chutes, conveyors, railroad and industrial cars.

Certain Minimum Properties of SAE 950 Steel as Furnished by the Mill

	Thickness or Diameter, inches				
Property[a]	Up to 0.0709, inclusive	0.0710 to 0.2299, inclusive	0.2300 to ½ inclusive	Over ½ to 1, inclusive	Over 1 to 2, inclusive
Minimum yield point, psi	50,000	50,000	50,000	47,000	45,000
Minimum tensile strength, psi	70,000	70,000	70,000	67,000	65,000
Elongation in 2 in., %	20	22	22	22	22

[a] For severe cold forming operations requiring greater ductility, relaxation of the yield point and tensile strength requirements is commonly negotiated between producer and consumer.

MATERIALS

Carbon Steels.—*SAE Steels 1006, 1008, 1010, 1015:* These steels are the lowest carbon steels of the plain carbon type, and are selected where cold formability is the primary requisite of the user. They are produced both as rimmed and killed steels. Rimmed steel is used for sheet, strip, rod, and wire where excellent surface finish or good drawing qualities are required, such as body and fender stock, hoods, lamps, oil pans, and other deep-drawn and -formed products. This steel is also used for cold-heading wire for tacks, and rivets and low carbon wire products. Killed steel (usually aluminum killed or special killed) is used for difficult stampings, or where nonaging properties are needed. Killed steels (usually silicon killed) should be used in preference to rimmed steel for forging or heat-treating applications.

These steels have relatively low tensile values and should not be selected where much strength is desired. Within the carbon range of the group, strength and hardness will rise with increases in carbon and/or with cold work, but such increases in strength are at the sacrifice of ductility or the ability to withstand cold deformation. Where cold rolled strip is used, the proper temper designation should be specified to obtain the desired properties.

With less than 0.15 carbon, the steels are susceptible to serious grain growth, causing brittleness, which may occur as the result of a combination of critical strain (from cold work) followed by heating to certain elevated temperatures. If cold-worked parts formed from these steels are to be later heated to temperatures in excess of 1100 degrees F, the user should exercise care to avoid or reduce cold working. When this condition develops, it can be overcome by heating the parts to a temperature well in excess of the upper critical point, or at least 1750 degrees F.

Steels in this group, being nearly pure iron or ferritic in structure, do not machine freely and should be avoided for cut screws and operations requiring broaching or smooth finish on turning. The machinability of bar, rod, and wire products is improved by cold drawing. Steels in this group are readily welded.

SAE 1016, 1017, 1018, 1019, 1020, 1021, 1022, 1023, 1024, 1025, 1026, 1027, 1030:

Steels in this group, due to the carbon range covered, have increased strength and hardness, and reduced cold formability compared to the lowest carbon group. For heat-treating purposes, they are known as carburizing or case hardening grades. When uniform response to heat treatment is required, or for forgings, killed steel is preferred; for other uses, semi-killed or rimmed steel may be indicated, depending on the combination of properties desired. Rimmed steels can ordinarily be supplied up to 0.25 carbon.

Selection of one of these steels for carburizing applications depends on the nature of the part, the properties desired, and the processing practice preferred. Increases in carbon give greater core hardness with a given quench, or permit the use of thicker sections. Increases in manganese improve the hardenability of both the core and case; in carbon steels this is the only change in composition that will increase case hardenability. The higher manganese variants also machine much better. For carburizing applications, SAE 1016, 1018, and 1019 are widely used for thin sections or water-quenched parts. SAE 1022 and 1024 are used for heavier sections or where oil quenching is desired, and SAE 1024 is sometimes used for such parts as transmission and rear axle gears. SAE 1027 is used for parts given a light case to obtain satisfactory core properties without drastic quenching. SAE 1025 and 1030, although not usually regarded as carburizing types, are sometimes used in this manner for larger sections or where greater core hardness is needed.

For cold-formed or -headed parts, the lowest manganese grades (SAE 1017, 1020, and 1025) offer the best formability at their carbon level. SAE 1020 is used for fan blades and some frame members, and SAE 1020 and 1025 are widely used for low-strength bolts. The next higher manganese types (SAE 1018, 1021, and 1026) provide increased strength.

All steels listed may be readily welded or brazed by the common commercial methods. SAE 1020 is frequently used for welded tubing. These steels are used for numerous forged parts, the lower-carbon grades where high strength is not essential. Forgings from the lower-carbon steels usually machine better in the as-forged condition without annealing, or after normalizing.

SAE 1030, 1033, 1034, 1035, 1036, 1038, 1039, 1040, 1041, 1042, 1043, 1045, 1046, 1049, 1050, 1052: These steels, of the medium-carbon type, are selected for uses where higher mechanical properties are needed and are frequently further hardened and strengthened by heat treatment or by cold work. These grades are ordinarily produced as killed steels.

Steels in this group are suitable for a wide variety of automotive-type applications. The particular carbon and manganese level selected is affected by a number of factors. Increases in the mechanical properties required in section thickness, or in depth of hardening, ordinarily indicate either higher carbon or manganese or both. The heat-treating practice preferred, particularly the quenching medium, has a great effect on the steel selected. In general, any of the grades over 0.30 carbon may be selectively hardened by induction or flame methods.

The lower-carbon and manganese steels in this group find usage for certain types of cold-formed parts. SAE 1030 is used for shift and brake levers. SAE 1034 and 1035 are used in the form of wire and rod for cold upsetting such as bolts, and SAE 1038 for bolts and studs. The parts cold-formed from these steels are usually heat-treated prior to use. Stampings are generally limited to flat parts or simple bends. The higher-carbon SAE 1038, 1040, and 1042 are frequently cold drawn to specified physical properties for use without heat treatment for some applications such as cylinder head studs.

Any of this group of steels may be used for forgings, the selection being governed by the section size and the physical properties desired after heat treatment. Thus, SAE 1030 and 1035 are used for shifter forks and many small forgings where moderate properties are desired, but the deeper-hardening SAE 1036 is used for more critical parts where a higher strength level and more uniformity are essential, such as some front suspension parts. Forgings such as connecting rods, steering arms, truck front axles, axle shafts, and tractor wheels are commonly made from the SAE 1038 to 1045 group. Larger forgings at similar strength levels need more carbon and perhaps more manganese. Examples are crankshafts made from SAE 1046 and 1052. These steels are also used for small forgings where high hardness after oil quenching is desired. Suitable heat treatment is necessary on forgings from this group to provide machinability. These steels are also widely used for parts machined from bar stock, the selection following an identical pattern to that described for forgings. They are used both with and without heat treatment, depending on the application and the level of properties needed. As a class, they are considered good for normal machining operations. It is also possible to weld these steels by most commercial methods, but precautions should be taken to avoid cracking from too rapid cooling.

SAE 1055, 1060, 1062, 1064, 1065, 1066, 1070, 1074, 1078, 1080, 1085, 1086, 1090, 1095: Steels in this group are of the high-carbon type, having more carbon than is required to achieve maximum as quenched hardness. They are used for applications where the higher carbon is needed to improve wear characteristics for cutting edges, to make springs, and for special purposes. Selection of a particular grade is affected by the nature of the part, its end use, and the manufacturing methods available.

In general, cold-forming methods are not practical on this group of steels, being limited to flat stampings and springs coiled from small-diameter wire. Practically all parts from these steels are heat treated before use, with some variations in heat-treating methods to obtain optimum properties for the particular use to which the steel is to be put.

Uses in the spring industry include SAE 1065 for pretempered wire and SAE 1066 for cushion springs of hard-drawn wire, SAE 1064 may be used for small washers and thin stamped parts, SAE 1074 for light flat springs formed from annealed stock, and SAE 1080 and 1085 for thicker flat springs. SAE 1085 is also used for heavier coil springs. Valve spring wire and music wire are special products.

Due to good wear properties when properly heat-treated, the high-carbon steels find wide usage in the farm implement industry. SAE 1070 has been used for plow beams, SAE 1074 for plow shares, and SAE 1078 for such parts as rake teeth, scrapers, cultivator shovels, and plow shares. SAE 1085 has been used for scraper blades, disks, and for spring tooth harrows. SAE 1086 and 1090 find use as mower and binder sections, twine holders, and knotter disks.

Free Cutting Steels.—*SAE 1111, 1112, 1113:* This class of steels is intended for those uses where easy machining is the primary requirement. They are characterized by a higher sulfur content than comparable carbon steels. This composition results in some sacrifice of cold-forming properties, weldability, and forging characteristics. In general, the uses are similar to those for carbon steels of similar carbon and manganese content.

These steels are commonly known as Bessemer screw stock, and are considered the best machining steels available, machinability improving within the group as sulfur increases. They are used for a wide variety of machined parts. Although of excellent strength in the cold-drawn condition, they have an unfavorable property of cold shortness and are not commonly used for vital parts. These steels may be cyanided or carburized, but when uniform response to heat-treating is necessary, open-hearth steels are recommended.

SAE 1109, 1114, 1115, 1116, 1117, 1118, 1119, 1120, 1126: Steels in this group are used where a combination of good machinability and more uniform response to heat treatment is needed. The lower-carbon varieties are used for small parts that are to be cyanided or carbonitrided. SAE 1116, 1117, 1118, and 1119 carry more manganese for better hardenability, permitting oil quenching after case-hardening heat treatments in many instances. The higher-carbon SAE 1120 and 1126 provide more core hardness when this is needed.

SAE 1132, 1137, 1138, 1140, 1141, 1144, 1145, 1146, 1151: This group of steels has characteristics comparable to carbon steels of the same carbon level, except for changes due to higher sulfur as noted previously. They are widely used for parts where large amounts of machining are necessary, or where threads, splines, or other contours present special problems with tooling. SAE 1137, for example, is widely used for nuts and bolts and studs with machined threads. The higher-manganese SAE 1132, 1137, 1141, and 1144 offer greater hardenability, the higher-carbon types being suitable for oil quenching for many parts. All these steels may be selectively hardened by induction or flame heating if desired.

Carburizing Grades of Alloy Steels.—*Properties of the Case:* The properties of carburized and hardened cases (surface layers) depend on the carbon and alloy content, the structure of the case, and the degree and distribution of residual stresses. The carbon content of the case depends on the details of the carburizing process, and the response of iron and the alloying elements present, to carburization. The original carbon content of the steel has little or no effect on the carbon content produced in the case. The hardenability of the case, therefore, depends on the alloy content of the steel and the final carbon content produced by carburizing, but not on the initial carbon content of the steel.

With complete carbide solution, the effect of alloying elements on the hardenability of the case is about the same as the effect of these elements on the hardenability of the core. As an exception to this statement, any element that inhibits carburizing may reduce the hardenability of the case. Some elements that raise the hardenability of the core may tend to produce more retained austenite and consequently somewhat lower hardness in the case.

Alloy steels are frequently used for case hardening because the required surface hardness can be obtained by moderate speeds of quenching. Slower quenching may mean less distortion than would be encountered with water quenching. It is usually desirable to select a steel that will attain a minimum surface hardness of 58 or 60 Rockwell C after carburizing and oil quenching. Where section sizes are large, a high-hardenability alloy steel may be necessary, whereas for medium and light sections, low-hardenability steels will suffice.

In general, the case-hardening alloy steels may be divided into two classes as far as the hardenability of the case is concerned. Only the general type of steel (SAE 3300–4100, etc.) is discussed. The original carbon content of the steel has no effect on the carbon content of the case, so the last two digits in the specification numbers are not meaningful as far as the case is concerned.

A) High-Hardenability Case: SAE 2500, 3300, 4300, 4800, 9300

As these are high-alloy steels, both the case and the core have high hardenability. They are used particularly for carburized parts having thick sections, such as bevel drive pinions and heavy gears. Good case properties can be obtained by oil quenching. These steels are likely to have retained austenite in the case after carburizing and quenching; consequently, special precautions or treatments, such as refrigeration, may be required.

B) Medium-Hardenability Case: SAE 1300, 2300, 4000, 4100, 4600, 5100, 8600, 8700

Carburized cases of these steels have medium hardenability, which means that their hardenability is intermediate between that of plain carbon steel and the higher-alloy carburizing steels discussed earlier. In general, these steels can be used for average-size case-hardened automotive parts such as gears, pinions, piston pins, ball studs, universal joint crosses, crankshafts, etc. Satisfactory case hardness is usually produced by oil quenching.

Core Properties: The core properties of case-hardened steels depend on both carbon and alloy content of the steel. Each of the general types of alloy case-hardening steel is usually made with two or more carbon contents to permit different hardenability in the core.

The most desirable hardness for the core depends on the design and functioning of the individual part. In general, where high compressive loads are encountered, relatively high core hardness is beneficial in supporting the case. Low core hardnesses may be desirable where great toughness is essential.

The case-hardening steels may be divided into three general classes, depending on hardenability of the core.

A) Low-Hardenability Core: SAE 4017, 4023, 4024, 4027,* 4028,* 4608, 4615, 4617,* 8615,* 8617*

B) Medium-Hardenability Core: SAE 1320, 2317, 2512, 2515,* 3115, 3120, 4032, 4119, 4317, 4620, 4621, 4812, 4815,* 5115, 5120, 8620, 8622, 8720, 9420

C) High-Hardenability Core: SAE 2517, 3310, 3316, 4320, 4817, 4820, 9310, 9315, 9317

Heat Treatments: In general, all the alloy carburizing steels are made with fine grain and most are suitable for direct quenching from the carburizing temperature. Several other types of heat treatment involving single and double quenching are also used for most of these steels.

Directly Hardenable Grades of Alloy Steels.—These steels may be considered in five groups on the basis of approximate mean carbon content of the SAE specification. In general, the last two figures of the specification agree with the mean carbon content. Consequently the heading "0.30 – 0.37 Mean Carbon Content of SAE Specification" includes steels such as SAE 1330, 3135, and 4137.

It is necessary to deviate from the above plan in the classification of the carbon molybdenum steels. When carbon molybdenum steels are used, it is customary to specify higher carbon content for any given application than would be specified for other alloy steels, due to the low alloy content of these steels. For example, SAE 4063 is used for the same applications as SAE 4140, 4145, and 5150. Consequently, in the following discussion, the car-

* Borderline classifications might be considered in the next higher hardenability group.

bon molybdenum steels have been shown in the groups where they belong on the basis of applications rather than carbon content.

Mean Carbon Content of SAE Specification	Common Applications
(a) 0.30–0.37 per cent	Heat-treated parts requiring moderate strength and great toughness.
(b) 0.40–0.42 per cent	Heat-treated parts requiring higher strength and good toughness.
(c) 0.45–0.50 per cent	Heat-treated parts requiring fairly high hardness and strength with moderate toughness.
(d) 0.50–0.62 per cent	Springs and hand tools.
(e) 1.02 per cent	Ball and roller bearings.

For the present discussion, steels of each carbon content are divided into two or three groups on the basis of hardenability. Transformation ranges and consequently heat-treating practices vary somewhat with different alloying elements even though the hardenability is not changed.

0.30–0.37 Mean Carbon Content of SAE Specification: These steels are frequently used for water-quenched parts of moderate section size and for oil-quenched parts of small section size. Typical applications of these steels are connecting rods, steering arms and steering knuckles, axle shafts, bolts, studs, screws, and other parts requiring strength and toughness where section size is small enough to permit the desired physical properties to be obtained with the customary heat treatment.

Steels falling in this classification may be subdivided into two groups on the basis of hardenability:

A) Low Hardenability: SAE 1330, 1335, 4037, 4042, 4130, 5130, 5132, 8630

B) Medium Hardenability: SAE 2330, 3130, 3135, 4137, 5135, 8632, 8635, 8637, 8735, 9437

0.40–0.42 Mean Carbon Content of SAE Specification: In general, these steels are used for medium and large size parts requiring high degree of strength and toughness. The choice of the proper steel depends on the section size and the mechanical properties that must be produced. The low and medium hardenabilty steels are used for average size automotive parts such as steering knuckles, axle shafts, propeller shafts, etc. The high hardenability steels are used particularly for large axles and shafts for large aircraft parts.

These steels are usually considered as oil quenching steels, although some large parts made of the low and medium hardenability classifications may be quenched in water under properly controlled conditions.

These steels may be divided into three groups on the basis of hardenability:

A) Low Hardenability: SAE 1340, 4047, 5140, 9440

B) Medium Hardenability: SAE 2340, 3140, 3141, 4053, 4063, 4140, 4640, 8640, 8641, 8642, 8740, 8742, 9442

C) High Hardenability: SAE 4340, 9840

0.45–0.50 Mean Carbon Content of SAE Specification: These steels are used primarily for gears and other parts requiring fairly high hardness as well as strength and toughness. Such parts are usually oil-quenched and a minimum of 90 per cent martensite in the as-quenched condition is desirable.

A) Low Hardenability: SAE 5045, 5046, 5145, 9747, 9763

B) Medium Hardenability: SAE 2345, 3145, 3150, 4145, 5147, 5150, 8645, 8647, 8650, 8745, 8747, 8750, 9445, 9845

C) High Hardenability: SAE 4150, 9850

0.50–0.63 Mean Carbon Content of SAE Specification: These steels are used primarily for springs and hand tools. The hardenability necessary depends on the thickness of the material and the quenching practice.

A) Medium hardenability: SAE 4068, 5150, 5152, 6150, 8650, 9254, 9255, 9260, 9261

B) High Hardenability: SAE 8653, 8655, 8660, 9262

1.02 Mean Carbon Content of SAE Specification—SAE 50100, 51100, 52100: These straight chromium electric furnace steels are used primarily for the races and balls or rollers of anti friction bearings. They are also used for other parts requiring high hardness and wear resistance. The compositions of the three steels are identical, except for a variation in chromium, with a corresponding variation in hardenability.

A) Low Hardenability: SAE 50100

B) Medium Hardenability: SAE 51100, 52100

Resulfurized Steel: Some of the alloy steels, SAE 4024, 4028, and 8641, are made resulfurized so as to give better machinability at a relatively high hardness. In general, increased sulfur results in decreased transverse ductility, notched impact toughness, and weldability.

Chromium Nickel Austenitic Steels (Not capable of heat treatment).—*SAE 30201:* This steel is an austenitic chromium–nickel–manganese stainless steel usually required in flat products. In the annealed condition, it exhibits higher strength values than the corresponding chromium–nickel stainless steel (SAE 30301). It is nonmagnetic in the annealed condition, but may be magnetic when cold-worked. SAE 30201 is used to obtain high strength by work-hardening and is well suited for corrosion-resistant structural members requiring high strength with low weight. It has excellent resistance to a wide variety of corrosive media, showing behavior comparable to stainless grade SAE 30301. It has high ductility and excellent forming properties. Owing to this steel's work-hardening rate and yield strength, tools for forming must be designed to allow for a higher spring back or recovery rate. It is used for automotive trim, automotive wheel covers, railroad passenger car bodies and structural members, and truck trailer bodies.

SAE 30202: Like its corresponding chromium–nickel stainless steel SAE 30302, this is a general-purpose stainless steel. It has excellent corrosion resistance and deep drawing qualities. It is nonhardenable by thermal treatments, but may be cold worked to high tensile strengths. In the annealed condition, it is nonmagnetic but slightly magnetic when cold-worked. Applications for this stainless steel are hub cap, railcar and truck trailer bodies, and spring wire.

SAE 30301: Capable of attaining high tensile strength and ductility by moderate or severe cold working. It is used largely in the cold-rolled or cold-drawn condition in the form of sheet, strip, and wire. Its corrosion resistance is good but not equal to SAE 30302.

SAE 30302: The most widely used of the general-purpose austenitic chromium–nickel stainless steels. It is used for deep drawing largely in the annealed condition. It can be worked to high tensile strengths but with slightly lower ductility than SAE 30301.

SAE 30303F: A free-machining steel recommended for the manufacture of parts produced on automatic screw machines. Caution must be used in forging this steel.

SAE 30304: Similar to SAE 30302 but somewhat superior in corrosion resistance and having superior welding properties for certain types of equipment.

SAE 30305: Similar to SAE 30304 but capable of lower hardness. Has greater ductility with slower work-hardening tendency.

SAE 30309: A steel with high heat-resisting qualities which is resistant to oxidation at temperatures up to about 1800 degrees F.

SAE 30310: This steel has the highest heat-resisting properties of any of the chromium nickel steels listed here and will resist oxidation at temperatures up to about 1900 degrees F.

SAE 30316: Recommended for use in parts where unusual resistance to chemical or salt water corrosion is necessary. It has superior creep strength at elevated temperatures.

SAE 30317: Similar to SAE 30316 but has the highest corrosion resistance of all these alloys in many environments.

SAE 30321: Recommended for use in the manufacture of welded structures where heat treatment after welding is not feasible. It is also recommended for use where temperatures up to 1600 degrees F are encountered in service.

SAE 30325: Used for such parts as heat control shafts.

SAE 30347: This steel is similar to SAE 30321. This niobium alloy is sometimes preferred to titanium because niobium is less likely to be lost in welding operations.

Stainless Chromium Irons and Steels.—*SAE 51409:* An 11 per cent chromium alloy developed, especially for automotive mufflers and tailpipes. Resistance to corrosion and oxidation is very similar to SAE 51410. It is nonhardenable and has good forming and welding characteristics. This alloy is recommended for mildly corrosive applications where surface appearance is not critical.

SAE 51410: A general-purpose stainless steel capable of heat treatment to show good physical properties. It is used for general stainless applications, both in the heat-treated and annealed condition but is not as resistant to corrosion as SAE 51430 in either the annealed or heat-treated condition.

SAE 51414: A corrosion and heat-resisting nickel-bearing chromium steel with somewhat better corrosion resistance than SAE 51410. It will attain slightly higher mechanical properties when heat-treated than SAE 51410. It is used in the form of tempered strip or wire, and in bars and forgings for heat-treated parts.

SAE 51416F: A free-machining grade for the manufacture of parts produced in automatic screw machines.

SAE 51420: This steel is capable of being heat-treated to a relatively high hardness. It will harden to a maximum of approximately 500 Brinell. Maximum corrosion resisting qualities exist only in the fully hardened condition. It is used for cutlery, hardened pump shafts, etc.

SAE 51420F: This is similar to SAE 51420 except for its free-machining properties.

SAE 51430: This high-chromium steel is not capable of heat treatment and is recommended for use in shallow parts requiring moderate draw. Corrosion and heat resistance are superior to SAE 51410.

SAE 51430F: This steel is similar to SAE 51430 except for its free-machining properties.

SAE 51431: This nickel-bearing chromium steel is designed for heat treatment to high mechanical properties. Its corrosion resistance is superior to other hardenable steels.

SAE 51440A: A hardenable chromium steel with greater quenched hardness than SAE 51420 and greater toughness than SAE 51440B and 51440C. Maximum corrosion resistance is obtained in the fully hardened and polished condition.

SAE 51440B: A hardenable chromium steel with greater quenched hardness than SAE 51440A. Maximum corrosion resistance is obtained in the fully hardened and polished condition. Capable of hardening to 50–60 Rockwell C depending on carbon content.

SAE 51440C: This steel has the greatest quenched hardness and wear resistance on heat treatment of any corrosion- or heat-resistant steel.

SAE 51440F: The same as SAE 51440C, except for its free-machining characteristics.

SAE 51442: A corrosion- and heat-resisting chromium steel with corrosion-resisting properties slightly better than SAE 51430 and with good scale resistance up to 1600 degrees F.

SAE 51446: A corrosion- and heat-resisting steel with maximum amount of chromium consistent with commercial malleability. Used principally for parts that must resist high temperatures in service without scaling. Resists oxidation up to 2000 degrees F.

SAE 51501: Used for its heat and corrosion resistance and good mechanical properties at temperatures up to approximately 1000 degrees F.

General Application of SAE Steels: These applications are intended as a general guide only since the selection may depend on the exact character of the service, cost of material, machinability when machining is required, or other factors. When more than one steel is recommended for a given application, information on the characteristics of each steel listed will be found in the section beginning on page 433.

Adapters, 1145
Agricultural steel, 1070, 1080
Aircraft forgings, 4140
Axles front or rear, 1040, 4140
Axle shafts, 1045, 2340, 2345, 3135, 3140, 3141, 4063, 4340
Ball-bearing races, 52100
Balls for ball bearings, 52100
Body stock for cars, rimmed*
Bolts and screws, 1035
Bolts
  anchor, 1040
  cold-headed, 4042
  connecting-rod, 3130
  heat-treated, 2330
  heavy-duty, 4815, 4820
  steering-arm, 3130
Brake levers, 1030, 1040
Bumper bars, 1085
Cams free-wheeling, 4615, 4620
Camshafts, 1020, 1040
Carburized parts, 1020, 1022, 1024, 1117, 1118, 1320, 2317, 2515, 3310, 3115, 3120, 4023, 4032
Chain pins transmission, 4320, 4815, 4820
Chains transmission, 3135, 3140
Clutch disks, 1060, 1070, 1085
Clutch springs, 1060
Coil springs, 4063
Cold-headed bolts, 4042
Cold-heading
  steel, 30905, 1070
  wire or rod, rimmed*, 1035
Cold-rolled steel, 1070
Connecting-rods, 1040, 3141
Connecting-rod bolts, 3130
Corrosion resisting, 51710, 30805
Covers transmission, rimmed*
Crankshafts, 1045, 1145, 3135, 3140, 3141
Crankshafts Diesel engine, 4340
Cushion springs, 1060
Cutlery stainless, 51335
Cylinder studs, 3130
Deep-drawing steel, rimmed*, 30905
Differential gears, 4023
Disks clutch, 1070, 1060
Ductile steel, 30905
Fan blades, 1020
Fatigue resisting 4340, 4640
Fender stock for cars, rimmed*
Forgings
  aircraft, 4140
  carbon steel, 1040, 1045
  heat-treated, 3240, 5140, 6150
  high-duty, 6150
  small or medium, 1035
  large, 1036
Free-cutting steel
  carbon, 1111, 1113
  chromium-nickel steel, 30615
  manganese steel, 1132, 1137
Gears
  carburized, 1320, 2317, 3115, 3120, 3310, 4119, 4125, 4320, 4615, 4620, 4815, 4820
  heat-treated, 2345
  car and truck, 4027, 4032
  cyanide-hardening, 5140
  differential, 4023
  high duty, 4640, 6150
  oil-hardening, 3145, 3150, 4340, 5150
  ring, 1045, 3115, 3120, 4119
  transmission, 3115, 3120, 4119
  truck and bus, 3310, 4320
Gear shift levers, 1030
Harrow disks, 1080
Hay-rake teeth, 1095
Key stock, 1030, 2330, 3130
Leaf springs, 1085, 9260
Levers
  brake, 1030, 1040
  gear shift, 1030
  heat-treated, 2330
Lock washers, 1060
Mower knives, 1085
Mower sections, 1070
Music wire, 1085
Nuts, 3130
  heat-treated, 2330
Oil pans automobile, rimmed*

* The "rimmed" and "killed" steels listed are in the SAE 1008, 1010, and 1015 group. See general description of these steels.

Numbering Systems for Metals and Alloys.—Several different numbering systems have been developed for metals and alloys by various trade associations, professional engineering societies, standards organizations, and by private industries for their own use. The numerical code used to identify the metal or alloy may or may not be related to a specification, which is a statement of the technical and commercial requirements that the product must meet. Numbering systems in use include those developed by the American Iron and Steel Institute (AISI), Society of Automotive Engineers (SAE), American Society for Testing and Materials (ASTM), American National Standards Institute (ANSI), Steel Founders Society of America, American Society of Mechanical Engineers (ASME), American Welding Society (AWS), Aluminum Association, Copper Development Association, U.S. Department of Defense (Military Specifications), and the General Accounting Office (Federal Specifications).

The Unified Numbering System (UNS) was developed through a joint effort of the ASTM and the SAE to provide a means of correlating the different numbering systems for metals and alloys that have a commercial standing. This system avoids the confusion caused when more than one identification number is used to specify the same material, or when the same number is assigned to two entirely different materials. It is important to understand that a UNS number is not a specification; it is an identification number for metals and alloys for which detailed specifications are provided elsewhere. UNS numbers are shown in Table 1; each number consists of a letter prefix followed by five digits. In some, the letter is suggestive of the family of metals identified by the series, such as A for aluminum and C for copper. Whenever possible, the numbers in the UNS groups contain numbering sequences taken directly from other systems to facilitate identification of the material; e.g., the corresponding UNS number for AISI 1020 steel is G10200. The UNS numbers corresponding to the commonly used AISI-SAE numbers that are used to identify plain carbon alloy and tool steels are given in Table 1.

Table 1. Unified Numbering System (UNS) for Metals and Alloys

UNS Series	Metal
A00001 to A99999	Aluminum and aluminum alloys
C00001 to C99999	Copper and copper alloys
D00001 to D99999	Specified mechanical property steels
E00001 to E99999	Rare earth and rare earth like metals and alloys
F00001 to F99999	Cast irons
G00001 to G99999	AISI and SAE carbon and alloy steels (except tool steels)
H00001 to H99999	AISI and SAE H-steels
J00001 to J99999	Cast steels (except tool steels)
K00001 to K99999	Miscellaneous steels and ferrous alloys
L00001 to L99999	Low-melting metals and alloys
M00001 to M99999	Miscellaneous nonferrous metals and alloys
N00001 to N99999	Nickel and nickel alloys
P00001 to P99999	Precious metals and alloys
R00001 to R99999	Reactive and refractory metals and alloys
S00001 to S99999	Heat and corrosion resistant (stainless) steels
T00001 to T99999	Tool steels, wrought and cast
W00001 to W99999	Welding filler metals
Z00001 to Z99999	Zinc and zinc alloys

Table 1. AISI and SAE Numbers and Their Corresponding UNS Numbers for Plain Carbon, Alloy, and Tool Steels

AISI-SAE Numbers	UNS Numbers	AISI-SAE Numbers	UNS Numbers	AISI-SAE Numbers	UNS Numbers	AISI-SAE Numbers	UNS Numbers
Plain Carbon Steels							
1005	G10050	1030	G10300	1070	G10700	1566	G15660
1006	G10060	1035	G10350	1078	G10780	1110	G11100
1008	G10080	1037	G10370	1080	G10800	1117	G11170
1010	G10100	1038	G10380	1084	G10840	1118	G11180
1012	G10120	1039	G10390	1086	G10860	1137	G11370
1015	G10150	1040	G10400	1090	G10900	1139	G11390
1016	G10160	1042	G10420	1095	G10950	1140	G11400
1017	G10170	1043	G10430	1513	G15130	1141	G11410
1018	G10180	1044	G10440	1522	G15220	1144	G11440
1019	G10190	1045	G10450	1524	G15240	1146	G11460
1020	G10200	1046	G10460	1526	G15260	1151	G11510
1021	G10210	1049	G10490	1527	G15270	1211	G12110
1022	G10220	1050	G10500	1541	G15410	1212	G12120
1023	G10230	1053	G10530	1548	G15480	1213	G12130
1025	G10250	1055	G10550	1551	G15510	1215	G12150
1026	G10260	1059	G10590	1552	G15520	12L14	G12144
1029	G10290	1060	G10600	1561	G15610	…	…
Alloy Steels							
1330	G13300	4150	G41500	5140	G51400	8642	G86420
1335	G13350	4161	G41610	5150	G51500	8645	G86450
1340	G13400	4320	G43200	5155	G51550	8655	G86550
1345	G13450	4340	G43400	5160	G51600	8720	G87200
4023	G40230	E4340	G43406	E51100	G51986	8740	G87400
4024	G40240	4615	G46150	E52100	G52986	8822	G88220
4027	G40270	4620	G46200	6118	G61180	9260	G92600
4028	G40280	4626	G46260	6150	G61500	50B44	G50441
4037	G40370	4720	G47200	8615	G86150	50B46	G50461
4047	G40470	4815	G48150	8617	G86170	50B50	G50501
4118	G41180	4817	G48170	8620	G86200	50B60	G50601
4130	G41300	4820	G48200	8622	G86220	51B60	G51601
4137	G41370	5117	G51170	8625	G86250	81B45	G81451
4140	G41400	5120	G51200	8627	G86270	94B17	G94171
4142	G41420	5130	G51300	8630	G86300	94B30	G94301
4145	G41450	5132	G51320	8637	G86370	…	…
4147	G41470	5135	G51350	8640	G86400	…	…
Tool Steels (AISI and UNS Only)							
M1	T11301	T6	T12006	A6	T30106	P4	T51604
M2	T11302	T8	T12008	A7	T30107	P5	T51605
M4	T11304	T15	T12015	A8	T30108	P6	T51606
M6	T11306	H10	T20810	A9	T30109	P20	T51620
M7	T11307	H11	T20811	A10	T30110	P21	T51621
M10	T11310	H12	T20812	D2	T30402	F1	T60601
M3-1	T11313	H13	T20813	D3	T30403	F2	T60602
M3-2	T11323	H14	T20814	D4	T30404	L2	T61202
M30	T11330	H19	T20819	D5	T30405	L3	T61203
M33	T11333	H21	T20821	D7	T30407	L6	T61206
M34	T11334	H22	T20822	O1	T31501	W1	T72301
M36	T11336	H23	T20823	O2	T31502	W2	T72302
M41	T11341	H24	T20824	O6	T31506	W5	T72305
M42	T11342	H25	T20825	O7	T31507	CA2	T90102
M43	T11343	H26	T20826	S1	T41901	CD2	T90402
M44	T11344	H41	T20841	S2	T41902	CD5	T90405
M46	T11346	H42	T20842	S4	T41904	CH12	T90812
M47	T11347	H43	T20843	S5	T41905	CH13	T90813
T1	T12001	A2	T30102	S6	T41906	CO1	T91501
T2	T12002	A3	T30103	S7	T41907	CS5	T91905
T4	T12004	A4	T30104	P2	T51602	…	…
T5	T12005	A5	T30105	P3	T51603	…	…

Standard Steel Numbering System.—The most widely used systems for identifying wrought carbon, low–alloy, and stainless steels are based on chemical composition, and are those of the American Iron and Steel Institute (AISI) and the Society of Automotive Engineers (SAE). These systems are almost identical, but they are carefully coordinated. The standard steels so designated have been developed cooperatively by producers and users and have been found through long experience to cover most of the wrought ferrous metals used in automotive vehicles and related equipment. These designations, however, are not specifications, and should not be used for purchasing unless accompanied by supplementary information necessary to describe commercially the product desired. Engineering societies, associations, and institutes whose members make, specify, or purchase steel products publish standard specifications, many of which have become well known and respected. The most comprehensive and widely used specifications are those published by the American Society for Testing and Materials (ASTM). The U.S. government and various companies also publish their own specification for steel products to serve their own special procurement needs. The Unified Numbering System (UNS) for metals and alloys is also used to designate steels (see pages 442 and 443).

The numerical designation system used by both AISI and SAE for wrought carbon, alloy, and stainless steels is summarized in Table .

Table 2. AISI-SAE System of Designating Carbon and Alloy Steels

AISI-SAE Designation[a]	Type of Steel and Nominal Alloy Content (%)
	Carbon Steels
10xx	Plain Carbon (Mn 1.00% max.)
11xx	Resulfurized
12xx	Resulfurized and Rephosphorized
15xx	Plain Carbon (Max. Mn range 1.00 to 1.65%)
	Manganese Steels
13xx	Mn 1.75
	Nickel Steels
23xx	Ni 3.50
25xx	Ni 5.00
	Nickel–Chromium Steels
31xx	Ni 1.25; Cr 0.65 and 0.80
32xx	Ni 1.75; Cr 1.07
33xx	Ni 3.50; Cr 1.50 and 1.57
34xx	Ni 3.00; Cr 0.77
	Molybdenum Steels
40xx	Mo 0.20 and 0.25
44xx	Mo 0.40 and 0.52
	Chromium–Molybdenum Steels
41xx	Cr 0.50, 0.80, and 0.95; Mo 0.12, 0.20, 0.25, and 0.30
	Nickel–Chromium–Molybdenum Steels
43xx	Ni 1.82; Cr 0.50 and 0.80; Mo 0.25
43BVxx	Ni 1.82; Cr 0.50; Mo 0.12 and 0.35; V 0.03 min.
47xx	Ni 1.05; Cr 0.45; Mo 0.20 and 0.35
81xx	Ni 0.30; Cr 0.40; Mo 0.12
86xx	Ni 0.55; Cr 0.50; Mo 0.20
87xx	Ni 0.55; Cr 0.50; Mo 0.25
88xx	Ni 0.55; Cr 0.50; Mo 0.35
93xx	Ni 3.25; Cr 1.20; Mo 0.12
94xx	Ni 0.45; Cr 0.40; Mo 0.12
97xx	Ni 0.55; Cr 0.20; Mo 0.20
98xx	Ni 1.00; Cr 0.80; Mo 0.25
	Nickel–Molybdenum Steels
46xx	Ni 0.85 and 1.82; Mo 0.20 and 0.25
48xx	Ni 3.50; Mo 0.25
	Chromium Steels

Table 2. AISI-SAE System of Designating Carbon and Alloy Steels

AISI-SAE Designation[a]		Type of Steel and Nominal Alloy Content (%)
50xx		Cr 0.27, 0.40, 0.50, and 0.65
51xx		Cr 0.80, 0.87, 0.92, 0.95, 1.00, and 1.05
50xxx		Cr 0.50; C 1.00 min.
51xxx		Cr 1.02; C 1.00 min.
52xxx		Cr 1.45; C 1.00 min.
		Chromium–Vanadium Steels
61xx		Cr 0.60, 0.80, and 0.95; V 0.10 and 0.15 min
		Tungsten–Chromium Steels
72xx		W 1.75; Cr 0.75
		Silicon–Manganese Steels
92xx		Si 1.40 and 2.00; Mn 0.65, 0.82, and 0.85; Cr 0.00 and 0.65
		High–Strength Low–Alloy Steels
9xx		Various SAE grades
xxBxx		B denotes boron steels
xxLxx		L denotes leaded steels
AISI	**SAE**	**Stainless Steels**
2xx	302xx	Chromium–Manganese–Nickel Steels
3xx	303xx	Chromium–Nickel Steels
4xx	514xx	Chromium Steels
5xx	515xx	Chromium Steels

[a] xx in the last two digits of the carbon and low–alloy designations (but not the stainless steels) indicates that the carbon content (in hundredths of a per cent) is to be inserted.

Classification of Tool Steels.—Steels for tools must satisfy a number of different, often conflicting requirements. The need for specific steel properties arising from widely varying applications has led to the development of many compositions of tool steels, each intended to meet a particular combination of applicational requirements. The resultant diversity of tool steels, their number being continually expanded by the addition of new developments, makes it extremely difficult for the user to select the type best suited to his needs, or to find equivalent alternatives for specific types available from particular sources.

As a cooperative industrial effort under the sponsorship of AISI and SAE, a tool classification system has been developed in which the commonly used tool steels are grouped into seven major categories. These categories, several of which contain more than a single group, are listed in the following with the letter symbols used for their identification. The individual types of tool steels within each category are identified by suffix numbers following the letter symbols.

Category Designation	Letter Symbol	Group Designation
High-Speed Tool Steels	M	Molybdenum types
	T	Tungsten types
Hot-Work Tool Steels	H1–H19	Chromium types
	H20–H39	Tungsten types
	H40–H59	Molybdenum types
Cold-Work Tool Steels	D	High-carbon, high-chromium types
	A	Medium-alloy, air-hardening types
	O	Oil-hardening types
Shock-Resisting Tool Steels	S	…
Mold Steels	P	…
Special-Purpose Tool Steels	L	Low-alloy types
	F	Carbon–tungsten types
Water-Hardening Tool Steels	W	…

Table 3. Quick Reference Guide for Tool Steel Selection

Application Areas	Tool Steel Categories and AISI Letter Symbol						
	High-Speed Tool Steels, M and T	Hot-Work Tool Steels, H	Cold-Work Tool Steels, D, A, and O	Shock-Resisting Tool Steels, S	Mold Steels, P	Special-Purpose Tool Steels, L and F	Water-Hardening Tool Steels, W
	Examples of Typical Applications						
Cutting Tools Single-point types (lathe, planer, boring) Milling cutters Drills Reamers Taps Threading dies Form cutters	General-purpose production tools: M2, T1 For increased abrasion resistance: M3, M4, and M10 Heavy-duty work calling for high hot hardness: T5, T15 Heavy-duty work calling for high abrasion resistance: M42, M44		Tools with keen edges (knives, razors) Tools for operations where no high-speed is involved, yet stability in heat treatment and substantial abrasion resistance are needed	Pipe cutter wheels			Uses that do not require hot hardness or high abrasion resistance. Examples with carbon content of applicable group: Taps (1.05/1.10% C) Reamers (1.10/1.15% C) Twist drills (1.20/1.25% C) Files (1.35/1.40% C)
Hot Forging Tools and Dies Dies and inserts Forging machine plungers and pierces	For combining hot hardness with high abrasion resistance: M2, T1	Dies for presses and hammers: H20, H21 For severe conditions over extended service periods: H22 to H26, also H43	Hot trimming dies: D2	Hot trimming dies Blacksmith tools Hot swaging dies			Smith's tools (1.65/0.70% C) Hot chisels (0.70/0.75% C) Drop forging dies (0.90/1.00% C) Applications limited to short-run production
Hot Extrusion Tools and Dies Extrusion dies and mandrels, Dummy blocks Valve extrusion tools	Brass extrusion dies: T1	Extrusion dies and dummy blocks: H20 to H26 For tools that are exposed to less heat: H10 to H19		Compression molding: S1			

Table 3. *(Continued)* **Quick Reference Guide for Tool Steel Selection**

Application Areas	Tool Steel Categories and AISI Letter Symbol						
	High-Speed Tool Steels, M and T	Hot-Work Tool Steels, H	Cold-Work Tool Steels, D, A, and O	Shock-Resisting Tool Steels, S	Mold Steels, P	Special-Purpose Tool Steels, L and F	Water-Hardening Tool Steels, W
	Examples of Typical Applications						
Cold-Forming Dies Bending, forming, drawing, and deep drawing dies and punches	Burnishing tools: M1, T1	Cold heading: die casting dies: H13	Drawing dies: 01 Coining tools: O1, D2 Forming and bending dies: A2 Thread rolling dies: D2	Hobbing and short-run applications: S1, S7 Rivet sets and rivet busters		Blanking, forming, and trimmer dies when toughness has precedence over abrasion resistance: L6	Cold-heading dies: W1 or W2 ($C \cong 1.00\%$) Bending dies: W1 ($C \cong 1.00\%$)
Shearing Tools Dies for piercing, punching, and trimming Shear blades	Special dies for cold and hot work: T1 For work requiring high abrasion resistance: M2, M3	For shearing knives: H11, H12 For severe hot shearing applications: M21, M25	Dies for medium runs: A2, A6 also O1 and O4 Dies for long runs: D2, D3 Trimming dies (also for hot trimming): A2	Cold and hot shear blades Hot punching and piercing tools Boilermaker's tools		Knives for work requiring high toughness: L6	Trimming dies (0.90/0.95% C) Cold blanking and punching dies (1.00% C)
Die Casting Dies and Plastics Molds		For zinc and lead: H11 For aluminum: H13 For brass: H21	A2 and A6 O1		Plastics molds: P2 to P4, and P20		
Structural Parts for Severe Service Conditions	Roller bearings for high-temperature environment: T1 Lathe centers: M2 and T1	For aircraft components (landing gear, arrester hooks, rocket cases): H11	Lathe centers: D2, D3 Arbors: O1 Bushings: A4 Gages: D2	Pawls Clutch parts		Spindles, clutch parts (where high toughness is needed): L6	Spring steel (1.10/1.15% C)
Battering Tools for Hand and Power Tool Use				Pneumatic chisels for cold work: S5 For higher performance: S7			For intermittent use: W1 (0.80% C)

Table 4. Molybdenum High-Speed Steels

Identifying Chemical Composition and Typical Heat-Treatment Data																			
	AISI Type	M1	M2	M3 Cl. 1	M3 Cl. 2	M4	M6	M7	M10	M30	M33	M34	M36	M41	M42	M43	M44	M46	M47
Identifying Chemical Elements in Per Cent	C	0.80	0.85; 1.00	1.05	1.20	1.30	0.80	1.00	0.85; 1.00	0.80	0.90	0.90	0.80	1.10	1.10	1.20	1.15	1.25	1.10
	W	1.50	6.00	6.00	6.00	5.50	4.00	1.75	…	2.00	1.50	2.00	6.00	6.75	1.50	2.75	5.25	2.00	1.50
	Mo	8.00	5.00	5.00	5.00	4.50	5.00	8.75	8.00	8.00	9.50	8.00	5.00	3.75	9.50	8.00	6.25	8.25	9.50
	Cr	4.00	4.00	4.00	4.00	4.00	4.00	4.00	4.00	4.00	4.00	4.00	4.00	4.25	3.75	3.75	4.25	4.00	3.75
	V	1.00	2.00	2.40	3.00	4.00	1.50	2.00	2.00	1.25	1.15	2.00	2.00	2.00	1.15	1.60	2.25	3.20	1.25
	Co	…	…	…	…	…	12.00	…	…	5.00	8.00	8.00	8.00	5.00	8.00	8.25	12.00	8.25	5.00
Heat-Treat. Data	Hardening Temperature Range, °F	2150–2225	2175–2225	2200–2250	2200–2250	2200–2250	2150–2200	2150–2225	2150–2225	2200–2250	2200–2250	2200–2250	2225–2275	2175–2220	2175–2210	2175–2220	2190–2240	2175–2225	2150–2200
	Tempering Temperature Range, °F	1000–1100	1000–1160	1000–1100	1000–1100	1000–1100	1000–1100	1000–1100	1000–1100	1000–1100	1000–1100	1000–1100	1000–1100	1000–1100	950–1100	950–1100	1000–1160	975–1050	975–1100
	Approx. Tempered Hardness, Rc	65–60	65–60	66–61	66–61	66–61	66–61	66–61	65–60	65–60	65–60	65–60	65–60	70–65	70–65	70–65	70–62	69–67	70–65
Relative Ratings of Properties (A = greatest to E = least)																			
Characteristics in Heat Treatment	Safety in Hardening	D	D	D	D	D	D	D	D	D	D	D	D	D	D	D	D	D	D
	Depth of Hardening	A	A	A	A	A	A	A	A	A	A	A	A	A	A	A	A	A	A
	Resistance to Decarburization	C	B	B	B	B	C	C	C	C	C	C	C	C	C	C	C	C	C
	Stability of Shape in Heat Treatment — Quenching Medium: Air or Salt	C	C	C	C	C	C	C	C	C	C	C	C	C	C	C	C	C	C
	Stability of Shape in Heat Treatment — Quenching Medium: Oil	D	D	D	D	D	D	D	D	D	D	D	D	D	D	D	D	D	D
Service Properties	Machinability	D	D	D	D/E	D	D	D	D	D	D	D	D	D	D	D	D	D	D
	Hot Hardness	B	B	B	B	B	A	B	B	A	A	A	A	A	A	A	A	A	A
	Wear Resistance	B	B	B	B	A	B	B	B	B	B	B	B	B	B	B	B	B	B
	Toughness	E	E	E	E	E	E	E	E	E	E	E	E	E	E	E	E	E	E

Table 5. Tungsten High-Speed Tool Steels

Identifying Chemical Composition and Typical Heat-Treatment Data									
	AISI Type		T1	T2	T4	T5	T6	T8	T15
Identifying Chemical Elements in Per Cent	C		0.75	0.80	0.75	0.80	0.80	0.75	1.50
	W		18.00	18.00	18.00	18.00	20.00	14.00	12.00
	Cr		4.00	4.00	4.00	4.00	4.50	4.00	4.00
	V		1.00	2.00	1.00	2.00	1.50	2.00	5.00
	Co		…	…	5.00	…	…	5.00	5.00
Heat-Treat. Data	Hardening Temperature Range, °F		2300–2375	2300–2375	2300–2375	2325–2375	2325–2375	2300–2375	2200–2300
	Tempering Temperature Range, °F		1000–1100	1000–1100	1000–1100	1000–1100	1000–1100	1000–1100	1000–1200
	Approx. Tempered Hardness, R_c		65–60	66–61	66–62	65–60	65–60	65–60	68–63
Relative Ratings of Properties (A = greatest to E = least)									
Characteristics in Heat Treatment	Safety in Hardening		C	C	D	D	D	D	D
	Depth of Hardening		A	A	A	A	A	A	A
	Resistance to Decarburization		A	A	B	C	C	B	B
	Stability of Shape in Heat Treatment — Quenching Medium	Air or Salt	C	C	C	C	C	C	C
		Oil	D	D	D	D	D	D	D
Service Properties	Machinability		D	D	D	D	D/E	D	D/E
	Hot Hardness		B	B	A	A	A	A	A
	Wear Resistance		B	B	B	B	B	B	A
	Toughness		E	E	E	E	E	E	E

Table 6. Hot-Work Tool Steels

AISI	Group	Chromium Types						Tungsten Types						Molybdenum Types		
	Identifying Chemical Composition and Typical Heat-Treatment Data															
	Type	H10	H11	H12	H13	H14	H19	H21	H22	H23	H24	H25	H26	H41	H42	H43
Identifying Chemical Elements in Per Cent	C	0.40	0.35	0.35	0.35	0.40	0.40	0.35	0.35	0.35	0.45	0.25	0.50	0.65	0.60	0.55
	W	…	…	1.50	…	5.00	4.25	9.00	11.00	12.00	15.00	15.00	18.00	1.50	6.00	…
	Mo	2.50	1.50	1.50	1.50	…	…	…	…	…	…	…	…	8.00	5.00	8.00
	Cr	3.25	5.00	5.00	5.00	5.00	4.25	3.50	2.00	12.00	3.00	4.00	4.00	4.00	4.00	4.00
	V	0.40	0.40	0.40	1.00	…	2.00	…	…	…	…	…	1.00	1.00	2.00	2.00
	Co	…	…	…	…	…	4.25	…	…	…	…	…	…	…	…	…
Heat-Treat. Data	Hardening Temperature Range, °F	1850–1900	1825–1875	1825–1875	1825–1900	1850–1950	2000–2200	2000–2200	2000–2200	2000–2300	2000–2250	2100–2300	2150–2300	2000–2175	2050–2225	2000–2175
	Tempering Temperature Range, °F	1000–1200	1000–1200	1000–1200	1000–1200	1100–1200	1000–1300	1100–1250	1100–1250	1200–1500	1050–1200	1050–1250	1050–1250	1050–1200	1050–1200	1050–1200
	Approx. Tempered Hardness, Rc	56–39	54–38	55–38	53–38	47–40	59–40	54–36	52–39	47–30	55–45	44–35	58–43	60–50	60–50	58–45
	Relative Ratings of Properties (A = greatest to D = least)															
Characteristics in Heat Treatment	Safety in Hardening	A	A	A	A	A	B	B	B	B	B	B	B	C	C	C
	Depth of Hardening	A	A	A	A	A	A	A	A	A	A	A	A	A	A	A
	Resistance to Decarburization	B	B	B	B	B	B	B	B	B	B	B	B	C	B	C
	Stability of Shape in Heat Treatment — Quenching Medium: Air or Salt	B	B	B	B	C	C	C	C	…	C	C	C	C	C	C
	Stability of Shape in Heat Treatment — Quenching Medium: Oil	…	…	…	…	…	D	D	D	D	D	D	D	D	D	D
Service Properties	Machinability	C/D	C/D	C/D	C/D	D	D	D	D	D	D	D	D	D	D	D
	Hot Hardness	C	C	C	C	C	C	C	C	B	B	B	B	B	B	B
	Wear Resistance	D	D	D	D	D	C/D	C/D	C/D	C/D	C	D	C	C	C	C
	Toughness	C	B	B	B	C	C	C	C	D	D	C	D	D	D	D

Table 7. Cold-Work Tool Steels

AISI	Group	High-Carbon, High-Chromium Types					Medium-Alloy, Air-Hardening Types								Oil-Hardening Types			
	Types	D2	D3	D4	D5	D7	A2	A3	A4	A6	A7	A8	A9	A10	O1	O2	O6	O7
Identifying Chemical Composition and Typical Heat-Treatment Data																		
Identifying Chemical Elements in Per Cent	C	1.50	2.25	2.25	1.50	2.35	1.00	1.25	1.00	0.70	2.25	0.55	0.50	1.35	0.90	0.90	1.45	1.20
	Mn	…	…	…	…	…	…	…	2.00	2.00	…	…	…	1.80	1.00	1.60	…	…
	Si	…	…	…	…	…	…	…	…	…	…	…	…	1.25	…	…	1.00	…
	W	…	…	…	…	…	…	…	…	…	1.00	1.25	…	…	0.50	…	…	1.75
	Mo	1.00	…	1.00	1.00	1.00	1.00	1.00	1.00	1.25	1.00	1.25	1.40	1.50	…	…	0.25	…
	Cr	12.00	12.00	12.00	12.00	12.00	5.00	5.00	1.00	1.00	5.25	5.00	5.00	…	0.50	…	…	0.75
	V	1.00	…	…	…	4.00	…	1.00	…	…	4.75	…	1.00	…	…	…	…	…
	Co	…	…	…	3.00	…	…	…	…	…	…	…	…	…	…	…	…	…
	Ni	…	…	…	…	…	…	…	…	…	…	…	1.50	1.80	…	…	…	…
Heat-Treatment Data	Hardening Temperature Range, °F	1800–1875	1700–1800	1775–1850	1800–1875	1850–1950	1700–1800	1750–1850	1500–1600	1525–1600	1750–1800	1800–1850	1800–1875	1450–1500	1450–1500	1400–1475	1450–1500	1550–1525
	Quenching Medium	Air	Oil	Air	Air	Air	Air	Air	Air	Air	Air	Air	Air	Air	Oil	Oil	Oil	Oil
	Tempering Temperature Range, °F	400–1000	400–1000	400–1000	400–1000	300–1000	350–1000	350–1000	350–800	300–800	300–1000	350–1100	950–1150	350–800	350–500	350–500	350–600	350–550
	Approx. Tempered Hardness, Rc	61–54	61–54	61–54	61–54	65–58	62–57	65–57	62–54	60–54	67–57	60–50	56–35	62–55	62–57	62–57	63–58	64–58
Relative Ratings of Properties (A = greatest to E = least)																		
Characteristics in Heat Treatment	Safety in Hardening	A	C	A	A	A	A	A	A	A	A	A	A	A	B	B	B	B
	Depth of Hardening	A	A	A	A	A	A	A	A	A	A	A	A	A	B	B	B	B
	Resistance to Decarburization	B	B	B	B	B	B	B	A/B	A/B	B	B	B	A/B	A	A	A	A
	Stability of Shape in Heat Treatment	A	B	A	A	A	A	A	A	A	A	A	A	A	B	B	B	B
Service Properties	Machinability	E	E	E	E	E	D	D	D/E	D/E	E	D	D	C/D	C	C	B	C
	Hot Hardness	C	C	C	C	C	C	C	D	D	C	C	C	D	E	E	E	E
	Water Resistance	B/C	B	B	B/C	A	C	B	C/D	C/D	A	C/D	C/D	C	D	D	D	D
	Toughness	E	E	E	E	E	D	D	D	D	E	C	C	D	D	D	D	C

Table 8. Shock-Resisting, Mold, and Special-Purpose Tool Steels

AISI	Category	Shock-Resisting Tool Steels				Mold Steels							Special-Purpose Tool Steels				
	Identifying Chemical Composition and Typical Heat-Treatment Data																
	Types	S1	S2	S5	S7	P2	P3	P4	P5	P6	P20	P21[a]	L2[b]	L3[b]	L6	F1	F2
Identifying Elements in Per Cent	C	0.50	0.50	0.55	0.50	0.07	0.10	0.07	0.10	0.10	0.35	0.20	0.50/ 1.10	1.00	0.70	1.00	1.25
	Mn	…	…	0.80	…	…	…	…	…	…	…	…	…	…	…	…	…
	Si	…	1.00	2.00	…	…	…	…	…	…	…	…	…	…	…	…	…
	W	2.50	…	…	…	…	…	…	…	…	…	…	…	…	…	1.25	3.50
	Mo	…	0.50	0.40	1.40	0.20	…	0.75	…	…	0.40	…	…	…	0.25	…	…
	Cr	1.50	…	…	3.25	2.00	0.60	5.00	2.25	1.50	1.25	…	1.00	1.50	0.75	…	…
	V	…	…	…	…	…	…	…	…	…	…	…	0.20	0.20	…	…	…
	Ni	…	…	…	…	0.50	1.25	…	…	3.50	…	4.00	…	…	1.50	…	…
Heat-Treat. Data	Hardening Temperature, °F	1650–1750	1550–1650	1600–1700	1700–1750	1525–1550[c]	1475–1525[c]	1775–1825[c]	1550–1600[c]	1450–1500[c]	1500–1600[c]	Soln. treat.	1550–1700	1500–1600	1450–1550	1450–1600	1450–1600
	Tempering Temp. Range, °F	400–1200	350–800	350–800	400–1150	350–500	350–500	350–900	350–500	350–450	900–1100	Aged	350–1000	350–600	350–1000	350–500	350–500
	Approx. Tempered Hardness, Rc	58–40	60–50	60–50	57–45	64–58[d]	64–58[d]	64–58[d]	64–58[d]	61–58[d]	37–28[d]	40–30	63–45	63–56	62–45	64–60	65–62
	Relative Ratings of Properties (A = greatest to E = least)																
Characteristics in Heat Treatment	Safety in Hardening	C	E	C	B/C	C	C	C	C	C	C	A	D	D	C	E	E
	Depth of Hardening	B	B	B	A	B[e]	B[e]	B[e]	B[e]	A[e]	B	A	B	B	B	C	C
	Resist. to Decarb.	B	C	C	B	A	A	A	A	A	A	A	A	A	A	A	A
	Stability of Shape in Heat Treatment — Quench. Med.: Air	…	…	…	A	…	…	B	…	B	C	A	…	…	…	…	…
	Stability of Shape in Heat Treatment — Quench. Med.: Oil	D	…	D	C	C	C	…	C	C	…	A	D	D	C	…	…
	Stability of Shape in Heat Treatment — Quench. Med.: Water[f]	…	E	…	…	…	…	…	E	…	…	…	E	E	…	E	E
Service Properties	Machinability	D	C/D	C/D	D	C/D	D	D/E	D	D	C/D	D	C	C	D	C	D
	Hot Hardness	D	E	E	C	E	E	D	E	E	E	D	E	E	E	E	E
	Wear Resistance	D/E	D/E	D/E	D/E	D	D	C	D	D	D/E	D	D/E	D	D	D	B/C
	Toughness	B	A	A	B	C	C	C	C	C	C	D	B	D	B	E	E

[a] Contains also about 1.20 per cent A1. Solution treated in hardening.

[b] Quenched in oil.

[c] After carburizing.

[d] Carburized case.

[e] Core hardenability.

[f] Sometimes brine is used.

Table 9. Classification, Approximate Compositions, and Properties Affecting Selection of Tool and Die Steels
(From SAE Recommended Practice)

Type of Tool Steel	Chemical Composition[a]								Non-warping Prop.	Safety in Hardening	Toughness	Depth of Hardening	Wear Resistance
	C	Mn	Si	Cr	V	W	Mo	Co					
Water Hardening													
0.80 Carbon	70–0.85	b	b	b	…	…	…	…	Poor	Fair	Good[c]	Shallow	Fair
0.90 Carbon	0.85–0.95	b	b	b	…	…	…	…	Poor	Fair	Good[c]	Shallow	Fair
1.00 Carbon	0.95–1.10	b	b	b	…	…	…	…	Poor	Fair	Good[c]	Shallow	Good
1.20 Carbon	1.10–1.30	b	b	b	…	…	…	…	Poor	Fair	Good[c]	Shallow	Good
0.90 Carbon–V	0.85–0.95	b	b	b	0.15–0.35	…	…	…	Poor	Fair	Good	Shallow	Fair
1.00 Carbon–V	0.95–1.10	b	b	b	0.15–0.35	…	…	…	Poor	Fair	Good	Shallow	Good
1.00 Carbon–VV	0.90–1.10	b	b	b	0.35–0.50	…	…	…	Poor	Fair	Good	Shallow	Good
Oil Hardening													
Low Manganese	0.90	1.20	0.25	0.50	0.20[d]	0.50	…	…	Good	Good	Fair	Deep	Good
High Manganese	0.90	1.60	0.25	0.35[d]	0.20[d]	…	0.30[d]	…	Good	Good	Fair	Deep	Good
High-Carbon, High-Chromium[e]	2.15	0.35	0.35	12.00	0.80[d]	0.75[d]	0.80[d]	…	Good	Good	Poor	Through	Best
Chromium	1.00	0.35	0.25	1.40	…	…	0.40	…	Fair	Good	Fair	Deep	Good
Molybdenum Graphitic	1.45	0.75	1.00	…	…	…	0.25	…	Fair	Good	Fair	Deep	Good
Nickel–Chromium[f]	0.75	0.70	0.25	0.85	0.25[d]	…	0.50[d]	…	Fair	Good	Fair	Deep	Fair
Air Hardening													
High-Carbon, High-Chromium	1.50	0.40	0.40	12.00	0.80[d]	…	0.90	0.60[d]	Best	Best	Fair	Through	Best
5 Per Cent Chromium	1.00	0.60	0.25	5.25	0.40[d]	…	1.10	…	Best	Best	Fair	Through	Good
High-Carbon, High-Chromium–Cobalt	1.50	0.40	0.40	12.00	0.80[d]	…	0.90	3.10	Best	Best	Fair	Through	Best
Shock-Resisting													
Chromium–Tungsten	0.50	0.25	0.35	1.40	0.20	2.25	0.40[d]	…	Fair	Good	Good	Deep	Fair
Silicon–Molybdenum	0.50	0.40	1.00	…	0.25[d]	…	0.50	…	Poor[g]	Poor[h]	Best	Deep	Fair
Silicon–Manganese	0.55	0.80	2.00	0.30[d]	0.25[d]	…	0.40[d]	…	Poor[g]	Poor[h]	Best	Deep	Fair
Hot Work													
Chromium–Molybdenum–Tungsten	0.35	0.30	1.00	5.00	0.25[d]	1.25	1.50	…	Good	Good	Good	Through	Fair
Chromium–Molybdenum–V	0.35	0.30	1.00	5.00	0.40	…	1.50	…	Good	Good	Good	Through	Fair
Chromium–Molybdenum–VV	0.35	0.30	1.00	5.00	0.90	…	1.50	…	Good	Good	Good	Through	Fair

Table 9. *(Continued)* **Classification, Approximate Compositions, and Properties Affecting Selection of Tool and Die Steels**
(From SAE Recommended Practice)

Type of Tool Steel	Chemical Composition[a]								Non-warping Prop.	Safety in Hardening	Toughness	Depth of Hardening	Wear Resistance
	C	Mn	Si	Cr	V	W	Mo	Co					
Tungsten	0.32	0.30	0.20	3.25	0.40	9.00	...	...	Good	Good	Good	Through	Fair
High Speed													
Tungsten, 18–4–1	0.70	0.30	0.30	4.10	1.10	18.00	...	...	Good	Good	Poor	Through	Good
Tungsten, 18–4–2	0.80	0.30	0.30	4.10	2.10	18.50	0.80	...	Good	Good	Poor	Through	Good
Tungsten, 18–4–3	1.05	0.30	0.30	4.10	3.25	18.50	0.70	...	Good	Good	Poor	Through	Best
Cobalt–Tungsten, 14–4–2–5	0.80	0.30	0.30	4.10	2.00	14.00	0.80	5.00	Good	Fair	Poor	Through	Good
Cobalt–Tungsten, 18–4–1–5	0.75	0.30	0.30	4.10	1.00	18.00	0.80	5.00	Good	Fair	Poor	Through	Good
Cobalt–Tungsten, 18–4–2–8	0.80	0.30	0.30	4.10	1.75	18.50	0.80	8.00	Good	Fair	Poor	Through	Good
Cobalt–Tungsten, 18–4–2–12	0.80	0.30	0.30	4.10	1.75	20.00	0.80	12.00	Good	Fair	Poor	Through	Good
Molybdenum, 8–2–1	0.80	0.30	0.30	4.00	1.15	1.50	8.50	...	Good	Fair	Poor	Through	Good
Molybdenum–Tungsten, 6–6–2	0.83	0.30	0.30	4.10	1.90	6.25	5.00	...	Good	Fair	Poor	Through	Good
Molybdenum–Tungsten, 6–6–3	1.15	0.30	0.30	4.10	3.25	5.75	5.25	...	Good	Fair	Poor	Through	Best
Molybdenum–Tungsten, 6–6–4	1.30	0.30	0.30	4.25	4.25	5.75	5.25	...	Good	Fair	Poor	Through	Best
Cobalt–Molybdenum–Tungsten, 6–6–2–8	0.85	0.30	0.30	4.10	2.00	6.00	5.00	8.00	Good	Fair	Poor	Through	Good

[a] C = carbon; Mn = manganese; Si = silicon; Cr = chromium; V = vanadium; W = tungsten; Mo = molybdenum; Co = cobalt.

[b] Carbon tool steels are usually available in four grades or qualities: *Special (Grade 1)*—The highest quality water-hardening carbon tool steel, controlled for hardenability, chemistry held to closest limits, and subject to rigid tests to ensure maximum uniformity in performance; *Extra (Grade 2)*—A high-quality water-hardening carbon tool steel, controlled for hardenability, subject to tests to ensure good service; *Standard (Grade 3)*—A good-quality water-hardening carbon tool steel, not controlled for hardenability, recommended for application where some latitude with respect to uniformity is permissible; *Commercial (Grade 4)*—A commercial-quality water-hardening carbon tool steel, not controlled for hardenability, not subject to special tests. On *special* and *extra* grades, limits on manganese, silicon, and chromium are not generally required if Shepherd hardenability limits are specified. For *standard* and *commercial* grades, limits are 0.35 max. each for Mn and Si; 0.15 max. Cr for standard; 0.20 max. Cr for commercial.

[c] Toughness decreases somewhat when increasing depth of hardening.

[d] Optional element. Steels have found satisfactory application either with or without the element present. In silicon–manganese steel listed under Shock-Resisting Steels, if chromium, vanadium, and molybdenum are not present, then hardenability will be affected.

[e] This steel may have 0.50 per cent nickel as an optional element. The steel has been found to give satisfactory application either with or without the element present.

[f] Approximate nickel content of this steel is 1.50 per cent.

[g] Poor when water quenched, fair when oil quenched.

[h] Poor when water quenched, good when oil quenched.

Table 10a. Typical Heat Treatments for SAE Carbon Steels (Carburizing Grades)

SAE No.	Normalize, Deg. F	Carburize, Deg. F	Cool[a]	Reheat, Deg. F	Cool[a]	2nd Reheat, Deg. F	Cool[a]	Temper,[b] Deg. F
1010 to 1022	…	1650–1700	A	…	…	…	…	250–400
	…	1650–1700	B	1400–1450	A	…	…	250–400
	…	1650–1700	C	1400–1450	A	…	…	250–400
	…	1650–1700	C	1650–1700	B	1400–1450	A	250–400
	…	1500–1650[cd]	B	…	…	…	…	Optional
	…	1350–1575[ed]	D	…	…	…	…	Optional
1024	1650–1750[f]	1650–1700	E	……	…	…	…	250–400
	…	1350–1575[ed]	D	…	…	…	…	Optional
1025 1026	…	1650–1700	A	…	…	…	…	250–400
	…	1500–1650[cd]	B	…	…	…	…	Optional
1027	…	1350–1575[ed]	D	…	…	…	…	Optional
1030	…	1500–1650[cd]	B	…	…	…	…	Optional
	…	1350–1575[ed]	D	…	…	…	…	Optional
1111 1112 1113	…	1500–1650[cd]	B	…	…	…	…	Optional
	…	1350–1575[ed]	D	…	…	…	…	Optional
1109 to 1120	…	1650–1700	A	…	…	…	…	250–400
	…	1650–1700	B	1400–1450	A	…	…	250–400
	…	1650–1700	C	1400–1450	A	…	…	250–400
	…	1650–1700	C	1650–1700	B	1400–1450	A	250–400
	…	1500–1650[cd]	B	…	…	…	…	Optional
	…	1350–1575[ed]	D	…	…	…	…	Optional
1126	…	1500–1650[cd]	B	…	…	…	…	Optional
	…	1350–1575[ed]	D	…	…	…	…	Optional

[a] Symbols: A = water or brine; B = water or oil; C = cool slowly; D = air or oil; E = oil; F = water, brine, or oil.

[b] Even where tempering temperatures are shown, tempering is not mandatory in many applications. Tempering is usually employed for partial stress relief and improves resistance to grinding cracks.

[c] Activated or cyanide baths.

[d] May be given refining heat as in other processes.

[e] Carbonitriding atmospheres

[f] Normalizing temperatures at least 50 deg. F above the carburizing temperature are sometimes recommended where minimum heat-treatment distortion is of vital importance.

Table 11a. Typical Heat Treatments for SAE Carbon Steels (Heat-Treating Grades)

SAE Number	Normalize, Deg. F		Anneal, Deg. F	Harden, Deg. F	Quench[a]	Temper, Deg. F
1025 & 1030	…		…	1575–1650	A	To Desired Hardness
1033 to 1035	…		…	1525–1575	B	
1036	1600–1700		…	1525–1575	B	
	…		…	1525–1575	B	
1038 to 1040	1600–1700		…	1525–1575	B	
	…		…	1525–1575	B	
1041	1600–1700	and/or	1400–1500	1475–1550	E	
1042 to 1050	1600–1700		…	1475–1550	B	
1052 & 1055	1550–1650	and/or	1400–1500	1475–1550	E	
1060 to 1074	1550–1650	and/or	1400–1500	1475–1550	E	
1078	…		1400–1500[a]	1450–1500	A	
1080 to 1090	1550–1650	and/or	1400–1500[a]	1450–1500	E[b]	
1095	…		1400–1500[a]	1450–1500	F	
	…		1400–1500[a]	1500–1600	E	
1132 & 1137	1600–1700	and/or	1400–1500	1525–1575	B	
1138 & 1140	…		…	1500–1550	B	
	1600–1700		…	1500–1550	B	
1141 & 1144	…		1400–1500	1475–1550	E	
	1600–1700		1400–1500	1475–1550	E	
1145 to 1151	…		…	1475–1550	B	
	1600–1700		…	1475–1550	B	

[a] Slow cooling produces a spheroidal structure in these high-carbon steels that is sometimes required for machining purposes.

[b] May be water- or brine-quenched by special techniques such as partial immersion or time quenched; otherwise they are subject to quench cracking.

Table 12a. Typical Heat Treatments for SAE Alloy Steels (Carburizing Grades)

SAE No.	Normalize[a]	Cycle Anneal[b]	Carburized, Deg. F	Cool[c]	Reheat, Deg. F	Cool[c]	Temper,[d] Deg. F
1320	yes	…	1650–1700	E	1400–1450[e]	E	250–350
	yes	…	1650–1700	E	1475–1525[f]	E	250–350
	yes	…	1650–1700	C	1400–1450[e]	E	250–350
	yes	…	1650–1700	C	1500–1550[f]	E	250–350
	yes	…	1650–1700	E[g]	…	…	250–350
	yes	…	1500–1650[h]	E	…	…	250–350
2317	yes	yes	1650–1700	E	1375–1425[e]	E	250–350
	yes	yes	1650–1700	E	1450–1500[f]	E	250–350
	yes	yes	1650–1700	C	1375–1425[e]	E	250–350
	yes	yes	1650–1700	C	1475–1525[f]	E	250–350
	yes	yes	1650–1700	E[g]	…	…	250–350
	yes	yes	1450–1650[h]	E	…	…	250–350
2512 to 2517	yes[i]	…	1650–1700	C	1325–1375[e]	E	250–350
	yes[i]	…	1650–1700	C	1425–1475[f]	E	250–350
3115 & 3120	yes	…	1650–1700	E	1400–1450[e]	E	250–350
	yes	…	1650–1700	E	1475–1525[f]	E	250–350
	yes	…	1650–1700	C	1400–1450[e]	E	250–350
	yes	…	1650–1700	C	1500–1550[f]	E	250–350
	yes	…	1650–1700	E[g]	…	…	250–350
	yes	……	1500–1650[h]	E	…	…	250–350
3310 & 3316	yes[i]	…	1650–1700	E	1400–1450[e]	E	250–350
	yes[i]	…	1650–1700	C	1475–1500[f]	E	250–350
4017 to 4032	yes	yes	1650–1700	E[g]	…	…	250–350
4119 & 4125	yes	…	1650–1700	E[g]	…	…	250–350
4317 & 4320, 4608 to 4621	yes	yes	1650–1700	E	1425–1475[e]	E	250–350
	yes	yes	1650–1700	E	1475–1527[f]	E	250–350
	yes	yes	1650–1700	C	1425–1475[e]	E	250–350
	yes	yes	1650–1700	C	1475–1525[f]	E	250–350
	yes	yes	1650–1700	E[g]	…	…	250–350
	yes	yes	1650–1700	E[g]	…	…	250–350
	yes	…	1500–1650[h]	E	…	…	250–350
4812 to 4820	yes[i]	yes	1650–1700	E	1375–1425[e]	E	250–350
	yes[i]	yes	1650–1700	E	1450–1500[f]	E	250–350
	yes[i]	yes	1650–1700	C	1375–1425[e]	E	250–350
	yes[i]	yes	1650–1700	C	1450–1500[f]	E	250–350
	…	…	1650–1700	E[g]	…	…	250–350
5115 & 5120	yes	…	1650–1700	E	1425–1475[e]	E	250–350
	yes	…	1650–1700	E	1500–1550[f]	E	250–350
	yes	…	1650–1700	C	1425–1475[e]	E	250–350
	yes	…	1650–1700	C	1500–1550[f]	E	250–350
	yes	…	1500–1650[h]	E	…	…	250–350
8615 to 8625, 8720	yes	yes	1650–1700	E	1475–1525[e]	E	250–350
	yes	yes	1650–1700	E	1525–1575[f]	E	250–350
	yes	yes	1650–1700	C	1475–1525[e]	E	250–350
	yes	yes	1650–1700	C	1525–1575[f]	E	250–350
	yes	yes	1650–1700	E[g]	…	…	250–350
	yes	yes	1500–1650[h]	E	…	…	250–350
9310 to 9317	yes[i]	…	1650–1700	E	1400–1450[e]	E	250–350
	yes[i]	…	1650–1700	C	1500–1525	E	250–350

[a] Normalizing temperatures should be not less than 50 deg. F higher than the carburizing temperature. Follow by air cooling.

[b] For cycle annealing, heat to normalizing temperature—hold for uniformity—cool rapidly to 1000–1250 deg. F; hold 1 to 3 hours, then air or furnace cool to obtain a structure suitable for machining and finishing.

[c] Symbols: C = cool slowly; E = oil.

[d] Tempering treatment is optional and is generally employed for partial stress relief and improved resistance to cracking from grinding operations.

[e] For use when case hardness only is paramount.

[f] For use when higher core hardness is desired.

[g] Treatment is for fine-grained steels only, when a second reheat is often unnecessary.

[h] Treatment is for activated or cyanide baths. Parts may be given refining heats as indicated for other heat-treating processes.

[i] After normalizing, reheat to temperatures of 1000–1200 deg. F and hold approximately 4 hours.

Table 13a. Typical Heat Treatments for SAE Alloy Steels
(Directly Hardenable Grades)

SAE No.	Normalize, Deg. F		Anneal, Deg. F	Harden, Deg. F	Quench[a]	Temper, Deg. F
1330	…		…	1525–1575	B	To desired hardness
	1600–1700	and/or	1500–1600	1525–1575	B	To desired hardness
1335 & 1340	…		…	1500–1550	E	To desired hardness
	1600–1700	and/or	1500–1600	1525–1575	E	To desired hardness
2330	…		…	1450–1500	E	To desired hardness
	1600–1700	and/or	1400–1500	1450–1500	E	To desired hardness
2340 & 2345	…		…	1425–1475	E	To desired hardness
	1600–1700	and/or	1400–1500	1425–1475	E	To desired hardness
3130	1600–1700		…	1500–1550	B	To desired hardness
3135 to 3141	…		…	1500–1550	E	To desired hardness
	1600–1700	and/or	1450–1550	1500–1550	E	To desired hardness
3145 & 3150	…		…	1500–1550	E	To desired hardness
	1600–1700	and/or	1400–1500	1500–1550	E	To desired hardness
4037 & 4042	…		1525–1575	1500–1575	E	Gears, 350–450 To desired hardness
4047 & 4053	…		1450–1550	1500–1575	E	To desired hardness
4063 & 4068	…		1450–1550	1475–1550	E	To desired hardness
4130	1600–1700	and/or	1450–1550	1600–1650	B	To desired hardness
4137 & 4140	1600–1700	and/or	1450–1550	1550–1600	E	To desired hardness
4145 & 4150	1600–1700	and/or	1450–1550	1500–1600	E	To desired hardness
4340	1600–1700	and draw	1100–1225	1475–1525	E	To desired hardness
4640	1600–1700	and/or	1450–1550	1450–1500	E	To desired hardness
	1600–1700	and/or	1450–1500	1450–1500	E	Gears, 350–450
5045 & 5046	1600–1700	and/or	1450–1550	1475–1500	E	250–300
5130 & 5132	1650–1750	and/or	1450–1550	1500–1550	G	To desired hardness
5135 to 5145	1650–1750	and/or	1450–1550	1500–1550	E	To desired hardness Gears, 350–400
5147 to 5152	1650–1750	and/or	1450–1550	1475–1550	E	To desired hardness Gears, 350–400
50100	…		1350–1450	1425–1475	H	To desired hardness
	…		1350–1450	1500–1600	E	To desired hardness
51100						
52100						
6150	1650–1750	and/or	1550–1650	1600–1650	E	To desired hardness
9254 to 9262	…		…	1500–1650	E	To desired hardness
8627 to 8632	1600–1700	and/or	1450–1550	1550–1650	B	To desired hardness
8635 to 8641	1600–1700	and/or	1450–1550	1525–1575	E	To desired hardness

Table 13a. *(Continued)* **Typical Heat Treatments for SAE Alloy Steels** (Directly Hardenable Grades)

SAE No.	Normalize, Deg. F		Anneal, Deg. F	Harden, Deg. F	Quench[a]	Temper, Deg. F
8642 to 8653	1600–1700	and/or	1450–1550	1500–1550	E	To desired hardness
8655 & 8660	1650–1750	and/or	1450–1550	1475–1550	E	To desired hardness
8735 & 8740	1600–1700	and/or	1450–1550	1525–1575	E	To desired hardness
8745 & 8750	1600–1700	and/or	1450–1500	1500–1550	E	To desired hardness
9437 & 9440	1600–1700	and/or	1450–1550	1550–1600	E	To desired hardness
9442 to 9747	1600–1700	and/or	1450–1550	1500–1600	E	To desired hardness
9840	1600–1700	and/or	1450–1550	1500–1550	E	To desired hardness
9845 & 9850	1600–1700	and/or	1450–1550	1500–1550	E	To desired hardness

[a] Symbols: B = water or oil; E = oil; G = water, caustic solution, or oil; H = water.

Table 14a. Typical Heat Treatments for SAE Alloy Steels
(Heat-Treating Grades—Chromium–Nickel Austenitic Steels)

SAE No.	Normalize	Anneal,[a] Deg. F	Harden, Deg. F	Quenching Medium	Temper
30301 to 30347	…	1800–2100	…	Water or Air	…

[a] Quench to produce full austenitic structure using water or air in accordance with thickness of section. Annealing temperatures given cover process and full annealing as used by industry, the lower end of the range being used for process annealing.

Table 15a. Typical Heat Treatments for SAE Alloy Steels
(Heat-Treating Grades—Chromium–Nickel Austenitic Steels)

SAE No.[a]	Normalize	Aub-critical Anneal, Deg. F	Full Anneal Deg. F	Harden Deg. F	Quenching Medium	Temper Deg. F
51410	…	1300–1350[b]	1550–1650[c]	…	Oil or air	To desired hardness
	…	…	…	1750–1850		
51414	…	1200–1250[b]	…	…	Oil or air	To desired hardness
	…	…	…	1750–1850		
51416	…	1300–1350[b]	1550–1650[c]	…	Oil or air	To desired hardness
	…	…	…	1750–1850		
51420 51420F	…	1350–1450[b]	1550–1650[c]	…	Oil or air	To desired hardness
	…	…	…	1800–1850		
51430	…	1400–1500[d]	…	…	…	…
51430F	…	1250–1500[d]	…	…	…	…
51431	…	1150–1225[b]	…	1800–1900	Oil or air	To desired hardness
51440A 51440B 51440C 51440F	…	1350–1440[b]	1550–1650[c]	1850–1950	Oil or air	To desired hardness
51442	…	1400–1500[d]	…	…	…	…
51446	…	1500–1650[d]	…	…	…	…
51501	…	1325–1375[b]	1525–1600[c]	1600–1700	Oil or air	To desired hardness

[a] Suffixes A, B, and C denote three types of steel differing in carbon content only. Suffix F denotes a free-machining steel.

[b] Usually air cooled, but may be furnace cooled.

[c] Cool slowly in furnace.

[d] Cool rapidly in air.

Brass, Bronze, Aluminum And Other Non-ferrous Alloys

Cast Brass and Bronze.—The following information on S.A.E. Standard Brass and Bronze Castings includes typical applications of the different alloys in the automotive industry, the composition in percentage, and physical properties based upon standard test bars cast to size with only a minimum amount of machining to remove the fin gate. Standard specimens of wrought material are taken parallel to the direction of rolling and all rods, bars and shapes are tested in full size when practicable.

Red Brass Castings — S.A.E. Standard No. 40.—Red brass is used for waterpump impellers, fittings for gasoline and oil lines, small bushings, small miscellaneous castings. This is a free-cutting brass with good casting and finished properties.

Composition of No. 40: Copper, 84 to 86; tin, 4 to 6; lead, 4 to 6; zinc, 4 to 6; iron, max., 0.25; nickel, max.,0.75; phosphorus, max., 0.05; aluminum, 0.00; sulphur, max., 0.05; antimony, max., 0.25; other impurities, max., 015 per cent.

Physical Properties: Tensile strength, 26,000 pounds per square inch; yield point. 12,000 pounds per square inch; elongation in 2 inches (or proportionate gage length), 15 per cent.

Yellow Brass Castings — S.A.E. Standard No. 41.—Yellow brass is used for radiator parts, fittings for water-cooling systems, battery terminals, miscellaneous castings. This alloy is intended for commercial castings when cheapness and good machining properties are essential.

Composition of No. 41: Copper, 62 to 67; lead, 1.50 to 3.50; tin, max., 1; iron, max., 0.75; nickel, max., 0.25; phosphorus, max.,0.03; aluminum, max., 0.30; sulphur, max., 0.05; antimony, max., 0.15; other impurities, max.,0.15 per cent; zinc, remainder.

Physical Properties: Tensile strength, 20,000 pounds per square inch; elongation in 2 inches (or proportionate gage length), 15 per cent.

Manganese Bronze Castings—S.A.E. Standard No. 43.—This alloy is intended for castings requiring strength and toughness. It is used for such automotive parts as gear-shifter forks; counters, spiders; brackets and similar fittings; parts for starting motors; landing-gear and tail-skid castings for airplanes.

Composition of No. 43: Copper, 55 to 60; zinc, 38 to 42; tin, max., 1.50; manganese, max., 3.50; aluminum, max., 1.50; iron, max., 2; lead, max., 0.40 per cent.

Physical Properties: Tensile strength, 65,000 pounds per square inch; elongation in 2 inches (or proportionate gage length), 25 per cent.

High Tensile Manganese Bronze Castings —S.A.E. Standard No. 430.—This alloy is intended for use in castings where high strength and toughness are required such as marine propellers, shafts and gears.

Composition of No. 430: Copper, 60 to 68; iron, 2 to 4; aluminum, 3 to 6; manganese, 2.5 to 5; tin, max., 0.50; lead, max., 0.20, and nickel, max., 0.50 per cent; zinc, remainder.

Physical Properties: This alloy is manufactured in two grades, distinguished by chemical composition: Grade A being in the lower, and Grade B in the higher range of manganese, aluminum and iron content. Tensile strength, Grade A, 90,000, and Grade B, 110,000 pounds per square inch; elongation in 2 inches, Grade A, 20, and Grade B, 12 per cent.

Cast Brass to be brazed— S.A.E. Standard No. 44.—This brass is used for water-pipe fittings which are to be brazed. It begins to melt at about 1830 degrees F. and is entirely melted at approximately 1870 degrees F. The alloy or spelter used for brazing must have a lower melting temperature. Silver solder may be used.

Composition of No. 44: Copper, 83 to 86; zinc, 14 to17 lead, max., 0.50; iron, max.,0.15 per cent.

Brazing Solder — S.A.E. Standard No. 45: This solder begins to melt at approximately 1560 degrees F. and is entirely melted at about 1600 degrees F. It may be used by melting in a crucible under a flux of borax, with or without the addition of boric acid. The part to be brazed is dipped into the melted solder. When used in powdered form, this solder, mixed with a flux, is applied to the material and then melted either by means of a brazing torch or by using a furnace.

Composition of No. 45: Copper, 48 to 52; lead, max., 0.50; iron, max., 0.10 per cent; zinc, remainder.

Hard Bronze Castings — S.A.E. Standard No. 62.— This is a strong general utility bronze suitable for severe working conditions and heavy pressures. Typical applications include gears; bearings; bushings for severe service; valve guides; valve-tappet guides; camshaft bearings; fuel pump, timer and distributor parts; connecting-rod bushings; piston-pins; rocker lever; steering sector and hinge bushings; starting-motor parts.

Composition of No. 62: Copper, 86 to 89; tin, 9 to 11; lead, max., 0.20; iron, max.,0.06; zinc,1 to 3 per cent.

Physical Properties: Tensile strength, 30,000 pounds per square inch; yield point, 25,000 pounds per square inch; elongation in 2 inches (or proportionate gage length), 14 per cent.

Leaded Gun Metal Castings — S.A.E. Standard No. 63.—This general-utility bronze combines strength with fair machining qualities. It is especially good for bushings subjected to heavy loads and severe working conditions. It is also used for fittings subjected to moderately high water or oil pressures.

Composition of No. 63: Copper, 86 to 89; tin, 9 to 11; phosphorus, max., 0.25; zinc and other impurities, max., 0.50; lead, 1 to 2.50 per cent.

Physical Properties: Tensile strength, 30,000 pounds per square inch; yield point, 12,000 pounds per square inch; elongation in 2 inches (or proportionate gage length), 10 per cent.

Phosphor Bronze Castings — S.A.E. Standard No. 64.—This alloy is excellent when anti-friction qualities are important and where resistance to wear and scuffing are desired. It is used for such parts as wrist-pins, piston-pins, valve rocker-arm bushings, fuel and water-pump bushings, steering-knuckle bushings, aircraft control bushings.

Properties of No. 64: Copper, 78.50 to 81.50; tin, 9 to 11; lead, 9 to 11; phosphorus, 0.05 to 0.25; zinc, max., 0.75; other impurities, max., 0.25 per cent.

Physical Properties: Tensile strength, 25,000 pounds per square inch; yield point, 12,000 pounds per square inch; elongation in 2 inches (or proportionate gage length), 8 per cent.

Phosphor Gear Bronze Castings — S.A.E. Standard No. 65.—This bronze is not used regularly but it may be employed for gears and worm wheels where the requirements are severe and a very hard bronze is necessary.

Properties of No. 65: Copper, 88 to 90; tin, 10 to 12; phosphorus, 0.10 to 0.30; nickel, max., 0.05; lead, zinc, and other impurities, max., 0.50 per cent.

Physical Properties: Tensile strength, 35,000 pounds per square inch; yield point, 20,000 pounds per square inch; elongation in 2 inches (or proportionate gage length), 10 per cent.

Bronze Backing for Lined Bearings — S.A.E. Standard No. 66.—This is an inexpensive but suitable alloy for bronze-backed bearings of connecting-rods or main engine bearings.

Composition: Copper, 83 to 86; tin, 4.50 to 6; lead, 8 to 10; zinc, max., 2; other impurities, max., 0.25 per cent.

Physical Properties: Tensile strength,25,000 pounds per square inch; yield point, 12,000 pounds per square inch; elongation in 2 inches, 8 percent.

Bronze Bearing Castings — S.A.E. Standard No. 660.— This composition is widely used for bronze bearings. Typical applications in the automotive industry include such parts as spring bushings, torque tube bushings, steering-knuckle bushings, piston-pin bushings, thrust washers, etc.

Composition of No. 660: Copper, 81 to 85; tin, 6.50 to 7.50; lead, 6 to 8; zinc, 2 to 4; iron, max., 0.20; antimony, max.,0.20; other impurities, max., 0.50 per cent.

Physical Properties: : Tensile strength, 30,000 pounds per square inch; yield point, 14,000 pounds per square inch; elongation in 2 inches, 18 per cent.

Cast Aluminum Bronze — S.A.E. Standard No. 68.—This alloy has considerable strength, resistance to corrosion, hardness equal to manganese bronze, and good bearing qualities under certain conditions. It is used for worm-wheels, gears, valve guides, valve seats, and forgings.

Composition of No. 68: Copper, (Grade A) 87 to 89, (Grade B) 89.50 to 90.50; aluminum, (Grade A) 7 to 9, (Grade B) 9.50 to 10.50; iron, (Grade A) 2.50 to 4, (Grade B) not over 1; tin, max., (Grade A) 0.5, (Grade B) 0.2; total other impurities, (Grade A), 1, (Grade B) 0.5 per cent.

Physical Properties: Tensile strength, (Grades A and B) as cast, 65,000 pounds per square inch; tensile strength, (Grade B) as heat-treated, quenched and drawn, 80,000 pounds per square inch; yield point, (Grades A and B) as cast, 25,000 pounds per square inch; yield point, (Grade B) as heat-treated, 50,000 pounds per square inch. Elongation in 2 inches, (Grade A) as cast, 20 per cent; (Grade B)15 per cent; (Grade B) as heat-treated, 4 per cent.

Wrought Copper and Copper Alloys

Brass Sheet and Strip — S.A.E. Standard No. 70.—There are two grades designated as 70*A* (Cartridge Brass) and 70*C* (Yellow Brass). Tempers range from quarter hard through extra spring. These are given in the accompanying table. The numbers following each temper designation in the table represent the amount of reduction in B. & S. gage numbers when the brass sheets are rolled. The greater the reduction, the harder the brass.

This alloy is used to make radiator cores and tanks in the automotive industry; bead chain, flashlight shells, socket and screw shells in the electrical industry; and eyelets, fasteners, springs and stampings in the hardware industry.

Composition of No. 70A: Copper, 68.5 to 71.5; lead, max., 0.07; iron, max., 0.05; zinc, remainder.

Composition of No. 70C: Copper, 64.0 to 68.5 lead, max., 0.15; iron, max., 0.05; zinc, remainder.

Mechanical Properties: Tensile strengths and Rockwell hardness numbers are given in the accompanying table.

Aluminum Bronze Rods, Bars, and Shapes — S.A.E. Standard No. 701.—This alloy is commonly used for bushings, gears, valve parts, bearings, sleeves, screws, pins, and fabricated sections. It is also used where strength at elevated temperatures, a low coefficient of friction against steel, or a combination of strength and corrosion resistance is required. Alloy grades are: 701*B*, 701*C* and 701*D*.

Composition of No. 701B: Copper, 80.0 to 93.0; aluminum, 6.5 to11.0; iron, max., 4.00; nickel, max., 1.00; manganese, max., 1.50; silicon, max., 2.25; tin, max., 0.60; zinc, max., 1.0; tellurium, max., 0.6; other elements, max., 0.50.

Composition of No. 701C: Copper, 78.0; aluminum, 9.0 to 11.0; iron, 2.0 to 4.0; nickel, 4.0 to 5.5; manganese, max., 1.50; silicon, max., 0.25; tin, max., 0.20; other elements, max. 0.50.

Composition of No. 701D: Copper, 88.0 to 92.5; aluminum, 6.0 to 8.0; iron, 1.5 to 3.5; other elements, max., 0.50.

Mechanical Properties: Minimum tensile strengths of the No. 701*B* alloy grade range from 70,000 to 80,000 psi, minimum yield strengths from 30,000 to 40,000 psi and minimum elongations in 2 inches from 12 to 9 per cent depending on the shape or size of rod or bar. Minimum tensile strengths of the No. 701*C* alloy grade range from 85,000 to 100,000 psi, minimum yield strengths from 42,500 to 50,000 psi and minimum elongations in 2 inches from 10 to 5 per cent depending on size.

This alloy must withstand cold bending without fracture through an angle of 120 degrees around a pin, the diameter of which is equal to twice the diameter of round rod or four times the thickness of bar or other shapes.

Table 16. Hardness and Ultimate Strength of No. 70 sheet Brass by Tempers

Temper of Brass Sheet (S.A.E. No. 70) and Equivalent Reduction in B.& S. Gage Numbers		Rockwell hardness numbers: *B* Scale $\frac{1}{16}''$ Ball 100 kg. Load		Rockwell hardness numbers: Superficial 30-*T* Scale $\frac{1}{16}''$ Ball 30kg. Load		Ultimate Strength Pounds per Square inch	
Temper	Gage Nos.	Min.	Max.	Min.	Max.	Min.	Max.
Grade A							
Quarter Hard	1	40	65	43	60	49,000	59,000
Half Hard	2	60	77	56	68	57,000	67,000
Three-Quarter Hard	3	72	82	65	72	64,000	74,000
Hard	4	79	86	70	74	71,000	81,000
Extra Hard	6	85	91	74	77	83,000	92,000
Spring	8	89	93	76	78	91,000	100,000
Extra Spring	10	91	95	77	79	95,000	104,000
Grade C							
Quarter Hard	1	40	65	43	60	49,000	59,000
Half Hard	2	57	74	54	66	55,000	65,000
Three-Quarter Hard	3	70	80	65	71	62,000	72,000
Hard	4	76	84	68	73	68,000	78,000
Extra Hard	6	83	89	73	76	79,000	89,000
Spring	8	87	92	75	78	86,000	95,000
Extra Spring	10	88	93	76	79	90,000	99,000

The hardness numbers equivalent to such temper designations as "quarter hard." "half hard," etc., vary over a wide range as shown by the table above. The hardness number represented by a given temper designation depends not only upon the kind of annealing and thickness of a given material, but may be affected decidedly by the composition or type of alloy. "Quarter hard" red brass sheet (S.A.E. No. 79). for example. may have a Rockwell hardness varying from 50 to 95 which differs considerably from the minimum and maximum numbers given in the table above opposite "quarter hard."

Hardness tests of the indentation type, such as Rockwell or Brinell, are generally used for thin materials; however, if the sheet is very thin, the test may be for comparison only with other sheets of the same composition and thickness. When the penetration is deep relative to the thickness, there may be an apparent decrease of hardness due to the flow or punching-through of the material because of lack of lateral support; however, when the penetration is even greater relative to thickness, there may be an apparent increase in hardness due to the pressure of the penetrator on the anvil of the instrument.

Copper Sheet and Strip — S.A.E. Standard No. 71.—This alloy is used for building fronts, roofing, radiators, chemical process equipment, rotating bands, and vats.

Composition of No. 71: Copper, min., 99.90 (plus silver). In one type of sheet used in the automotive industry 6 to 10 troy ounces of silver may be added to one ton (avoirdupois) of copper. This is sufficient to raise the recrystallization temperature appreciably.

Mechanical Properties: Minimum tensile strengths range from 30,000 to 52,000 psi depending on temper. Generally the higher the strength the harder the temper.

Free Cutting Brass Rod — S.A.E. Standard No. 72.—This alloy is used for small screw machine parts, pins, nuts, plugs, screws, valve discs and caps.

Composition of No. 72: Copper, 60.0 to 63.0; lead, 2.5 to 3.7; iron, max., 0.35; other elements, max., 0.50; zinc, remainder.

Mechanical Properties: In the soft temper the minimum tensile strength ranges from 40,000 to 48,000 psi, the minimum yield strength from 15,000 to 20,000 psi and the minimum elongation in 2 inches from 25 to 15 per cent as the thickness decreases down from over 2 inches. In the half hard temper the minimum tensile strength ranges from 45,000 to 57,000 psi, the minimum yield strength ranges from 15,000 to 25,000 psi and the minimum elongation in 2 inches from 20 to 7 per cent as the size decreases down from over 2 inches. In the hard temper the minimum tensile strength is 80,000 psi and the minimum yield strength 45,000 psi for thicknesses of $\frac{1}{8}$ to $\frac{3}{16}$ inch. The minimum tensile strength is 70,000 psi, the minimum yield strength is 35,000 psi and the minimum elongation in 2 inches is 4 percent for thicknesses over $\frac{3}{16}$ to $\frac{5}{16}$ inch.

Naval Brass Rods, Bars, Forgings, and Shapes — S.A.E. No. 73.—This material is intended for use where brass rod that is stronger, tougher, and more corrosion resistant than commercial brass rod is required. Uses include forgings, water pump and propeller shafts, studs and nuts, bushings, turnbuckle barrels, adjusting stud ends, and screw machine parts.

Composition of No. 73: Copper, 59.0 to 62.0; tin, 0.50 to 1.00; lead, max., 0.20; iron, max., 0.10; other elements, max., 0.10; zinc, remainder.

Mechanical Properties: Rods and bars in the soft temper have a minimum tensile strength ranging from 50,000 to 54,000 psi, a minimum yield strength of 20,000 psi and a minimum elongation in 4 times the diameter or thickness of 30 per cent as the size decreases. Rods and bars in the half hard temper have a minimum tensile strength ranging from 54,000 to 60,000 psi, a minimum yield strength ranging from 22,000 to 27,000 psi and a minimum elongation in 4 times the diameter or thickness ranging from 30 to 22 per cent as the size decreases. Rods and bars in the hard temper have a minimum tensile strength ranging from 54,000 to 67,000 psi, a minimum yield strength ranging from 22,000 to 45,000 psi and a minimum elongation in 4 times the diameter or thickness ranging from 30 to 13 per cent as the size decreases.

Seamless Brass Tubes— S.A.E. Standard No. 74.—The alloys comprising these tubes are identified by the letters *A*, *B*, *C*, and *D*. Nos. 74*A* and 74*D* are used for condenser and heat exchanger tubes and flexible hose. Nos. 74*B* and 74*C* are general purpose materials used for water pipe radiator and ornamental work. The tubes may be formed, bent, upset, squeezed, swaged, flared, roll threaded and knurled.

Composition of No. 74A (Munts Metal): Copper, 59.0 to 63.0; lead, max., 0.30; iron, max., 0.07; zinc, remainder.

Composition of No. 74B (Yellow Brass): Copper, 65.0 to 68.0; lead, 0.20 to 0.80; iron, max., 0.07; zinc, remainder.

Composition of No. 74C (Cartridge Brass): Copper, 68.5 to 71.5; lead, max., 0.07; iron, max., 0.05; zinc, remainder.

Composition of No. 74D (Red Brass, 85%): Copper, 84.0 to 86.0; lead, max., 0.06; iron, max., 0.05; zinc, remainder.

Mechanical Properties of No. 74A: This tube in drawn temper exhibits a minimum tensile strength of 54,000 psi. Common tempers of this tube include light annealed, drawn general purpose, and hard drawn.

Mechanical Properties of Nos. 74B and 74C: These tubes in drawn temper exhibit a minimum tensile strength of 54,000 psi. In hard temper they exhibit a minimum tensile strength of 66,000 psi. Common tempers of these tubes include drawn general purpose and hard drawn.

Mechanical Properties of Nos. 74D: This tube in light, drawn, and hard tempers exhibits minimum tensile strengths of 44,000 44,000 and 57,000 psi, respectively. Common tempers of this tube include light, drawn general purpose, and hard.

Copper Tubes — S.A.E. Standard No. 75.—These tubes which contain a minimum of 99.90 per cent deoxidized copper are used for general engineering purposes, including gasoline, hydraulic and oil lines.

Mechanical Properties: In the light drawn temper the minimum tensile strength is 36,000 psi and the maximum tensile strength is 47,000 psi. In the drawn general purpose temper the minimum tensile strength is 36,000 psi and in the hard drawn temper (applying to tubes up to 1 inch outside diameter, inclusive, with wall thicknesses from 0.020 to 0.120 inch; tubes over 1 to 2 inches outside diameter, inclusive with wall thicknesses from 0.035 to 0.180 inch; and tubes over 2 to 4 inches outside diameter with wall thicknesses from 0.060 to 0.250 inch) the minimum tensile strength is 45,000 psi.

Phosphor Bronze Sheet and Strip — S.A.E. Standard No. 77.—Typical uses for this sheet and strip include springs, switch parts, sleeve bushings, clutch discs, diaphragms, fuse clips, and fasteners. There are two grades of this alloy, 77*A* and 77*C*. Six tempers are applied to this alloy, namely, soft, half hard, hard, extra hard, spring and extra spring.

Composition of No. 77A: Tin, 3.5 to 5.8; phosphorus, 0.03 to 0.35; lead, max., 0.05; iron, max., 0.10; zinc, max., 0.30, antimony, max., 0.05; copper, tin, and phosphorus, min., 99.50.

Composition of No. 77C: Tin, 7.0 to 9.0; phosphorus, 0.03 to 0.35; lead, max., 0.05; iron, max., 0.10; zinc, max., 0.20; antimony, max., 0.01; copper, tin, and phosphorus, min., 99.50.

Mechanical Properties: The minimum tensile strength of the No. 77*A* alloy ranges from 40,000 to 96,000 psi as the temper ranges from soft to extra spring. The minimum tensile strength of the No. 77*C* alloy ranges from 53,000 to 110,000 psi as the temper ranges from soft to extra spring.

Red Brass and Low Brass Sheet and Strip — S.A.E. Standard No. 79.—There are two grades designated as 79*A* (Red Brass, 85 per cent) and 79*B* (Low Brass, 80 per cent). Common tempers of No. 79*A* strip are quarter hard, half hard, extra hard, and spring. Common temper of No. 79*A* sheet is half hard. Common tempers of No. 79*B* strip are quarter hard, half hard, hard, and spring. Typical uses include weather strip, trim, conduit, sockets, fasteners, radiator cores and costume jewelry.

Composition of No. 79A: Copper, 84.0 to 86.0; lead, 0.05; iron, 0.05; zinc, remainder.

Composition of No. 79B: Copper, 78.5 to 81.5; lead, 0.05; iron, 0.05; zinc, remainder.

Mechanical Properties: Minimum tensile strengths of the 79*A* alloy range from 44,000 to 82,000 psi as the temper ranges from quarter hard to extra spring and minimum tensile strengths of the 79*B* alloy range from 48,000 to 89,000 psi as the temper ranges from quarter hard to extra spring.

Brass Wire — S.A.E. Standard No. 80.—This wire is used for making springs, locking wire, rivets, screws, and for wrapping turnbuckles. There are two grades, 80*A* and 80*B*.

Composition of No. 80A: Copper, 68.5 to 71.5; lead, max., 0.07; iron, max., 0.05; zinc remainder.

Composition of No. 80B: Copper, 63.0 to 68.5; lead, max., 0.10; iron, max., 0.05; zinc remainder.

Mechanical Properties: Minimum tensile strengths of the 80A and SoB alloys range from 50,000 to 120,000 psi as tempers range from eighth hard to spring.

Phosphor Bronze Wire and Rod — S.A.E. Standard No. 81.—This alloy is used for springs, switch parts, fasteners, and cotter pins. It should withstand being bent cold through an angle of 120 degrees without fracture, around a pin with a. diameter twice the diameter of the wire.

Composition of No. 81: Tin, 3.50 to 5.80; phosphorus, 0.03 to 0.35; lead, max., 0.05; iron, max., 0.10; zinc, max., 0.30; copper, tin, and phosphorus, min., 99.50.

Mechanical Properties: Minimum tensile strengths of hard drawn wire in coils range from145,000 to 105,000 as the wire diameter ranges from 0.025 to 0.500 inch. Minimum tensile strengths of spring temper rods range from 125,000 to 90,000 psi as the rod diameter ranges from 0.025 to 0.500 inch.

Annealed Copper Wire — S.A.E. Standard No. 83.— This wire is used primarily for electrical purposes but it is also used for metal spraying and copper brazing. No composition limits are specified for this wire but the copper should be of such quality and purity that when drawn and annealed should exhibit the mechanical properties (maximum tensile strength and minimum elongation) and electrical characteristics called for in the standard. Its electrical resistivity should not exceed 875.20 ohms per mile-lb. (100 per cent electrical conductivity IACS, International Annealed Copper Standard) at a temperature of 20 degrees C.

Mechanical Properties: Maximum tensile strengths of annealed wire range from 36,000 to 38,000 psi for wire diameters ranging from 0.4600 down to over 0.0201 inch. Minimum elongations in 10 inches of annealed wire range from 15 to 35 percent as the wire diameter ranges from over 0.0030 to 0.4600 inch.

Brass Forgings — S.A.E. Standard No. 88.—Typical uses for this alloy are forgings and pressings of all kinds.

Composition of No. 88: Copper, 58.0 to 61.0; lead, 1.50 to 2.50; iron, max., 0.30; other elements, max., 0.50; zinc, remainder.

Mechanical Properties: Hot-pressed forgings made of this alloy should have tensile strengths ranging from 45,000 to 60,000 psi and elongations in 2 inches ranging from 25 to 60 per cent.

Copper-Silicon and Beryllium Copper Alloys

Everdur.—This copper-silicon alloy is available in five slightly different nominal compositions for applications which require high strength, good fabricating and fusing qualities, immunity to rust, free-machining and a corrosion resistance equivalent to copper. The following table gives the nominal compositions and tensile strengths, yield strengths, and per cent elongations for various tempers and forms.

Table 17. Nominal Composition and Properties of Everdur

Desig. No.	Nominal Composition[a]					Temper[b]	Strength, Thousands of Pounds per Square Inch		Percent Elongation
	Cu	Si	Mn	Pb	Al		Tensile	Yield	
1010	95.80	3.10	1.10	…	…	A HRA CRA CRHH CRH H	52 50 52 71 70 to 85	15 18 18 40 60 38 to 50	35[c] 40 35 10 3 17 to 8[c]

Desig. No.	Nominal Composition[a]					Temper[b]	Strength, Thousands of Pounds per Square Inch		Percent Elongation
	Cu	Si	Mn	Pb	Al		Tensile	Yield	
1015	98.25	1.50	0.25	…	…	AP HP XHB	38 50 75 to 85	10 40 45 to 55	35 8 to 6[c]
1012	95.60	3.00	5.00	0.40	…	A H	52 85	15 50	35[c] 13 to 8[c]
1000	94.90	4.00	1.10	…	…	AC	45	…	15
1014	90.75	2.00	…	…	7.25	A	75 to 90	37.5 to 45	12 to 9[c]

[a] The following chemical symbols are used: Cu for copper, Si for silicon, Mn for manganese, Pb for lead, and Al for aluminum.

[b] Symbols used are: HRA for hot-rolled and annealed tank plates; CRA for cold-rolled sheets and strips; CRHH for cold-rolled half hard strips; and CRH for cold-rolled hard strips. For round, square, hexagonal, and octagonal rods: A for annealed; H for hard; and XHB for extra-hard bolt temper (in coils for cold-heading). For pipe and tube: AP for annealed; and HP for hard. For castings: AC for as cast.

[c] Per cent elongation in 4 times the diameter or thickness of the specimen. All other values are percent elongation in 2 inches.

Designation numbers are those of The American Brass Co.

The values given for the tensile strength, yield strength and elongation are all minimum values. Where ranges are shown, the first values given are for the largest diameter or largest size specimens. Yield strength values were determined at 0.50 per cent elongation under load.

Uses: (1010) Hot-rolled-and-annealed plates for unfired pressure vessels, and rods for hot forging, hot upsetting, and machining. (1015) Cold-headed-and-roll threaded bolts and cold-drawn seamless tubes for electrical metallic tubing and rigid conduit. (1012) Screw machine products. (1000) Castings. (1014) Hot forgings and for free machining applications; not for cold working or welding.

Beryllium Copper.— These alloys which contain copper, beryllium, cobalt and in the case of one alloy, silver, fall into two groups. One group whose beryllium content is greater than one per cent is characterized by its high strength and hardness and the other, whose beryllium content is less than 1 per cent, by its high electrical and thermal conductivity. The alloys have many applications in the electrical and aircraft industries or wherever strength, corrosion resistance, conductivity, non-magnetic and non-sparking properties are essential. Beryllium copper is obtainable in the form of strips, rods and bars, wire, platers bars, billets, tubes, and casting ingots.

Composition end Physical Properties: The accompanying table lists some of the more common wrought and casting alloys and gives some of their physical properties.

Table 18. Nominal Composition and Properties of Beryllium Copper Alloys

Alloy[a]	Composition[b] Be	Co	Ag	Cu	Form	Temper[c]	Tensile Strength, Thousands of Lbs. per Sq. In.	% Elong. in 2 In.	Rockwell Hardness, Scale-Range
	Per Cent								
25	1.80 to 2.05	0.18 to 0.30	…	Bal.	Strip	A H HT	60 to 78 100 to 120 190 to 215	35 to 60 2 to 7 1 to 3	B-45 to 78 B-96 to 102 C-40 to 45
					Rod	A AT	60 to 85 165 to 190	35 to 60 4 to 10	B-45 to 85 C-36 to 41
					Wire	A AT	58 to78 165 to 190	35 to 55 3 to 8	… …
165	1.60 to 1.80	0.18 to 0.30	…	Bal.	Strip	A H HT	60 to 78 100 to 120 180 to 200	35 to 60 2 to 7 1 to 3	B-45 to 78 B-96 to 102 C-39 to 41

Table 18. *(Continued)* **Nominal Composition and Properties of Beryllium Copper**

Alloy[a]	Composition[b] Be	Co	Ag	Cu	Form	Temper[c]	Tensile Strength, Thousands of Lbs. per Sq. In.	% Elong. in 2 In.	Rockwell Hardness, Scale-Range
	Per Cent								
10	0.40 to 0.70	2.35 to 2.70	...	Bal.	Strip	A	38 to 55	20 to 35	B-20 to 45
						H	70 to 85	5 to 8	B-70 to 8o
						HT	110 to 130	5 to 12	B-95 to 102
					Rod	A	38 to 55	20 to 35	B-20 to 45
						AT	100 to 120	10 to 25	B-92 to I00
50	0.25 to 0.50	1.40 to 1.70	0.90 to 1.10	Bal.	Rod	A	38 to 55	20 to 35	B-20 to 45
						AT	100 to 120	10 to 25	B-92 to 100
20C	1.90 to 2.15	0.35 to 0.65	...	Bal.	...	AT	150 to 175	1 to 3	C-38 to 45
275C	2.50 to 2.75	0.35 to 0.65	...	Bal.	...	AT	140 to165	1 to 2	C-42 to 48
10C	0.45 to 0.75	2.35 to 2.70	...	Bal.	...	AT	100 to 120	5 to 12	B-92 to 103

[a] Alloys with number designations are wrought alloys and those with number and letter designations are casting alloys. Designations are those of The Beryllium Corp.

[b] Chemical symbols are used to designate the constituent metals: Be, beryllium; Co. cobalt; Ag, silver; Cu, copper.

[c] Temper and condition symbol designations: A, solution annealed; H, hard; HT, heat treated from hard; AT, heat treated from solution annealed.

Aluminum and Aluminum Alloys

Pure aluminum is a silver-white metal characterized by a slightly bluish cast. It has a specific gravity of 2.70, resists the corrosive effects of many chemicals, and has a malleability approaching that of gold. When alloyed with other metals, numerous properties are obtained that make these alloys useful over a wide range of applications.

Aluminum alloys are light in weight compared with steel, brass, nickel, or copper; can be fabricated by all common processes; are available in a wide range of sizes, shapes, and forms; resist corrosion; readily accept a wide range of surface finishes; have good electrical and thermal conductivities; and are highly reflective to both heat and light.

Characteristics of Aluminum and Aluminum Alloys.—Aluminum and its alloys lose part of their strength at elevated temperatures, although some alloys retain good strength at temperatures from 400 to 500 degrees F. At subzero temperatures, however, their strength increases without loss of ductility so that aluminum is a particularly useful metal for low-temperature applications.

When aluminum surfaces are exposed to the atmosphere, a thin invisible oxide skin forms immediately that protects the metal from further oxidation. This self-protecting characteristic gives aluminum its high resistance to corrosion. Unless exposed to some substance or condition that destroys this protective oxide coating, the metal remains protected against corrosion. Aluminum is highly resistant to weathering, even in industrial atmospheres. It is also corrosion resistant to many acids. Alkalis are among the few substances that attack the oxide skin and therefore are corrosive to aluminum. Although the metal can safely be used in the presence of certain mild alkalis with the aid of inhibitors, in general, direct contact with alkaline substances should be avoided. Direct contact with certain other metals should be avoided in the presence of an electrolyte; otherwise, galvanic corrosion of the aluminum may take place in the contact area. Where other metals must be fastened to aluminum, the use of a bituminous paint coating or insulating tape is recommended.

Aluminum is one of the two common metals having an electrical conductivity high enough for use as an electric conductor. The conductivity of electric-conductor (EC) grade is about 62 per cent that of the International Annealed Copper Standard. Because aluminum has less than one-third the specific gravity of copper, however, a pound of aluminum will go almost twice as far as a pound of copper when used as a conductor. Alloying lowers the conductivity somewhat so that wherever possible the EC grade is used in electric conductor applications. However, aluminum takes a set, which often results in loosening of screwed connectors, leading to arcing and fires. Special clamping designs are therefore required when aluminum is used for electrical wiring, especially in buildings.

Aluminum has nonsparking and nonmagnetic characteristics that make the metal useful for electrical shielding purposes such as in bus bar housings or enclosures for other electrical equipment and for use around inflammable or explosive substances.

Aluminum can be cast by any method known. It can be rolled to any desired thickness down to foil thinner than paper and in sheet form can be stamped, drawn, spun, or roll-formed. The metal also may be hammered or forged. Aluminum wire, drawn from rolled rod, may be stranded into cable of any desired size and type. The metal may be extruded into a variety of shapes. It may be turned, milled, bored, or otherwise machined in equipment often operating at their maximum speeds. Aluminum rod and bar may readily be employed in the high-speed manufacture of parts made on automatic screw-machine.

Almost any method of joining is applicable to aluminum—riveting, welding, or brazing. A wide variety of mechanical aluminum fasteners simplifies the assembly of many products. Resin bonding of aluminum parts has been successfully employed, particularly in aircraft components.

For the majority of applications, aluminum needs no protective coating. Mechanical finishes such as polishing, sandblasting, or wire brushing meet the majority of needs. When additional protection is desired, chemical, electrochemical, and paint finishes are all used. Vitreous enamels have been developed for aluminum, and the metal may also be electroplated.

Temper Designations for Aluminum Alloys.—The temper designation system adopted by the Aluminum Association and used in industry pertains to all forms of wrought and cast aluminum and aluminum alloys except ingot. It is based on the sequences of basic treatments used to produce the various tempers. The temper designation follows the alloy designation, being separated by a dash.

Basic temper designations consist of letters. Subdivisions of the basic tempers, where required, are indicated by one or more digits following the letter. These digits designate specific sequences of basic treatments, but only operations recognized as significantly influencing the characteristics of the product are indicated. Should some other variation of the same sequence of basic operations be applied to the same alloy, resulting in different characteristics, then additional digits are added.

The basic temper designations and subdivisions are as follows:

–F, as fabricated: Applies to products that acquire some temper from shaping processes not having special control over the amount of strain-hardening or thermal treatment. For wrought products, there are no mechanical property limits.

–O, annealed, recrystallized (wrought products only): Applies to the softest temper of wrought products.

–H, strain-hardened (wrought products only): Applies to products that have their strength increased by strain-hardening with or without supplementary thermal treatments to produce partial softening.

The *–H* is always followed by two or more digits. The first digit indicates the specific combination of basic operations, as follows:

–H1, strain-hardened only: Applies to products that are strain-hardened to obtain the desired mechanical properties without supplementary thermal treatment. The number following this designation indicates the degree of strain-hardening.

–H2, strain-hardened and then partially annealed: Applies to products that are strain-hardened more than the desired final amount and then reduced in strength to the desired level by partial annealing. For alloys that age-soften at room temperature, the –H2 tempers have approximately the same ultimate strength as the corresponding –H3 tempers. For other alloys, the –H2 tempers have approximately the same ultimate strengths as the corresponding –H1 tempers and slightly higher elongations.The number following this designation indicates the degree of strain-hardening remaining after the product has been partially annealed.

–H3, strain-hardened and then stabilized: Applies to products which are strain-hardened and then stabilized by a low-temperature heating to slightly lower their strength and increase ductility. This designation applies only to the magnesium-containing alloys that, unless stabilized, gradually age-soften at room temperature.The number following this designation indicates the degree of strain-hardening remaining after the product has been strain-hardened a specific amount and then stabilized.

The second digit following the designations –H1, –H2, and –H3 indicates the final degree of strain-hardening. Numeral 8 has been assigned to indicate tempers having a final degree of strain-hardening equivalent to that resulting from approximately 75 per cent reduction of area. Tempers between –O (annealed) and 8 (full hard) are designated by numerals 1 through 7. Material having an ultimate strength about midway between that of the –O temper and that of the 8 temper is designated by the numeral 4 (half hard); between –O and 4 by the numeral 2 (quarter hard); and between 4 and 8 by the numeral 6 (three-quarter hard). (*Note:* For two-digit –H tempers whose second figure is odd, the standard limits for ultimate strength are exactly midway between those for the adjacent two-digit –H tempers whose second figures are even.) Numeral 9 designates extra-hard tempers.

The third digit, when used, indicates a variation of a two-digit –H temper, and is used when the degree of control of temper or the mechanical properties are different from but close to those for the two-digit –H temper designation to which it is added. (*Note:* The minimum ultimate strength of a three-digit –H temper is at least as close to that of the corresponding two-digit –H temper as it is to the adjacent two-digit –H tempers.) Numerals 1 through 9 may be arbitrarily assigned and registered with the Aluminum Association for an alloy and product to indicate a specific degree of control of temper or specific mechanical property limits. Zero has been assigned to indicate degrees of control of temper or mechanical property limits negotiated between the manufacturer and purchaser that are not used widely enough to justify registration with the Aluminum Association.

The following three-digit –H temper designations have been assigned for wrought products in all alloys:

–H111: Applies to products that are strain-hardened less than the amount required for a controlled H11 temper.

–H112: Applies to products that acquire some temper from shaping processes not having special control over the amount of strain-hardening or thermal treatment, but for which there are mechanical property limits, or mechanical property testing is required.

The following three-digit H temper designations have been assigned for wrought products in alloys containing more than a normal 4 per cent magnesium.

–H311: Applies to products that are strain-hardened less than the amount required for a controlled H31 temper.

–H321: Applies to products that are strain-hardened less than the amount required for a controlled H32 temper.

–H323: Applies to products that are specially fabricated to have acceptable resistance to stress-corrosion cracking.

–H343: Applies to products that are specially fabricated to have acceptable resistance to stress-corrosion cracking.

The following three-digit –H temper designations have been assigned for

Patterned or Embossed Sheet	Fabricated Form
–H114	–O temper
–H124, –H224, –H324	–H11, –H21, –H31 temper, respectively
–H134, –H234, –H334	–H12, –H22, –H32 temper, respectively
–H144, –H244, –H344	–H13, –H23, –H33 temper, respectively
–H154, –H254, –H354	–H14, –H24, –H34 temper, respectively
–H164, –H264, –H364	–H15, –H25, –H35 temper, respectively
–H174, –H274, –H374	–H16, –H26, –H36 temper, respectively
–H184, –H284, –H384	–H17, –H27, –H37 temper, respectively
–H194, –H294, –H394	–H18, –H28, –H38 temper, respectively
–H195, –H395	–H19, –H39 temper, respectively

–W, solution heat-treated: An unstable temper applicable only to alloys that spontaneously age at room temperature after solution heat treatment. This designation is specific only when the period of natural aging is indicated.

–T, thermally treated to produce stable tempers other than –F, –O, or –H: Applies to products that are thermally treated, with or without supplementary strain-hardening, to produce stable tempers.The –T is always followed by one or more digits. Numerals 2 through 10 have been assigned to indicate specific sequences of basic treatments, as follows:

–T1, naturally aged to a substantially stable condition: Applies to products for which the rate of cooling from an elevated temperature-shaping process, such as casting or extrusion, is such that their strength is increased by room-temperature aging.

–T2, annealed (cast products only): Designates a type of annealing treatment used to improve ductility and increase dimensional stability of castings.

–T3, solution heat-treated and then cold-worked: Applies to products that are cold-worked to improve strength, or in which the effect of cold work in flattening or straightening is recognized in applicable specifications.

–T4, solution heat-treated and naturally aged to a substantially stable condition: Applies to products that are not cold-worked after solution heat treatment, or in which the effect of cold work in flattening or straightening may not be recognized in applicable specifications.

–T5, artificially aged only: Applies to products that are artificially aged after an elevated-temperature rapid-cool fabrication process, such as casting or extrusion, to improve mechanical properties or dimensional stability, or both.

–T6, solution heat-treated and then artificially aged: Applies to products that are not cold-worked after solution heat-treatment, or in which the effect of cold work in flattening or straightening may not be recognized in applicable specifications.

–T7, solution heat-treated and then stabilized: Applies to products that are stabilized to carry them beyond the point of maximum hardness, providing control of growth or residual stress or both.

–T8, solution heat-treated, cold-worked, and then artificially aged: Applies to products that are cold-worked to improve strength, or in which the effect of cold work in flattening or straightening is recognized in applicable specifications.

–T9, solution heat-treated, artificially aged, and then cold-worked: Applies to products that are cold-worked to improve strength.

–T10, artificially aged and then cold-worked: Applies to products that are artificially aged after an elevated-temperature rapid-cool fabrication process, such as casting or extrusion, and then cold-worked to improve strength.

Additional digits may be added to designations –T1 through –T10 to indicate a variation in treatment that significantly alters the characteristics of the product. These may be arbitrarily assigned and registered with The Aluminum Association for an alloy and product to indicate a specific treatment or specific mechanical property limits.

These additional digits have been assigned for wrought products in all alloys:

–T_51, stress-relieved by stretching: Applies to products that are stress-relieved by stretching the following amounts after solution heat-treatment:

Plate	1½ to 3 per cent permanent set
Rod, Bar and Shapes	1 to 3 per cent permanent set
Drawn tube	0.5 to 3 per cent permanent set

Applies directly to plate and rolled or cold-finished rod and bar.

These products receive no further straightening after stretching.

Applies to extruded rod and bar shapes and tube when designated as follows:

–T_510 : Products that receive no further straightening after stretching.

–T_511: Products that receive minor straightening after stretching to comply with standard tolerances.

–T_52, stress-relieved by compressing: Applies to products that are stress-relieved by compressing after solution heat-treatment, to produce a nominal permanent set of 2½ per cent.

–T__54, stress-relieved by combined stretching and compressing: applies to die forgings that are stress relieved by restriking cold in the finish die.

The following two-digit –T temper designations have been assigned for wrought products in all alloys:

–T42: Applies to products solution heat-treated and naturally aged that attain mechanical properties different from those of the –T4 temper.

–T62: Applies to products solution heat-treated and artificially aged that attain mechanical properties different from those of the –T6 temper.

Aluminum Casting Alloy Designation Systems.—Aluminum casting alloys are listed in many specifications of various standardizing agencies. The numbering systems used by each differ and are not always correlatable. Casting alloys are available from producers who use a commercial numbering system and this numbering system is the one used in the tables of aluminum casting alloys given further along in this section.

A system of four-digit numerical designations for wrought aluminum and wrought aluminum alloys was adopted by the Aluminum Association in 1954. This system is used by the commercial producers and is similar to the one used by the SAE; the difference being the addition of two prefix letters.

The first digit of the designation identifies the alloy type: 1) indicating an aluminum of 99.00 per cent or greater purity; 2) copper; 3) manganese; 4) silicon; 5) magnesium; 6) magnesium and silicon; 7) zinc; 8) some element other than those aforementioned; and 9) unused (not assigned at present).

If the second digit in the designation is zero, it indicates that there is no special control on individual impurities; integers 1 through 9 indicate special control on one or more individual impurities.

In the 1000 series group for aluminum of 99.00 per cent or greater purity, the last two of the four digits indicate to the nearest hundredth the amount of aluminum above 99.00 per cent. Thus designation 1030 indicates 99.30 per cent minimum aluminum. In the 2000 to 8000 series groups the last two of the four digits have no significance but are used to identify different alloys in the group. At the time of adoption of this designation system most of the existing commercial designation numbers were used for these last two digits, as for example, 14S became 2014, 3S became 3003, and 75S became 7075. When new alloys are developed and are commercially used these last two digits are assigned consecutively beginning with –01, skipping any numbers previously assigned at the time of initial adoption.

Experimental alloys are also designated in accordance with this system but they are indicated by the prefix X. The prefix is dropped upon standardization.

Table 19 lists the nominal composition of commonly used aluminum casting alloys. Table 19 shows the product forms and nominal compositions of common wrought aluminum alloys.

Heat-treatability of Wrought Aluminum Alloys.—In high-purity form, aluminum is soft and ductile. Most commercial uses, however, require greater strength than pure aluminum affords. This extra strength is achieved in aluminum first by the addition of other elements to produce various alloys, which singly or in combination impart strength to the metal. Further strengthening is possible by means that classify the alloys roughly into two categories, non-heat-treatable and heat-treatable.

Non-heat-treatable alloys: The initial strength of alloys in this group depends upon the hardening effect of elements such as manganese, silicon, iron and magnesium, singly or in various combinations. The non-heat-treatable alloys are usually designated, therefore, in the 1000, 3000, 4000, or 5000 series. These alloys are work-hardenable, so further strengthening is made possible by various degrees of cold working, denoted by the "H" series of tempers. Alloys containing appreciable amounts of magnesium when supplied in strain-hardened tempers are usually given a final elevated-temperature treatment called *stabilizing* for property stability.

Heat-treatable alloys: The initial strength of alloys in this group is enhanced by the addition of alloying elements such as copper, magnesium, zinc, and silicon. These elements singly or in various combinations show increasing solid solubility in aluminum with increasing temperature, so it is possible to subject them to thermal treatments that will impart pronounced strengthening.

The first step, called *heat-treatment* or *solution heat-treatment,* is an elevated-temperature process designed to put the soluble element in solid solution. This step is followed by rapid quenching, usually in water, which momentarily "freezes" the structure and for a short time renders the alloy very workable. Some fabricators retain this more workable structure by storing the alloys at below freezing temperatures until they can be formed. At room or elevated temperatures the alloys are not stable after quenching, however, and precipitation of the constituents from the supersaturated solution begins. After a period of several days at room temperature, termed *aging* or *room-temperature precipitation,* the alloy is considerably stronger. Many alloys approach a stable condition at room temperature, but some alloys, particularly those containing magnesium and silicon or magnesium and zinc, continue to age-harden for long periods of time at room temperature.

Heating for a controlled time at slightly elevated temperatures provides even further strengthening and properties are stabilized. This process is called *artificial aging* or *precipitation hardening.* By application of the proper combination of solution heat-treatment, quenching, cold working and artificial aging, the highest strengths are obtained.

Table 19. Nominal Compositions (in per cent) of Common Aluminum Casting Alloys (AA/ANSI)

Alloy Designation)	Product	Si	Fe	Cu	Mn	Mg	Cr	Ni	Zn	Ti	Others	
											Each	Total
201.0	S	0.10	0.15	4.0–5.2	0.20–0.50	0.15–0.55	…	…	…	0.15–0.35	0.05[a]	0.10
204.0	S&P	0.20	0.35	4.2–5.0	0.10	0.15–0.35	…	0.05	0.10	0.15–0.30	0.05[b]	0.15
208.0	S&P	2.5–3.5	1.2	3.5–4.5	0.50	0.10	…	0.35	1.0	0.25	…	0.50
222.0	S&P	2.0	1.5	9.2–10.7	0.50	0.15–0.35	…	0.50	0.8	0.25	…	0.35
242.0	S&P	0.7	1.0	3.5–4.5	0.35	1.2–1.8	0.25	1.7–2.3	0.35	0.25	0.05	0.15
295.0	S	0.7–1.5	1.0	4.0–5.0	0.35	0.03	…	…	0.35	0.25	0.05	0.15
308.0	P	5.0–6.0	1.0	4.0–5.0	0.50	0.10	…	…	1.0	0.25	…	0.50
319.0	S&P	5.5–6.5	1.0	3.0–4.0	0.50	0.10	…	0.35	1.0	0.25	…	0.50
328.0	S	7.5–8.5	1.0	1.0–2.0	0.20–0.6	0.20–0.6	0.35	0.25	1.5	0.25	…	0.50
332.0	P	8.5–10.5	1.2	2.0–4.0	0.50	0.50–1.5	…	0.50	1.0	0.25	…	0.50
333.0	P	8.0–10.0	1.0	3.0–4.0	0.50	0.05–0.50	…	0.50	1.0	0.25	…	0.50
336.0	P	11.0–13.0	1.2	0.50–1.5	0.35	0.7–1.3	…	2.0–3.0	0.35	0.25	0.05	…
355.0	S&P	4.5–5.5	0.6[c]	1.0–1.5	0.50[c]	0.40–0.6	0.25	…	0.35	0.25	0.05	0.15
C355.0	S&P	4.5–5.5	0.20	1.0–1.5	0.10	0.40–0.6	…	…	0.10	0.20	0.05	0.15
356.0	S&P	6.5–7.5	0.6[c]	0.25	0.35[c]	0.20–0.45	…	…	0.35	0.25	0.05	0.15
356.0	S&P	6.5–7.5	0.20	0.20	0.10	0.25–0.45	…	…	0.10	0.20	0.05	0.15
357.0	S&P	6.5–7.5	0.15	0.05	0.03	0.45–0.6	…	…	0.05	0.20	0.05	0.15
A357.0	S&P	6.5–7.5	0.20	0.20	0.10	0.40–0.7	…	…	0.10	0.04–0.20	0.05[d]	0.15
443.0	S&P	4.5–6.0	0.8	0.6	0.50	0.05	0.25	…	0.50	0.25	…	0.35
B443.0	S&P	4.5–6.0	0.8	0.15	0.35	0.05	…	…	0.35	0.25	0.05	0.15
A444.0	P	6.5–7.5	0.20	0.10	0.10	0.05	…	…	0.10	0.20	0.05	0.15
512.0	S	1.4–2.2	0.6	0.35	0.8	3.5–4.5	0.25	…	0.35	0.25	0.05	0.15
513.0	P	0.30	0.40	0.10	0.30	3.5–4.5	…	…	1.4–2.2	0.20	0.05	0.15
514.0	S	0.35	0.50	0.15	0.35	3.5–4.5	…	…	0.15	0.25	0.05	0.15
520.0	S	0.25	0.30	0.25	0.15	9.5–10.6	…	…	0.15	0.25	0.05	0.15
705.0	S&P	0.20	0.8	0.20	0.40–0.6	1.4–1.8	0.20–0.40	…	2.7–3.3	0.25	0.05	0.15
707.0	S&P	0.20	0.8	0.20	0.40–0.6	1.8–2.4	0.20–0.40	…	4.0–4.5	0.25	0.05	0.15
710.0	S	0.15	0.50	0.35–0.65	0.05	0.6–0.8	…	…	6.0–7.0	0.25	0.05	0.15
711.0	P	0.30	0.7–1.4	0.35–0.65	0.05	0.25–0.45	…	…	6.0–7.0	0.20	0.05	0.15
712.0	S	0.30	0.50	0.25	0.10	0.50–0.65	0.40–0.6	…	5.0–6.5	0.15–0.25	0.05	0.20
850.0	S&P	0.7	0.7	0.7–1.3	0.10	0.10	…	0.7–1.3	…	0.20	—[e]	0.30
851.0	S&P	2.0–3.0	0.7	0.7–1.3	0.10	0.10	…	0.30–0.7	…	0.20	—[e]	0.30

[a] Also contains 0.40–1.0 per cent silver.

[b] Also contains 0.05 max. per cent tin.

[c] If iron exceeds 0.45 per cent, manganese content should not be less than one-half the iron content.

[d] Also contains 0.04-0.07 per cent beryllium.

[e] Also contains 5.5–7.0 per cent tin.

S = sand cast; P = permanent mold cast. The sum of those "Others" metallic elements 0.010 per cent or more each, expressed to the second decimal before determining the sum. *Source:* Standards for Aluminum Sand and Permanent Mold Castings. Courtesy of the Aluminum Association.

Clad Aluminum Alloys.—The heat-treatable alloys in which copper or zinc are major alloying constituents are less resistant to corrosive attack than the majority of non-heat-treatable alloys. To increase the corrosion resistance of these alloys in sheet and plate form they are often clad with high-purity aluminum, a low magnesium-silicon alloy, or an alloy containing 1 per cent zinc. The cladding, usually from 2½ to 5 per cent of the total thickness on each side, not only protects the composite due to its own inherently excellent corrosion resistance but also exerts a galvanic effect that further protects the core material.

Special composites may be obtained such as clad non-heat-treatable alloys for extra corrosion protection, for brazing purposes, or for special surface finishes. Some alloys in wire and tubular form are clad for similar reasons and on an experimental basis extrusions also have been clad.

Characteristics of Principal Aluminum Alloy Series Groups.—*1000 series:* These alloys are characterized by high corrosion resistance, high thermal and electrical conductivity, low mechanical properties and good workability. Moderate increases in strength may be obtained by strain-hardening. Iron and silicon are the major impurities.

2000 series: Copper is the principal alloying element in this group. These alloys require solution heat-treatment to obtain optimum properties; in the heat-treated condition mechanical properties are similar to, and sometimes exceed, those of mild steel. In some instances artificial aging is employed to further increase the mechanical properties. This treatment materially increases yield strength, with attendant loss in elongation; its effect on tensile (ultimate) strength is not as great. The alloys in the 2000 series do not have as good corrosion resistance as most other aluminum alloys and under certain conditions they may be subject to intergranular corrosion. Therefore, these alloys in the form of sheet are usually clad with a high-purity alloy or a magnesium-silicon alloy of the 6000 series which provides galvanic protection to the core material and thus greatly increases resistance to corrosion. Alloy 2024 is perhaps the best known and most widely used aircraft alloy.

3000 series: Manganese is the major alloying element of alloys in this group, which are generally non-heat-treatable. Because only a limited percentage of manganese, up to about 1.5 per cent, can be effectively added to aluminum, it is used as a major element in only a few instances. One of these, however, is the popular 3003, used for moderate-strength applications requiring good workability.

4000 series: The major alloying element of this group is silicon, which can be added in sufficient quantities to cause substantial lowering of the melting point without producing brittleness in the resulting alloys. For these reasons aluminum-silicon alloys are used in welding wire and as brazing alloys where a lower melting point than that of the parent metal is required. Most alloys in this series are non-heat-treatable, but when used in welding heat-treatable alloys they will pick up some of the alloying constituents of the latter and so respond to heat-treatment to a limited extent. The alloys containing appreciable amounts of silicon become dark gray when anodic oxide finishes are applied, and hence are in demand for architectural applications.

5000 series: Magnesium is one of the most effective and widely used alloying elements for aluminum. When it is used as the major alloying element or with manganese, the result is a moderate to high strength non-heat-treatable alloy. Magnesium is considerably more effective than manganese as a hardener, about 0.8 per cent magnesium being equal to 1.25 per cent manganese, and it can be added in considerably higher quantities. Alloys in this series possess good welding characteristics and good resistance to corrosion in marine atmospheres. However, certain limitations should be placed on the amount of cold work and the safe operating temperatures permissible for the higher magnesium content alloys (over about 3½ per cent for operating temperatures over about 150 deg. F) to avoid susceptibility to stress corrosion.

6000 series: Alloys in this group contain silicon and magnesium in approximate proportions to form magnesium silicide, thus making them capable of being heat-treated. The major alloy in this series is 6061, one of the most versatile of the heat-treatable alloys. Though less strong than most of the 2000 or 7000 alloys, the magnesium-silicon (or magnesium-silicide) alloys possess good formability and corrosion resistance, with medium strength. Alloys in this heat-treatable group may be formed in the –T4 temper (solution heat-treated but not artificially aged) and then reach full –T6 properties by artificial aging.

7000 series: Zinc is the major alloying element in this group, and when coupled with a smaller percentage of magnesium, results in heat-treatable alloys of very high strength. Other elements such as copper and chromium are usually added in small quantities. A notable member of this group is 7075, which is among the highest strength aluminum alloys available and is used in air-frame structures and for highly stressed parts.

S.A.E. Cast Magnesium Alloys

S.A.E. Standard No. 50 Alloy.—This alloy is used for most commercial applications. It is used in the "as cast," "heat treated," or "heat treated and aged" condition as may be required.

Composition of No. 50: Aluminum, 5.3 to 6.7; manganese, min., 0.15; zinc, 2.5 to 3.5; silicon, max., 0.5; copper, max., 0.05; nickel, max., 0.03; other impurities, max., 0.3 per cent and the remainder, magnesium.

Physical Properties: For sand castings as in the "as cast," "heat treated" and "heat treated and aged" conditions the minimum tensile strengths are respectively: 24,000, 30,000 and 32,000 pounds per square inch; the minimum yield strengths are respectively: 10,000, 10,000 and 16,000 pounds per square inch and the elongations in 2 inches are respectively: 4, 6 and 2 per cent.

S.A.E. Standard No. 500 Alloy.—This is a sand casting alloy to be used particularly where maximum pressure tightness is required. It may be used in the "as cast," "heat treated" or "heat treated and aged" condition as may be required.

Composition of No. 500: Aluminum 8.3 to 9.7; manganese, min., 0.10; zinc, 1.7 to 2.3; silicon, max., 0.5; copper, max., 0.05; nickel, max., 0.03; other impurities, max., 0.3 per cent and the remainder, magnesium.

Physical Properties: For sand castings in the "as cast," "heat treated" and "heat treated and aged" conditions, the minimum tensile strengths are respectively:20,000, 30,000 and 32,000 pounds per square inch; the yield strengths are respectively: 10,000, 10,000 and 17,000 pounds per square inch and the elongations in 2 inches are respectively; 1, 6 and 1 per cent.

S.A.E. Wrought Magnesium Alloys

S.A.E. Standard No.51 Alloy.— This alloy is used where maximum salt water resistance and weldability are desired. It is used in the annealed temper for applications requiring maximum formability, such as aircraft tanks and wheel fairings.

Composition of No. 51: Manganese, min., 1.20; silicon, max., 0.3; copper, max., 0.05; nickel, max., 0.03 other impurities, max., 0.3 per cent and the remainder, magnesium.

Physical Properties: Standard tensile test specimens machined from plate or sheet stock in thicknesses between 0.016 inch and 0.025 inch have a minimum tensile strength of 32,000 pounds per square inch in the hard rolled temper, a maximum tensile strength of 35,000 pounds per square inch in the annealed temper and an elongation in 2 inches of 4 per cent in the hard rolled temper and 12 per cent in the annealed temper.

S.A.E. Standard No. 510 Alloy.—This alloy is generally used where moderate formability and mechanical properties are required.

Composition of No. 510: Aluminum, 3.3 to 4.7; manganese, min., 0.20; zinc, max., 0.3; silicon, max., 0.5; copper, max., 0.05; nickel, max., 0.03; other impurities, max., 0.3 per cent and the remainder, magnesium.

Physical Properties: Standard tensile test specimens machined from plate or sheet stock in thicknesses between 0.16 and 0.125 inch have a tensile strength of 36,000 pounds per square inch, minimum in the hard rolled temper and 38,000 pounds per square inch, maximum in the annealed temper; a yield strength of 25,000 pounds per square inch in the hard rolled temper and an elongation in 2inches of 4 per cent in the hard rolled temper and 10 per cent in the annealed temper.

S.A.E. Standard No. 511 Alloy.—This alloy is used where high mechanical properties are required. It is available in the hard rolled and annealed tempers

Composition of No. 511: Aluminum 5.8 to 7.2; manganese, min., 0.15; zinc, max., 0.3; silicon, max., 0.5; copper, max., 0.05; nickel, 0.03; other impurities, 0.3 per cent and the remainder, magnesium.

Physical Properties: Standard tension test specimens machined from plate or sheet stock in thicknesses between 0.016 inch and 0.125 inch have tensile strength of 39,000 pounds per square inch, minimum in the hard rolled temper and 42,000 pounds per square inch, maximum in the annealed temper; a yield strength of 28,000 pounds per square inch in the hard rolled temper and an elongation in 2 inches of 3 per cent in the hard rolled temper and 10 per cent in the annealed temper.

S.A.E. Standard No. 52 Alloy.—This is a general purpose alloy with moderate strength and fair weldability. It is especially suited for the production of thin wall tubing and other sections requiring good extrusion characteristics.

Composition of No. 52: Aluminum, 2.4 to 3.0; manganese, min., 0.20; zinc, 0.7 to 1.3 silicon, max., 0.5; copper, max., 0.05; nickel, max., 0.03; other impurities, max., 0.3 per cent and the remainder magnesium.

Physical Properties: Standard test specimens machined from solid bar stock and structural shapes have a minimum tensile strength of 37,000 pounds per square inch in extruded bars up to 1½ inches and 34,000 pounds per square inch in structural shapes; a yield strength of 25,000 pounds per square inch in the former and 17,000 pounds per square inch in the latter and an elongation in 2 inches of 12 per cent in the former and 10 per cent in the latter.

S.A.E. Standard No. 520 Alloy.—This alloy is used for extruded bars, rods and shapes with good strength and fair weldability.

Composition of No. 520: Aluminum, 5.8 to 7.2; manganese, min., 0.15; zinc, 0.4 to 1.0; silicon, max., 0.5 iron, max., 0.05; nickel, max., 0.03; other impurities, max., 0.3 per cent and the remainder, magnesium.

Physical Properties: Standard tension test specimens machined from solid bar stock and structural shapes have a minimum tensile strength of 40,000 pounds per square inch in extruded bars up to 1½ inches and 38,000 pounds per square inch in structural shapes; a yield strength of 26,000 pounds per square inch in the former and 23,000 pounds per square inch in the latter and an elongation in 2 inches of 12 per cent in the former and 10 per cent in the latter.

S.A.E. Standard No. 522 Alloy.—This is an extrusion alloy used for applications requiring maximum weldability.

Composition of No. 522: Manganese, min. 1.2; silicon, max. 0.3 copper, max. O.05 nickel, max. 0.03; and calcium, 0.3 per cent; remainder, magnesium.

Physical Properties: Tensile strength is 30,000 pounds per square inch for extruded bars, ¼ inch to 1½ inches; 29,000 pounds per square inch for structural shapes and 28,000

pounds per square inch for hollow shapes. Elongation in 2 inches is 3 per cent for extruded bars, ¼ inch to 1½ inches and 2 per cent for structural and hollow shapes.

S.A.E.Standard Nos. 53, 531, 532 and 533 Alloys.—These are forging alloys. Nos. 53 and 533 are suitable for hammer forging. The former has somewhat better physical properties but the latter may be readily welded and contains no tin. No. 533 may also be press forged. Hammer forgings are normally more economical than press forgings but can only be used for applications involving moderate stresses. Press forging alloys Nos. 531 and 532 are used in applications involving higher stresses. No. 532 is stronger than No. 531 but more difficult to forge and is usually employed only for comparatively simple forgings requiring highest physical properties.

Composition of No. 53: Aluminum, 3.0 to 4.0; manganese, min. 0.2 zinc, max. 0.3; silicon, max. 0.3 copper, max. 0.05; nickel, max. 0.005; iron, max. 0.005 and tin, 4.0 to 6.0 per cent; remainder, magnesium.

Composition of No. 531: Aluminum, 5.8 to 7.2; manganese, minimum 0.15 zinc, 0.4 to 1.5; silicon, maximum 0.3; copper, max. 0.05; nickel, maximum 0.005 and iron, maximum 0.005 per cent; remainder, magnesium.

Composition of No. 532 : Aluminum, 7.8 to 9.2 manganese, minimum 0.12; zinc, 0.2 to 0.8; silicon, maximum 0.3; copper, maximum 0.05; nickel, maximum 0.005 and iron, maximum 0.005 per cent; remainder, magnesium.

Composition of No. 533: Manganese, minimum 1.2; silicon, maximum 0.3; copper, maximum 0.05 and nickel, maximum 0.03 per cent; remainder, magnesium.

Physical Properties: In the as forged condition, No. 53 has a minimum tensile strength of 36,000; No. 531, 38,000; No. 532, 42,000; and No. 533, 30,000 pounds per square inch. Nos. 53 and 531 have a yield strength of 22,000; No. 532, 26,000; and No. 533, 18,000 pounds per square inch. No. 53 has a minimum elongation in 2 inches of 7; No. 531, 6; No. 532, 5; and No. 533, 3 per cent.

Nickel and Nickel Alloys

Nickel.—Nickel is noted for its corrosion resistance, good electrical conductivity and high heat-transfer properties. It is used to fabricate process equipment for handling pure foods and drugs, electrical contact parts, and radio and X-ray tub elements.

Approximate Composition: (Commercially pure wrought nickel:) Nickel (including cobalt), 99.4; copper, 0.1; iron, 0.15 manganese, 0.25; silicon, 0.05; carbon, 0.05 and sulphur, 0.005. (Cast nickel:) Nickel, 97.0; copper, 0.3; iron, manganese, 0.5 silicon, 1.6; and carbon, 0.5.

Average Physical Properties: Wrought nickel in the annealed, hot-rolled, cold drawn, and hard temper cold-rolled conditions exhibits yield strengths (o. 2 percent offset) of 20,000, 25,000, 70,000, and 95,000 pounds per square inch, respectively tensile strengths of 70,000, 75,000, 95,000, and 105,000 pounds per square inch, respectively; elongations in 2 inches of 40, 40, 25, and 5 per cent, respectively; and Brinell hardness of 100, 110, 170, and 210, respectively.

Low-Carbon Nickel.—A special type of nickel that is corrosion resistant an has a high ductility and heat resistance. It lends itself well to spinning and cold coining or forging and is used in the manufacture of tubing and molds for the beverage and food industries.

Approximate Composition: Nickel, 99.4 copper, 0.05; iron, 0.1; silicon, 0.15; manganese, 0.2; carbon, 0.01; and sulphur, 0.005.

Average Physical Properties: Annealed low-carbon nickel exhibits a yield strength (0.2 per cent offset) of 15,000 pounds per square inch, a tensile strength of 60,000 pounds per square inch, an elongation in 2 inches of 50 per cent and a Brinell hardness of 90.

Duranickel: This age-hardenable alloy has good spring and low-sparking properties and is slightly magnetic after heat treatment. Items such as corrosion-resistant paper machine shaker springs, diaphragms, and extrusion dies for plastics are made from it.

Approximate Composition: Nickel, 93.7 copper, 0.05; iron, 0.35; aluminum, 4.4; silicon, 0.5; manganese, 0.3; carbon, 0. 17; and sulphur, 0.005.

Average Physical Properties: : In the hot-rolled, hot-rolled and age-hardened, cold-drawn, and cold-drawn and age-hardened conditions this alloy exhibits yield strengths (0.2 per cent offset) of 50,000, 130,000 90,000, and 135,000 pounds per square inch, respectively; tensile strengths of 105,000 170,000, 120,000, and 175,000 pounds per square inch, respectively; elongations in 2 inches of 35, 15, 25, and 15 per cent, respectively; and Brinell hardnesses of 180, 320, 220, and 340, respectively.

Monel.—This general purpose alloy is corrosion-resistant, strong, tough and has a silvery-white color. It is used for making abrasion- and heat-resistant valves and pump parts, propeller shafts, laundry machines, chemical processing equipment, etc.

Approximate Composition: Nickel, 67; copper, 30; iron, 1.4; silicon, 0.1; manganese, 1; carbon, 0. 15; and sulphur 0.01.

Average Physical Properties: Wrought Monel in the annealed, hot-rolled, cold-drawn, and hard temper cold-rolled conditions exhibits yield strengths (0.2 per cent offset) of 35,000, 50,000, 80,000, and 100,000 pounds per square inch, respectively; tensile strengths of 75,000, 90,000, 100,000, and 110,000 pounds per square inch, respectively; elongations in 2 inches of 40, 35, 25, and 5 per cent, respectively; and Brinell hardnesses of 125, 150, 190, and 240, respectively.

"R" Monel.—This free-cutting, corrosion resistant alloy is used for automatic screw machine products such as bolts, screws and precision parts.

Approximate Composition: Nickel, 67; copper, 30; iron, 1.4; silicon, 0.05; manganese, 1; carbon, 0.15; and sulphur, 0.035.

Average Physical Properties: In the hot-rolled and cold-drawn conditions this alloy exhibits yield strengths (0.2 per cent offset) of 45,000 and 75,000 pounds per square inch, respectively; tensile strengths of 85,000 and 90,000 pounds per square inch, respectively; elongations in 2 inches of 35, and 25 per cent, respectively; and Brinell hardnesses of 145 and 180, respectively.

"K" Monel.—This strong and hard alloy, comparable to heat-treated alloy steel, is age-hardenable, non-magnetic and has low-sparking properties. It is used for corrosive applications where the material is to be machined or formed, then age hardened. Pump and valve parts, scrapers, and instrument parts are made from this alloy.

Approximate Composition: Nickel, 66; copper, 29; iron, 0.9; aluminum, 2.75; silicon, 0.5 manganese, 0.75; carbon, 0.15; and sulphur, 0.005.

Average Physical Properties: In the hot-rolled, hot-rolled and age-hardened, colddrawn, and cold-drawn and age-hardened conditions the alloy exhibits yield strengths (0.2 per cent offset) of 45,000, 110,000, 85,000, and 115,000 pounds per square inch, respectively; tensile strengths of 100,000, 150,000 115,000, and 155,000 pounds per square inch, respectively; elongations in 2 inches of 40, 25, 25, and 20 per cent, respectively; and Brinell hardnesses of160, 280, 210, and 290, respectively.

"KR" Monel.—This strong, hard, age-hardenable and non-magnetic alloy is more readily machinable than "K" Monel. It is used for making valve stems, small parts for pumps, and screw machine products requiring an age-hardening material that is corrosion-resistant.

Approximate Composition: Nickel, 66; copper, 29; iron, 0.9; aluminum, 2.75; silicon, 0.5; manganese, 0.75; carbon, 0.28; and sulphur, 0.005.

Average Physical Properties: Essentially the same as "K" Monel.

"S" Monel.—This extra hard casting alloy is non-galling, corrosion-resisting, non-magnetic, age-hardenable and has low-sparking properties. It is used for gall-resistant pump and valve parts which have to withstand high temperatures, corrosive chemicals and severe abrasion.

Approximate Composition: Nickel, 63; copper, 30; iron, 2; silicon, 4 manganese, 0.75; carbon, 0.1; and sulphur, 0.015.

Average Physical Properties: In the annealed sand-cast, as-cast sand-cast, and age-hardened sand-cast conditions it exhibits yield strengths (0.2 per cent offset) of 70,000, 100,000, and 100,000 pounds per square inch, respectively; tensile strengths of 90,000, 130,000, and 130,000 pounds per square inch, respectively; elongations in 2 inches of 3, 2, and 2 per cent, respectively; and Brinell hardnesses of 275, 320, and 350, respectively.

"H" Monel.—An extra hard casting alloy with good ductility, intermediate strength and hardness that is used for pumps, impellers and steam nozzles.

Approximate Composition: Nickel, 63;Copper, 31 iron, 2; silicon, 3; manganese, 0.75; carbon, 0.1; and sulphur, 0.015.

Average Physical Properties: In the as-cast sand-cast condition this alloy exhibits a yield strength (0.2 per cent offset) of 60,000 pounds per square inch, a tensile strength of 100,000 pounds per square inch, an elongation in 2 inches of 15 per cent and a Brinell hardness of 210.

Inconel.—This heat resistant alloy retains its strength at high heats, resists oxidation and corrosion, has a high creep strength and is non-magnetic. It is used for high temperature applications (up to 2000 degrees F.) such as engine exhaust manifolds and furnace and heat treating equipment. Springs operating at temperatures up to 700 degrees F. are also made from it.

Approximate Composition: Nickel, 76; copper, 0.20; iron, 7.5; chromium, 15.5; silicon, 0.25; manganese, 0.25; carbon, 0.08; and sulphur, 0.007.

Physical Properties: Wrought Inconel in the annealed, hot-rolled, cold-drawn, and hard temper cold-rolled conditions exhibits yield strengths (0.2 per cent offset) of 35,000, 60,000, 90,000, and 110,000 pounds per square inch, respectively; tensile strengths of 85,000, 100,000 110,000, and 135,000 pounds per square inch, respectively; elongations in 2 inches of 45, 35, 20, and per cent, respectively; and Brinell hardnesses of 150, 180, 200, and 260, respectively.

Inconel "X" .—This alloy has a low creep rate, is age-hardenable and non-magnetic, resists oxidation and exhibits a high strength at elevated temperatures. Uses include the making of bolts and turbine rotors used at temperatures up to 1500 degrees F., aviation brake drum springs and relief valve and turbine springs with low load-loss or relaxation for temperatures up to 1000 degrees F.

Approximate Composition: Nickel, 73; copper, 0.2 maximum; iron, 7; chromium, 15; aluminum, 0.7; silicon, 0.4; manganese, 0.5; carbon, 0.04; sulphur, 0.007; columbium, 1; and titanium, 2.5.

Average Physical Properties: Wrought Inconel "X" in the annealed and agehardened hot-rolled conditions exhibits yield strengths (0.2 per cent offset) of 50,000 and 120,000 pounds per square inch, respectively; tensile strengths of 115,000 and 180,000 pounds per square inch, respectively; elongations in 2 inches of 50 and 25 percent, respectively; and brinell hardness of 200 and 360, respectively.

Titanium and Titanium Alloys

Titanium.—This metal is used in its commercially pure state and in alloy form (being alloyed with manganese or ferrochromium) for applications requiring a metal with properties of light weight, high strength, and good temperature- and corrosion-resistance. Titanium and its alloys weigh approximately 44 per cent less than stainless or alloy steels, are equal or greater in yield and ultimate tensile strength than structural alloys in common use, withstand temperatures up to 800 degrees F. and higher temperatures up to 2000 degrees F. for short periods and are resistant to the corrosive effects of salt water and many acids, alkalis and other chemicals. It is available in the form of plates, sheets, strip, forgings, ingots, bars, rods, and wire.

Table 20. Properties of Titanium and Titanium Alloys

Type[a]	Yield Strength, Pounds per Sq. In.	Ult. Tensile Strength, Pounds per Sq. In.	Percent Elongation in 2 Inch.
Titanium, pure	40,000 to 85,000	60,000 to 110,000	30 to 20
3% Ferrochromium or 4% Manganese	75,000 to 110,000	100,000 to 125,000	20 to 15
6% Ferrochromium	110,000 to 125,000	120,000 to 155,000	18 to 10
7% Manganese	120,000 to 160,000	130,000 to 170,000	18 to 7

[a] The percentage of chief alloying elements is given, except for the first entry which is commercially pure titanium.

Composition and Properties: The accompanying table gives the nominal compositions, yield strengths, tensile strengths and elongations of titanium and some of its alloys.

Powder Metallurgy

Powder metallurgy is a process whereby metal parts in large quantities can be made by the compressing and sintering of various powdered metals such as brass, bronze, aluminum, and iron. Compressing of the metal powder into the shape of the part to be made is done by accurately formed dies and punches in special types of hydraulic or mechanical presses. The "green" compressed pieces are then sintered in an atmosphere controlled furnace at high temperatures, causing the metal powder to be bonded together into a solid mass. A subsequent sizing or pressing operation and supplementary heat treatments may also be employed. The physical properties of the final product are usually comparable to those of cast or wrought products of the same composition. Using closely controlled conditions, steel of high hardness and tensile strength has also been made by this process.

Any desired porosity from 5 to 50 per cent can be obtained in the final product. Large quantities of porous bronze and iron bearings, which are impregnated with oil for self-lubrication, have been made by this process. Other porous powder metal products are used for filtering liquids and gases. Where continuous porosity is desired in the final product, the voids between particles are kept connected or open by mixing one per cent of zinc stearate or other finely powdered metallic soap throughout the metal powder before briquetting and then boiling this out in a low temperature baking before the piece is sintered.

The dense type of powdered metal products include refractory metal wire and sheet, cemented carbide tools, and electrical contact materials (products which could not be made as satisfactorily by other processes) and gears or other complex shapes which might also have been made by die casting or the precise machining of wrought or cast metal.

Advantages of Powder Metallurgy.—Parts requiring irregular curves, eccentrics, radial projections, or recesses often can be produced only by powder metallurgy. Parts that require irregular holes, keyways, flat sides, splines or square holes that are not easily machined, can usually be made by this process. Tapered holes and counter-bores are easily produced. Axial projections can be formed but the permissible size depends on the extent to which the powder will flow into the die recesses. Projections not more than one-quarter the length of the part are practicable. Slots, grooves, blind holes, and recesses of varied depths are also obtainable.

Limiting Factors in Powdered Metal Process.—The number and variety of shapes that may be obtained are limited by lack of plastic flow of powders, i.e., the difficulty with which they can be made to flow around corners. Tolerances in diameter usually cannot be held closer than 0.001 inch and tolerances in length are limited to 0.005 inch. This difference in diameter and length tolerances may be due to the elasticity of the powder and spring of the press.

Factors Affecting Design of Briquetting Tools.—High-speed steel is recommended for dies and punches and oil-hardening steel for strippers and knock-outs. One manufacturer specifies dimensional tolerances of 0.0002 inch and super-finished surfaces for these tools. Because of the high pressures employed and the abrasive character of certain refractory materials used in some powdered metal composition, there is frequently a tendency toward severe wear of dies and punches. In such instances, carbide inserts, chrome plating, or highly resistant die steels are employed. With regard to the shape of the die, corner radii, fillets, and bevels should be used to avoid sharp corners. Feather edges, threads, and reentrant angles are usually impracticable. The making of punches and dies is particularly exacting because allowances must be made for changes in dimensions due to growth after pressing and shrinkage or growth during sintering.

HARDNESS TESTING

Testing the Hardness of Metals

Brinell Hardness Test.—The Brinell test for determining the hardness of metallic materials consists in applying a known load to the surface of the material to be tested through a hardened steel ball of known diameter. The diameter of the resulting permanent impression in the metal is measured and the Brinell Hardness Number (BHN) is then calculated from the following formula in which D = diameter of ball in millimeters, d = measured diameter at the rim of the impression in millimeters, and P = applied load in kilograms.

$$\text{BHN} = \frac{\text{load on indenting tool in kilograms}}{\text{surface area of indentation in sq. mm.}} = \frac{P}{\frac{\pi D}{2}(D - \sqrt{D^2 - d^2})}$$

If the steel ball were not deformed under the applied load and if the impression were truly spherical, then the preceding formula would be a general one, and any combination of applied load and size of ball could be used. The impression, however, is not quite a spherical surface because there must always be some deformation of the steel ball and some recovery of form of the metal in the impression; hence, for a standard Brinell test, the size and characteristics of the ball and the magnitude of the applied load must be standardized. In the standard Brinell test, a ball 10 millimeters in diameter and a load of 3000, 1500, or 500 kilograms is used. It is desirable, although not mandatory, that the test load be of such magnitude that the diameter of the impression be in the range of 2.50 to 4.75 millimeters. The following test loads and approximate Brinell numbers for this range of impression diameters are: 3000 kg, 160 to 600 BHN; 1500 kg, 80 to 300 BHN; 500 kg, 26 to 100 BHN. In making a Brinell test, the load should be applied steadily and without a jerk for at least 15 seconds for iron and steel, and at least 30 seconds in testing other metals. A minimum period of 2 minutes, for example, has been recommended for magnesium and magnesium alloys. (For the softer metals, loads of 250, 125, or 100 kg are sometimes used.)

According to the American Society for Testing and Materials Standard E10-66, a steel ball may be used on material having a BHN not over 450, a Hultgren ball on material not over 500, or a carbide ball on material not over 630. The Brinell hardness test is not recommended for material having a BHN over 630.

Rockwell Hardness Test.—The Rockwell hardness tester is essentially a machine that measures hardness by determining the depth of penetration of a penetrator into the specimen under certain fixed conditions of test. The penetrator may be either a steel ball or a diamond spheroconical penetrator. The hardness number is related to the depth of indentation and the number is higher the harder the material. A minor load of 10 kg is first applied, causing an initial penetration; the dial is set at zero on the black-figure scale, and the major load is applied. This major load is customarily 60 or 100 kg when a steel ball is used as a penetrator, but other loads may be used when necessary. The ball penetrator is $\frac{1}{16}$ inch in diameter normally, but other penetrators of larger diameter, such as $\frac{1}{8}$ inch, may be employed for soft metals. When a diamond spheroconical penetrator is employed, the load usually is 150 kg. Experience decides the best combination of load and penetrator for use. After the major load is applied and removed, according to standard procedure, the reading is taken while the minor load is still applied.

The Rockwell Hardness Scales.—The various Rockwell scales and their applications are shown in the following table. The type of penetrator and load used with each are shown in Tables and, which give comparative hardness values for different hardness scales.

Scale	Testing Application
A	For tungsten carbide and other extremely hard materials. Also for thin, hard sheets.
B	For materials of medium hardness such as low- and medium-carbon steels in the annealed condition.
C	For materials harder than Rockwell B-100.
D	Where a somewhat lighter load is desired than on the C scale, as on case hardened pieces.
E	For very soft materials such as bearing metals.
F	Same as the E scale but using a $1/16$-inch ball.
G	For metals harder than tested on the B scale.
H & K	For softer metals.
15–N; 30–N; 45–N	Where a shallow impression or a small area is desired. For hardened steel and hard alloys.
15–T; 30–T; 45-T	Where a shallow impression or a small area is desired for materials softer than hardened steel.

Shore's Scleroscope.—The scleroscope is an instrument that measures the hardness of the work in terms of elasticity. A diamond-tipped hammer is allowed to drop from a known height on the metal to be tested. As this hammer strikes the metal, it rebounds, and the harder the metal, the greater the rebound. The extreme height of the rebound is recorded, and an average of a number of readings taken on a single piece will give a good indication of the hardness of the work. The surface smoothness of the work affects the reading of the instrument. The readings are also affected by the contour and mass of the work and the depth of the case, in carburized work, the soft core of light-depth carburizing, pack-hardening, or cyanide hardening, absorbing the force of the hammer fall and decreasing the rebound. The hammer weighs about 40 grains, the height of the rebound of hardened steel is in the neighborhood of 100 on the scale, or about $6\frac{1}{4}$ inches, and the total fall is about 10 inches or 255 millimeters.

Vickers Hardness Test.—The Vickers test is similar in principle to the Brinell test. The standard Vickers penetrator is a square-based diamond pyramid having an included point angle of 136 degrees. The numerical value of the hardness number equals the applied load in kilograms divided by the area of the pyramidal impression: A smooth, firmly supported, flat surface is required. The load, which usually is applied for 30 seconds, may be 5, 10, 20, 30, 50, or 120 kilograms. The 50-kilogram load is the most usual. The hardness number is based upon the diagonal length of the square impression. The Vickers test is considered to be very accurate, and may be applied to thin sheets as well as to larger sections with proper load regulation.

Knoop Hardness Numbers.—The Knoop hardness test is applicable to extremely thin metal, plated surfaces, exceptionally hard and brittle materials, very shallow carburized or nitrided surfaces, or whenever the applied load must be kept below 3600 grams. The Knoop indentor is a diamond ground to an elongated pyramidal form and it produces an indentation having long and short diagonals with a ratio of approximately 7 to 1. The longitudinal angle of the indentor is 172 degrees, 30 minutes, and the transverse angle 130 degrees. The Tukon Tester in which the Knoop indentor is used is fully automatic under electronic control. The Knoop hardness number equals the load in kilograms divided by the projected area of indentation in square millimeters. The indentation number corresponding to the long diagonal and for a given load may be determined from a table computed for a theoretically perfect indentor. The load, which may be varied from 25 to 3600 grams, is applied for a definite period and always normal to the surface tested. Lapped plane surfaces free from scratches are required.

Monotron Hardness Indicator.—With this instrument, a diamond-ball impressor point ¾ mm in diameter is forced into the material to a depth of 9⁄5000 inch and the pressure required to produce this constant impression indicates the hardness. One of two dials shows the pressure in kilograms and pounds, and the other shows the depth of the impression in millimeters and inches. Readings in Brinell numbers may be obtained by means of a scale designated as *M*–1.

Keep's Test.—With this apparatus, a standard steel drill is caused to make a definite number of revolutions while it is pressed with standard force against the specimen to be tested. The hardness is automatically recorded on a diagram on which a dead soft material gives a horizontal line, and a material as hard as the drill itself gives a vertical line, intermediate hardness being represented by the corresponding angle between 0 and 90 degrees.

Comparison of Hardness Scales.—Tables and show comparisons of various hardness scales. All such tables are based on the assumption that the metal tested is homogeneous to a depth several times that of the indentation. To the extent that the metal being tested is not homogeneous, errors are introduced because different loads and different shapes of penetrators meet the resistance of metal of varying hardness, depending on the depth of indentation. Another source of error is introduced in comparing the hardness of different materials as measured on different hardness scales. This error arises from the fact that in any hardness test, metal that is severely cold-worked actually supports the penetrator, and different metals, different alloys, and different analyses of the same type of alloy have different cold-working properties. In spite of the possible inaccuracies introduced by such factors, it is of considerable value to be able to compare hardness values in a general way.

The data shown in Table are for hardness measurements of unhardened steel, steel of soft temper, grey and malleable cast iron, and most nonferrous metals. Again these hardness comparisons are not as accurate for annealed metals of high Rockwell B hardness such as austenitic stainless steel, nickel and high nickel alloys, and cold-worked metals of low B-scale hardness such as aluminum and the softer alloys.

The data shown in Table are based on extensive tests on carbon and alloy steels mostly in the heat-treated condition, but have been found to be reliable on constructional alloy steels and tool steels in the as-forged, annealed, normalized, quenched, and tempered conditions, providing they are homogeneous. These hardness comparisons are not as accurate for special alloys such as high manganese steel, 18–8 stainless steel and other austenitic steels, nickel-base alloys, constructional alloy steels, and nickel-base alloys in the cold-worked condition.

Table 21. Comparative Hardness Scales for Steel

Rockwell C-Scale Hardness Number	Diamond Pyramid Hardness Number Vickers	Brinell Hardness Number 10-mm Ball, 3000-kgf Load			Rockwell Hardness Number		Rockwell Superficial Hardness Number Superficial Diam. Penetrator			Shore Scleroscope Hardness Number
		Standard Ball	Hultgren Ball	Tungsten Carbide Ball	A-Scale 60-kgf Load Diam. Penetrator	D-Scale 100-kgf Load Diam. Penetrator	15-N Scale 15-kgf Load	30-N Scale 30-kgf Load	45-N Scale 45-kgf Load	
68	**940**	…	…	…	**85.6**	**76.9**	**93.2**	**84.4**	**75.4**	97
67	**900**	…	…	…	**85.0**	**76.1**	**92.9**	**83.6**	**74.2**	95
66	**865**	…	…	…	**84.5**	**75.4**	**92.5**	**82.8**	**73.3**	92
65	**832**	…	…	**739**	**83.9**	**74.5**	**92.2**	**81.9**	**72.0**	91
64	**800**	…	…	**722**	**83.4**	**73.8**	**91.8**	**81.1**	**71.0**	88
63	**772**	…	…	**705**	**82.8**	**73.0**	**91.4**	**80.1**	**69.9**	87
62	**746**	…	…	**688**	**82.3**	**72.2**	**91.1**	**79.3**	**68.8**	85

Table 21. *(Continued)* **Comparative Hardness Scales for Steel**

Rockwell C-Scale Hardness Number	Diamond Pyramid Hardness Number Vickers	Brinell Hardness Number 10-mm Ball, 3000-kgf Load			Rockwell Hardness Number		Rockwell Superficial Hardness Number Superficial Diam. Penetrator			Shore Sclero-scope Hard-ness Num-ber
		Standard Ball	Hultgren Ball	Tungsten Carbide Ball	A-Scale 60-kgf Load Diam. Penetra-tor	D-Scale 100-kgf Load Diam. Penetra-tor	15-N Scale 15-kgf Load	30-N Scale 30-kgf Load	45-N Scale 45-kgf Load	
61	**720**	…	…	**670**	**81.8**	**71.5**	**90.7**	**78.4**	**67.7**	83
60	**697**	…	**613**	**654**	**81.2**	**70.7**	**90.2**	**77.5**	**66.6**	81
59	**674**	…	**599**	**634**	**80.7**	**69.9**	**89.8**	**76.6**	**65.5**	80
58	**653**	…	**587**	**615**	**80.1**	**69.2**	**89.3**	**75.7**	**64.3**	78
57	**633**	…	**575**	**595**	**79.6**	**68.5**	**88.9**	**74.8**	**63.2**	76
56	**613**	…	**561**	**577**	**79.0**	**67.7**	**88.3**	**73.9**	**62.0**	75
55	**595**	…	**546**	**560**	**78.5**	**66.9**	**87.9**	**73.0**	**60.9**	74
54	**577**	…	**534**	**543**	**78.0**	**66.1**	**87.4**	**72.0**	**59.8**	72
53	**560**	…	**519**	**525**	**77.4**	**65.4**	**86.9**	**71.2**	**58.6**	71
52	**544**	**500**	**508**	**512**	**76.8**	**64.6**	**86.4**	**70.2**	**57.4**	69
51	**528**	**487**	**494**	**496**	**76.3**	**63.8**	**85.9**	**69.4**	**56.1**	68
50	**513**	**475**	**481**	**481**	**75.9**	**63.1**	**85.5**	**68.5**	**55.0**	67
49	**498**	**464**	**469**	**469**	**75.2**	**62.1**	**85.0**	**67.6**	**53.8**	66
48	**484**	**451**	**455**	**455**	**74.7**	**61.4**	**84.5**	**66.7**	**52.5**	64
47	**471**	**442**	**443**	**443**	**74.1**	**60.8**	**83.9**	**65.8**	**51.4**	63
46	**458**	**432**	**432**	**432**	**73.6**	**60.0**	**83.5**	**64.8**	**50.3**	62
45	**446**	**421**	**421**	**421**	**73.1**	**59.2**	**83.0**	**64.0**	**49.0**	60
44	**434**	**409**	**409**	**409**	**72.5**	**58.5**	**82.5**	**63.1**	**47.8**	58
43	**423**	**400**	**400**	**400**	**72.0**	**57.7**	**82.0**	**62.2**	**46.7**	57
42	**412**	**390**	**390**	**390**	**71.5**	**56.9**	**81.5**	**61.3**	**45.5**	56
41	**402**	**381**	**381**	**381**	**70.9**	**56.2**	**80.9**	**60.4**	**44.3**	55
40	**392**	**371**	**371**	**371**	**70.4**	**55.4**	**80.4**	**59.5**	**43.1**	54
39	**382**	**362**	**362**	**362**	**69.9**	**54.6**	**79.9**	**58.6**	**41.9**	52
38	**372**	**353**	**353**	**353**	**69.4**	**53.8**	**79.4**	**57.7**	**40.8**	51
37	**363**	**344**	**344**	**344**	**68.9**	**53.1**	**78.8**	**56.8**	**39.6**	50
36	**354**	**336**	**336**	**336**	**68.4**	**52.3**	**78.3**	**55.9**	**38.4**	49
35	**345**	**327**	**327**	**327**	**67.9**	**51.5**	**77.7**	**55.0**	**37.2**	48
34	**336**	**319**	**319**	**319**	**67.4**	**50.8**	**77.2**	**54.2**	**36.1**	47
33	**327**	**311**	**311**	**311**	**66.8**	**50.0**	**76.6**	**53.3**	**34.9**	46
32	**318**	**301**	**301**	**301**	**66.3**	**49.2**	**76.1**	**52.1**	**33.7**	44
31	**310**	**294**	**294**	**294**	**65.8**	**48.4**	**75.6**	**51.3**	**32.5**	43
30	**302**	**286**	**286**	**286**	**65.3**	**47.7**	**75.0**	**50.4**	**31.3**	42
29	**294**	**279**	**279**	**279**	**64.7**	**47.0**	**74.5**	**49.5**	**30.1**	41
28	**286**	**271**	**271**	**271**	**64.3**	**46.1**	**73.9**	**48.6**	**28.9**	41
27	**279**	**264**	**264**	**264**	**63.8**	**45.2**	**73.3**	**47.7**	**27.8**	40
26	**272**	**258**	**258**	**258**	**63.3**	**44.6**	**72.8**	**46.8**	**26.7**	38
25	**266**	**253**	**253**	**253**	**62.8**	**43.8**	**72.2**	**45.9**	**25.5**	38
24	**260**	**247**	**247**	**247**	**62.4**	**43.1**	**71.6**	**45.0**	**24.3**	37
23	**254**	**243**	**243**	**243**	**62.0**	**42.1**	**71.0**	**44.0**	**23.1**	36

Table 21. *(Continued)* **Comparative Hardness Scales for Steel**

		Brinell Hardness Number 10-mm Ball, 3000-kgf Load			Rockwell Hardness Number		Rockwell Superficial Hardness Number Superficial Diam. Penetrator			
Rockwell C-Scale Hardness Number	Diamond Pyramid Hardness Number Vickers	Standard Ball	Hultgren Ball	Tungsten Carbide Ball	A-Scale 60-kgf Load Diam. Penetra-tor	D-Scale 100-kgf Load Diam. Penetra-tor	15-N Scale 15-kgf Load	30-N Scale 30-kgf Load	45-N Scale 45-kgf Load	Shore Sclero-scope Hard-ness Num-ber
22	**248**	**237**	**237**	**237**	**61.5**	**41.6**	**70.5**	**43.2**	**22.0**	35
21	**243**	**231**	**231**	**231**	**61.0**	**40.9**	**69.9**	**42.3**	**20.7**	35
20	**238**	**226**	**226**	**226**	**60.5**	**40.1**	**69.4**	**41.5**	**19.6**	34
(18)	230	219	219	219	…	…	…	…	…	33
(16)	222	212	212	212	…	…	…	…	…	32
(14)	213	203	203	203	…	…	…	…	…	31
(12)	204	194	194	194	…	…	…	…	…	29
(10)	196	187	187	187	…	…	…	…	…	28
(8)	188	179	179	179	…	…	…	…	…	27
(6)	180	171	171	171	…	…	…	…	…	26
(4)	173	165	165	165	…	…	…	…	…	25
(2)	166	158	158	158	…	…	…	…	…	24
(0)	160	152	152	152	…	…	…	…	…	24

Note: The values in this table shown in **boldface** type correspond to those shown in American Society for Testing and Materials Specification E140-67.

Values in ( ) are beyond the normal range and are given for information only.

Turner's Sclerometer.—In making this test a weighted diamond point is drawn, once forward and once backward, over the smooth surface of the material to be tested. The hardness number is the weight in grams required to produce a standard scratch.

Mohs's Hardness Scale.—Hardness, in general, is determined by what is known as Mohs's scale, a standard for hardness that is applied mainly to nonmetallic elements and minerals. In this hardness scale, there are ten degrees or steps, each designated by a mineral, the difference in hardness of the different steps being determined by the fact that any member in the series will scratch any of the preceding members.

This scale is as follows: 1) talc; 2) gypsum; 3) calcite; 4) fluor spar; 5) apatite; 6) orthoclase; 7) quartz; 8) topaz; 9) sapphire or corundum; and 10) diamond.

These minerals, arbitrarily selected as standards, are successively harder, from talc, the softest of all minerals, to diamond, the hardest. This scale, which is now universally used for nonmetallic minerals, is not applied to metals.

Relation Between Hardness and Tensile Strength.—The approximate relationship between the hardness and tensile strength is shown by the following formula:

Tensile strength = $Bhn \times 515$ (for Brinell numbers up to 175).

Tensile strength = $Bhn \times 490$ (for Brinell numbers larger than 175).

The above formulas give the tensile strength in pounds per square inch for steels. These approximate relationships between hardness and tensile strength do not apply to nonferrous metals with the possible exception of certain aluminum alloys.

Table 22. Comparative Hardness Scales for Unhardened Steel, Soft-Temper Steel, Grey and Malleable Cast Iron, and Nonferrous Alloys

Rockwell Hardness Number			Rockwell Superficial Hardness Number			Rockwell Hardness Number			Brinell Hardness Number	
Rockwell B scale 1/16″ Ball Penetrator 100-kg Load	Rockwell F scale 1/16″ Ball Penetrator 60-kg Load	Rockwell G scale 1/16″ Ball Penetrator 150-kg Load	Rockwell Superficial 15-T scale 1/16″ Ball Penetrator	Rockwell Superficial 30-T scale 1/16″ Ball Penetrator	Rockwell Superficial 45-T scale 1/16″ Ball Penetrator	Rockwell E scale 1/8″ Ball Penetrator 100-kg Load	Rockwell K scale 1/8″ Ball Penetrator 150-kg Load	Rockwell A scale "Brale" Penetrator 60-kg Load	Brinell Scale 10-mm Standard Ball 500-kg Load	Brinell Scale 10-mm Standard Ball 3000-kg Load
100	…	82.5	93.0	82.0	72.0	…	…	61.5	201	240
99	…	81.0	92.5	81.5	71.0	…	…	61.0	195	234
98	…	79.0	…	81.0	70.0	…	…	60.0	189	228
97	…	77.5	92.0	80.5	69.0	…	…	59.5	184	222
96	…	76.0	…	80.0	68.0	…	…	59.0	179	216
95	…	74.0	91.5	79.0	67.0	…	…	58.0	175	210
94	…	72.5	…	78.5	66.0	…	…	57.5	171	205
93	…	71.0	91.0	78.0	65.5	…	…	57.0	167	200
92	…	69.0	90.5	77.5	64.5	…	100	56.5	163	195
91	…	67.5	…	77.0	63.5	…	99.5	56.0	160	190
90	…	66.0	90.0	76.0	62.5	…	98.5	55.5	157	185
89	…	64.0	89.5	75.5	61.5	…	98.0	55.0	154	180
88	…	62.5	…	75.0	60.5	…	97.0	54.0	151	176
87	…	61.0	89.0	74.5	59.5	…	96.5	53.5	148	172
86	…	59.0	88.5	74.0	58.5	…	95.5	53.0	145	169
85	…	57.5	…	73.5	58.0	…	94.5	52.5	142	165
84	…	56.0	88.0	73.0	57.0	…	94.0	52.0	140	162
83	…	54.0	87.5	72.0	56.0	…	93.0	51.0	137	159
82	…	52.5	…	71.5	55.0	…	92.0	50.5	135	156
81	…	51.0	87.0	71.0	54.0	…	91.0	50.0	133	153
80	…	49.0	86.5	70.0	53.0	…	90.5	49.5	130	150
79	…	47.5	…	69.5	52.0	…	89.5	49.0	128	147
78	…	46.0	86.0	69.0	51.0	…	88.5	48.5	126	144
77	…	44.0	85.5	68.0	50.0	…	88.0	48.0	124	141
76	…	42.5	…	67.5	49.0	…	87.0	47.0	122	139
75	99.5	41.0	85.0	67.0	48.5	…	86.0	46.5	120	137
74	99.0	39.0	…	66.0	47.5	…	85.0	46.0	118	135
73	98.5	37.5	84.5	65.5	46.5	…	84.5	45.5	116	132
72	98.0	36.0	84.0	65.0	45.5	…	83.5	45.0	114	130
71	97.5	34.5	…	64.0	44.5	100	82.5	44.5	112	127
70	97.0	32.5	83.5	63.5	43.5	99.5	81.5	44.0	110	125
69	96.0	31.0	83.0	62.5	42.5	99.0	81.0	43.5	109	123
68	95.5	29.5	…	62.0	41.5	98.0	80.0	43.0	107	121
67	95.0	28.0	82.5	61.5	40.5	97.5	79.0	42.5	106	119
66	94.5	26.5	82.0	60.5	39.5	97.0	78.0	42.0	104	117
65	94.0	25.0	…	60.0	38.5	96.0	77.5	…	102	116
64	93.5	23.5	81.5	59.5	37.5	95.5	76.5	41.5	101	114
63	93.0	22.0	81.0	58.5	36.5	95.0	75.5	41.0	99	112
62	92.0	20.5	…	58.0	35.5	94.5	74.5	40.5	98	110
61	91.5	19.0	80.5	57.0	34.5	93.5	74.0	40.0	96	108
60	91.0	17.5	…	56.5	33.5	93.0	73.0	39.5	95	107
59	90.5	16.0	80.0	56.0	32.0	92.5	72.0	39.0	94	106
58	90.0	14.5	79.5	55.0	31.0	92.0	…	71.0	38.5	92
57	89.5	13.0	…	54.5	30.0	91.0	…	70.5	38.0	91
56	89.0	11.5	79.0	54.0	29.0	90.5	…	69.5	…	90
55	88.0	10.0	78.5	53.0	28.0	90.0	…	68.5	37.5	89
54	87.5	8.5	…	52.5	27.0	89.5	…	68.0	37.0	87
53	87.0	7.0	78.0	51.5	26.0	89.0	…	67.0	36.5	86
52	86.5	5.5	77.5	51.0	25.0	88.0	…	66.0	36.0	85
51	86.0	4.0	…	50.5	24.0	87.5	…	65.0	35.5	84
50	85.5	2.5	77.0	49.5	23.0	87.0	…	64.5	35.0	83
50	85.5	2.5	77.0	49.5	23.0	87.0	…	64.5	35.0	83
49	85.0	1.0	76.5	49.0	22.0	86.5	…	63.5	…	82

Table 22. *(Continued)* **Comparative Hardness Scales for Unhardened Steel, Soft-Temper Steel, Grey and Malleable Cast Iron, and Nonferrous Alloys**

Rockwell Hardness Number			Rockwell Superficial Hardness Number			Rockwell Hardness Number			Brinell Hardness Number	
Rockwell B scale 1/16" Ball Penetrator 100-kg Load	Rockwell F scale 1/16" Ball Penetrator 60-kg Load	Rockwell G scale 1/16" Ball Penetrator 150-kg Load	Rockwell Superficial 15-T scale 1/16" Ball Penetrator	Rockwell Superficial 30-T scale 1/16" Ball Penetrator	Rockwell Superficial 45-T scale 1/16" Ball Penetrator	Rockwell E scale 1/8" Ball Penetrator 100-kg Load	Rockwell K scale 1/8" Ball Penetrator 150-kg Load	Rockwell A scale "Brale" Penetrator 60-kg Load	Brinell Scale 10-mm Standard Ball 500-kg Load	Brinell Scale 10-mm Standard Ball 3000-kg Load
48	84.5	…	…	48.5	20.5	85.5	…	62.5	34.5	81
47	84.0	…	76.0	47.5	19.5	85.0	…	61.5	34.0	80
46	83.0	…	75.5	47.0	18.5	84.5	…	61.0	33.5	…
45	82.5	…	…	46.0	17.5	84.0	…	60.0	33.0	79
44	82.0	…	75.0	45.5	16.5	83.5	…	59.0	32.5	78
43	81.5	…	74.5	45.0	15.5	82.5	…	58.0	32.0	77
42	81.0	…	…	44.0	14.5	82.0	…	57.5	31.5	76
41	80.5	…	74.0	43.5	13.5	81.5	…	56.5	31.0	75
40	79.5	…	73.5	43.0	12.5	81.0	…	55.5	…	…
39	79.0	…	…	42.0	11.0	80.0	…	54.5	30.5	74
38	78.5	…	73.0	41.5	10.0	79.5	…	54.0	30.0	73
37	78.0	…	72.5	40.5	9.0	79.0	…	53.0	29.5	72
36	77.5	…	…	40.0	8.0	78.5	100	52.0	29.0	…
35	77.0	…	72.0	39.5	7.0	78.0	99.5	51.5	28.5	71
34	76.5	…	71.5	38.5	6.0	77.0	99.0	50.5	28.0	70
33	75.5	…	…	38.0	5.0	76.5	…	49.5	…	69
32	75.0	…	71.0	37.5	4.0	76.0	98.5	48.5	27.5	…
31	74.5	…	…	36.5	3.0	75.5	98.0	48.0	27.0	68
30	74.0	…	70.5	36.0	2.0	75.0	…	47.0	26.5	67
29	73.5	…	70.0	35.5	1.0	74.0	97.5	46.0	26.0	…
28	73.0	…	…	34.5	…	73.5	97.0	45.0	25.5	66
27	72.5	…	69.5	34.0	…	73.0	96.5	44.5	25.0	…
26	72.0	…	69.0	33.0	…	72.5	…	43.5	24.5	65
25	71.0	…	…	32.5	…	72.0	96.0	42.5	…	64
24	70.5	…	68.5	32.0	…	71.0	95.5	41.5	24.0	…
23	70.0	…	68.0	31.0	…	70.5	…	41.0	23.5	63
22	69.5	…	…	30.5	…	70.0	95.0	40.0	23.0	…
21	69.0	…	67.5	29.5	…	69.5	94.5	39.0	22.5	62
20	68.5	…	…	29.0	…	68.5	…	38.0	22.0	…
19	68.0	…	67.0	28.5	…	68.0	94.0	37.5	21.5	61
18	67.0	…	66.5	27.5	…	67.5	93.5	36.5	…	…
17	66.5	…	…	27.0	…	67.0	93.0	35.5	21.0	60
16	66.0	…	66.0	26.0	…	66.5	…	35.0	20.5	…
15	65.5	…	65.5	25.5	…	65.5	92.5	34.0	20.0	59
14	65.0	…	…	25.0	…	65.0	92.0	33.0	…	…
13	64.5	…	65.0	24.0	…	64.5	…	32.0	…	58
12	64.0	…	64.5	23.5	…	64.0	91.5	31.5	…	…
11	63.5	…	…	23.0	…	63.5	91.0	30.5	…	…
10	63.0	…	64.0	22.0	…	62.5	90.5	29.5	…	57
9	62.0	…	…	21.5	…	62.0	…	29.0	…	…
8	61.5	…	63.5	20.5	…	61.5	90.0	28.0	…	…
7	61.0	…	63.0	20.0	…	61.0	89.5	27.0	…	56
6	60.5	…	…	19.5	…	60.5	…	26.0	…	…
5	60.0	…	62.5	18.5	…	60.0	89.0	25.5	…	55
4	59.5	…	62.0	18.0	…	59.0	88.5	24.5	…	…
3	59.0	…	…	17.0	…	58.5	88.0	23.5	…	…
2	58.0	…	61.5	16.5	…	58.0	…	23.0	…	54
1	57.5	…	61.0	16.0	…	57.5	87.5	22.0	…	…
0	57.0	…	…	15.0	…	57.0	87.0	21.0	…	53

Not applicable to annealed metals of high B-scale hardness such as austenitic stainless steels, nickel and high-nickel alloys nor to cold-worked metals of low B-scale hardness such as aluminum and the softer alloys.

(Compiled by Wilson Mechanical Instrument Co.)

Durometer Tests.—The durometer is a portable hardness tester for measuring hardness of rubber, plastics, and some soft metals. The instrument is designed to apply pressure to the specimen and the hardness is read from a scale while the pressure is maintained. Various scales can be used by changing the indentor and the load applied.

Creep.—Continuing changes in dimensions of a stressed material over time is called creep, and it varies with different materials and periods under stress, also with temperature. Creep tests may take some time as it is necessary to apply a constant tensile load to a specimen under a selected temperature. Measurements are taken to record the resulting elongation at time periods sufficiently long for a relationship to be established. The data are then plotted as elongation against time. The load is applied to the specimen only after it has reached the testing temperature, and causes an initial elastic elongation that includes some plastic deformation if the load is above the proportional limit for the material.

Some combinations of stress and temperature may cause failure of the specimen. Others show initial high rates of deformation, followed by decreasing, then constant, rates over long periods. Generally testing times to arrive at the constant rate of deformation are over 1000 hours.

Creep Rupture.—Tests for creep rupture are similar to creep tests but are prolonged until the specimen fails. Further data to be obtained from these tests include time to rupture, amount of elongation, and reduction of area. Stress-rupture tests are performed without measuring the elongation, so that no strain data are recorded, time to failure, elongation and reduction of area being sufficient. Sometimes, a V-notch is cut in the specimen to allow measurement of notch sensitivity under the testing conditions.

Identifying Metals.—When it is necessary to sort materials, several rough methods may be used without elaborate chemical analysis. The most obvious of these is by using a magnet to pick out those materials that contain magnetic elements. To differentiate various levels of carbon and other elements in a steel bar, hold the bar in contact with a grinding wheel and observe the sparks. With high levels of carbon, for instance, sparks are produced that appear to split into several bright tracers. Patterns produced by several other elements, including small amounts of aluminum and titanium, for instance, can be identified with the aid of Data Sheet 13, issued by the American Society for Metals (ASM), Metals Park, OH.

Stress Analysis.—Stresses, deflections, strains, and loads may be determined by application of strain gages or lacquers to the surface of a part, then applying loads simulating those to be encountered in service. Strain gages are commercially available in a variety of configurations and are usually cemented to the part surface. The strain gages are then calibrated by application of a known moment, load, torque, or pressure. The electrical characteristics of the strain gages change in proportion to the amount of strain, and the magnitude of changes in these characteristics under loads to be applied in service indicate changes caused by stress in the shape of the components being tested.

Lacquers are compounded especially for stress analysis and are applied to the entire part surface. When the part is loaded, and the lacquer is viewed under light of specific wavelength, stresses are indicated by color shading in the lacquer. The presence and intensity of the strains can then be identified and measured on the part(s) or on photographs of the setup. From such images, it is possible to determine the need for thicker walls, strengthening ribs and other modifications to component design that will enable the part to withstand stresses in service.

Most of these tests have been standardized by the American Society for Testing and Materials (ASTM), and are published in their *Book of Standards* in separate sections for metals, plastics, rubber, and wood. Many of the test methods are also adopted by the American National Standards Institute (ANSI).

INDEX

A

B

C

D

E

F

G

H

I

J—K

L

M

N—O

P

Q—R

S

T

U—Z

NOTES